Atomically Precise Nanochemistry

Atomically Precise Nanochemistry

Edited by

Rongchao Jin
Carnegie Mellon University
USA

De-en Jiang
Vanderbilt University
USA

WILEY

Registered Office
John Wiley & Sons Ltd, The Atrium, Southern Gate, Chichester, West Sussex, PO19 8SQ, UK

For details of our global editorial offices, customer services, and more information about Wiley products, visit us at www.wiley.com.

Wiley also publishes its books in a variety of electronic formats and by print-on-demand. Some content that appears in standard print versions of this book may not be available in other formats.

Library of Congress Cataloging-in-Publication Data
Names: Jin, Rongchao, editor. | Jiang, De-en, 1975– editor.
Title: Atomically precise nanochemistry / edited by Rongchao Jin and De-en Jiang.
Description: Hoboken, NJ : Wiley, 2023. | Includes bibliographical references and index.
Identifiers: LCCN 2022051975 (print) | LCCN 2022051976 (ebook) | ISBN 9781119788645 (cloth) | ISBN 9781119788683 (adobe pdf) | ISBN 9781119788652 (epub)
Subjects: LCSH: Nanochemistry.
Classification: LCC QC176.8.N35 A88 2023 (print) | LCC QC176.8.N35 (ebook) | DDC 620/.5–dc23/eng20230119
LC record available at https://lccn.loc.gov/2022051975
LC ebook record available at https://lccn.loc.gov/2022051976

Cover Design: Wiley
Cover Image: Courtesy of Dr. Yingwei Li.

Set in 9.5/12.5pt STIXTwoText by Straive, Pondicherry, India

Contents

List of Contributors

Osman M. Bakr
KAUST Catalysis Center (KCC)
Division of Physical Sciences and Engineering
King Abdullah University of Science
and Technology (KAUST)
Dhuwal, Saudi Arabia

Sourav Biswas
School of Chemistry
Indian Institute of Science Education
and Research Thiruvananthapuram
Kerala, India

Alexander I. Boldyrev
Department of Chemistry and Biochemistry
Utah State University
Logan, USA

Guido Bussoli
Department of Industrial Chemistry
"Toso Montanari"
University of Bologna
Bologna, Italy

Yitao Cao
Department of Chemical and Biomolecular
Engineering
National University of Singapore
Singapore

Cristiana Cesari
Department of Industrial Chemistry
"Toso Montanari"
University of Bologna
Bologna, Italy

Ning Chen
College of Chemistry
Chemical Engineering
and Materials Science
and State Key Laboratory of Radiation Medicine
and Protection
Soochow University, Suzhou, Jiangsu, China

Tiankai Chen
Department of Chemical and Biomolecular
Engineering
National University of Singapore
Singapore

Woojun Choi
Department of Chemistry and Medical Chemistry
Yonsei University
Wonju, Gangwon, Republic of Korea

Chunwei Dong
KAUST Catalysis Center (KCC)
Division of Physical Sciences
and Engineering
King Abdullah University of Science
and Technology (KAUST)
Thuwal, Saudi Arabia

Cristina Femoni
Department of Industrial Chemistry
"Toso Montanari"
University of Bologna
Bologna, Italy

Mei-Yan Gao
State Key Laboratory of Structural Chemistry
Fujian Institute of Research on the
Structure of Matter
Chinese Academy of Sciences
China

and

Department of Chemical Sciences
Bernal Institute, University of Limerick
Limerick, Limerick County
Republic of Ireland

Wanmiao Gu
Key Laboratory of Materials Physics
Anhui Key Laboratory of Nanomaterials
and Nanotechnology
CAS Center for Excellence in Nanoscience
Institute of Solid State Physics
Chinese Academy of Sciences
Hefei, China

Wenhua Han
Department of Physics, School of Physical
Science and Technology
Ningbo University
Ningbo, China

Hironobu Hayashi
Division of Materials Science
Nara Institute of Science and Technology
Japan

Trevor W. Hayton
Department of Chemistry and Biochemistry
University of California Santa Barbara
Santa Barbara, USA

Maria C. Iapalucci
Department of Industrial Chemistry
"Toso Montanari"
University of Bologna
Bologna, Italy

De-en Jiang
Department of Chemical and Biomolecular
Engineering
Vanderbilt University
Nashville, USA

Hongjie Jiang
College of Chemistry
Chemical Engineering and Materials Science
and State Key Laboratory of Radiation Medicine
and Protection
Soochow University, Suzhou, Jiangsu, China

Rongchao Jin
Department of Chemistry
Carnegie Mellon University
Pittsburgh, USA

Yongsung Jo
Department of Chemistry
Yonsei University
Seoul, Republic of Korea

Dongil Lee
Department of Chemistry
Yonsei University
Seoul, Republic of Korea

Pengye Liu
Department of Physics, School of Physical
Science and Technology
Ningbo University
Ningbo, China

Sukhendu Mandal
School of Chemistry
Indian Institute of Science Education
and Research Thiruvananthapuram
Kerala, India

Alvaro Muñoz-Castro
Grupo de Química Inorgánica y Materiales
Moleculares
Facultad de Ingeniería
Universidad Autonoma de Chile
El Llano Subercaseaux, Santiago, Chile

Saidkhodzha Nematulloev
KAUST Catalysis Center (KCC)
Division of Physical Sciences and Engineering
King Abdullah University of Science
and Technology (KAUST)
Thuwal, Saudi Arabia

Jiawei Qiu
College of Chemistry
Chemical Engineering and Materials Science
and State Key Laboratory of Radiation Medicine
and Protection
Soochow University, Suzhou, Jiangsu, China

Silvia Ruggieri
Laboratory of Luminescent Materials
Department of Biotechnology
University of Verona
Verona, Italy

Hoeun Seong
Department of Chemistry
Yonsei University
Seoul, Republic of Korea

Di Sun
School of Chemistry and Chemical Engineering
State Key Laboratory of Crystal Materials
Shandong University, Ji'nan, China

Fang Sun
School of Chemistry and Chemical Engineering
Chongqing Key Laboratory of Theoretical
and Computational Chemistry
Chongqing University, Chongqing, China

Panpan Sun
School of Chemistry and Chemical Engineering
State Key Laboratory of Crystal Materials
Shandong University, Ji'nan, China

Zhong-Ming Sun
State Key Laboratory of Elemento-Organic
Chemistry School of Materials Science
and Engineering Nankai University
Tianjin, China

Qing Tang
School of Chemistry and Chemical Engineering
Chongqing Key Laboratory of Theoretical
and Computational Chemistry
Chongqing University, Chongqing, China

Nikolay V. Tkachenko
Department of Chemistry and Biochemistry
Utah State University
Logan, USA

Zhao-Yang Wang
Henan Key Laboratory of Crystalline Molecular
Functional Materials
Henan International Joint Laboratory of Tumor
Theranostical Cluster Materials
Green Catalysis Center
and College of Chemistry
Zhengzhou University, Zhengzhou, China

Zhikun Wu
Key Laboratory of Materials Physics
Anhui Key Laboratory of Nanomaterials
and Nanotechnology
CAS Center for Excellence in Nanoscience
Institute of Solid State Physics
Chinese Academy of Sciences
Hefei, China

Xin Xia
School of Chemistry and Chemical Engineering
State Key Laboratory of Crystal Materials
Shandong University, Ji'nan, China

Yunpeng Xia
College of Chemistry
Chemical Engineering and Materials Science
and State Key Laboratory of Radiation Medicine
and Protection
Soochow University, Suzhou, Jiangsu, China

Jianping Xie
Joint School of National University of Singapore
and Tianjin University
International Campus of Tianjin University
Fujian, China

and

Department of Chemical and Biomolecular
Engineering
National University of Singapore
Singapore

Wen Wu Xu
Department of Physics, School of Physical
Science and Technology
Ningbo University
Ningbo, China

Yu-He Xu
State Key Laboratory of Elemento-Organic
Chemistry
School of Materials Science and Engineering
Nankai University, China

Hiroko Yamada
Division of Materials Science
Nara Institute of Science and Technology
Japan

Qiaofeng Yao
Joint School of National University of Singapore
and Tianjin University
International Campus of Tianjin University
Fujian, China

and

Department of Chemical and Biomolecular
Engineering
National University of Singapore
Singapore

Yang-Rong Yao
College of Chemistry
Chemical Engineering and Materials Science
and State Key Laboratory of Radiation Medicine
and Protection
Soochow University, Suzhou, Jiangsu, China

and

Department of Materials Science and
Engineering
University of Science and Technology of China
Hefei, China

Peng Yuan
KAUST Catalysis Center (KCC)
Division of Physical Sciences and Engineering
King Abdullah University of Science and
Technology (KAUST)
Thuwal, Saudi Arabia

Stefano Zacchini
Department of Industrial Chemistry
"Toso Montanari"
University of Bologna
Bologna, Italy

Shuang-Quan Zang
Henan Key Laboratory of Crystalline Molecular
Functional Materials
Henan International Joint Laboratory of Tumor
Theranostical Cluster Materials
Green Catalysis Center
and College of Chemistry
Zhengzhou University, Zhengzhou China

Jian Zhang
State Key Laboratory of Structural Chemistry
Fujian Institute of Research on the
Structure of Matter
Chinese Academy of Sciences
Fujian, China

Lei Zhang
State Key Laboratory of Structural Chemistry
Fujian Institute of Research on the
Structure of Matter
Chinese Academy of Sciences
Fujian, China

and

Institute of Modern Optics
College of Electronic Information and
Optical Engineering
Nankai University, Tianjin, China

Lihao Zheng
College of Chemistry
Chemical Engineering and Materials Science
and State Key Laboratory of Radiation Medicine
and Protection
Soochow University, Suzhou, Jiangsu, China

Manman Zhou
Department of Chemistry and Centre for Atomic
Engineering of Advanced Materials
Key Laboratory of Structure and Functional
Regulation of Hybrid Materials of Ministry
of Education
Anhui Province Key Laboratory of Chemistry
for Inorganic/Organic Hybrid Functionalized
Materials
Anhui University, Anhui, China

Manzhou Zhu
Department of Chemistry and Centre for Atomic
Engineering of Advanced Materials
Key Laboratory of Structure and Functional
Regulation of Hybrid Materials of Ministry
of Education
Anhui Province Key Laboratory of Chemistry
for Inorganic/Organic Hybrid Functionalized
Materials
Anhui University, Anhui, China

Preface

Chemists have long been motivated to create atomically precise nanoclusters, not only for addressing some fundamental issues that were not possible to tackle with imprecise nanoparticles but also to provide new opportunities for applications such as catalysis, optics, and biomedicine. Given the breadth of the book, De-en and I decided to invite a number of experts who are working on various types of atomically precise nanoclusters. We thank all the experts for their warm support of the book and timely completion of the chapters. Due to space limitations, we must apologize to some colleagues for missing their excellent work that could not be included in this book.

This book comprises 16 chapters. Chapter 1 provides an introduction to atomically precise nanochemistry. Chapters 2 to 10 cover atomically precise metal nanoclusters, such as Au, Ag, Cu, Ni, Rh, and the doped/alloyed nanoclusters, as well as the electrocatalytic application in CO_2 reduction and water splitting. Endohedral metallofullerenes, graphene nanoribbons, Zintl clusters, and Ti-oxo nanoclusters are discussed in Chapters 11 to 14, respectively. Finally, Chapters 15 and 16 are devoted to the assembly of nanoclusters (such as Au, Ag, and Cu), including the crystalline assembly and the use of nanoclusters as nodes for constructing special types of metal-organic frameworks, as well as the sensing and other applications. The atomic-level control in the synthesis, the new types of structures, and the physical/chemical properties of nanoclusters are illustrated in various chapters. This book contains not only experimental contributions but also theoretical insights into the atomic and electronic structures, as well as the catalytic mechanisms. We expect this book to be suitable for graduate and undergraduate students, researchers, and industry practitioners.

Overall, the concept of atomic precision is expected to have a major impact on future nanoscience research and other areas. We hope that atomically precise nanochemistry will serve as a hub for the unification of various research areas in which precision materials are being created and studied. In future research, exquisite nanochemistry will surely bring exciting opportunities in both fundamental research and practical applications. Progress in various types of atomically precise nanoclusters and the hybridization of two or more types of nanoclusters, as well as the assembly of nanoclusters into meso- and macroscopic functional materials, will lead to more exciting frontiers.

Rongchao Jin and De-en Jiang
October 2022

1

Introduction to Atomically Precise Nanochemistry

Rongchao Jin

Department of Chemistry, Carnegie Mellon University, Pittsburgh, PA, 15213, USA

1.1 Why Atomically Precise Nanochemistry?

Since the beginning of the twenty-first century, nanoscience has made significant progress [1–8], especially in the creation of a variety of nanostructures with size and shape control and the discovery of new phenomena at the nanoscale. The rapid progress of nanoscience heavily relies on the synthetic breakthroughs [1–7, 9]. In terms of chemical synthesis, Mother Nature is indeed the master, evidenced by, for example, the creation of giant molecules such as DNA and proteins from atomic building blocks, and even the buildup of complex photosynthesis machinery, all being at the level of atomic precision. While there is still a long way to go for chemists to be on par with Mother Nature, we expect that nanochemistry will make giant leaps in the near future toward controlling nanostructures with atomic precision [10].

Beside the nanoscience field, the pursuit of atomic precision is also of critical importance in other areas (Scheme 1.1), including inorganic cluster chemistry, gas phase cluster science and solid-state materials science, as well as supramolecular chemistry. In moving toward larger sizes and building up complex architectures (Figure 1.1), precise control over size and structure will certainly become more challenging, yet highly exciting and rewarding [11].

1.1.1 Motivations from Nanoscience Research

Over the past two decades, significant advances have been made in controlling the size, shape, crystallinity, and composition of many types of nanoparticles, including metal nanoparticles (e.g. Au, Ag, Pt, Pd, Rh), semiconductor nanocrystals (e.g. CdSe, InP), and magnetic nanoparticles (e.g. Fe, Co, Ni), all with high monodispersity (e.g. size distributions of 5–10%). In addition, several shape-controlled nanostructures have been successfully developed, [1–9] such as the nanorods/nanowires (1D) and nanoprisms (2D). Research on the anisotropic nanostructures (both 1D and 2D) has significantly expanded the fundamental understanding of the new physicochemical properties of nanostructures enabled by shape control. Based on the new properties observed, a wide range of applications, such as sensing, catalysis, optics, and electronics, have been designed. New frontiers keep emerging, which has greatly pushed the frontier of nanoscience research.

Despite the impressive progress in nanoscience, some issues still exist. Below we briefly discuss a few issues that hamper nanoscience research from going deeper. By pursuing atomic precision,

Atomically Precise Nanochemistry, First Edition. Edited by Rongchao Jin and De-en Jiang.

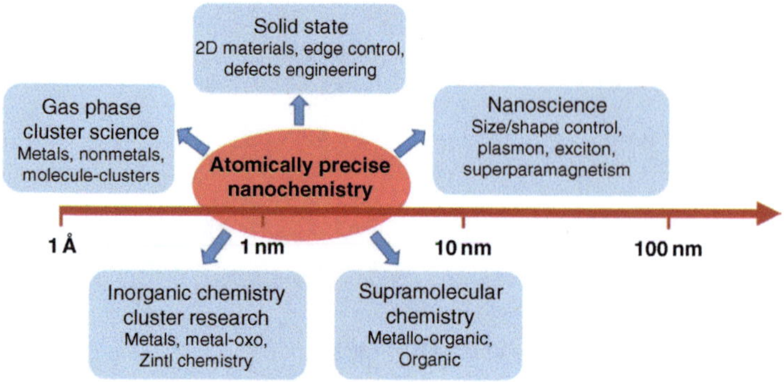

Scheme 1.1 Atomically precise nanochemistry as a "hub" for nanoscience, inorganic cluster chemistry, gas phase cluster science, and other areas.

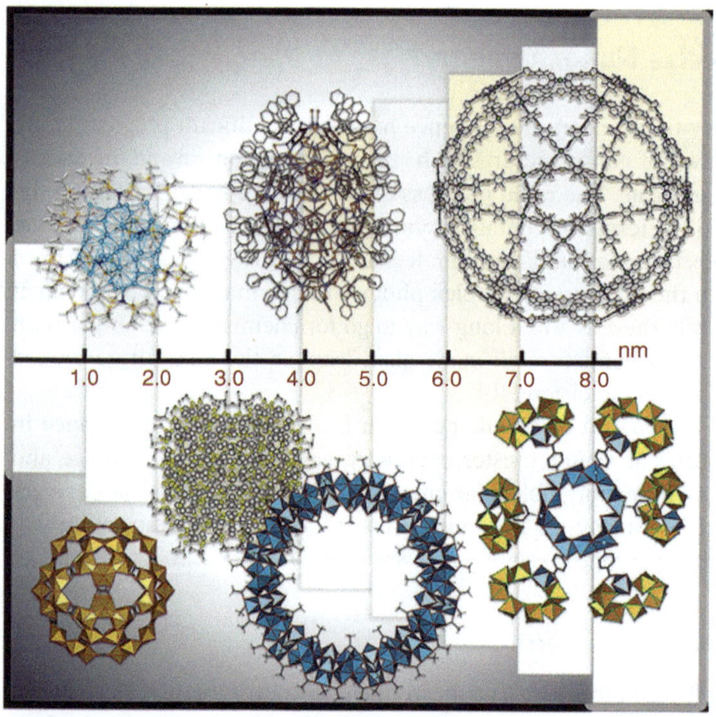

Figure 1.1 Examples of giant structures of nanoclusters and molecular architectures determined by X-ray crystallography. *Source:* Reprinted with permission from [11]. © 2021 American Chemical Society.

those issues can be overcome or at least alleviated, which will enable nanoscience research to reach a new level.

First of all, the polydispersity of nanoparticles (NPs) has long been an issue in nanoscience research. The synthesis of NPs tends to produce particles with a polydispersity (more or less). When the polydispersity is controlled to be better than 15% (e.g. 10 ± 1.5 nm), such NPs are typically called *monodisperse*. Although highly monodisperse NPs (e.g. polydispersity down to 5%) have also been made in some cases (Figure 1.2a/b), these NPs are still not of the same size at the *atomic* scale. In other words, no two NPs are the same! Therefore, a major dream of nanochemists has long been to synthesize truly uniform NPs (i.e. atomically precise NPs). This dream has now

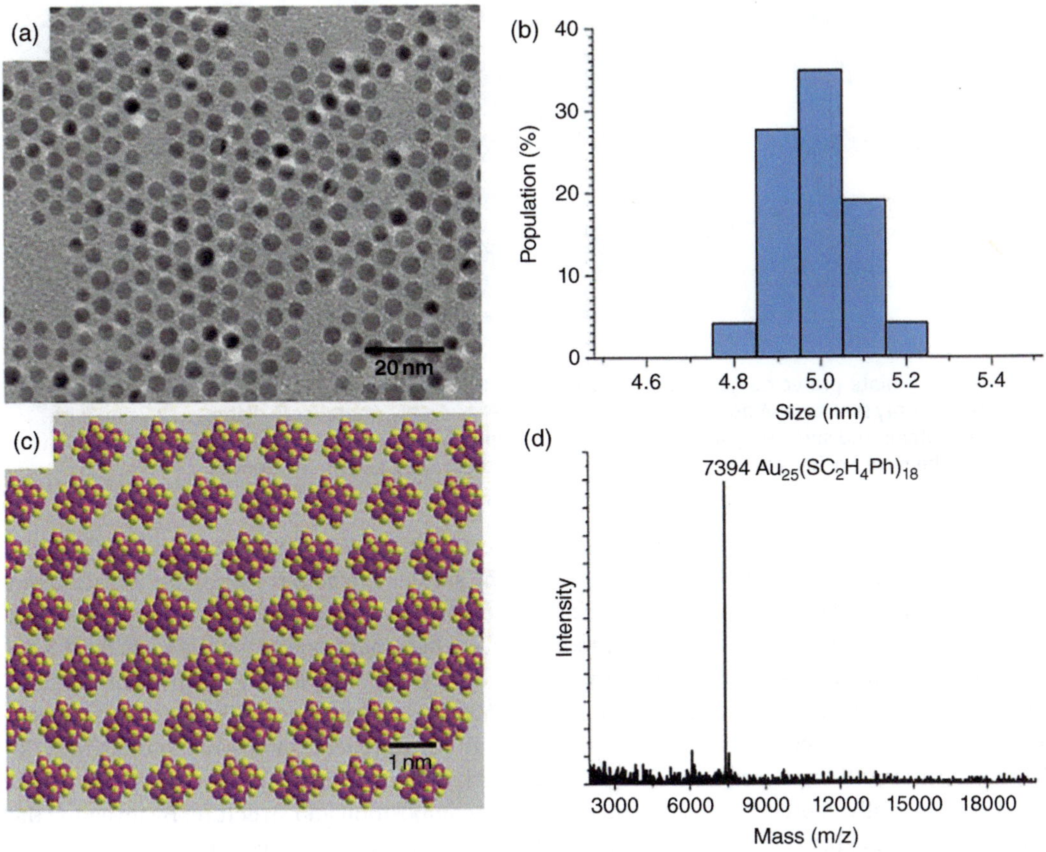

Figure 1.2 Comparison between regular nanoparticles and atomically precise nanoclusters. (a) Monodisperse nanoparticles (average: ~5 nm, standard deviation: ~5%) imaged by transmission electron microscopy; (b) Typical size distribution (e.g. 5 ± 0.3 nm diameter); (c) Atomically precise $Au_{25}(SC_2H_4Ph)_{18}$ nanoclusters (1 nm metal core diameter, $-C_2H_4Ph$ groups are omitted for clarity) assembled in a single crystal; (d) Mass spectrometry characterization of $Au_{25}(SC_2H_4Ph)_{18}$. *Source:* Reprinted with permission from [12]. © 2014 Springer.

been partially realized in the case of ultrasmall gold NPs of 1–3 nm in size (Figure 1.2c/d) and silver as well [10]. Such atomically precise NPs are often termed as nanoclusters (NCs) in order to differentiate them from the regular NPs. The success in obtaining atomically precise NCs is critically important, because they can serve as models for nanochemists to gain fundamental understanding of some important issues that were previously not feasible to tackle, such as the nanoparticle isomerism, the origin of surface plasmon resonance (SPR), and chemical bonding evolution with size [10].

Second, the surface of NPs often remains poorly controlled in the synthesis and thus poorly understood. Questions on the precise composition of the surface adsorbates (i.e. stabilizers) and how the stabilizers are adsorbed on the particle surface are generally not known for most nanoparticle samples. While transmission electron microscopy (TEM) and various spectroscopy techniques are powerful in characterizing NPs, they often cannot reveal the true composition and bonding structure of the surface. For solution phase NPs, the surface includes the organic stabilizers and the interface to the inorganic core. This organic–inorganic interface (Figure 1.3a) is unfortunately very tough to study, [12, 13] because TEM is ineffective in imaging the particle surface (Figure 1.3b)

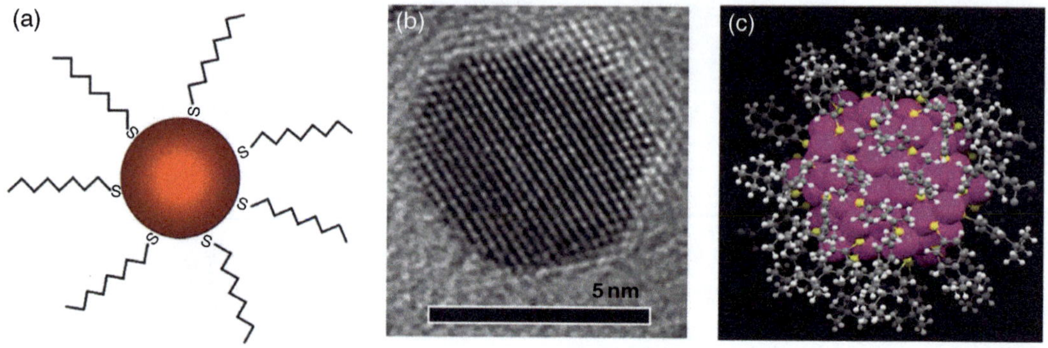

Figure 1.3 Thiolate-protected gold nanoparticles. (a) Cartoon, (b) High resolution TEM image (ligands invisible), (c) X-ray structure of atomically precise $Au_{246}(SPh\text{-}p\text{-}CH_3)_{80}$ (metal core diameter: 2.2 nm) with both metal atoms and surface ligands visible. *Source:* Adapted with permission from [10, 13]. © 2016 American Chemical Society and American Association for the Advancement of Science.

due to insufficient electron scattering by organics (e.g. light atoms of S, C, N, O, and H). While scanning probe microscopy [14] can readily reveal the surface molecules via van der Waals forces or tunneling currents, unfortunately it is incapable of showing the underlying interface between the inorganic core and surface ligands. In molecular science, many spectroscopic tools have been developed and are very powerful for analyzing molecules, such as nuclear magnetic resonance (NMR), infrared (IR), and Raman scattering, but when they are used for NPs, the polydispersity and heterogeneity of regular NPs make it very difficult (or unreliable) to correlate the NMR/IR/Raman signals with imprecise NPs. Thus, well-defined NPs are critically needed for fundamental research, especially in order to understand the surface composition and structure. By revealing the interfacial bonding between the stabilizers and the underlying inorganic core through X-ray crystallography (XRC) (Figure 1.3c), many fundamental questions could be addressed, such as the nature of active sites in nanocatalysis, charge transfer and catalytic mechanisms, photoluminescence blinking, surface magnetism, self-assembled monolayer (SAM) structure, and nanoparticle assembly mechanisms [10, 13].

Third, conventional NPs often possess various defects in the interior and/or on the surface, which are highly detrimental to many physical and chemical properties, including the stability of NPs, photoluminescence, and charge transport. How to eliminate those defects? Is it possible to create perfect NPs? These questions are of paramount importance in nanoscience research. Thus, new chemistry should be developed to attain atomically precise NPs. Even more exciting is to develop capabilities of tailoring or engineering the NP surface for specific applications, e.g. catalysis and biomedicine. These tasks call for the atomically precise NCs.

Fourth, the mechanisms for shape-controlled synthesis of nanoparticles are still not well understood, such as the nucleation and growth mechanisms. For example, small nanoprisms (<10 nm) were observed as nuclei in the photo- or plasmon-induced transformation of spherical Ag NPs to nanoprisms, [1, 9] but thus far it is not yet clear how the transformation occurs during the nucleation stage [15, 16], nor how the nuclei subsequently grow to larger nanoprisms (e.g. atom by atom in the classical mechanism, particle edge-to-edge fusion [9], or other unknown processes). Considering the important roles of stabilizers, facet-selective binding of stabilizers (or ligands) is often invoked to explain the formation of nonspherical (=anisotropic) nanostructures, but many details are still unknown yet, including how the ligands are bonded to the surface atoms on specific facets such as {111} and {100} of nanoparticles, and what the bonding geometry is (e.g. terminal vs. bridging bonds). To understand the mechanism for shape control and relevant issues, atomic-level information for both the nucleation and growth processes is critically needed, in which atomically precise NCs may provide useful hints.

Last but not least, the applications of NPs such as nanocatalysis and nanomedicine require the precise knowledge of the total structure of the NP, not just the core size/shape from TEM analysis. In nanocatalysis, the adsorption of reactant(s) occurs on the surface; thus, without knowing the surface composition and structure, it would not be possible to understand the catalytic mechanism and provide the design principles for efficient catalysts [17]. The same is true in nanomedicine [8]. Upon the nanoparticles (e.g. carrying DNA or drug compounds) being injected into an animal or human body, these nanoparticles first experience interactions with proteins in blood, glutathione in cells, and other types of biomolecules. Knowing the surface of nanoparticles will be critically important and can lead to a fundamental understanding of the biodistribution and pharmacokinetics [18, 19]; hence, precision medicine can be designed.

Beside the research efforts toward atomically precise metal NCs, it is worth noting that the pursuit of atomic precision in semiconductor quantum dot is also emerging [20–22]. In quantum dots, the strong quantum confinement makes the properties extremely sensitive to size variation, thus, precise control over quantum dot size and structural uniformity is very pressing. Efforts toward such goals are underway [22, 23]. In the case of magnetic NPs, the surface disorder often results in electron spin randomization (so-called spin canting), hence, a lower magnetization than the bulk value, but in certain cases an enhanced magnetization was observed [24, 25]. To unravel the mysteries, a precise understanding of the surface atomic structure should be first obtained [26].

To address all the above important issues and many other fundamental ones in nanoscience, atomically well-defined NPs are critically needed. Nanoscientists are thus strongly motivated to develop atomically precise nanochemistry.

1.1.2 Motivations from Inorganic Chemistry Research

The cluster state is also of wide interest in the inorganic chemistry field. Understanding the chemical bonding pattern in the cluster state and the pattern evolution with increasing size has long been a central task. Since the early twentieth century, research in inorganic chemistry has led to the production of metal clusters (typically in ligand-protected form) and also boron clusters, Zintl clusters, and metal-oxo nanoclusters. There is strong interest in pushing up the size of such clusters.

Boron cluster research was pioneered by Stock in the 1910s [27] and has significantly expanded the concept of chemical bonding [28]. In contrast to classical covalent bonds in which two electrons are shared by two atoms or centers (denoted "2c-2e," so-called Lewis pair [29]), multicenter bonding such as 3c-2e [30] was introduced for boron clusters. In addition, the concept of three-dimensional (3D) aromaticity was also put forth [31, 32], which is in contrast with the 2D case, i.e. Hückel aromaticity. Such aromaticity rules have been widely applied in later research in different areas.

The Zintl cluster research was initiated by Zintl in the 1930s [33], such as As_7^{3-} and Sb_7^{3-} in liquid ammonia. A striking feature is that Zintl clusters are bare (without the need of ligand protection in solution phase) [34]. Larger sizes are such as $[As@Ni_{12}@As_{20}]^{3-}$, [35] $[Au_8Pb_{33}]^{6-}$, and $[Au_{12}Pb_{44}]^{8-}$ [36]. Understanding the electronic structure of Zintl clusters and the size evolution is of particular interest [34, 37]. Several theories, such as the Zintl-Klemm concept [38], were developed in early work. The Wade-Mingos electron counting rules from the polyhedral borane clusters were later introduced into the Zintl cluster field. The concepts of aromaticity and multicenter delocalized bonding [32, 39], as well as the superatom electronic picture, [37d, 40] have all been applied to Zintl clusters.

Another line of inorganic cluster chemistry involves the metal-oxo chemistry or polyoxometalate clusters (POM), which also features atomic precision [41]. The POMs are closely relevant to the corresponding metal oxide in terms of composition and structure, [42] thus, atomically precise POMs can serve as molecular or nanoscale models for metal oxides, such as Ti-oxo NCs for modeling TiO_2 and Ce-oxo for CeO_2. Using the Ti-oxo cluster (TOC) system as an example, the inorganic Ti—O core is coordinated by organic or inorganic ligands. In early work, a hydrolysis product of titanium tetraethoxide was identified by XRC to be $[Ti_7O_4(OEt)_{20}]$ in 1967 [43]. Since then, many TOCs have been reported, such as Ti_{42}, Ti_{44}, and Ti_{52} [44–46]. Their structures can be controlled, such as spherical cages, rings or wheels, and many types of dense structures [41]. In addition to monometallic TOCs, there are also bimetallic ones with doped ions being main group ions, transition metal ions, and lanthanide ions [41]. The incorporation of Ag clusters into TOCs is quite appealing, such as the Ag_{14}@Ti_{16}-oxo nanocluster [47].

Finally, metal coordination-based supramolecular chemistry has also moved toward the nanoscale by creating ever-larger nanostructures, such as wheels/rings, cages, and polyhedrons, as well as tubular architectures [11]. The construction of large and exquisite molecular architectures calls for new development of nanochemistry [48].

1.1.3 Motivations from Gas Phase Cluster Research

The field of gas phase cluster science has moved from small sizes (several to a dozen atoms in earlier research) to larger sizes in the current efforts [49]. Since the 1960s, intense research on gas phase clusters has been carried out, including inert-gas-element clusters (e.g. xenon clusters, Xe_n), metal clusters (e.g. Na_n clusters), carbon clusters (C_n), water clusters ($[H_2O]_n$), and many other types [50]. The Xe_n cluster research led to an important insight into the structural stability of clusters, i.e. the formation of icosahedral packing structures, such as the one-shelled Xe_{13}, two-shelled Xe_{55}, and three-shelled Xe_{147} [51]. These are called the *geometric magic numbers,* and the atomic shell closing endows high stability to the clusters.

In the 1980s, research on gas phase Na_n clusters led to another important concept, that is, the electron shell closing of $1S\,1P\,2S\,1D\ldots$, with notable sizes of Na_8, Na_{34}, Na_{58}, and Na_{92} (so-called superatoms because of the atomic-like orbitals in such clusters) [52]. Much work has also been done on gas phase gold clusters, [53, 54] and theoretical computations on the 2D to 3D structure transition (Figure 1.4) [55, 56]. To investigate the evolution of the superatom electronic shell picture, larger sized NCs should be pursued. As the size boundary moves up, the cluster science indeed merges with nanoscience. Preparation of clusters in gas phase with hundreds of atoms per core has been pursued in recent years, [49] although major effort is still required in the pursuit of atomic precision.

Carbon clusters (or fullerenes) are one of the significant areas in gas phase cluster research [57, 58]. Fullerenes are typically produced using the electric arc discharge or thermal chemical vapor deposition (CVD) method. From the initial discovery of C_{60} in 1985, a variety of sizes have been reported, with smaller ones such as C_{20}, C_{24}, C_{28}, C_{32}, C_{36} and larger ones such as C_{70}, C_{80}, C_{82}, C_{108} and even up to C_{540} [59]. The size-dependent structure, electronic, and optical properties of C_n are under active investigation, and the cage structure of C_n clusters is particularly appealing for encapsulating atoms, molecules, and clusters of few-atom in size, forming endohedral clusters [60]. New additions to the nanocarbon family also include the atomically precise graphene nanoribbons (GNR) [61] and some other forms.

1.1.4 Motivations from Other Areas

The concept of atomic precision is also critically important in solid-state materials. A variety of solid-state materials are being intensively pursued in the field of materials science, such as the low dimensional materials (2D van der Waals materials, stacked 2D materials, etc.) and quantum

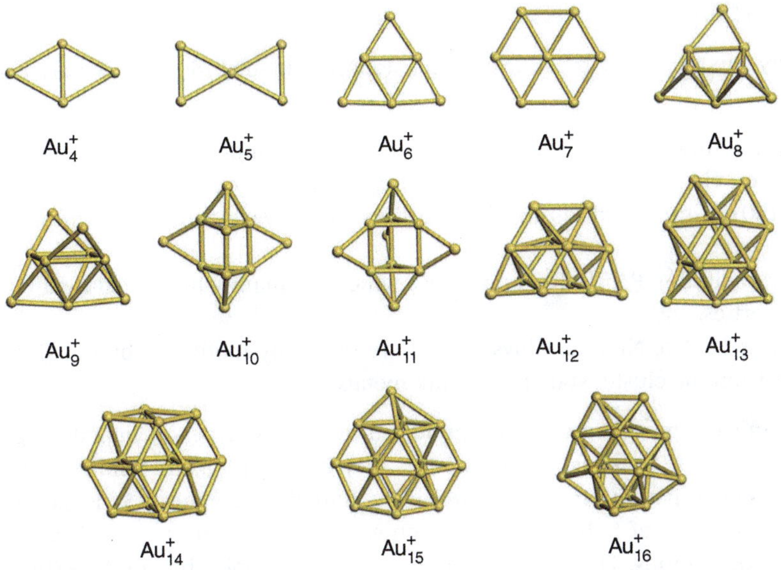

Figure 1.4 Computed structures of bare cationic Au_n^+ clusters and 2D to 3D structural transition. *Source:* Reprinted with permission from [55]. © 2019 Royal Society of Chemistry.

materials. Precise control over the edge and topology is of vital importance for controlling the functionality, and heteroatom doping (e.g. doped WS_2) also calls for atomic precision strategies [62].

In the semiconductor industry, silicon chips have moved toward the few-nm regime, in which precise control of the film quality and interface has become a major issue in order to prevent electrical leak in the few-nm transistors [63]. This calls for atomically precise fabrication strategies, which is vital in advancing toward the 1-nm transistor technology.

1.2 Types of Nanoclusters Covered in This Book

Intense efforts in the past years have led to success in synthesizing Au, Ag, Cu, Pt, Pd, Rh, Al, Fe, Co, and Ni nanoclusters, as well as alloys. The obtained atomically precise nanoclusters can serve as new models or platforms for solving many issues of structure and bonding evolution, and thus enable significant progress in nanoscience research.

This book is intended to provide an overview of the progress in atomically precise nanochemistry with a focus on inorganic nanostructures, including metals (e.g. Au, Ag, Cu, Ni, Rh), Zintl clusters, Ti-oxo nanoclusters, and carbon clusters, as well as catalytic application and assembly of NCs into functional materials. On a note, other types (e.g. Pd_n, [64] Pt_n, [65] Al_n [66] and cerium-oxo NCs [42], metallo-rings and wheels [11], supramolecular cages/polyhedrons/capsules [11, 48], and DNA-templated Au_n and Ag_n NCs [67, 68]) are not covered due to the limitations of the book. Semiconductor NCs [20–23] are also not covered.

Among the chapters, atomically precise metal NCs (Au, Ag, Cu, Ni, Rh) and their catalytic applications are discussed in Chapters 2–10. Some other types of NCs, including endohedral metallofullerenes (EMF), GNRs, Zintl clusters, and Ti-oxo nanoclusters, are presented in Chapters 11 to 14. Finally, Chapters 15 and 16 are devoted to the assembly of metal NCs (Au, Ag, Cu), including their incorporation into metal organic frameworks (MOFs) toward the fabrication of assembled functional materials.

1.2.1 Atomically Precise Metal Nanoclusters (Au, Ag, Cu, Ni, Rh)

In recent years, atomically precise metal NCs have been intensely pursued. Such NCs may be divided into three types:

i) *Coinage metals,* including Au, Ag, Cu, and alloys, which are primarily motivated by the elegant optical properties of such elements in the form of NCs (*c.f.* their larger counterparts – plasmonic NPs in the 3–100 nm range exhibiting distinct SPRs in the visible and near-infrared wavelength range),
ii) *Platinum group metals,* including Pd, Pt, Rh, and alloys, which are mainly motivated by their excellent catalytic properties, and
iii) *Magnetic metals,* such as Fe, Co, Ni, and alloys, which are primarily motivated by the size-dependent magnetism from the cluster state to the bulk metals.

Among the types of atomically precise metal NCs protected by various types of ligands (e.g. phosphines [69, 70], thiolates [71, 72], alkynyls [73], carbenes [74], halides [75], and Stibine [76]), the gold-thiolate (Au-SR) system is perhaps the most extensively studied one (Figure 1.5). Its larger counterparts (i.e. plasmonic Au NPs of 5–100 nm) have been widely studied in the past decades and explored in numerous applications. The success in atomically precise gold-thiolate NCs (tens to hundreds of gold atoms per core) will help explain the regular NPs in terms of the latter's surface structure, assembly, and catalytic mechanism, as well as many other fundamental aspects [10, 77].

In this book, Chapter 2 by Xie's group summarizes the progress in total synthesis of gold NCs, including the control over size, structure, composition and ligand engineering, and Chapter 3 by Wu's group reviews the advances in the synthesis, characterization, and application of gold NCs with a focus on the categorization of structural series. The structures of metal NCs constitute the basis for understanding the functionality and stability of NCs. A theoretical perspective on the structure evolution and structural design of $Au_n(SR)_m$ NCs is provided by Xu and coworkers (see Chapter 4). Among the applications of Au NCs, catalysis (e.g. electrocatalytic water splitting and CO_2 reduction) is summarized in Chapter 5 by Lee's group, and theoretical simulations on the electrocatalysis mechanisms are given in Chapter 6 by Tang's and Jiang's groups.

While a large body of research focuses on gold NCs, there has also been substantial progress in Ag NCs (reviewed in Chapter 7 by Zhu's group), Cu NCs (Chapter 8 by Bakr's group), Ni NCs (Chapter 9 by Hayton's group), and Rh and its alloy NCs (Chapter 10 by Femoni's group). Some selected crystal structures of these metal NCs are shown in Figure 1.6.

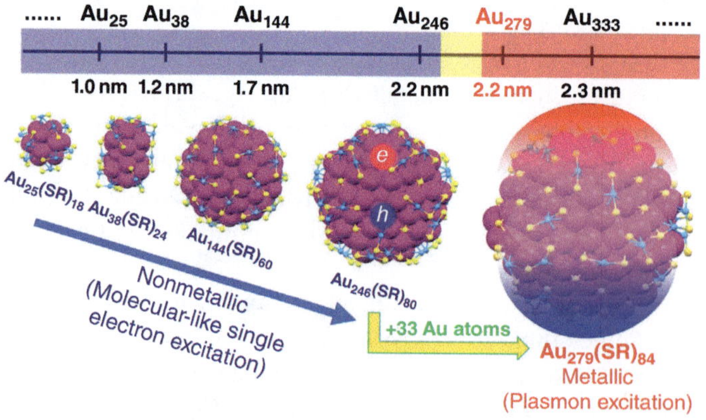

Figure 1.5 Atomically precise gold nanoclusters with the transition from nonmetallic Au_{246} to metallic Au_{279}. *Source:* Reprinted with permission from [77]. © 2021 The Authors (Jin and Higaki).

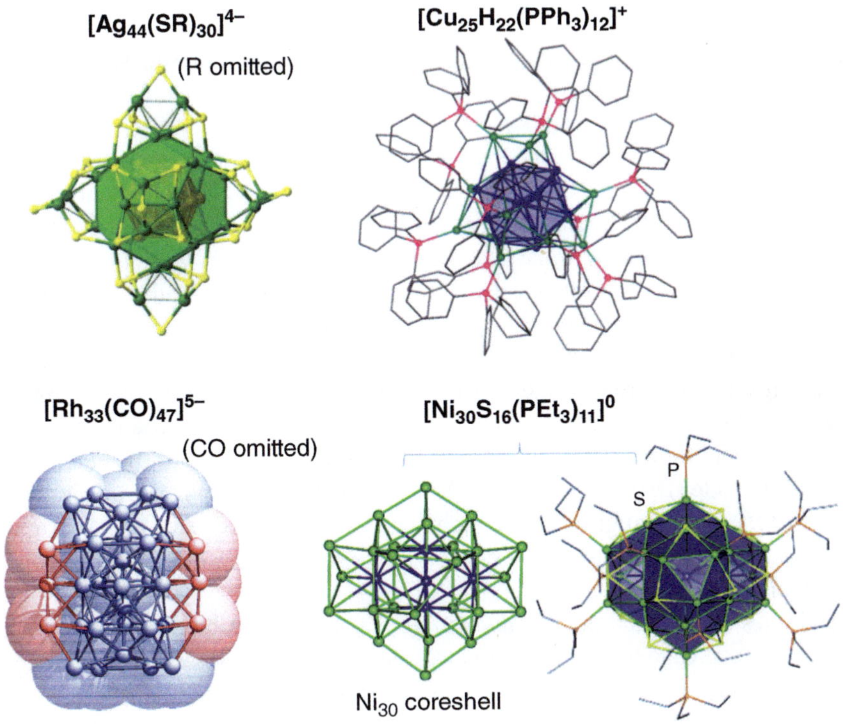

Figure 1.6 Examples of atomically precise Ag, Cu, Rh, and Ni nanoclusters with metal—metal bonding. *Source:* Adapted from Chapters 7–10.

The types of metal NCs that are discussed in Chapters 2–10 contain a certain number of delocalized valence electrons of metal atoms, such as 2e, 8e, 18e, 20e, 34e, 40e, 58e, 70e, and 92e. These so-called electronic magic numbers are closely relevant to the stability and properties of the NCs. Unlike the zero valent Au, Ag, Pt, Pd, and Rh NCs – which can even be in negative (e-rich) states for Rh atoms [78], Cu NCs tend to be in Cu(I) state or have low numbers of core valence electrons, such as 2e in the NCs of $[Cu_{25}H_{22}(PPh_3)_{12}]^+$ [79], $[Cu_{13}\{S_2CN^nBu_2\}_6(CCPh)_4]^+$ [80], $Cu_{14}(C_2B_{10}H_{10}S_2)_6(CH_3CN)_8$ [81], and $[Cu_{61}(S^tBu)_{26}S_6Cl_6H_{14}]^+$ [82]. However, upon introducing hetereometal dopants, the valence electron number could be raised, such as the 10e $[Pt_2Cu_{34}(PET)_{22}Cl_4]^{2-}$ [83a] and the 67e $[Cu_{43}Al_{12}](Cp^*)_{12}$ [83b].

Historically, chemists have long been interested in counting the valence electrons, such as the organic aromaticity rule of $4n + 2$ (π electrons) put forth by Hückel in 1931, and the Wade-Mingos electron counting rules for boron and metal clusters [38, 39] developed in the 1970s, the superatom electron counts [52] in the 1980s, and so on. More discussions on the counting of cluster valence electrons are provided in Section 1.3.2.

On the other hand, there are also many Au(I), Ag(I), and Cu(I) NCs that have no free valence electrons, such as the anion-templated Cu(I) and Ag(I) NCs. While these systems are not the focus of this book, some Cu(I) and Ag(I) NCs are discussed in Chapter 8 and 16, respectively. In Chapter 8, Dong et al. provides a very systematic summary of the syntheses of Cu(I) and Cu(0) NCs, whereas Chapter 16 by Wang and Zang focuses on the assembly of Ag(I) NCs for framework materials, and Chapter 15 by Mandal's and Sun's groups summarizes the assembly of Au and Ag NCs into 3D and 2D crystals.

1.2.2 Endohedral Fullerenes and Graphene Nanoribbons

Nanocarbon has many types, including the fullerenes (C_n), EMFs, carbon nanotubes, graphene quantum dots, and nanoribbons. Some of them are in atomically precise forms, such as fullerenes and endohedral fullerenes, as well as certain GNRs.

Since the discovery of C_{60}, carbon clusters have long been of major interest to chemists, physicists, and materials scientists owing to the unique cage structure, superconductivity, and other unprecedented functionalities [57–60]. In this book, Chapter 11 is devoted to the endohedral fullerenes (Figure 1.7). Chen's group has summarized recent progress in EMFs with a monomer, dimer, and even cluster enclosed in the carbon cage, forming metallofullerenes and clusterfullerenes, and the new properties are discussed, such as optical absorption, electrochemical, and magnetic properties, as well as the chemical reactivity. Charge transfer from the enclosed metal atom to the fullerene cage may result in different isomeric fullerene structures. In the case of two metal atoms enclosed, metal—metal interactions (including bonding and non-bonding) lead to interesting phenomena. The inclusion of more atoms gives rise to clusterfullerenes, such as the enclosing of metal nitride (e.g. Sc_3N), oxide (Dy_2O), sulfide (Sc_2S), carbide (M_2C_2, where M = metal), and carbonitride (Sc_3CN) clusters. The configuration of the enclosed few-atom cluster may be rigid (e.g. M_3N) or flexible (e.g. M_2C_2), and the enclosed cluster may also transfer a certain number of valence electrons to the fullerene cage.

Fullerenes and doped ones exhibit rich optical absorption peaks in the visible to near-infrared (NIR) range. The different metal-cage interaction and electron transfers greatly affect the energy distribution of molecular orbitals of endohedral fullerenes and thus their optical absorption spectra. There are also rich electrochemical properties of endohedral fullerenes, and both the first oxidation and first reduction potentials can be largely varied by the metal atoms in the cage.

In addition to the zero-dimensional (0D) C_n clusters with spherical or spheroidal shape, atomically precise 2D GNRs are also of major interest (Figure 1.8). Chapter 12 by Yamada and Hayashi summarizes some recent progress in the on-surface synthesis of GNRs and bandgap engineering toward transistor applications.

1.2.3 Zintl Clusters

Zintl cluster research constitutes an important branch in the field of inorganic chemistry. Previous work focused on the main group elements such as Ge, Pb, and Sb. Recent efforts have also investigated transition metals (Figure 1.9), for example, Dehnen's group carried out a series of work on endohedral Zintl clusters such as $[Th@Bi_{12}]^{4-}$, $[Co@Sn_6Sb_6]^{3-}$, and double-metal centered $[Co_2@Sn_5Sb_7]^{3-}$ [34]. Sun's group recently obtained large sized $[Au_8Pb_{33}]^{6-}$ and $[Au_{12}Pb_{44}]^{8-}$ with gold kernels and lead shells [36]. Chapter 13 by Sun and coworkers have summarized some recent

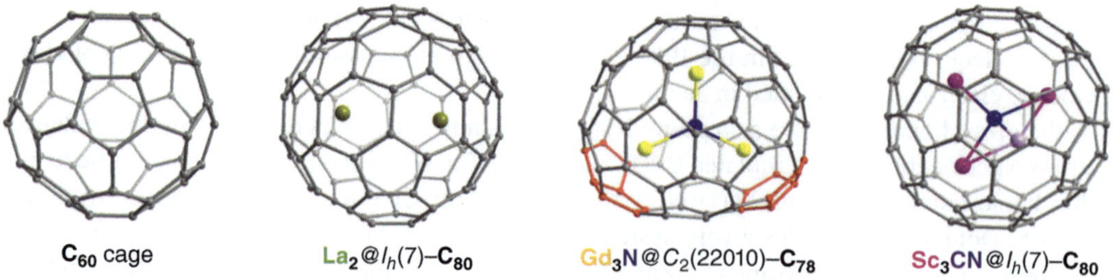

Figure 1.7 Structures of C_{60} and some endohedral fullerenes (the letter colors code the elements). *Source:* Adapted from Chapter 11.

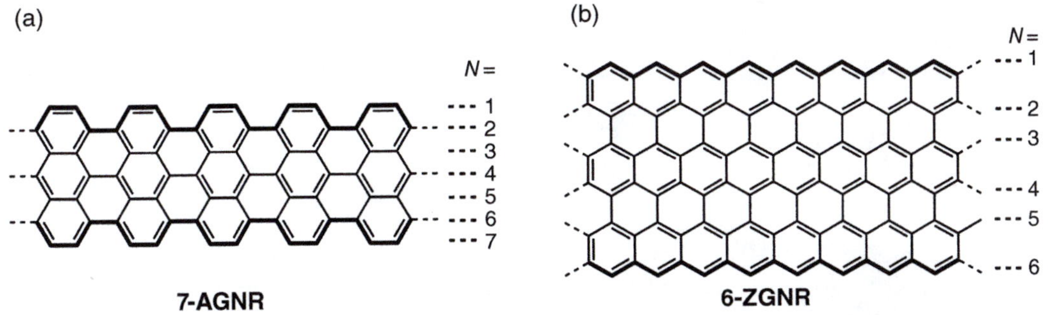

Figure 1.8 Types of graphene nanoribbons (GNRs), (a) Armchair type (e.g. 7-AGNR), (b) Zigzag type (e.g. 6-ZGNR. *Source:* Adapted from Chapter 12).

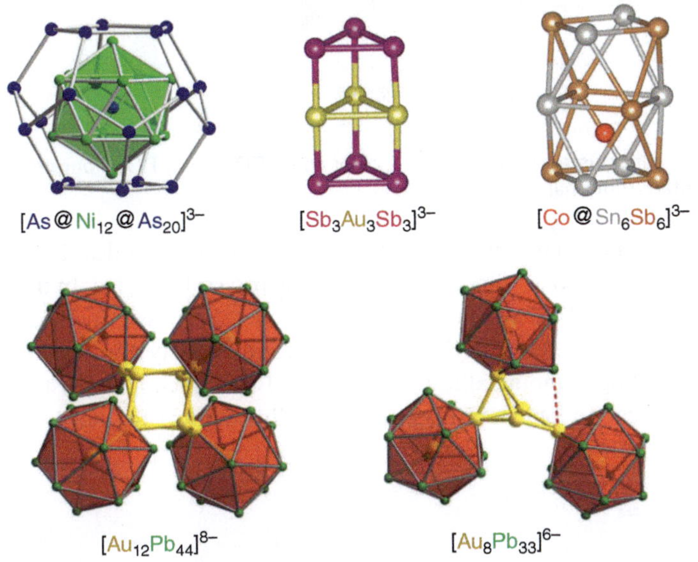

Figure 1.9 Structures of some Zintl clusters (the letter colors code the elements). *Source:* Adapted from Chapter 13.

progress in Zintl clusters of group 15 elements (i.e. pnictogens), including homo- and heterometallic clusters. The anionic clusters can be readily isolated from the solutions of corresponding binary Zintl phase composed of pnictogens and alkali metal salt. The introductions of group 13 or 14 elements or transition metals such as Au and Ag, leads to rich configurations for binary and ternary clusters, which provide new systems for studying the geometric and electronic properties, as well as the bonding pattern. Extensive theoretical work on Zintl clusters has been carried out. The adaptive natural density partitioning method (AdNDP) [84] has been applied to both Zintl clusters and ligand-protected metal NCs. A comparison of their electronic structures would benefit a deeper understanding of both types of NCs.

1.2.4 Metal-Oxo Nanoclusters

Metal-oxide nanoparticles such as TiO_2 and CeO_2 have also received wide interest owing to their importance in catalysis [17]. Research on the cluster form, i.e. metal-oxo few-atom clusters and nanoclusters, provides atomically precise models for understanding the surface structure and

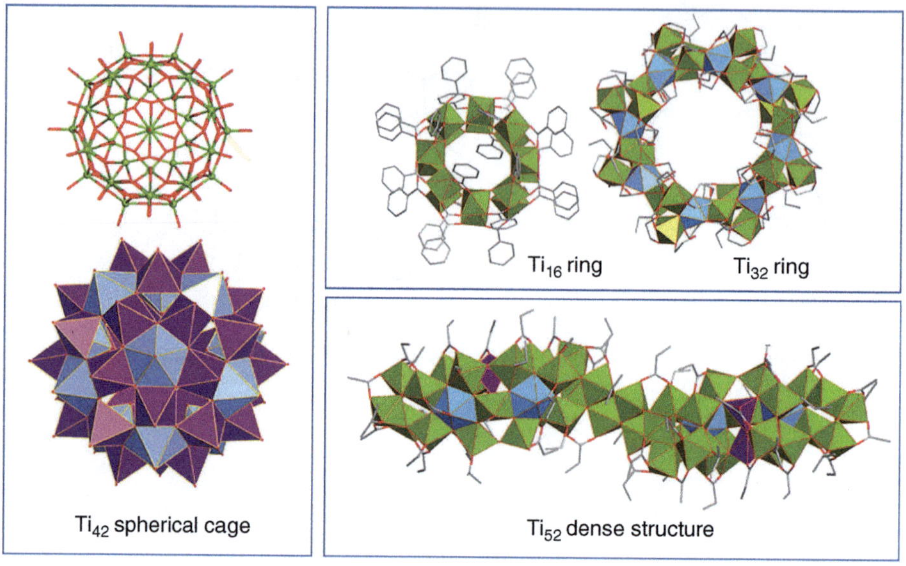

Figure 1.10 Structures of some Ti-oxo nanoclusters (denoted as Ti_n, color code: TiO_4 = red; TiO_5 = purple, TiO_6 = green, TiO_7 = blue). *Source:* Adapted from Chapter 14.

coordination bonding that are involved in regular nanoparticles of the corresponding oxides. There are many types of metal-oxo nanoclusters, such as Ti-oxo, Mo-oxo, and W-oxo systems [11]. In this book, Chapter 14 by Gao et al. illustrates the TOC system and the doped ones. Impressive progress has been made on the study of TOCs in its relatively short history [41, 43], and a variety of structures have been reported (Figure 1.10), including homometallic and bimetallic TOCs. Such nanoclusters are quite promising in solar energy utilization for catalysis as well as nonlinear optics [41, 45, 47].

1.3 Some Fundamental Aspects

Many types of NCs are being pursued in different communities, but there has been little communication between these communities. This book brings together these different research areas under the umbrella of "atomic precision" for potential stimulations. There are indeed some common themes or pursuits among the different areas, such as the size dependent structure evolution, the superatom electronic structure, structural and bonding patterns, and the doping strategy for tailoring the functionality. Among them, the superatom concept has been pursued across different research communities such as the metal clusters [56], semiconductor heterostructures [85], EMFs [60], and Zintl clusters [40] owing to their atom-like orbitals and valency. In this section, we select some general themes for further discussions.

1.3.1 Synthesis and Crystallization

Atomically precise NCs require precise control over the size, structure, and composition. It is worthy of a comment on the gas phase versus solution phase synthesis. In gas phase preparations, clusters are formed by condensation of atomic vapor; thus, no chemical treatment such as etching or size-focusing [10] could be performed in order to achieve uniformity. Thus the gas phase products are typically subject to mass selection by mass spectrometry (MS). In contrast, solution phase synthesis of clusters and nanoclusters use metal salts or complexes as precursors, which are reduced by a reductant such as $NaBH_4$. During the reaction process, conditions such as temperature and the ligand-to-metal ratio can be controlled to obtain the target product. The subsequent purification by solvent extraction or chromatography isolation can give rise to a pure product, then single crystal growth for structure determination by XRC.

Here we illustrate the solution phase synthesis of atomically precise gold NCs, whose characterization involves both MS and XRC. The MS includes matrix-assisted laser desorption ionization (MALDI) and electrospray ionization (ESI) methods. Among the synthetic methods, the size-focusing methodology and the ligand exchange-induced size/structure transformation (LEIST) methodology have been established [86, 87], which have allowed for the synthesis of a series of $Au_n(SR)_m$ NCs. For example, the 2.2 nm $Au_{246}(p\text{-MBT})_{80}$ (where, *p*-MBT = SPh-*p*-Me) nanocluster was synthesized by a two-step "size-focusing" method [13]. In the first step, a polydisperse product in the range of ~10 k to ~70 k Dalton was made by $NaBH_4$ reduction of polymeric Au(I)SR precursors. The second step involved a size-focusing process: in the presence of excess thiol and under heating conditions, the initially polydisperse NCs were gradually converged to the most stable product of 56 kDa mass (Figure 1.11) due to the latter's high stability under the applied chemical conditions.

Unlike regular NPs that are represented by their average size, atomically precise NCs are reported by their precise formulas, resembling the case of molecules. Thus, the same sized NCs, no matter how they were made (e.g. by different methods or laboratories), must possess the same formula. Thus, there is no concern for size heterogeneity. The NCs are also of molecular purity (e.g. >99%) [72], resembling molecular compounds, and they adopt the same structure (except the case of isomerism). Among the characterization methods, MS and XRC play vital roles.

Both MALDI- and ESI-MS methods are widely used in nanocluster analysis. In MALDI analysis, some ligands are often knocked off, hence, the precise mass of NCs is not obtainable by MALDI, but the wide range scan can provide important information on the sample purity. Compared to the

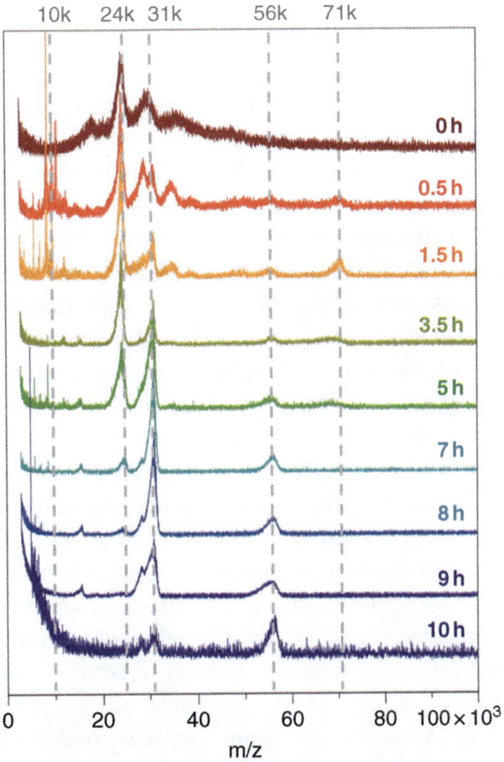

Figure 1.11 Size-focusing synthesis of $Au_{246}(p\text{-MBT})_{80}$ monitored by MALDI-MS. During the size-focusing process, the initially polydisperse product gradually converges to the most stable 56 k Dalton species corresponding to $Au_{246}(p\text{-MBT})_{80}$. *Source:* Reprinted with permission from [13].

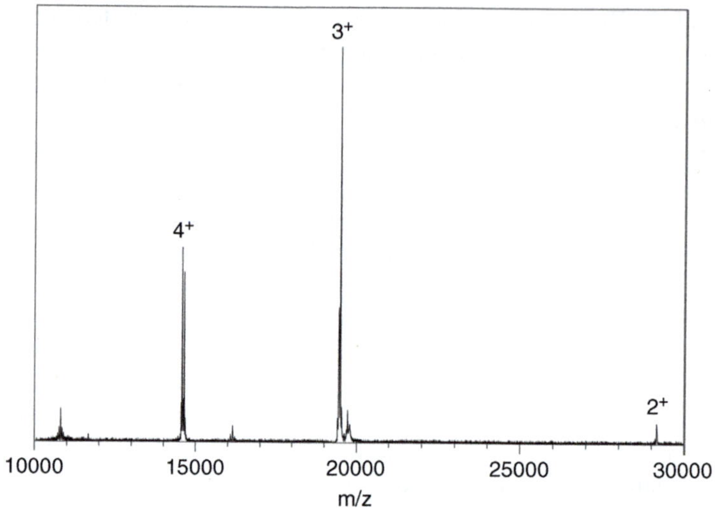

Figure 1.12 ESI-MS analysis of Au_{246}(*p*-MBT)$_{80}$. The peaks correspond to the 4^+, 3^+, and 2^+ ion sets of Au_{246}(*p*-MBT)$_{80}$ (MW = 58 309). *Source:* Reprinted with permission from [13]. © 2016 American Association for the Advancement of Science.

MALDI, ESI is a much softer ionization technique and is thus more suitable for the formula determination. Based on the precise mass determined by ESI, one can easily deduce the formula. For example, the above 56 kDa product was deduced to be Au_{246}(*p*-MBT)$_{80}$ (Figure 1.12). In this analysis, in-source ionization led to the 2+, 3+, and 4+ molecular ion peaks (note: the 1+ peak is beyond the ESI detector's mass range, up to m/z 40 k only, whereas MALDI can go to a much higher range). With the molecular mass (MW = 58 309) determined, one can deduce the number of gold atoms (n) and ligands (m) by applying conditions of (i) $n > m$ (this is owing to the metal—metal bond formation in NCs, whereas the $Au^{I}SR$ complex features $n = m$), and (ii) accurate match between the formula mass and the ESI-determined one. Therefore, high precision ESI is indispensable in order to precisely determine the cluster formula. The as-determined formula should also be confirmed by XRC analysis. When both analyses are consistent, one can be confident about the formula, especially for large sizes (>100 gold atoms).

For structural analysis, the most reliable and popular method is the single-crystal X-ray diffraction analysis (i.e. XRC) [11]. Typically, the single crystal growth is performed by diffusion of an antisolvent (e.g. acetonitrile) into a solution of the NCs (e.g. in toluene). The total structures of NCs reveal not only the core (i.e. the arrangements of metal atoms) but also the surface structure (i.e. the arrangements of ligands and bonding between the ligands and the core); see an example in Figure 1.13. Intense research over the past years has led to the successful syntheses and structure determinations of a series of sizes of Au-SR NCs (see Chapter 3).

In addition to size control, ligand engineering is also important because the ligands play some critical roles in photoluminescence, catalysis, and biomedicine [10, 18]. Xie's group [88] reported the incorporation of two or multiple types of thiolates into the ligand shell of Au_{25}, such as the thiolates containing carboxylic, hydroxyl, or amine groups (Figure 1.14a). To obtaining a ligand shell with positive charges, Ishida et al. synthesized $Au_{25}(SR^+)_{18}$ with cationic ligands such as (11-mercaptoundecyl)-N,N,N-trimethylammonium, HS-$(CH_2)_{11}N(CH_3)_3^+$, and their recent work has also demonstrated a surface reaction route for obtaining such fully cationized NCs (Figure 1.14b) [89a]. Clickable azide-functionalized-Au_{25} was reported by Gunawardene et al. [89b]. Such NCs provide new opportunities for surface reactions. Xie's group carried out impressive work to obtain insights into the nanocluster growth mechanisms in aqueous systems by

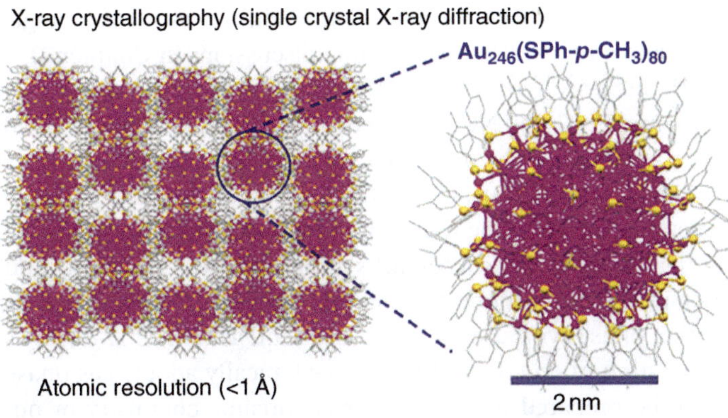

Figure 1.13 Crystallization of atomically precise nanoclusters and total structure determination by X-ray crystallography. *Source:* Redrawn from the cif file [13] of $Au_{246}(SPh\text{-}p\text{-}CH_3)_{80}$.

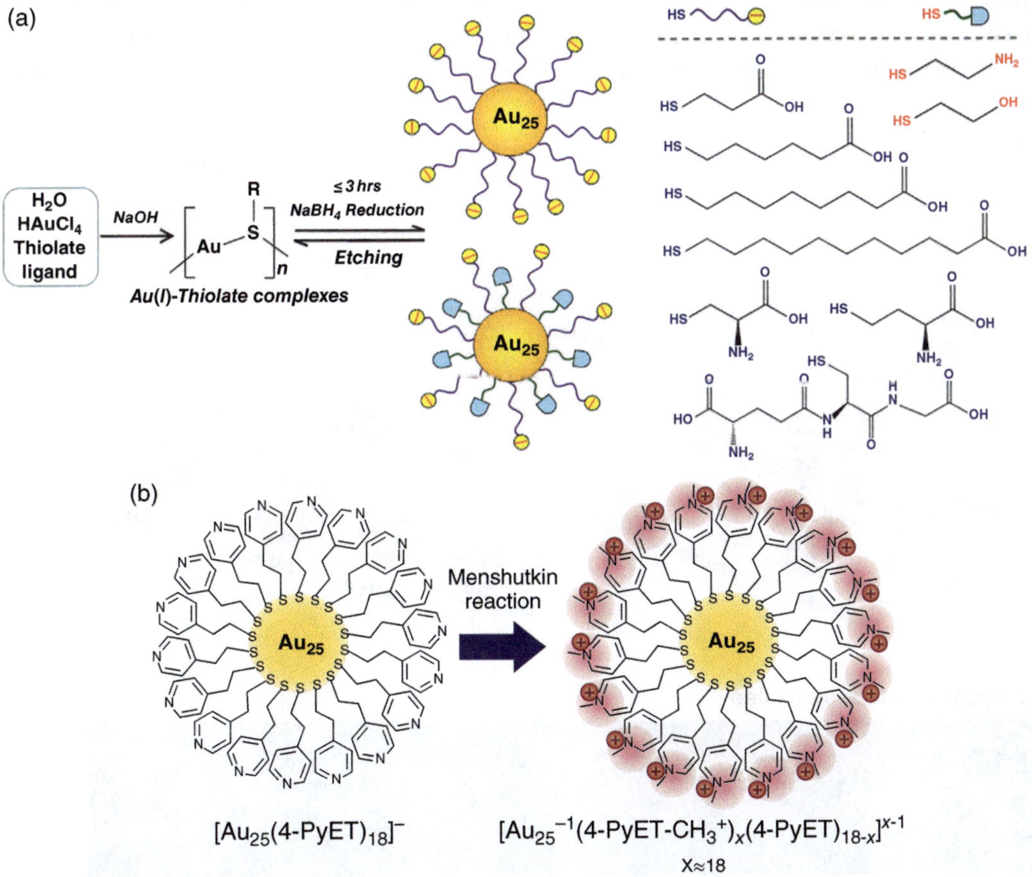

Figure 1.14 Ligand engineering for the $Au_{25}(SR)_{18}$ nanocluster, (a) through NaOH-assisted $NaBH_4$ reduction method, and (b) through surface reaction for transformation into fully cationized NCs. *Source:* Panel A adapted from Chapter 2; panel B, reprinted with permission from [89a]. © 2021 American Chemical Society.

ESI-MS analysis, including the details of $Au_{25}(SR)_{18}$ formation, intercluster metal exchange, surface motif exchange, isomerization, and size conversion (see more discussions in Chapter 2).

While noble metal NCs (Au, Ag, Rh, etc.) are typically synthesized by chemical reduction of their salts or complexes, the cases of Cu and Fe/Co/Ni NCs with zero valence (i.e. the metal—metal bond formation) become challenging and have only met with very limited success. Hayton's group [79] earlier reported $[Cu_{25}H_{22}(PPh_3)_{12}]^+$ (with partial Cu(0)) and recently also tackled the Ni nanoclusters. While the $Ni^{II}(SR)_2$ precursor is hard to be reduced, the use of $[Ni(1,5\text{-cod})_2]$ as the Ni source and PEt_3 as a supporting ligand led to the successful synthesis of $Ni_{30}S_{16}(PEt_3)_{11}$ and structural characterization [90], which provides new opportunities for pursuing atomically precise magnetic NCs. Compared to the solution phase syntheses of metal and Zintl as well as Ti-oxo NCs (Chapters 2–10, 13, and 14), the preparation of endohedral fullerenes typically adopts gas phase methods (Chapter 11), and GNRs can be prepared by solution phase organic chemistry or on-surface synthesis, both at atomic precision (Chapter 12).

1.3.2 Structural and Bonding Patterns

Since much work has been reported on gold NCs, here we focus on the gold-thiolate system. Based on the reported X-ray structures of gold NCs, there are several major types of core structures [10], which are the icosahedral (I_h), decahedral (D_h, including the Marks and Ino types, denoted as m-D_h and i-D_h), face-centered cubic (fcc), tetrahedral Au_4 network, hexagonal close packed (hcp), body-centered cubic (bcc), and mixed phases (Figure 1.15a). It is noteworthy that bulk gold is fcc;

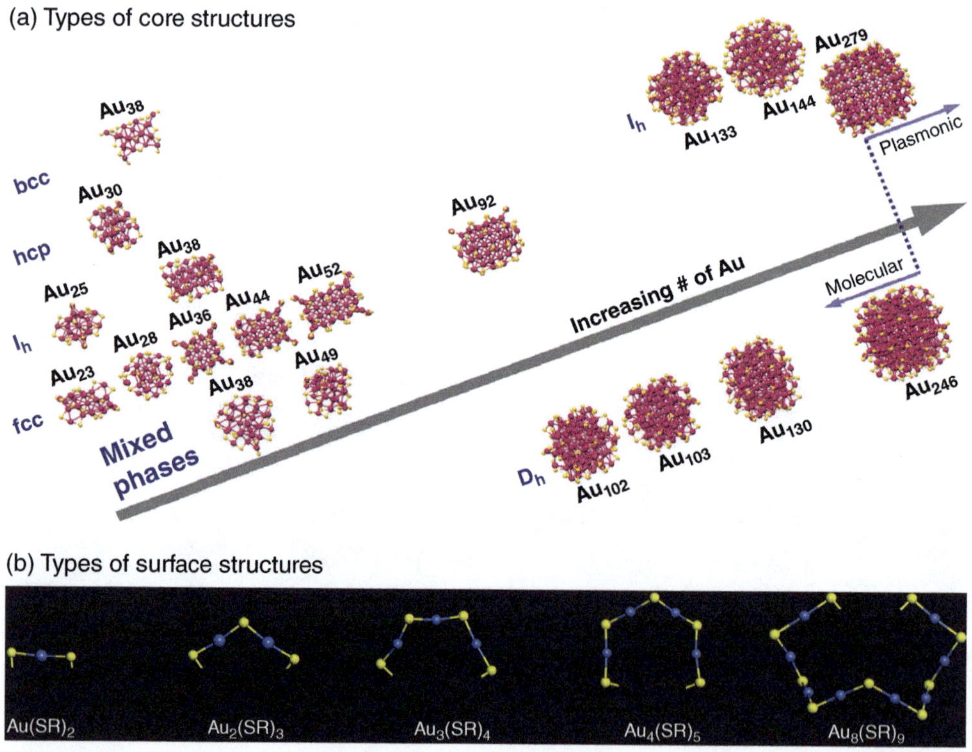

Figure 1.15 Types of core structures (a) and surface-protecting motifs (b) in $Au_n(SR)_m$ nanoclusters. *Source:* Panel A: Reprinted with permission from [91]. © 2020 American Chemical Society. Panel. B: adapted from Chapter 3.

thus, the observations of hcp and bcc phases in gold NCs are quite intriguing. Different from the fcc/hcp/bcc types that are crystallographic (i.e. possessing translational symmetry), other structural types such as I_h and D_h as well as the Au_4 network type are noncrystallographic because the presence of five-fold symmetry or other factors destroy the translational symmetry, but they still possess molecular symmetry, which endows high stability to such noncrystallographic NCs. It is noteworthy that Wu's group reported quite a number of mixed-phase NCs [92], including Au_{46}(*m*-MBT)$_{26}$, Au_{48}(*m*-MBT)$_{26}$, $Au_{49}(DMBT)_{27}$, and $Au_{67}(SCH_2Ph)_{35}$ (more discussions in Chapter 3).

Another important aspect pertains to the surface structures that cannot be mapped out by electron microscopy but can be readily mapped out by XRC analysis. A traditional view on the nanoparticle structure is a polyhedral core (or truncated one) with a protecting layer of organic ligands. But, surprisingly, the X-ray structures of $Au_n(SR)_m$ revealed that some gold atoms are partitioned into the surface-protecting layer, forming unique staple-like motifs (Figure 1.15b) such as the monomer, dimer, trimer, and tetramer, as well as the plain bridging thiolate (–S(R)–, without any gold atom incorporated). The *x*-mer staples incorporate an *x* number of gold atoms and can be expressed as $Au_x(SR)_{x+1}$, with $x = 1–4$ [10]. An octameric staple motif, $Au_8(SR)_9$, has also been found in the $Au_{21}(S\text{-}Adm)_{15}$ structure [93]. In small sizes of $Au_n(SR)_m$, ring-like motifs are also present, including the octameric $Au_8(SR)_8$ ring (as opposed to the octameric *staple*) in Au_{20}(S-Ph-*p*-^{t}Bu)$_{16}$ [94a] and hexameric $Au_6(SR)_6$ ring in $Au_{28}(S\text{-}Adm)_{22}$ [94b].

The growth pattern of NCs reveals the evolution of structure with increasing size. While early research only reported some scattered sizes without mutual relationships, later work indeed revealed intriguing growth patterns, including 1D, 2D, and box-type addition growth. In these patterns, the fundamental building blocks are the Au_4 tetrahedron, fcc-Ag_{14} cube, and M_{13} icosahedron (or cuboctahedron), as well as M_5 decahedron, where M = Au or Ag. Below we illustrate a few series of NCs.

One-dimensional periodic growth of NCs was first discovered in the series that is composed of $Au_{28}(TBBT)_{20}$, $Au_{36}(TBBT)_{24}$, $Au_{44}(TBBT)_{28}$, and $Au_{52}(TBBT)_{32}$, where TBBT = 4-*tert*-butylbenzenethiolate (SPh-^{t}Bu) [95]. In this series (Figure 1.16), a uniform increment of $Au_8(SR)_4$ was found, and X-ray structures revealed a double helical extension of Au_4 units. Theoretically, groups of Pei, Gao, and Zeng provided insights on this series and also predicted some longer 1D NCs such as $Au_{60}(SR)_{36}$, $Au_{68}(SR)_{40}$, and $Au_{76}(SR)_{44}$, [96, 97] but no experimental success yet. The HOMO-LUMO gap shrinks with increasing size (i.e. length) but does not reduce to zero, rather, it converges to 0.7 eV [96]. On a note, Wu and coworkers [98] also found $Au_{56}(TBBT)_{34}$ with a helical growth in relation with $Au_{52}(TBBT)_{32}$.

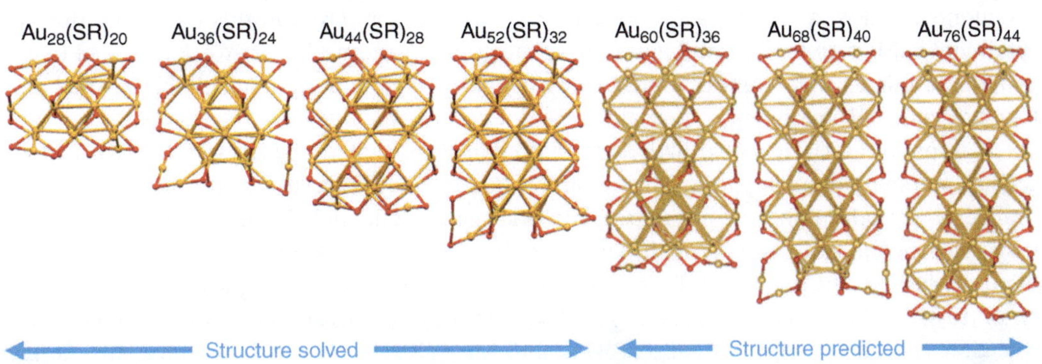

Figure 1.16 One dimensional periodic growth of $Au_{8n+4}(SR)_{4n+8}$ nanoclusters (n = 3, 4, 5, 6, etc.). Color labels: golden = Au, red = S, carbon tails omitted for clarity. *Source:* Reprinted with permission from [96]. © 2016 Royal Society of Chemistry.

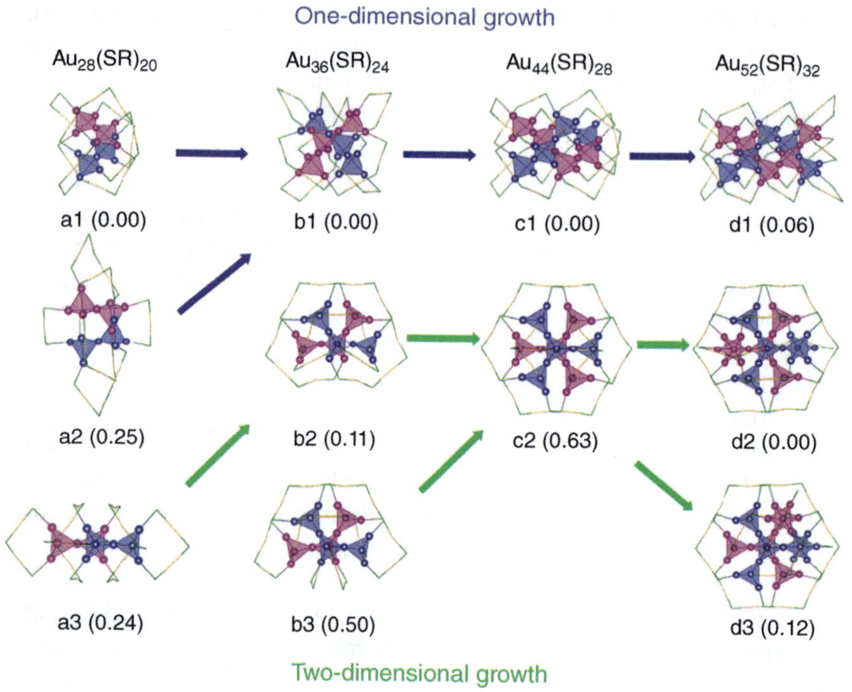

Figure 1.17 Relation between 1D growth (blue arrows) and 2D growth (green arrows) of $Au_{8n+4}(SR)_{4n+8}$ nanoclusters. Each vertical column shows the isomers for the corresponding size (e.g. a1, a2, and a3 for the $Au_{28}(SR)_{20}$ formula; the relative energy (eV) for each isomer is shown in parentheses). *Source:* Reprinted with permission from [99]. © 2020 The Authors (Liu et al.)

Based on the Au_{28} to Au_{52} 1D growth pattern, Gao et al. further predicted a 2D growth pattern (Figure 1.17, green arrows) and it was experimentally verified [99]. The pattern shows that starting from the a3 isomer of $Au_{28}(SR)_{20}$, b2 of $Au_{36}(SR)_{24}$ can be obtained via a 2D growth. Experiment by Zhu's group indeed obtained the b2 isomer using 2,4-dimethylbenzenethiol (DMBT) as the ligand via a two-step size-focusing method. Both b1 and b2 isomers were produced in one pot, but chromatography isolated them, and the structure of b2 was determined to be the same as predicted [99]. Based on the superatom network model, AdNDP analysis on the $Au_{20}{}^{8+}$ kernels of b1 and b2 isomers showed that the 12 valence electrons are equally distributed on the six Au_4 tetrahedra (2e superatoms), that is, a type of 4c-2e bond in each tetrahedron. Overall, this combined theory/experiment work is quite insightful in revealing the growth patterns. Future work may lead to the synthesis of other isomers under each formula (Figure 1.17) by designing suitable ligands and other experimental conditions.

In addition to the 1D helical growth and the 2D pattern, Chapter 3 by Gu and Wu provided an excellent summary of other patterns based on the Au_4 and Au_3 units such as the quasi-linear Au_4 tetrahedron growth by sharing one vertex, Au_3-additive growth, and Kekulé-like ring growth, as well as icosahedral Au_{13} based growth. They also discussed the series of NCs with "homo-kernel hetero staples," i.e. the same kernel can be protected by different surface staple motifs, such as the pair of $Au_{28}(TBBT)_{20}$ and $Au_{28}(S\text{-}c\text{-}C_6H_{11})_{20}$ NCs with the same Au_{14} kernel but different motifs for protection, as well as the pair of $Au_{44}(TBBT)_{28}$ and $Au_{42}(TBBT)_{26}$ NCs with the same Au_{26} kernel but different *numbers* of dimeric and monomeric motifs (though identical types) [100]. These results motivated Xu et al. [101] to develop a new method for theoretical structural design of $Au_n(SR)_m$ NCs by redistributing the staple motifs on the known Au kernel structures, hence, a

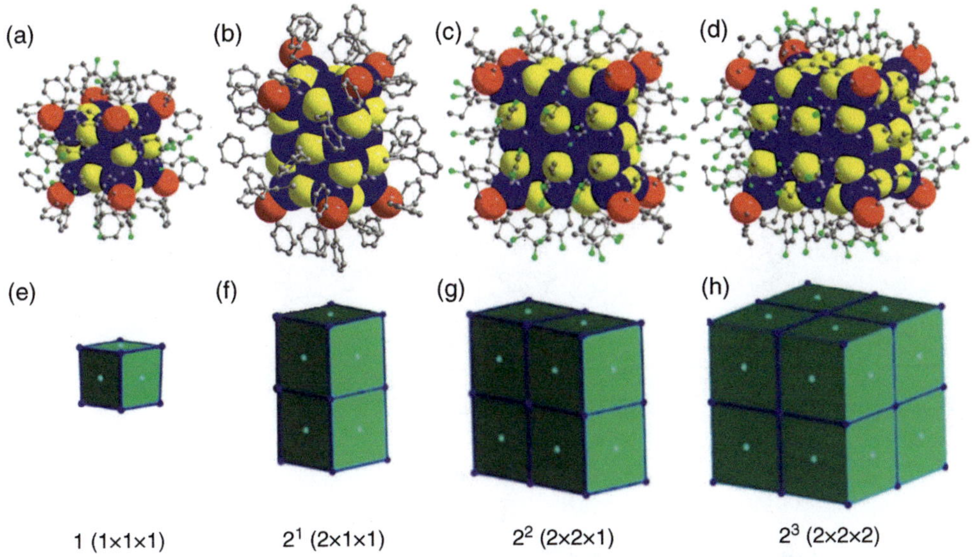

Figure 1.18 "Box"-addition growth pattern in a fcc series of Ag_n NCs: (a) $[Ag_{14}(SPhF_2)_{12}(PPh_3)_8]$ (denoted 1), (b) $[Ag_{23}(SC_2H_4Ph)_{18}(PPh_3)_8]$ (2^1), (c) $[Ag_{38}(SPhF_2)_{26}(P\text{-}^nBu_3)_8]$ (2^2), (d) $[Ag_{63}(SPhF_2)_{36}(P\text{-}^nBu_3)_8]^+$ (2^3), and their corresponding models (e–h) depicted as the idealized fcc box addition sequence. Color codes: blue and light blue sphere = Ag; red sphere = P; yellow sphere = S; green = F; gray = C. *Source:* Redrawn from [103] and [104]. © 2017 American Chemical Society and 2018 The Authors (Liu et al.).

prediction of new stable $Au_n(SR)_m$ nanoclusters without the need of the demanding global-minimum search (see Chapter 4). Chapter 4 (Xu et al.) summarizes an additional series of periodic NC growth. It is also worth noting that Pei's group reported a "gold-atom insertion, thiolate-group elimination" rule for the evolution from $Au_{28}(SR)_{20}$ to $Au_{29}(SR)_{19}$ to $Au_{30}(SR)_{18}$, [102a] and a 2D growth pattern for the evolution from $Au_{44}(SR)_{28}$ to $Au_{68}(SR)_{36}$ to $Au_{92}(SR)_{44}$ [102b]. These are experimentally supported [71c, d].

In the Ag_n research, groups of Zheng [103] and Li [104] identified a box-addition type of growth pattern in the series composed of Ag_{14}, Ag_{23}, Ag_{38}, and Ag_{63} (Figure 1.18). These NCs feature the fcc Ag_{14} unit and are protected by binary ligands (thiolates on the facets and eight phosphines on the corners). Zheng's group also reported a series of $[Au_{12+n}Cu_{32}(SR)_{30+n}]^{4-}$ ($n = 0, 2, 4, 6$) with surface motif exchange while retaining the kernel structure ($Au_{12}@Cu_{20}$) [105].

To understand the structure evolution and unify the reported $Au_n(SR)_m$ structures, groups of Xu, Gao, and Zeng have made great efforts. A grand unified model (GUM) was established in 2016 [106], which offers insights into structure stability and patterns, as well as predictions of new NCs. This model was inspired by the quark model in particle physics, in which a proton or neutron is composed of three quarks, as well as the tetraquark containing two quarks and two antiquarks, whereas in gold NCs, the elementary blocks are the triangular Au_3^+ (analogous to the proton/neutron) and the Au_4^{2+} (analogous to the tetraquark), each block holding two valence electrons. The gold atoms in each elementary block may be of three "flavors," Au(1e), Au(0.5e) and Au(0e), in other words, three different valence-electron states. The icosahedral Au_{13} can be viewed as a secondary building block constructed from the Au_3 to Au_4 elementary ones. Here, it is worth noting that the valence electrons refer to those that are not bonded (or localized) by surface ligands, rather, being in metal—metal bonds. Given the delocalization nature of metal—metal bonds, some authors may use the term of free valence electrons, but the term "free" should not be mistaken as the free electrons in the metallic or plasmonic state, because in NCs of smaller than $Au_{279}(SR)_{84}$, the strong quantum

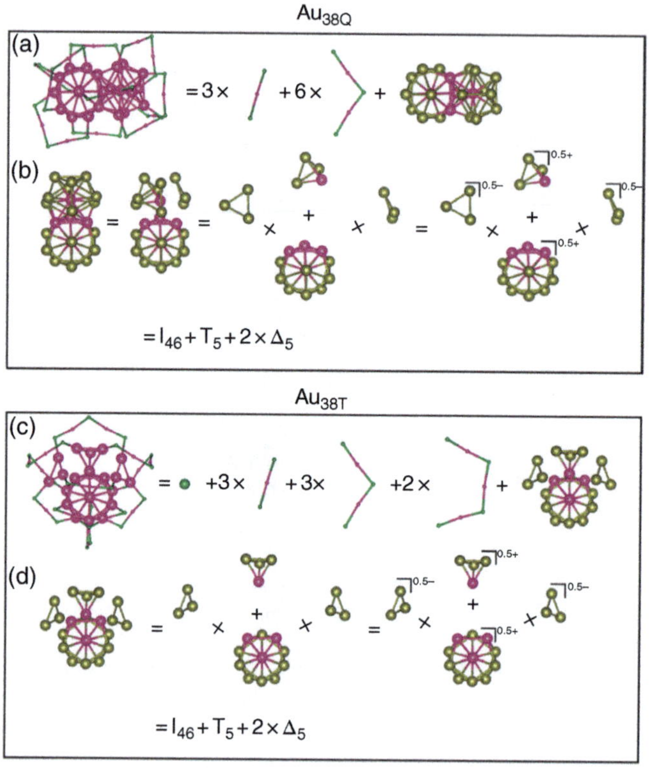

Figure 1.19 The GUM-guided structural dissection of Au_{38Q} (a and b) and its isomer Au_{38T} (c and d). Color code: magenta and dark yellow represent Au atoms with 1e and 0.5e valence states, respectively; dark green represents S atom. R groups are omitted for clarity. Reprinted with permission from [110]. © 2018 American Chemical Society.

confinement effect leads to discrete orbitals on which the delocalized valence electrons occupy; they are not free electrons; [107–109] otherwise, gapless excitations (i.e. plasmons, as opposed to excitons) would occur in sizes below $Au_{279}(SR)_{84}$ [107]. With the GUM, a plethora of complex and seemingly unrelated structures of gold NCs have been deciphered, and the structure evolution was understood [106]. Using $Au_{38}(SR)_{24}$ as an example, its structure can be dissected as shown in Figure 1.19, and the core isomerism can also be rationalized [110]. The stabilities of NCs originate in the local stability of each building block, and the growth of NCs can be viewed as a sequential addition of the building blocks, as opposed to the traditional view of addition of individual atoms.

The stability of $Au_n(SR)_m$ NCs has also been studied in terms of thermodynamics. Based on the experimental structures, Mpourmpakis and coworkers reported a thermodynamic stability model for the $Au_n(SR)_m$ NCs, [111] in which they identified that when the binding energy (BE) of shell-to-core equals the cohesive energy (CE) of the core, stable NCs result (see the "parity line" in Figure 1.20). This energy balance is indeed maintained over a large size range of Au NCs from the molecular (e.g. Au_{18}, with large $E_g \sim 1.7$ eV) to the metallic regime (Au_{279}, E_g of meV supposedly), even extendable to the bimetallic NCs [111].

The particular stability of certain sizes of NCs is interesting, which typically originates from two major factors, the geometric and electronic stabilities [51, 52]. The discussions in this section have illustrated the types of atomic structures of NCs; high-symmetry structures typically lead to high stability. Here we further discuss the electronic factor, which contributes to the stability and affects the geometric structure (e.g. by the Jahn-Teller effect in superatomic 1*P* state of bimetallic M_{25} and

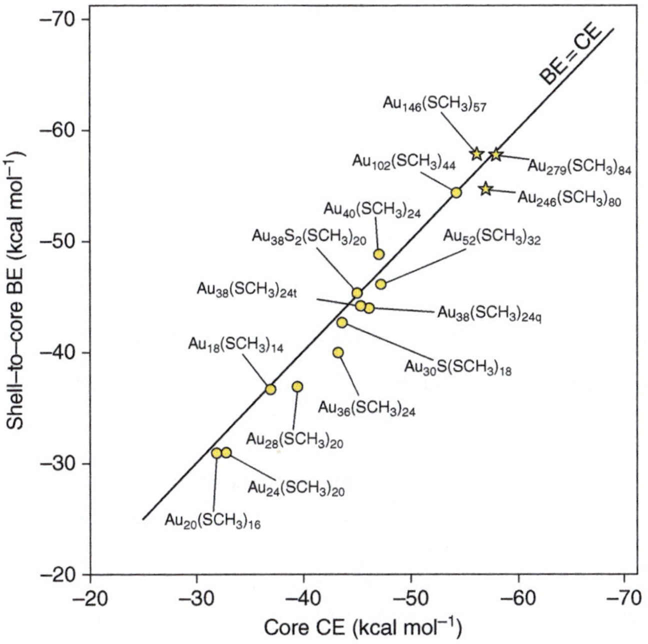

Figure 1.20 The origin of thermodynamic stability of $Au_n(SR)_m$ nanoclusters (BE: binding energy of shell to core; CE: cohesive energy of the core). *Source:* Reprinted with permission from [111b]. Copyright 2019 Royal Society of Chemistry.

homogold Au_{25} NCs, [112] in contrast with the Jahn-Teller effect of atomic *d* state in transition metal complexes). The electronic structure of NCs is also closely relevant to the bonding pattern.

It may be worthwhile to start with the electronic structure of elements in the periodic table. The inert gases (e.g. He, Ne, Ar, Kr, Xe) are chemically inert due to electron shell closing, and the stability numbers are 2e, 10e, 18e, 36e, and 54e. Shortly after the elucidation of the electronic structure for elements in the 1910s–1920s, Sidgwick [113] developed an electron counting scheme for metal complexes – for example, $Ni(CO)_4$ possesses a total of 36e, that is, 28e (Ni) + 2e × 4 (CO) = 36e, hence, it is stable because of the same electron number as that of Kr. The octet rule and the 18-electron rule are well known in inorganic chemistry for structure and reactivity predictions. Both rules describe the tendency of the central metal atom (or ion) to achieve the noble gas electron configuration in the valence shell, i.e. $ns^2\ np^6$ or $ns^2\ np^6\ (n-1)d^{10}$ (here, n is the valence shell number). Complexes with unfilled $(n-1)d$ orbitals can bind more ligands and exhibit higher reactivity.

The concept of chemical bonding is of central importance in chemistry and the cluster research has significantly expanded the concept. Here we briefly outline the evolution from conventional bonds to new types in boron and metal clusters. In the 1910s, Lewis discussed the covalent bond as a pair of electrons shared by two atoms ("2c-2e"), where, "c" = center [29]. But this concept was later challenged by boron clusters due to boron's electron-deficient nature. New concepts such as 3c-2e, multi-center bonds and 3D aromaticity were put forth then. Based on the structures of boron clusters, Williams systematically described the deltahedral structural pattern in 1971 and classified the boron clusters as three categories (i.e. closo-, nido- or arachno-) according to their formulas [114]. Wade [115] explored the concept of skeletal electron pair (SEP) for cluster bonding and rationalizing the shape of borane clusters. For example, the icosahedral $B_{12}H_{12}^{2-}$ cluster (50e total) with high stability possesses 13 SEPs (or 26e by the $2n+2$ rule, that is, 2e × 12 [B atoms] + 2 = 26e), which are highly delocalized on the icosahedral shell. Such σ-aromatic bonding is regarded as spherical aromaticity [116], similar to the 2D case (e.g. π aromatic bonding in benzene). Mingos [117] also introduced the SEP concept into gold-phosphine clusters. The relationship between the structure and electron count in gold clusters was understood in terms of the overlap of the radially directed s/d_{z^2} hybrids on each gold atom [118]. Several rules were summarized (called Wade's or Wade-Mingos rules), including:

i) The *closo*-type of clusters with n vertices has $\mathbf{2n+2}$ skeletal electrons (or $n+1$ SEPs).
ii) The *nido*-type, which is derived from the closo parent by removing one vertex, should have $2n+4$ electrons.
iii) The *arachno*-type should possess $2n+6$ electrons since this type has 2 missing vertices.

The Wade-Mingos rules – which rationalized the valence electron counting and structure of polyhedral boranes and metal-phosphine clusters – have been introduced into broader types of metal clusters and Zintl chemistry as well. For example, according to the $\mathbf{14n+2}$ rule, an icosahedron gives rise to 170e (i.e. $14\times12+2=170$e, where n the number of vertices for an icosahedron, excluding the central atom). For metal carbonyl clusters such as the $[Rh_{12}E(CO)_{27}]^{q-}$ (where E = Sn with $q=4$, and Sb/Bi with $q=3$), one obtains 170e by counting Rh 9e, 4e for Sn (or 5e for Sb/Bi), 2e for each CO, and the charge q [78]. In Zintl clusters, cage clusters of group 14 also fit well into the rule of $\mathbf{4n+2}$ (*Note:* Main group elements have no d electrons; hence, $10n$ less in the electron count), [34, 37] such as closo-Ge_{10}^{2-} and Pb_{12}^{2-}. The Pb_{12}^{2-} cage possesses $4n+2=50$e, considering that Pb is not a transition metal and has no d electrons for metal—metal bonding, hence, 12×10e = 120e less than the 170e count for shell closing in transition metal clusters [78].

With the concepts of aromaticity and multicenter delocalized bonding, the closo-type Ge_{10}^{2-} and Pb_{12}^{2-} can be considered as 3D aromatic. In endohedral lead-based Zintl clusters, the $4n+2$ rule also explains the particular stability of the 60e Zintl clusters of $[M@Pb_{12}]^{q-}$ (M = Ni, Pd, Pt, $q=2$; M = Co, Rh, Ir, $q=3$) and the 58e $[Mn@Pb_{12}]^{3-}$, in which both the M (d^{10}) and the closo-$[Pb_{12}]^{2-}$ cage acquires closed-shell configurations. The 62e $[Au@Pb_{12}]^{3-}$ is also in line with the $4n+2$ rule. In the analysis of bonding pattern of Zintl clusters, for example, Sn_5^{6-} can be understood as classical Lewis 2c-2e bonds and three delocalized 5c-2e bonds (Figure 1.21), which is indeed similar to the case of organic $C_5H_5^-$. The prismatic tri-layered $[Sb_3Au_3Sb_3]^{3-}$ can be considered as π aromatic,

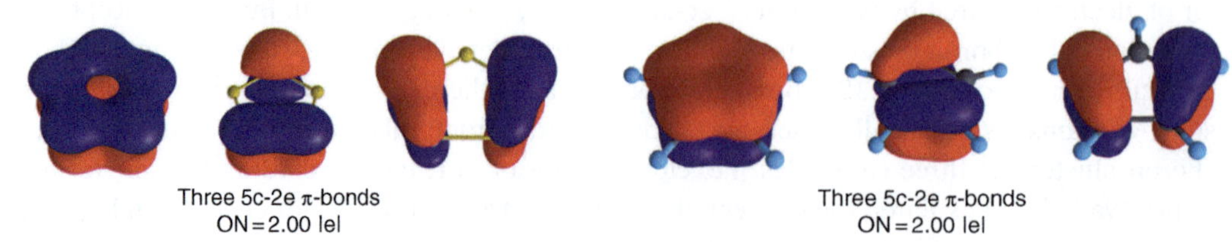

Figure 1.21 Bonding pattern of (a) Sn_5^{6-} and (b) $C_5H_5^-$ involving multicenter-2e bonds (computed by the adaptive natural density partitioning (AdNDP) algorithm). *Source:* Adapted from Chapter 13.

and the top or bottom Sb_3 triangle achieves a total of 15 valence electrons and involves 3c-3e π-bonds (three of such bonds) within the layer. More discussions can be found in Chapter 13.

In the 1980s, research on gas phase Na_n clusters led to another important concept now called *superatoms,* that is, the particularly abundant Na_8, Na_{34}, Na_{58}, and Na_{92} sizes, which adopt electron shell structures of closed forms, such as $1S^2\ 1P^6$ (8e total) for Na_8 [52]. This picture was later confirmed in extensive research on other types of metal clusters [50]. The electron configuration of superatomic clusters leads to the metal core valence electrons filling as in the following sequence [119]:

$$1S^2\ 1P^6\ 1D^{10}\ 2S^2\ 1F^{14}\ 2P^6\ 1G^{18}\ 2D^{10}\ 1H^{22}\ 3S^2\ 2F^{14}\ 1I^{26}\ 3P^6\ 1J^{30}\ 2G^{18}\ 3D^{10}\ 4S^2\ 1K^{34}\ 2H^{22}\ \ldots$$

Note that the sequence was constructed based on an electron in a spherical potential without considering any electronic correlation or other effects; thus, the sequence of higher levels may differ from other theoretical treatments [50, 56], such as the jellium model or with the exchange-correlation incorporated. Different treatments lead to different degeneracy of high-energy orbitals such as *I*, *J* and *K*, hence, various magic numbers of electron shell closing, but the lower numbers such as 2e, 8e, 18e, . . . are quite robust regardless of different treatments. With the orbital sequence, electron configurations for 34e, 40e, 58e, etc. can be constructed readily. Castleman and coworkers carried out much work on the reactivity of gas phase Al_n superatoms, such as the highly reactive 39e (open-shell) Al_{13} and the insert Al_{13}^- of 40e (shell closing) [120].

A modified superatom model has been introduced into the ligand-protected metal NCs [121]. In consideration of the fact that some ligands (e.g. thiolate) form 2c-2e bonds with the metal core atom(s), some valence electrons of metal atoms are localized and these do not participate in the delocalized metal—metal bonding. The number of metal valence electrons being localized by each type of ligands is listed in Table 1.1, in which ligands such as phosphines do not consume metal electrons due to dative bonding with metal atoms. With the superatom model, the electron configuration of some NCs can be rationalized; for instance, $Au_{25}(SR)_{18}^-$ [121, 122] possesses 25(Au $6s^1$) – 18(SR) + 1 (charge) = 8e, hence, a $1S^2\ 1P^6$ configuration and accordingly high stability due

Table 1.1 Localization of metal core valence electron(s) by surface ligands in the superatom model. Note that the electron counts for ligands differ in various models.

Ligand	Number of electrons localized by ligand
Phosphine (PPh_3, PR_3)	0
Carbonyl (CO)	0
Thiolate or thiolato (-SR)	1
Dithiolate, e.g. 1,3-benzenedithiolate (-S-CR_2-S-)	2
Dithiocarbamate (S_2C-NR_2)	1
Dithiophosphate ($S_2P(OR)_2$)	1
Halide (−Cl, −Br)	1
Alkynl (CCPh, CCR)	1
Amine ($-NH_2$, −NHR)	0
Carboxylic group (COO−)	1
NHC (nitrogen-heterocyclic carbene)	0
Hydride (H^-)	1
S (bare), Se (bare)	2

to shell closing. Similarly, $Au_{102}(SR)_{44}$ [123] possesses 58e with a configuration of $1S^2\ 1P^6\ 1D^{10}\ 2S^2\ 1F^{14}\ 2P^6\ 1G^{18}$, hence, high stability by shell closing. Overall, the stability of $Au_{25}(SR)_{18}^-$ and $Au_{102}(SR)_{44}$ was well explained by electron shell closing (plus their high symmetry geometric structures of spherical shape [122, 123]), but some other $Au_n(SR)_m$ NCs could not be explained by the superatom picture, for example, $Au_{38}(SR)_{24}$ with 14e [124].

The superatom model was also used in Zintl clusters [37d]; for example, a celebrated Zintl cluster is the fullerene-like $[As@Ni_{12}@As_{20}]^{3-}$ cluster; its geometric structure consists of an As-centered Ni_{12} icosahedron which is enclosed in an exterior As_{20} pentagonal dodecahedral shell [35]. Theoretically, a charge distribution analysis revealed a pattern of $[\{As@Ni_{12}\}^{3-}@\{As_{20}\}^0]$, with the kernel possessing superatomic $1S^21P^62S^2$ electron configuration [40].

It is also of interest to discuss the trio NCs of 8e, among which the $Au_{25}(SR)_{18}^-$ and $Au_{28}(SR)_{20}$ both possess 8e and can be viewed as $1S^21P^6$ (Figure 1.22), but $Au_{23}(SR)_{16}^-$ (also 8e) is not the case, and its electron configuration should be viewed as $1S^21P^41D^2$ (Figure 1.22) [125]. In addition, the neutral $Au_{25}(SR)_{18}^0$ should be $1S^21P^5$, hence, being open shell and thus magnetic [126]. Unlike common radicals being short lived, $Au_{25}(SR)_{18}^0$ exhibits high stability and longevity and is similar to the anionic form, which came as a surprise [126]. The facile conversion of $Au_{25}(SR)_{18}^-$ to $Au_{25}(SR)_{18}^0$ [127] implies that the $1P^6$ electrons are not as tightly bound as expected, which is different from the case of atoms (e.g. Ne atoms being highly insert due to $2s^22p^6$ electron shell closing and tight binding). In addition, the case of spherical $Au_{133}(SR)_{52}$ is 81e, rather than the expected 70e or 92e for shell closing; the $1H$ open shell implies magnetism, which has indeed been confirmed by experiment [128]. Similar to $Au_{25}(SR)_{18}^0$ (7e), the $Au_{133}(SR)_{52}^0$ "radical" is also very stable [128].

To go beyond the superatom model, Yang and Chen [129] put forth a "superatom network" model in which Au_4 (2e) or Au_{13} (8e) superatoms serve as building blocks for constructing superatom-based supermolecules such as $Au_{24}(SR)_{20}$ and $Au_{38}(SR)_{24}$. The fcc series (Au_{28}, Au_{36}, Au_{44}, Au_{52}) cannot be explained by the superatom model, but the superatom-network model (with Au_4^{2+} units) indeed explains this series well [96, 97, 130]. The supermolecule picture also explains the 2D pattern of Au_4 units [99, 125, 130]. The discovered new bonding patterns are quite intriguing.

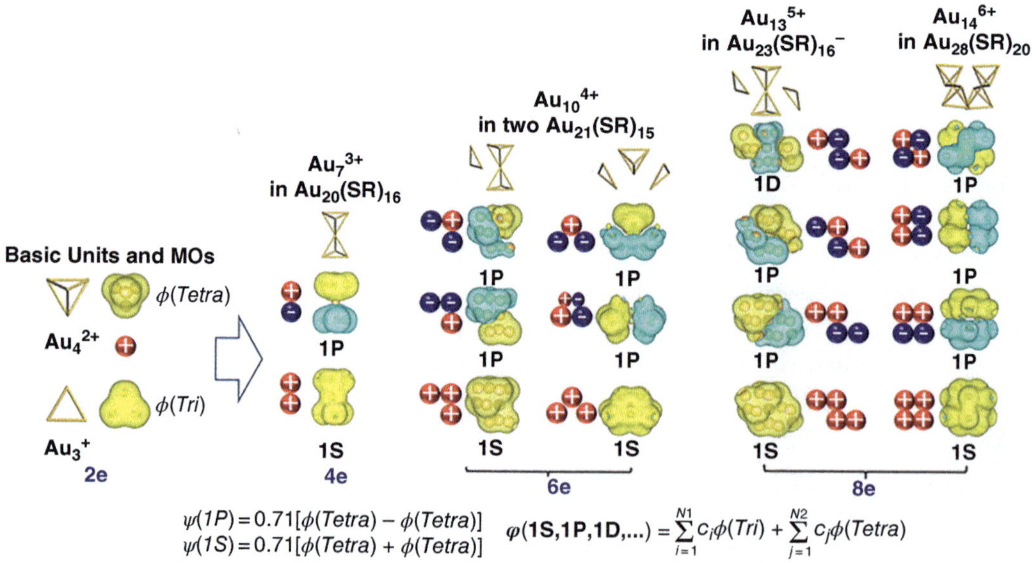

Figure 1.22 Electron density diagrams of the valence MOs of the Au cores in 2e, 4e, 6e, and 8e clusters. *Source:* Reprinted with permission from [125]. © 2019 American Chemical Society.

New types of bonding have also been found in the endohedrally doped fullerenes. For example, $Dy_2@C_{80}$ shows the presence of a single-electron Dy—Dy bond (2c-1e), that is, Dy-e-Dy. In dimetallofullerenes, the nature of metal—metal bonding inside a cage could be fundamentally different from the case of isolated M—M, because the enclosed case depends on the cage size and structure. More discussions can be found in Chapter 11. It is worth noting that the endohedral Zintl clusters containing a metal dimer, such as $[Pd_2@Ge_{18}]^{4-}$, [37b] is of interest for a comparison with the dimetallofullerenes for insights into the bonding and electronic properties. Such crossing-community efforts should be fostered in future work, which is indeed one of our motivations for organizing this book by bringing together various communities under the common umbrella of "atomic precision."

Overall, the cluster research on boron, Zintl, and metal NCs has significantly extended the chemical bonding concept. Beyond the classical 2c-2e bonding, many new types (e.g. 2c-1e, *n*c-2e, and *n*c-*m*e) have been put forth. Spherical aromaticity has successfully explained the stability of cage structures. The discovered 1D and 2D bonding patterns in $Au_n(SR)_m$ NCs are quite intriguing. The development of electron-counting rules [118, 131] may accommodate broader types of clusters. In future work, new structures are expected to further push the concept of bonding in areas of atomically precise metal NCs (homo- and hetero-metallic ones), endohedrally doped clusterfullerenes, and Zintl clusters (homo- and hetero-metallic types). The unusual bonding behavior will also lead to new functionalities of such materials.

1.3.3 Transition from Nonmetallic to Metallic State: Emergence of Plasmon

The advent of atomically precise nanochemistry opens exciting opportunities for addressing some longstanding questions, such as the nonmetal-to-metal transition over the size evolution. The transition in bulk states of elements in the periodic table is relatively straightforward, that is, bandgap $E_g = 0$ for metals and nonzero for nonmetals. However, nanoclusters are finite systems and the E_g value exhibits size dependency. Then, questions arise: At what size does the metallic state emerge? Would the nonmetal-to-metal transition be smooth or abrupt? Given the difficulties in measuring millielectronvolt (meV) gaps [132], a more convenient method is to probe the size at which plasma emerges in a metal nanocluster. When the plasma state is formed in a particle, a collective excitation of free electrons will give rise to SPRs, which can be spectroscopically probed. For spherical or nearly spherical Au nanoparticles, a single SPR peak is exhibited. Next, we briefly discuss some progress in addressing the transition from nonmetallic to metallic state in gold NCs.

Small NCs exhibit discrete electronic energy levels, such as the case of Au_{25} (Figure 1.23), which is caused by ultrasmall size or quantization effect. With increasing size, the energy level spacing becomes smaller and eventually the HOMO-LUMO gap should disappear; accordingly, the metallic state is formed. Recent success in attaining Au_n NCs over the size regime of hundreds of atoms has permitted us to address this fundamental question.

Here we illustrate two critical sizes, $Au_{246}(SR)_{80}$ and $Au_{279}(SR)_{84}$, both being ~2.2 nm core diameter. The electronic properties are closely related to the particle shape, thus, their structures were first analyzed by XRC. It was revealed that Au_{246} is nearly spherical, [13] with metal core diameter of 2.2 nm and overall diameter (including the ligand shell) of 3.3 nm. The $Au_{246}(SR)_{80}$ structure can be divided into four shells. The kernel is a three-shell Au_{116} Ino decahedron (*i*-D_h) that exposes {111} facets at the top/bottom poles and {100} facets at the waist (Figure 1.24a). The fourth shell is composed of 90 gold atoms (Figure 1.24b, magenta), which rounds up the particle into a sphere. The surface layer (Au—S interface) contains various types of motifs, including the plain bridging —S—, gold-containing monomeric —S—Au—S— staples, and dimeric —S—Au—S—Au—S— staples (Figure 1.24c), all of which adapt to the facets and grooves on the Au_{206} core. Hence, the Au_{206} core

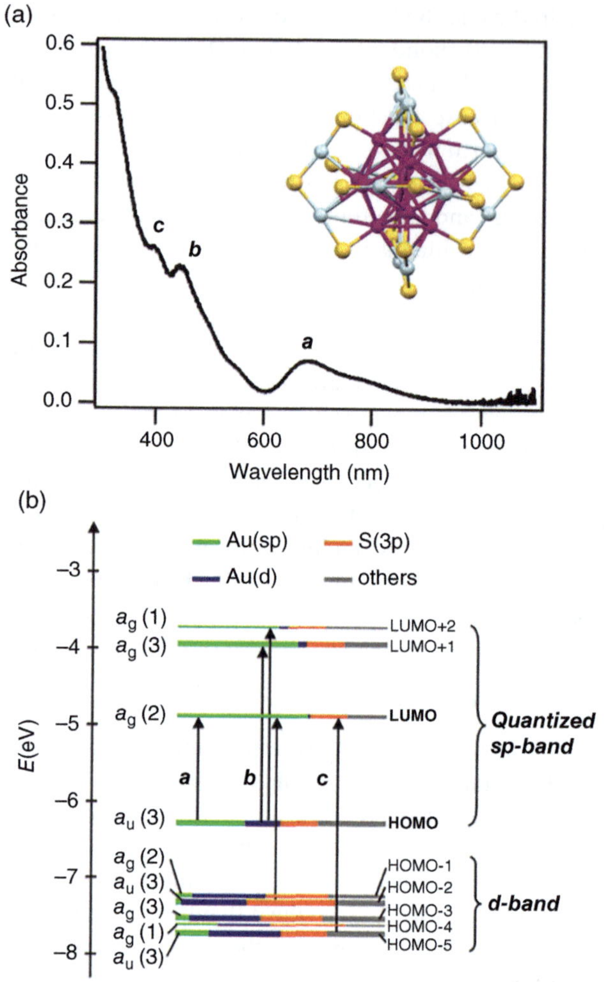

Figure 1.23 Quantization of the electronic structure in $[Au_{25}(SR)_{18}]^-$. (a) UV–vis absorption spectrum of Au_{25} (inset: the crystal structure, carbon tails are omitted for clarity). Color codes: magenta = Au core, blue = surface Au, yellow = S. (b) Kohn–Sham orbital level diagram of Au_{25}. *Source:* Adapted with permission from [122a]. © 2008 American Chemical Society.

is well protected. In the carbon tail layer, C—H···π interactions lead to the formation of rotational and parallel patterns of surface –R groups (Figure 1.24d).

The $Au_{279}(TBBT)_{84}$ nanocluster (where, TBBT = 4-*tert*-butylbenzenethiolate) was synthesized by a LEIST method using $Au_{333}(TBBM)_{79}$ (where, TBBM = 4-*tert*-butylbenzyl mercaptan) as the starting material, followed by reaction with excess TBBT thiol at 80 °C [108]. ESI-MS determined the formula to be $Au_{279}(TBBT)_{84}$ (obs. MW: 68838.4 g/mol, cal.: 68836.7 g/mol), Figure 1.25a. XRC analysis (Figure 1.25b) revealed that $Au_{279}(TBBT)_{84}$ adopts a fcc structure [108], as opposed to the decahedral structure of $Au_{246}(SR)_{80}$. The kernel of $Au_{279}(SR)_{84}$ is an Au_{147} cuboctahedron, and the next layer is composed of 102 gold-atoms in the forms of {100} and {111} facets. The as-formed Au_{249} core is then protected by 18 monomeric staples, 6 dimeric staples, and 30 simple bridging thiolates. On a note, Dass's group also reported this structure [109a]. Overall, Au_{279} is nearly spherical.

At first glance, Au_{246} would be thought to be metallic since its size is quite large. But, interestingly, Au_{246} is nonmetallic (or nonplasmonic), evidenced by multiple peaks at 400, 460, and 600 nm

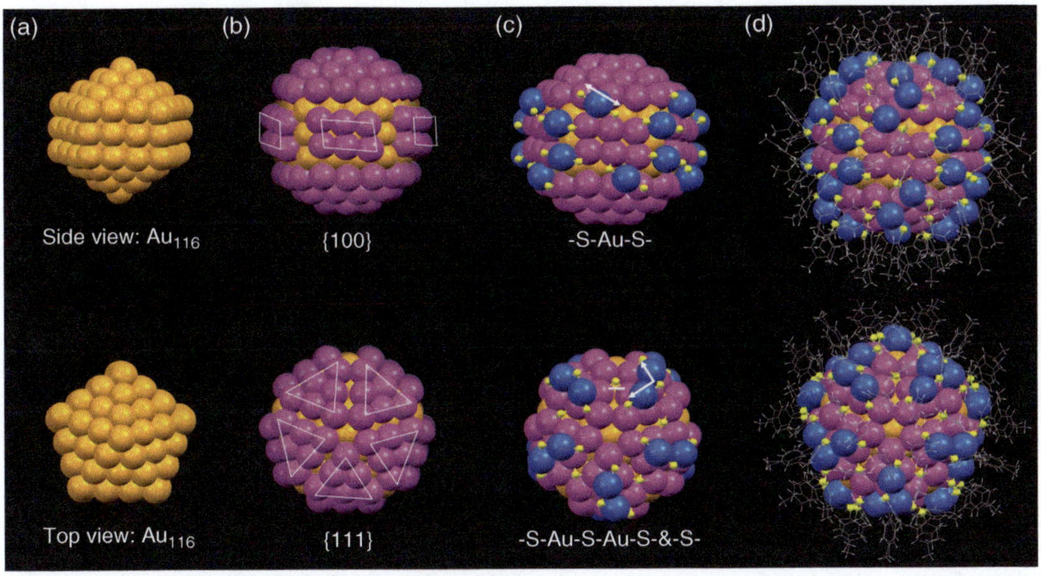

Figure 1.24 X-ray crystal structure of the 2.2 nm $Au_{246}(SR)_{80}$ nanocluster. (a) Au_{116} *i*-D_h kernel, upper panel: side view; lower panel: top view. (b) Transition layer (colored in magenta). (c) Gold-sulfur interfacial structure containing 20 —S—Au—S— staple motifs (top), 10 —S—Au—S—Au—S— staple motifs, and 10 —S— bridging motifs (bottom). (d) Surface carbon-tail layer and overall structure. *Source:* Reprinted with permission from [13]. © 2016 American Association for the Advancement of Science.

(Figure 1.26a) [107]. If Au_{246} were metallic, it would exhibit a single SPR due to its spherical shape. In contrast, Au_{279} exhibits a single peak (i.e. nascent plasmon resonance) at 506 nm (Figure 1.26a) [108]. Similar features are also shown in transient absorption spectra by femtosecond time-resolved spectroscopy (Figure 1.26b) [107, 108]. The observation of a sharp transition (within a merely 33-atom increment) from nonmetallic Au_{246} to metallic Au_{279} is remarkable, which sheds light on the long outstanding questions – at what size the plasmon resonance is born and whether the transition occurs smoothly or abruptly.

Femtosecond spectroscopy further revealed that the photoexcited Au_{246} relaxes within ~10 ps for the majority of the signals, but a weak, long decay component was also found, which decays to zero in the following 200 ps [107]. Three decay components (time-constants) were found by global fitting of the data (i.e. 1.9 ps, 4.6 ps, and 78 ps), with the 1.9 ps ascribed to the internal conversion from higher to lower excited state, the 4.6 ps component to the relaxation from the lowest excited state to the ground state, and the 78 ps to the hot ground state relaxation. The nonmetallicity of Au_{246} indicates an ultrasmall band gap (not measurable but supposedly of k_BT level); such a small gap is very challenging to measure experimentally. In contrast, the electron dynamics of Au_{279} shows that the majority of the signals disappear within ~5 ps (the fitted lifetime $\tau = 1.23$ ps), but the signals do not completely decay to zero even at 100 ps, indicating an additional decay component of long lifetime. The fast process is attributed to the electron–phonon coupling (time constant $\tau_{e\text{-}ph} = 1.23$ ps), and the long-lived one involves phonon–phonon coupling [108].

Further studies on the electron dynamics of Au_{246} and Au_{279} were performed under different pump fluences (Figure 1.27). Although Au_{246} is large in size, it exhibits laser power-independent dynamics (Figure 1.27a), which is similar to the features of molecules, hence, nonmetallic [107]. In contrast, the relaxation dynamics in Au_{279} exhibits a dependence on the laser fluence (Figure 1.27b), indicating that Au_{279} has a plasma-like electronic structure, i.e. a free electron gas is formed [108]. With the observed power dependent electron–phonon coupling in Au_{279}, the

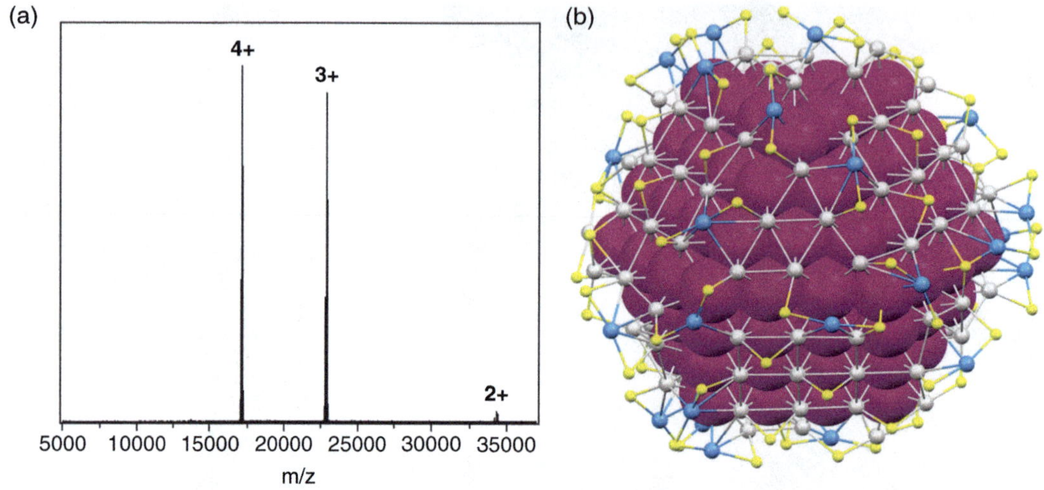

Figure 1.25 (a) ESI-MS analysis of $Au_{279}(TBBT)_{84}$ (in-source ionized peaks of 4+, 3+, and 2+ sets). (b) Crystal structure of $Au_{279}S_{84}$ by X-ray crystallography (–R groups not resolved due to insufficient crystal quality). Yellow = S; other colors = Au. *Source:* Reprinted with permission from [108]. © 2018 American Chemical Society.

intrinsic τ_{e-ph} was obtained via extrapolation to zero pump-power, which is 0.88 ps (Figure 1.27c), comparable to the previous work on plasmonic NPs in which the τ_{e-ph} decreases as the size becomes smaller (e.g. ~0.6 ps for ~2.2 nm Au NPs) [133]. Fitting of the Au_{246} data shows a decay time constant of 4.6 ps independence of laser fluence (Figure 1.27c) [107]. Similar results on the Au_{279} dynamics were obtained by Sakthivel et al. [109b]

Regarding the electronic state and optical spectral evolution of $Au_n(SR)_m$ NCs, Figure 1.28 shows a series of NCs from Au_{25} to Au_{333}. Below the size of Au_{279}, the reported NCs exhibit multiple peaks and thus are of excitonic nature, whereas Au_{279} and above exhibit nascent SPRs [10, 134–136]. The series from Au_{25} to Au_{333} implies a physical picture that the excitons in NCs larger than Au_{246} start to "communicate" and develop a coherence, and the concerted excitonic transitions are generated, manifested as a plasmon excitation in Au_{279} (Figure 1.28) [134]. The development of such a coherence and oscillator's energy "focusing" into the SPR energy (~2.45 eV) is intriguing (comparing the Au_{246} and Au_{279} spectra in Figure 1.28). Beyond the Au_{246} to Au_{279} transition, Dass et al. reported Au_{191} with interesting electron dynamics [135] and Higaki et al. found interesting phonon dynamics in Au_{333} [136]. More details on large-sized NCs [137–139] remain to be revealed in future research. For example, there is potentially some dependence on the ligand type. Wang's group set out to tackle alkynyl-protected Au_n NCs such as $Au_{156}(CCR)_{60}$ and performed femtosecond spectroscopic analysis on the ultrafast electron dynamics [137]. Larger sizes are still required to be synthesized in order to probe the nonmetal-to-metal transition in Au-alkynyl NCs. Yi et al. [139] investigated the aqueous NCs for the nonmetal to metal transition. The shape dependence for the transition [77] should also be investigated in future research.

Regarding the Ag_n NCs, work by Yang et al. [140] and Ma et al. [141] reported the crystal structures of Ag_{374} and Ag_{307}, respectively, both being decahedral structures and plasmonic (single peak at 465 nm). There seems a trend that Ag_n NCs give birth to the plasmon state at a much smaller size than the Au_n system, as indicated by Ag_{206}, Ag_{210}, and Ag_{211} NCs all displaying one dominant peak at ~465 nm [142, 143], but Ag_{155} and smaller ones all show multiple peaks [144]. Thus, the transition seems to occur between Ag_{155} and Ag_{206} (c.f. Au_{246} to Au_{279} for the gold system), but no detailed probing such as electron dynamics and phonon dynamics has been reported yet on such Ag_n NCs.

Figure 1.26 (a) Steady-state optical absorption spectra of Au_{246} and Au_{279} in dichloromethane solutions. (b) Transient absorption spectra of Au_{246} and Au_{279} at 0.3 ps after pump. *Source:* Reprinted with permission from [108]. © 2018 American Chemical Society.

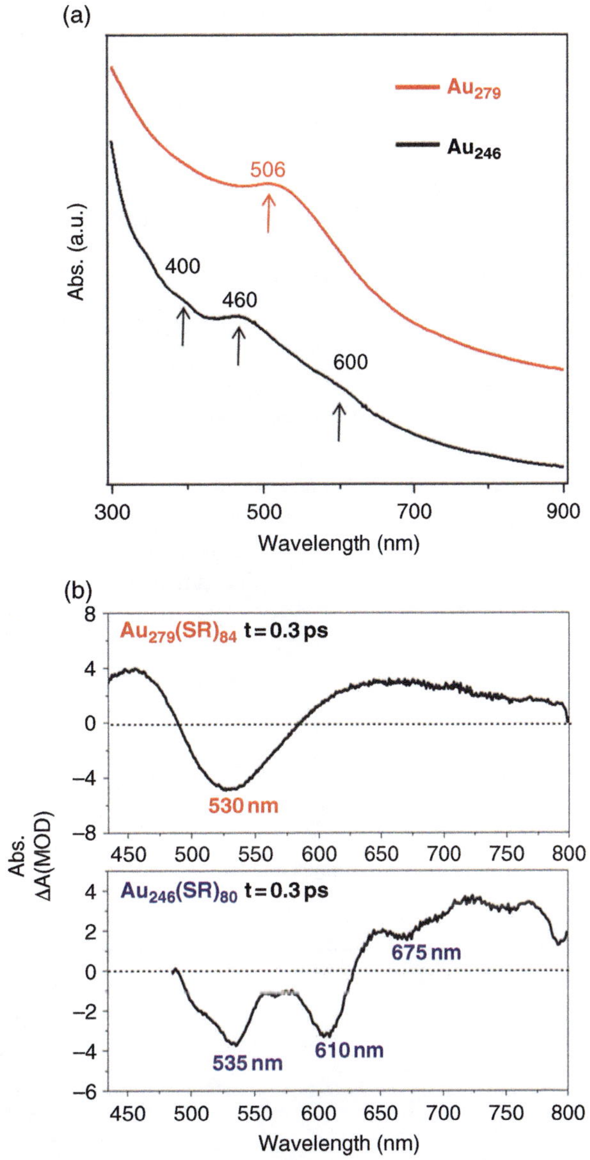

Future work may push the limit of synthesis and crystallization of atomically precise NCs. Giant NCs (>500 metal atoms) are particularly appealing in terms of bridging up with regular nanoparticles with extended facets, and the giant NCs are expected to reveal size-dependent patterns in both the core and the surface. The structural rules to be learned will provide important implications about the mysterious surface (i.e. Au—S interface) of regular plasmonic nanoparticles. New breakthroughs are needed in both the synthesis (i.e. achieving atomic precision for giant sizes) and the crystallization of such NCs for total structure determination by XRC.

1.3.4 Transition from Metal Complexes to the Cluster State: Emergence of Core

The nonmetal-to-metal transition discussed above pertains to the quantized NCs and plasmonic NPs. On the other hand, the transition from Au(I) complexes to the cluster state (i.e. metal—metal bond formation) is also of interest. No delocalized valence electrons are present in Au(I) complexes

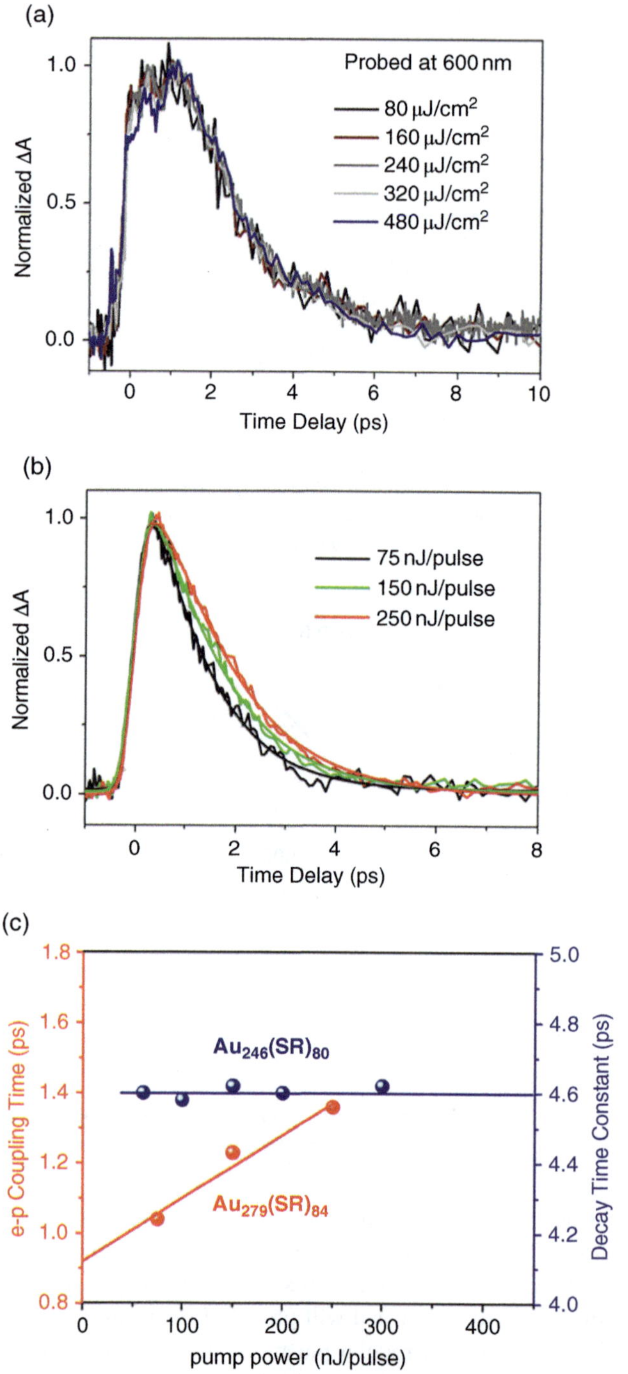

Figure 1.27 (a) Au_{246} normalized decay kinetics at 600 nm as a function of laser fluence with 470-nm pump. (b) Au_{279} normalized decay kinetics at 530 nm as a function of laser fluence with 360-nm pump. (c) The extracted time-constant of Au_{246} (in blue, 470 nm pump) and the $\tau_{e\text{-}ph}$ of Au_{279} (in red, 360 nm pump) as a function of pump fluence. *Source:* Adapted with permission from [107, 108]. © 2017 Wiley VCH and 2018 American Chemical Society.

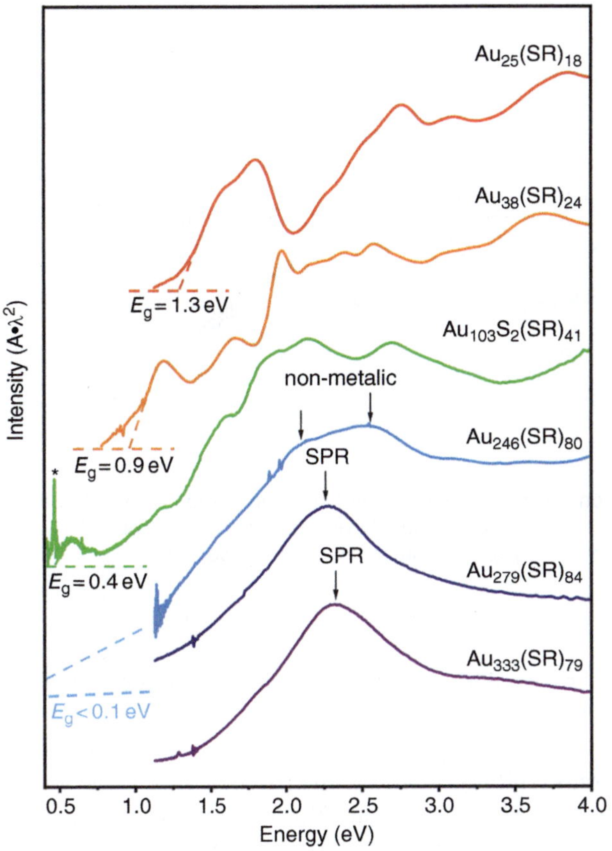

Figure 1.28 Evolution of E_g and optical absorption spectra with size increasing from Au_{25} to Au_{333}. In the Au_{103} spectrum, the asterisk at ~0.4 eV is the solvent vibrational overtone absorption peak. Solvents: CH_2Cl_2 for the UV–vis range, and CCl_4 for the UV–vis–NIR range. Temperature: r.t. *Source:* Reprinted with permission from [134]. © 2022 Royal Society of Chemistry.

such as $Au_{12}(SR)_{12}$, thus, only metal–ligand charge transfer signals can be observed, [145] whereas $Au_{15}(SR)_{13}$ starts to possess 2e, indicating Au—Au bond formation or the core emergence in the cluster [146, 147]. Thus, there is an interesting transition in terms of the electronic structure, and this may also lead to structural rearrangements due to emergence of Au—Au bond (as opposed to sole Au—S bonds in complexes) and delocalized electrons in the core (e.g. Au_4 in the $Au_{15}(SR)_{13}$).

In recent work, Zhu's group [148] used cyclohexanethiol and diphophine (1,4-bis[diphenylphosphino]butane, abbrev. L^4) to convert Au_{16}(S-Adm)$_{12}$ (S-Adm = 1-adamantanethiolate) to Au_{14}(S-*c*-C_6H_{11})$_{10}L^4$ via the LEIST method [86c], and further down to $[Au_{13}(\text{S-Adm})_8(L^4)_2]^+$. XRC analysis identified a planar Au_5 "bowtie" in the core of $[Au_{13}(\text{S-Adm})_8(L^4)_2]^+$, which is quite interesting as it links to the gas phase few-atom clusters (see the bowtie Au_5 in Figure 1.4) that were theoretically computed to be planar structures until the size of ~13 atoms (the onset of 3D structure) [55, 56]. DFT simulations by Pei and coworkers [148] revealed evolution pathways from the $Au_{12}(SR)_{12}$ complex to the Au_{13} nanocluster, in which a "Au_4S_4 to Au_5" nucleation process was identified (Figure 1.29). DFT also showed that the HOMO–LUMO gap decreases dramatically from 2.95 eV of $Au_{12}(SR)_{12}$ to 1.91 eV of Au_{13}, but there is no distinct change from Au_{13} to Au_{14} (E_g = 1.87 eV). This work provides some hints for the nucleation of cluster core with Au—Au bond formation from the Au(I)-SR complex.

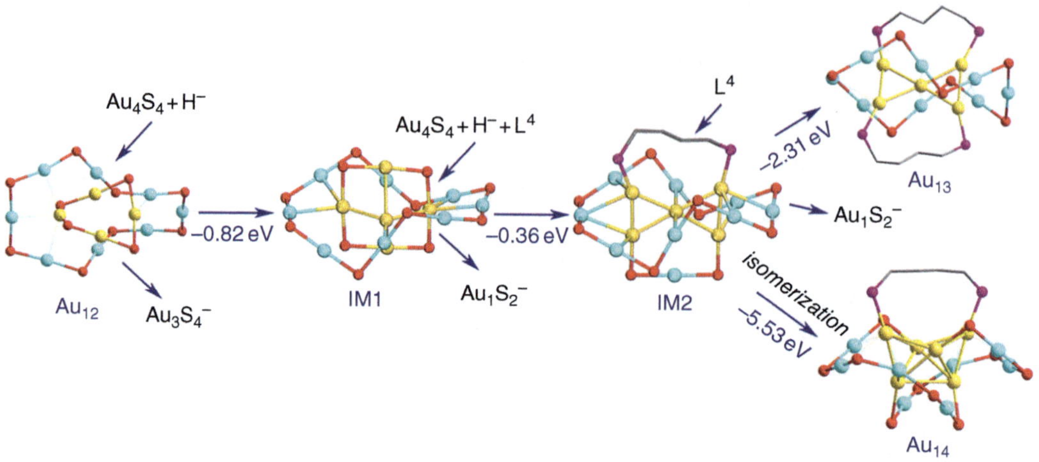

Figure 1.29 Crystal structure-based simulations of the nucleation (the formation of the first Au—Au bond) from the $Au_{12}(SR)_{12}$ complex to the Au_{13} cluster (Color labels: yellow, blue = Au; red = S; magenta = P; gray = C). *Source:* Reprinted with permission from [148]. © 2021 American Chemical Society.

1.3.5 Doping and Alloying

Doping and alloying are important strategies for tailoring the functionality of materials, and have been widely used in research on metal NCs, [149–153] Zintl clusters, [34–36, 154, 155] Ti-oxo nanoclusters, [47] endohedral fullerenes, [156] and many other systems [120]. For the case of doping into the non-empty NCs, the heteroatoms typically substitute the parent atoms, while for the cages, dopants typically go to the central region. Below, we briefly illustrate some examples using metal NCs as examples, including some distinct effects from doping or alloying.

In $Au_n(SR)_m$ NCs, Negishi's group reported single Pd doping into the Au_{25} NC to produce $Pd_1Au_{24}(SR)_{18}$ [157]. Qian et al. [158] found that single Pt atom doping into Au_{25} led to drastic changes to the optical spectrum, which was due to the Pt perturbation to the cluster electronic structure revealed by DFT simulations by Jiang et al. [158] In recent work, specific doping by several metal elements into the nanocluster has also been realized [149, 150]. Various types of dopants such as Pt, Pd, Ag, Cu, Ir, Cd, and Hg can be doped into gold NCs (Scheme 1.2).

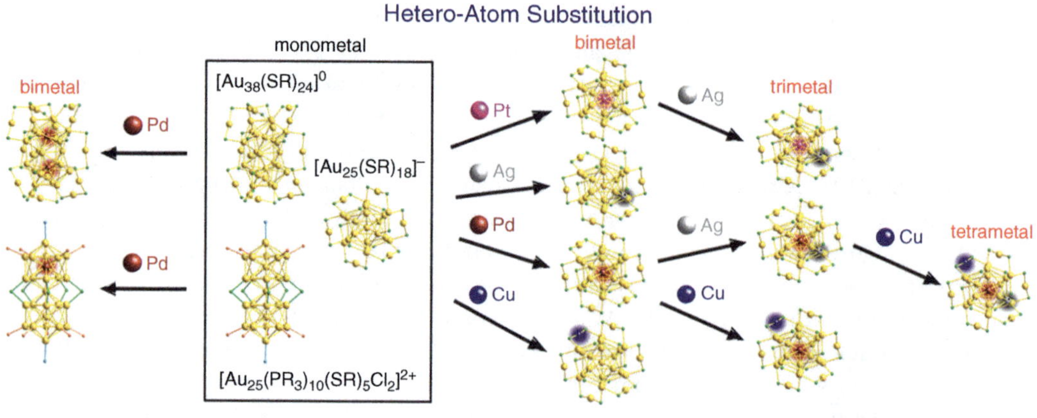

Scheme 1.2 Heteroatom doping of gold NCs by Pt, Pd, Ag, and Cu. *Source:* Reprinted with permission from [150]. © 2018 American Chemical Society.

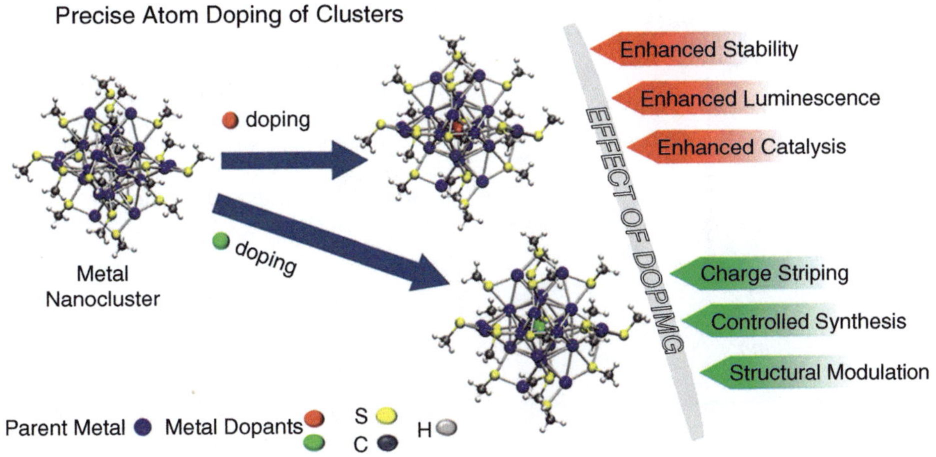

Scheme 1.3 Heteroatom doping in metal nanoclusters. *Source:* Reprinted with permission from [151]. © 2018 American Chemical Society.

With respect to the effects of heterometal doping (Scheme 1.3), the Pt and Pd doping typically leads to higher stability of the NCs, [157, 158] while Ag and Cu doping does not [149–151]. The effects of doping on catalysis have been reported by many groups. For example, Qian et al. [158] found an increased conversion in styrene oxidation reaction on Pt_1Au_{24} compared to Au_{25}. Tsukuda's group found a significant improvement in aerobic oxidation of benzyl alcohol over Pd_1Au_{24} compared to Au_{25} [159]. Lee's group reported a large enhancement of electrocatalytic hydrogen evolution on Pt_1Au_{24} compared to the Au_{25} [160]. Doping can also lead to PL enhancement [149, 151], for example, the case of $[Au_{12}Ag_{13}(SR)_5(PPh_3)_{10}X_2]^{2+}$ compared to $[Au_{25}(SR)_5(PPh_3)_{10}X_2]^{2+}$, and $Au@Ag_{24}(SR)_{18}^-$ compared to $Ag_{25}(SR)_{18}^-$, but not the case of $Pd@Ag_{24}(SR)_{18}^{2-}$. Liu's group also reported PL enhancement in copper clusters after doping, such as $[Au@Cu_{12}(S_2CN\text{-}^nBu_2)_6(CCPh)_4]^+$ [153].

In addition to specific doping with one or more atoms, alloying with a nonspecific number (x, typically a range) of heterometal atoms has also been reported, such as $Ag_xAu_{25-x}(SR)_{18}$, [150] in which case it would be highly desirable to separate the components. Significantly, Negishi's group [161] successfully isolated the $Ag_xAu_{25-x}(SR)_{18}$ alloys according to the x value, and their work even demonstrated the isolation of isomers for specific x. Figure 1.30 shows the x components (0–4) by high-resolution chromatography isolation, including $[Au_{25}(SC_4H_9)_{18}]^0$ (i), $[Au_{24}Ag(SC_4H_9)_{18}]^0$ (ii), $[Au_{23}Ag_2(SC_4H_9)_{18}]^0$ (iii), $[Au_{22}Ag_3(SC_4H_9)_{18}]^0$ (v), and $[Au_{21}Ag_4(SC_4H_9)_{18}]^0$.

Doping was also extensively reported in Ag, Cu, Rh, and Ni nanoclusters. For example, rhodium carbonyl NCs have been incorporated with Ge, Sn, Sb, Bi, and Au [78]. Nickel NCs can be doped with Sb and many other elements (e.g. Pt, Pd, and Au). In Zintl clusters, fullerenes, and Ti-oxo systems, significant amounts of work have also been carried out in doping with various elements. Readers are referred to Chapters 2–14 for more details.

1.3.6 Redox and Magnetism

The molecule-like nature of metal NCs and fullerenes makes them exhibit rich redox properties and. Electrochemical measurements [162] such as cyclic voltammetry (CV), differential pulse voltammetry (DPV) and square wave voltammetry (SWV) have been carried out on metal NCs (Chapters 5 and 10) and endohedral fullerenes (Chapter 11).

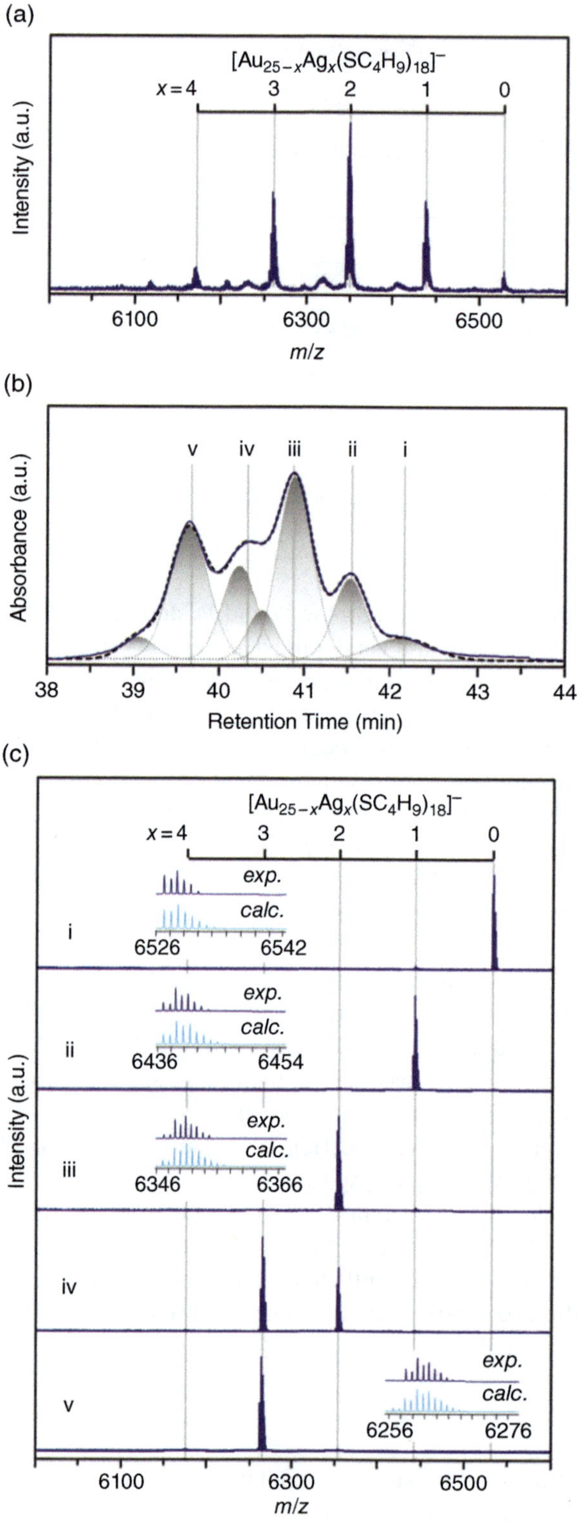

Figure 1.30 High-resolution separation of $Au_{25-x}Ag_x(SR)_{18}$ (x = 0–4) into $[Au_{25}(SC_4H_9)_{18}]^0$ (i), $[Au_{24}Ag(SC_4H_9)_{18}]^0$ (ii), $[Au_{23}Ag_2(SC_4H_9)_{18}]^0$ (iii), $[Au_{22}Ag_3(SC_4H_9)_{18}]^0$ (iv), and $[Au_{21}Ag_4(SC_4H_9)_{18}]^0$ (v). (a, c, d) Mass spectrometry analysis, (b, e) chromatography analysis. *Source:* Reprinted with permission from [161]. Copyright 2018 American Chemical Society.

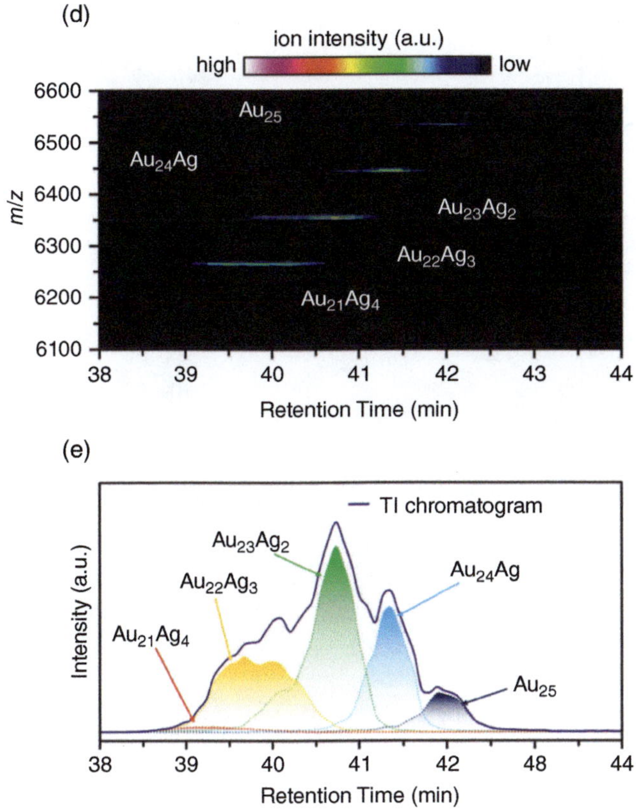

Figure 1.30 (Continued)

Figure 1.31 illustrates the SWVs of $Au_n(SR)_m$ NCs spanning the nonmetallic to metallic regime [163]. The HOMO-LUMO gap can be determined by the electrochemical gap (between O1 and R1 in Figure 1.31, here O means oxidation and R for reduction) after subtracting the charging energy (typically 0.2–0.3 eV by taking the O1 and O2 difference). Detailed analyses can also lead to the construction of electron orbital diagrams of NCs [162, 164]. More discussions on the redox properties of metal-thiolate NCs (e.g. Au, Ag, and doped ones) can be found in Chapter 5.

The redox properties of NCs are closely relevant to the magnetism in NCs [165, 166]. For example, the anionic $Au_{25}(SR)_{18}^-$ (counter ion: $[N(C_8H_{17})_4]^+$) is diamagnetic, but the charge neutral Au_{25} (7e) exhibits paramagnetism, and quantification by electron paramagnetic resonance (EPR) spectroscopic analysis revealed one spin (unpaired electron) per particle [126]. Regarding the large sizes of $Au_n(SR)_m$ NCs (e.g. $n > 100$), Zeng et al. [128] reported magnetism of $[Au_{133}(SR)_{52}]^0$ (81e) by EPR, while the oxidized product, $[Au_{133}(SR)_{52}]^+$, showed no EPR signal. The EPR signals of Au_{25} and Au_{133} exhibited different symmetry. In recent work, Wang and coworkers found paramagnetism in the $Au_{99}(CCR)_{40}$ NC (59e) by SQUID measurement [167]. The effect of NC assembly on the magnetism was reported by Maran's group, in which various coupling effects were observed, such as antiferromagnetism in crystals [166].

Several Ag_n NCs were also reported to be EPR active, but all the signals showed no large splitting as in gold NCs, with g values of open-shelled Ag NCs being very close to 2 [104]. For bimetallic NCs, Zhu's group [168] reported the charge state variability of $[Au_{18}Cu_{32}(SPhCl)_{36}]^{2-/3-}$ and the EPR signal of the 3- state (17e) with S = ½ and g = (2.02, 2.35). Tsukuda's group also reported the EPR spectrum of $PtAu_{24}(SR)_{18}^-$ (7e) while the zero charge state is nonmagnetic [169]. Li et al. [165]

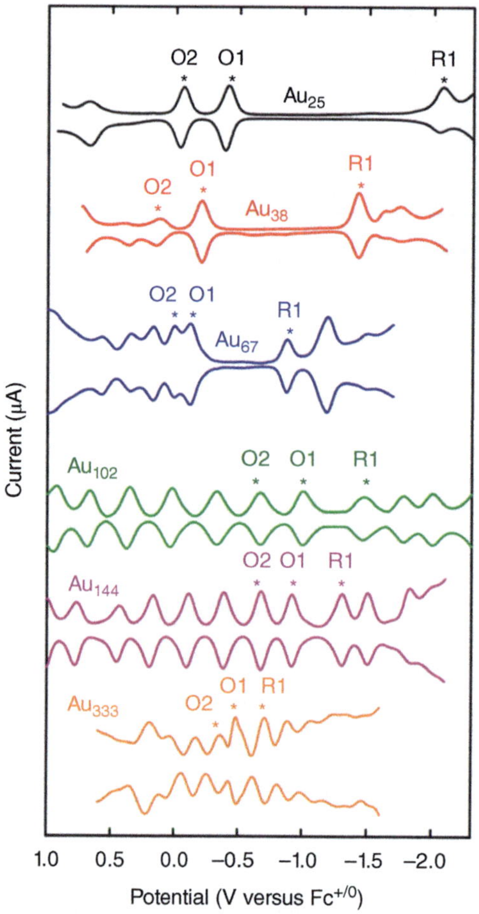

Figure 1.31 SWVs of a series of $Au_n(SR)_m$ nanoclusters in solution (O1: the first oxidation wave, R1: the first reduction wave). *Source:* Adapted from Chapter 5.

found that silver doping into $Au_{25}(SCH_2CH_2Ph)_{18}{}^0$ leads to a linear decrease in axial splitting with increasing Ag doping (Figure 1.32), which is attributed to the smaller spin-orbit coupling of silver compared to gold. The ligands (aromatic vs. nonaromatic ligands) also have a small influence on the EPR signals [170].

In the case of magnetic metals (Fe, Co, Ni), the $Ni_{30}S_{16}(PEt_3)_{11}$ NC obtained by Hayton's group [90] features a compact "metal-like" core with a high degree of Ni—Ni bonding. Interestingly, it has an open-shell electronic structure, hence, magnetic. Variable temperature magnetic susceptibility (χ_T) measured by SQUID magnetometry found that Ni_{30} possesses a manifold of closely spaced electronic states near the HOMO–LUMO gap [90]. Specifically, the field induces a mixing of a triplet ground state with a low-lying quintet excited state. The closely spaced states near the HOMO–LUMO gap should originate from the metal—metal bonding in the core. The magnetization vs. temperature curves of Ni_{30} showed no magnetic blocking in zero-field-cooled (ZFC) and field-cooled (FC) down to 1.8 K, neither any hysteresis when cycling magnetization vs. applied field. More discussions can be found in Chapter 9.

Overall, an odd number of delocalized valence electrons in the metal core should indicate that the NC be magnetic due to open-shell electronic structure, which is in resemblance with molecules. Thus, electron counting offers a quick way for the identification of magnetic NCs. On the

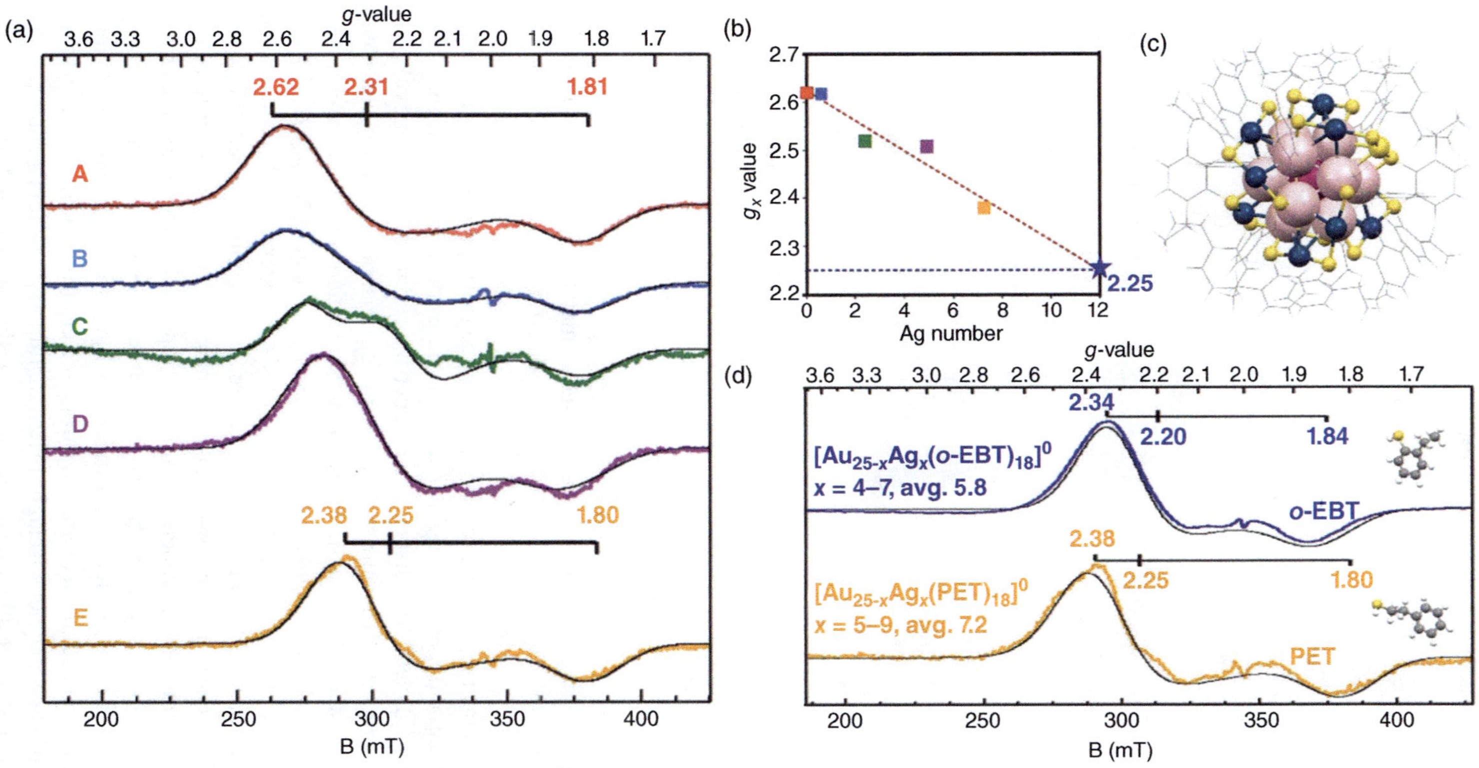

Figure 1.32 (a) X-band (9.645 GHz) EPR spectra measured at 16 K of $[Au_{25-x}Ag_x(SC_2H_4Ph)_{18}]^0$ with Ag dopant x = (A) 0, (B) 0–2, (C) 0–6, (D) 3–7, and (E) 5–9; microwave power 0.2 mW. (b) Plot of g_x vs the average Ag dopant number in $Au_{25-x}Ag_x$ (or the $Au_{12-x}Ag_x$ shell since the center and exterior staple motifs contain gold exclusively) with maximum x possibly being 12 and the corresponding linear fit. (c) Crystal structure of $[Au_{25-x}Ag_x(SPh\text{-}o\text{-}C_2H_5)_{18}]^0$ (avg. x = 5.8). (d) Comparison of the X-band EPR spectra of $[Au_{25-x}Ag_x(SPh\text{-}o\text{-}C_2H_5)_{18}]^0$ (x = 4–7, avg. 5.8) with g = (2.34, 2.25, 1.84) and $[Au_{25-x}Ag_x(SC_2H_4Ph)_{18}]^0$ (x = 5–9, avg. 7.2) with g = (2.38, 2.25, 1.80). Color codes: magenta = central Au; pink = icosahedral shell Au/Ag; blue = surface motif Au; yellow = S; gray = C, white = H. *Source:* Reproduced with permission from [170]. © 2020 American Chemical Society.

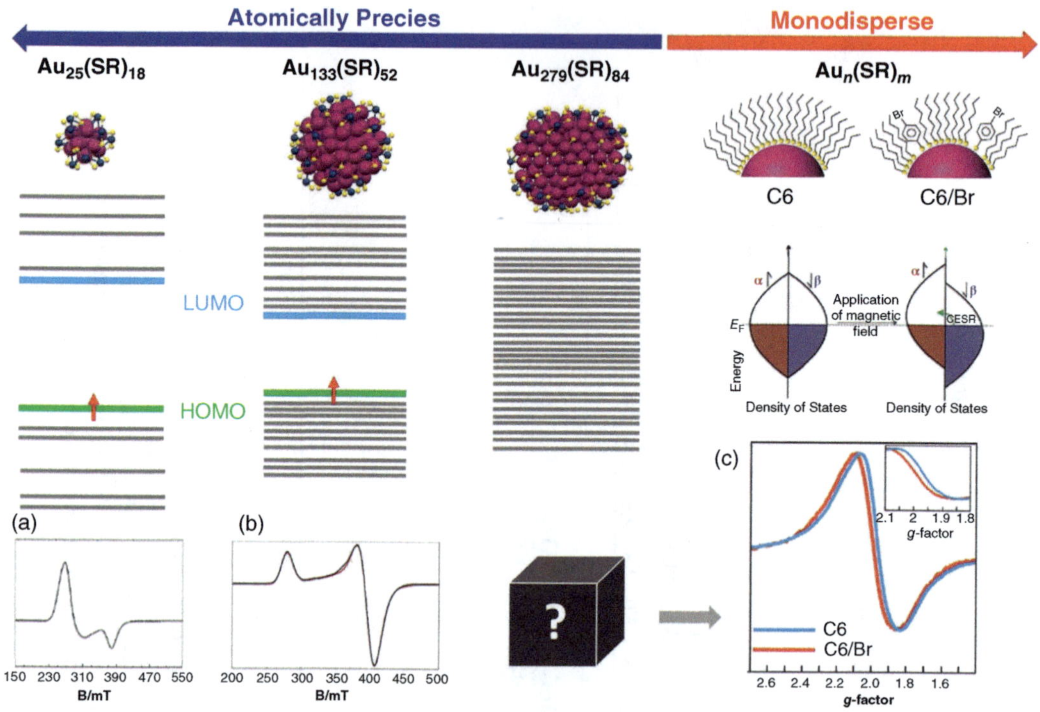

Scheme 1.4 Gold NCs with schematic energy states and corresponding EPR spectra of (a) $Au_{25}(SR)_{18}$, (b) $Au_{133}(SR)_{52}$, and $Au_{279}(SR)_{84}$ (its EPR spectrum unknown yet). (c) Schematic of Pauli paramagnetism of Au-SR NPs protected by (left) *n*-hexanethiolate (C6 for short), and (right) partially modified with 4-bromobenzenthiolate (C6/Br), and corresponding CESR spectra. Color codes: magenta = kernel Au; blue = motif Au; yellow = S. *Source:* Panel (a) Reprint with permission from [126]. © 2009 American Chemical Society. Panel (b) Reprinted with permission from [128]. © 2019 Royal Society of Chemistry. Panel (c) Reproduced with permission from [171]. © 2015 Wiley-VCH.

other hand, no strong ferromagnetism in atomically precise metal NCs has been found thus far, which requires strong spin-orbit coupling. Nevertheless, antiferromagnetism and low-temperature ferromagnetism were observed by Maran et al. in microcrystals assembled from $Au_{25}(SR)_{18}{}^{0}$ NCs [166]. Future work may map out the size evolution of magnetic properties (Scheme 1.4) and reveal a deeper fundamental understanding, such as the surface effect on magnetism.

Other than the open-shell metal NCs that exhibit magnetism, endohedral fullerenes also give rise to interesting magnetic properties. In contrast to no strong ferromagnetic coupling observed yet in atomically precise Au, Ag, and Ni NCs, quite many endohedral fullerenes indeed exhibit hysteresis behavior (so-called single molecule magnets, SMM); for example, $Dy_2@I_h(7)\text{-}C_{80}(CH_2Ph)$ showed a magnetization loop below 21 K (Figure 1.33), with the blocking temperature (T_B) determined to be 21.9 K [172]. Of note, the observed intra-molecular ferromagnetic coupling [156] differs from the above case of $Au_{25}(SR)_{18}{}^{0}$ crystal that involves inter-molecular antiferromagnetic or ferromagnetic coupling [166]. In endohedrally doped fullerenes, the observed SMM behavior originates from the strong ferromagnetic coupling between the two Dy ions through the single-electron metal—metal bond (2c-1e). Many other magnetic endohedral fullerenes have been reported (see Chapter 11). Generally, when the encapsulated metal atoms have unpaired electrons or transfer an odd number of electrons to the carbon cage, magnetism can be observed. Important factors include the encapsulated metal atoms (ions) and the resulted structure, as well as the cage size and structure. By tailoring such factors, strong SMMs with broader hysteresis, longer relaxation times, and higher blocking temperatures (e.g. room temperature) may be created in future research.

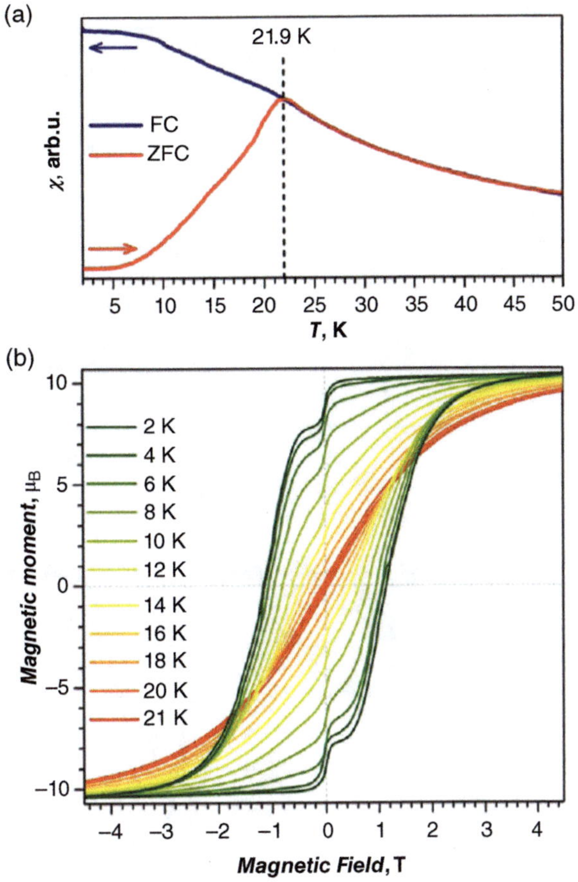

Figure 1.33 Magnetization curve (a) and magnetic hysteresis (b) of $Dy_2@I_h(7)$-$C_{80}(CH_2Ph)$ in which the cage is derivatized by $-CH_2Ph$. The blocking temperatures (T_B) was determined to be 21.9 K. *Source:* Adapted from Chapter 11.

1.3.7 Energy Gap Engineering

The vast majority of atomically precise NCs possess a distinct gap in their electronic structure due to their ultrasmall size. Therefore, by controlling the size, one can readily control the HOMO-LUMO gap or bandgap (denoted E_g for both). For example, the $Au_n(SR)_m$ NCs exhibit E_g from ~3 eV for $Au^I(SR)$ complexes to 2.5 eV for $Au_{15}(SR)_{13}$, 1.3 eV for $Au_{25}(SR)_{18}^-$, 0.9 eV for $Au_{38}(SR)_{24}$, 0.4 eV for $Au_{102}(SR)_{44}$, 0.17 eV for $Au_{144}(SR)_{60}$, [163] and the milli-eV range for $Au_{246}(SR)_{80}$ and $Au_{279}(SR)_{84}$ [134]. The $Au_{279}(SR)_{84}$ and larger sizes such as $Au_{333}(SR)_{79}$ exhibit nascent plasmon resonances; thus, these can be viewed as "gapless." The gapless excitations of SPRs imply the emergence of free conduction electrons, [134] which is in contrast to the single-electron excitations (excitons, i.e. bound electron–hole pairs) in smaller NCs [122].

The bandgap engineering is particularly important in GNRs research, because one of the potential applications of GNRs is in the transistors, which requires an optimum bandgap of ~1 eV [173]. GNRs with large width have extremely small (meV) gaps; thus, one should control the width within 1 nm (Figure 1.34a).

Here we also briefly discuss the E_g control of Ti-oxo NCs, which are inherently large gap materials (note: bulk TiO_2 E_g = 3.2 eV and quantum confinement leads to even larger gaps), but, interestingly, Gao et al. [174] demonstrated TOCs with low bandgaps (1.5 eV) in recent work (Figure 1.34b) and such TOCs appear black due to panchromatic absorption. The attainment of TOCs bandgap control from 3.5 to 1.5 eV is of vital importance in photocatalysis applications. The small bandgap also enhances the third-order nonlinear optical properties [174].

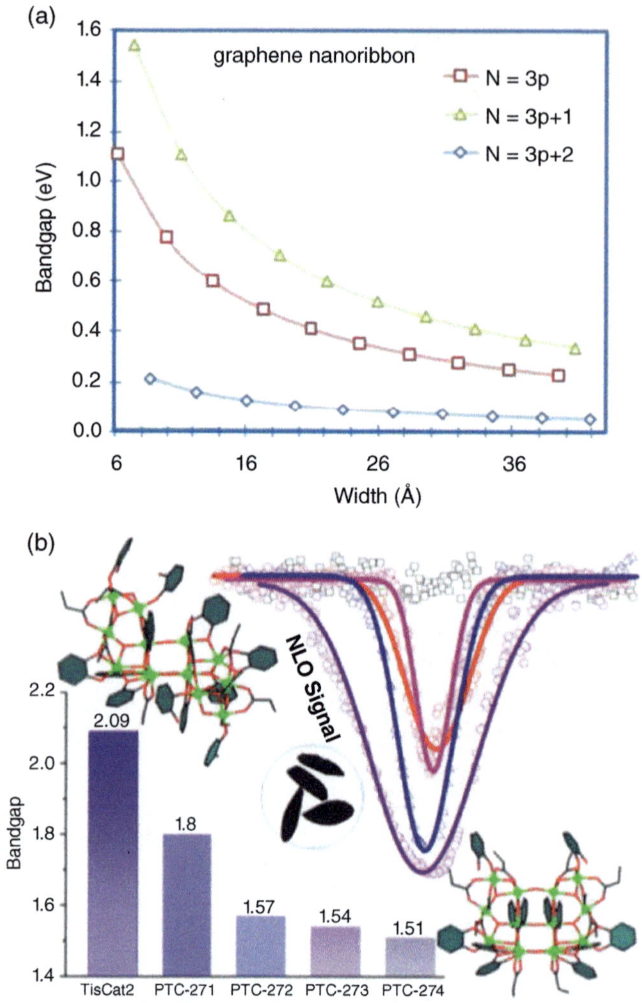

Figure 1.34 Bandgap engineering of GNRs (a) and TOCs (b). *Source:* Panel A: Redrawn from [173]. © 2014 American Scientific Publishers; Panel B: Reprinted with permission from [174]. © 2022 American Chemical Society.

1.3.8 Assembly of Atomically Precise Nanoclusters

Assembly is one of the major topics for molecular and nanoscale materials [175, 176]. The precise building blocks such as NCs are akin to atoms in a solid-state compound and provide exciting opportunities for assembly into diverse types of superstructures (Scheme 1.5) [177–180]. Such structures could be utilized as hosts for drug delivery and in the design of sensing devices and construction of hierarchical architectures. The self-assembly of NCs can be driven by covalent bonding or noncovalent forces [177–181], as well as charge transfer-induced electronic coupling [175]. During the self-assembly process, the system's energy is reduced. Assembly by noncovalent forces is more common than covalent bonding, and the nanocluster surfaces can interact via hydrogen bonding (e.g. the presence of –COOH or –OH groups on ligands [180]), electrostatic interactions (e.g. charges on the surface or core), $\pi\cdots\pi$ stacking (e.g. benzene groups), van der Waals interactions, dipole–dipole interactions, C—H$\cdots\pi$ interactions, and so on [13, 177, 178, 181].

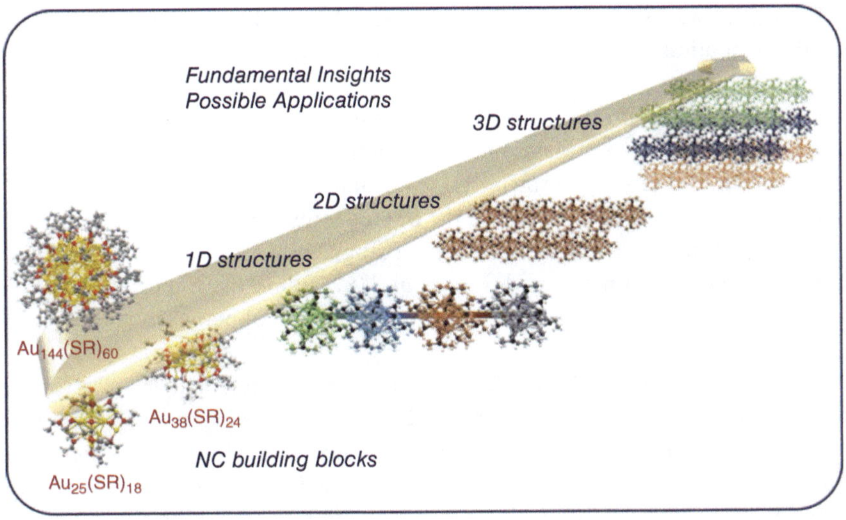

Scheme 1.5 Assembly of atomically precise nanocluster building blocks into 1D, 2D, and 3D meso- or macro-structures. *Source:* Reprinted with permission from [177a]. © 2020 Wiley VCH.

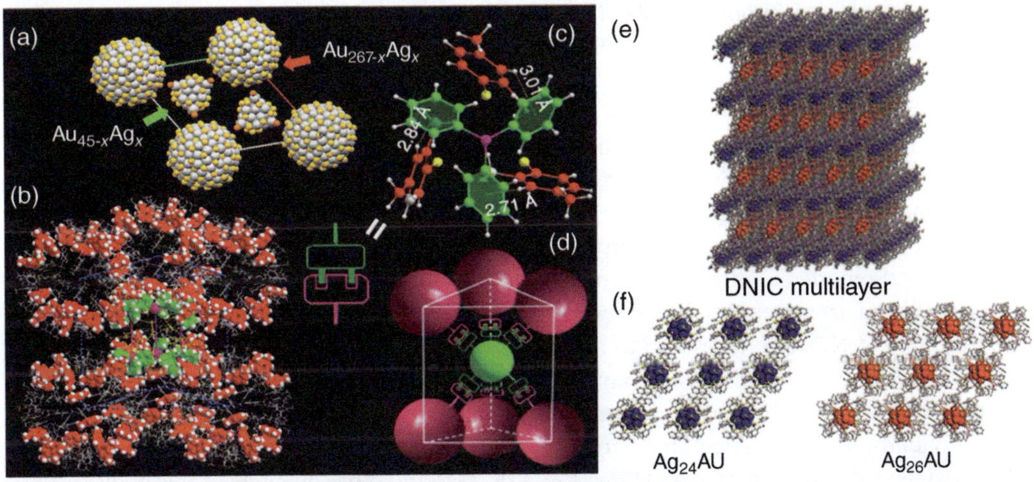

Figure 1.35 Co-assembly of two types of NCs into single crystals. (a–d) co-crystal structure of $Au_{267-x}Ag_x(SR)_{80}$ and $Au_{45-x}Ag_x(SR)_{27}(PPh_3)_6$ NCs into a hexagonal unit cell; (e–f) co-assembly of $[Ag_{24}Au(SR)_{18}]^-$ and $[Ag_{26}Au(SR)_{18}(PPh_3)_6]^+$ into a layer-by-layer crystalline structure. *Source:* Reprinted with permission from [91]. © 2020 American Chemical Society. Original work in [183, 184].

Both noncrystalline and crystalline superstructures have been reported, with the former type being such as 1D fibril mesostructures of micron sizes in diameter [182] and the latter type being single crystals. While the single crystals typically involve one type of NCs, some co-assembly examples have been reported, [183, 184] which involve two types of NCs (Figure 1.35).

Xie's group obtained shape-controlled assembly of $[Ag_{44}(p\text{-MBA})_{30}]^{4-}$ NCs into well-defined supercrystals (regular octahedra vs. concave octahedral from the same NCs) via adjusting the interparticle electrostatic interactions with ironic strength and solvent polarity [185]. Assembly of NCs onto plasmonic Au nanorods was realized by Pradeep et al. [179], in which they first functionalized plasmonic Au nanorods with *p*-MBA via ligand exchange, then the *p*-MBA functionalized

GNRs were mixed with $[Ag_{44}(p\text{-MBA})_{30}]^{4-}$ (counterion: Na^+) in DMF, and the hydrogen bonding between ligands led to the formation of an octahedral shape for the assemblies (as opposed to the starting rod shape).

With respect to the assembly-induced effects on the properties of NCs, here we illustrate the transport properties, magnetic coupling, luminescence enhancement, and mechanical properties.

The transport properties (e.g. the electron transport) of 1D and 2D superstructures are quite appealing. For example, Galchenko et al. [186] reported the photoconductivity in the films of Au_{25}-rod NCs. Li et al. [182] pursued the electrical conductivity of single crystals of various Au_{21} NCs and identified the important roles of counterions. Fetzer et al. [187] found the importance of the structural order in charge transport. An anisotropic effect (in plane vs. out of plane) in crystals was found by Yuan et al. [188].

In magnetic studies, Roy et al. assembled $Ni_9Te_6(PEt_3)_8$ with fullerenes, in which electron transfer to the fullerenes led to the formation of a rock-salt–related structure and a magnetically ordered phase was observed at low temperature [175b]. Maran and coworkers [189] found that the $[Au_{25}(SBu)_{18}{}^0]_n$ 1D-polymers are stabilized by inter-cluster aurophilic interaction with proper orientation via a twist-and-lock mechanism. Interestingly, the formation of the aurophilic bonds is only made possible when the ligands are long enough (starting from R = n-butyl) while shorter -SPr and -SEt ligands on Au_{25} just cannot form the polymeric wire. Heterometal doping may destroy such 1D assembly, e.g. the case of doped $Au_{24}Pt(SBu)_{18}$. These results imply that electronic factors may also play an important role. The 1D $[Au_{25}(SBu)_{18}{}^0]_n$ polymers exhibit antiferromagnetism at low temperature owing to the spin exchange interaction along the chain, as revealed by EPR analysis and DFT calculations [189].

The assembled NCs may produce new or enhanced functionality such as stronger photoluminescence [181]. For example, Zhu's group found a 2.4X enhancement of PL in crystals compared to amorphous assemblies [190]. Furthermore, Pradeep's group reported the effect of crystal type on the PL enhancement, for example, the cubic crystal of Ag_{29} NCs was found to exhibit stronger PL than the trigonal crystal (Figure 1.36), also with a red-shift of ~30 nm in the emission peak [191]. Zang et al. [192] reported a mirror image relationship of circularly polarized luminescence (CPL) signals from chiral arrangement of silver clusters in a crystalline state.

The assembly of NCs also gives rise to new opportunities for studying the mechanical properties [180]. For example, Bigioni's group found that the crystals of the hydrogen bond-assisted assembly of $Ag_{44}(SR)_{32}{}^{4-}$ NCs exhibited a response to hydrostatic compression, manifested in anomalous pressure softening and correlated chiral rotation of the NCs. For more discussions on this topic, readers are referred to several reviews [175a, 177, 178, 181] and other chapters in this book.

Chapters 15 by Mandal's and Sun's groups, and Chapter 16 by Zang's group provide excellent summaries on the topics of assembly of NCs, including the use of NCs as nodes for the construction of special types of metal–organic frameworks with linkers such as bipyridine, dicarboxyl, and multidentate POMs. Chiral crystals with CPL signals can be constructed [181].

Overall, the assembly of NCs will lead to not only new opportunities in fundamental research of understanding the assembly mechanisms, [13, 177] but also exciting new materials for applications in chemical sensing, optics, and many other fields [175, 178, 181].

1.4 Some Applications

The atomically precise NCs and their assembled materials have found a variety of applications, including catalysis, sensing, optics, solar energy conversion, and biomedicine [17, 193–200]. Here we briefly illustrate a few.

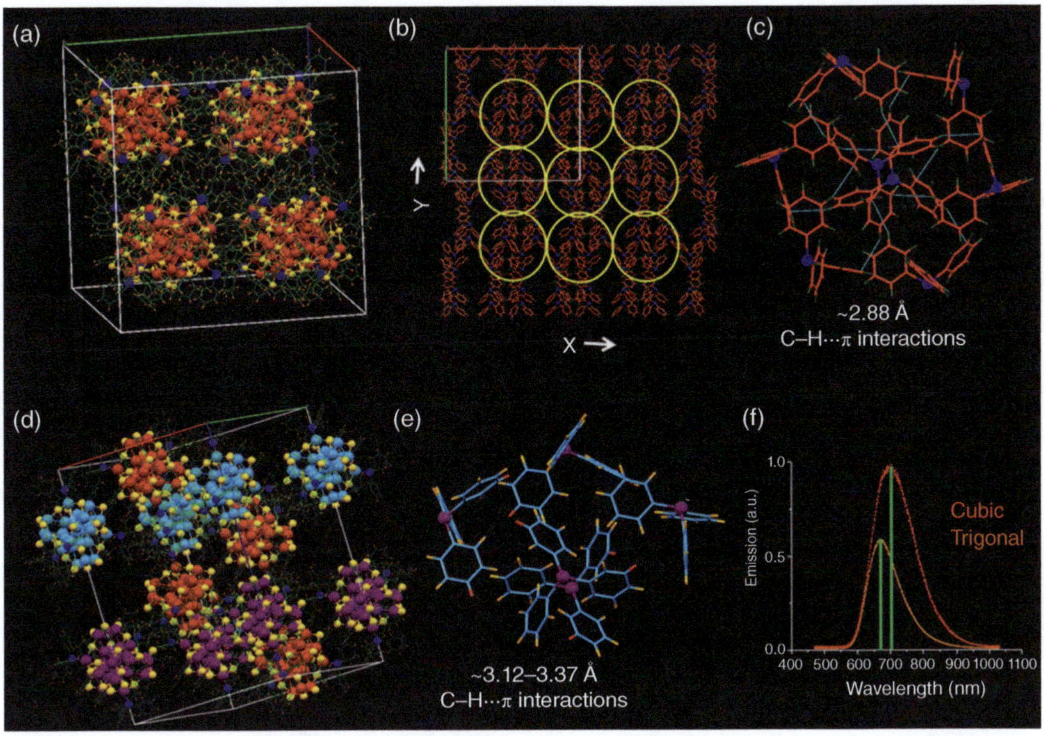

Figure 1.36 Crystalline assembly for enhancing photoluminescence from NCs. (a) Cubic unit cell of $[Ag_{29}(BDT)_{12}(PPh_3)_4]^{3-}$ NCs. (b) Packing of PPh_3 in the assemblies. (c) Strong C—H···π interaction mode (T-shaped) between the PPh_3 ligands. (d) Trigonal unit cell of the Ag_{29} NCs. (e) C—H···π interaction mode in the trigonal unit cell. (f) Emission spectra of single crystals with the cubic and trigonal lattices, respectively. *Source:* Reprinted with permission from [192a/b]. © 2018 Royal Society of Chemistry and 2019 American Chemical Society.

1.4.1 Chemical and Biological Sensing

Sensing by the use of metal NCs [194] has been extensively reported, including the detection of

i) toxic ions (e.g. Hg^{2+}, Cd^{2+}, Ag^+, Cu^{2+}, As^{3+}, S^{2-}, NO_2^-, CN^-),
ii) important molecules (e.g. 2,4,6-trinitrotoluene (TNT), H_2, H_2O_2),
iii) biomolecules (e.g. glucose, adenosine triphosphate, dopamine, uric acid, amino acids, cholesterol),
iv) intracellular pH, temperature, and reactive oxygen species (ROS),
v) DNA and protein biomarkers.

Wu and coworkers [201a] reported the use of $Au_{25}(SG)_{18}^-$ as a probe to detect toxic ions and attained a sensitivity of ~200 nM for Ag^+ detection. Pradeep's group [201b] utilized $Au_{15}(SG)_{13}$ for Cu^{2+} detection and obtained an excellent visual sensitivity of 1 ppm. The assemblies of NCs are also quite promising in sensing applications (see Chapter 16 by Zang's group).

The sensing applications primarily rely on the luminescence properties of NCs, including both photoluminescence [202] and electrochemiluminescence [203]. Generally, the luminescence originates from the opening of a distinct HOMO-LUMO gap in metal NCs [204]. For example, the $Au_{25}(SR)_{18}^-$ nanocluster exhibits red to near-IR luminescence [205]. Although the majority of metal NCs are not strong in emission (quantum yield QY often below a few percent), there are a few examples of high QYs, such as $Au_{12}Ag_{13}(PPh_3)_{10}(SR)_5X_2$ (QY ~ 50%) [202]. Some strategies have been devised to enhance the luminescence of those weakly emitting NCs, such as heterometal doping into the metal core, aggregation-induced enhancement, crystallization-induced enhancement, ligand engineering, etc. [202, 204] New types of NCs with strong emission are also being designed in current research.

1.4.2 Biomedical Imaging, Drug Delivery, and Therapy

Gold NCs are particularly attractive for bioimaging, photoacoustic imaging, and cancer therapy [199, 200, 206]. Compared to the regular Au NPs (e.g. 13 nm) that have been extensively explored for biomedical applications [8], gold NCs possess several distinct advantages (Scheme 1.6), [206] including:

i) their ultrasmall sizes (subnanometer to 2 nm core diameters), which facilitates cellular entry, even crossing the blood-brain-barrier (BBB) [8, 206];
ii) atomic uniformity, which eliminates the heterogeneity in biological interactions [18, 19];
iii) NIR photoluminescence, which allows for deeper penetration of biological tissue and minimum photo-damage to biological samples, as well as low auto-luminescence from background in the living system [200a].

Bioimaging based on glutathione-protected $Au_{25}(SG)_{18}$ luminescence or photoacoustic effect was reported by Zheng et al. [207a] The groups of Xie and Zhang [200a] utilized $Au_{25}(SR)_{18}$ for NIR imaging of brain blood flow and found significant differences between healthy and injured brain, which allows one to distinguish the lipopolysaccharides-induced brain injury and stroke *in vivo*. NIR imaging was also used to monitor in real-time cancer metastasis [200a]. Overall, luminescent metal NCs will play an important role in diagnosis and treatment of various diseases, [206] not just cancers and tumors.

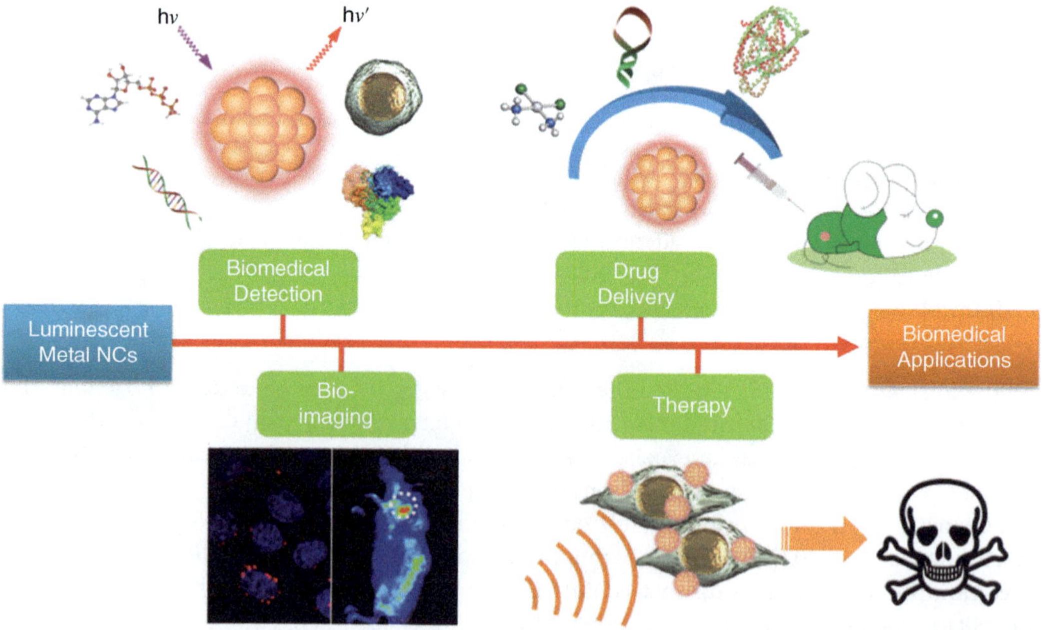

Scheme 1.6 Biomedical applications of luminescent metal NCs. *Source:* Reprinted with permission from [206]. © 2019 Tsinghua University Press and Springer.

In renal clearance of NCs, Zheng and coworkers discovered surprising behavior of Au_{10-12}, Au_{15}, Au_{18}, and Au_{25} NCs (all protected by glutathione) [207b]. Specifically, a merely few-atom decrease in size than Au_{25} was found to result in four- to ninefold reduction in renal clearance efficiency in the early elimination stage. This unique nano–bio interaction in the subnanometer regime is also manifested in the slowdown of the extravasation of Au NCs from normal blood vessels; hence, such NCs can be utilized to enhance their targeting to cancer tissues through an enhanced permeability and retention effect. This atomically precise size effect of NCs opens a new pathway for the development of nanomedicines for diseases associated with glycocalyx dysfunction [207b]. The use of NCs for drug delivery also hold promise in other diseases.

Jia et al. explored gold NCs as a new type of radiosensitizers [200b]. When irradiated by X-rays, Au NCs produce reactive oxygen species, resulting in irreversible cancer cell apoptosis. In the work of Jia et al., they constructed an atomically precise Au_8-levonorgestrel nanocluster and used it as a radiosensitizer for enhancing cancer therapy. The stable isotope of ^{197}Au in gold NCs can also be converted into radioactive ^{198}Au and ^{199}Au radionuclides, which are excellent β^- emitters for radionuclide therapy applications.

In addition to metal NCs, fullerenes doped with Gd and surface derivatized to be water-soluble also show great potential in cancer treatment, and encapsulating of radioactive isotopes could find applications in nuclear medicine.

1.4.3 Antibacteria

The antibacterial functionality of metal NCs (Au, Ag) is quite attractive and has recently been reviewed by Xie and coworkers [208a]. Briefly, the core size of metal NCs was found to not affect the antibacterial ability, but the composition of both the core and the surface ligands greatly influences the antibacterial efficacy [208a].

The assemblies of NCs are also promising in antibacterial applications (see Chapter 16); for example, the silver cluster assemblies can activate O_2 to various ROS, including singlet oxygen ($^1O_2$), superoxide anion (O_2^-), and hydrogen peroxide (H_2O_2), and accordingly exhibit an outstanding photocatalytic inactivation of superbacteria [208b]. The ROS capability of both Ag and Au NCs [208c] may enable their other applications in biology and biomedicine.

1.4.4 Solar Energy Conversion

Fullerenes (e.g. C_{60} derivatives) were widely utilized as electron acceptors in organic or polymer solar cells, albeit in recent years fullerenes have been substituted by organic counterparts that achieve higher solar power conversion efficiencies.

Metal NCs (e.g. Au, Ag) have recently been explored for solar cell applications (Scheme 1.7) owing to their strong absorption of visible and near infrared light. Bang and coworkers have recently reviewed the topic of $Au_n(SR)_m$ NCs for solar energy conversion [198a]. It remains to understand the photosensitization loss in NCs under practical operating conditions. In future research, mechanistic studies toward a better understanding of the photoelectrochemical behavior of metal NCs should be carried out, which will greatly help the establishment of new design principles [198].

1.4.5 Catalysis

Metal NCs are particularly appealing for applications in catalysis, including thermo-, photo-, and electro-catalysis [17, 193]. Compared to conventional NPs in which the size dispersity and surface

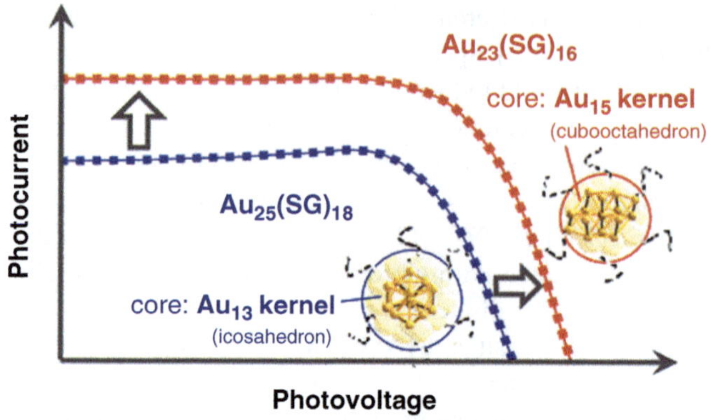

Scheme 1.7 Gold NCs as sensitizers for TiO_2-based photoelectrochemical solar cells. *Source:* Reprinted with permission from [198b]. © 2021 American Chemical Society.

uncertainties complicate the catalytic research, atomically precise NCs provide well-defined models for identifying the active-site and understanding the mechanism at the atomic/molecular level [209]. For example, $Au_{25}(SR)_{18}$ can activate O_2 at room temperature by one-electron transfer from the cluster to O_2, [210] and $Au_{22}(L^8)_6$ (where, L^8 represent diphosphine with an eight-carbon-chain spacer) can hold up to six hydrogen atoms when it activates H_2 [211]. Among the Ag and Cu NCs containing hydride (H^-), a few were reported to be quite useful in catalysis [212]. In contrast, the presence of hydrides in gold nanoclusters is very rare but quite common in Cu and Ag NCs, mainly due to the different electrochemical potentials of Cu, Ag, and Au [79]. In recent work, Tsukuda's group found the presence of H^- in small gold phoshine clusters such as HAu_9, $HPdAu_8$, and $HPdAu_{10}$ (all protected by ligands) [213].

Ligand-protected NCs may not be highly active in catalyzing certain reactions, thus ligand-off NCs are often pursued [193d], which typically exhibit much higher catalytic reactivity but also result in instability and aggregation since there is no ligand protection. In this regard, partially ligand-off NCs could be more promising (Scheme 1.8), in which one or few surface sites become bare and can thus activate reactants while the metal core structure can be retained [17]. Based on the X-ray structures of NCs, DFT simulations can play important roles in revealing the catalytic mechanism [209, 212, 214].

In organic transformation reactions, gold NCs and heterometal-doped ones have been explored for chemoselective hydrogenation of C=O and N=O against C=C, semihydrogenation of C≡C,

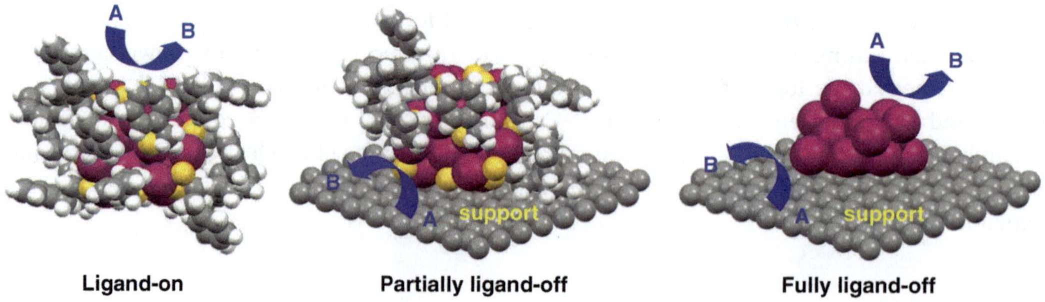

Scheme 1.8 Utilizing atomically precise $Au_n(SR)_m$ nanoclusters for catalysis, including the ligand-on $Au_n(SR)_m$, partially ligand-off $Au_n(SR)_{m-x}$/support, and fully ligand-off Au_n/support catalysts. *Source:* Reprinted with permission from [17]. © 2021 American Chemical Society.

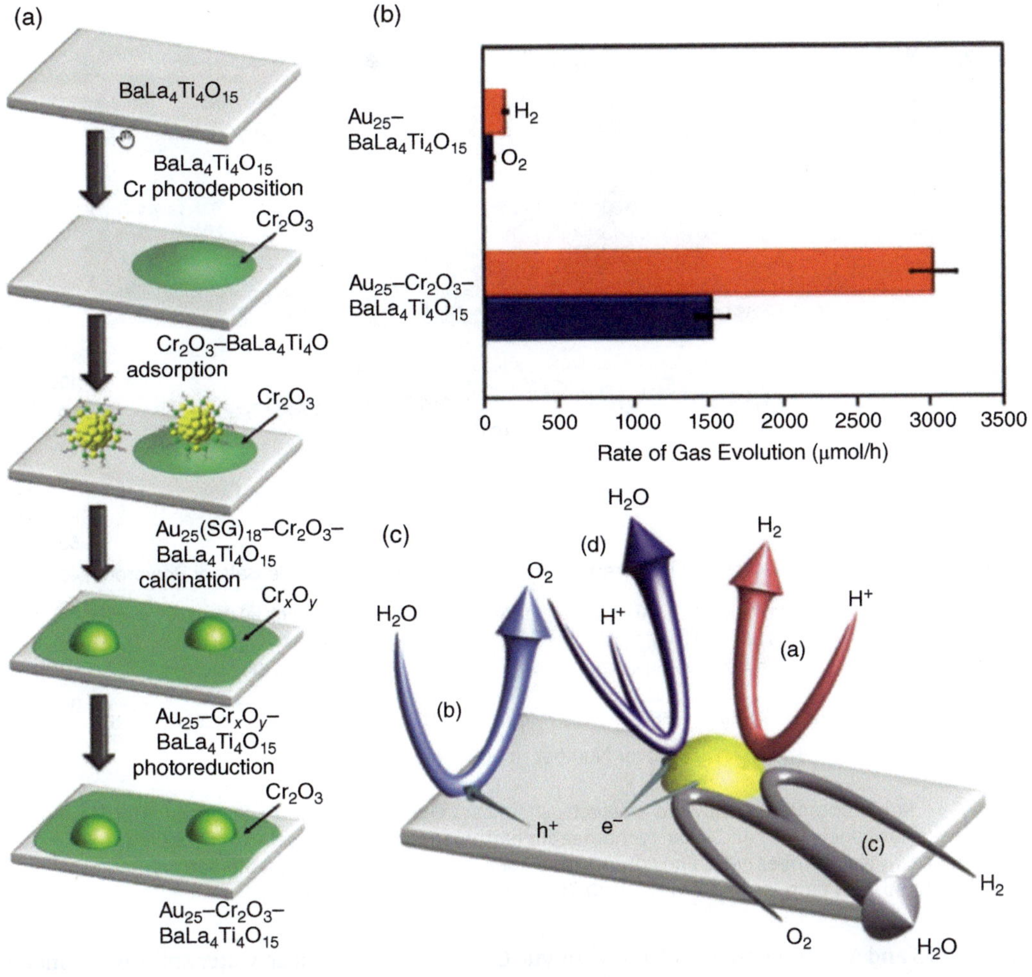

Figure 1.37 Photocatalytic water splitting on metal NCs. (a) Catalyst preparation, (b) Comparison of photocatatlytic activity, (c) Schematic diagram of reactions involved. *Source:* Reprinted with permission from [215a]. © 2018 American Chemical Society.

selective oxidation of alcohol, styrene oxidation and glucose oxidation, carbon–carbon coupling reactions, and many other reactions [17, 193].

In energy catalysis, examples include photocatalytic H_2 generation (e.g. Au_{25}, Ag_{44}), [215] electrocatalytic H_2 evolution [160] and CO_2 reduction, [216, 217] as well as O_2 reduction [218] in fuel cell reactions. For example, Negishi et al. [215] demonstrated photocatalytic water splitting to H_2 and O_2 with high efficiency on Au_{25}-Cr_2O_3-$BaLa_4Ti_4O_{15}$ by suppressing both the backward reaction and Au_{25} aggregation via surface Cr_2O_3 deposition (Figure 1.37). The catalyst showed high activity and stability.

Metal NCs are also promising for electrocatalytic CO_2 reduction [216, 217]. Kauffman et al. [219] explored $Au_{25}(SR)_{18}$ as a renewably powered electrocatalyst for CO_2 reduction to CO with reaction rates between 400 and 800 L of CO_2 per gram catalytic metal per hour and product selectivities of 80–95%. The electrochemical system was coupled with inexpensive consumer grade solar panel or a solar-rechargeable battery (Figure 1.38). Recent work by Li et al. reported single-atom tailoring of gold NCs for enhancing the catalytic activity and electrochemical stability [209]. The well-defined NCs allow one to map out the electrocatalytic active sites [219–221].

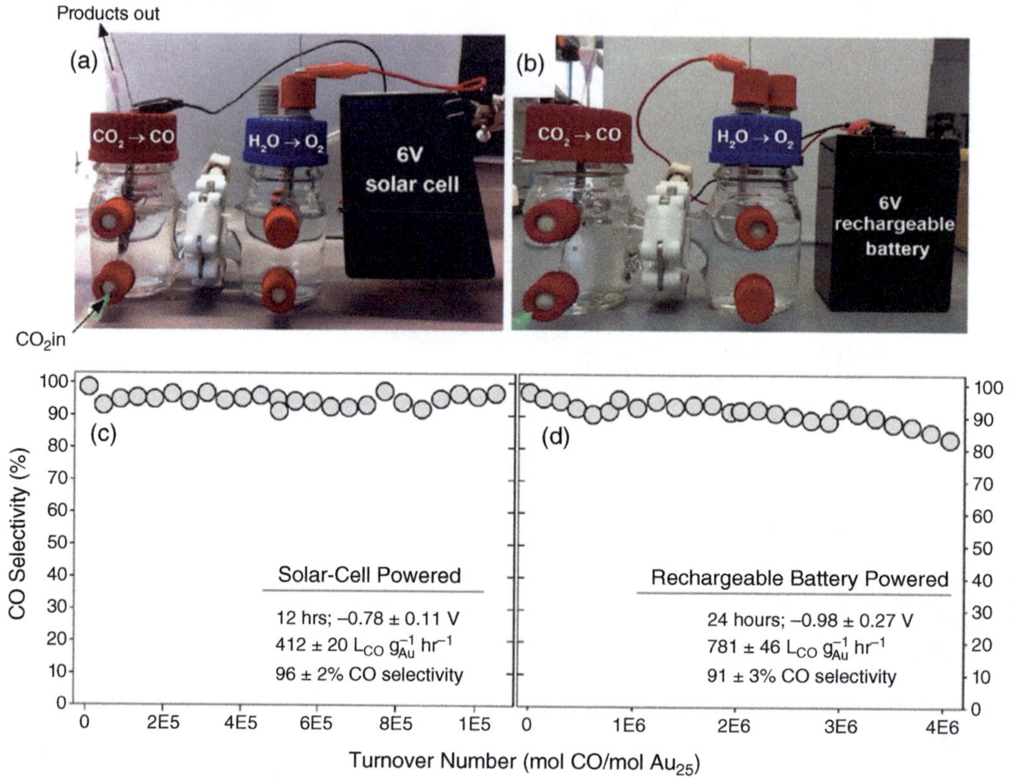

Figure 1.38 (a/b) Photographs showing an electrochemical CO_2 reactor powered by easily-affordable inexpensive solar panels (left) and solar-rechargeable battery (right). (c/d) CO selectivity of CO_2RR as a function of turnover number (when powered with a solar cell or solar-rechargeable battery). *Source:* Reprinted with permission from [219]. © 2015 American Chemical Society.

Chapters 5 and 6 summarize the electrocatalytic CO_2 reduction and also water splitting on metal NCs (e.g. Au, Ag, and Cu-hydrido NCs). The available X-ray structures of NCs enable theoretical modeling for mechanistic understanding (see Chapter 6), including the active site and its possible dynamics, the reactant adsorption configuration, the intermediates, and rate-determining step, as well as the transition states. Atomic-level tailoring (e.g. kernel and shell) has been carried out for enhancing the catalytic activity, selectivity, and durability [17, 193]. The NC-incorporated MOFs have also been utilized for catalysis (see Chapter 16); for instance, Ag_{27}-MOF was demonstrated to be a high-performance catalyst for CO_2-cycloadditions of N-propargylamines.

Enzyme-like catalysis by atomically precise NCs have been demonstrated, such as horseradish peroxidise [222], and *in vivo* biocatalysis has recently been reported by Zhang et al. [223]. With atomically precise NCs, we believe that more exciting progress will be achieved in the development of highly active and selective catalysts for biologically important reactions, such as glutathione peroxidase and superoxide dismutase-like enzymatic activities. By combining with other functionalities of NCs, multi-modal probes can be designed for simultaneous biocatalysis and bioimaging (or therapy) based on the NCs.

With respect to metal-oxo nanoclusters, TOCs are important photocatalysts as their bulk oxide TiO_2. In addition, TOCs may also be utilized as well-defined catalytic supports. The tunable bandgap of TOCs by metal/non-metal doping and ligand modification [174] makes such NCs appealing for visible-light photocatalysis (e.g. by the black TOCs). The TOCs have indeed been applied in photocatalytic H_2 production and CO_2 photoreduction.

1.5 Concluding Remarks

There are many applications beyond those discussed above, including optics (e.g. utilizing the nonlinear optical properties of metal NCs [195–197] and metal-oxo NCs [174]), electronics (e.g. single electron/single particle transistors, [175] 2D transistors with GNRs [61]), optoelectronics (e.g. light-emitting diodes), [202] chiral recognition and separation (e.g. using chiral NCs) [181], magnetic devices (e.g. endohedral metallofullerenes [60] and superatom-assembled materials [175]), etc. New applications also remain to be visioned in future research.

Overall, atomically precise nanochemistry serves as a "hub" for many areas, and the precision materials created by such exquisite chemistry will bring about exciting opportunities in both fundamental research and practical applications, and the concept of atomic precision is expected to have major impact in future nanoscience and other areas.

Acknowledgment

I am grateful to my former and current students, postdoctoral researchers, visiting scholars, and collaborators for their significant contributions to atomically precise nanocluster research. I also thank the funding support from the U.S. National Science Foundation (DMR 1808675), AFOSR, the Camille and Henry Dreyfus Foundation, and DOE.

References

1 Jin, R., Cao, Y.W., Mirkin, C.A. et al. (2001). *Science* 294: 1901.
2 Skrabalak, S.E., Au, L., Li, X., and Xia, Y. (2007). *Nat. Protoc.* 2: 2182.
3 Lohse, S.E. and Murphy, C.J. (2013). *Chem. Mater.* 25: 1250.
4 Narayanan, R. and El-Sayed, M.A. (2004). *J. Phys. Chem. B* 108: 5726.
5 Murray, C.B., Kagan, C.R., and Bawendi, M.G. (2000). *Annu. Rev. Mater. Sci.* 30: 545.
6 Peng, X. (2009). *Nano Res.* 2: 425.
7 Hu, J., Ouyang, M., Yang, P., and Lieber, C.M. (1999). *Nature* 399: 48.
8 Rosi, N.L. and Mirkin, C.A. (2005). *Chem. Rev.* 105: 1547–1562.
9 Jin, R., Charles Cao, Y., Hao, E. et al. (2003). *Nature* 425: 487.
10 Jin, R., Zeng, C., Zhou, M., and Chen, Y. (2016). *Chem. Rev.* 116: 10346.
11 Virovets, A.V., Peresypkina, E., and Scheer, M. (2021). *Chem. Rev.* 121: 14485.
12 Zeng, C. and Jin, R. (2014). *Gold Nanoclusters: Size-Controlled Synthesis and Crystal Structures, in "Gold Clusters, Colloids and Nanoparticles I"* (ed. D.M.P. Mingos), 87–115. Springer.
13 Zeng, C., Chen, Y., Kirschbaum, K. et al. (2016). *Science* 354: 1580.
14 Jackson, A.M., Hu, Y., Silva, P.J., and Stellacci, F. (2006). *J. Am. Chem. Soc.* 128: 11135.
15 Xue, C., Métraux, G.S., Millstone, J.E., and Mirkin, C.A. (2008). *J. Am. Chem. Soc.* 130: 8337.
16 Millstone, J.E., Hurst, S.J., Métraux, G.S. et al. (2008). *Small* 5: 646.
17 Jin, R., Li, G., Sharma, S. et al. (2021). *Chem. Rev.* 121: 567.
18 Wong, O.A., Hansen, R.J., Ni, T.W. et al. (2013). *Nanoscale* 5: 10525.
19 Marjomäki, V., Lahtinen, T., Martikainen, M. et al. (2014). *Proc. Natl Acad. Sci. USA* 111: 1277.
20 Cossairt, B.M. and Owen, J.S. (2011). *Chem. Mater.* 23: 3114.
21 Zhang, J., Feng, P., Bu, X., and Wu, T. (2022). *Natl. Sci. Rev.* 9: nwab076.
22 Beecher, A.N., Dziatko, R.A., Steigerwald, M.L. et al. (2016). *J. Am. Chem. Soc.* 138: 16754.
23 He, L., Luan, C., Rowell, N. et al. (2021). *Chem. Res.* 54: 776.

24 Ma, Z., Mohapatra, J., Wei, K. et al. (2021). *Chem. Rev.* https://doi.org/10.1021/acs.chemrev.1c00860.

25 Respaud, M., Broto, J.M., Rakoto, H. et al. (1998). *Phys. Rev. B* 57: 2925.

26 Chang, H., Bootharaju, M.S., Lee, S. et al. (2021). *Bull. Korean Chem. Soc.* 42: 1386.

27 Stock, A. (1933). *Hydrides of Boron and Silicon*. Ithaca, New York: Cornell Univ. Press.

28 Fox, M.A. and Wade, K. (2003). *Pure Appl. Chem.* 75: 1315.

29 Lewis, G.N. (1916). *J. Am. Chem. Soc.* 38: 762.

30 Longuet-Higgins, H.C. (1949). *J. Chim. Phys.* 46: 275.

31 Lipscomb, W.N. (1959). *Adv. Inorg. Chem. Radiochem.* 1: 117.

32 Wade, K. (1971). *J. Chem. Soc. D* 792.

33 Zintl, E. and Dullenkopf, W. (1932). *Z. Phys, Chem. Abt. B* 16: 183.

34 McGrady, J.E., Weigend, F., and Dehnen, S. (2022). *Chem. Soc. Rev.* 51: 628.

35 Moses, M.J., Fettinger, J.C., and Eichhorn, B.W. (2003). *Science* 300: 778.

36 Shu, C.C., Morgan, H.W.T., Qiao, L. et al. (2020). *Nat. Commun.* 11: 3477.

37 a) Liu, C., Popov, I.A., Chen, Z. et al. (2018). *Chem.–A Eur. J.* 24: 14583. b) Goicoechea, J.M. and Sevov, S.C. (2005). *J. Am. Chem. Soc.* 127: 7676. c) Stegmaier, S. and Fässler, T.F. (2011). *J. Am. Chem. Soc.* 133: 19758. d) King, R.B. and Silaghi-Dumitrescu, I. (2008). *Dalton Trans.* 6083.

38 Nesper, R. (2014). *Z. Anorg. Allg. Chem.* 640: 2639.

39 Mingos, D.M.P. (1984). *Acc. Chem. Res.* 17: 311.

40 Carey, D.M., Morales-Verdejo, C., and Muñoz-Castro, A. (2015). *Chem. Phys. Lett.* 638: 99.

41 Schubert, U. (2021). *Chem. Eur. J.* 27: 11239.

42 Mitchell, K.J., Abboud, K.A., and Christou, G. (2017). *Nat. Commun.* 8: 1445.

43 Watenpaugh, K. and Caughlan, C.N. (1967). *Chem. Commun.* 76.

44 Gao, M.Y., Wang, F., Gu, Z.G. et al. (2016). *J. Am. Chem. Soc.* 138: 2556.

45 Gao, M.Y., Zhang, L., and Zhang, J. (2019). *Chem. Eur. J.* 25: 1.

46 Fang, W.H., Zhang, L., and Zhang, J. (2016). *J. Am. Chem. Soc.* 138: 7480.

47 Fan, X., Yuan, F., Li, D. et al. (2021). *Angew. Chem. Int. Ed.* 60: 12949.

48 Chakrabarty, R., Mukherjee, P.S., and Stang, P.J. (2011). *Chem. Rev.* 111: 6810.

49 Plant, S.R., Cao, L., and Palmer, R.E. (2014). *J. Am. Chem. Soc.* 136: 7559.

50 Alonso, J.A. (2005). *Structure and Properties of Atomic Nanoclusters*. Imperial College Press.

51 Echt, O., Sattler, K., and Rechnagel, E. (1981). *Phys. Rev. Lett.* 47: 1121.

52 Knight, W.D., Clemenger, K., de Heer, W.A. et al. (1984). *Phys. Rev. Lett.* 52: 2141.

53 Li, J., Li, X., Zhai, H.J., and Wang, L.S. (2003). *Science* 299: 864.

54 Gruene, P., Rayner, D.M., Redlich, B. et al. (2008). *Science* 321: 674.

55 Wu, P., Liu, Q., and Chen, G. (2019). *RSC Adv.* 9: 20989.

56 a) Häkkinen, H. (2008). *Chem. Soc. Rev.* 37: 1847–1859. b) Martin, T.P., Bergmann, T., Goehlich, H., and Lange, T. (1991). *J. Phys. Chem.* 95: 6421.

57 Kroto, H.W., Heat, J.R., O'Brian, S.C. et al. (1985). *Nature* 318: 162.

58 Krätschmer, W., Lamb, L.D., Fostiropoulos, K., and Huffman, D.R. (1990). *Nature* 347: 354.

59 Thilgen, C. and Diederich, F. (2006). *Chem. Rev.* 106: 5049.

60 a) Cai, W., Chen, C.H., Chen, N., and Echegoyen, L. (2019). *Acc. Chem. Res.* 52: 1824. b) Guo, T., Diener, M.D., Chai, Y. et al. (1992). *Science* 257: 1661.

61 Yamaguchi, J., Hayashi, H., Jippo, H. et al. (2020). *Commun. Mater.* 1: 36.

62 Cochrane, K.A., Lee, J.H., Kastl, C. et al. (2021). *Nat. Commun.* 12: 7287.

63 Gao, X.J, Mendez, J.P., Lu, T.M. et al. (2021). *International Conference on Simulation of Semiconductor Processes and Devices (SISPAD 2021)*, 102–106. https://sispad2021.org/.

64 Tran, N.T., Powell, D.R., and Dahl, L.F. (2000). *Angew. Chem. Int. Ed.* 39: 4121.

65 Batchelor, L.K., Berti, B., Cesari, C. et al. (2018). *Dalton Trans.* 47: 4467.

66 Köhnlein, H., Purath, A., Klemp, C. et al. (2001). *Inorg. Chem.* 40: 4830.

67 Chen, Y., Phipps, M.L., Werner, J.H. et al. (2018). *Acc. Chem. Res.* 51: 2756.

68 a) Gonzàlez-Rosell, A., Cerretani, C., Mastracco, P. et al. (2021). *Nanoscale Adv.* 3: 1230. b) Cerretani, C., Kanazawa, H., Vosch, T., and Kondo, J. (2019). *Angew. Chem. Int. Edit.* 58: 17153. c) Huard, D.J.E., Demissie, A., Kim, D. et al. (2018). *J. Am. Chem. Soc.* 141: 11465.

69 Konishi, K. (2014). Phosphine-coordinated pure-gold clusters: diverse geometrical structures and unique optical properties/responses. In: *Gold Clusters, Colloids and Nanoparticles I* (ed. D.M.P. Mingos), 49–86. Springer.

70 a) Kenzler, S. and Schnepf, A. (2021). *Chem. Sci.* 12: 3116. b) Adnan, R.H., Madridejos, J.M.L., Alotabi, A.S. et al. (2022). *Adv. Sci.* 9: 2105692.

71 a) Higaki, T., Li, Q., Zhou, M. et al. (2018). *Acc. Chem. Res.* 51: 2764. b) Zhou, M., Higaki, T., Hu, G. et al. (2019). *Science* 364: 279. c) Li, Y., Zhou, M., Song, Y. et al. (2021). *Nature* 594: 380. d) Zeng, C., Liu, C., Chen, Y. et al. (2016). *J. Am. Chem. Soc.* 138: 8710.

72 a) Qian, H., Zhu, M., Gayathri, C. et al. (2011). *ACS Nano* 5: 8935. b) Dainese, T., Agrachev, M., Antonello, S. et al. (2018). *Chem. Sci.* 9: 8796.

73 Lei, Z., Wan, X.K., Yuan, S.F. et al. (2018). *Acc. Chem. Res.* 51: 2465.

74 a) Kulkarni, V.K., Khiarak, B.N., Takano, S. et al. (2022). *J. Am. Chem. Soc.* 144: 9000. b) Shen, H., Wu, Q., Malola, S. et al. (2022). *J. Am. Chem. Soc.* 144: 10844.

75 a) Yang, S., Chai, J., Song, Y. et al. (2015). *J. Am. Chem. Soc.* 137: 10033. b) Jiang, D.E. and Walter, M. (2012). *Nanoscale* 4: 4234.

76 a) Li, Y.Z., Ganguly, R., Hong, K.Y. et al. (2018). *Chem. Sci.* 9: 8723. b) Patty, J.B., Havenridge, S., Tietje-Mckinney, D. et al. (2021). *J. Am. Chem. Soc.* 144: 478.

77 Jin, R. and Higaki, T. (2021). *Commun. Chem.* 4: 28.

78 Femoni, C., Iapalucci, M.C., Ruggieri, S., and Zacchini, S. (2018). *Acc. Chem. Res.* 51: 2748.

79 Nguyen, T.A., Jones, Z.R., Goldsmith, B.R. et al. (2015). *J. Am. Chem. Soc.* 137: 13319.

80 Chakrahari, K.K., Liao, J.H., Kahlal, S. et al. (2016). *Angew. Chem. Int. Ed.* 55: 14704.

81 Li, Y.L., Wang, J., Luo, P. et al. (2019). *Adv. Sci.* 6: 1900833.

82 Ghosh, A., Huang, R.W., Alamer, B. et al. (2019). *ACS Mater. Lett.* 1: 297.

83 a) Lee, S., Bootharaju, M.S., Deng, G. et al. (2021). *J. Am. Chem. Soc.* 143: 12100. b) Wessing, J., Ganesamoorthy, C., Kahlal, S. et al. (2018). *Angew. Chem. Int. Ed.* 57: 14630.

84 Zubarev, D.Y. and Boldyrev, A.I. (2008). *Phys. Chem. Chem. Phys.* 10: 5207.

85 Shinohara, H. (2000). *Rep. Prog. Phys.* 63: 843.

86 a) Qian, H., Zhu, M., Wu, Z., and Jin, R. (2012). *Acc. Chem. Res.* 45: 1470. b) Wu, Z., MacDonald, M.A., Chen, J. et al. (2011). *J. Am. Chem. Soc.* 133: 9670. c) Zeng, C., Chen, Y., Das, A., and Jin, R. (2015). *J. Phys. Chem. Lett.* 6: 2976.

87 a) Gratious, S., Nair, A.S., Mukherjee, S. et al. (2021). *J. Phys. Chem. Lett.* 12: 10987. b) Mukherjee, S., Jayakumar, D., and Mandal, S. (2021). *J. Phys. Chem. C* 125: 12149.

88 Yao, Q., Chen, T., Yuan, X., and Xie, J. (2018). *Acc. Chem. Res.* 51: 1338.

89 a) Narita, K., Ishida, Y., Nukui, S. et al. (2021). *J. Phys. Chem. Lett.* 12: 11761. b) Gunawardene, P.N., Corrigan, J.F., and Workentin, M.S. (2019). *J. Am. Chem. Soc.* 141: 11781.

90 Touchton, A.J., Wu, G., and Hayton, T.W. (2022). *Chem. Sci.* 13: 5171.

91 Li, Y. and Jin, R. (2020). *J. Am. Chem. Soc.* 142: 13627.

92 Zhou, Y., Liao, L., Zhuang, S. et al. (2021). *Angew. Chem. Int. Ed.* 60: 8668.

93 Chen, S., Xiong, L., Wang, S. et al. (2016). *J. Am. Chem. Soc.* 138: 10754.

94 a) Zeng, C., Liu, C., Chen, Y. et al. (2014). *J. Am. Chem. Soc.* 136: 11922. b) Dong, J., Gan, Z., Gu, W. et al. (2021). *Angew. Chem. Int. Ed.* 60: 17932.

95 Zeng, C., Chen, Y., Iida, K. et al. (2016). *J. Am. Chem. Soc.* 138: 3950.

96 Ma, Z., Wang, P., and Pei, Y. (2016). *Nanoscale* 8: 17044.

97 Xu, W.W., Li, Y., Gao, Y., and Zeng, X.C. (2016). *Nanoscale* 8: 7396.

98 Liao, L., Wang, C., Zhuang, S. et al. (2020). *Angew. Chem. Int. Ed.* 59: 731.

99 Liu, X., Xu, W.W., Huang, X. et al. (2020). *Nat. Commun.* 11: 3349.

100 a) Chen, Y., Liu, C., Tang, Q. et al. (2016). *J. Am. Chem. Soc.* 138: 1482. b) Zhuang, S., Liao, L., Zhao, Y. et al. (2018). *Chem. Sci.* 9: 2437.

101 Yang, L., He, Q., Han, W. et al. (2021). *Chem. Phys. Lett.* 785: 139133.

102 a) Xiong, L., Peng, B., Ma, Z. et al. (2017). *Nanoscale* 9: 2895. b) Wang, P., Sun, X., Liu, X. et al. (2017). *J. Phys. Chem. Lett.* 8: 1248.

103 Yang, H., Yan, J., Wang, Y. et al. (2017). *J. Am. Chem. Soc.* 139: 31.

104 Liu, C., Li, T., Abroshan, H. et al. (2018). *Nat. Commun.* 9: 744.

105 Yang, H., Wang, Y., Yan, J. et al. (2014). *J. Am. Chem. Soc.* 136: 7197.

106 Xu, W.W., Zhu, B., Zeng, X.C., and Gao, Y. (2016). *Nat. Commun.* 7: 13574.

107 Zhou, M., Zeng, C., Song, Y. et al. (2017). *Angew. Chem.* 129: 16475.

108 Higaki, T., Zhou, M., Lambright, K.J. et al. (2018). *J. Am. Chem. Soc.* 140: 5691.

109 a) Sakthivel, N.A., Theivendran, S., Ganeshraj, V. et al. (2017). *J. Am. Chem. Soc.* 139: 15450. b) Sakthivel, N.A., Stener, M., Sementa, L. et al. (2018). *J. Phys. Chem. Lett.* 9: 1295.

110 Xu, W.W., Zeng, X.C., and Gao, Y. (2018). *Acc. Chem. Res.* 51: 2739.

111 a) Taylor, M.G. and Mpourmpakis, G. (2017). *Nat. Commun.* 8: 15988. b) Cowan, M.J. and Mpourmpakis, G. (2019). *Nanoscale Adv.* 1: 184.

112 a) Kwak, K., Tang, Q., Kim, M. et al. (2015). *J. Am. Chem. Soc.* 137: 10833. b) Tofanelli, M.A., Salorinne, K., Ni, T.W. et al. (2016). *Chem. Sci.* 1882: 7.

113 Sidgwick, N.V. (1929). *The Electronic Theory of Valency*. London: Oxford University Press.

114 Williams, R.E. (1971). *Inorg. Chem.* 10: 210.

115 Wade, K. (1976). *Adv. Inorg. Chem. Radiochem.* 18: 1.

116 a) King, R.B. (2001). *Chem. Rev.* 101: 1119. b) Boldyrev, A.I. and Wang, L.S. (2005). *Chem. Rev.* 105: 3716. c) von Schleyer, P.R. and Jiao, H.J. (1996). *Pure Appl. Chem.* 68: 209.

117 Mingos, D.M.P. (1976). *J. Chem. Soc. Dalton Trans.* 13: 1163.

118 Mingos, D.M.P. (2015). *Dalton Trans.* 44: 6680.

119 Qian, H., Zhu, Y., and Jin, R. (2012). *Proc. Natl Acad. Sci. USA* 109: 696.

120 a) Castleman, A.W. Jr. and Khanna, S.N. (2009). *J. Phys. Chem. C* 113: 2664. b) Luo, Z., Castleman, A.W. Jr., and Khanna, S.N. (2016). *Chem. Rev.* 116: 14456. c) Jena, P. and Sun, Q. (2018). *Chem. Rev.* 118: 5755.d Zhao, J., Du, Q., Zhou, S., and Kumar, V. (2020). *Chem. Rev.* 120: 9021.

121 a) Akola, J., Walter, M., Whetten, R.L. et al. (2008). *J. Am. Chem. Soc.* 130: 3756. b) Walter, M., Akola, J., Lopez-Acevedo, O. et al. (2008). *Proc. Natl Acad. Sci. USA* 105: 9157.

122 a) Zhu, M., Aikens, C.M., Hollander, F.J. et al. (2008). *J. Am. Chem. Soc.* 130: 5883. b) Aikens, C.M. (2010). *J. Phys. Chem. Lett.* 1: 2594. c) Heaven, M.W., Dass, A., White, P.S. et al. (2008). *J. Am. Chem. Soc.* 130: 3754.

123 Jadzinsky, P.D., Calero, G., Ackerson, C.J. et al. (2007). *Science* 318: 430.

124 Qian, H., Eckenhoff, W.T., Zhu, Y. et al. (2010). *J. Am. Chem. Soc.* 132: 8280.

125 Pei, Y., Wang, P., Ma, Z., and Xiong, L. (2019). *Acc. Chem. Res.* 52: 23.

126 Zhu, M., Aikens, C.M., Hendrich, M.P. et al. (2009). *J. Am. Chem. Soc.* 131: 2490.

127 Zhu, M., Eckenhoff, W.T., Pintauer, T., and Jin, R. (2008). *J. Phys. Chem. C* 112: 14221.

128 Zeng, C., Weitz, A., Withers, G. et al. (2019). *Chem. Sci.* 10: 9684.

129 a) Cheng, L., Yuan, Y., Zhang, X., and Yang, J. (2013). *Angew. Chem. Int. Ed.* 52: 9035–9039. b) Cheng, L., Ren, C., Zhang, X., and Yang, J. (2013). *Nanoscale* 5: 1475.

130 Zeng, C., Chen, Y., Liu, C. et al. (2015). *Sci. Adv.* 1: e1500425.

131 a) Teo, B.K. and Yang, S.Y. (2015). *J. Clust. Sci.* 26: 1923–1941. b) Jemmis, E.D., Balakrishnarajan, M.M., and Pancharatna, P.D. (2001). *J. Am. Chem. Soc.* 123: 4313.

132 Zhou, M., Du, X., Wang, H., and Jin, R. (2021). *ACS Nano* 15: 13980.

133 Maioli, P., Stoll, T., Sauceda, H.E. et al. (2018). *Nano Lett.* 18: 6842.

134 Du, X., Liu, Z., Higaki, T. et al. (2022). *Chem. Sci.* 13: 1925.

135 Sakthivel, N.A., Shabaninezhad, M., Sementa, L. et al. (2020). *J. Am. Chem. Soc.* 142: 15799.

136 Higaki, T., Zhou, M., He, G. et al. (2019). *Proc. Natl Acad. Sci. USA* 116: 13215.

137 Hu, F., Guan, Z.J., Yang, G. et al. (2021). *J. Am. Chem. Soc.* 143: 17059.

138 Negishi, Y., Nakazaki, T., Malola, S. et al. (2015). *J. Am. Chem. Soc.* 137: 1206.

139 Yi, C., Zheng, H., Tvedte, L.M. et al. (2015). *J. Phys. Chem. C* 119: 6307.

140 Yang, H., Wang, Y., Chen, X. et al. (2016). *Nat. Commun.* 7: 12809.

141 Ma, M.X., Ma, X.L., Liang, G.M. et al. (2021). *J. Am. Chem. Soc.* 143: 13731.

142 Yan, J., Zhang, J., Chen, X. et al. (2018). *Natl. Sci. Rev.* 5: 694.

143 Liu, J.Y., Alkan, F., Wang, Z. et al. (2018). *Angew. Chem.* 131: 201.

144 a Wang, Z., Alkan, F., Aikens, C.M. et al. (2022). *Angew. Chem. Int. Ed.* 61: e202206742.b Song, Y., Lambright, K., Zhou, M. et al. (2018). *ACS Nano* 12: 9318.

145 Zhou, M. and Jin, R. (2021). *Annu. Rev. Phys. Chem.* 72: 121.

146 Tlahuice-Flores, A., Jose-Yacamán, M., and Whetten, R.L. (2013). *Phys. Chem. Chem. Phys.* 15: 19557.

147 Jiang, D., Overbury, S.H., and Dai, S. (2013). *J. Am. Chem. Soc.* 135: 8786.

148 Li, Q., Huang, B., Yang, S. et al. (2021). *J. Am. Chem. Soc.* 143: 15224.

149 Kang, X., Li, Y., Zhu, M., and Jin, R. (2020). *Chem. Soc. Rev.* 49: 6443.

150 Hossain, S., Niihori, Y., Nair, L.V. et al. (2018). *Acc. Chem. Res.* 51: 3114.

151 Ghosh, A., Mohammed, O.F., and Bakr, O.M. (2018). *Acc. Chem. Res.* 51: 3094.

152 a) Gan, Z., Xia, N., and Wu, Z. (2018). *Acc. Chem. Res.* 51: 2774. b) Krishnadas, K.R., Baksi, A., Ghosh, A. et al. (2017). *Acc. Chem. Res.* 50: 1988.

153 Sharma, S., Chakrahari, K.K., Saillard, J.Y., and Liu, C.W. (2018). *Acc. Chem. Res.* 51: 2475.

154 Esenturk, E.N., Fettinger, J., and Eichhorn, B. (2006). *J. Am. Chem. Soc.* 128: 9178.

155 Spiekermann, A., Hoffmann, S.D., Kraus, F., and Fässler, T.F. (2007). *Angew. Chem. Int. Ed.* 46: 1638.

156 Popov, A.A., Yang, S., and Dunsch, L. (2013). *Chem. Rev.* 113: 5989.

157 Negishi, Y., Kurashige, W., Niihori, Y. et al. (2010). *Phys. Chem. Chem. Phys.* 12: 6219.

158 Qian, H., Jiang, D., Li, G. et al. (2012). *J. Am. Chem. Soc.* 134: 16159.

159 Xie, S., Tsunoyama, H., Kurashige, W. et al. (2012). *ACS Catal.* 2: 1519.

160 Kwak, K., Choi, W., Tang, Q. et al. (2017). *Nat. Commun.* 8: 14723.

161 Niihori, Y., Koyama, Y., Watanabe, S. et al. (2018). *J. Phys. Chem. Lett.* 9: 4930.

162 Murray, R.W. (2008). *Chem. Rev.* 108: 2688.

163 a) Kwak, K., Thanthirige, V.D., Pyo, K. et al. (2017). *J. Phys. Chem. Lett.* 8: 4898. b) Chen, S., Higaki, T., Ma, H. et al. (2020). *ACS Nano* 14: 16781.

164 Tang, Z., Robinson, D.A., Bokossa, N. et al. (2011). *J. Am. Chem. Soc.* 133: 16037.

165 Li, Y. and Jin, R. (2021). *J. Phys. Chem. C* 125: 15773.

166 Agrachev, M., Ruzzi, M., Venzo, A., and Maran, F. (2019). *Acc. Chem. Res.* 52: 44.

167 Li, J.J., Liu, Z., Guan, Z.J. et al. (2022). *J. Am. Chem. Soc.* 144: 690.

168 Li, Q., Chai, J., Yang, S. et al. (2021). *Small* 17: 1907114.

169 Suyama, M., Takano, S., Nakamura, T., and Tsukuda, T. (2019). *J. Am. Chem. Soc.* 141: 14048.

170 Li, Y., Biswas, S., Luo, T.Y. et al. (2020). *Chem. Mater.* 32: 9238.

171 Cirri, A., Silakov, A., and Lear, B.J. (2015). *Angew. Chem. Int. Ed.* 54: 11750.

172 Liu, F., Krylov, D.S., Spree, L. et al. (2017). *Nat. Commun.* 8: 16098.

173 Tayo, B.O. (2014). *Mater. Focus* 3: 248.

174 Gao, M.Y., Wang, Z., Li, Q.H. et al. (2022). *J. Am. Chem. Soc.* 144: 8153.

175 a) Doud, E.A., Voevodin, A., Hochuli, T.J. et al. (2020). *Nat. Rev. Mater.* 5: 371. b) Roy, X., Lee, C.H., Crowther, A.C. et al. (2013). *Science* 341: 157.

176 Qian, E.A., Wixtrom, A.I., Axtell, J.C. et al. (2017). *Nat. Chem.* 9: 333.

177 a Bonacchi, S., Antonello, S., Dainese, T., and Maran, F. (2020). *Chem.–Eur. J.* 27: 30. b) Sugi, K.S., Maier, A., and Scheele, M. (2022). *Chem. Commun.* 58: 6998.

178 Rival, J.V., Mymoona, P., Lakshmi, K.M. et al. (2021). *Small* 17: 2005718.

179 Chakraborty, A., Fernandez, A.C., Som, A. et al. (2018). *Angew. Chem.* 130: 6632.

180 Yoon, B., Luedtke, W.D., Barnett, R.N. et al. (2014). *Nat. Mater.* 13: 807.

181 Jin, Y., Zhang, C., Dong, X.Y. et al. (2021). *Chem. Soc. Rev.* 50: 2297.

182 Li, Q., Russell, J.C., Luo, T.Y. et al. (2018). *Nat. Commun.* 9: 3871.

183 Yan, J., Malola, S., Hu, C. et al. (2018). *Nat. Commun.* 9: 3357.

184 He, L., Gan, Z., Xia, N. et al. (2019). *Angew. Chem. Int. Ed.* 58: 9897.

185 Yao, Q., Yu, Y., Yuan, X. et al. (2015). *Angew. Chem. Int. Ed.* 54: 184.

186 Galchenko, M., Black, A., Heymann, L., and Klinke, C. (2019). *Adv. Mater.* 31: 1900684.

187 Fetzer, F., Maier, A., Hodas, M. et al. (2020). *Nat. Commun.* 11: 6188.

188 Yuan, P., Zhang, R., Selenius, E. et al. (2020). *Nat. Commun.* 11: 2229.

189 De Nardi, M., Antonello, S., Jiang, D. et al. (2014). *ACS Nano* 8: 8505.

190 Chen, T., Yang, S., Chai, J. et al. (2017). *Sci. Adv.* 3: e1700956.

191 a) Nag, A., Chakraborty, P., Bodiuzzaman, M. et al. (2018). *Nanoscale* 10: 9851. b) Chakraborty, P., Nag, A., Chakraborty, A., and Pradeep, T. (2019). *Acc. Chem. Res.* 52: 2.

192 Han, Z., Dong, X.Y., Luo, P. et al. (2020). *Sci. Adv.* 6: eaay0107.

193 a) Du, Y., Sheng, H., Astruc, D., and Zhu, M. (2020). *Chem. Rev.* 120: 526. b) Fang, J., Zhang, B., Yao, Q. et al. (2016). *Coord. Chem. Rev.* 322: 1. c) Aiken, J.D. III, Lin, Y., and Finke, R.G. (1996). *J. Mol. Catal. A* 114: 29. d) Sudheeshkumar, V., Shivhare, A., and Scott, R.W.J. (2017). *Cat. Sci. Technol.* 7: 272.

194 Chakraborty, I. and Pradeep, T. (2017). *Chem. Rev.* 117: 8208.

195 Olesiak-Banska, J., Waszkielewicz, M., Obstarczyk, P., and Samoc, M. (2019). *Chem. Soc. Rev.* 48: 4087.

196 a) Ramakrishna, G., Varnavski, O., Kim, J. et al. (2008). *J. Am. Chem. Soc.* 130: 5032. b) Yousefalizadeh, G., Ahmadi, S., Mosey, N.J., and Stamplecoskie, K.G. (2021). *Nanoscale* 13: 242.

197 a) Bonacic-Koutecký, V. and Antoine, R. (2019). *Nanoscale* 11: 12436. b) Basu, S., Fakhouri, H., Moulin, C. et al. (2021). *Nanoscale* 13: 4439.

198 a) Khan, R., Naveen, M.H., and Bang, J.H. (2021). *ACS Energy Lett.* 6: 2713. b) Lee, J., Naveen, M.H., Park, J. et al. (2021). *ACS Energy Lett.* 6: 2305.

199 Du, B., Yu, M., and Zheng, J. (2018). *Nat. Rev. Mater.* 3: 358.

200 a) Liu, H., Hong, G., Luo, Z. et al. (2019). *Adv. Mater.* 31: 1901015. b) Jia, T.T., Yang, G., Mo, S.J. et al. (2019). *ACS Nano* 13: 8320.

201 a) Wu, Z., Wang, M., Yang, J. et al. (2012). *Small* 8: 2028. b) George, A., Shibu, E.S., Maliyekkal, S.M. et al. (2012). *ACS Appl. Mater. Interfaces* 4: 639.

202 a) Chen, T., Lin, H., Cao, Y. et al. (2022). *Adv. Mater.* 34: 2103918. b) Wang, S., Meng, X., Das, A. et al. (2014). *Angew. Chem.* 126: 2408.

203 a) Hesari, M. and Ding, Z. (2017). *Acc. Chem. Res.* 50: 218. b) Chen, S., Ma, H., Padelford, J.W. et al. (2019). *J. Am. Chem. Soc.* 141: 9603.

204 a) Zheng, J., Zhou, C., Yu, M., and Liu, J. (2012). *Nanoscale* 4: 4073. b) Maity, S., Bain, D., and Patra, A. (2019). *J. Phys. Chem. C* 123: 2506.

205 a) Zhou, M. and Song, Y. (2021). *J. Phys. Chem. Lett.* 12: 1514. b) Li, Q., Zeman, C.J. IV, Schatz, G.C., and Gu, X.W. (2021). *ACS Nano* 15: 16095.

206 Su, Y., Xue, T., Liu, Y. et al. (2019). *Nano Res.* 12: 1251.

207 a) Jiang, X., Du, B., Tang, S. et al. (2019). *Angew. Chem.* 131: 6055. b) Du, B., Jiang, X., Das, A. et al. (2017). *Nat. Nanotechnol.* 12: 1096.

208 a Zheng, K. and Xie, J. (2021). *Acc. Chem. Res.* 2: 1104–1116. b) Cao, M., Wang, S., Hu, J.H. et al. (2021). *Adv. Sci.* 9: 2103721. c) Kawamura, K., Hikosou, D., Inui, A. et al. (2019). *J. Phys. Chem. C* 123: 26644.

209 Li, S., Nagarajan, A.V., Alfonso, D.R. et al. (2021). *Angew. Chem. Int. Ed.* 60: 6351.

210 Li, G. and Jin, R. (2013). *Acc. Chem. Res.* 46: 1749.

211 Zhang, Q.F., Chen, X., and Wang, L.S. (2018). *Acc. Chem. Res.* 51: 2159.

212 Tang, Q., Hu, G., Fung, V., and Jiang, D. (2018). *Acc. Chem. Res.* 51: 2793.

213 Takano, S., Hasegawa, S., Suyama, M., and Tsukuda, T. (2018). *Acc. Chem. Res.* 51: 3074.

214 Sun, C., Mammen, N., Kaappa, S. et al. (2019). *ACS Nano* 13: 5975.

215 a) Kurashige, W., Kumazawa, R., Ishii, D. et al. (2018). *J. Phys. Chem. C* 122: 13669. b) Wang, Y., Liu, X., Wang, Q. et al. (2020). *Angew. Chem. Int. Ed.* 59: 7748.

216 Kauffman, D.R., Alfonso, D., Matranga, C. et al. (2012). *J. Am. Chem. Soc.* 134: 10237.

217 a) Cai, X., Li, G., Hu, W., and Zhu, Y. (2022). *ACS Catal.* 12: 10638. b) Lin, X., Ma, W., Sun, K. et al. (2021). *J. Phys. Chem. Lett.* 12: 552. c) Qin, L., Sun, F., Ma, X. et al. (2021). *Angew. Chem.* 133: 26340.

218 Sumnner, L., Sakthivel, N.A., Schrock, H. et al. (2018). *J. Phys. Chem. C* 122: 24809.

219 Kauffman, D.R., Thakkar, J., Siva, R. et al. (2015). *ACS Appl. Mater. Interfaces* 7: 15626.

220 Nagarajan, A.V., Loevlie, D.J., Cowan, M.J., and Mpourmpakis, G. (2022). *Curr. Opin. Chem. Eng.* 36: 100784.

221 Seong, H., Efremov, V., Park, G. et al. (2021). *Angew. Chem. Int. Ed.* 60: 14563.

222 Zhao, Y., Zhuang, S., Liao, L. et al. (2019). *J. Am. Chem. Soc.* 142: 973.

223 Liu, H., Li, Y., Sun, S. et al. (2021). *Nat. Commun.* 12: 114.

2

Total Synthesis of Thiolate-Protected Noble Metal Nanoclusters

Qiaofeng Yao[1,2], Yitao Cao[2], Tiankai Chen[2], and Jianping Xie[1,2]

[1] *Joint School of National University of Singapore and Tianjin University, International Campus of Tianjin University, Binhai New City, Fujian, China 350207*
[2] *Department of Chemical and Biomolecular Engineering, National University of Singapore, 4 Engineering Drive 4, Singapore Singapore 117585*

2.1 Introduction

Since 1828, total synthesis of organic and biomolecules has continuously advanced the understanding of their structure-property relationships quantitatively, which has led to rational design and fabrication of a vast library of organic/biomolecules of desired attributes with atomic precision [1].

A good example of such a molecule is penicillin, which is a widely used antibacterial agent, whose successful clinic use in the World War II has saved millions of lives. Although penicillin was first extracted from *Penicillium* fungus, the development of its total synthesis chemistry by Sheehan et al. in 1957 has allowed rational structural derivatization for broadening their spectrum of activity [2]. Total synthesis of organic and biomolecules has been made possible largely because of the wealth of known chemistry of step-by-step reactions along the synthetic routes of the target molecules. As nanoscience enters the era of atomic precision, the rational design of functional nanomaterials at the atomic level has been extensively pursued in the past years. Although global efforts have made commendable achievements in this realm, there are still many long-lasting puzzles in the synthesis, functionalization, and decomposition of metal nanoparticles, which largely restricts mankind's capability of fine modulating metal nanomaterials. Therefore, revealing the finest step-by-step reaction map along the synthesis, functionalization, and decomposition routes of metal nanoparticles is essential for decoding the mysteries in nanoscience. However, due to the vast challenge in producing conventional metal nanoparticles at the atomic precision, the development of total synthesis chemistry in metal nanoparticles is largely lagged in comparison to that in organic chemistry.

Thiolate-protected noble metal (e.g. Au and Ag) nanoclusters (NCs) are a family of ultra-small particles with a core size of 3 nm or lower, representing the missing link between discrete metal atoms and metal nanocrystals (>3 nm) [3]. They can be produced with molecular purity (e.g. molecular formula of $[M_m(SR)_n]^q$, where m, n, and q are the number of metal atoms (M), thiolate ligands (SR), and net charge per cluster, respectively). Recent advances in X-ray crystallography reveal that the packing structures of metal atoms and protecting ligands are highly tunable in metal NCs, not necessarily following the crystalline packing adopted by larger NPs and bulk

Atomically Precise Nanochemistry, First Edition. Edited by Rongchao Jin and De-en Jiang.

metals [4]. It has now become known that $[M_m(SR)_n]^q$ NCs can be described by a "divide-and-protect" scheme, [5] where an M(0) core is capped by a shell of M(I)-SR protecting motifs. In addition, due to the strong quantum confinement effect in this size regime, metal NCs exhibit many size-dependent molecule-like physical and chemical properties (e.g. HOMO-LUMO transition [4m, 6], strong fluorescence [7], intrinsic chirality [8], and enhanced catalytic activity [9]), which have promoted their applications in various fields (e.g. biomedicine [7e, 10], clean energy [11], green catalysis [12], and environmental remediation [13]). Taking the molecule-like properties and structures together, metal nanoclusters can be regarded as "metallic molecules," and the atomic-level size/structure dependency of cluster properties makes these metallic molecules an ideal platform for unraveling the total synthesis chemistry of metal nanoparticles at the atomic precision.

In this chapter, we employ thiolate-protected noble metal NCs as paradigm particles and exemplify the state-of-the-art development of total synthesis chemistry of metal nanoparticles at the molecular and atomic levels. The discussion is based on the core-shell structure of metal NCs, including not only the methodologies delivering the atomic precision in customizing property-dictating attributes (e.g. size, composition, and atomic packing structure) but also the atom-level chemistry governing these tailorabilities. Specifically, we start our discussion with a concise summary on the engineering chemistry of the cluster size, followed by the principles governing the composition customization of metal NCs. Subsequently, recent discoveries on the isomerization of metal NCs are depicted as a means for independently tailoring the atomic packing structure of metal NCs. Besides the growth and functionalization chemistry, we also exemplify the decomposition chemistry of metal NCs in the presence of etchant (e.g. thiol), which is crucial in stabilization and application of metal NCs. At the end of this chapter, we present our perspectives on the further development of total synthesis chemistry of metal nanoparticles.

2.2 Size Engineering of Metal Nanoclusters

Due to its molecule-like nature, metal NCs exhibit highly size-dependent properties. Metal NCs with only a few atoms' difference may show distinct properties, and thus behave differently in their applications [3c, 4a, g, 14]. There are a number of reports comparing the properties of different-sized metal NCs, as well as their performance in various fields of applications like catalysis [15]. For example, the dominant optical absorption peaks of four glutathione-protected Au NCs – namely $Au_{10-12}(SG)_{10-12}$, $Au_{15}(SG)_{13}$, $Au_{18}(SG)_{14}$, and $[Au_{25}(SG)_{18}]^-$ (H-SG = glutathione) – are located at 330/370, 375/415, 560/620, and 675/805 nm, respectively [16]. Their differences in the light absorption efficiency, together with the differences in the quantum yield for electron transfer, have resulted in the situation whereby $Au_{18}(SG)_{14}$ is the best photosensitizer for light harvesting among the four NCs [17]. Therefore, the development of efficient size engineering protocols, especially at atomic precision, is critical to the metal NC/nanoparticle research [18].

2.2.1 Size Engineering by Reduction-Growth Strategy

Employing size engineering protocols at the reduction-growth stage is facile and efficient for producing thermodynamically controlled metal NC sizes. A typical synthesis method for metal NCs is modified from the Brust-Schiffrin method [19]. Metal salts (e.g. $HAuCl_4$ or $AgNO_3$) is mixed with thiol ligands in specific solvent to yield M(I)-SR complexes (M = Au or Ag), followed by the introduction of a reducing agent (e.g. $NaBH_4$ or CO). The reduction-growth process is a fast and vigorous process at the beginning; after that, thermodynamically controlled size focusing reactions will take place. In the size-focusing reactions, the metal NCs of various sizes will be converted to the thermodynamically stable one, thus achieving atomic precision in the synthesis of metal NCs [20]. Hence, size engineering of metal NCs can be achieved by controlling the states of precursors (i.e. M(I)-SR complexes) and reduction environment in the synthesis. By changing the type of ligands, the precursor state (and their tendency of being reduced) may vary. It was reported that different-sized Ag NCs – including Ag_{9-15}, Ag_{25}, and Ag_{44} – can be obtained by using glutathione (GSH), 6-mercaptohexanoic acid (MHA), and *p*-mercaptobenzoic acid (*p*-MBA) as ligands, respectively, while keeping other experimental conditions largely unchanged [21].

Alternatively, pH of the solution in the synthesis regulates both the precursor state and the reducing power of the reducing agent. In the synthesis of GSH-protected Au NCs by CO reduction, $Au_{10-12}(SG)_{10-12}$, $Au_{15}(SG)_{13}$, $Au_{18}(SG)_{14}$, and $[Au_{25}(SG)_{18}]^-$ can be obtained at atomic precision by only tuning the pH of the solution to 7, 9, 10, and 11, respectively [16]. The strategy of size engineering by controlling the pH of solution works well with other ligands (e.g. 3-mercaptopropanic acid [MPA]).

Despite the achievements made in engineering the precursor state and reduction environment, it is still crucial to understand the governing principles in reduction-growth of metal NCs at the molecular level to achieve the size engineering of metal NCs. One useful and powerful way of mapping out the possible reaction pathway is via spectroscopic identification of important reaction intermediates. In the past decade, electrospray ionization mass spectrometry (ESI-MS) has been widely applied to characterize metal NCs and their reaction intermediates [22]. One of the keys to the spectroscopic identification of reaction intermediates is to slow down the reaction kinetics so that the lifetime of the reaction intermediates is prolonged. It can be potentially achieved by replacing the strong reducing agents with a mild one (e.g. CO) [23]. In 2014, our group reported the tracing of stable intermediates in the CO-mediated growth of $[Au_{25}(m\text{-MBA})_{18}]^-$ NCs (*m*-MBA = *meta*-mercaptobenzoic acid) [24]. With the application of ESI-MS, 27 complex or NC species have been identified in the CO-mediated reduction-growth process, as shown in Figure 2.1. All the identified intermediate species have even-numbered valence electron count ($N^* = m-n-q$) ranging from 2 to 10, suggesting that the reduction is a $2e^-$ hopping process. Furthermore, time-dependent evolution of the intermediate species can be studied. The reduction-growth process is comprised of two stages, including kinetically controlled growth and thermodynamically con trolled size focusing. In the first stage, a narrow size distribution of NCs is formed upon reduction, while in the second stage, they slowly transform into atomically precise Au_{25} NCs. Taking the formation and consumption of these intermediate species into consideration, the balanced reactions during the reduction-growth synthesis of Au_{25} NCs can be proposed. Overall, these reactions fall into four categories, which are reduction, isoelectric addition, disproportionation, and comproportionation reactions.

Although the CO-mediated reduction-growth can be monitored by quenching the reaction to observe the intermediates, similar intermediate-tracking process remained challenging for $NaBH_4$-mediated reduction-growth until real-time ESI-MS techniques were developed for metal NCs. It requires the reaction to be moderately slow, but more importantly, the introduced amount of small inorganic ions should be minimized. As such, it can be achieved by reducing the amount of $NaBH_4$ from great excess level to stoichiometric level, as shown in Figure 2.2 [25]. From the analysis on the intermediate species, the $NaBH_4$-mediated reduction-growth also follows a $2e^-$ hopping process, although the sizes of the intermediate species slightly differ from those observed in CO-mediated reduction.

Moreover, from the identification of intermediates and the byproducts, it can be calculated that production of one $[Au_{25}(p\text{-MBA})_{18}]^-$ (*p*-MBA = *para*-mercaptobenzoic acid) NC requires 32 Au atoms from the precursors and 8 electrons from the reducing agents. It can be further proved by the

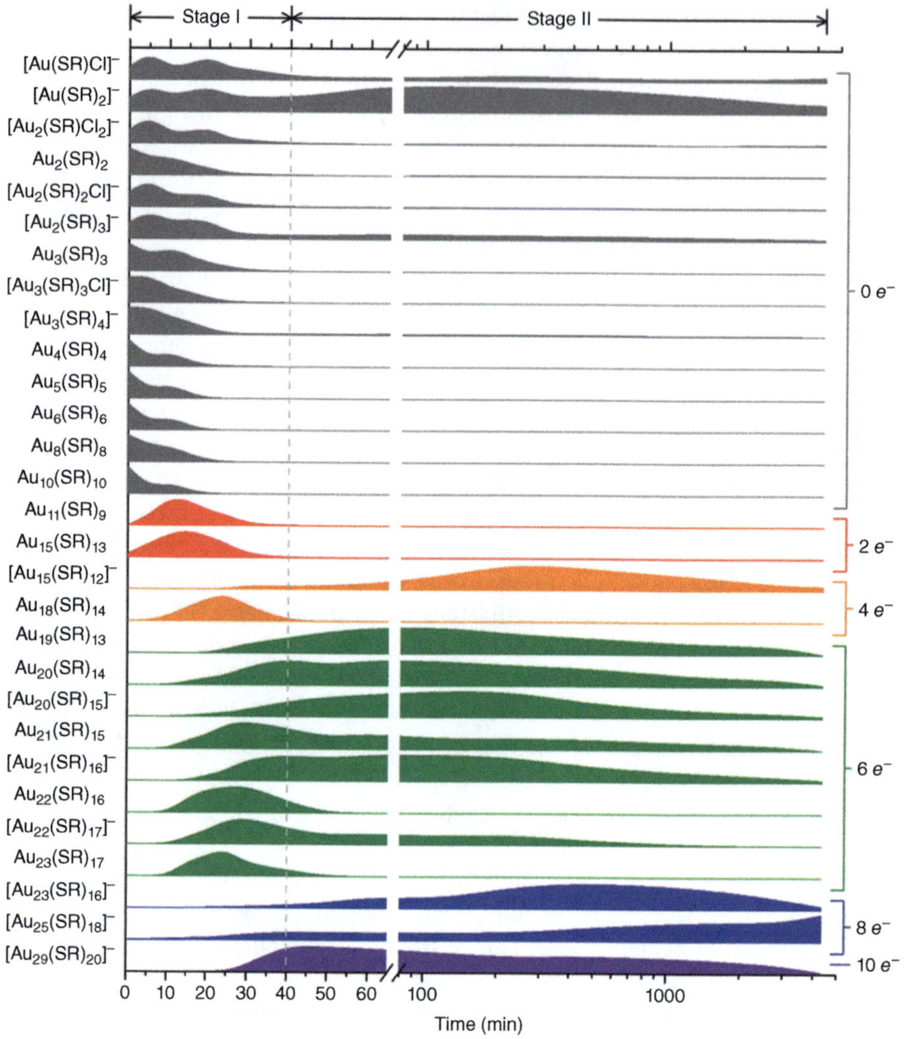

Figure 2.1 Time-dependent abundance of the Au(I) complexes and Au NCs identified by ESI-MS in the CO-mediated synthesis of atomically precise $[Au_{25}(m\text{-MBA})_{18}]^-$. *Source:* Reproduced with permission [24]. Copyright 2014, American Chemical Society.

successful synthesis of $[Au_{25}(p\text{-MBA})_{18}]^-$ NCs with high yield and purity with only stoichiometric amount of reactants were supplied (stoichiometric synthesis). In this synthesis, the amount of reducing agent is calculated to be equivalent to providing 8 electrons per 32 Au atoms in the precursors. Inspired by the stoichiometry and mechanism uncovered, controlling the stoichiometry of the reactants can be an efficient process for size engineering of metal NCs. Similar strategies have been discovered and applied in organic chemistry at atomic precisions for decades, like in the halogenation of alkynes, increasing the moles of halogens will lead the product from haloalkenes to haloalkanes.

It has also been demonstrated that metal NCs that are larger in size (e.g. $Au_{38}(p\text{-MBA})_{24}$ and $[Au_{44}(p\text{-MBA})_{26}]^{2-}$) will form when the amount of reducing agent ($NaBH_4$) increase from the amount used in the stoichiometric synthesis of $[Au_{25}(p\text{-MBA})_{18}]^-$, [25] and smaller intermediates will form when the amount of reducing agent decreases [26].

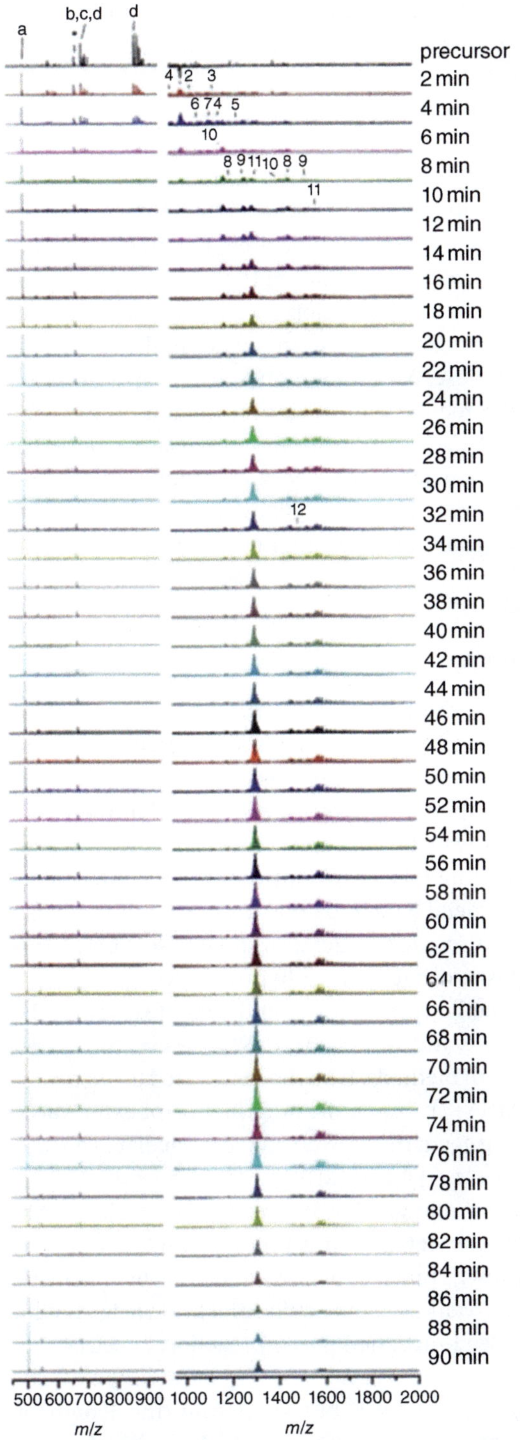

No.	0 e⁻ Complexes
a	$[Au(SR)_2]^-$
b	$Au_4(SR)_4$
c	$Au_6(SR)_6$
d	$Au_{10}(SR)_{10}$
No.	**2 e⁻ NCs**
1	$Au_{15}(SR)_{13}$
No.	**4 e⁻ NCs**
2	$[Au_{16}(SR)_{13}]^-$
3	$[Au_{17}(SR)_{14}]^-$
4	$Au_{18}(SR)_{14}$
No.	**6 e⁻ NCs**
5	$Au_{20}(SR)_{14}$
6	$Au_{21}(SR)_{15}$
7	$[Au_{21}(SR)_{16}]^-$
8	$Au_{23}(SR)_{17}$
9	$Au_{24}(SR)_{18}$
No.	**8 e⁻ NCs**
10	$[Au_{23}(SR)_{16}]^-$
11	$[Au_{25}(SR)_{18}]^-$
No.	**10 e⁻ NCs**
12	$[Au_{29}(SR)_{20}]^-$

Figure 2.2 Real-time ESI mass spectra of Au(I) complexes and Au NCs in the stoichiometric synthesis of $[Au_{25}(p\text{-MBA})_{18}]^-$. *Source:* Reproduced with permission [25]. Copyright 2018, American Chemical Society.

In addition to Au NCs, efforts have been devoted to understanding the reduction-growth of Ag NCs to realize their size engineering at atomic precision. In a $NaBH_4$-mediated reduction-growth of Ag NCs, where the product contains $[Ag_{17}(SPh\text{-}tBu)_{12}]^{3-}$ and $[Ag_{44}(SPh\text{-}tBu)_{30}]^{4-}$ (SPh-*t*Bu = 4-tertbutylbenzenethiolate), it was found by ESI-MS that the two NCs are formed through two distinct pathways, featuring precursors and intermediates containing different numbers of Ag atoms [27]. The pathway leading to $[Ag_{17}(SPh\text{-}tBu)_{12}]^{3-}$ features intermediate species with 22–34 Ag atoms, and the other pathway features intermediate species with 35–57 Ag atoms. Therefore, dedicated control on the intermediate size can be useful in the size engineering of Ag NCs.

Besides ESI-MS, other techniques can also generate useful information that may guide the size engineering of metal NCs. Matrix-assisted laser desorption/ionization (MALDI) is another soft ionization technique other than ESI that can be used for capturing intermediates in the reduction-growth process of metal NCs [28]. The community also uses X-ray techniques like X-ray-absorption fine structure (XAFS) to gain some insights into the reduction-growth process and propose the principles governing the size engineering of metal NCs [29]. Furthermore, polyacrylamide gel electrophoresis (PAGE), which can simultaneously separate metal NCs of different sizes and provide insights on their sizes, has been applied to separate the intermediates during the growth of GSH-protected Ag NCs [30]. A sequential growth model has been proposed for the reduction-growth of these Ag NCs.

2.2.2 Size Engineering by Size Conversion Strategy

Other than the controls applied in the reduction-growth of metal NCs, the size engineering of metal NCs can also be achieved by one-to-one size conversion of metal NCs. The size conversion of metal NCs can be induced by several ways. First, introducing a stimulus of reduction (reducing agent) may further reduce the metal NCs and convert them into larger sizes. Second, introducing a stimulus of oxidation (oxidizing agent) may oxidize the metal NCs and convert them into smaller sizes. This process is typically known as the etching of metals. Lastly, introducing a nonoxidative or reductive stimulus may also lead the conversion of metal NCs into other sizes, via processes like ligand exchange. The etching of metal NCs will be discussed in detail in a later section of this chapter; here we focus on how the size engineering is realized by the other two ways.

Seed-mediated growth is one of the most reliable methods in controlling the size (as well as the morphology) of nanoparticles in nanoscience. Likewise, the size of metal NCs can be tuned and controlled at atomic precision with seed-mediated growth methods. It has been reported that $[Au_{25}(p\text{-}MBA)_{18}]^-$ can grow into atomically precise $[Au_{44}(p\text{-}MBA)_{26}]^{2-}$ upon the introduction of extra Au(I)-thiolate complexes and CO [31]. CO serves as a reducing agent that powers the seed-mediated growth from an $8e^-$ NC (i.e. $[Au_{25}(p\text{-}MBA)_{18}]^-$) into a $20e^-$ one (i.e. $[Au_{44}(p\text{-}MBA)_{26}]^{2-}$).

Moreover, this seed-mediated growth process has been studied at the molecular level. ESI-MS coupled with several other techniques were applied to understand this process. It was found that two different pathways exist in the seed-mediated growth process, which are LaMer-like pathway and aggregative pathway, as shown in Figure 2.3. As a result, the molecule-level knowledge on the size growth pathway makes it possible to drive the seed-mediated growth reaction to selectively produce some intermediate cluster sizes of interest.

Starting from the same metal NC $[Au_{25}(p\text{-}MBA)_{18}]^-$ as seed, but shortening the length of Au(I)-SR complexes introduced and decreasing the reducing power of CO (by lowering the pH), another metal NC species $Au_{38}(p\text{-}MBA)_{24}$ can be synthesized other than $[Au_{44}(p\text{-}MBA)_{26}]^{2-}$. This is because such engineering in the seed-mediated growth process favors the LaMer-like process and $Au_{38}(p\text{-}MBA)_{24}$ is a critical intermediate species in this process, thus realizing the kinetic trapping of this NC species.

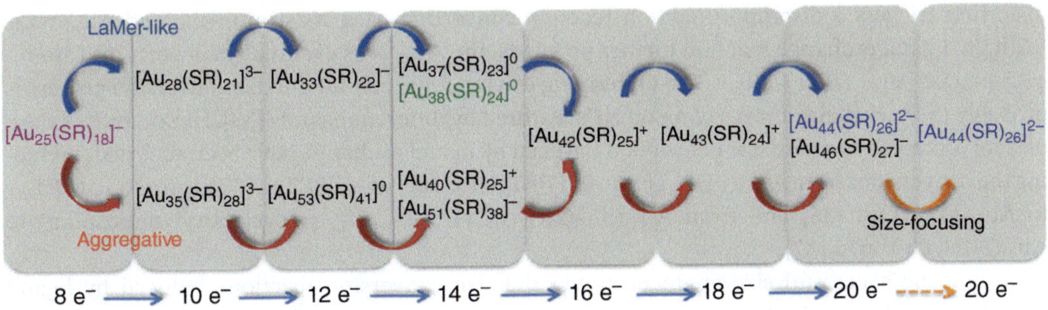

Figure 2.3 Schematic illustration of the growth pathways from $[Au_{25}(SR)_{18}]^-$ to $[Au_{44}(SR)_{26}]^{2-}$ (SR = *p*-mercaptobenzoic acid), including the LaMer-like pathway and the aggregative pathway. *Source:* Reproduced with permission [31]. Copyright 2017, the Authors, published by Springer Nature.

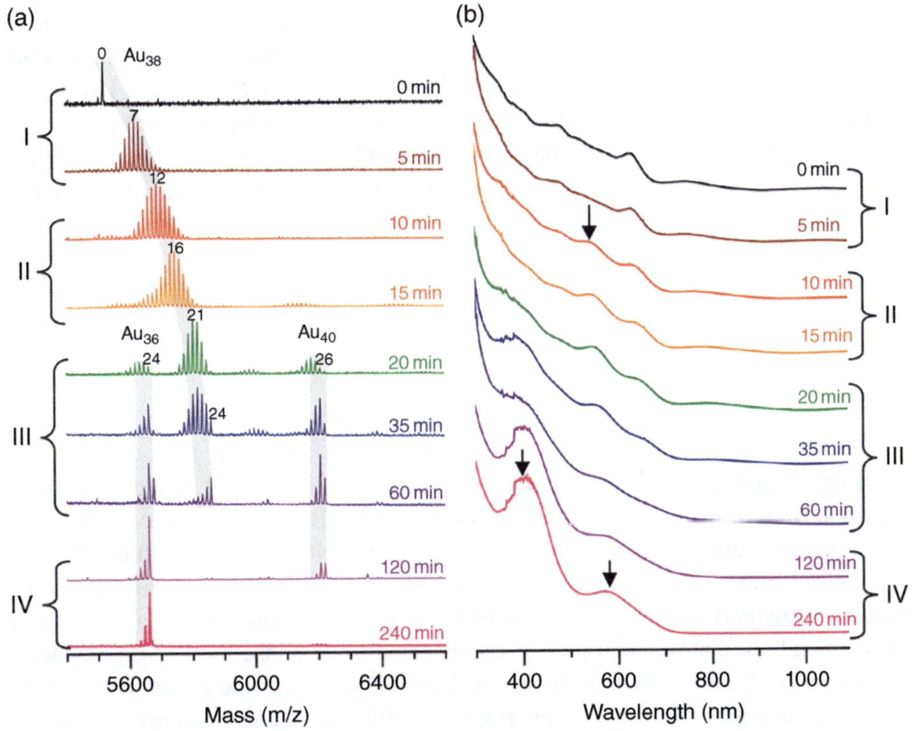

Figure 2.4 Time-course (a) ESI-MS and (b) UV-vis absorption spectra of the ligand exchange induced size conversion process from $Au_{38}(PET)_{24}$ to $Au_{36}(TBBT)_{24}$. *Source:* Reproduced with permission [32]. Copyright 2013, American Chemical Society.

Size conversion of metal NCs may also occur without deliberately introducing redox reactions. One good example is that the metal NCs of one size may convert to another via ligand exchange reactions. Ligand exchange induced size conversion was achieved by adding 4-*tert*-butylbenzenethiol (TBBT) into $Au_{38}(PET)_{24}$ NCs (PET = 2-phenylethanethiol) [32]. As shown in Figure 2.4, the introduced TBBT first replaces up to 12 PET ligands on the $Au_{38}(SR)_{24}$ NCs at the early stage of this reaction. In the subsequent stage, more PET ligands (>12 per NC) are replaced by the introduced TBBT. With the extensive replacement of PET by TBBT, structural distortion of the metal core start to take place in $Au_{38}(SR)_{24}$ NCs because of the bulkier hydrocarbon tails of TBBT. This structural distortion triggers a disproportionation reaction converting Au_{38} NCs into Au_{40} and Au_{36} NCs. With the ligand exchange reaction further proceeds, the Au_{40} NCs eventually converts into atomically precise $Au_{36}(TBBT)_{24}$ NCs. The underlying driving force behind the size conversion process is probably the varied magic sizes of metal NCs featured by different ligands [33]. Hence, more examples for atomically precise size conversion induced by ligand exchange have been reported, including the conversions from $Au_{25}(PET)_{18}$ to $Au_{28}(TBBT)_{20}$ [34] or $Au_{20}(TBBT)_{16}$ [35], from $Au_{144}(PET)_{60}$ to $Au_{133}(TBBT)_{52}$, [36] and from $Au_{23}(S\text{-}c\text{-}C_6H_{11})_{16}$ (H-S-c-C_6H_{11} = cyclohexyl mercaptan) to $Au_{24}(SCH_2Ph\text{-}t\text{Bu})_{20}$ [37].

Despite the successful size engineering via the size conversion reactions induced by ligand exchange reactions, it is also important to develop synthetic method for customizing cluster size while maintaining the type of ligands unchanged. This is because the ligand also plays an important part in dictating the recognition, self-assembly, and application chemistry of metal NCs. Owing to worldwide efforts devoted to synthetic chemistry in the past years, ligand-maintained size conversion has been made possible via a couple of delicate methods. For example, in the presence of light and oxygen, $[Au_{23}(S\text{-}c\text{-}C_6H_{11})_{16}]^-$ NCs can grow into $Au_{28}(S\text{-}c\text{-}C_6H_{11})_{20}$, which is larger in size but same in the type of ligands [38]. Mechanistic studies suggest that the conversion process requires photons possessing energy higher than the HOMO-LUMO gap of $[Au_{23}(S\text{-}c\text{-}C_6H_{11})_{16}]^-$. The excited NCs decompose into smaller NCs like $Au_{18}(S\text{-}c\text{-}C_6H_{11})_{14}$, which later react with Au(I)-(S-c-C_6H_{11}) complexes and reassemble into $Au_{28}(S\text{-}c\text{-}C_6H_{11})_{20}$. Interestingly, the conversion of these two NCs can also be induced by an oxidation process by introducing H_2O_2 to $[Au_{23}(S\text{-}c\text{-}C_6H_{11})_{16}]^-$ [39]. Although oxygen is involved in the conversion process, these two NCs are identical in the valence electron count (both being 8 e^- NC). These findings suggest that the size engineering of metal NCs at atomic precisions can be realized via one-to-one size conversion initiated by various stimuli.

2.3 Composition Engineering of Metal Nanoclusters

Besides the size of cluster, the composition of metal NCs is equally important in dictating their properties. The composition of metal NCs include the metal composition and ligand composition, where the former dominates the HOMO-LUMO structure of metal NCs, [4m] while the latter not only influences the electronic structure of metal NCs but also determines the recognition, assembly, and application chemistry of metal NCs. Due to their pivotal importance in fundamental and applied research of metal NCs, atomically precise engineering to compose metal NCs has been consistently pursued since the very dawn of cluster research. The worldwide efforts have established a series of methodologies toward customizing the metal and ligand compositions at the atomic precision. These methods include co-reduction or in-situ reduction, metal/ligand/surface motif exchange, and inter-cluster reaction.

2.3.1 Metal Composition Engineering

Precisely alloying metal NCs is a long-lasting pursue in cluster research, which requires control of the alloying content and alloying site at the atomic level. The controls of alloying content and alloying site are both highly element dependent. Taking the most investigated $[Au_{25}(SR)_{18}]^-$ as an example, decades of development have allowed Au atoms substituted by many other metal elements like Ag, Cu, Pt, Pd, Cd, and Hg, where monodisperse $[Au_{24}M(SR)_{18}]^{0/-1}$ NCs are yielded with M = Pt, Pd, Cd, and Hg [40], while a mixture of $[Au_{25-x}M_x(SR)_{18}]^-$ NCs are preferred with M = Ag and Cu [41]. This should be in principle attributed to the varied chemical environments of

Au atoms in the $[Au_{25}(SR)_{18}]^-$ NCs. $[Au_{25}(SR)_{18}]^-$ consists of a centered-icosahedral Au_{13} core and 6 SR-$[Au(I)\text{-}SR]_2$ surface motifs [41l, m]. Such core-shell structure of $Au@Au_{12}@6[Au(SR)_2]$ gives rise to three categories of Au sites: 1 central Au atom, 12 Au atoms in the apex sites of Au_{13} core, and 12 Au atoms in the Au(I)-SR motifs. The X-ray structure analysis results indicate that Ag heteroatoms preferentially substitute the Au atoms in the apex sites of M_{13} core [41a], while the Pt heteroatoms favor the central site in the M_{13} core [42]. This can well corroborate the monodisperse and polydisperse alloying contents in the cases of M = Pt and M = Ag, respectively. However, it should also be mentioned that Cd heteroatom is observed in both the central and apex sites of M_{13} core [43], and Hg heteroatom is witnessed in the M(I)-SR motifs [44]. As there are 12 equivalent Au sites in both apex sites of Au_{13} core and Au(I)-SR motifs, the monodisperse $[Au_{24}M(SR)_{18}]^0$ (M = Cd and Hg) may be a product of kinetics [43b].

The most common way to produce alloy NCs is co-reduction of M(I)-SR complexes. By co-reduction of Au(I)-$(SC_{12}H_{25})$ and Ag(I)-$(SC_{12}H_{25})$ complexes (H-$SC_{12}H_{25}$ = 1-dodecanethiol), Negishi et al. synthesized a series of bimetallic $[Au_{25-x}Ag_x(SR)_{18}]^-$ NCs with x ranging from 0 to 11 [45]. Although it is extremely difficult to narrow the x value down to a defined number, the authors reported that the x value can be controlled in a narrow range with its average value tunable via the feeding ratio of Au(I)/Ag(I). Similar tunability of alloying content is also reported in water-soluble $[Au_{25-x}Ag_x(SR)_{18}]^-$ NCs. Dou et al. employed a NaOH-assisted $NaBH_4$ reduction method to produce a variety of $[Au_{25-x}Ag_x(MHA)_{18}]^-$ with x = 0–11, where the average x was tailored by the ratio of feeding Au(I)/Ag(I) (Figure 2.5) [46]. It should be noted that the marked stability of $[Au_{25}(SR)_{18}]^-$ often preserves it as a byproduct in the synthesis of bimetallic $[Au_{25-x}M_x(SR)_{18}]^-$ via the co-reduction method [45, 46]. This remains true in the co-reduction synthesis of $Au_{24}Pt(SR)_{18}$ [47]. Qian et al. reduced combined Au(III) and Pt(II) by $NaBH_4$ in the presence of

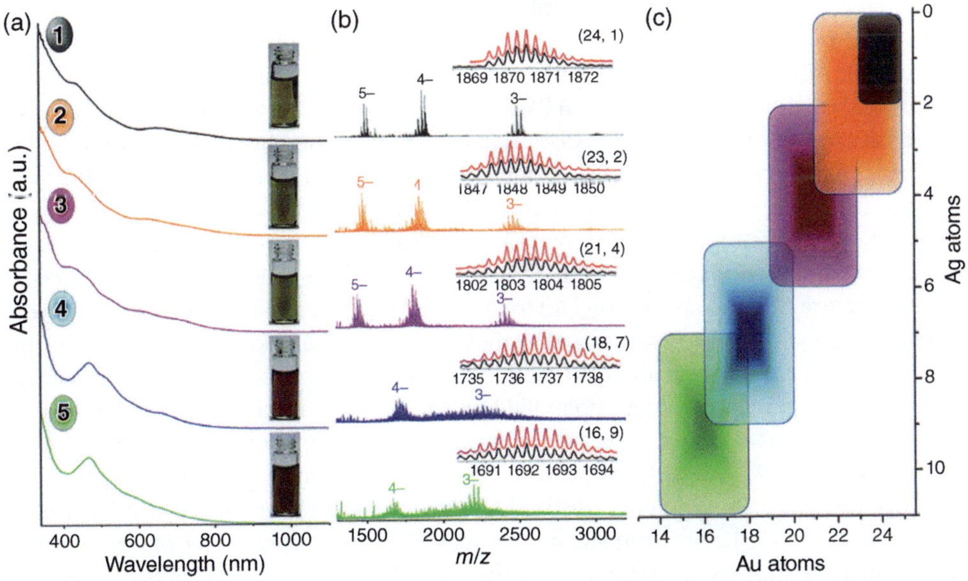

Figure 2.5 (a) UV-vis absorption spectra, (b) ESI-MS spectra, and (c) metal atom distribution of $[Au_{25-x}Ag_x(SR)_{18}]^-$ NCs synthesized by co-reduction of Au(I)-SR and Ag(I)-SR complexes at varied feeding molar ratios of Au(I)/Ag(I): 1/24 (NC-1), 22/3 (NC-2), 18/7 (NC-3), 16/9 (NC-4), and 14/11 (NC-5). The insets in (a) are the digital photos of corresponding alloy NCs, while those in (b) show experimental (black lines) and simulated (red lines) isotope patterns of $Au_{25-x}Ag_x(SR)_{18}$ NCs carrying 4- charges. The parentheses indicate (25−x, x) of individual cluster species. *Source:* Reproduced with permission [46], Copyright 2014, Royal Society of Chemistry.

excess 1-phenylethanethiol (H-SC_2H_4Ph), giving rise to a mixture of $[Au_{25}(SC_2H_4Ph)_{18}]^-$ and $Au_{24}Pt(SC_2H_4Ph)_{18}$. To eliminate the undesired $[Au_{25}(SC_2H_4Ph)_{18}]^-$, the authors added H_2O_2 to oxidatively decompose $[Au_{25}(SC_2H_4Ph)_{18}]^-$, leaving alone monodisperse $Au_{24}Pt(SC_2H_4Ph)_{18}$ as the final product. Similar co-reduction methods have been used for the synthesis of many other-sized metal NCs, including $Au_{38-x}Ag_x(SR)_{24}$, [48] $[Ag_{28}Pt(S_2R)_{12}(PPh_3)_4]^{4-}$ (S_2R denotes dithiolate ligand), [49], and $Au_{144-x}Cu_x(SR)_{60}$ [50].

In addition to co-reduction of M(I)-SR complexes of mixed metals, alloy NCs can be facilely synthesized by the NC-templated methods, where presynthesized metal NCs are employed as template to react with hetero-metal precursors in diverse forms like metal salt, M(I)-SR complexes, and metal NCs. As the most straightforward approach, the synthesis of alloy NCs can be made possible by the galvanic replacement reactions between parental metal NCs and hetero-metal salt. Bootharaju et al. developed a galvanic replacement method for producing monodisperse $[Ag_{24}Au(SR)_{18}]^-$ NCs [51]. $[Ag_{25}(SR)_{18}]^-$ NCs were employed as parental NCs and allowed to react with Au(I) salt ($AuClPPh_3$ in this case), where the galvanic replacement reaction between the Ag(0) core of $[Ag_{25}(SR)_{18}]^-$ and $AuClPPh_3$ gave rise to high-quality $[Ag_{24}Au(SR)_{18}]^-$ NCs. The X-ray crystallography analysis on as-formed $[Ag_{24}Au(SR)_{18}]^-$ NCs manifests a M-S framework similar to those of pure $[Ag_{25}(SR)_{18}]^-$ or $[Au_{25}(SR)_{18}]^-$ NCs, while the Au heteroatom is suggested to replace the central Ag atom of parental $[Ag_{25}(SR)_{18}]^-$ NCs. Although the detailed formation mechanism of $[Ag_{24}Au(SR)_{18}]^-$ NCs is unknown, the authors proposed that the galvanic replacement reaction occurred initially at the apex site of icosahedral Ag_{13} core, followed by the intra-cluster diffusion of Au(0) heteroatom from the apex site to the central site of M_{13} icosahedron.

In sharp contrast to galvanic replacement reactions, anti-galvanic replacement reaction is a recent discovery in the synthetic chemistry of metal NCs. It represents a reaction in which the M(0) core of less reactive metal is oxidatively replaced by ions of more reactive metal. For example, Yao et al. prepared monodisperse $Au_{24}Cd(SR)_{18}$ NCs by reacting preformed $[Au_{25}(SR)_{18}]^-$ with Cd^{2+} in a acetonitrile solution [43b]. This is similar to the production of monodisperse $Au_{24}Hg(SR)_{18}$ by the anti-galvanic reaction between parental $[Au_{25}(SR)_{18}]^-$ with Hg^{2+}, [44] but in sharp contrast to the reactions of $[Au_{25}(SR)_{18}]^-$ with Zn^{2+}, Co^{2+}, and Ni^{2+} documented by the same group [43b]. In the reactions of $[Au_{25}(SR)_{18}]^-$ with Zn^{2+}, Co^{2+}, and Ni^{2+}, the oxidization of anionic $[Au_{25}(SR)_{18}]^-$ into neutral $Au_{25}(SR)_{18}$ (in the cases of Co^{2+} and Ni^{2+}) or cationic $[Au_{25}(SR)_{18}]^+$ (in the case of Zn^{2+}) was observed rather than the metal replacement reaction by heteroatoms. The authors attributed the successful incorporation of Cd heteroatom into the $[Au_{25}(SR)_{18}]^-$ to the similar size and electronic structure Cd (and apparently Hg) to those of Au [43b]. The molecular monodispersity of $Au_{24}M(SR)_{18}$ (M = Cd or Hg) synthesized by the anti-galvanic reactions should, however, be attributed to the limited reducing capability of the parental NCs, which can reduce only one equivalent of Cd^{2+} or Hg^{2+} and incorporate it into individual NC. More intriguingly, $Au_{24}Cd(SR)_{18}$ can be transferred into $Au_{24}Hg(SR)_{18}$ by mixing the former with excess Hg^{2+}, while the reverse conversion from $Au_{24}Hg(SR)_{18}$ to $Au_{24}Cd(SR)_{18}$ is proved difficult [43b]. X-ray crystallography data from the same group also suggests different substituting sites of heteroatom in $Au_{24}Cd(SR)_{18}$ and $Au_{24}Hg(SR)_{18}$, where the Cd heteroatom sits on the apex site of the icosahedral M_{13} core of the former and the Hg heteroatom is accommodated in the M(I)-SR protecting motifs of the latter. This observation indicates that, in the reaction of $Au_{24}Cd(SR)_{18}$ with Hg^{2+}, the replacement of Cd by Hg may first occur at the apex site of M_{13} core, followed by the diffusion of Hg heteroatom from the M_{13} core to the M(I)-SR motif. Similar anti-galvanic reaction between $[Au_{25}(SR)_{18}]^-$ and Ag^+ was also investigated by Choi et al. based on combined electrochemical and spectroscopic means [52]. The author speculated that bimetallic $[Au_{25-x}Ag_x(SR)_{18}]^q$ NCs were formed via a different addition-dissociation mechanism, featuring formation of $Au_{25}Ag(SR)_{18}$ intermediate adduct.

In addition to metal salts, preformed metal NCs can react with M(I)-SR complexes for the production of alloy NCs. Our group employed $[Ag_{44}(p\text{-MBA})_{30}]^{4-}$ as template NCs to react with Au(I)-SR′ complexes, in which SR′ can be identical to or different from *p*-MBA [53]. The reaction gives rise to monodisperse $[Ag_{44-x}Au_x(SR)_{30}]^{4-}$ NCs, whose *x* values exhibit marked dependence on the feeding dosage of Au(I)-SR complexes. With increasing dosage of Au(I)-SR complexes, the *x* values ascend and reach a maximum at 12, corroborating to the documented superior stability of $[Ag_{32}Au_{12}(SR)_{30}]^{4-}$. The alloying process was monitored by delicate ESI-MS analyses, indicating the reaction follows Eq. (2.1):

$$\begin{aligned}&\left[Ag_{44}\left(SR\right)_{30}\right]^{4-}+x\left[Au_2\left(SR'\right)_2Cl\right]^{-}\rightarrow\\&\quad\left[Ag_{44-x}Au_x\left(SR\right)_{30-2x}\left(SR'\right)_{2x}\right]^{4-}+x\left[Au\left(SR\right)_2\right]^{-}+x\,AgCl\left(x=1-12\right)\end{aligned}\qquad(2.1)$$

Based on the stoichiometry detailed above and the known crystal structure of $[Ag_{44}(SR)_{30}]^{4-}$, we proposed a surface motif exchange mechanism for the alloying reaction between parental $[Ag_{44}(SR)_{30}]^{4-}$ NCs and $Au_2(SR')_2Cl$ complexes. As shown in Figure 2.6, association of $Au_2(SR')_2Cl$ to a surface Ag(I) atom of $[Ag_{44}(SR)_{30}]^{4-}$ can induce the destruction of Ag—S bond and formation of Ag—Cl bond on this particular Ag(I) atom. Subsequent dissociation of AgCl and $Au(SR)_2$ from the NC adduct is able to produce $[Ag_{43}Au(SR)_{28}(SR')_2]^{4-}$. Upon sufficient supply of $Au_2(SR')_2Cl$ complexes, the surface motif exchange reaction can proceed to its completion on the surface of $[Ag_{44}(SR)_{30}]^{4-}$, where all 12 surface Ag(I) atoms can be replaced by Au(I) heteroatoms. In addition to $[Ag_{44}(SR)_{30}]^{4-}$ NCs, many other mono-metallic (e.g. $[Au_{25}(SR)_{18}]^-$ and $Au_{38}(SR)_{24}$) [43a, 54] and bimetallic (e.g. $[Au_{25-x}Ag_x(SR)_{18}]^-$) [55] NCs have been employed as template NCs in the reaction with M(I)-SR complexes, giving rise to alloy NCs preserving the M-S framework of the parental NCs.

A recent discovery in the NC-templated synthesis is the intra-cluster heteroatom diffusion, as the original substitution site of heteroatom may be different from its energetically favorable site in the alloy NCs. For example, Zheng et al. synthesized $[Ag_{24}Au(MHA)_{18}]^-$ NCs by reacting parental

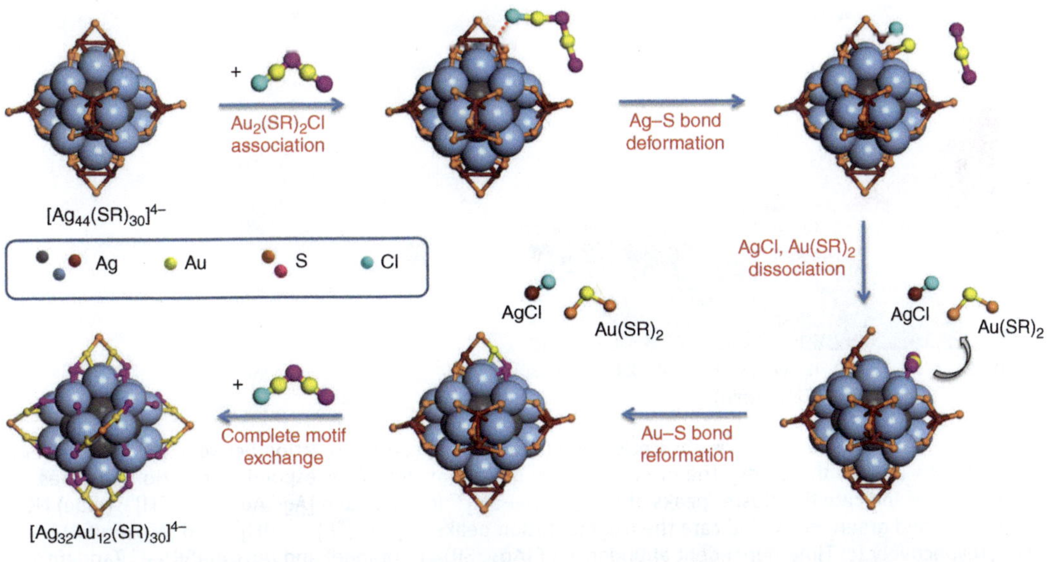

Figure 2.6 Schematic illustration showing the proposed surface motif exchange mechanism for the alloying reaction between $[Ag_{44}(SR)_{30}]^{4-}$ and $Au_2(SR)Cl$. *Source:* Reproduced with permission [53]. Copyright 2017, the Authors, published by Springer Nature.

$[Ag_{25}(MHA)_{18}]^-$ NCs with Au(I)-(MHA) complexes with a molar ratio of $Ag_{25}/Au(I) = 1/1$ [56]. The combined UV-vis absorption and ESI-MS spectra indicate that the Au heteroatom sits in the center of icosahedral M_{13} core in the thermodynamically stable $[Ag_{24}Au(MHA)_{18}]^-$ NCs. In order to furnish more details about the dynamics of the alloy reaction, the authors monitored the alloying reaction by time-course UV-vis absorption, ESI-MS, and tandem MS (MS/MS) (Figure 2.7). The combined optical and mass spectra data reveals an intra-cluster diffusion mechanism for the structure evolution of $[Ag_{24}Au(MHA)_{18}]^-$ NCs. Specifically, the reaction of $[Ag_{25}(MHA)_{18}]^-$ NCs with Au(I)-(MHA) complexes would first incorporate a Au heteroatom into the surface motifs of the $[Ag_{25}(MHA)_{18}]^-$, yielding $[Ag_{24}Au(MHA)_{18}]^-$. With the reaction proceeding, the Au heteroatom would gradually migrate from the surface motifs, via the apex sites of the icosahedral M_{13} core, and finally to the center of the icosahedral M_{13} core. It should be reminded that the intra-cluster diffusion has long been speculated as an important mechanism for the structure evolution of alloy NCs (see above discussion for some examples) [43b, 51], and this work provides for the first time the molecule-level evidence on the inward diffusion of heteroatoms.

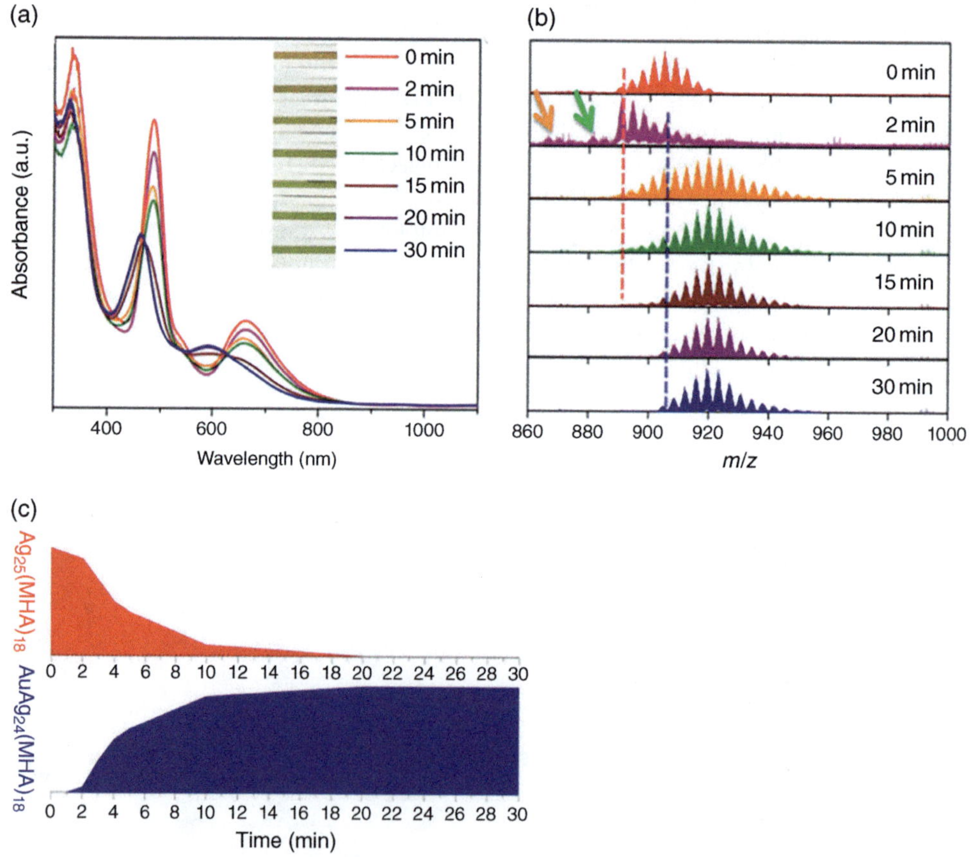

Figure 2.7 Time-course (a) UV-vis absorption and (b) ESI-MS spectra of reaction between $[Ag_{25}(SR)_{18}]^-$ and Au(I)-SR (SR = MHA in this study). The insets in (a) are digital images of corresponding reaction mixtures. The dash lines indicate the cluster peaks of $[Ag_{25}(SR)_{18}-5H]^{6-}$ (orange) and $[Ag_{24}Au(SR)_{18}-5H]^{6-}$ (blue) NCs. The orange and green arrows indicate the fragmentation peaks of $[Ag_{25}(SR)_{17} - 4H]^{6-}$ and $[Ag_{24}Au(SR)_{17} - 4H]^{6-}$, respectively. (c) Time-dependent abundance of $[Ag_{25}(SR)_{18}]^-$ (orange) and $[Ag_{24}Au(SR)_{18}]^-$. Tandem mass spectra of (d) $[Ag_{25}(SR)_{18}-5H]^{6-}$ and (e) $[Ag_{24}Au(SR)_{18}-5H]^{6-}$ NCs recorded at the collision energies of 2 and 6 eV at varied reaction time. (f) Time-dependent abundance of each fragment ions captured in (d) and (e). *Source:* Reproduced with permission [56]. Copyright 2019, American Chemical Society.

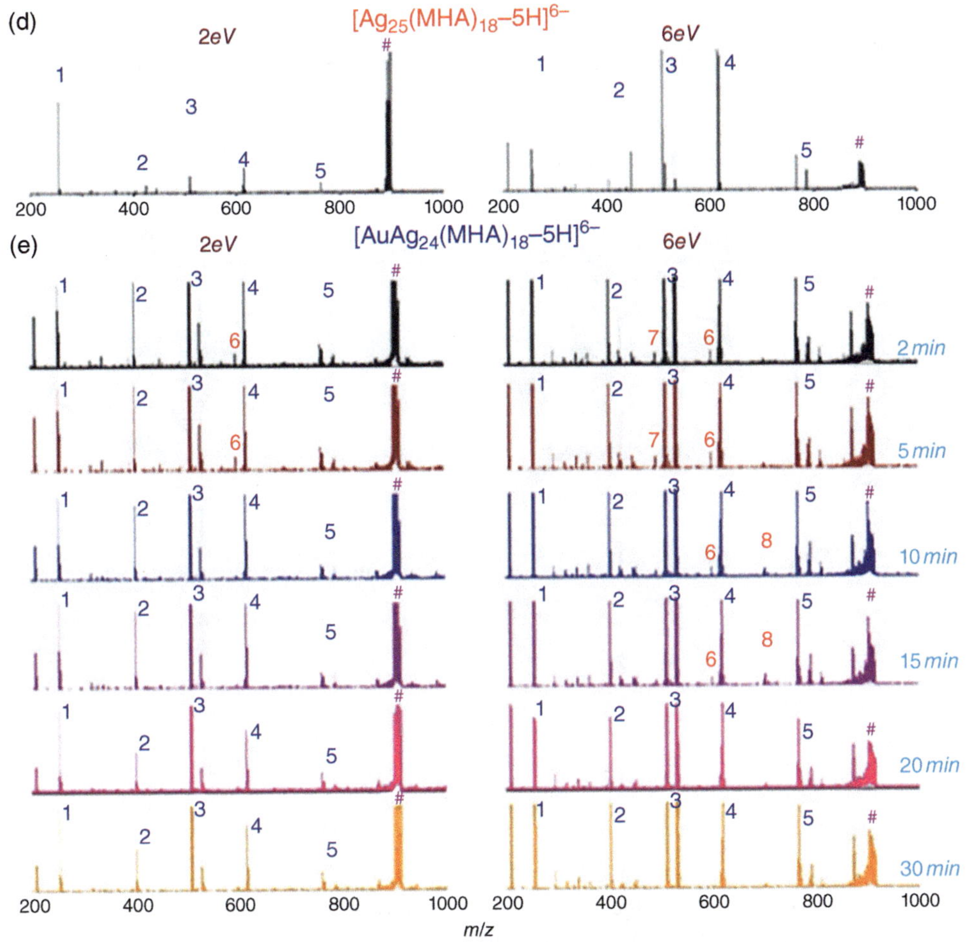

Figure 2.7 (Continued)

The alloying reaction can also be made possible via inter-cluster reactions, which usually lead to exchange the M(I)-SR moieties on the surface of parental NCs. As a demonstrative example, Krishnadas et al. mixed $[Au_{25}(PET)_{18}]^-$ and $[Ag_{25}(DMBT)_{18}]^-$ (DMBT = 2,4-dimethylbenzenethiol) NCs in dichloromethane (DCM) with a molar ratio of Ag_{25}/Au_{25} = 0.3/1.0 [57]. The reaction between these two NCs yielded a mixture of $[Au_{25-x}Ag_x(SR)_{18}]^-$ with x = 0–25 (Figure 2.8a–c), suggesting the inter-cluster reaction between $[Au_{25}(SR)_{18}]^-$ and $[Ag_{25}(SR)_{18}]^-$ NCs could lead to metal or M(I)-SR moiety exchange between the parental NCs, but largely preserve their M-S framework unaltered. More intriguingly, the authors captured an important intermediate, $[Ag_{25}Au_{25}(SR)_{36}]^{2-}$ by ESI-MS in the inter-cluster reaction, implying the inter-cluster reaction was most probably initiated by dimerization of parental NCs. Figure 2.8d depicts a structural model of $[Ag_{25}Au_{25}(SR)_{36}]^{2-}$ optimized by density functional theory (DFT) calculation, in which an inter-cluster Ag—S bond is witnessed. It should be noted that the inter-cluster reaction was also observed in $Au_{38-x}Ag_x(SR)_{24}$ NCs. Niihori et al. synthesized $Au_{38-x}Ag_x(S\text{-}C_4H_9)_{24}$ (H-S-C_4H_9 = 1-butanethiol) NCs by reacting $Au_{38}(S\text{-}C_4H_9)_{24}$ with Ag(I)-(S-C_4H_9) complexes [54]. Interestingly, the retention time profile of as-synthesized $Au_{38-x}Ag_x(S\text{-}C_4H_9)_{24}$ NCs in the high-performance liquid chromatography (HPLC) was observed changing with incubation time, after their ESI-MS spectra remain unaltered. This can only be made possible by inter-cluster reaction between $Au_{38-x}Ag_x(S\text{-}C_4H_9)_{24}$

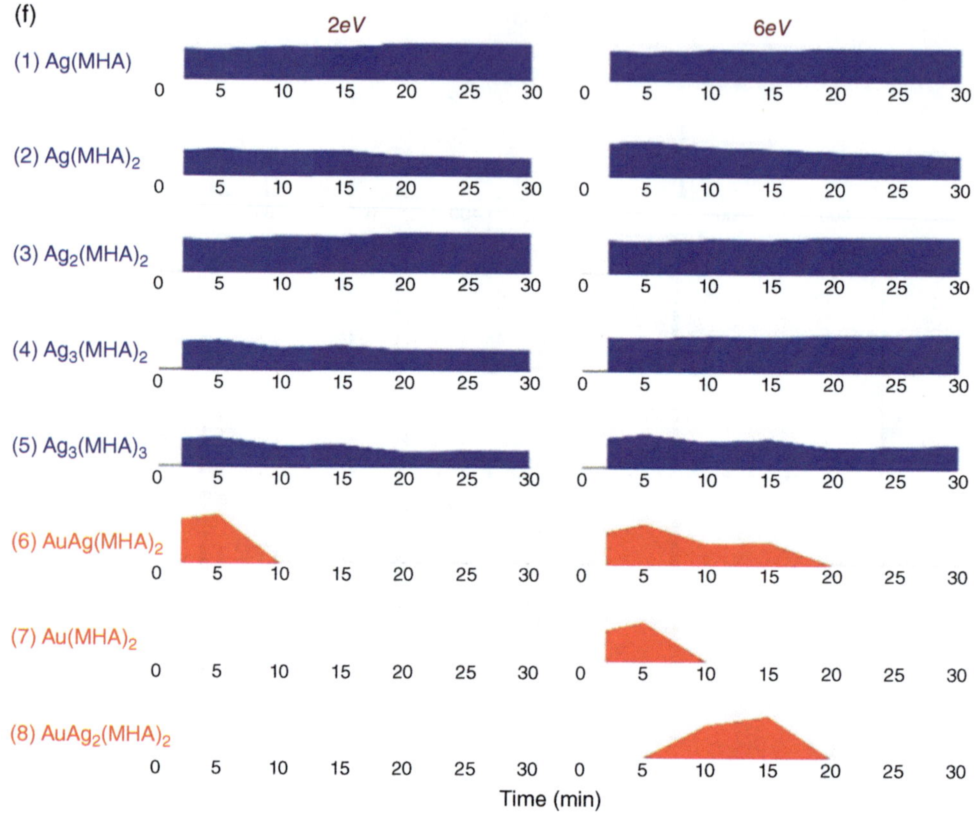

Figure 2.7 (Continued)

NCs in the HPLC column, causing changes in the distribution of x value of the alloy NCs. Similar inter-cluster reactions have also been documented for other cluster pairs like $[Au_{25}(SR)_{18}]^-$ vs. $[Ag_{44}(SR)_{30}]^{4-}$ [58] and $[Au_{25}(SR)_{18}]^-$ vs. $MAg_{28}(BDT)_{12}(PPh_3)_4$ (BDT = 1,3-benzenedithiol; M = Ni, Pd, or Pt) [59].

2.3.2 Ligand Composition Engineering

In comparison to the metal composition control, the control of the ligand composition of metal NCs is less explored. The organic ligand shell is the primary interface of metal NCs interacting with their surroundings, largely determining the stability, recognition, self-assembly, and application chemistry of metal NCs. The composition of ligand shell can now be tailored by a variety of methods quite similar to those used in metal composition engineering. These methods include in-situ reduction, ligand/surface motif exchange, and inter-cluster reaction.

Thiolate ligands with designed functionality can be allocated into the protection shell of metal NCs via in-situ reduction of corresponding M(I)-SR complexes. For example, Yuan et al. devised a NaOH-assisted $NaBH_4$ reduction method for the gram-scale synthesis of mono-, bi-, and tri-thiolate-protected Au_{25} NCs [60]. By introducing a proper amount of NaOH into the aqueous reaction mixture containing Au(I)-SR complexes and $NaBH_4$, the reducing capability of $NaBH_4$ can be largely moderated while the etching capability of SR ligands is significantly enhanced. The growth rate (determined by the reducing capability) and etching rate of freshly formed Au(0) core can thus be balanced by this way, conducive to the formation of thermodynamically stable $[Au_{25}(SR)_{18}]^-$ NCs in a short period (within three hours).

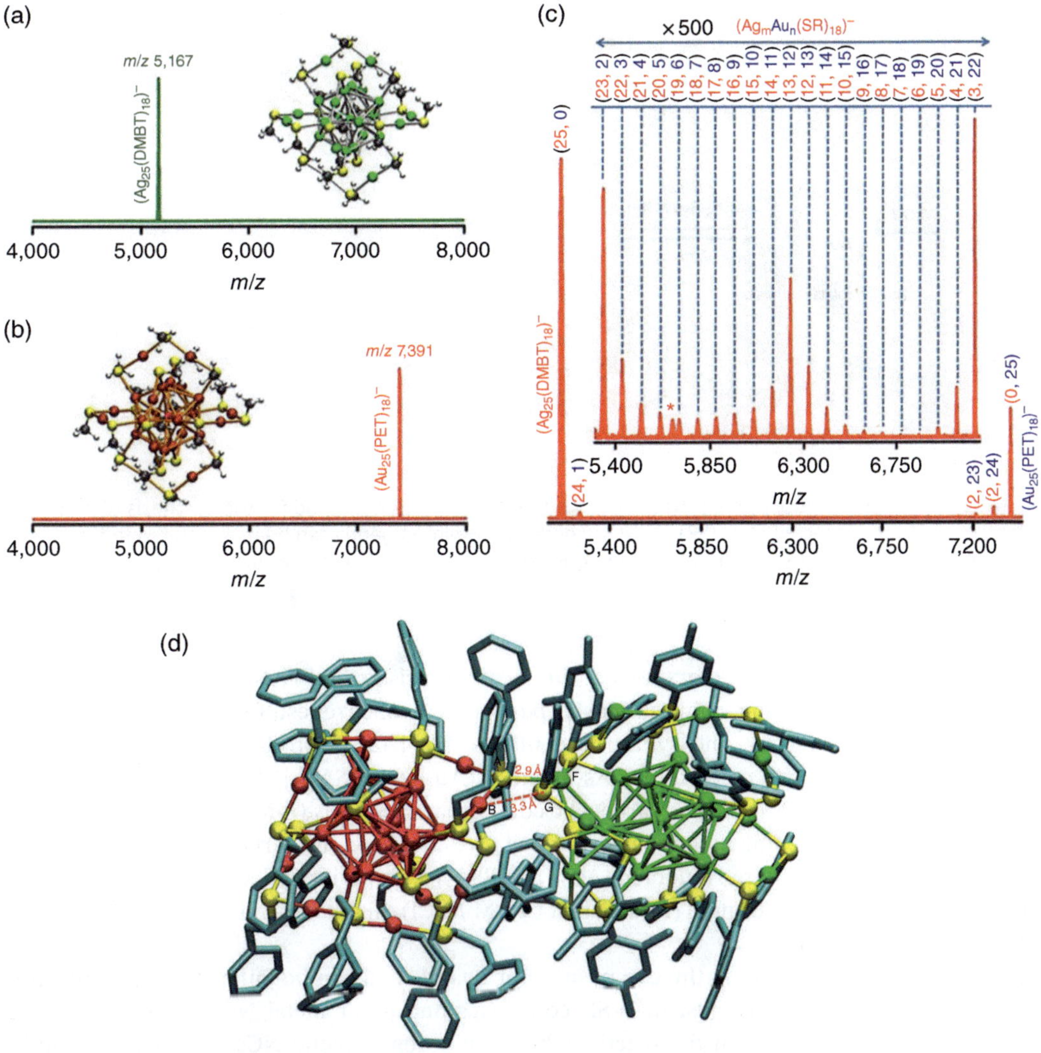

Figure 2.8 ESI-MS spectra of (a) $[Ag_{25}(DMBT)_{18}]^-$ and (b) $[Au_{25}(PET)_{18}]^-$, and (c) their reaction product $[Ag_{25-x}Au_x(SR)_{18}]^-$, where DMBT and PET are 2,4-dimethylbenzenethiol and 2-phenylethanethiol, respectively. (d) DFT calculations optimized structure of $[Ag_{25}Au_{25}(DMBT)_{18}(PET)_{18}]^{2-}$ dimer. The insets in (a) and (b) are corresponding structure models of clusters, where the SR ligands are simplified as SCH_3. Color code: red, Au; green, Ag; yellow, S; gray/cyan, C; and light gray, H. *Source:* Reproduced with permission [57]. Copyright 2016, the Authors, published by Springer Nature.

The cluster growth environment established by NaOH-assisted $NaBH_4$ method is quite robust against the altered SR ligands, where mono-, bi-, and tri-thiolate-protected Au_{25} NCs were synthesized with a variety of hydrophilic ligands containing carboxylic, hydroxyl, or amine groups (detailed in Figure 2.9), simply by mixing designed thiolate ligands with $HAuCl_4$ before $NaBH_4$ reduction. Of note, atomically precise metal NCs with highly positive surface charge have been long pursued in the community due to their potential biomedical applications.

Following the successful attempts of Yuan et al. in incorporating amine-containing thiolate ligands into the protection shell of $[Au_{25}(SR)_{18}]^-$, Ishida et al. managed to produce $[Au_{25}(SR)_{18}]^{0/-}$ with cationic (11-mercaptoundecyl)-N,N,N-trimethylammonium (H-S-$(CH_2)_{11}N(CH_3)_3^+$) as the sole protecting ligand, where a similar NaOH-assisted $NaBH_4$ method was used [61]. In addition to the high density of positive surface charge, high density of negative surface charge has also been

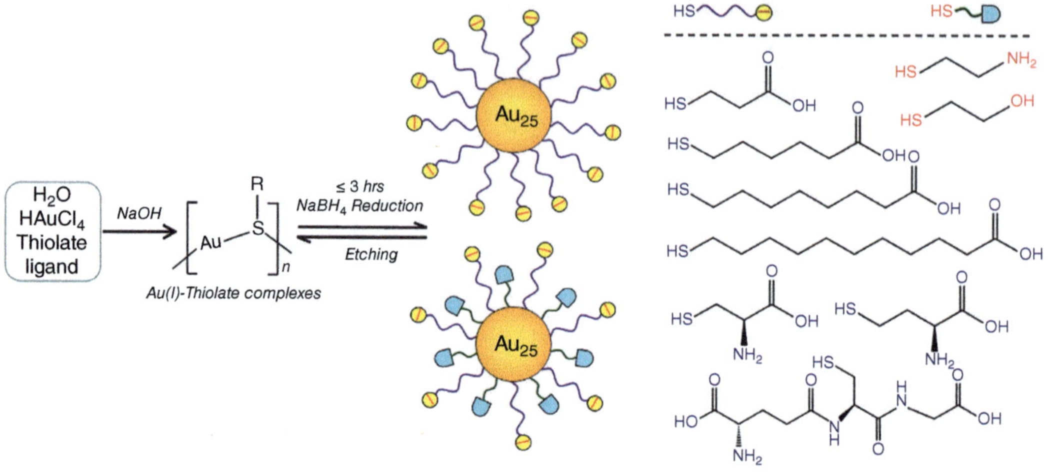

Figure 2.9 Schematic illustration of NaOH-assisted $NaBH_4$ reduction method for the synthesis of mono-, bi-, and tri-thiolate-protected $[Au_{25}(SR)_{18}]^-$ NCs, where the thiolate ligands can be incorporated in the protection shell of Au_{25} NCs are shown on the right. *Source:* Reproduced with permission [60]. Copyright 2014, Wiley VCH.

pursued in recent synthetic explorations of atomically precise metal NCs. Instead of commonly used carboxylic thiol, Zhang et al. employed *para*-mercaptobenzenesulfonic acid (*p*-MBSA) as model sulfonic thiol ligand, and synthesized ultra-stable $[Au_{25}(p\text{-MBSA})_{18}]^-$ NCs by the CO-mediated reduction-growth method [62]. As-obtained $[Au_{25}(p\text{-MBSA})_{18}]^-$ NCs exhibit consistently more negative ζ-potential than their carboxylic counterpart (i.e. $[Au_{25}(p\text{-MBA})_{18}]^-$ NCs) in a wide pH range of 2–11, evidencing higher density of negative charge on the surface of the former. Such high density of negative charge rendered $[Au_{25}(p\text{-MBSA})_{18}]^-$ superior molecular and colloidal stability against strong ionic strength (established by 2 M NaCl), high temperature (up to 80 °C), and extreme pH (from 3 to 10).

Besides the in-situ reduction, the composition of ligand shell can be tailored by reacting template metal NCs with SR ligands, M(I)-SR complexes and other metal NCs. Ligand exchange approaches depend literally on the reactions between parental metal NCs and foreign ligands, which can, however, yield different ligand composition on cluster surface according to the structures of foreign ligands and parental NCs. Early ligand exchange explorations were largely motivated by incorporating extra functionalities into the protection shell of metal NCs. For example, Tracy et al. performed ligand exchange reaction between $[Au_{25}(PET)_{18}]^-$ and thiolated poly(ethylene glycol) (S-PEG), producing $[Au_{25}(PET)_{18-x}(S\text{-PEG})_x]^-$ (x is up to 13) NCs [63]. Due to the high affinity of PEG moiety to the alkaline metal cations (e.g. Na^+, K^+, Rb^+, and Cs^+), the ligand substitution by S-PEG made $[Au_{25}(PET)_{18-x}(S\text{-PEG})_x]^-$ NCs easier to be ionized in the ESI process, facilitating the determination of their accurate mass and thus formulae by ESI-MS. Other than heterogenous thiolate-protected metal NCs, ligand exchange reaction can also proceed to completion and produce ligand completely exchanged metal NCs. AbdulHalim et al. devised a two-phase method for complete ligand exchange of $[Ag_{44}(MNBA)_{30}]^{4-}$ (MNBA = 5-mercapto-2-nitrobenzoic acid) with 4-fluorothiophenol (4-NTP) [64]. The $[Ag_{44}(MNBA)_{30}]^{4-}$ NCs were originally synthesized and accommodated in aqueous phase, followed by introducing a DCM/ethanol (1/1 v/v) solution of 4-NTP and tetraphenylphosphonium bromide. By vigorous stirring for <30 seconds, the majority (>94%) of Ag_{44} NCs can be transferred from aqueous phase into organic phase. The combined UV-vis absorption spectroscopy, ESI-MS, ^{13}C nuclear magnetic resonance (NMR) spectroscopy, and element analysis verify that the phase-transferred NCs are size and quality uncompromised $[Ag_{44}(4\text{-NTP})_{30}]^{4-}$ NCs. This two-phase ligand exchange approach is reasonably versatile and can be used to exchange with other aromatic thiolate ligands like 4-nitrothiophenol (4-NTP) and 2-naphthalenethiol (2-NT).

In addition to thiolate ligands, several other thiolate-containing compounds like M(I)-SR complexes and thiolate-protected metal NCs have been deployed as source of foreign ligands to react with parental metal NCs. As already discussed, $[Ag_{44}(SR)_{30}]^{4-}$ NCs can react with Au(I)-SR′ complexes via a surface motif exchange mechanism following the Eq. (2.1) [53]. By using 4-NTP as the foreign thiolate ligands (SR′) in the Au(I)-SR′ complexes, 4-NTP could be incorporated into the protecting shell of $[Ag_{44}(p\text{-MBA})_{30}]^{4-}$, yielding $[Ag_{44-x}Au_x(p\text{-MBA})_{30-2x}(4\text{-NTP})_{2x}]^{4-}$ ($x = 1$–12) NCs. Similar surface motif exchange reaction can also occur between $[Au_{25}(p\text{-MBA})_{18}]^-$ and Au(I)-(4-NTP) complexes, selectively furnishing the cluster surface with 4-NTP [65]. The inter-cluster reaction is another important method of current interest for engineering the ligand composition of metal NCs. Krishnadas et al. performed inter-cluster reaction between $[Au_{25}(PET)_{18}]^-$ and $[Ag_{44}(4\text{-FTP})_{30}]^{4-}$ in DCM, generating $[Au_{25-x}Ag_x(PET)_{18-y}(4\text{-FTP})_y]^-$ within 15 minutes [58]. The authors carried out detailed ESI-MS and MALDI-MS analysis on the inter-cluster reactions under varied ratios of Au_{25} and Ag_{44} NCs. The mass spectral data suggests that the foreign 4-FTP ligands are immobilized on the surface of Au_{25} NCs prominently via Ag-(4-FTP) and Au-(PET) exchange. More intriguingly, the same group also investigated the inter-cluster reaction between $[Au_{25}(PET)_{18}]^-$ and $MAg_{28}(BDT)_{12}(PPh_3)_4$ (M = Ni, Pd, or Pt), where only Au—Ag exchange but no ligand exchange was observed between the parental NCs [59]. This may be due to the distinctly different coordination habits and hydrocarbon tails of BDT (bidentate) and PET (monodentate) ligands.

Although remarkable progress has been made in terms of engineering the ligand composition of metal NCs, it remains as one of the greatest challenges to control the ligand composition at the single molecule level. Our group tried to tackle this challenging problem by reversible adding and removing a single ligand from the protecting shell of $[Au_{25}(MHA)_{18}]^-$ [66]. As shown in Figure 2.10a,b, the $[Au_{25}(MHA)_{18}]^-$ was gradually transferred into $[Au_{25}(MHA)_{19}]^-$ in the presence of excess MHA ligand and dissolved molecular O_2 in an aqueous solution of pH = ~9.0. On the basis of extensive ESI-MS, tandem MS, and ^{1}H-NMR analyses, we concluded that the ligand addition reaction of $[Au_{25}(MHA)_{18}]^-$ occurred via an oxidative pathway, where $[Au_{25}(MHA)_{18}]^-$ was oxidized by O_2 to $[Au_{25}(MHA)_{18}]^0$, and finally to $[Au_{25}(MHA)_{18}]^+$. Subsequent isoelectronic addition of one free MHA ligand to $[Au_{25}(MHA)_{18}]^+$ led to the formation of $Au_{25}(MHA)_{19}$. The overall oxidative nature of the ligand addition reaction implies that the reaction may be reversed in the presence of sufficiently strong reducing agent. This hypothesis has been experimentally verified, where introducing CO into the aqueous solution of $[Au_{25}(MHA)_{19}]^-$ at pH = ~12 converted the cluster into $[Au_{25}(MHA)_{18}]^-$. This addition and removal cycle of the single MHA ligand can be repeated for at least three cycles without any observable degradation of cluster quality (Figure 2.10c,d).

2.4 Structure Engineering of Metal Nanoclusters

The physiochemical properties of metal NCs not only rely on the number and types of metal atoms they contain (i.e. the size and composition) but also highly depend on the atomic-level arrangements of these metal atoms (i.e. the structure). Taking Au NCs as an example, in bulk Au and crystalline Au nanoparticles (>3 nm), the Au atoms normally adopt face-centered cubic (FCC) close packing mode, which is, however, not necessarily true for Au NCs. There are far more atomic packing modes in ultrasmall Au NCs, including FCC, hexagonal close packing (HCP), body-centered cubic, decahedra, icosahedra, multi-tetrahedral, etc. [67] Different atomic packing

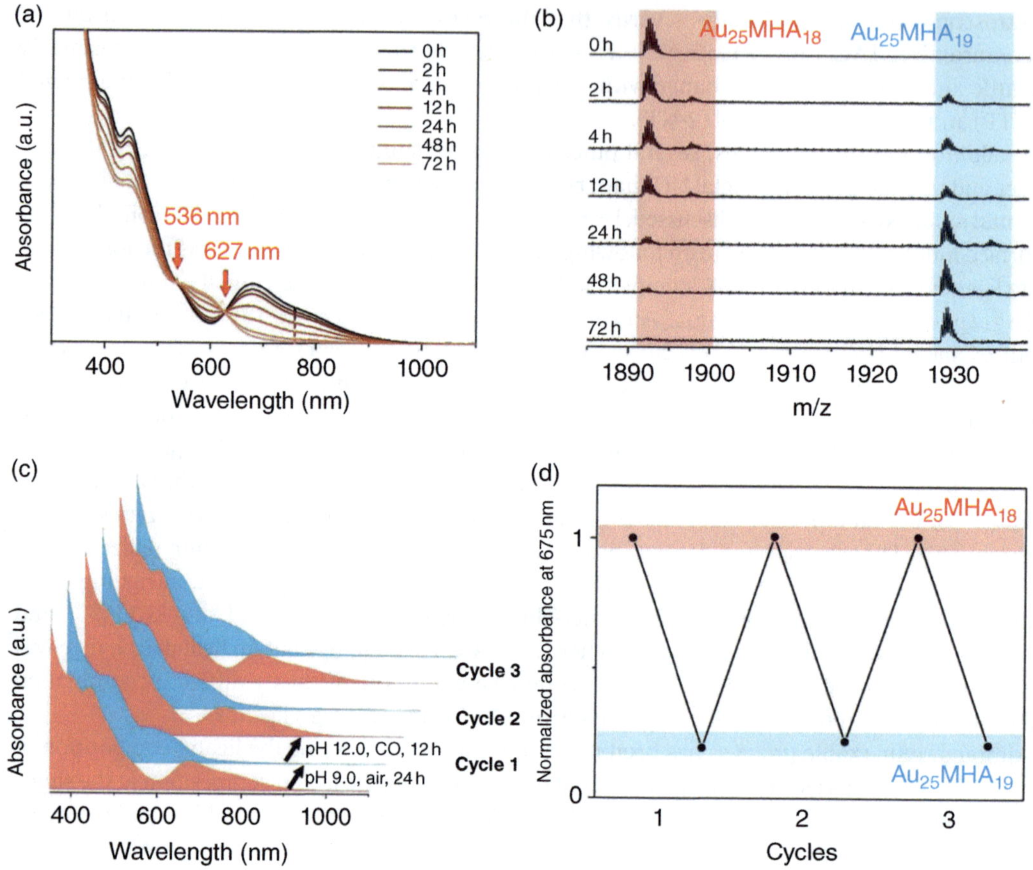

Figure 2.10 Time-course (a) UV-vis absorption and (b) ESI-MS spectra showing the conversion from $[Au_{25}(MHA)_{18}]^-$ to $Au_{25}(MHA)_{19}$. (c) UV-vis absorption spectra and (d) the normalized characteristic absorbance of $[Au_{25}(MHA)_{18}]^-$ at 675 nm recorded in the reversible conversion between $[Au_{25}(MHA)_{18}]^-$ and $Au_{25}(MHA)_{19}$ for three cycles. *Source:* Reproduced with permission [66]. Copyright 2020, the Authors, published by Springer Nature.

structures of Au NCs play a crucial role in determining their properties, such as the energy levels. The atomic structure of metal NCs usually changes simultaneously with their size, which means metal NCs with given size usually have an optimized structure corresponding to such size. As we have already discussed the size engineering of metal NCs above, here in this section we will focus on the strategy independently customizing the cluster structure without changing their size.

The size-preserved structure engineering is an emerging topic in cluster research community because if one can independently customize the structure of metal NCs without changing their size, one should be able to establish a more precise and reliable correlation between their properties and structures. As metal NCs can be regarded as metallic molecules, this kind of structure engineering can be referred to as isomerism, a basic concept in molecular science. In an isomerization process, molecules possessing identical formula change their atom arrangements in space. The isomerism has been widely spotted in organic compounds as a primary source of their structural complexity and property diversity. Thus, isomerization is a promising strategy that can greatly diversify the types and functions of metal NCs.

Unlike the bulk metals in which we can use physical parameters such as pressure and temperature to control the crystal phase, the structure engineering of metal NCs normally uses the surface strategies because of their extremely small size and large specific surface area. The most straightforward surface strategy, of course, is the surface ligand exchange. However, ligand exchange changes the molecular formula of metal NCs though keeping the numbers of metal atoms and ligands unchanged; thus, we term this kind of structure engineering as pseudo-isomerization. In the following content of this section, we will concisely discuss the pseudo-isomerization with ligand exchange and the "genuine" isomerization without changing the ligands.

2.4.1 Pseudo-Isomerization

Though the surface ligand exchange has been widely used in transformation of metal NC species, most of the ligand exchange induced transformation reactions involve size conversions, [33] while the example of ligand exchange induced pseudo-isomerization is still limited. A representative example of pseudo-isomerization reaction is the inter-conversion between two pseudo-isomers of $Au_{28}(SR)_{20}$ (Figure 2.11) [68]. The two pseudo-isomers, $Au_{28}(S\text{-}c\text{-}C_6H_{11})_{20}$ and $Au_{28}(SPh\text{-}t Bu)_{20}$, can convert into each other by reacting with corresponding free thiol ligand at 80 °C. This example unambiguously shows the importance of surface ligand in dictating the structure of metal NCs. It has also been shown that the two pseudo-isomers have different catalytic reactivities. However, the pseudo-isomerization of metal NCs makes it nontrivial (if not impossible) to decouple the effects of ligand and atomic arrangement on NC properties, which calls for more delicate structure engineering strategies.

2.4.2 Isomerization

Pairs of isomers with identical molecular formulae have been discovered in a variety of metal NCs of different sizes. For example, the first pair of structural isomers of metal NCs is found in $Au_{38}(SR)_{24}$ by X-ray crystallography [14e]. The two isomers, Au_{38T} and Au_{38Q}, can be synthesized separately by different protocols. More importantly, the metastable isomer, Au_{38T}, can transform into the more stable one, Au_{38Q}, at 50 °C in toluene within 36 hours. Apparently, it is a thermodynamically preferred spontaneous transformation process, where the reverse transition from Au_{38Q} to Au_{38T} is largely prohibited. It indicates that the structural reconstruction involved in the isomerization reaction is most likely to occur through breaking/regenerating chemical bonds, whose energy barrier is too higher to make the reversed structure transition possible.

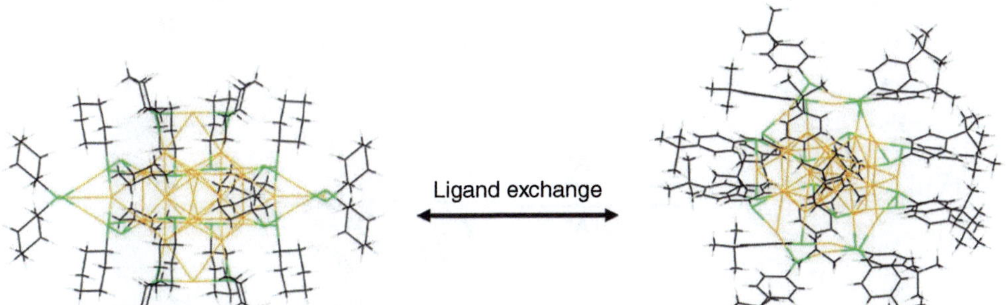

Figure 2.11 Schematic illustration of ligand-exchange-induced pseudo-isomerization of $Au_{28}(SR)_{20}$. Color code: yellow, Au; green, S; gray, C. *Source:* Reproduced with permission [68]. Copyright 2016, American Chemical Society.

The reversible isomerization process is more interesting because "reversible," in most cases, means "controllable." It has been shown that the two structural isomers of $Au_{13}Ag_{12}$ NCs are stable and spontaneously inter-convertible at desired temperatures [69]. Another inter-conversion process between cluster isomers was observed in Au_{28} NCs, whose isomerization process involved reconstruction of their surface motif in response to different solvent environments [70]. Both cases mentioned above involve breaking/regenerating of chemical bonds in the conversion process, and their molecule-level driving forces remain largely unknown. Isomerization processes have also been observed in extreme conditions – such as electron beam irradiation and cluster-cluster collision in gas phase, which are out of the scope of this Chapter [71].

Recently, our group reported a weak supramolecular interaction induced reversible isomerization mechanism for hydrophilic metal NCs. It has been shown that in the hydrophilic $[Au_{25}(p\text{-}MBA)_{18}]^-$ NCs, the coupling/decoupling of the inter-molecular interaction between the *p*-MBA ligand and a cationic surfactant, cetyltrimethylammonium (CTA^+) cation, can drive the reversible isomerization reaction [72]. As shown in Figure 2.12, after CTA^+ adsorption, the isomer in water, **$(Au_{25})_R$** (reddish-brown in color, the classical isomer), will be transformed into a new isomer, **$(Au_{25})_G$** (dark green in color, the newly discovered isomer in the solution phase). When **$(Au_{25})_G$** is transferred to methanol solution and dissociates the CTA^+ ions, a reverse transformation can be realized. Owing to the high controllability on the reversible isomerization process, fundamentals of this isomerization reactions have also been investigated. Both forward and backward isomerization reactions follow the first-order reaction kinetics (Figure 2.13), and the activation energies are calculated to be 1.16 and 1.17 eV, respectively. The transition state lifetime of ~10^{-14} s and the isosbestic points in the absorption spectra indicate that there is no observable intermediate species formed during the isomerization. The first-order reaction kinetics, low activation energies (~1 eV), and the absence of reaction intermediates collaboratively suggest that this is a typical inorganic NC isomerization reaction, which can be realized by displacive reconfiguration of metal atoms without breaking any chemical bonds [73]. The displacive reconfiguration process has also been confirmed by DFT calculations (Figure 2.14), which show that the two isomers are topologically connected via a simple rotation of the gold core. In this process, all the atoms in the NC simultaneously change their relative positions while maintaining the binding coordination with each other. The estimated activation barrier for such displacive reconfiguration is only about 0.6 eV.

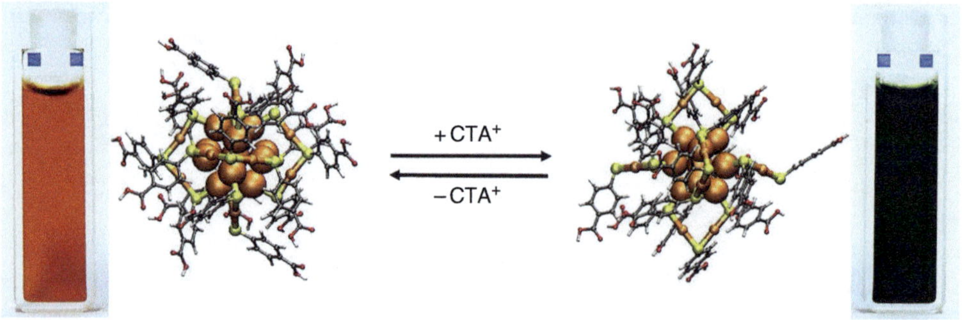

Figure 2.12 Schematic illustration of reversible isomerization reactions of $[Au_{25}(p\text{-}MBA)_{18}]^-$ induced by intermolecular interaction with CTA^+. The insets are digital images of corresponding structural isomers of $[Au_{25}(p\text{-}MBA)_{18}]^-$. Color code: orange, Au; yellow, S; gray, C; red, O. *Source:* Reproduced with permission [72]. Copyright 2021, Elsevier Inc.

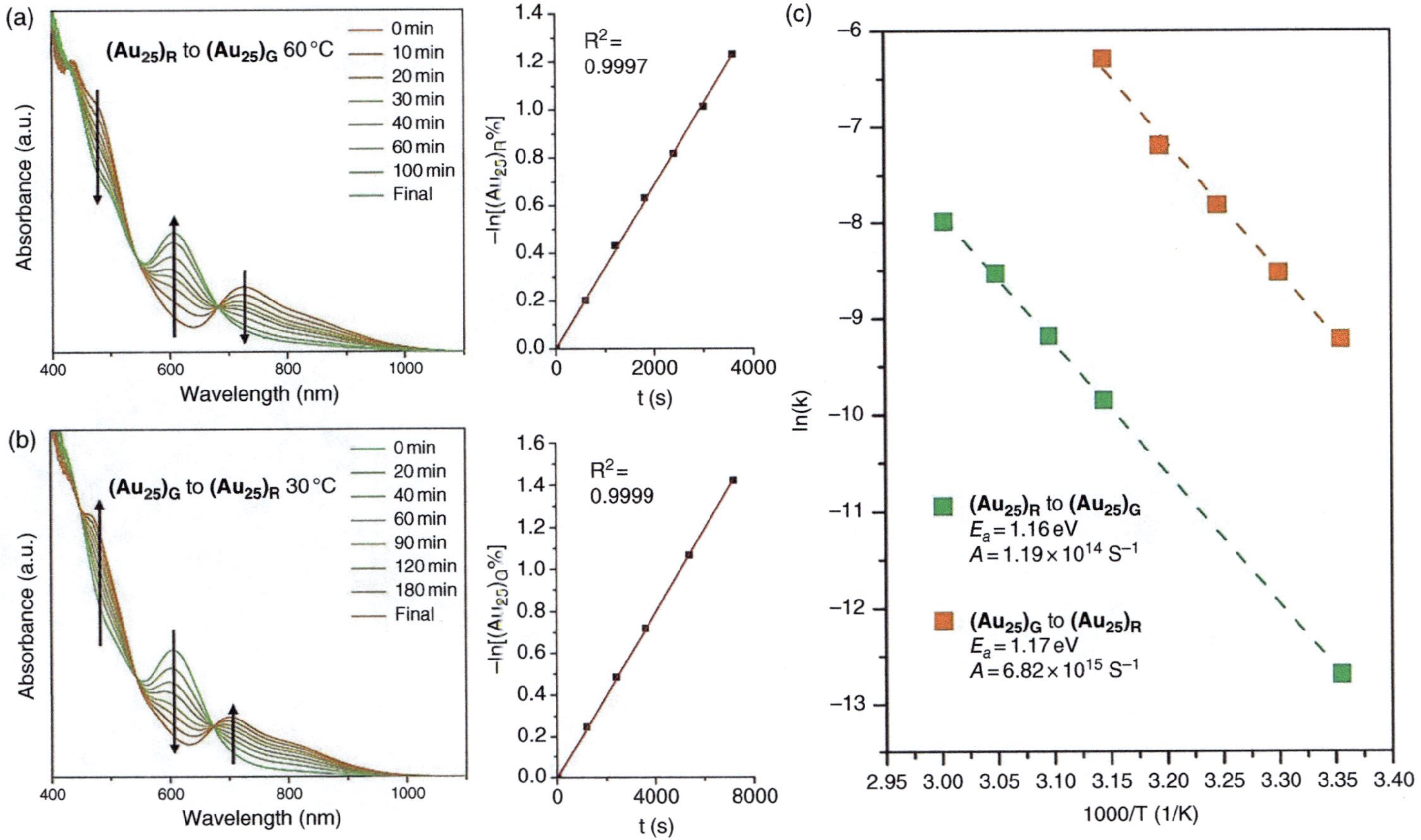

Figure 2.13 (a) Forward and (b) reverse reaction kinetics of isomerization from $(Au_{25})_R$ to $(Au_{25})_G$. (c) Arrhenius plots for the inter-conversion processes and their linear fittings (dash lines). E_a is activation energy, and A denotes Arrhenius pre-exponential factor. *Source:* Reproduced with permission [72]. Copyright 2021, Elsevier Inc.

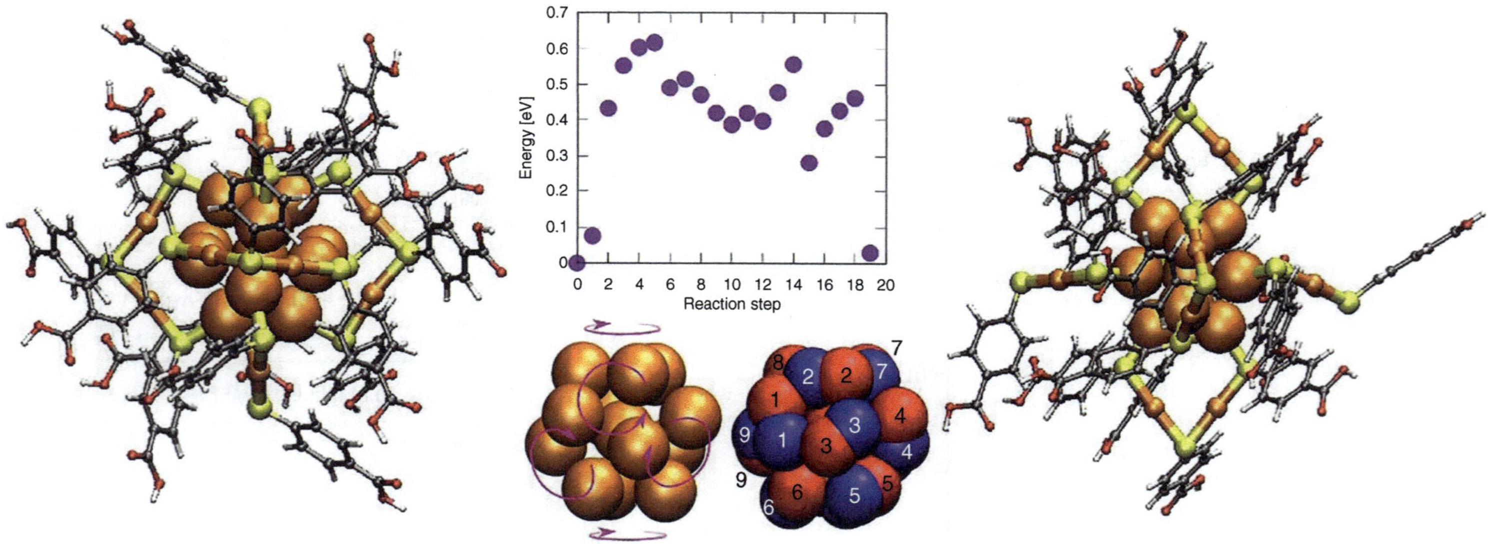

Figure 2.14 Topologically connected structures of the two structural isomers of $[Au_{25}(p\text{-MBA})_{18}]^-$ and their transformation mode. Color code: orange, Au; yellow, S; gray, C; red, O. *Source:* Reproduced with permission [72]. Copyright 2021, Elsevier Inc.

2.5 Top-Down Etching Reaction of Metal Nanoclusters

The reduction-growth reaction represents the bottom-up process to produce metal NCs, and the etching reaction describing the top-down process is also an important aspect in the study of atomically precise nanochemistry. In our daily life, metal etching is one of the most common chemical reactions. Etching becomes more prominent in particles of nanoscale size due to their small size and large specific surface area. Etching of metal NCs is one of the most critical chemical reactions dictating their synthesis, storage, and applications. On one hand, a considerably good resistance of metal NCs against the etching reaction is a prerequisite to their storage and usage. On the other hand, etching is a governing reaction in the size-focusing process in the wet-chemical synthesis to dictate the size distribution and purity of metal NCs. Therefore, understanding the underlying chemistry of etching will greatly benefit the synthesis, storage, and application of metal NCs.

Etching of metal NCs normally is an oxidation process in which metal NCs give their electrons to an etchant and degrade into smaller species (top-down process). This process can be understood using the valence electron count (i.e. N^*) of metal NCs. In a reductive growth reaction, the N^* increases, while in an etching reaction of metal NCs, on the contrary, the N^* decreases. In thiolate-protected Au NCs, the crucial role of excess thiol ligands as etchants to drive the size-focusing process has long been recognized. It helps convert the polydisperse NCs generated to thermodynamically stable ones, like $[Au_{25}(SR)_{18}]^-$.

The etching capability of free thiol ligands has been shown to closely relate to their reactivity toward oxygen (O_2), which can generate thiol radicals [74]. The radicals generated by reaction between O_2 and thiol, as an etchant, will cleave the surface of metastable metal NCs, and oxidize the Au(0) core, followed by drawing out the core Au(0) atoms to initiate the top-down conversion process. The top-down conversion process will finally produce thermodynamically stable metal NCs or magic-sized metal NCs (Figure 2.15).

A molecular insight into the etching reactions of metal NCs was made possible by taking advantage of the advanced ESI-MS techniques [75]. A real-time ESI-MS test was conducted on the etching reaction of water-soluble $[Au_{25}(MHA)_{18}]^-$ NCs in the presence of excess thiol ligand to map out all the observable intermediates and elementary reactions. The precursor, $[Au_{25}(MHA)_{18}]^-$, was carefully processed by diverse techniques such as ultrafiltration and polyacrylamide gel electrophoresis (PAGE) to make the signals of the NC species readily captured by ESI-MS without further purifications, thus enabling a continuous real-time tracking of the whole etching reaction in 30 days. Twenty different intermediate species were captured (Figure 2.16a). By systematically analyzing the formation and consumption process of each species, the entire oxidative etching process of $[Au_{25}(MHA)_{18}]^-$ was mapped out (Figure 2.16b). It is suggested that the decrease of N^* follows a 2 e^- hopping mode, stepwise from 8 of $[Au_{25}(MHA)_{18}]^-$ to 6, 4, 2, and 0, indicating that at least from the aspect of N^*, oxidative etching is a reverse process of reductive growth. A more detailed analysis of the reaction process indicates that the etching reaction can be divided into two stages: an initial decomposition process and a subsequent recombination process (Figure 2.16c). In Stage I, $[Au_{25}(MHA)_{18}]^-$ is decomposed by thiol radicals (generated from reaction with O_2) to produce a series of species with smaller sizes and N^*, such as $[Au_{20}(MHA)_{15}]^-$ ($N^* = 6$), $[Au_{21}(MHA)_{16}]^-$ ($N^* = 6$), $Au_{18}(MHA)_{14}$ ($N^* = 4$), $Au_{15}(MHA)_{13}$ ($N^* = 2$), together with short Au(I)-SR complexes ($[Au(MHA)_2]^-$ and $[Au_2(MHA)_3]^-$, $N^* = 0$) as a byproduct. These reactions fulfill the features of a radical-initiated top-down process. Moreover, new features that have never been revealed before were observed in the next stage.

In Stage II, the early formed small NC species would slow down the top-down etching reaction so they recombine with the short Au(I)-SR complexes to produce isoelectric NCs (i.e. NCs with the same N^*) through isoelectric addition. For example, $Au_{18}(MHA)_{14}$ will react with Au(I)-(MHA) complexes to generate $Au_{20}(MHA)_{16}$, $Au_{22}(MHA)_{18}$, and $Au_{24}(MHA)_{20}$. The most interesting

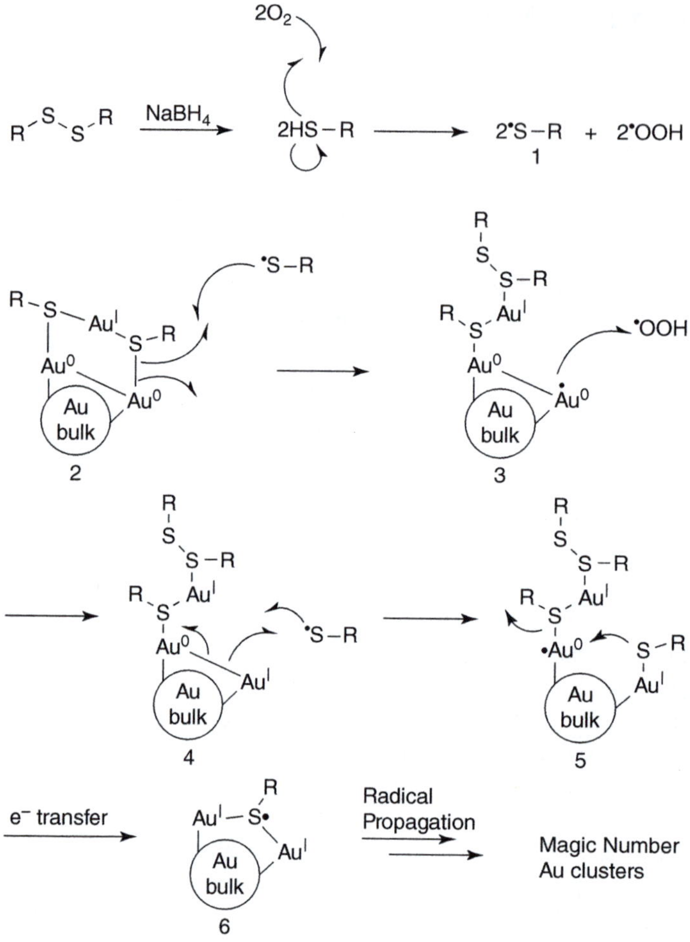

Figure 2.15 Proposed etching reaction of metal NCs mediated by O_2 and thiol. *Source:* Reproduced with permission [74]. Copyright 2015, Wiley VCH.

information that Stage II can tell is that the size of NC species does not necessarily decrease in the oxidative environment. The NC species produced through the isoelectric addition reaction have a larger size and higher ligand-to-metal ratio. Most of these NC species are unique in the oxidative etching reactions. They have never been observed in the reduction-growth process of $[Au_{25}(SR)_{18}]^-$, where NC species with relatively lower ligand-to-metal ratios are often observed (Figure 2.17). This data suggests that the oxidative etching of metal NCs is not necessarily the reverse process of the reductive growth. The unique species formed in an oxidative environment (e.g. $Au_{20}(SR)_{16}$, $Au_{22}(SR)_{18}$, and $Au_{24}(SR)_{20}$) are more resistant to the further oxidative etching reaction and survive as final products in the long-time reaction. These species are expected to have lengthened motifs and rigidified ligand shell, which will provide better protection to the Au(0) core.

2.6 Conclusion and Outlooks

In summary, the global efforts devoted into the cluster chemistry have made thiolate-protected metal NCs increasingly available with their size, composition, and atomic packing structure customizable at the unprecedented atomic level. In the most used reduction-growth scheme of metal

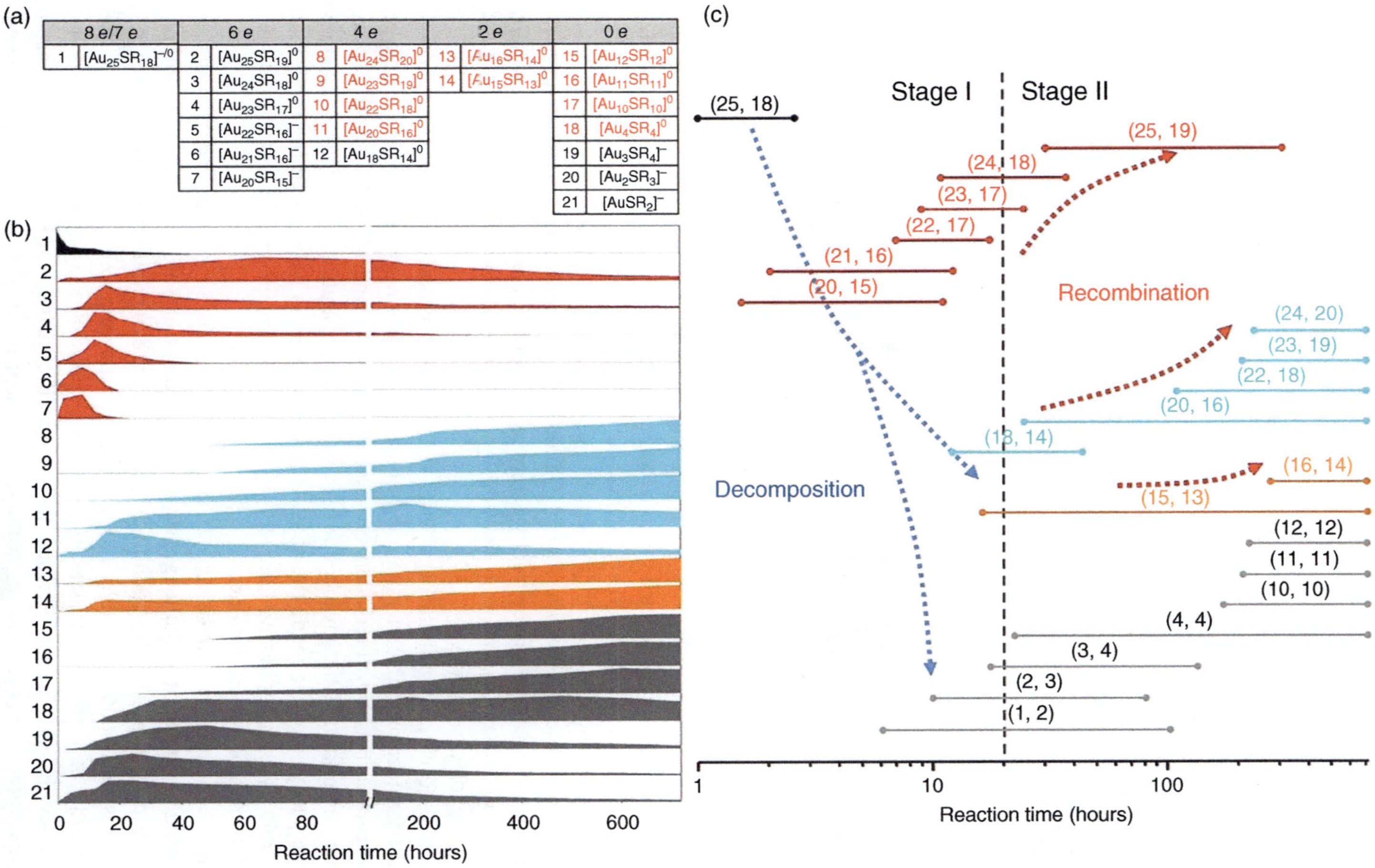

Figure 2.16 Oxidative etching reaction process of $[Au_{25}(MHA)_{18}]^-$ with excess thiolate ligands in 30 days. (a) Species detected in reaction process. (b) Normalized ESI-MS spectral intensity profiles of the 21 species identified throughout the etching process. (c) Segments representing the full width at half maximum (FWHM) of the normalized ESI-MS spectral intensity profiles. Species with different N^* were marked with different colors: black, N^* = 7 and 8; red, N^* = 6; cyan, N^* = 4; orange, N^* = 2; and gray, N^* = 0. *Source:* Reproduced with permission [75]. Copyright 2021, the Authors, published by Springer Nature.

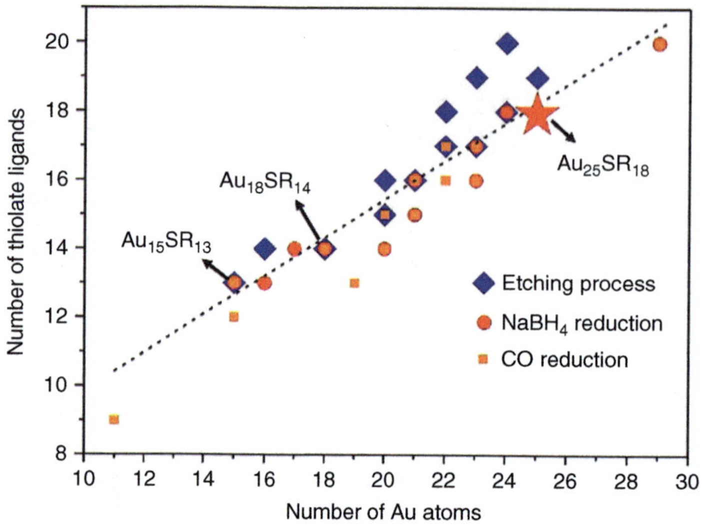

Figure 2.17 Comparison of the NC species detected during the etching (blue square) and reduction-growth (using $NaBH_4$ (red circle) or CO (orange square) as reducing agent) of water-soluble $[Au_{25}(SR)_{18}]^-$. The black dash line is an eye-guide for comparing the relative positions of the NC species. *Source:* Reproduced with permission [75]. Copyright 2021, the Authors, published by Springer Nature.

NC synthesis, the size and composition of metal NCs can be tailored by the type and states of M(I)-SR complexes and the reducing power of the reducing agents. In addition to the reduction-growth approaches, the size and composition engineering of metal NCs can be made possible via templated-NC methods, where pre-formed metal NCs are employed as the templates to react with varied species including reducing/oxidizing agents, metal salts, M(I)-SR complexes, and other metal NCs. In addition to the size and composition of metal NCs, the atomic packing structure has risen as another property-dictating attribute which should be independently (to both size and composition) controlled. This can be preferentially achieved by isomerization of metal NCs, where genuine reversible isomerization has been realized by week inter-molecular interactions of SR ligands with hydrophobic cations (e.g. CTA^+).

Based on the well-developed synthetic chemistry of atomically precise metal NCs, their decomposition chemistry under oxidative etching conditions has also been explored, where a unique recombination reaction between the oxidatively etched metal NCs and M(I)-SR complexes was identified. More importantly, the advances in the synthesis, separation, and characterization chemistry in the past years have made the above-mentioned process traceable at the molecular and/or even atomic levels, mapping out their step-by-step reaction maps reminiscent of total synthesis routes of diverse organic molecules.

Despite the above-mentioned achievements in the development of total synthesis chemistry of metal NCs, there still exist essential challenges requiring additional efforts from worldwide groups. In the size engineering of metal NCs, a pending question of fundamental interest is whether the size of metal NCs is continuously tunable. This question is attracting recent attention due to the increasing availability of stable cluster sizes in a wide size spectrum ranging from a few to hundreds of metal atoms. Besides the size diversity, the site-specific engineering strategy for the metal and ligand composition is of core interest. In the development of such site-specific alloying and ligand engineering methodology, the molecular and atomic dynamics of thiolate ligands and metal atoms, respectively, should be carefully considered, where intra-cluster metal atom diffusion and thiolate ligand migration may further challenge site-specific composition engineering chemistry.

When the topic comes to the atomic packing structure engineering, more supramolecular interactions should be investigated in terms of their feasibility in inducing the reversible isomerization process of metal NCs. Also, the importance of isomerization in determining the structural diversity of metal NCs should be explored. Last but not least, the decomposition chemistry of metal NCs should also be established in the presence of etchants other than free thiols, which may pave another top-down strategy for engineering the metal NCs at the atomic precision.

To wrap up, the advances of total synthesis chemistry of thiolate-protected metal NCs not only vastly diversified the cluster library for fundamental and applied research but also revealed molecule-level clues for understanding the growth/etching, alloying/surface engineering, and isomerization reactions of metal NCs. Therefore, the total synthesis exploration constitutes a necessary first step for promoting the nanochemistry research to the unprecedented atomic resolution. We believe the methodological and mechanistic achievements systematized in this chapter can greatly add to the acceptance of metal NCs in various application scenarios.

Contributions

Q.Y., Y.C., and T.C. contributed equally to this chapter under the guidance of J.X.

References

1 Nicolaou, K.C., Vourloumis, D., Winssinger, N., and Baran, P.S. (2000). *Angew. Chem. Int. Ed.* 39: 44–122.

2 a) Sheehan, J.C. and Henery-Logan, K.R. (1957). *J. Am. Chem. Soc.* 79: 1262–1263. b) Rolinson, G.N. and Geddes, A.M. (2007). *Int. J. Antimicrob. Ag.* 29: 3–8.

3 a) Jin, R., Zeng, C., Zhou, M., and Chen, Y. (2016). *Chem. Rev.* 116: 10346–10413. b) Yao, Q., Yuan, X., Chen, T. et al. (2018). *Adv. Mater.* 30: 1802751. c) Chakraborty, I. and Pradeep, T. (2017). *Chem. Rev.* 117: 8208–8271. d) Du, Y., Sheng, H., Astruc, D., and Zhu, M. (2019). *Chem. Rev.* 120: 526–622.

4 a) Jadzinsky, P.D., Calero, G., Ackerson, C.J. et al. (2007). *Science* 318: 430–433. b) Zeng, C., Chen, Y., Kirschbaum, K. et al. (2016). *Science* 354: 1580–1584. c) Zeng, C., Qian, H., Li, T. et al. (2012). *Angew. Chem. Int. Ed.* 51: 13114–13118. d) Yan, N., Xia, N., Liao, L. et al. (2018). *Sci. Adv.* 4: eaat7259. e) Chen, S., Wang, S., Zhong, J. et al. (2015). *Angew. Chem. Int. Ed.* 54: 3145–3149. f) Yang, H., Wang, Y., Huang, H. et al. (2013). *Nat. Commun.* 4: 2422. g) Desireddy, A., Conn, B.E., Guo, J. et al. (2013). *Nature* 501: 399–402. h) Joshi, C.P., Bootharaju, M.S., Alhilaly, M.J., and Bakr, O.M. (2015). *J. Am. Chem. Soc.* 137: 11578–11581. i) Sakthivel, N.A., Theivendran, S., Ganeshraj, V. et al. (2017). *J. Am. Chem. Soc.* 139: 15450–15459. j) Dainese, T., Antonello, S., Gascón, J.A. et al. (2014). *ACS Nano* 8: 3904–3912. k) Tofanelli, M.A., Salorinne, K., Ni, T.W. et al. (2016). *Chem. Sci.* 7: 1882–1890. l) Heaven, M.W., Dass, A., White, P.S. et al. (2008). *J. Am. Chem. Soc.* 130: 3754–3755. m) Zhu, M., Aikens, C.M., Hollander, F.J. et al. (2008). *J. Am. Chem. Soc.* 130: 5883–5885.

5 Häkkinen, H., Walter, M., and Grönbeck, H. (2006). *J. Phys. Chem. B* 110: 9927–9931.

6 Walter, M., Akola, J., Lopez-Acevedo, O. et al. (2008). *Proc. Natl. Acad. Sci. U. S. A.* 105: 9157–9162.

7 a) Luo, Z., Yuan, X., Yu, Y. et al. (2012). *J. Am. Chem. Soc.* 134: 16662–16670. b) Wu, Z., Yao, Q., Chai, O.J.H. et al. (2020). *Angew. Chem. Int. Ed.* 59: 9934–9939. c) Li, Y., Galli, G., and Gygi, F.O. (2008). *ACS Nano* 2: 1896–1902. d) Yau, S.H., Varnavski, O., and Goodson, T. (2013). *Acc. Chem. Res.* 46: 1506–1516. e) Liu, H., Hong, G., Luo, Z. et al. (2019). *Adv. Mater.* 31: 1901015.

8 a) Knoppe, S. and Bürgi, T. (2014). *Acc. Chem. Res.* 47: 1318–1326. b) Zhang, M.-M., Li, K., and Zang, S.-Q. (2020). *Adv. Opt. Mater.* 8: 1902152. c) Dolamic, I., Knoppe, S., Dass, A., and Bürgi, T. (2012). *Nat. Commun.* 3: 798.

9 Yamazoe, S., Koyasu, K., and Tsukuda, T. (2013). *Acc. Chem. Res.* 47: 816–824.

10 a) Shang, L., Stockmar, F., Azadfar, N., and Nienhaus, G.U. (2013). *Angew. Chem. Int. Ed.* 52: 11154–11157. b) Chen, Y., Li, L., Gong, L. et al. (2019). *Adv. Funct. Mater.* 29: 1806945. c) Du, B., Yu, M., and Zheng, J. (2018). *Nat. Rev. Mater.* 3: 358–374. d) Zhang, J., Fu, Y., Conroy, C.V. et al. (2012). *J. Phys. Chem. C* 116: 26561–26569.

11 a) Zhang, H., Liu, H., Tian, Z. et al. (2018). *Nat. Nanotechnol.* 13: 900–905. b) Chen, Y.-S. and Kamat, P.V. (2014). *J. Am. Chem. Soc.* 136: 6075–6082. c) Chen, Y.-S., Choi, H., and Kamat, P.V. (2013). *J. Am. Chem. Soc.* 135: 8822–8825.

12 a) Zhang, Y., Song, P., Chen, T. et al. (2018). *Proc. Natl. Acad. Sci. U. S. A.* 115: 10588–10593. b) Zhuang, S., Chen, D., Liao, L. et al. (2020). *Angew. Chem. Int. Ed.* 59: 3073–3077. c) Li, G., Sui, X., Cai, X. et al. (2021). *Angew. Chem. Int. Ed.* 60: 10573–10576.

13 a) Liu, F., Liu, Y., Yao, Q. et al. (2020). *Environ. Sci. Technol.* 54: 5913–5921. b) Chen, P.-C., Periasamy, A.P., Harroun, S.G. et al. (2016). *Coord. Chem. Rev.* 320-321: 129–138.

14 a) Kang, X., Li, Y., Zhu, M., and Jin, R. (2020). *Chem. Soc. Rev.* 49: 6443–6514. b) Cerretani, C., Kanazawa, H., Vosch, T., and Kondo, J. (2019). *Angew. Chem. Int. Edit.* 58: 17153–17157. c) Guan, Z., Zeng, J., Yuan, S. et al. (2018). *Angew. Chem. Int. Edit.* 57: 5703–5707. d) Crasto, D., Malola, S., Brosofsky, G. et al. (2014). *J. Am. Chem. Soc.* 136: 5000–5005. e) Tian, S., Li, Y., Li, M. et al. (2015). *Nat. Commun.* 6: 8667.

15 a) Chen, W. and Chen, S. (2009). *Angew. Chem. Int. Edit.* 48: 4386–4389. b) Li, W., Liu, C., Abroshan, H. et al. (2016). *J. Phys. Chem. C* 120: 10261–10267. c) Duchesne, P.N., Li, Z., Deming, C.P. et al. (2018). *Nat. Mater.* 17: 1033–1039. d) Kwak, K., Choi, W., Tang, Q. et al. (2017). *Nat. Commun.* 8: 14723. e) Zhao, S., Jin, R., Abroshan, H. et al. (2017). *J. Am. Chem. Soc.* 139: 1077–1080. f) Abbas, M.A., Kim, T.Y., Lee, S.U. et al. (2016). *J. Am. Chem. Soc.* 138: 390–401. g) Song, X., Zhu, W., Ge, X. et al. (2021). *Angew. Chem. Int. Edit.* 60: 1306–1312. h) Du, B., Jiang, X., Das, A. et al. (2017). *Nat. Nanotechnol.* 12: 1096–1102. i) Xie, J., Zheng, Y., and Ying, J. (2010). *Chem. Commun.* 46: 961–963.

16 Yu, Y., Chen, X., Yao, Q. et al. (2013). *Chem. Mater.* 25: 946–952.

17 Stamplecoskie, K.G. and Kamat, P.V. (2014). *J. Am. Chem. Soc.* 136: 11093–11099.

18 a) Jiang, D.-e. (2013). *Nanoscale* 5: 7149–7160. b) Häkkinen, H. (2008). *Chem. Soc. Rev.* 37: 1847–1859.

19 a) Brust, M., Walker, M., Bethell, D. et al. (1994). *J. Chem. Soc. Chem. Commun.* 801–802. b) Kim, B.H., Hackett, M.J., Park, J., and Hyeon, T. (2014). *Chem. Mater.* 26: 59–71. c) Uehara, A., Booth, S.G., Chang, S.Y. et al. (2015). *J. Am. Chem. Soc.* 137: 15135–15144.

20 Jin, R., Qian, H., Wu, Z. et al. (2010). *J. Phys. Chem. Lett.* 1: 2903–2910.

21 Zheng, K., Yuan, X., and Xie, J. (2017). *Chem. Commun.* 53: 9697–9700.

22 Chen, T., Yao, Q., Nasaruddin, R.R., and Xie, J.P. (2019). *Angew. Chem. Int. Edit.* 58: 11967–11977.

23 Chen, T. and Xie, J. (2016). *Chem. Rec.* 16: 1761–1771.

24 Luo, Z., Nachammai, V., Zhang, B. et al. (2014). *J. Am. Chem. Soc.* 136: 10577–10580.

25 Chen, T., Fung, V., Yao, Q. et al. (2018). *J. Am. Chem. Soc.* 140: 11370–11377.

26 Chen, T., Yao, Q., Cao, Y., and Xie, J. (2020). *Cell Rep. Phys. Sci.* 1: 100206.

27 Cao, Y., Guo, J., Shi, R. et al. (2018). *Nat. Commun.* 9: 2379.

28 Dharmaratne, A.C., Krick, T., and Dass, A. (2009). *J. Am. Chem. Soc.* 131: 13604–13605.

29 Ishida, R., Hayashi, S., Yamazoe, S. et al. (2017). *J. Phys. Chem. Lett.* 8: 2368–2372.

30 Zaker, Y., Ashenfelter, B.A., Bhattarai, B. et al. (2021). *Small* 17: 20202238.

31 Yao, Q., Yuan, X., Fung, V. et al. (2017). *Nat. Commun.* 8: 927.

32 Zeng, C., Liu, C., Pei, Y., and Jin, R. (2013). *ACS Nano* 7: 6138–6145.

33 Zeng, C., Chen, Y., Das, A., and Jin, R. (2015). *J. Phys. Chem. Lett.* 6: 2976–2986.

34 Zeng, C., Li, T., Das, A. et al. (2013). *J. Am. Chem. Soc.* 135: 10011–10013.

35 Zeng, C., Liu, C., Chen, Y. et al. (2014). *J. Am. Chem. Soc.* 136: 11922–11925.

36 Zeng, C., Chen, Y., Kirschbaum, K. et al. (2015). *Sci. Adv.* 1: e1500045.

37 Das, A., Li, T., Li, G. et al. (2014). *Nanoscale* 6: 6458–6462.

38 Tang, L., Kang, X., Wang, S., and Zhu, M. (2019). *Langmuir* 35: 12350–12355.

39 Higaki, T., Liu, C., Chen, Y. et al. (2017). *J. Phys. Chem. Lett.* 8: 866–870.

40 a) Suyama, M., Takano, S., Nakamura, T., and Tsukuda, T. (2019). *J. Am. Chem. Soc.* 141: 14048–14051. b) Tian, S., Liao, L., Yuan, J. et al. (2016). *Chem. Commun.* 52: 9873–9876. c) Hossain, S., Imai, Y., Suzuki, D. et al. (2019). *Nanoscale* 11: 22089–22098.

41 a) Kumara, C., Aikens, C.M., and Dass, A. (2014). *J. Phys. Chem. Lett.* 5: 461–466. b) Negishi, Y., Munakata, K., Ohgake, W., and Nobusada, K. (2012). *J. Phys. Chem. Lett.* 3: 2209–2214.

42 Christensen, S.L., MacDonald, M.A., Chatt, A. et al. (2012). *J. Phys. Chem. C* 116: 26932–26937.

43 a) Wang, S., Song, Y., Jin, S. et al. (2015). *J. Am. Chem. Soc.* 137: 4018–4021. b) Yao, C., Lin, Y.-J., Yuan, J. et al. (2015). *J. Am. Chem. Soc.* 137: 15350–15353.

44 Liao, L., Zhou, S., Dai, Y. et al. (2015). *J. Am. Chem. Soc.* 137: 9511–9514.

45 Negishi, Y., Iwai, T., and Ide, M. (2010). *Chem. Commun.* 46: 4713–4715.

46 Dou, X., Yuan, X., Yao, Q. et al. (2014). *Chem. Commun.* 50: 7459–7462.

47 Qian, H., Jiang, D.-e., Li, G. et al. (2012). *J. Am. Chem. Soc.* 134: 16159–16162.

48 Kumara, C. and Dass, A. (2012). *Nanoscale* 4: 4084–4086.

49 Bootharaju, M., Kozlov, S., Cao, Z. et al. (2017). *Nanoscale* 9: 9529–9536.

50 Dharmaratne, A.C. and Dass, A. (2014). *Chem. Commun.* 50: 1722–1724.

51 Bootharaju, M.S., Joshi, C.P., Parida, M.R. et al. (2015). *Angew. Chem. Int. Ed.* 55: 922–926.

52 Choi, J.-P., Fields-Zinna, C.A., Stiles, R.L. et al. (2010). *J. Phys. Chem. C* 114: 15890–15896.

53 Yao, Q., Feng, Y., Fung, V. et al. (2017). *Nat. Commun.* 8: 1555.

54 Niihori, Y., Hashimoto, S., Koyama, Y. et al. (2019). *J. Phys. Chem. C* 123: 13324–13329.

55 Yang, S., Wang, S., Jin, S. et al. (2015). *Nanoscale* 7: 10005–10007.

56 Zheng, K., Fung, V., Yuan, X. et al. (2019). *J. Am. Chem. Soc.* 141: 18977–18983.

57 Krishnadas, K.R., Baksi, A., Ghosh, A. et al. (2016). *Nat. Commun.* 7: 13447.

58 Krishnadas, K.R., Ghosh, A., Baksi, A. et al. (2016). *J. Am. Chem. Soc.* 138: 140–148.

59 Khatun, E., Chakraborty, P., Jacob, B.R. et al. (2019). *Chem. Mater.* 32: 611–619.

60 Yuan, X., Zhang, B., Luo, Z. et al. (2014). *Angew. Chem. Int. Ed.* 53: 4623–4627.

61 Ishida, Y., Narita, K., Yonezawa, T., and Whetten, R.L. (2016). *J. Phys. Chem. Lett.* 7, 3718–7, 3722.

62 Zhang, B., Wu, Z., Cao, Y. et al. (2021). *J. Phys. Chem. C* 125: 489–497.

63 Tracy, J.B., Kalyuzhny, G., Crowe, M.C. et al. (2007). *J. Am. Chem. Soc.* 129: 6706–6707.

64 AbdulHalim, L.G., Kothalawala, N., Sinatra, L. et al. (2014). *J. Am. Chem. Soc.* 136: 15865–15868.

65 Yao, Q., Fung, V., Sun, C. et al. (1979). *Nat. Commun.* 2018: 9.

66 Cao, Y., Fung, V., Yao, Q. et al. (2020). *Nat. Commun.* 11: 5498.

67 Li, Y., Zhou, M., and Jin, R. (2021). *Adv. Mater.* 33: e2006591.

68 Chen, Y., Liu, C., Tang, Q. et al. (2016). *J. Am. Chem. Soc.* 138: 1482–1485.

69 Qin, Z., Zhang, J., Wan, C. et al. (2020). *Nat. Commun.* 11, 6019.

70 Xia, N., Yuan, J., Liao, L. et al. (2020). *J. Am. Chem. Soc.* 142: 12140–12145.

71 a) Takahata, R., Yamazoe, S., Maehara, Y. et al. (2020). *J. Phys. Chem. C* 124: 6907–6912. b) Kalenius, E., Malola, S., Matus, M.F. et al. (2021). *J. Am. Chem. Soc.* 143: 1273–1277.

72 Cao, Y., Malola, S., Matus, M.F. et al. (2021). *Chem* 7: 2227–2244.

73 Williamson, C.B., Nevers, D.R., Nelson, A. et al. (2019). *Science* 363: 731–735.

74 Dreier, T.A. and Ackerson, C.J. (2015). *Angew. Chem. Int. Ed.* 54: 9249–9252.

75 Cao, Y., Liu, T., Chen, T. et al. (2021). *Nat. Commun.* 12: 3212.

3

Thiolated Gold Nanoclusters with Well-Defined Compositions and Structures

Wanmiao Gu and Zhikun Wu

Key Laboratory of Materials Physics, Anhui Key Laboratory of Nanomaterials and Nanotechnology, CAS Center for Excellence in Nanoscience, Institute of Solid State Physics, Chinese Academy of Sciences, Hefei, 230031, P.R.China

3.1 Introduction

Metal nanocrystals have attracted extensive attention, especially for researchers in materials science, chemistry, biology, and physics due to their distinctive optical, electrical and magnetic properties and potential applications in fields such as molecular recognition, biomedicine, catalysis, and environment monitoring [1, 2]. After more than a century of efforts, today's scientists have accomplished excellent control of particle size, composition, morphology, and surface properties [3, 4]. By improving the dispersion range during the synthesis, nanocrystals can have extremely high uniformity at the nanometer level (e.g. size distribution ~5%). Nevertheless, it is difficult to have identical nanoparticles (NPs) at the atomic level. In addition, the polydispersity and other factors (e.g. large mass) hinder some state-of-the-art characterizations, such as mass spectrometry (MS) and single crystal X-ray diffraction (SCXC). Modern electron microscopy can be used to identify the arrangements of metal atoms in nanocrystals, but the organic stabilizers and the interface bonding between organic stabilizers and underlying metal atoms are not clear [5, 6]. The size heterogeneity and structural imprecision of nanocrystals pose obstacles to the deep understanding of many fundamental properties of NPs. Nanoscientists have been struggling to prepare monodisperse NPs with precise surface structures, and the emergence of nanocluster materials provides opportunities for resolving this problem, at least in the ultrasmall size regime.

Metal nanoclusters (NCs) are ultrasmall-sized NPs (1 to sub-3 nm core diameter) and stable entities composed of 10s–100s of metal atoms in the size range [7, 8]. Compared with conventional nanocrystals, metal NCs show some attractive advantages, such as atomic monodispersity and precise compositions/structures. From a structural point of view, the NCs typically display a core-shell pattern—that is, the metal cores are enclosed by external organic ligand shells to stabilize the NCs and prevent aggregation (Figure 3.1) [9]. Therefore, metal NCs with well-defined compositions and structures can be expressed by $[M_n(L)_m]^q$ (where M and L denote the metal atom and the organic ligand, respectively, n is the exact number of metal atoms, m refers to the number of protecting ligands, and q is the net charge of NCs).

The emergence of NCs not only realizes the researchers' dream to prepare truly uniform particles at the atomic level, but also provides a basis for exploring the evolution from metal complexes to conventional metal nanocrystals. Due to the quantum size effect, NCs show intrinsic properties that distinguish them from conventional nanocrystals, such as molecular-like structures [10], discrete

Atomically Precise Nanochemistry, First Edition. Edited by Rongchao Jin and De-en Jiang.

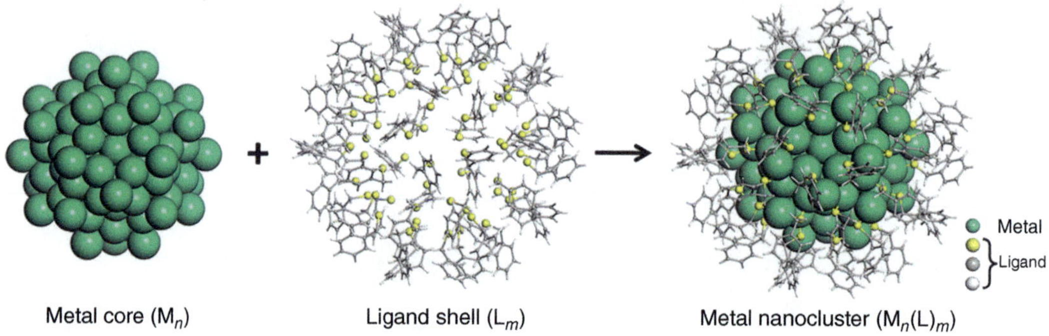

Figure 3.1 Structure of the metal NCs.

electronic energy levels [11], photoluminescence (PL) [12], and optical chirality [13], which make the NCs have wide application potentials in biological detection [14–16], catalysis [17, 18], and other fields [19, 20]. Studies show that the changes of any atomic position and number in NCs may significantly affect their physical and chemical properties [21, 22]. Benefiting from their precisely characterized compositions and structures, NCs can serve as ideal models of nanocrystals for investigating the mechanism of structure (or composition)-dependent properties at the atomic level [23–25].

To date, with the continuous progress of wet chemistry, various metal NCs with atomic precision have been successfully prepared and isolated, including gold [10, 26], silver [27, 28], and copper [29, 30] NCs. Here we mainly discuss the gold nanoclusters (abbreviated as Au NCs) with the most abundant species, which can be dated back to the 1960s. In 1969, the first phosphine-protected Au NC (i.e. $Au_{11}(SCN)_3(PPh_3)_7$, PPh_3 = triphenylphosphine) with precise composition and structure was characterized [31]. Afterward, many phosphine-stabilized Au NCs such as $Au_{55}(PPh_3)_{12}Cl_6$ [32] and $[Au_{39}(PPh_3)_{14}Cl_6]Cl_2$ [33] were synthesized and identified. Although some remarkable advances in Au NCs have been made, the stability had been the focus problem in this period. Until 1994, Brust et al. employed thiolates (SR) as surface ligands to prepare stable Au NCs by utilizing strong coordination between SR and Au atoms [34]. Compared with the Au—P bond of phosphine-protected Au NCs, the Au—S bond of thiolated Au NCs is more robust. After that, various synthesis and purification methods have been continuously developed or improved [35, 36], laying the foundation for the prosperity of cluster research. Furthermore, the advances in SCXC and synchrotron radiation technology have accelerated the exploration of nanocluster structures [26, 37].

In this chapter, we focus on the thiolated Au NCs (abbreviated as $Au_n(SR)_m$) with well-defined compositions and structures. First, the synthesis, purification, and characterization of NCs are summarized. Second, the anatomy of structures is illustrated. Based on the precise size and structure, the optical absorption, luminescence, chirality, magnetism, and various applications of $Au_n(SR)_m$ are illustrated. Finally, some important issues in the future research of NCs are proposed. It is worth noting that most of the structures of $Au_n(SR)_m$ involved in this chapter are experimentally obtained, and only a small part of the structures are theoretically calculated.

3.2 Synthesis, Purification, and Characterization of Gold Nanoclusters

3.2.1 Synthesis

Preparation of monodisperse and stable Au NCs with definite compositions has always been an important goal of cluster research. Inspired by the self-assembled monolayer of thiolate on metal surface, Brust and Schiffrin et al. selected -SR as a protective ligand and used $NaBH_4$ to reduce

Au(I)-SR precursor in a two-phase solvent system, thus preparing thermodynamically stable, 1–3 nm thiolated Au NCs [34, 38]. Considering the outstanding contribution to the preparation of Au NCs, this method is also called Brust-Schiffrin two-phase synthesis method. Up to now, many research groups devoted tremendous efforts in the NCs field, and proposed various preparation methods such as gold salt (complex) reduction [39, 40], ligand induction [41], and anti-galvanic reaction [42], providing the basis for the vigorous development of NCs. In this section, we review general synthesis strategies, facile synthesis methods, and common synthesis parameters that influence the structures and properties of NCs, basing on the results of the literature.

3.2.1.1 Synthesis Strategy

Due to the lack of understanding of synthesis mechanism and cluster chemistry, rational designing and synthesizing of Au NCs remain a challenge. Recently, researchers have focused on the fundamentals of the underlying reaction process to provide some guidance for cluster synthesis. For example, Zhu et al. improved the Brust-Schiffrin method and revealed the importance of some kinetic factors such as the stirring speed [43]. By adjusting the reaction temperature and stirring speed, for example, the yield of $Au_{25}(SR)_{18}$ clusters was greatly improved. Subsequently, Wu et al. proposed the strategy of "kinetic control and thermodynamic selection" for the preparation of atomically mono-dispersed nanoclusters illustrated by the synthesis of the Au_{19} cluster [44]. As shown in Figure 3.2, the initial size distribution of crude NCs is mainly controlled by adjusting the reduction speed. Then, by controlling the thermodynamic factors such as aging time, the unstable components in the polydispersed clusters can be decomposed or transformed into stable products. This strategy is also realized by other groups. For example, Xie and coworkers demonstrated the utility of such a strategy by monitoring the formation of water-soluble Au_{25} cluster using mass spectrometry [45]. They used NaOH-mediated $NaBH_4$ reduction method to modulate the formation kinetics and thermodynamics of Au NCs. Specifically, the added NaOH reduces the reduction ability of $NaBH_4$ and accelerates the etching ability of free thiolate, therefore resulting in the formation of a balanced reversible reaction and the accelerating of the thermodynamical selection of Au_{25}. Although various methods have been developed, the formation process of mono-dispersed Au NCs can all be interpreted by kinetic control and thermodynamic selection. In this chapter, we discuss several major synthesis methods under the rule of "kinetic control and thermodynamic selection" strategy.

3.2.1.2 Gold Salt (Complex) Reduction Method

This method is actually the above-mentioned Brust-Schiffrin method or its modified one, whose features are that the gold precursor is gold salt (complex) and thus a reduction process is necessary for the preparation of Au NCs.

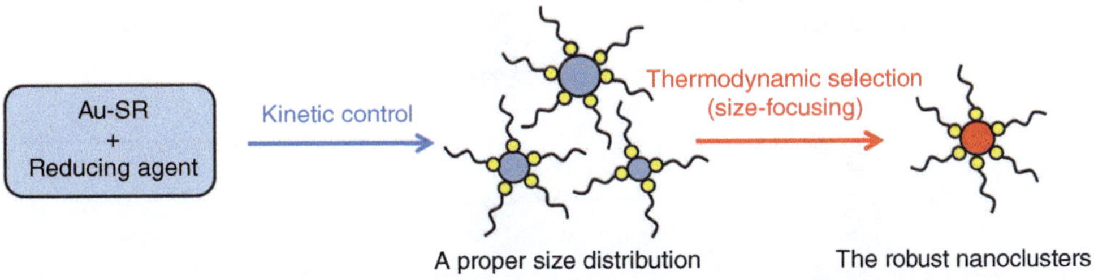

Figure 3.2 Illustration of the synthesis strategy of "kinetic control and thermodynamic selection". *Source:* Redrawn from Ref. [44] with permission from American Chemical Society, copyright 2011.

Adjustable experimental parameters for this method include the ligand, temperature, reducing reagent, assistant reagent, and others. Due to differences in steric hindrance, electronic structure, and bonding mode of ligands, the size, and structure of the final products can be affected by the protective ligands. Generally, cyclohexanthiol (HS-*c*-C_6H_{11}) and adamantanethiol (HS-Adm) ligands with relatively strong steric repulsion tend to result in small-sized NCs, such as Au_{18}(S-*c*-$C_6H_{11})_{14}$ [46], Au_{23}(S-*c*-$C_6H_{11})_{16}^-$ [40], Au_{28}(S-*c*-$C_6H_{11})_{20}$ [47], $Au_{16}(S\text{-}Adm)_{12}$ [48], $Au_{21}(S\text{-}Adm)_{15}$ [49] and $Au_{30}(S\text{-}Adm)_{18}$ [50]. The 4-*tert*-butylbenzenelthiolate(HSPh-^{t}Bu)-protected Au NCs obtained by the reduction of gold salts (complexes) prefer the face-centered cubic(fcc)-packing structure, such as $Au_{44}(SPh\text{-}^tBu)_{28}$ [51], $Au_{52}(SPh\text{-}^tBu)_{32}$ [52], $Au_{56}(SPh\text{-}^tBu)_{34}$ [53], and $Au_{92}(SPh\text{-}^tBu)_{44}$ [54, 55], while phenylethanethiolated (HSC_2H_4Ph) Au NCs prepared by the similar method mainly adopt the icosahedral core-shell structure, like $Au_{25}(SC_2H_4Ph)_{18}$ [10], $Au_{38}(SC_2H_4Ph)_{24}$ [56], and $Au_{144}(SC_2H_4Ph)_{60}$ [9, 57].

Strong reducing agents (such as $NaBH_4$ [43]) can provide fast reaction kinetics for the synthesis of NCs, while weak reducing agents (such as borane *tert*-butylamine complex [44], CO [58], and $NaBH_3CN$ [59]) are expected to slow down the reduction kinetics, resulting in different products. For instance, under the otherwise similar conditions, the borane *tert*-butylamine complex and $NaBH_4$ as reducing agents can led to the major $Au_{19}(SC_2H_4Ph)_{13}$ [44] and $Au_{25}(SC_2H_4Ph)_{18}$ [43] product, respectively. Yu et al. used CO as a mild reducing agent, and controlled the size of Au NCs by changing the solution pH value [58], as shown in Figure 3.3. The same group also adjusted the synthesis of Au_{25} cluster by controlling the amount of the reducing agent [60].

Recently, researchers also tried to add acids, bases, foreign metal ions and other assistant reagents to influence the reaction process and the final products. For examples, Zhuang et al. developed an acid-induction method for successfully preparing a medium-sized $Au_{52}(SC_2H_4Ph)_{32}$ cluster [61]. The addition of acid not only increases the hydrolysis of $NaBH_4$ by enhancing its reduction ability but also reduces the reaction reactivity of thiolate by weakening the interaction between Au and thiolate. Shortly afterward, $Au_{42}(SPh\text{-}^tBu)_{26}$ [62], $Au_{44}(SPh\text{-}^tBu)_{26}$ [63], and $Au_{48}(SPh\text{-}^tBu)_{28}$ [63] clusters were obtained by the acid-assisted method. Zhuang et al. also reported an example of cluster preparation by using foreign Cd ion as assistant reagent [64]. The addition of Cd may affect the kinetics and thermodynamics in the producing of non-fcc-structured $Au_{42}(SPh\text{-}^tBu)_{26}$ by forming unstable Au/Cd intermediates, by adjusting the reducing ability of $NaBH_4$, by affecting the etching rate of thiol, and so on.

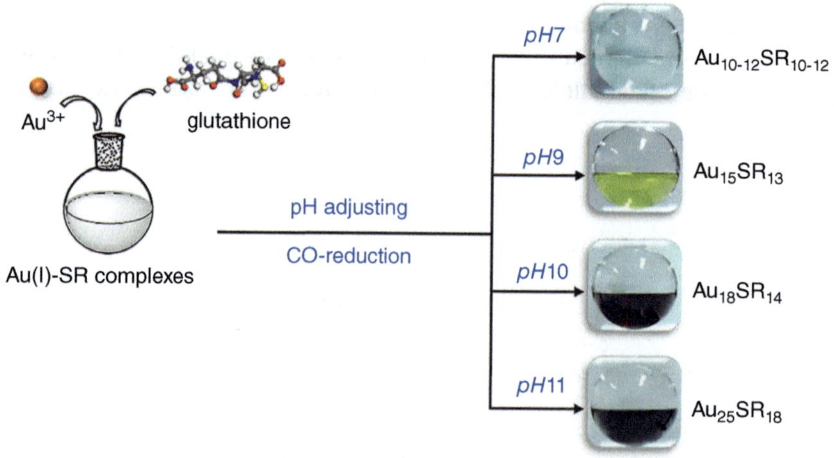

Figure 3.3 Synthesis of different sized Au NCs using the CO reduction method by adjusting pH. *Source:* Reproduced with permission from Ref. [58]. Copyright 2013 American Chemical Society.

3.2.1.3 Ligand Induction Method

In this method, the precursor is the monodispersed or polydispersed Au NCs, which react with the incoming ligands, thus resulting in the gold nanocluster products. Ligand induction synthesis usually includes ligand exchange, structure transformation, and size-focusing steps. Studies show that the type of ligand has a great influence on cluster products. For example, Jin group and Wu group successfully transformed $Au_{38}(SC_2H_4Ph)_{24}$ into $Au_{36}(SPh\text{-}^tBu)_{24}$ and $Au_{60}S_6(SCH_2Ph)_{36}$ clusters by using HSPh-^{t}Bu and benzyl mercaptan ($HSCH_2Ph$) as the induction ligand [65, 66], respectively. Another example is the ligand exchange of phenylethanethiolated $Au_{25}(SC_2H_4Ph)_{18}$. $Au_{28}(SPh\text{-}^tBu)_{20}$ was prepared after mixing $Au_{25}(SC_2H_4Ph)_{18}$ with excess HSPh-^{t}Bu thiol [67], while the $Au_{24}(SCH_2Ph)_{20}$ cluster was obtained by the exchange reaction of $Au_{25}(SC_2H_4Ph)_{18}$ with excess benzyl mercaptan ($HSCH_2Ph$) [68]. In addition, Chen et al. selected methyl phenylthiols (MBT) with different substitution position to explore the ligand effect on the size of Au NCs [69]. By etching polydispersed Au NCs with para-, meta- and ortho-MBT, $Au_{130}(p\text{-MBT})_{50}$, $Au_{104}(m\text{-MBT})_{41}$, and $Au_{40}(o\text{-MBT})_{24}$ were prepared, respectively (Figure 3.4).

Temperature plays an important role in the induction reaction since some transformations require thermal activation. For examples, Zeng et al. took the multi-sized $Au_x(SPh\text{-}^tBu)_y$ as precursors and added excessive HSPh-^{t}Bu at 60 °C and 80 °C, preparing $Au_{44}(SPh\text{-}^tBu)_{28}$ [51] and $Au_{52}(SPh\text{-}^tBu)_{32}$ [52], respectively; Wu et al. successfully obtained $Au_{92}(SPh\text{-}^tBu)_{44}$ [55] and $Au_{56}(SPh\text{-}^tBu)_{24}$ [53] clusters by treating multisized $Au_x(SPh\text{-}^tBu)_y$ with excessive HSPh-^{t}Bu under 65 °C and 50 °C, respectively. The induction time is also one of the most important parameters. For example, Liao et al. reacted the multisized $Au_x(DMBT)_y$ precursors with excess 2,4-dimethylbenzenethiolate (DMBT) at 40 °C, and obtained $Au_{44}(DMBT)_{26}$ cluster after etching for 20 hours [70], while a novel $Au_{49}(DMBT)_{27}$ cluster was synthesized by shortening the etching time to only 18 hours [71].

3.2.1.4 Anti-Galvanic Reaction Method

In 2012, Wu revealed an unprecedented phenomenon; that is, relatively noble metal nanoparticles can reduce relatively active metal ions (e.g. silver nanoparticle can reduce copper ions), which is

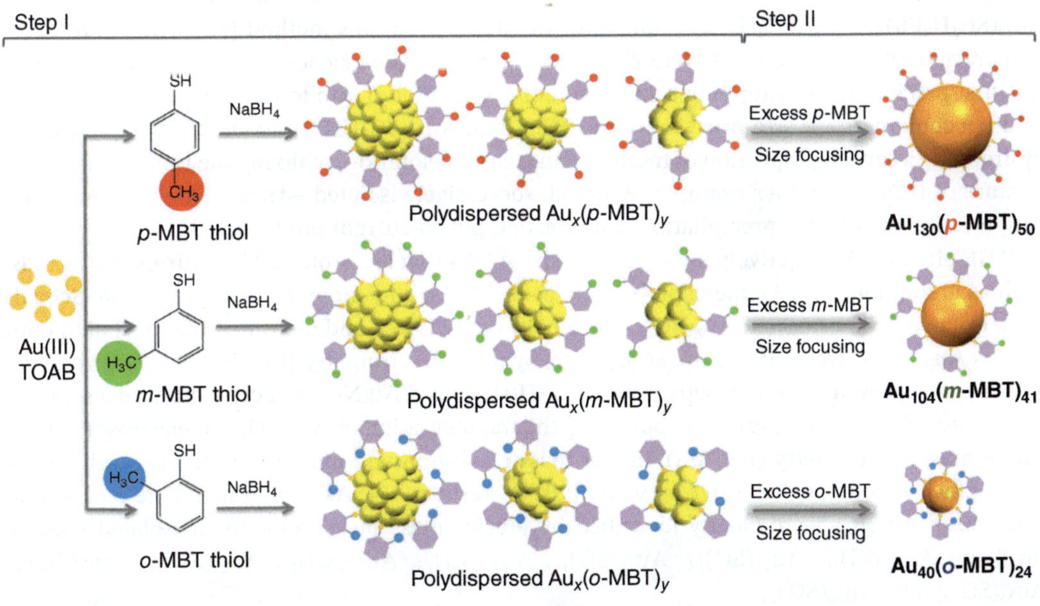

Figure 3.4 Synthesis of different magic-sized NCs by exploring three isomeric thiol ligands: *p*-MBT, *m*-MBT, and *o*-MBT. *Source:* Reproduced with permission from Ref. [69]. Copyright 2015 American Chemical Society.

termed anti-galvanic reduction (AGR) [42] and provides a facile and powerful method for the synthesis of metal nanoclusters, especially for the synthesis of iso-atom doped metal NCs. Until now, a lot of alloy nanoclusters have been synthesized by this method, including $Au_{25}Ag_2(SR)_{18}$ [72], $Au_{24}Cd(SR)_{18}$ [73], $Au_{24}Hg(SR)_{18}$ [73], $Au_{20}Cd_4(SH)(SR)_{19}$ [74], and $Au_{47}Cd_2(SR)_{31}$ [75]. Au NCs precursors are the main reactants in the reaction and thus certainly play a crucial role in AGR. Zhu and Zhang et al. investigated the anti-galvanic reaction between $Au_{23}(S\text{-}c\text{-}C_6H_{11})_{16}^-$ (or $Au_{34}(S\text{-}c\text{-}C_6H_{11})_{22}$) and $Cd(S\text{-}c\text{-}C_6H_{11})_2$ complex and obtained different products: $Au_{20}Cd_4(SH)(S\text{-}c\text{-}C_6H_{11})_{19}$ [74] for Au $Au_{23}(S\text{-}c\text{-}C_6H_{11})_{16}^-$ and $Au_{26}Cd_4(S\text{-}c\text{-}C_6H_{11})_{22}$ [76] for $Au_{34}(S\text{-}c\text{-}C_6H_{11})_{22}$. The type of ion precursor was extensively revealed to influence both AGR kinetics and thermodynamics. For example, $Au_{24}Cd(SC_2H_4Ph)_{18}$ [73, 77] and $[Au_{13}Cd_2(PPh_3)(SC_2H_4Ph)_6(NO_3)_2]_2Cd(NO_3)_4$ [78] clusters were prepared by reacting $Au_{25}(SC_2H_4Ph)_{18}^-$ with Cd-SC_2H_4Ph complex and $Cd(PPh_3)_2(NO_3)_2$, respectively. Another example is the reaction of $Au_{23}(S\text{-}c\text{-}C_6H_{11})_{16}^-$ with different Cd precursors [74]. When the $Cd(NO_3)_2$ and $Cd(S\text{-}c\text{-}C_6H_{11})_2$ complexes were treated with $Au_{23}(S\text{-}c\text{-}C_6H_{11})_{16}^-$ clusters, the $Au_{28}(S\text{-}c\text{-}C_6H_{11})_{20}$ and $Au_{20}Cd_4(SH)(S\text{-}c\text{-}C_6H_{11})_{19}$ clusters were obtained, respectively.

3.2.2 Isolation and Purification

In some cases, it is difficult to obtain monodisperse Au NCs by simple protocols (e.g. recrystallization) from the initial products. The polydispersity of NCs will bring great interference to the composition determination, structure characterization, and subsequent property research. Therefore, it is necessary to separate the crude products and obtain monodisperse nanoclusters. The purification methods of Au NCs mainly involves solvent fractionated precipitation, polyacrylamide gel electrophoresis (PAGE) [39, 79], preparative thin layer chromatography (PTLC) [36], recrystallization [80], size exclusion chromatography (SEC) [35], and silica gel column chromatography [81].

Fractionated precipitation is an effective and simple separation method based on the size-dependent solubility difference in mixed solvents. When a poorly soluble solvent is added to the NCs mixture, the relatively poorly soluble Au NCs will be preferentially precipitated out. For example, Nimmala et al. isolated the $Au_{67}(SC_2H_4Ph)_{35}$ cluster from the mixture of $Au_{25}(SC_2H_4Ph)_{18}$, $Au_{38}(SC_2H_4Ph)_{24}$, $Au_{67}(SC_2H_4Ph)_{35}$, and $Au_{102}(SC_2H_4Ph)_{44}$ by this method [82]. The entire isolation process can be tracked by MS, as shown in Figure 3.5. The heaviest $Au_{102}(SC_2H_4Ph)_{44}$ cluster was first precipitated by dropping a small amount of methanol into tetrahydrofuran (or toluene) solution containing the size-mixed clusters. After that, $Au_{67}(SC_2H_4Ph)_{35}$ could be precipitated out by dropping a certain proportion of methanol into the left solution, realizing the isolation from the mixture of NCs. In another example, Wu et al. successively isolated ~4 nm NPs, ~2 nm NPs, and Au_{25} NCs by fractionated precipitation from the one-pot co-current products [83].

PAGE often works effectively for separating those NCs that are protected by hydrophilic ligands, such as glutathione (GSH), mercaptobenzoic acid (MBA), and captopril (Capt). In 1998, Schaaff et al. isolated GSH-protected $Au_n(SG)_m$ clusters by using PAGE and identified the most abundant species $Au_{28}(SG)_{16}$ (revised to $Au_{25}(SG)_{18}$ later) by mass spectrometry [84]. In 2005, Negishi et al. effectively improved the PAGE separation of GSH-stabilized Au NCs and characterized a series of Au_n clusters ($n = 10$–39) [39]. By suppressing the fragmentation of $Au_n(SG)_m$ in electrospray ionization mass spectrometry (ESI-MS), higher mass resolution and more accurate mass calibration were achieved compared to their previous ESI-MS results [79]. As shown in Figure 3.6, several small-sized Au NCs were clearly identified by precise mass measurements of isolated species, involving $Au_{10}(SG)_{10}$, $Au_{15}(SG)_{13}$, $Au_{18}(SG)_{14}$, $Au_{22}(SG)_{16}$, $Au_{22}(SG)_{17}$, $Au_{25}(SG)_{18}$, $Au_{29}(SG)_{20}$, $Au_{33}(SG)_{22}$, and $Au_{39}(SG)_{24}$.

The separation mechanism of PTLC is similar to that of silica gel column chromatography. Au NCs of different polarity possess different adsorption capacity on silica gel; that is, the NCs with

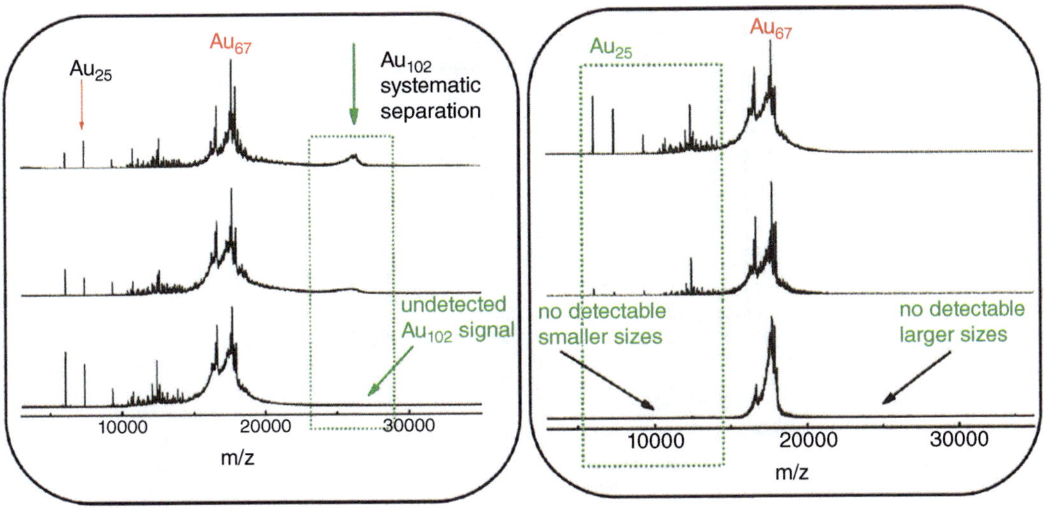

Figure 3.5 $Au_{67}(SR)_{35}$ purification separating larger cluster (left image) and smaller clusters (right image). *Source:* Adapted with permission from Ref. [82]. Copyright 2013 American Chemical Society.

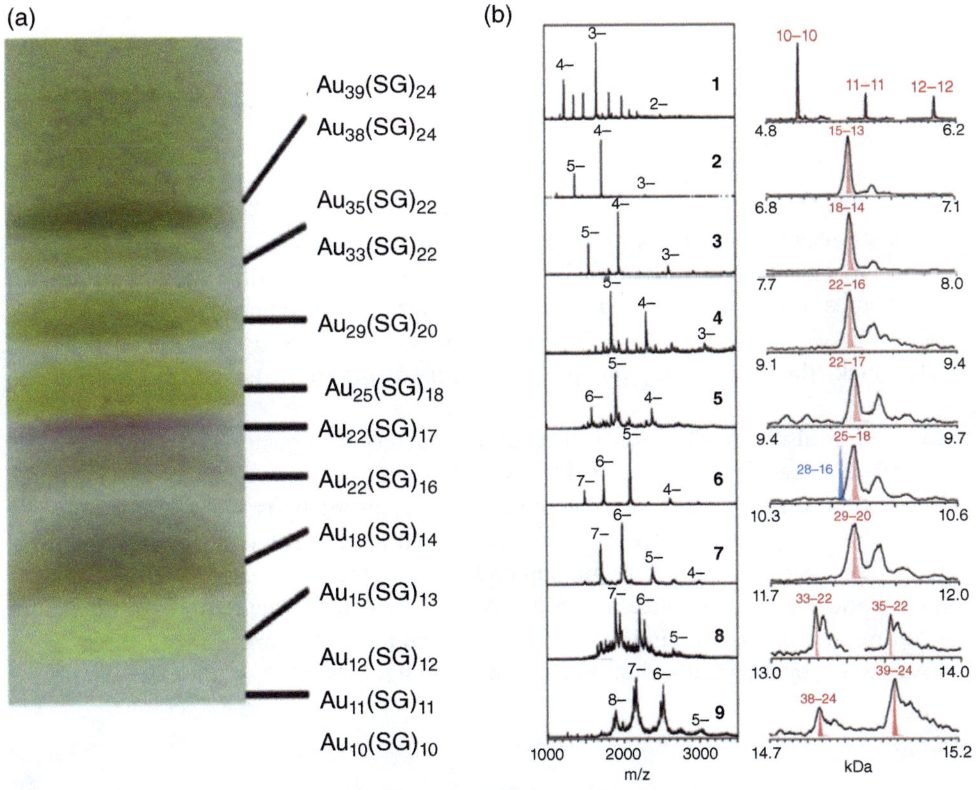

Figure 3.6 (a) Separation of $Au_n(SG)_m$ clusters by PAGE. (b) Low-resolution MS of the fractionated Au:SG clusters (left). The MS spectra reproduced from the most intense peaks in the high-resolution spectra (right). *Source:* Adapted with permission from Ref. [39]. Copyright 2005 American Chemical Society.

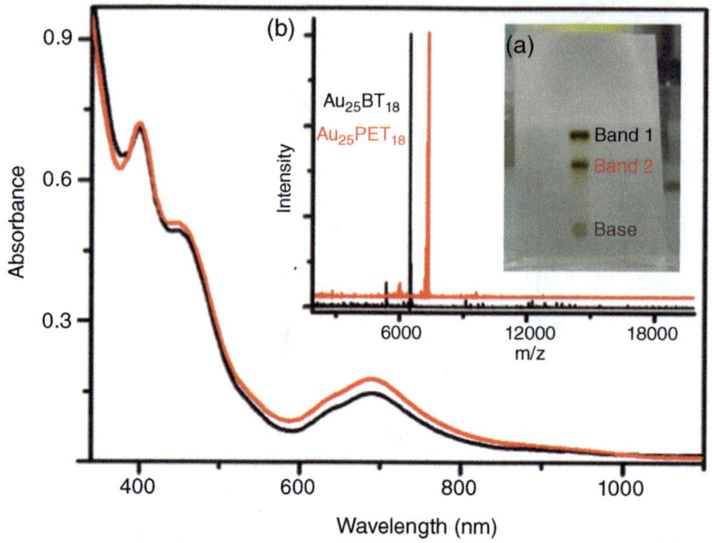

Figure 3.7 UV-vis spectra of PTLC separated materials: (a) Photograph of the PTLC plate used for cluster separation. Bands 1 ($Au_{25}BT_{18}$) and 2 ($Au_{25}PET_{18}$) are two separated NCs. (b) MS data of bands 1 (black trace) and 2 (red trace), respectively. Note, BT = butanethiolate, C_4H_9S-; PET = phenylethanethiolate, PhC_2H_4S-. *Source:* Reproduced with permission from Ref. [36]. Copyright 2014 American Chemical Society.

large polarity have strong interaction with silica gel and retain for a long time, whereas those clusters with small polarity have weak interactions with silica gel and thus short retention times. The NCs with different polarities and sizes can reach different positions of chromatoplate, making the separation efficient and visual.

In 2014, Li et al. reported the recycling of $Au_{25}(SC_2H_4Ph)_{18}$ by column chromatography packed by silica gel [85]. One year later, Ghosh et al. applied PTLC technology to cluster separation [36]. This separation strategy successfully applied to binary mixtures of Au_{25} clusters with different protecting ligands (Figure 3.7), as well as the mixtures of Au_{25} and Au_{144} with the same ligand but different cluster cores. Tian et al. separated the first pair of structural isomers ($Au_{38}(SC_2H_4Ph)_{24-T}$ and $Au_{38}(SC_2H_4Ph)_{24-Q}$) in Au NCs by PTLC [86]. Furthermore, this group adopted PTLC to quantitatively track the transformation process from $Au_{44}(SPh\text{-}^tBu)_{28}$ to $Au_{36}(SPh\text{-}^tBu)_{24}$ cluster [41].

Recrystallization is also a purification process, which refers to repeated crystallization. For example, Schaaff et al. recrystallized the 29 kDa Au NCs (later determined as $Au_{144}(SR)_{60}$) and obtained microcrystals from slow precipitation out of a dilute solvent–nonsolvent interaction [80]. Specifically, the cluster was precipitated from a concentrated toluene solution by the addition of excess ethanol; after a period of time at room temperature, the precipitate was filtered, then redissolved in toluene and precipitated twice. The repeated steps can ensure the removal of unreacted thiolates, resulting in a fine crystalline black powder. In addition, Wu and coworkers obtained high-purity $Au_{25}(SG)_{18}$ by recrystallization in water-methanol solution [87, 88].

3.2.3 Characterization

The full/partial composition of Au NCs can be determined by, for example, mass spectrometry, thermogravimetric analysis, XPS photoelectron spectroscopy, and nuclear magnetic resonance (NMR). Mass spectrometry can provide the molecular weight of the NCs and thus one can infer the atomic composition [40, 51]. Thermogravimetric analysis as an auxiliary tool [86, 89] can obtain the ratio of metal to organic ligand through the weight loss ratio of clusters, thereby verifying the molecular formula of NCs. XPS can provide the basic information of all elements except H in clusters [86], including element species, element atomic ratio, and chemical bonds. The existence of counterions in NCs can be determined by the element types. The accuracy of molecular formula of clusters can also be judged by comparing the atomic ratios of each element. NMR can be used to analyze chemical environments for organic ligands on NCs [90].

For the determination of the total structure, the most reliable approach is SCXC or synchrotron radiation [26, 91]. The prerequisite of experimental structure analysis is to grow high-quality single crystals that meet analytical standards. The obtained crystal data can also provide the complete formula of NCs and solve many fundamental issues [10, 40].

3.3 Structures of Gold Nanoclusters

In the past two decades, researchers have carried out extensive and substantial research in the field of Au NCs. The obtained atomically precise NCs present wide sizes and diverse structures. To facilitate the study and application, researchers classify Au NCs from different views and attributes. Au NCs can be classified by the surface ligand, such as thiolate-, phosphine-, alkyne- [92, 93], selenolate- [94, 95], carbine- [96, 97], and mix-ligands-protected Au NCs [98]. Although the NCs size varies only in the range of 1–3 nm, NCs can be further categorized into three types based on the number of Au atoms: small-sized range (i.e. less than 40 gold atoms); medium-sized (i.e. ~40 to ~100 gold atoms) [99]; and large-sized NCs (i.e. more than 100 gold atoms).

Crystal structure analysis is an indispensable tool for deeply understanding the structure of these nanoclusters. The metal framework structure and spatial arrangement of the organic stabilizers around the metal kernel assist in unrevealing the structure-property relationships and fundamental issues involved in their fabrication. Although high-resolution transmission electron microscopy (TEM) is powerful to characterize the morphology of NPs and has indeed become an indispensable tool in nanoscience, its use for the characterization of Au NCs with atomic resolution is unfortunately limited. Given the momentum transfer, thermal effect, and ligand stripping, the strong electron beam of TEM may damage the small NCs. In recent years, some progress has been made in TEM analysis of NCs using low-dose electron beams [100, 101]. Nevertheless, the method only maps out the inorganic core but not the metal-ligand interface nor the surface ligands. The overall structures of NCs in these reports were often aided by further theoretical simulations [102]. Crystallization of NCs can be achieved owing to the high molecular purity of NCs. Experimental tools such as SCXC and synchrotron radiation can be readily utilized for structural characterization of NCs, directly providing the information of the Au kernel, Au-ligand interface structure, and outer organic ligand structure. Studies of NCs show that no matter what the composition of Au_nL_m is, they mostly adopt a structural mode; that is, an Au core protected by simple organic ligands L_x ($x = 1, 2, 3, 4, \ldots$), staple-like Au_xL_{x+1} motifs, or ring-like Au_xL_x motifs [8]. This mode has been widely adopted by researchers and has gradually become a universal style to analyze NCs structure.

Based on the packing modes of Au kernel atoms, Au NCs can be divided into fcc or quasi-fcc [52, 89], body-centered cubic (bcc) [103], hexagonal close-paced (hcp) [46, 104], icosahedral [10], decahedral [26] structures, etc. Nonetheless, such a structural classification can only describe part of Au NCs kernel structures. The Au atoms of core can also be divided into tetrahedron, octahedron, decahedron, icosahedron, and other basic structural units. Based on these structural units, researchers proposed different growth modes, including tetrahedron-Au_4-based growth mode [51, 52], polyhedron-fusion mode [51, 67], layer-by-layer mode [51, 54], and shell-by-shell mode [105–107]. The small building blocks extend to the larger structure, thus completing the kernel structural classification of most of the reported Au NCs. In this section, we provide a summary of the known thiolate-protected $Au_n(SR)_m$ structures, mainly involving the experimental crystal structures resolved by X-ray crystallography.

3.3.1 Kernel Structures of $Au_n(SR)_m$

Atomically precise ultrasmall-sized Au NCs retain the original state of the Au—Au bonds, not only providing an important benchmark for basic research on the origin of condensed matter, but also helping to elucidating the structure evolution from Au(I)-complexes to Au NCs. To bridge the gaps between Au complexes and NCs, Li et al. develops a novel ligand-induced kernel-tailoring method to prepare the smallest Au NCs [108]. Starting from small-size NCs, three novel transition-size NCs including $[Au_{13}(S\text{-}Adm)_8(dppb)_2](BPh_4)$, $Au_{14}(S\text{-}c\text{-}C_6H_{11})_{10}(dppb)$, and $Au_{16}(S\text{-}c\text{-}C_6H_{11})_{11}(dppbz)$ (dppb = $Ph_2P(CH_2)_4PPh_2$, dppbz = $Ph_2P(C_6H_4)PPh_2$) are obtained by controlled clipping on the surface and kernel of initial NCs (Figure 3.8). Structural analysis shows that the smallest $[Au_{13}(S\text{-}Adm)_8(dppb)_2](BPh_4)$ cluster has a unique near-planar Au_5 inner core enclosed by a $Au_8(SR)_8$ ring-type outer shell. The Au_{13} cluster is highly correlated in structures with the ring-in-ring $Au_{12}(SR)_{12}$ complex, which exhibits a $Au_4(SR)_4$ inner ring surrounded by a $Au_8(SR)_8$ outer ring. Combining the structures of the $Au_{12}(SR)_{12}$ complex and $Au_{18}(S\text{-}c\text{-}C_6H_{11})_{14}$ cluster, the group revealed the original nucleation process in the Au NCs; that is, a $Au_4(SR)_4$ ring to a planar Au_5 inner core. Furthermore, structural analysis also showed that the $Au_{14}(S\text{-}c\text{-}C_6H_{11})_{10}(dppb)$ and $Au_{16}(S\text{-}c\text{-}C_6H_{11})_{11}(dppbz)$ both contain tetrahedron building blocks. The Au_6 kernel of $Au_{14}(S\text{-}c\text{-}C_6H_{11})_{10}(dppb)$ contains a pair of tetrahedrons that share a common edge, while the Au_7 kernel of $Au_{16}(S\text{-}c\text{-}C_6H_{11})_{11}(dppbz)$ consists of two Au_4 tetrahedrons combined by vertex-sharing.

3.3.2 Kernels Based on Tetrahedral Au_4 Units

The tetrahedral-based geometry is common in $Au_n(SR)_m$ clusters, such as $Au_{20}(SPh\text{-}^tBu)_{16}$ [110], $Au_{24}(SCH_2Ph\text{-}^tBu)_{20}$ [111], and $Au_{28}(SPh\text{-}^tBu)_{20}$ [67]. As illustrated in Figure 3.9a, 4-*tert*-butylbenzenethiol-protected $Au_{20}(SPh\text{-}^tBu)_{16}$ cluster features a vertex-sharing bi-tetrahedral Au_7 kernel and a big ring-type $Au_8(SR)_8$ motif. The 4-*tert*-butylphenylmethanethiolate-capped $Au_{24}(SCH_2Ph^tBu)_{20}$ exhibits a prolate-shape Au_8 kernel, which can be viewed as two Au_4 tetrahedrons joined together without sharing any Au atoms (Figure 3.9b). The four-tetrahedral Au_{14} kernel in $Au_{28}(SPh\text{-}^tBu)_{20}$ is made up of Au_7 double helixes and is protected by four $Au_2(SR)_3$ dimeric staples and two $Au_3(SR)_4$ staples (Figure 3.9c).

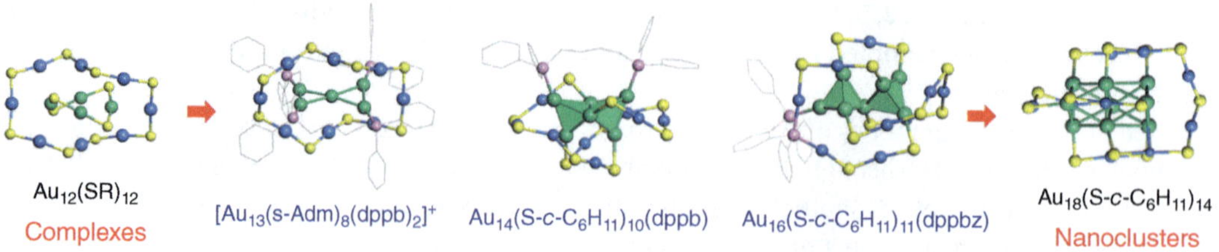

Figure 3.8 Structural evolution of transition size Au NCs. Color labels: yellow = S, lilac = P, gray = C, others = Au. *Source:* Structures are redrawn from the cifs presented in Refs. [104, 108, 109].

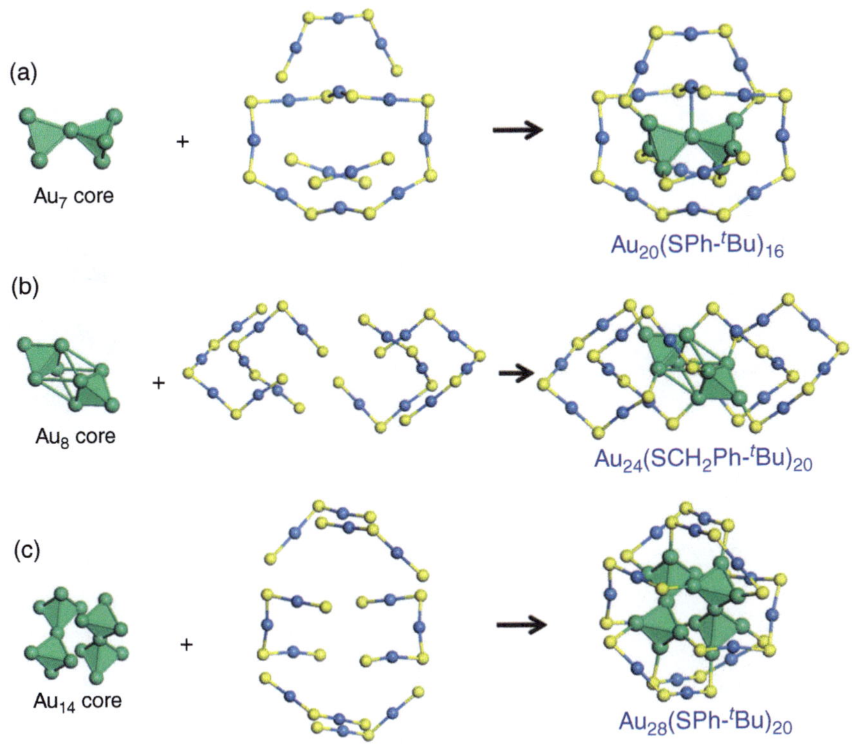

Figure 3.9 Anatomy of three Au NCs structures: (a) $Au_{20}(SPh\text{-}^tBu)_{16}$; (b) $Au_{24}(SCH_2Ph\text{-}^tBu)_{20}$; (c) $Au_{28}(SPh\text{-}^tBu)_{20}$. Color labels: yellow = S, others = Au. *Source:* Structures are redrawn from the cifs presented in Refs. [67, 110, 111].

In addition to these cases, there are a series of $Au_n(SR)_m$ clusters with tetrahedrons as building units. Five types of kernel evolution patterns associated with Au_4 tetrahedron blocks have been proposed for these NCs [51, 112–114], as categorized below:

1) Quasi-linear Au_4 tetrahedral growth by sharing one vertex, as shown in the Au cores of $Au_{15}(SR)_{13}$ (DFT-structure) [115], $Au_{20}(SPh\text{-}^tBu)_{16}$ [110], and $Au_{28}(SCH_2Ph\text{-}^tBu)_{22}$ [114] clusters (Figure 3.10a). The Au kernel grows by successively adding a Au_4 tetrahedron via vertex sharing, resulting in two tetrahedrons for $Au_{20}(SPh\text{-}^tBu)_{16}$ and three tetrahedrons for $Au_{28}(SCH_2Ph\text{-}^tBu)_{22}$.
2) The growth of triangular-Au_3 unit at one side or both sides of tetrahedral-Au_4 unit (Figure 3.10b) or bi-tetrahedral-Au_7 units (Figure 3.10c) or Au_7 double helixes (Figure 3.10d). For examples, the Au_7 kernel of $Au_{16}(S\text{-}Adm)_{12}$ can be divided into a triangular-Au_3 plus a tetrahedral-Au_4 unit [48]; the Au_{10} kernel of $Au_{21}(S\text{-}Adm)_{15}$ comprises a tetrahedral-Au_4 unit and two symmetric triangular-Au_3 units [49]; in $Au_{21}(S\text{-}^tBu)_{15}$, Au_{10} kernel can be viewed as a triangular-Au_3 plus a bi-tetrahedral-Au_7 unit [112]; the Au_{13} kernel of $Au_{23}(S\text{-}c\text{-}C_6H_{11})_{16}^-$ consists of a bi-tetrahedral Au_7 unit and two anti-symmetric triangular-Au_3 units [40]; the kernel of $Au_{29}(S\text{-}Adm)_{19}$ and $Au_{30}(S\text{-}Adm)_{18}$ are composed of Au_7 double-helixes and one or two triangular-Au_3 units [116], respectively.
3) Double-helical growth along one dimension, as presented in Figure 3.10e. In 2015, Zeng et al. resolved the structure of $Au_{52}(SPh\text{-}^tBu)_{32}$ cluster [52], which features a double-helical Au_{32} kernel composed of 10 tetrahedral-Au_4 units. Within each helix, five tetrahedrons are connected by vertex sharing. Besides, such double-helical structures can also be observed in the

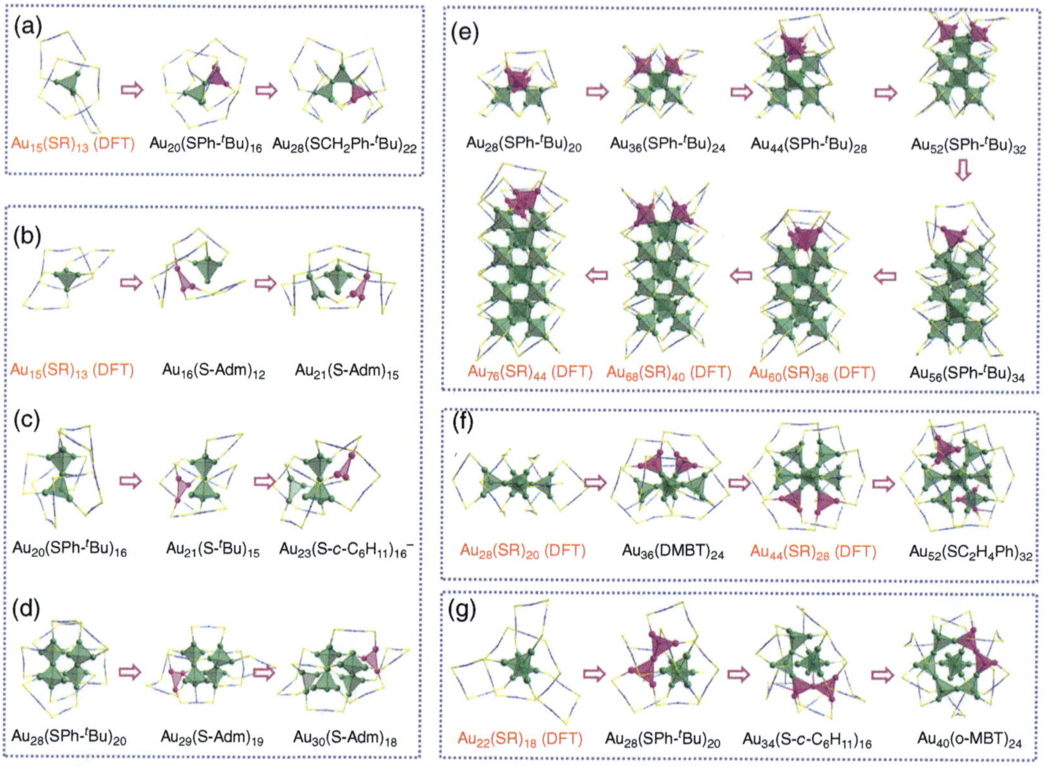

Figure 3.10 Kernel growth patterns of $Au_n(SR)_m$ based on tetrahedral Au_4 units: (a) quasi-linear growth of Au_4 tetrahedron; (b–d) symmetric or anti-symmetric growth of triangular-Au_3 unit; (e) one-dimensional growth; (f) two-dimensional growth; (g) Kekulé-like ring growth. Color labels: yellow = S, others = Au.

Au_{14} core of $Au_{28}(SPh\text{-}^tBu)_{20}$ [67], Au_{20} kernel of $Au_{36}(SPh\text{-}^tBu)_{24}$ [89], and Au_{26} kernel of $Au_{44}(SPh\text{-}^tBu)_{28}$ [51]. The underlying Au kernel in the sequence of $Au_{8n+4}(SR)_{4n+8}$ (n = 3–6) clusters grows by successively adding two tetrahedrons to the bottom of the double helixes, suggesting their one-dimensional growth pattern. Based on the structural evolution pathway, the one-dimensional core structure of longer $Au_{60}(SR)_{36}$ [117], $Au_{68}(SR)_{40}$ [117], $Au_{76}(SR)_{44}$ [118, 119] clusters were obtained by theoretical calculations. Recently, the atomic structure of $Au_{56}(SPh\text{-}^tBu)_{34}$ was successfully characterized by SCXC [53], which revealed that it almost obeys such a one-dimensional structural evolution. It is worth noting that $Au_{56}(SPh\text{-}^tBu)_{34}$ contains a double-helical Au_{35} core, within which one helix has one more tetrahedral-Au_4 unit than the other helix.

4) Two-dimensional growth with Au_4 tetrahedrons. With the rapid increase in $Au_n(SR)_m$ cluster library, structural isomerism with the same composition but different structures has become increasingly common. Inspired by the isomeric structures of the predicted $Au_{28}(SR)_{20}$ [120] and experimental $Au_{52}(SC_2H_4Ph)_{32}$ [61], Liu et al. predicted two new isomeric structures of $Au_{36}(SR)_{24}$ and $Au_{44}(SR)_{28}$ [113]. They also proposed a two-dimensional growth mode of the $Au_{8n+4}(SR)_{4n+8}$ (n = 3–6) cluster with Au_4 tetrahedrons as the blocks (Figure 3.10f). Among them, an $Au_{36}(SR)_{24}$ isomer was confirmed by experiment and its crystal structure was in good agreement with the predicted result.
5) Kekulé-like ring growth along the central Au_7 bi-tetrahedron, as shown in the $Au_{22}(SR)_{18}$ (DFT-structure) [121], $Au_{28}(SPh\text{-}^tBu)_{20}$ [67], $Au_{34}(S\text{-}c\text{-}C_6H_{11})_{16}$ [122], and $Au_{40}(o\text{-MBT})_{24}$ clusters [52] (Figure 3.10g).

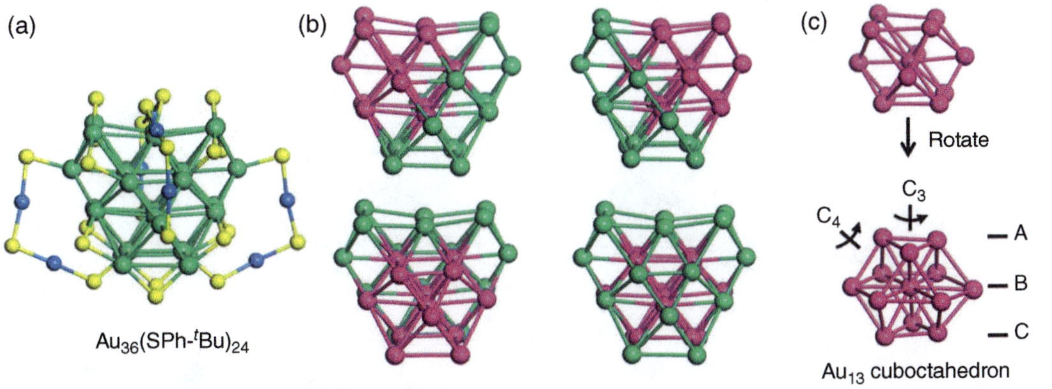

Figure 3.11 (a) Crystal structure of $Au_{36}(SPh\text{-}^tBu)_{24}$; (b) Highlight of the four cuboctahedra in the fcc-Au_{28} kernel; (c) Au_{13} cuboctahedron structure (A/B/C fcc sequence). Color labels: yellow = S, others = Au. *Source:* Structures are redrawn from the cif presented in Ref. [89].

3.3.2.1 Kernels in fcc Structure

With the extensive exploration of SR ligand types, Zeng et al. first resolved an fcc-type core structure in $Au_{36}(SPh\text{-}^tBu)_{24}$ (Figure 3.11a) [89]. The fcc-Au_{28} core in $Au_{36}(SPh\text{-}^tBu)_{24}$ is composed of four interpenetrating Au_{13} cuboctahedrons (Figure 3.11b). Such cuboctahedral Au_{13} units can be regarded as a fragment of FCC structure. As illustrated in Figure 3.11c, the 3 : 7 : 3 atom layer sequence can be recognized by observing the cubic octahedron along the C_3 axis, which is consistent with the A:B:C stacking sequence of FCC structure. Ever since the fcc-structured $Au_{36}(SPh\text{-}^tBu)_{24}$, numerous $Au_n(SR)_m$ with fcc or quasi-fcc structures have been solved by SCXC. Considering that the boundary between the kernel Au atoms and the staple Au atoms is not well defined, Zeng et al. provided three alternative anatomical views for fcc-structured $Au_{8n+4}(SR)_{4n+8}$ (n = 3–6) clusters based on the Au—Au bond length [51]. In addition to the tetrahedral helixes growth pattern mentioned above, other views are cuboctahedron interpenetration and layer-by-layer ones. As shown in Figure 3.12a, the core structure of $Au_{28}(SPh\text{-}^tBu)_{20}$, $Au_{36}(SPh\text{-}^tBu)_{24}$, and $Au_{44}(SPh^tBu)_{28}$, $Au_{52}(SPh\text{-}^tBu)_{32}$ can be viewed as the multiple interpenetrating cuboctahedrons, and the number of cuboctahedrons is 2, 4, 6, and 8, respectively. As more Au atoms are incorporated into the kernel of $Au_{8n+4}(SR)_{4n+8}$ (n = 3–6) clusters, these four clusters exhibit 4×4×3, 4×4×4, 4×4×5, 4×4×6 layer tetragonal rod-like structures, respectively (Figure 3.12b).

The cuboctahedron interpenetration and layer-by-layer growth modes have gradually become common structural perspectives to analyze the fcc-structured clusters [54, 122].

3.3.2.2 Kernels Arranged in hcp and bcc Fashions

In addition to the fcc-structure, some rare structures (e.g. hcp and bcc) have been observed. In 2015, both the Jin group and Zhu group resolved the structures of cyclohexanethiol-capped $Au_{18}(S\text{-}c\text{-}C_6H_{11})_{14}$ cluster [46, 104]. As shown in Figure 3.13a, the cluster with four free valence electrons contains face-fused bi-octahedral Au_9 kernel. The Au_9 kernel structure can also be viewed as A-B-A three layers of Au_3 planes arranged in an hcp manner, and is protected by a shell consisting of one $Au_4(SR)_5$ tetramer, one $Au_2(SR)_3$ dimer, and three $Au(SR)_2$ monomers.

Note that there is also a rotationally threefold symmetrical, hcp-structured $Au_{30}(S\text{-}Adm)_{18}$ cluster (Figure 3.13b) [50]. The Au_{18} kernel is composed of four Au layers (i.e. Au_3-Au_6-Au_6-Au_3) in an A-B-A-B arrangement, and is capped by six dimeric $Au_2(SR)_3$ staples. The novel structure of $Au_{30}(S\text{-}Adm)_{18}$ is distinctly different from the fcc-structured $Au_{30}(SR)_{18}$ [116, 123]. From another

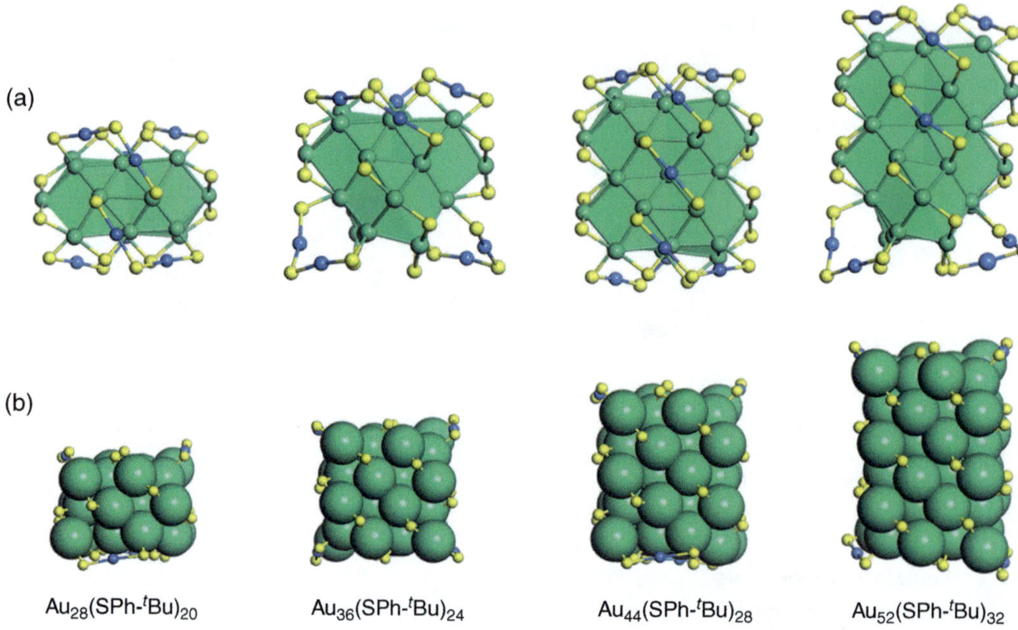

Figure 3.12 Two views of the growth pattern: (a) cuboctahedron interpenetration; (b) anisotropic growth of the fcc lattice. Color labels: yellow = S, others = Au. *Source:* Adapted with permission from Ref. [51]. Copyright 2016 American Chemical Society.

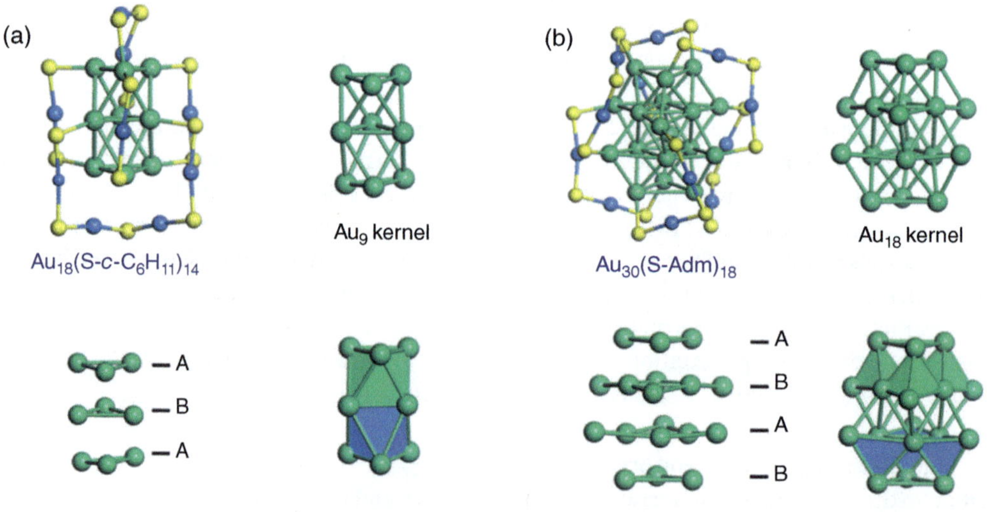

Figure 3.13 Structures of hcp in (a) $Au_{18}(S\text{-}c\text{-}C_6H_{11})_{14}$ and (b) $Au_{30}(S\text{-}Adm)_{18}$. Color labels: yellow = S, others = Au. *Source:* Structures are redrawn from the cifs presented in Refs. [46, 50, 104].

perspective, the Au_{18} kernel of $Au_{30}(S\text{-}Adm)_{18}$ can be regarded as an assembly of six Au_4 tetrahedrons. The $Au_{30}(S\text{-}Adm)_{18}$ has an anomalous solubility (i.e. only soluble in benzene but not in other common solvents), which also reminds researchers whether some clusters were ignored due to their insolubility.

The first bcc-structured $Au_n(SR)_m$ (i.e. $Au_{38}S_2(S\text{-}Adm)_{20}$) was synthesized by a size-focusing strategy (Figure 3.14) [103]. The bcc structure is in striking contrast with the fcc-structured bulk

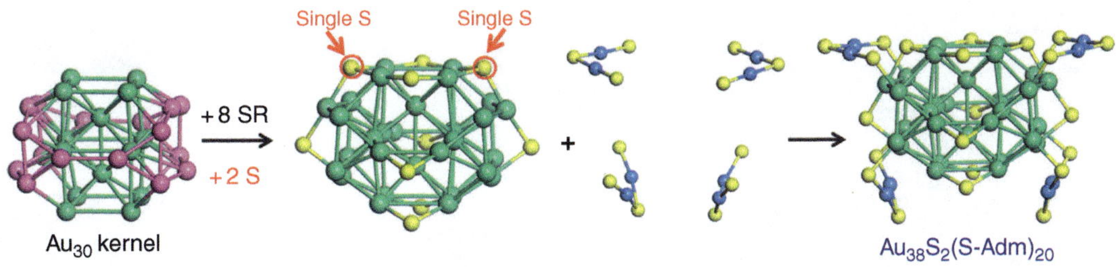

Figure 3.14 Structure of the bcc $Au_{38}S_2(S\text{-}Adm)_{20}$. Color labels: yellow = S, others = Au. *Source:* Structures are redrawn from the cif presented in Ref. [103].

gold and conventional Au NPs, as well as the bi-icosahedral structure of the same size $Au_{38}(SC_2H_4Ph)_{24-Q}$ [56]. The central part of the $Au_{38}S_2(S\text{-}Adm)_{20}$ structure consists of two bcc cubes stacked vertically, forming a Au_{14} unit. The remaining 16 Au atoms extend the Au_{14} inner core according to the bcc structural rule. The bcc-Au_{30} core is protected by four dimeric $Au_2(SR)_3$ staples, eight simple bridging SR. Notably, two bare sulfur atoms in the outer layer are bonded to the Au atoms in a tripodal manner (highlighted by the red arrows).

3.3.2.3 Kernels in Mirror Symmetry and Dual-Packing (fcc and non-fcc)

Zhuang et al. adopted an ion-induction method and prepared a novel $Au_{42}(SPh\text{-}^tBu)_{26}$ with a twist mirror symmetry structure (Figure 3.15) [64]. Different from the previously reported $Au_{42}(SR)_{26}$ (SR = HS-*c*-C_6H_{11} or HSPh-^{t}Bu) cluster with fcc structure [62, 122], the novel $Au_{42}(SPh\text{-}^tBu)_{26}$ exhibits a non-fcc kernel packing. The non-fcc Au_{26} kernel can be split into three parts: Au_9, Au_8, and Au_9 units. The bottom Au_9 unit is mirror symmetric with the upper Au_9 unit after 85° counterclockwise rotation along the longitudinal axis. The middle Au_8 unit shows a concave quadrilateral interface. The Au_{26} kernel of $Au_{42}(SPh\text{-}^tBu)_{26}$ is capped by four $Au_3(SR)_4$, four $Au(SR)_2$ staples, and two SR units.

Recently, several mix-structured $Au_n(SR)_m$ clusters have been reported, including $Au_{46}(m\text{-}MBT)_{26}$ [21], $Au_{48}(m\text{-}MBT)_{26}$ [21], $Au_{49}(DMBT)_{27}$ [71], and $Au_{67}(SCH_2Ph)_{35}$ [124]. $Au_{49}(DMBT)_{27}$ as the first $Au_n(SR)_m$ cluster with dual-packed (fcc and non-fcc) Au_{34} kernel (Figure 3.16a), is chiral, and the chirality originates from not only the staples but also the quasi-fcc-Au_{21} and the non-fcc-Au_{13} units. The Au_{34} kernel is stabilized by an exterior shell, including six $Au_2(SR)_3$ dimers,

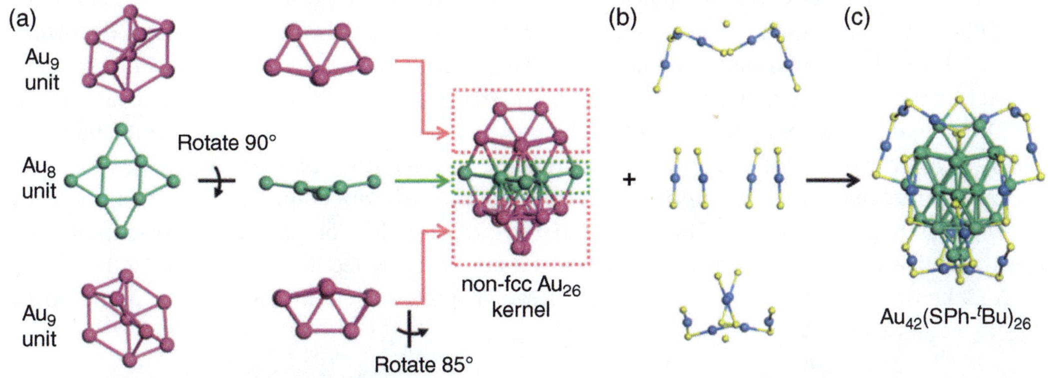

Figure 3.15 Anatomy of the structure of non-fcc-structured $Au_{42}(SPh\text{-}^tBu)_{26}$: (a) kernel structure; (b) staple structures; (c) total structure. Color labels: yellow = S, others = Au. *Source:* Adapted with permission from Ref. [64]. Copyright 2019 Wiley-VCH.

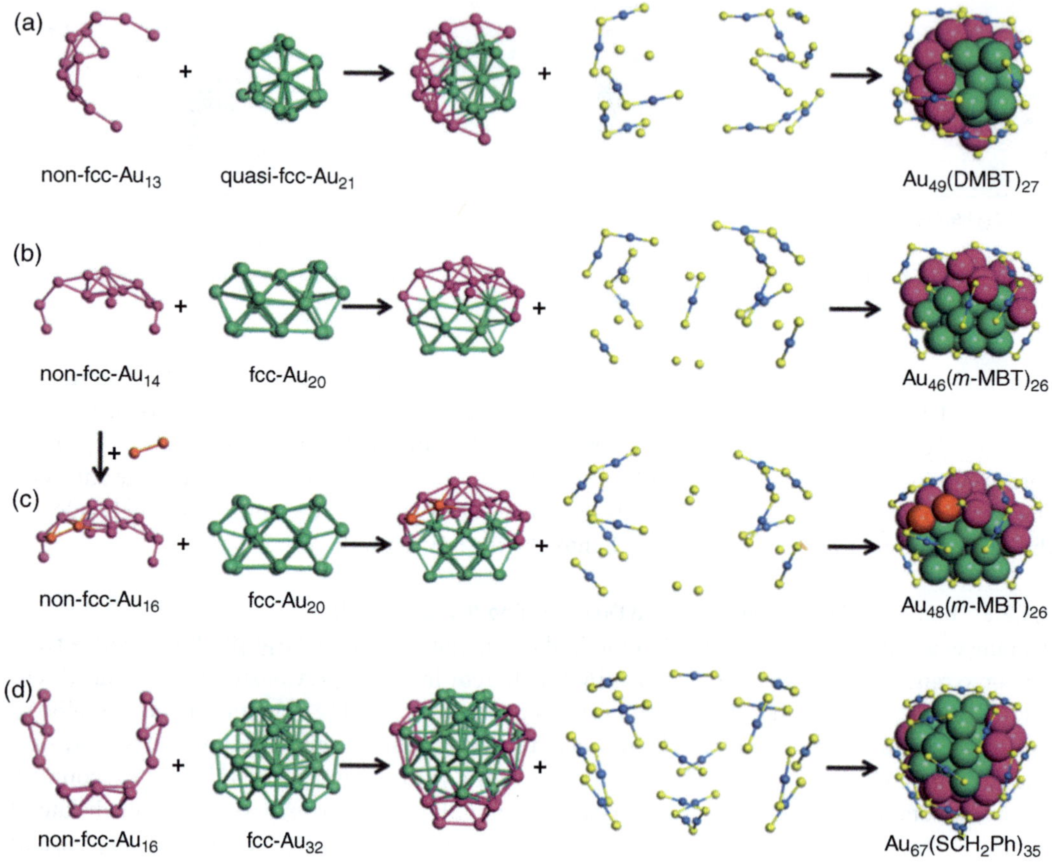

Figure 3.16 Anatomy of the structures of (a) $Au_{49}(DMBT)_{27}$; (b) $Au_{46}(m\text{-}MBT)_{26}$; (c) $Au_{48}(m\text{-}MBT)_{26}$; and (d) $Au_{67}(SCH_2Ph)_{35}$. Color labels: yellow = S, others = Au. *Source:* Structures are redrawn from the cifs presented in Refs. [21, 71, 124].

three $Au(SR)_2$ monomers, and three SR ligands. The dual-structured $Au_{48}(m\text{-}MBT)_{26}$ was prepared by modifying the conventional two-step method, and its Au_{36} kernel contains one non-fcc-Au_{16} cap and one rod-shaped fcc-Au_{20} unit. Notably, such a Au_{20} unit bears a resemblance to the kernel of $Au_{28}(SPh\text{-}^tBu)_{20}$, consisting of two interpenetrating cuboctahedrons with slight distortion (Figure 3.16b). The transformation process from $Au_{48}(m\text{-}MBT)_{26}$ to $Au_{46}(m\text{-}MBT)_{26}$ can be accomplished by applying a simple thermal treatment. The structure of $Au_{46}(m\text{-}MBT)_{26}$ was also resolved by SCXC (Figure 3.16c), which reveals that there is only a difference of two gold atoms between two NCs. After the loss of two Au atoms from the non-fcc-Au_{16} unit of $Au_{48}(m\text{-}MBT)_{26}$, the remained kernel maintains its original framework, and the exterior staples are also retained with only slight sliding and rotation. The $Au_{67}(SCH_2Ph)_{35}$ cluster also contains a mix-structured Au_{48} kernel, which is composed of one fcc-like Au_{32} unit and one non-fcc-like Au_{16} unit (Figure 3.16d). The Au_{48} kernel is protected by one $Au_4(SR)_5$ tetramer staple with a unique conformation and 15 $Au(SR)_2$ monomer staples.

3.3.2.4 Kernels Based on Icosahedral Au_{13} Unit

As early as 1981, a complete centered icosahedral Au_{13} kernel was found in a phosphine-coordinated Au_{13} cluster [125]. After that, numerous Au NCs based on icosahedral-Au_{13} unit have been

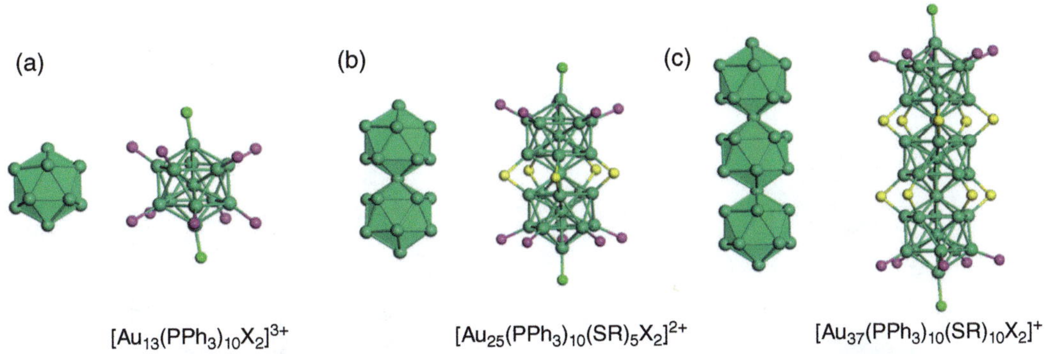

Figure 3.17 Structures of (a) $[Au_{13}(PPh_3)_{10}X_2]^{3+}$; (b) $[Au_{25}(PPh_3)_{10}(SR)_5X_2]^{2+}$; and (c) $[Au_{37}(PPh_3)_{10}(SR)_{10}X_2]^{+}$ clusters. Color labels: yellow = S, purple = P, others = Au. *Source:* Structures are redrawn from the cifs presented in Refs. [98, 125, 126].

reported. For example, Shichibu and Jin et al. resolved the atomic structures of two multiple-ligands-stabilized $[Au_{25}(PPh_3)_{10}(SR)_5X_2]^{2+}$ [126] and $[Au_{37}(PPh_3)_{10}(SR)_{10}X_2]^{+}$ (X = Cl/Br) [98] clusters, both of which contain a rod-like kernel assembled from two and three icosahedral Au_{13} units in a linear manner (Figure 3.17), respectively. The evolution from the mono-icosahedral $[Au_{13}(PPh_3)_{10}X_2]^{3+}$ (Figure 3.17a) to the vertex-sharing bi-icosahedral $[Au_{25}(PPh_3)_{10}(SR)_5X_2]^{2+}$ (Figure 3.17b) and to the vertex-sharing tri-icosahedral $[Au_{37}(PPh_3)_{10}(SR)_{10}X_2]^{+}$ (Figure 3.17c) cluster points to the possibility of achieving even longer rod NCs based on the assembly of icosahedral building blocks.

As shown in Figure 3.18, the icosahedral Au_{13} unit as the kernel building blocks not only exists in the phosphine-protected Au NCs, but also in other types of Au NCs. Examples of the icosahedral $Au_n(SR)_m$ clusters are $Au_{25}(SC_2H_4Ph)_{18}^-$ [10], the pair of structurally isomeric $Au_{38}(SC_2H_4Ph)_{24}$ clusters (denoted as $Au_{38}(SC_2H_4Ph)_{24-T}$ [86] and $Au_{38}(SC_2H_4Ph)_{24-Q}$ [56], where T and Q are the surname initial of the first author in the two works), $Au_{40}(S\text{-}Adm)_{22}$ [90], $Au_{44}(SR)_{26}$ [63, 70] (SR = HSPh-^{t}Bu or 2,4-DMBT), and $Au_{48}(SPh\text{-}^tBu)_{28}$ [63], all of which contain icosahedral Au_{13} units in their kernel structures. The $Au_{25}(SC_2H_4Ph)_{18}^-$ (counterion: $[N(C_6H_{17})_4]^+$) is the second thiolate-protected Au NCs after $Au_{102}(SR)_{44}$ with complete crystal structure characterized by SCXC. $Au_{25}(SC_2H_4Ph)_{18}^-$ possesses a centered icosahedral Au_{13} core, which is capped by an exterior shell composed of six dimeric $Au_2(SR)_3$ staples (Figure 3.18a). Both $Au_{38}(SC_2H_4Ph)_{24-T}$ and $Au_{40}(S\text{-}Adm)_{22}$ clusters can be regarded as the derived structures of one Au_{13} icosahedron unit (Figure 3.18b,c). The Au_{29} kernel of $Au_{40}(S\text{-}Adm)_{22}$ composes of one Au_{13} icosahedron and one Au_{16} cap, which is capped by an exterior shell, including six $Au(SR)_2$ staple, one $Au_2(SR)_3$ staple, one $Au_3(SR)_4$ staple, and three SR ligands. Similarly, the Au_{23} kernel of $Au_{38}(SC_2H_4Ph)_{24-T}$ consists of one Au_{13} icosahedron and one Au_{10} cap, and the surface layer contains two $Au_3(SR)_4$ staple units, three $Au_2(SR)_3$ staple units, three $Au(SR)_2$ staple, and one bridging SR ligand.

$Au_{38}(SC_2H_4Ph)_{24-T}$ is distinctly different from the earlier reported $Au_{38}(SC_2H_4Ph)_{24-Q}$. $Au_{38}(SC_2H_4Ph)_{24-T}$ and $Au_{38}(SC_2H_4Ph)_{24-Q}$ are the first pair of structural isomers in the NCs as revealed by SCXC. The anatomy of the $Au_{38}(SC_2H_4Ph)_{24-Q}$ structure starts with a face-fused bi-icosahedral Au_{23} core, which is protected by six $Au_2(SR)_3$ staples and three $Au(SR)_2$ staples (Figure 3.18d). The core structure of $Au_{48}(SPh\text{-}^tBu)_{28}$ is similar to that of the $Au_{44}(SR)_{26}$ cluster, which contains the same bi-icosahedroal Au_{23} unit capped with a similar bottom cap in the kernel (Figure 3.18e,f). Studies show that $Au_{38}(SC_2H_4Ph)_{24-Q}$, $Au_{44}(SR)_{26}$ and $Au_{48}(SPh\text{-}^tBu)_{28}$ clusters not only have the same core unit (i.e. Au_{23} bi-icosahedron) but also have some similar properties (e.g. electrochemical properties). They are comparable to the homologues in organic chemistry.

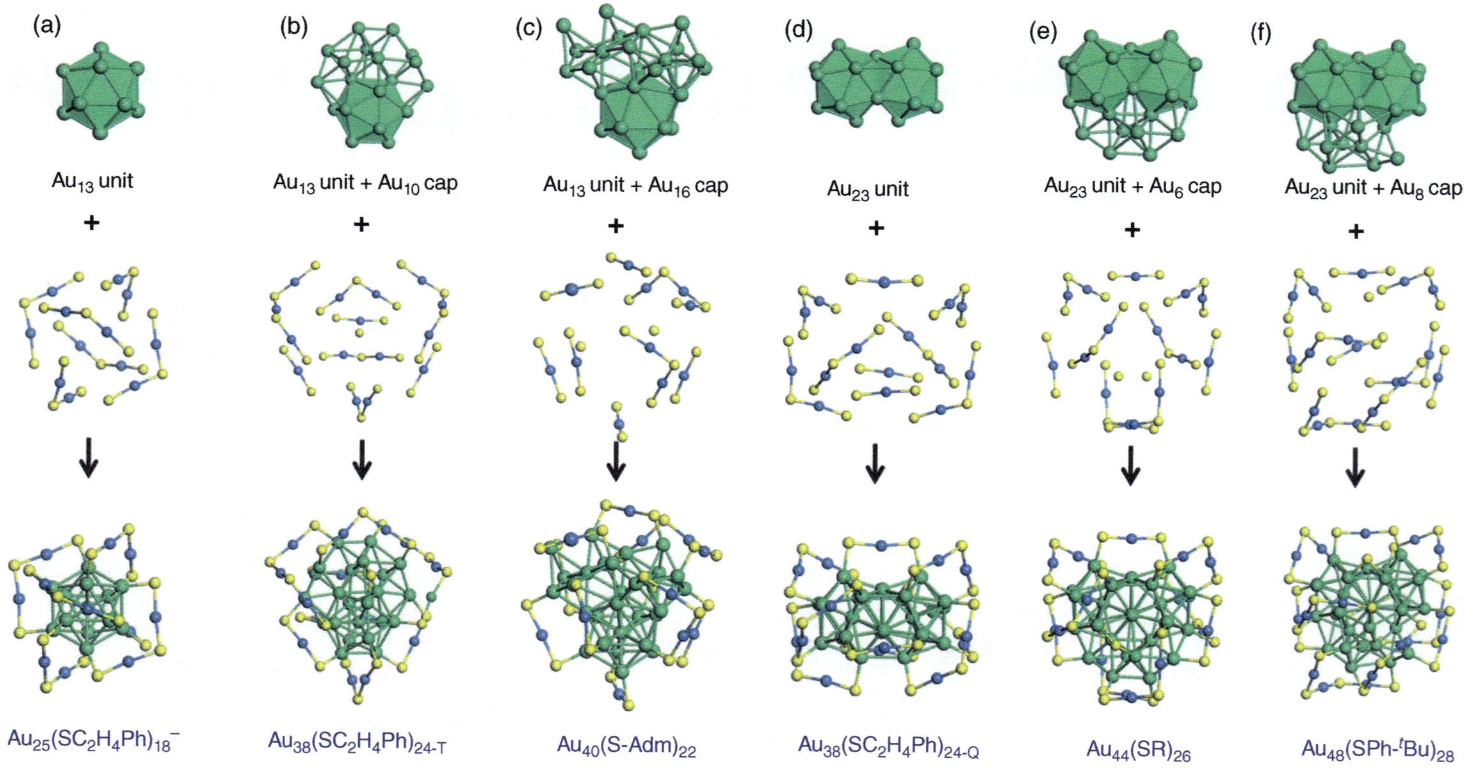

Figure 3.18 Anatomy of the structures of (a) $Au_{25}(SC_2H_4Ph)_{18}^-$; (b) $Au_{38}(SC_2H_4Ph)_{24-T}$; (c) Au_{40}(S-Adm)$_{22}$; (d) $Au_{38}(SC_2H_4Ph)_{24-Q}$; (e) $Au_{44}(SR)_{26}$; and (f) Au_{48}(SPh-^{t}Bu)$_{28}$. Color labels: yellow = S, others = Au. *Source:* Structures are redrawn from the cifs presented in Refs. [10, 56, 63, 70, 86, 90].

Therefore, the concept of kernel homology was introduced by Wu's group to describe the phenomenon of metal NCs sharing the identical "functional groups", which accordingly leads to similar properties.

3.3.2.5 Kernels with Multiple Shells

Multishell kernel growth belongs to the three-dimensional growth mode, which leads to isotropic expansion of the Au kernel structure. This mode is widely used in structural analysis of large-sized $Au_n(SR)_m$ clusters (i.e. clusters with more than 100 Au atoms), including $Au_{102}(p\text{-MBA})_{44}$ [26] (*p*-mercaptobenzoic acid), $Au_{103}S_2(S\text{-Nap})_{41}$ [127] (HS-Nap = naphthalenethiol), $Au_{130}(p\text{-MBT})_{50}$ [128], $Au_{133}(SPh\text{-}^tBu)_{52}$ [107, 129], $Au_{144}(SCH_2Ph)_{60}$ [9], $Au_{191}(SPh\text{-}^tBu)_{66}$ [106], $Au_{246}(p\text{-MBT})_{80}$ [130], and $Au_{279}(SPh\text{-}^tBu)_{84}$ [105].

In 2007, Jadzinsky et al. reported the X-ray structure determination of a *p*-MBA protected $Au_{102}(p\text{-MBA})_{44}$ [26], which was the first elucidated structure of $Au_n(SR)_m$ clusters. The groundbreaking work in X-ray crystallography of $Au_{102}(p\text{-MBA})_{44}$ provided clarity and significant insight into the structural assembly of Au NCs and surface protection modes. As shown in Figure 3.19a, the decahedral Au_7 inner core with D_{5h} symmetry is wrapped by a second shell of Au_{32} units, forming a two-shell Au_{39} Ino decahedron. The third shell (Au_{40}) consisting of Au_{15}, Au_{10} and Au_{15} units

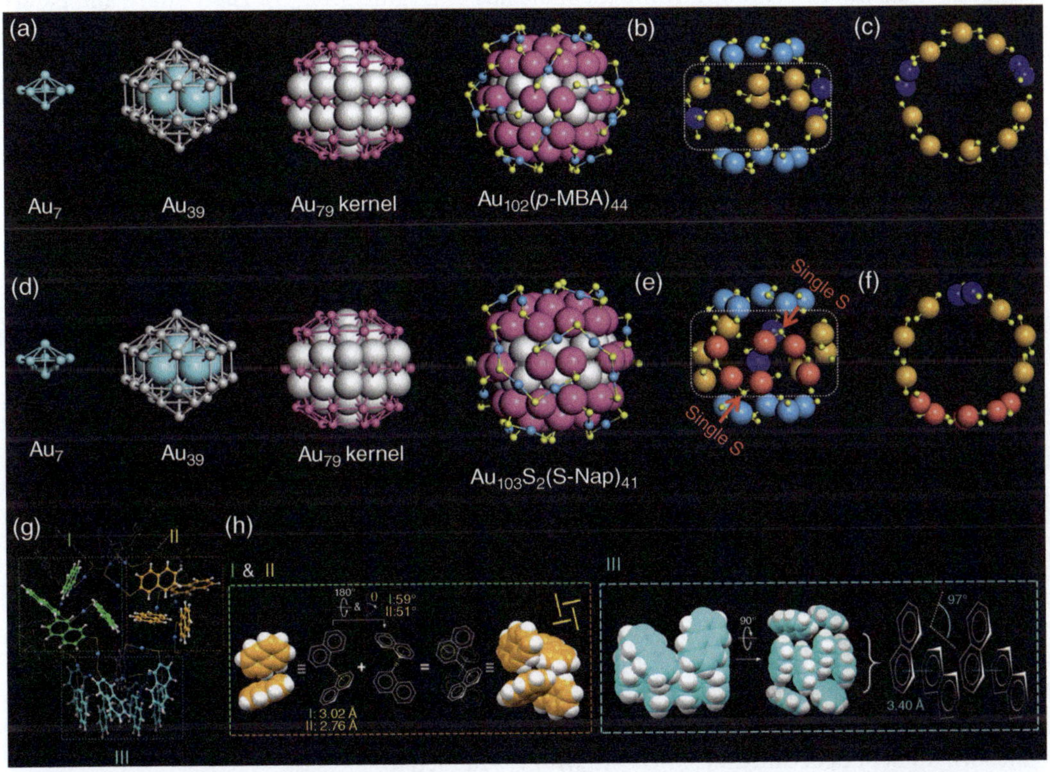

Figure 3.19 Anatomy of the structures of (a) $Au_{102}(p\text{-MBA})_{44}$ and (d) $Au_{103}S_2(S\text{-Nap})_{41}$. Side views of Au—S staple motifs of (b) $Au_{102}(p\text{-MBA})_{44}$ and (e) $Au_{103}S_2(S\text{-Nap})_{41}$. Top views of the staple motifs at waist position of (c) $Au_{102}(p\text{-MBA})_{44}$ and (f) $Au_{103}S_2(S\text{-Nap})_{41}$. (g) Highlight of the intracluster ligand-ligand interactions in the surface staple motifs of $Au_{103}S_2(S\text{-Nap})_{41}$. (h) Anatomy of the tetrameric herringbone structure via C—H⋯π interactions (left) and the parallel tetramer via staggered π⋯π interactions (right). *Source:* (a–f) Structures of $Au_{102}(p\text{-MBA})_{44}$ and $Au_{103}S_2(S\text{-Nap})_{41}$ are redrawn from the cifs presented in Refs. [26, 127]. (g, h) Reproduced with permission from Ref. [127]. Copyright 2017 American Chemical Society.

covers the top, waist, and bottom of the decahedron, maintaining C_5 symmetry and forming an Au_{79} Marks decahedron. The Au_{79} kernel is protected by 19 monomeric $Au(SR)_2$ staples and 2 $Au_2(SR)_3$ dimers (Figure 3.19b). The discrete nature of the $Au_{102}(p\text{-MBA})_{44}$ may be explained by the closing of a 58-electron shell.

Higaki et al. determined the structure of naphthalenethiol-protected $Au_{103}S_2(S\text{-Nap})_{41}$ with the same valence electrons and Au_{79} kernel as those in the previously reported $Au_{102}(p\text{-MBA})_{44}$ (Figure 3.19d) [127], which indicates the robustness of the decahedral structure as well as the 58-electron configuration, albeit the naphthalene ligand is much bulkier than the *p*-MBA ligand. The top and bottom protecting patterns of the $Au_{103}S_2(S\text{-Nap})_{41}$ and $Au_{102}(p\text{-MBA})_{44}$ clusters are the similar five $Au(SR)_2$ staples in C_5 symmetry. The difference between the two clusters lies in the waist protection pattern (Figure 3.19b,e): the $Au_{102}(p\text{-MBA})_{44}$ possesses two $Au_2(SR)_3$ staples and nine $Au(SR)_2$ staples (Figure 3.19c), while $Au_{103}S_2(S\text{-Nap})_{41}$ contains one $Au_2(SR)_3$ staples, six $Au(SR)_2$ staples, and two unique trimeric-like $Au_3S^*(SR)_3$ staples (i.e. -SR-Au-S*-Au-SR-Au-SR-, Figure 3.19f). The S* atom bears no carbon tail and the two S* atoms are highlighted by the red arrows (Figure 3.19e). Given the consideration that no external sulfide source was added during the synthesis, the two single S* should come from the thiol via S—C bond breaking. Furthermore, the surface ligands of the $Au_{103}S_2(S\text{-Nap})_{41}$ cluster exhibit some interesting and aesthetic patterns by the edge-to-face C—H···π interactions (termed as T-shape interactions) and parallel π···π interactions of bulk naphthalene groups within the nanocluster (Figure 3.19g,h). The T-shape and parallel interactions had not been previously reported, indicating the major role of ligands in constructing the surface structure.

The structure of $Au_{130}(p\text{-MBT})_{50}$ can also be dissected by a four-shell pattern (Figure 3.20a) [128]. The decahedral Au_{13} inner kernel is enclosed by an Au_{42} shell, assembling into an Ino decahedral Au_{55} unit with two shells. The third shell contains three parts, Au_{15}, Au_{20}, and Au_{15} units, covering the top, waist, and bottom of the decahedron, respectively. The three-shelled Au_{105} kernel exhibits quasi-D_{5h} symmetry, resembling a pentagonal barrel, which is protected by 25 monomeric $Au(SR)_2$ staple motifs (Figure 3.20b). Notably, the 25 monomers of $Au_{130}(p\text{-MBT})_{50}$ cluster self-assemble into five “ripple-like” stripe patterns with different radii, which are distributed parallel to each other on the surface of the Au_{105} kernel, as shown in Figure 3.20a. Every five $Au(SR)_2$ staples are aligned in a circle, forming a pentagonal ripple (Figure 3.20b). The structural anatomy shows that there is a close relationship between $Au_{130}(p\text{-MBT})_{50}$ and the $Au_{102}(p\text{-MBA})_{44}$. Both clusters can be included in the decahedron family, and $Au_{130}(\text{p-MBT})_{50}$ can be viewed as an elongated version of the $Au_{102}(p\text{-MBA})_{44}$. Such an extension creates an additional layer of footprints in the third shell, making it possible to form an ordered surface pattern of -S-Au-S- ripples in $Au_{130}(p\text{-MBT})_{50}$.

The pentagonal ripple also appears in $Au_{144}(SCH_2Ph)_{60}$ cluster [9]. To stabilize the Au_{114} core of $Au_{144}(SCH_2Ph)_{60}$, 30 $Au(SR)_2$ staples are distributed on the surface of the Au core (Figure 3.20e). Among them, 20 monomers form four pentagonal ripples, and the remaining 10 monomers form an almost planar 10-membered ring structure attached to the waist of the core (Figure 3.20f). The kernel of $Au_{144}(SCH_2Ph)_{60}$ with shell-by-shell growth pattern is based on a hollow icosahedral Au_{12} unit (Figure 3.20e). The second shell is composed of 42 Au atoms, forming the Au_{54} Mackay icosahedron with a hollow Au_{12} unit. The third Au_{60} shell encloses the Au_{54} icosahedron, leading to the formation of dodecahedral Au_{114} core with I_h symmetry.

The $Au_{133}(SPh\text{-}^tBu)_{52}$ has a similar size and formula with $Au_{130}(p\text{-MBT})_{50}$, but their structures are completely different. The kernel structure of $Au_{133}(SPh\text{-}^tBu)_{52}$ cluster is related to that of $Au_{144}(SCH_2Ph)_{60}$ but slightly different in the first and third shells (Figure 3.20c). As shown in Figure 3.20, in contrast to the hollow Au_{12} inner kernel of $Au_{144}(SCH_2Ph)_{60}$, $Au_{133}(SPh\text{-}^tBu)_{52}$ starts with a center-filled Au_{13} icosahedron [107, 129]. Compared to the rhombicosidodecahedral Au_{60} shell of $Au_{144}(SCH_2Ph)_{60}$, the third shell of $Au_{133}(SPh\text{-}^tBu)_{52}$ is incomplete and consists of only 52 gold atoms. The fourth shell contains 26 monomeric $Au(SR)_2$ staple motifs,

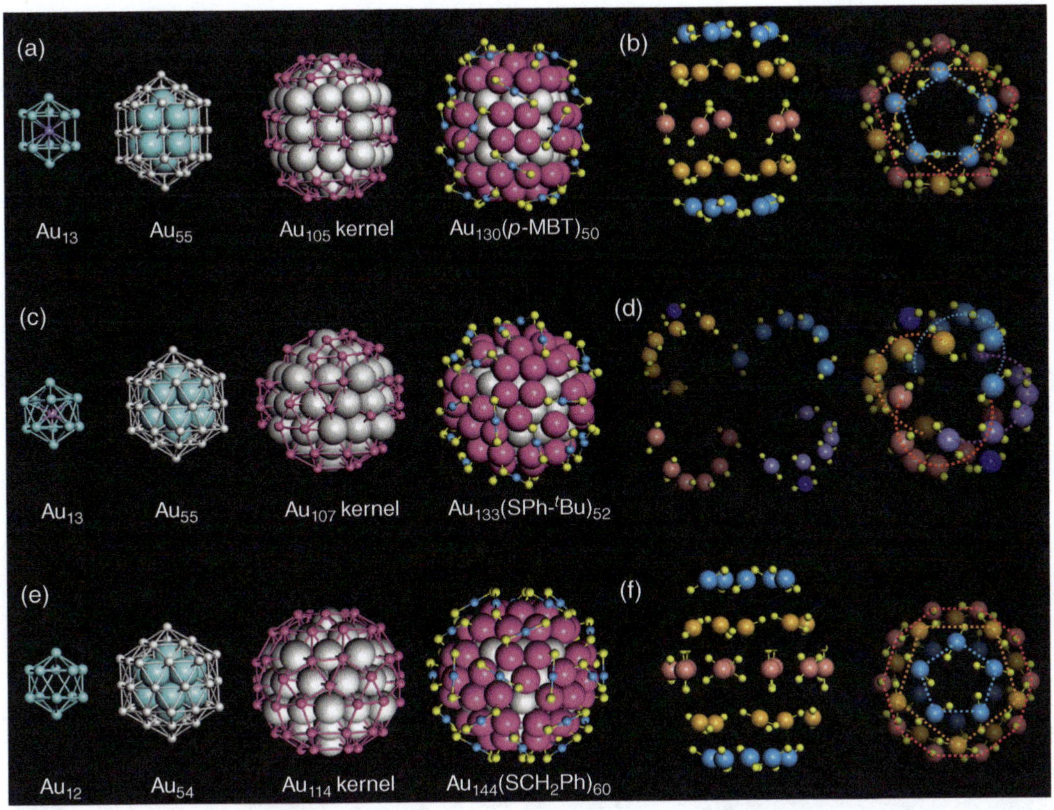

Figure 3.20 Anatomy of the structures of (a) $Au_{130}(p\text{-MBT})_{50}$, (c) $Au_{133}(SPh\text{-}^tBu)_{52}$, and (e) $Au_{144}(SCH_2Ph)_{60}$. Side and top views of Au—S staple motifs of the (b) $Au_{130}(p\text{-MBT})_{50}$; (d) $Au_{133}(SPh\text{-}^tBu)_{52}$; and (f) $Au_{144}(SCH_2Ph)_{60}$. Color labels: yellow = S, others = Au. *Source:* Structures are redrawn from the cifs presented in Refs. [9, 107, 128, 129].

which cap on the surface of the Au_{107} kernel with aesthetic patterns (Figure 3.20d). Among them, 24 monomeric $Au(SR)_2$ staple motifs self-assemble into four "helical stripes" and each stripe comprises six $Au(SR)_2$ staples that are parallel to each other and form a ladder-like helix. These four helices are attached to the globular Au_{107} kernel and coil up like a four-stranded rope. By further examining the arrangement of the carbon tails of the thiolates on the spherical surface, it was surprising to find that the carbon tails of $Au_{133}(SPh\text{-}^tBu)_{52}$ tend to self-assemble into multiple swirls, rather than the spiral stripe pattern that would be expected from the underlying Au—S staple patterns. As shown in Figure 3.21, each swirl is composed of four rotatively arranged phenyl rings.

Recently, Sakthivel et al. reported an anisotropic, monotwinned $Au_{191}(SPh\text{-}^tBu)_{66}$ cluster with an approximate dimension of 3.4×3.2 nm [106]. $Au_{191}(SPh\text{-}^tBu)_{66}$ contains an multi-shelled Au_{155} kernel with D_{3h} symmetry, arranged as $Au_3@Au_{23}@Au_{63}@Au_{66}$ (Figure 3.22a). The three-shelled Au_{89} inner core exhibits 14 faceted polyhedral structure including 8 [111] and 6 [100] facets, which provides sites for the fourth shell of 66 Au atoms. The Au_{66} shell comprises 6-atom triangular [111] facets, 5-atom trapezoidal [111] facets, and 4-atom square [100] facets. The final ligand shell consists of 24 monomeric $Au(SR)_2$ staples and 6 dimeric $Au_2(SR)_3$ staples. These protecting staple motifs can be divided into four types (Figure 3.22b). Two 6-atom triangular facets at the poles of the Au_{155} kernel are protected by 12 $Au(SR)_2$ staples. At the equatorial position of the Au_{155} kernel, 6 $Au(SR)_2$ staples are used to fix the two square [100] facets of the upper and lower parts of the body. On the body of the Au_{155} kernel, 6 $Au(SR)_2$ staples are anchored to the square [100] facets

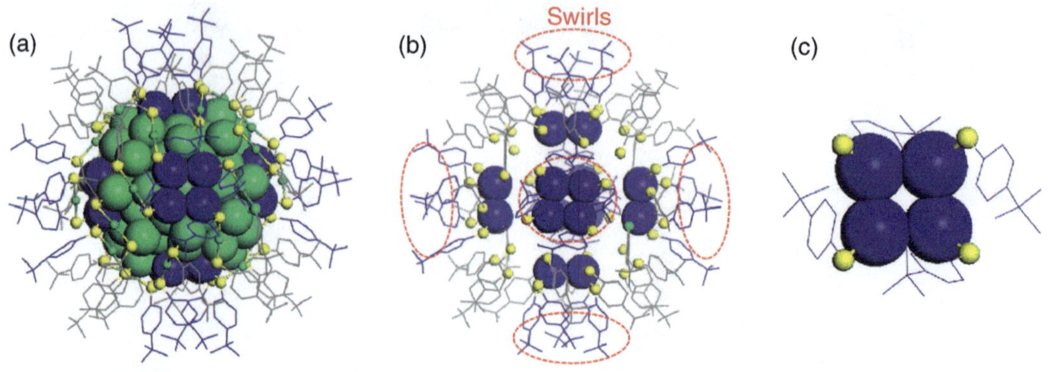

Figure 3.21 Self-assembly of the carbon tails of the SPh-^{t}Bu ligands. (a) Structure of $Au_{133}(SPh\text{-}^tBu)_{52}$; (b) The rotative arrangement of phenyl rings results in the formation of fourfold swirls; (c) Carbon-tail swirls on the Au_4 square unit. Color labels: yellow = S. *Source:* Structures are redrawn from the cifs presented in Refs. [129].

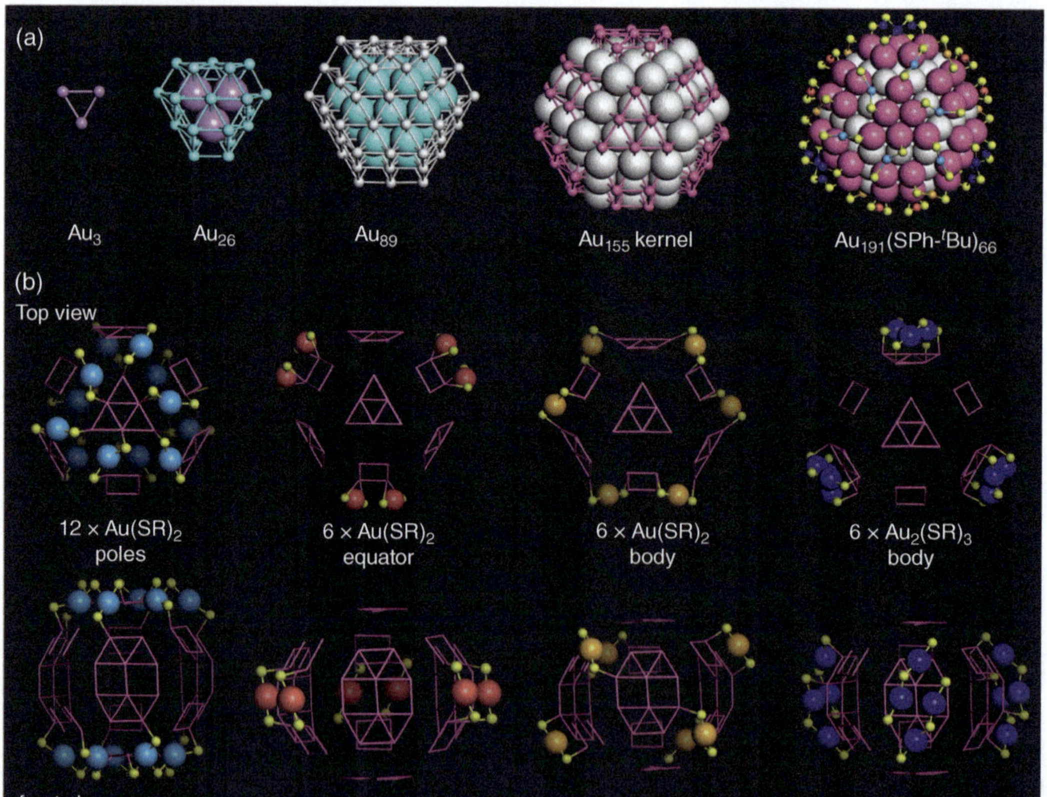

Figure 3.22 (a) X-ray crystal structure of the $Au_{191}(SPh\text{-}^tBu)_{66}$ cluster; (b) Different views of the staple motif assembly on $Au_{191}(SPh\text{-}^tBu)_{66}$. Color labels: yellow = S, others = Au. *Source:* Adapted with permission from Ref. [106]. Copyright 2020 American Chemical Society.

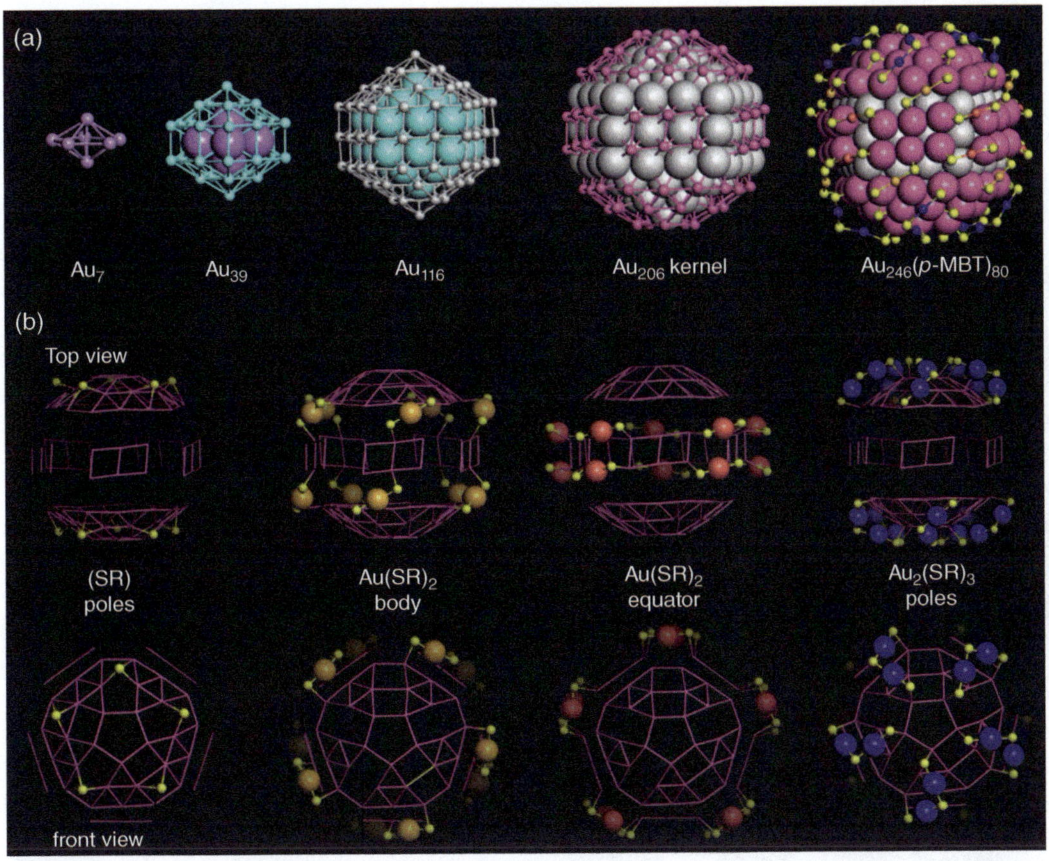

Figure 3.23 (a) Crystal structure of Au_{246}(*p*-MBT)$_{80}$ cluster; (b) Different views of the staple motif assembly on Au_{246}(*p*-MBT)$_{80}$. Color labels: yellow = S, others = Au. *Source:* Structures are redrawn from the cif presented in Ref. [130].

and trapezoidal [111] facets. The remaining 6 dimeric $Au_2(SR)_3$ staples are solely located on the trapezoidal [111] facets of the Au_{155} kernel.

In 2016, Zeng et al. grew high-quality single crystals and successfully elucidated the complete structure of a giant Au_{246}(*p*-MBT)$_{80}$ cluster via SCXC [130]. The cluster shows an overall diameter (including the ligand shell) of ~3.3 nm. As presented in Figure 3.23a, the innermost core of Au_{246}-(*p*-MBT)$_{80}$ is a three-shell Au_{116} Ino decahedron. The fourth Au_{90} shell is composed of two hemispherical Au_{30} units and six rectangular Au_6 units, covering the poles and equatorial position of the decahedron, respectively. The Au_{90} shell is a transition shell, which "sphericizes" the Au_{116} Ino decahedron, and provides locations for anchoring surface protecting motifs (Figure 3.23b). At the pole sites, the two hemispherical Au_{30} units of the fourth shell are capped by five simple bridging thiolates and five dimeric $Au_2(SR)_3$ staples. At the equatorial position, six rectangular Au_6 units of the fourth shell are covered by 10 monomeric $Au(SR)_2$ staple motifs. The remaining 10 monomeric $Au(SR)_2$ staples are anchored to the grooves between the hemispherical Au_{30} and rectangular Au_6 units. Although each part follows different assembling rules, the Au_{246}(*p*-MBT)$_{80}$ always maintains the fivefold symmetry. In addition, the SR ligands are highly ordered and self-assembled into two different surface types on the Au_{246}(*p*-MBT)$_{80}$ (Figure 3.24). At the poles, 25 thiolates are rotationally aligned into four pentagonal circles. Each circle has the same "latitude" and rotational direction. This pattern is termed as α-rotation (Figure 3.24a). At the waist, six of thiolates are arranged

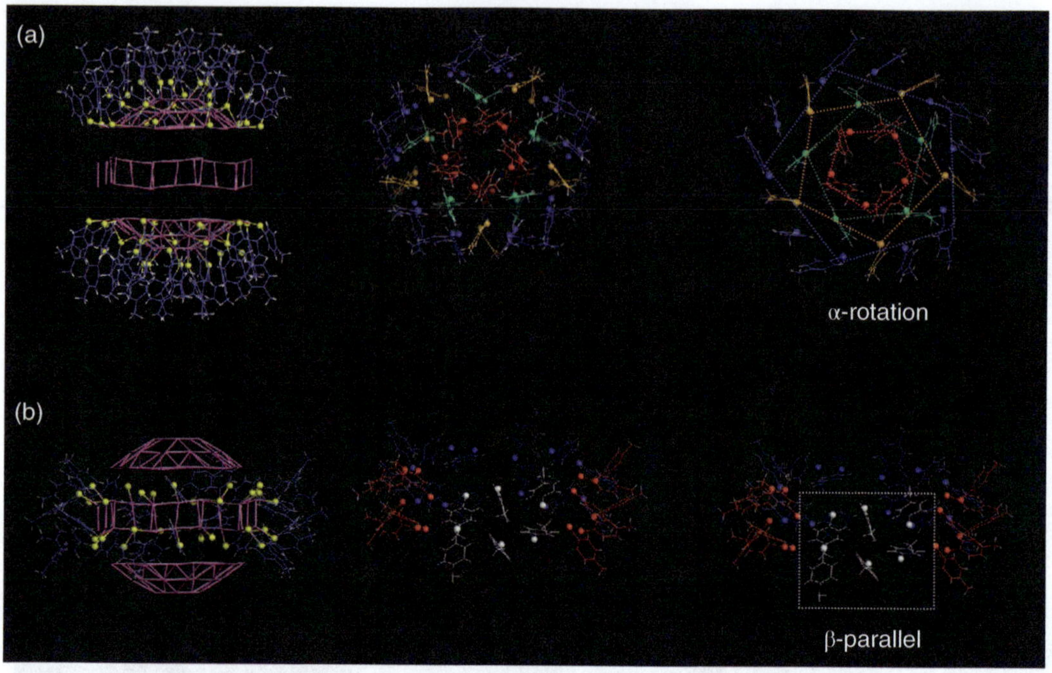

Figure 3.24 Self-assembled surface patterns of the ligands on the Au_{246}(*p*-MBT)$_{80}$. (a) Rotational packing of ligands at the pole site of the Au_{246}(*p*-MBT)$_{80}$; (b) Parallel packing of ligands at the waist of the Au_{246}(*p*-MBT)$_{80}$. *Source:* Structures are redrawn from the cif presented in Ref. [130].

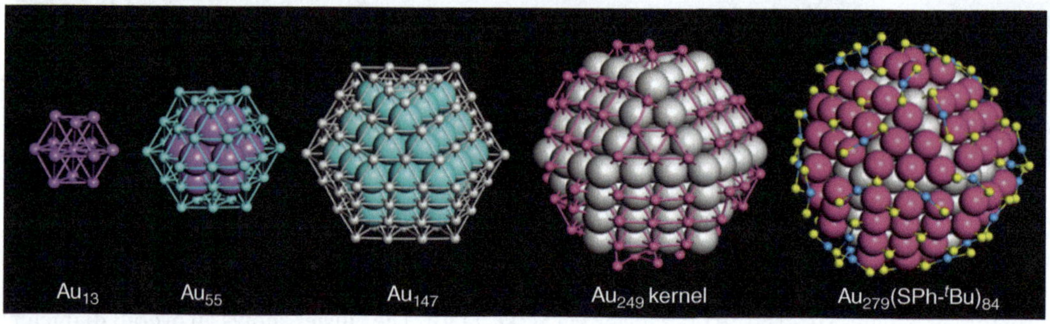

Figure 3.25 Anatomy of the structures of Au_{279}(SPh-^{t}Bu)$_{84}$. Color labels: yellow = S, others = Au. *Source:* Structures are redrawn from the cif presented in Ref. [105].

into three alternating parallel pairs to form a parallel pattern, which is called as β-parallel (Figure 3.24b). In total there are five such β-parallel patterns.

The Au_{279}(SPh-^{t}Bu)$_{84}$ is the largest thiolate-protected Au cluster with precise atomic composition and structure thus far [105]. The X-ray diffraction structure of $Au_{279}(SR)_{84}$ shows that the nanocrystal has a core-shell structure containing a truncated octahedral core with bulk fcc-like arrangement (Figure 3.25). The core structure follows the four-shell growth pattern Au_{13}@Au_{42}@Au_{92}@Au_{102}, starting with a centered Au_{13} cuboctahedral inner core, enclosed by a second Au_{42} cuboctahedral shell and a third Au_{92} cuboctahedral shell, and then wrapped by Au_{102} shell formed by Au_{54} truncated octahedron and Au_{48} distorted truncated cube. The Au_{249} kernel of Au_{279}(SPh-^{t}Bu)$_{84}$ are protected by a ligand shell with three types of staple motifs, namely 30 bridging thiolates, 18 monomeric $Au(SR)_2$ staples, and 6 dimeric $Au_2(SR)_3$ staples.

3.3.3 Protecting Surface Motifs of $Au_n(SR)_m$ Clusters

3.3.3.1 Staple-like $Au_x(SR)_{x+1}$ (x = 1, 2, 3, 4, 8) motifs

In addition to the kernel structure, the "staple" structure on the surface should be paid attention to, which plays a critical role in stabilizing the Au NCs. Given the high specific surface areas of NCs, their surface-protecting motifs may also determine some physical and chemical properties of the clusters to a certain extent. The simple SR ligand and oligomeric $Au_x(SR)_{x+1}$ (x = 1, 2, 3, 4) staple-like motifs are the most common surface structures in the $Au_n(SR)_m$ clusters. As shown in Figure 3.26, the $Au(SR)_2$ monomer with a linear-shaped manner (i.e. -SR-Au-SR-), "V-shaped" $Au_2(SR)_3$ dimer (i.e. -SR-Au-SR-Au-SR-), trimeric $Au_3(SR)_4$ staples (i.e. -SR-Au-SR-Au-SR-Au-SR-), and tetrameric $Au_4(SR)_5$ staple (i.e. -SR-Au-SR-Au-SR-Au-SR-Au-SR-) with a chair-like conformation were first experimentally identified in the $Au_{102}(p\text{-MBA})_{44}$ [26], $Au_{25}(SC_2H_4Ph)_{18}^-$ [10], $Au_{23}(S\text{-}c\text{-}C_6H_{11})_{16}^-$ [40], and $Au_{24}(SCH_2Ph\text{-}^tBu)_{20}$ clusters [111], respectively. Moreover, Chen et al. reported a long $Au_8(SR)_9$ staple motif in $Au_{21}(S\text{-Adm})_{15}$ cluster [49]. All the S-Au-S and Au-S-Au angles in these motifs are ca. 180° and 100°, respectively. According to the reported "staple" types in $Au_n(SR)_m$ clusters, the following conclusions can be drawn: (i) the kernel of the ultrasmall-sized NCs may contain multiple curved faces and usually requires longer $Au_3(SR)_4$ or $Au_4(SR)_5$ motifs to adapt to the kernel structure; (ii) the longer "staples" such as $Au_3(SR)_4$ or $Au_4(SR)_5$ motifs mostly exist at the edge or corner of NCs; and (iii) the cores of large-sized spherical NCs are usually stabilized by shorter $Au(SR)_2$ or $Au_2(SR)_3$ motifs.

3.3.3.2 Ring-like $Au_x(SR)_x$ (x = 4, 5, 6, 8) Motifs

Recently, Xu et al. resolved the ring-in-ring structure of Au(I)-thiolate $Au_{10}(SR)_{10}$ cluster, which consists of a pair of interlocked $Au_5(SR)_5$ ring with a nearly planar conformation [109] (Figure 3.27). It was found that the $Au_{10}(SR)_{10}$ cluster in alkane solvents can transform to a novel ring-in-ring $Au_{12}(SR)_{12}$ cluster. In the $Au_{12}(SR)_{12}$, the $Au_4(SR)_4$ inner ring and $Au_8(SR)_8$ outer ring adopt a butterfly-type and double-helical-like conformations, respectively. Dissolution of $Au_{12}(SR)_{12}$ in CH_2Cl_2 or $CHCl_3$ can reverse the transformation process and converts $Au_{12}(SR)_{12}$ back to $Au_{10}(SR)_{10}$. In early work, Zeng et al. identified the surface structure of a giant "ring" $Au_8(SR)_8$ motif in the $Au_{20}(SPh\text{-}^tBu)_{16}$ cluster [110]. Remarkably, the octameric ring motif in $Au_{20}(SPh\text{-}^tBu)_{16}$ adopts a chair conformation, rather than a double-helical-like structure. In $Au_{28}(SCH_2Ph\text{-}^tBu)_{22}$, two $Au_6(SR)_6$ hexamer rings with chair-like conformation cap the inner core [114].

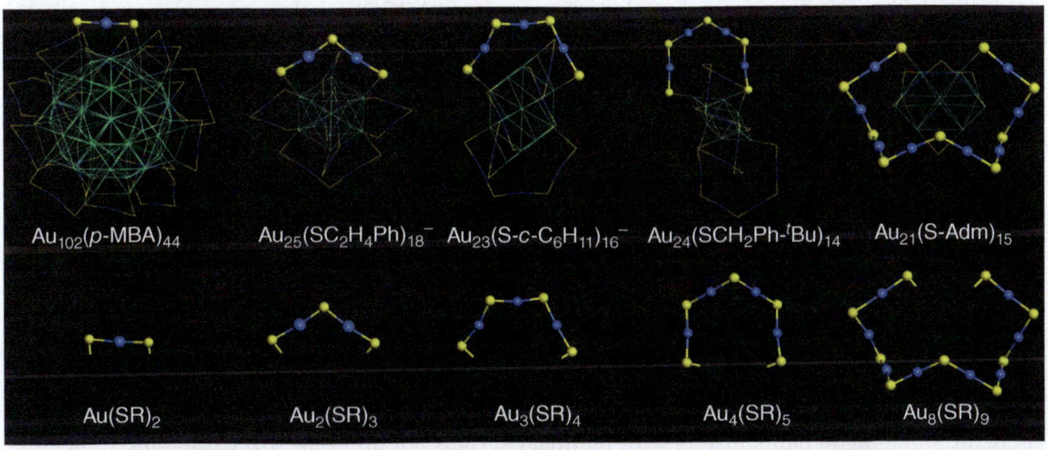

Figure 3.26 Different staple-like $Au_x(SR)_{x+1}$ motifs. Color labels: yellow = S, others = Au.

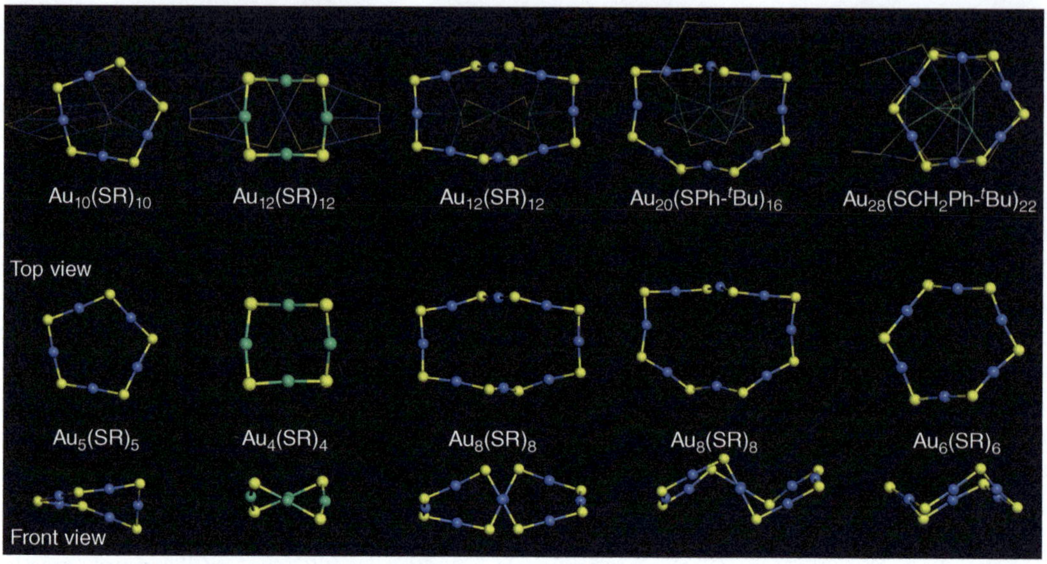

Figure 3.27 Different ring-like $Au_x(SR)_x$ motifs. Color labels: yellow = S, others = Au.

The double ring structures could endow $Au_{28}(SCH_2Ph\text{-}^tBu)_{22}$ with high photoluminescence, probably due to the enhanced charge transfer between the bi-ring and the kernel.

3.3.3.3 Giant $Au_{20}S_3(SR)_{18}$ and $Au_{23}S_4(SR)_{18}$ Staple Motifs

Gan et al. synthesized the $Au_{60}S_6(SCH_2Ph)_{36}$ cluster by a thermally induced ligand exchange reaction of $Au_{38}(SC_2H_4Ph)_{38-Q}$ with excess phenylmethanethiol [65]. SCXC revealed that $Au_{60}S_6(SCH_2Ph)_{36}$ consists of a fcc-like Au_{20} kernel protected by a pair of giant $Au_{20}S_3(SR)_{18}$ staple motifs (Figure 3.28a). One $Au_{20}S_3(SR)_{18}$ staple caps the Au_{20} kernel by five common terminal μ_2-S atoms (binding to one -R group and two Au atoms) and three bridging μ_4-S atoms (every μ_4-S binds to one kernel Au atom and three staple Au atoms, highlighted in red circle and arrows in Figure 3.28a), and the other $Au_{20}S_3(SR)_{18}$ staple binds to the core in the same manner after rotating 180°. The bare μ_4-S atoms should be derived from thiolate, which may undergo S—C bond cleavage under heating during the ligand-exchange-induced structural transformation. More interestingly, $Au_{60}S_6(SCH_2Ph)_{36}$ adopts a special stacking sequence of "ABCDEF" along the close-packed [001] direction with 60° rotation between two neighboring clusters. The new crystallographic stacking was named the 6H left-handed helical (6HLH) arrangement, which is the fourth packing mode, revealed long after the third closest crystallographic packing (4H) (discovered in 1979). Subsequently, this group introduced a single sulfur atom into $Au_{60}S_6(SCH_2Ph)_{36}$ and obtained the $Au_{60}S_7(SCH_2Ph)_{36}$ cluster by treating $Au_{60}S_6(SCH_2Ph)_{36}$ at 100 °C overnight with excessive phenylmethanethiol [131]. As presented in Figure 3.28b, $Au_{60}S_7(SCH_2Ph)_{36}$ consists of a fcc Au_{17} kernel, protected by one $Au_{20}S_3(SR)_{18}$ and one $Au_{23}S_4(SR)_{18}$ motif. The Au_{17} core in $Au_{60}S_7(SCH_2Ph)_{36}$ can be regarded as the Au_{20} core in $Au_{60}S_6(SCH_2Ph)_{36}$ losing three local Au atoms. Notably, the $Au_{23}S_4(SR)_{18}$ motif as the largest motif found in metal NCs, can be regarded as the adduct of $Au_{20}S_4(SR)_{18}$ with tetrahedral Au_3S.

3.3.3.4 Homo-Kernel Hetero-Staples

There are some metal nanoclusters with identical kernels but different surface staples, indicating the existence of homo-kernel-hetero-staples phenomenon. Jones et al. reported the crystal

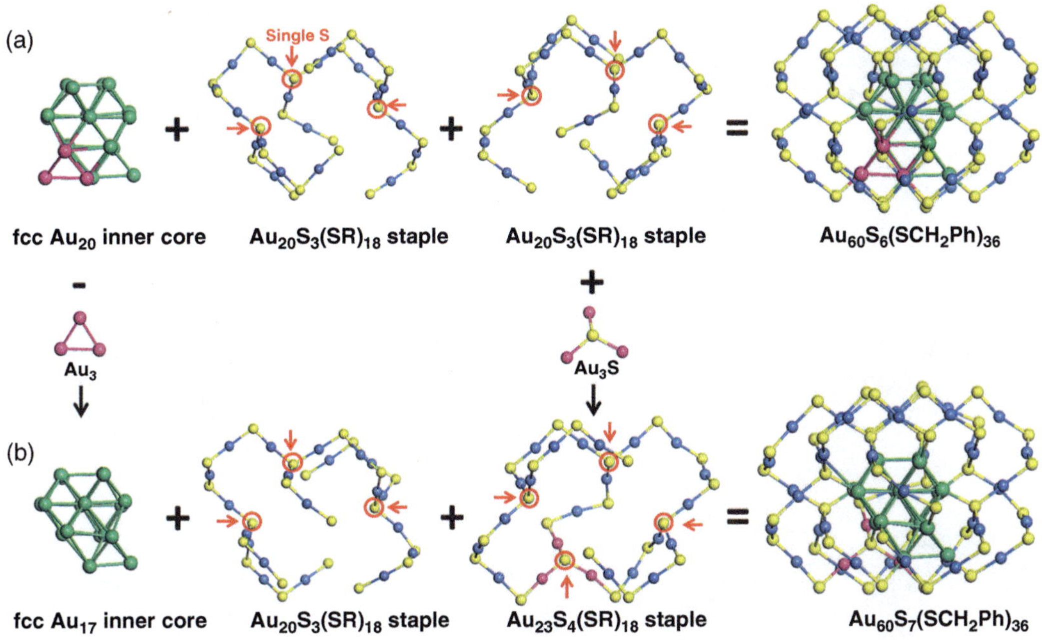

Figure 3.28 Anatomy of the structures of (a) $Au_{60}S_6(SCH_2Ph)_{36}$; and (b) $Au_{60}S_7(SCH_2Ph)_{36}$. Color labels: yellow = S, others = Au. *Source:* Structures are redrawn from the cifs presented in Refs. [65, 131].

structure of $Au_{21}S(S\text{-}Adm)_{15}$ cluster with single sulfur coordination [132], which is different from but related closely to $Au_{21}(S\text{-}Adm)_{15}$ [49]. As shown in Figure 3.29a,b, the Au_{10} kernel of $Au_{21}S(S\text{-}Adm)_{15}$ is identical to the core of $Au_{21}(S\text{-}Adm)_{15}$, both of which are composed of two octahedral Au_6 units sharing one edge. The Au_{10} kernel of $Au_{21}(S\text{-}Adm)_{15}$ is protected by four different motif structures, including a very long $Au_8(SR)_9$ staple, one $Au_2(SR)_3$ dimeric staple, one $Au(SR)_2$ monomeric staple, and one simple SR ligand (Figure 3.29a). By comparing the outer structure, it was found that one extra μ_3-coordinated sulfur atom (i.e. binding to three Au atoms) in $Au_{21}S(S\text{-}Adm)_{15}$ bears no carbon tail and caused a slight distortion of the outer structure (Figure 3.29b).

As presented in Figure 3.29c,d, the slight distortion of the outer layers caused by the additional sulfur atoms was also observed in $Au_{30}(S\text{-}^tBu)_{18}$ and $Au_{30}S(S\text{-}^tBu)_{18}$ clusters [123, 133]. The Au_{22} cores of both clusters can be described structurally as a rod-like Au_{20} bicuboctahedron face-capped by two Au atoms. The rod-like Au_{20} unit consists of two distorted interpenetrating Au_{13} cuboctahedra, which bears a strong resemblance to the fcc-Au_{20} core of $Au_{28}(SPh\text{-}^tBu)_{20}$. In the $Au_{30}(S\text{-}^tBu)_{18}$ cluster without single μ_3-sulfur atom, two $Au_3(SR)_4$ staples, two $Au(SR)_2$ staples, and six SR ligands cap the surface of the Au_{20} core (Figure 3.29c). The most distinct difference between these two Au_{30} clusters lies in the arrangement of the outer staples. The extra sulfur atom in $Au_{30}S(S\text{-}^tBu)_{18}$ slightly shifts the position of the remaining staple patterns (Figure 3.29d).

Zeng et al. reported a pair of $Au_{28}(SR)_{20}$ clusters with the similar rod-shaped fcc-Au_{20} kernel but different Au—S interface structure, denoted as $Au_{28}(SPh\text{-}^tBu)_{20}$ [67] and $Au_{28}(S\text{-}c\text{-}C_6H_{11})_{20}$ [47], respectively. As depicted in Figure 3.30a,b, the Au_{20} kernel of $Au_{28}(SPh\text{-}^tBu)_{20}$ is stabilized by four $Au_2(SR)_3$ staples and eight bridging SR ligands; whereas the Au_{20} kernel of $Au_{28}(S\text{-}c\text{-}C_6H_{11})_{20}$ is capped by two $Au_3(SR)_4$ staples, two $Au(SR)_2$ staples and eight bridging SR ligands. Moreover, Gan et al. synthesized a novel $Au_{32}(S\text{-}Adm)_{24}$, which possesses a similar fcc-Au_{20} kernel [134]. The Au—S interface of $Au_{32}(S\text{-}Adm)_{24}$ is composed of four trimetric $Au_3(SR)_4$ staple motifs and eight bridging SR ligands (Figure 3.30c).

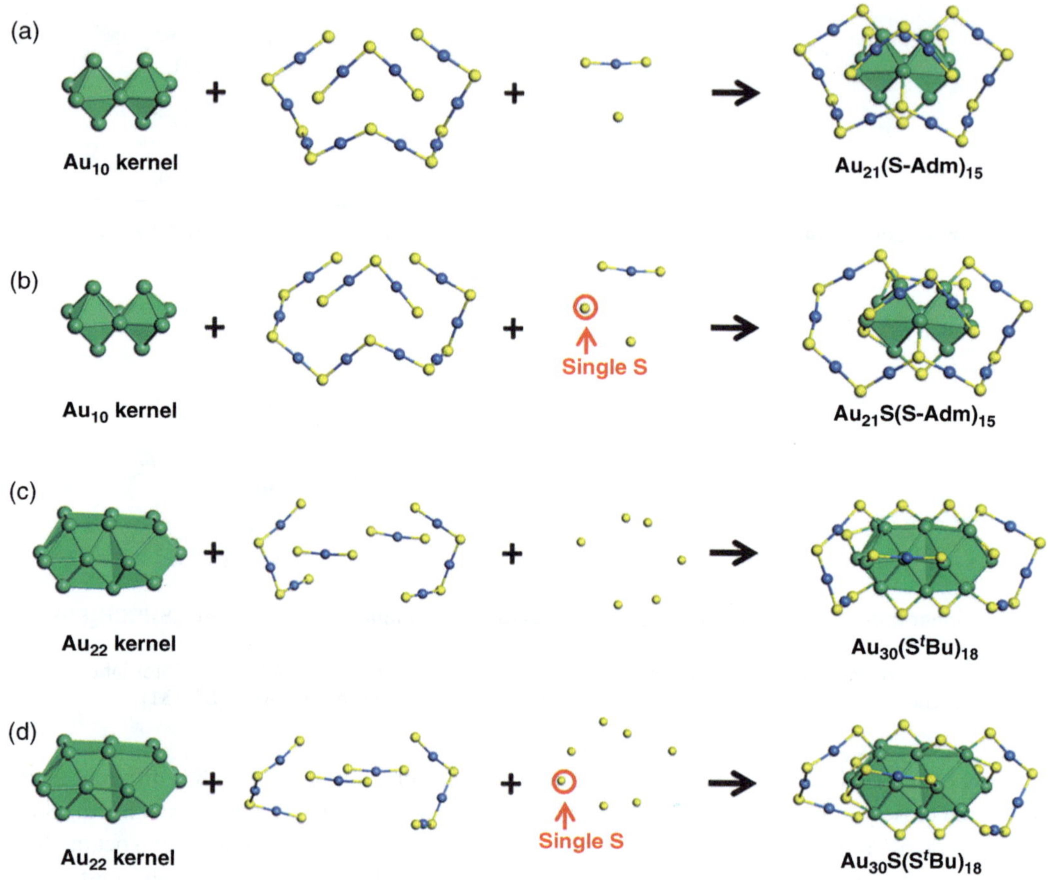

Figure 3.29 Anatomy of the structures of (a) $Au_{21}(S\text{-}Adm)_{15}$; (b) $Au_{21}S(S\text{-}Adm)_{15}$; (c) $Au_{30}(S\text{-}^tBu)_{18}$; and (d) $Au_{30}S(S\text{-}^tBu)_{18}$ [133]. Color labels: yellow = S, others = Au. *Source:* Structures are redrawn from the cifs presented in Refs. [49, 123, 132].

The homo-kernel-hetero-staples phenomenon also occurs in $Au_{42}(SR)_{26}$ (SR = S-*c*-C_6H_{11} or SPh-^{t}Bu) [62, 122] and $Au_{44}(SPh\text{-}^tBu)_{28}$ [51] clusters. As presented in Figure 3.30d,e, the Au_{34} kernel of $Au_{42}(SR)_{26}$ is composed of six interpenetrating Au_{13} cuboctahedrons, and it is identical to the kernel of $Au_{44}(SPh\text{-}^tBu)_{28}$. Although the two clusters have identical fcc-Au_{34} kernels, they possess different surface patterns. The Au_{34} kernel of $Au_{44}(SPh\text{-}^tBu)_{28}$ is protected by four $Au_2(SR)_3$ dimers at the top and bottom and two $Au(SR)_2$ monomers at the waist. In contrast, there are two $Au_2(SR)_3$ dimers at the top and bottom of $Au_{42}(SR)_{26}$ kernel and four $Au(SR)_2$ monomers at the waist. Of note, there are also 12 bridging SR ligands for every cluster.

Gan et al. recently successfully separated two types of crystals (rectangular vs. needle-like) of $Au_{60}S_8(SCH_2Ph)_{36}$ ($Au_{60}S_{8r}$ and $Au_{60}S_{8n}$ for short; r and n represent the crystal types rectangle and need-like, respectively) [135], which have different NC conformations and arrangements, as determined by SCXC. As shown in Figure 3.31a,c, the two $Au_{60}S_8(SCH_2Ph)_{36}$ clusters with identical formula have no obvious differences in the framework structure, which is composed of a Au_{14} kernel protected by a pair of $Au_{23}S_4(SR)_{18}$ staples. The Au_{14} kernel can be viewed as one Au_7 double-helixes, and the $Au_{23}S_4(SR)_{18}$ staple can convert to the other one by rotating 180° along the C_2 symmetry axis. The difference between the two $Au_{60}S_8(SCH_2Ph)_{36}$ clusters lies in the assembly patterns of phenylmethanethiolates on the cluster surfaces (Figure 3.31b,d). In other words, the two $Au_{60}S_8(SCH_2Ph)_{36}$ clusters are a pair of conformational isomers.

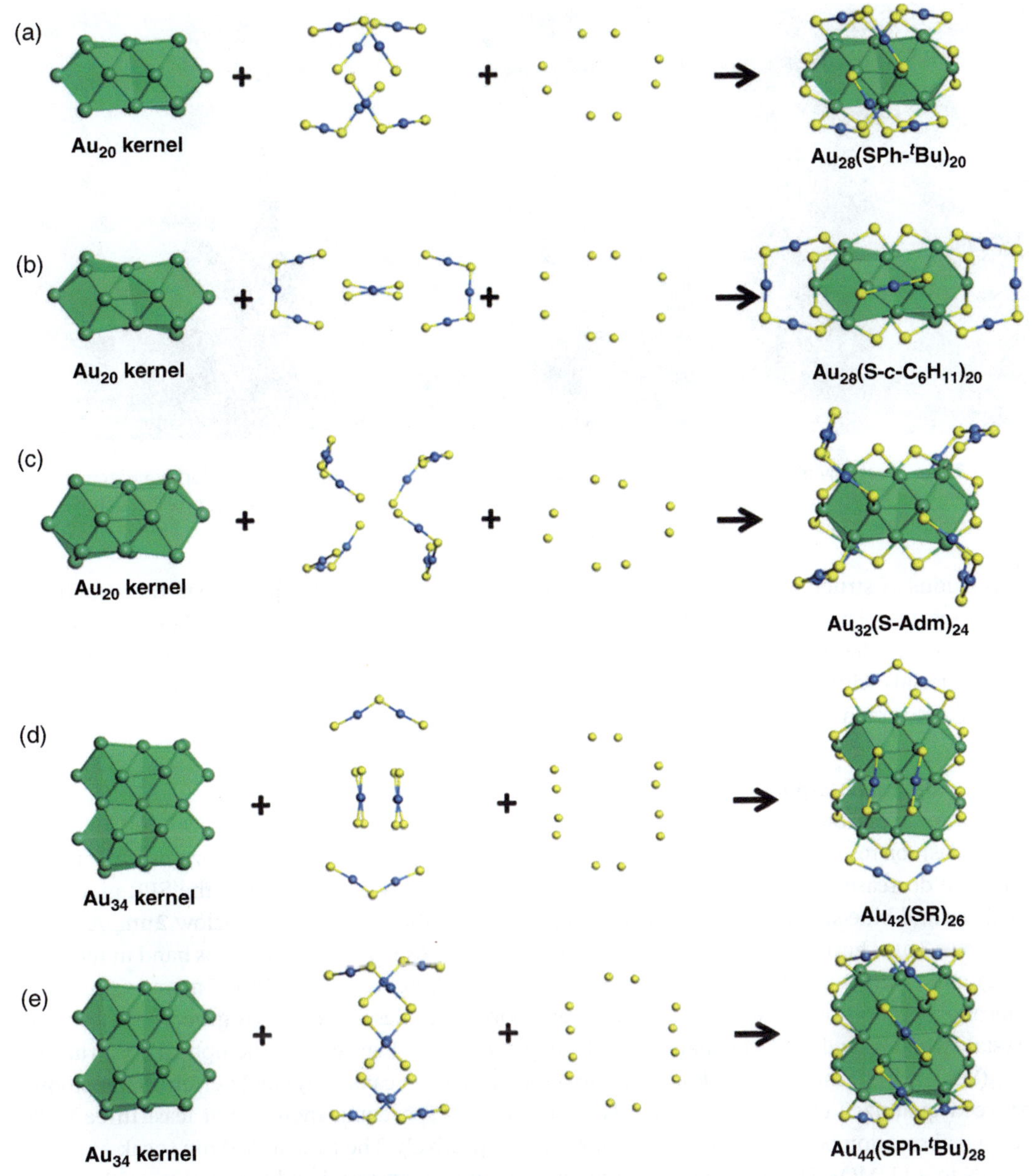

Figure 3.30 Anatomy of the structures of (a) $Au_{28}(SPh\text{-}^tBu)_{20}$; (b) $Au_{28}(S\text{-}c\text{-}C_6H_{11})_{20}$, (c) $Au_{32}(S\text{-}Adm)_{24}$; (d) $Au_{42}(SR)_{26}$; and (e) $Au_{44}(SPh\text{-}^tBu)_{28}$. Color labels: yellow = S, others = Au. *Source:* Structures are redrawn from the cifs presented in Refs. [47, 51, 62, 67, 122, 134].

3.4 Properties and Applications

3.4.1 Properties

Au NCs with atomically precise structures serve as an important link between Au complexes and plasmonic Au NPs for understanding the evolution rules from Au complexes to plasmonic Au NPs. Compared with large-sized Au NPs, Au NCs exhibit discrete electronic states and molecule-like behaviors due to the strong quantum-size effect, which provide opportunities to study precise

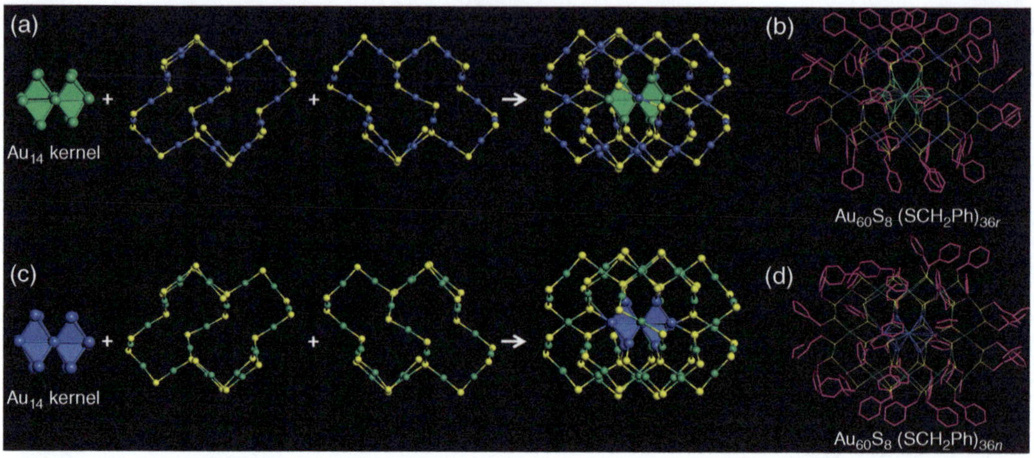

Figure 3.31 Anatomy of the $Au_{60}S_8(SCH_2Ph)_{36}$ structures of (a) $Au_{60}S_{8r}$ and (c) $Au_{60}S_{8n}$. Surface patterns of the ligands on (b) $Au_{60}S_{8r}$ and (d) $Au_{60}S_{8n}$. Color labels: yellow = S, red = C, others = Au. *Source:* Structures are redrawn from the cifs presented in Ref. [135].

correlations of structure- properties at the atomic level. The relationship between the structures and properties plays important roles in the field of clusters, because they not only help researchers to understand the origins of the properties but also provide certain guidance for the designs of nanomaterials with expected properties. In this section, we mainly review the optical properties (optical absorption and photoluminescence), magnetism, and chirality of thiolate-protected Au NCs.

3.4.1.1 Optical Absorption

Conventional plasmonic Au NPs of spherical shape (diameter from ~3 nm to ~100 nm) exhibit a single plasmon resonance peak (SPR). The SPR properties of spherical Au NPs are size-dependent [8]. With the decrease of particle size, the metallic state gradually fades out, and the SPR gradually weakens and blue shifts, and eventually disappears when the particle size is below 2 nm. Au NCs (less than 3 nm) have discrete electron energy levels (as opposite to the continuous band in metallic state), and exhibit molecular-like electronic transitions. Taking $Au_{25}(SC_2H_4Ph)_{18}^-$ as an example, its optical absorption properties and electron transitions were described by Jin group based on the crystal structure and DFT calculations [11]. As presented in Figure 3.32, the optical spectrum of $Au_{25}(SC_2H_4Ph)_{18}^-$ shows multiple absorption peaks, which are caused by single-electron transitions between quantized electron energy levels. In the UV-vis spectrum, there are at least three well-defined absorption peaks at 670, 450, and 400 nm, respectively. The peak at 670 nm (peak a) corresponds to a LUMO←HOMO transition, which is essentially an intraband (sp←sp) transition. The peak at 450 nm (peak b) arises from mixed intraband (sp←sp) and interband (sp←d) transitions. The peak at 400 nm (peak c) arises principally from an interband transition (sp←d). Notably, HOMO, LUMO, and other orbitals (including HOMO-1 and LUMO-1) consist almost entirely of electron orbitals in the Au_{13} core. Therefore, the absorption peak of $Au_{25}(SC_2H_4Ph)_{18}^-$ at 670 nm is completely derived from the transition caused by the Au_{13} core. Furthermore, HOMO displays a two-lobe distribution similar to the *p*-like orbital of an atom, while LUMO exhibits *d*-like properties. Thus, $Au_{25}(SC_2H_4Ph)_{18}^-$ can be regarded as a superatom with a valence electron number of 8e.

Recent advances in optical properties of clusters also provide insights into the transition from exciton state to plasmonic state, which has always been a central question in nanoscience. Zhou et al. investigated the evolution of the optical properties by comparing the steady-state absorption, transient absorption (TA), as well as carrier dynamics of a series of $Au_n(SR)_m$ clusters [136].

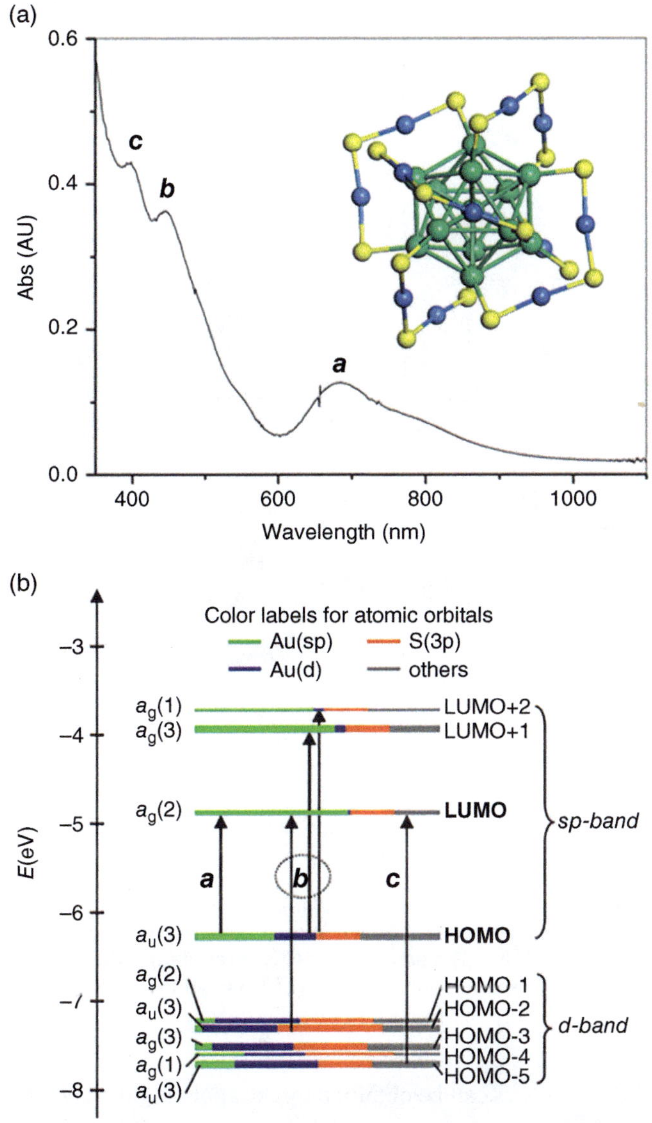

Figure 3.32 (a) Absorption spectrum and structure (inset) of $Au_{25}(SC_2H_4Ph)_{18}^-$; (b) Kohn-Sham orbital energy level diagram for a model compound $Au_{25}(SH)_{18}^-$. *Source:* Adapted with permission from Ref. [11]. Copyright 2008 American Chemical Society.

Figure 3.33a shows UV-Vis absorption of eight small-sized Au NCs. Multiple absorption peaks are observed in all these smaller Au NCs. TA measures the difference between the absorption of the excited state and that of the ground state (i.e. ΔA, Figure 3.33b). Figure 3.33c compares the carrier lifetimes of these Au NCs from the TA measurements. Except for $Au_{38}(SR)_{24}$, which exhibits a short lifetime of 4 ns, all other sizes display long carrier lifetimes and very subtle differences. This suggests that the size does not play an important role in the carrier lifetimes of these Au NCs. In the NCs smaller than $Au_{52}(SR)_{32}$, it is concluded that the structure, rather than size, determines the optical properties. As the size increases, quantum confinement becomes weaker, resulting in less pronounced absorption peaks due to crowdedness of electronic states. Three medium-sized clusters ($Au_{64}(SR)_{32}$, $Au_{92}(SR)_{44}$, $Au_{99}(SR)_{42}$) were selected to compare their optical properties

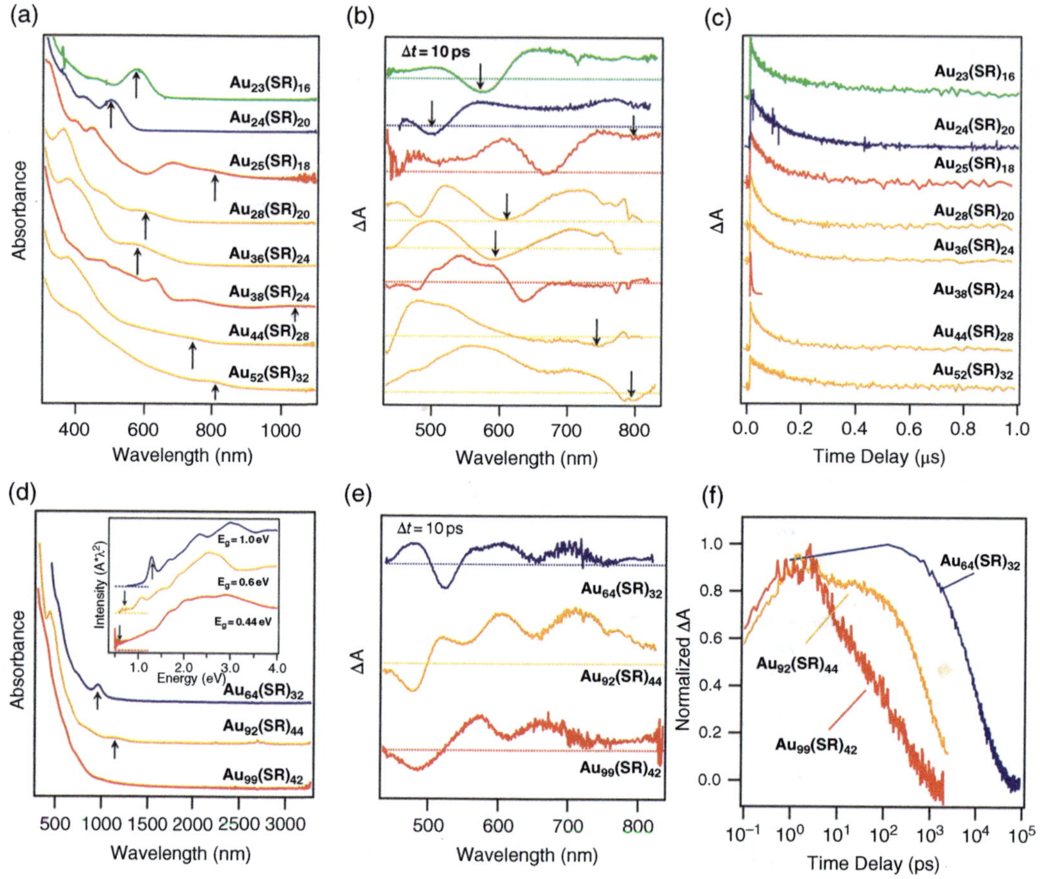

Figure 3.33 UV-vis-NIR absorption spectra of a series of (a) small-sized and (d) medium-sized NCs; TA spectra probed at 10 ps with 360 nm excitation: (b) small-sized NCs, (e) medium-sized NCs; Kinetic decays obtained from time-resolved TA spectra: (c) small-sized NCs, (f) medium-sized NCs. Inset of (d) is the absorption spectra in energy scale. *Source:* Adapted with permission from Ref. [136]. Copyright 2019 American Chemical Society.

(Figure 3.33d–f). The optical bandgap (E_g) of Au NCs can be obtained by extrapolating the absorbance to zero. From the evolution in E_g (inset of Figure 3.33d) and carrier dynamics, it is found that the size begins to play an important role in the optical properties of medium-sized Au NCs, in contrast to the previously discussed ultrasmall sizes, but the UV-vis-NIR absorption spectra are still largely determined by their atomic structures. Therefore, in the medium size range, both structure and size determine the optical performance of Au NCs.

For the Au NCs with size larger than Au_{99}, the effect of structure on optical properties is weakened and the optical properties show size dependence regardless of the structure types (Figure 3.34). The UV-vis-NIR and TA spectra of Au_{103} to Au_{246} clusters all show similar profiles (Figure 3.34a,b). From their relaxation dynamics (Figure 3.34c), it was found that the decay becomes faster as the size increases. The power dependence of these Au NCs was also tested, and it was found that their TA decays were independent of the pumping power. The above results reveal that Au_{103} to Au_{246} clusters are still in molecular state. Notably, the giant $Au_{246}(SR)_{80}$ (2.2 nm core diameter) is the largest nonmetallic Au NCs reported so far; whereas $Au_{279}(SR)_{84}$ is the smallest Au nanocrystal to exhibit metallic behavior. As shown in Figure 3.34d, Au NCs larger than Au_{279} all show a SPR peak in UV-vis-NIR spectra and a broad negative ground-state bleaching (GSB) in the TA spectra, which are typical observations in plasmonic Au NPs.

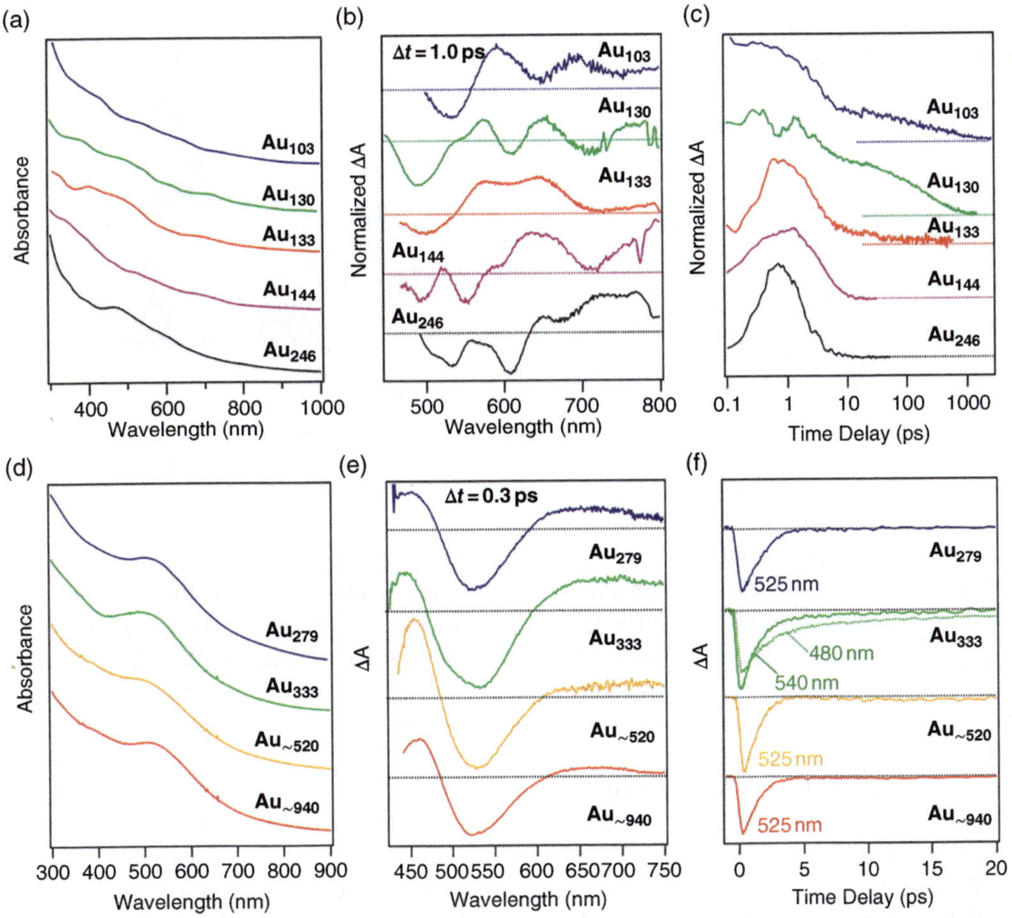

Figure 3.34 UV-vis-NIR absorption spectra of (a) Au_{103} to Au_{246} and (d) Au_{279} to Au_{940} clusters. TA spectra of these NCs at time delay: (b) 1 ps and (e) 0.3 ps; (c) Kinetic decays probed at selected wavelengths in Au_{103} to Au_{246} clusters; (f) Kinetic traces probed at the maximum of the SPR of Au_{279} to Au_{940} clusters. *Source:* Reproduced with permission from Ref. [136]. Copyright 2019 American Chemical Society.

3.4.1.2 Photoluminescence

With the rapid development of research on metal NCs, the PL properties have attracted increasing interest. The NCs lay a good foundation for an in-depth understanding of their PL properties owing to their precise compositions and structures. On the other hand, the NCs have the advantages of size-sensitive properties, rich yet tailorable surface chemistry, and many water-soluble NCs also have excellent biocompatibility, showing broad application prospects in the biomedical fields, such as drug delivery and biological imaging. Although most reported atomically precise $Au_n(SR)_m$ NCs exhibit low PL quantum yield (QY < 1%) in normal solvents and ambient environments, some clusters indeed show medium quantum yields (QY = 5–10%), including $Au_{18}(SG)_{14}$ [59], $Au_{22}(SG)_{18}$ [137], $Au_{24}(SCH_2Ph\text{-}^tBu)_{20}$ [111], $Au_{28}(SCH_2Ph\text{-}^tBu)_{22}$ [114], $Au_{21}(S\text{-}Adm)_{15}$ [138], and $Au_{38}S_2(S\text{-}Adm)_{20}$ [138]. The water-soluble $Au_{18}(SG)_{14}$ [59] and $Au_{22}(SG)_{18}$ [137] cluster both showed obvious red emission under UV illumination. $Au_{18}(SG)_{14}$ exhibits a maximum emission wavelengths at 745 nm with a QY of 5.3% (Figure 3.35), whereas $Au_{22}(SG)_{18}$ luminesces intensely at 665 nm with a QY of 8%. Given the difficulties with crystallization of water-soluble clusters, the crystal structures of the two clusters were not obtained at that time. Later, Chen et al. prepared $Au_{18}(S\text{-}c\text{-}C_6H_{11})_{14}$ using the $Au_{18}(SG)_{14}$ as precursor by reacting $Au_{18}(SG)_{14}$ with HS-c-C_6H_{11} [104],

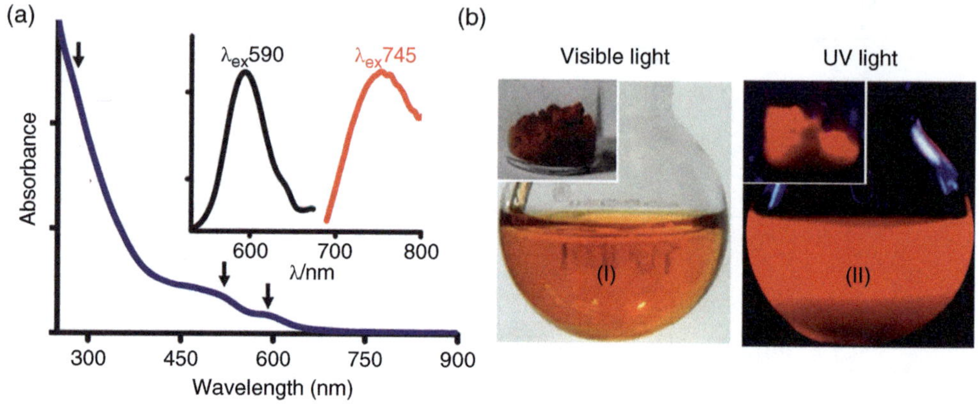

Figure 3.35 (a) UV-vis-NIR absorption spectrum of $Au_{18}(SG)_{14}$. Inset: PL spectrum of the $Au_{18}(SG)_{14}$ in water. (b) Photographs of the $Au_{18}(SG)_{14}$ aqueous solution and solid powder (inset) in visible light and UV light. *Source:* Reproduced with permission from Ref. [59]. Copyright 2012 American Chemical Society.

and successfully determined the X-ray structure of $Au_{18}(S\text{-}c\text{-}C_6H_{11})_{14}$. Han et al. synthesized alkynyl-protected $Au_{22}(C{\equiv}C^tBu)_{18}$ with methanol solution of NaOH as the reducing agent and resolved its structure by SCXC [139]. The QY of fluorescent $Au_{24}(SCH_2Ph\text{-}^tBu)_{20}$ and $Au_{28}(SCH_2Ph\text{-}^tBu)_{22}$ are 3.0% and 5.1% [114], respectively. Gan et al. attributed the intense PL of $Au_{24}(SR)_{20}$ to its structural features [68]; that is, two pairs of interlocked tetrameric motifs increase the rigidity of the cluster and reduce the nonradiative loss by vibrations, and the interactions between the kernel and thiolate motifs enhance electron transfer from ligand to kernel through the Au—S bonds. $Au_{28}(SCH_2Ph\text{-}^tBu)_{22}$ has the relatively high QY [114]. The tri-tetrahedron kernel with two bridging thiolates and two hexamer rings form a rigid, compact package, which can efficiently reduce non-radiation loss and thus enhance the emission. In addition, different from the $Au_{24}(SCH_2Ph\text{-}^tBu)_{20}$ cluster, $Au_{28}(SCH_2Ph\text{-}^tBu)_{22}$ cluster might realize charge transfer process through Au—Au bond. $Au_{28}(SCH_2Ph\text{-}^tBu)_{22}$ has a tight bi-ring structure and shows a fast electron transfer rate from the bi-ring to the kernel via compact Au—Au bonds, which also benefits the emission. In $Au_{38}S_2(S\text{-}Adm)_{20}$, a bright NIR PL at 900 nm with QY up to 15% was found, which is the highest reported QY among hydrophobic $Au_n(SR)_m$ clusters [138]. The relatively lower QY for $Au_{21}(S\text{-}Adm)_{15}$ (4%) compared to that of $Au_{38}S_2(S\text{-}Adm)_{20}$ is attributed to the lowest-lying excited state being symmetry-disallowed, as evidenced by the pressure-dependent unexpected shift of the absorption spectra compared with that of PL [138]. On the basis of the structures of the two NCs, this work revealed that the Au—S lock rings encircling the Au kernel and the surface sulfur lock atoms bridging the kernel units (e.g. tetrahedra, icosahedra, etc.) can immobilize the Au kernel and suppress the nonradiative motions, which significantly enhance the QY. The results also indicate that weak band-edge transition and abundant excited-state transitions/transformations in Au NCs are the two fundamental origins for the common PL characteristics (e.g. large Stokes shift and long PL lifetime).

Elucidating the luminescence mechanism of NCs and enhancing their luminescence QY have become two hot topics in the NCs field. Many research groups have done lots of works on the luminescence mechanism of Au NCs. Here, the $Au_{25}(SR)_{18}^-$ NC is taken as an example. The luminescence study of $Au_{25}(SR)_{18}^-$ started ever since its structure determination. $Au_{25}(SR)_{18}^-$ comprises an icosahedral Au_{13} core protected by six dimeric $Au_2(SR)_3$ staple motifs (also called semi-ring motifs).

A previous study by Link et al. observed two emission bands at around 1.5 and 1.15 eV in $Au_{25}(SG)_{18}^-$ (originally mis-assigned as $Au_{28}(SG)_{16}$) and proposed two different origins [140]: the visible emission (i.e. high energy band) corresponds to fluorescence, originating from the sp←d interband transition; while the NIR emission (i.e. low energy band) is phosphorescence, corresponding to the sp←sp intraband transition (Figure 3.36a–c). Shortly after, work by Lee et al.

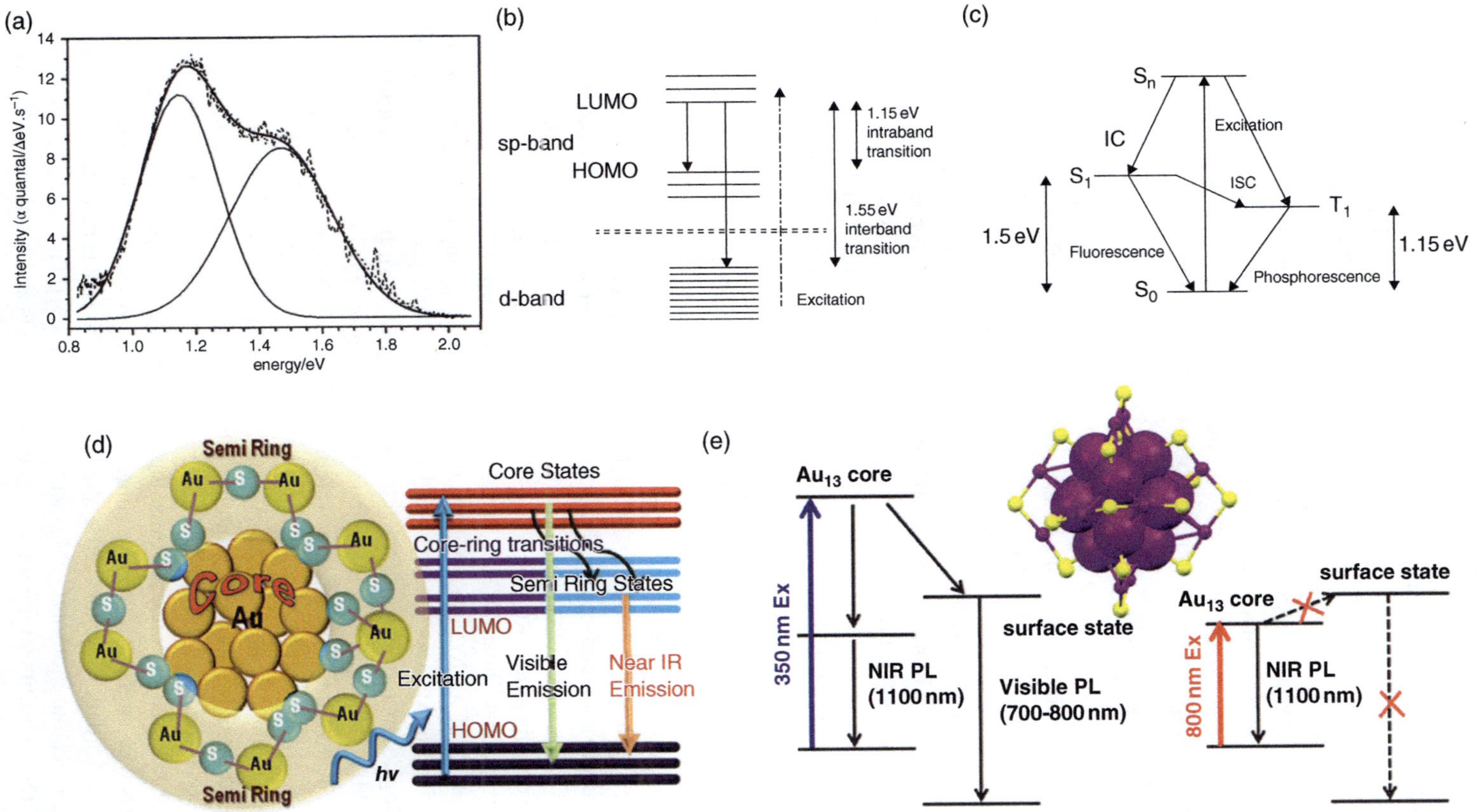

Figure 3.36 (a) Luminescence spectrum of $Au_{25}(SG)_{18}^-$ NCs. The solid line represents a fit of the experimental data to a sum of two Gaussian curves. (b, c) Two models for the origin of the luminescence in $Au_{25}(SG)_{18}^-$ NCs. (d) The relaxation pathways in $Au_{25}(SR)_{18}^-$ NCs showing the core states and staple states. (e) Mechanism of emissions in $Au_{25}(SR)_{18}^-$ NCs. *Source:* (a–c) Adapted with permission from Ref. [140]. Copyright 2002 American Chemical Society. (d) Reproduced with permission from Ref. [141]. Copyright 2010 American Chemical Society. (e) Reproduced with permission from Ref. [142]. Copyright 2021 American Chemical Society.

reported a broad emission band around 1000 nm for the $Au_{25}(SC_2H_4Ph)_{18}$ NCs, which was resolvable into 902 nm (1.38 eV) and 1025 nm (1.2 eV) [143]. The 1.38 eV emission matched well with the HOMO-LUMO energy gap of 1.33 eV and was assigned as the gap luminescence, while the 1.2 eV emission was considered to be a sub-gap energy luminescence.

Miller et al. studied the excited state relaxation dynamics of $Au_{25}(SC_2H_4Ph)_{18}^-$ by femtosecond laser spectroscopy [144]. When the 450 and 680 nm excitation wavelengths (assigned to ninefold degenerate HOMO←LUMO+1 and sixfold degenerate HOMO←LUMO excitations, respectively) were used, the system relaxation was not detected until the lowest fluorescence state (about 1000 nm). They observed time constants greater than 4 ps at a detection wavelength of 725–800 nm that they attributed to emission from staple state to the ground state. Moreover, an 80 cm^{-1} vibration localized at the Au_{13} core reflects the strong vibronic coupling of the delocalized Au—Au bond stretching vibrations.

Wu and Jin et al. studies the PL of $[Au_{25}(SR)_{18}]^q$ (SR = SC_2H_4Ph, $SC_{12}H_{25}$, SC_6H_{13}, where q is the charge state of NCs) [145], and reported that the visible PL at around 750 nm can be enhanced by increasing the ligand's capability of donating charges or increasing the electropositivity of the metal core (increasing q from −1 to +2). The observed fluorescence was attributed to charge transfer from ligands to metal core (i.e. LMNCT) through the Au—S bonds. Devadas et al. identified that the NIR luminescence (700–800 nm) of $Au_{25}(SR)_{18}$ (SR = SC_6H_{13} or SG) arises from the surface state emission, and the visible luminescence (~500 nm) is from the Au_{13} core states using ultrafast luminescence measurements (Figure 3.36d) [141]. Subsequently, Weerawardene et al., based on the theoretical calculations, proved that the PL of $Au_{25}(SR)_{18}^-$ originated from the electronic states of the core-based orbitals rather than charge-transfer states or the staple-based states [146].

Recently, on the basis of time-resolved emission and nanosecond transient absorption spectroscopy analysis, Zhou and Song proposed a model that can well explain the emission bands of $Au_{25}(L)_{18}^-$ (L = SC_2H_4Ph, SG, S-Nap, SePh) in the visible to near-infrared region [142]. It is found that the 700–800 nm visible PL bands of four $Au_{25}(SR)_{18}^-$ all arise from the surface states (i.e. the staple motifs) while the 1000–1200 nm NIR PL originates from the Au_{13} metal core state (Figure 3.36e).

Despite these research advances in Au_{25} NCs, there is no universal model yet that can well explain the luminescence mechanism. The luminescence has alternately been assigned to intraband and interband transition, fluorescence and phosphorescence, surface or staple states, or ligand-metal charge transfer states. Researchers have come to different conclusions in the origin of luminescence, possibly due to the difference in the protecting ligands of NCs.

In view of the weak PL of Au NCs, researchers also developed various strategies to improve the PL QY, such as capping the Au core surface with different types of ligands [68, 145], adjusting the size [21, 53], or doping homogold NCs with other metal atoms [147, 148], and adopting aggregation-induced emission (AIE) [149, 150]. Wu et al. found that the ligands with electron-rich atoms or groups can greatly improve the luminescence intensity of $Au_{25}(SR)_{18}$ clusters [145]. The similar surface-regulated PL enhancement was also reported by the Millstone [151] and Tsukuda groups [152]. Xie et al. discovered the AIE properties in Au(I)-SR complexes system (Figure 3.37) [149]. The nonluminescent oligomer Au(I)-SR complexes can emit strong fluorescence after aggregation induced by solvents or cations. The intensity and color of the luminescence largely depend on the degree of concentration. Based on this AIE property, Xie et al. prepared luminescent Au NCs with high QY (15%) by controlling the aggregation of Au(I)-SR complexes onto in-situ generated Au(0) cores to form core-shell clusters. The AIE feature can be attributed to a restriction of intramolecular motions (RIM) mechanism [153, 154]. The aggregation of Au(I)-SR motifs can increase the restraints on the intramolecular vibrations and rotations, reducing nonradiative relaxation of the excited states to enhance the luminescence. In addition to AIE, crystalization-induced luminescence enhancement (CIE) was also reported in Au—Ag bimetallic

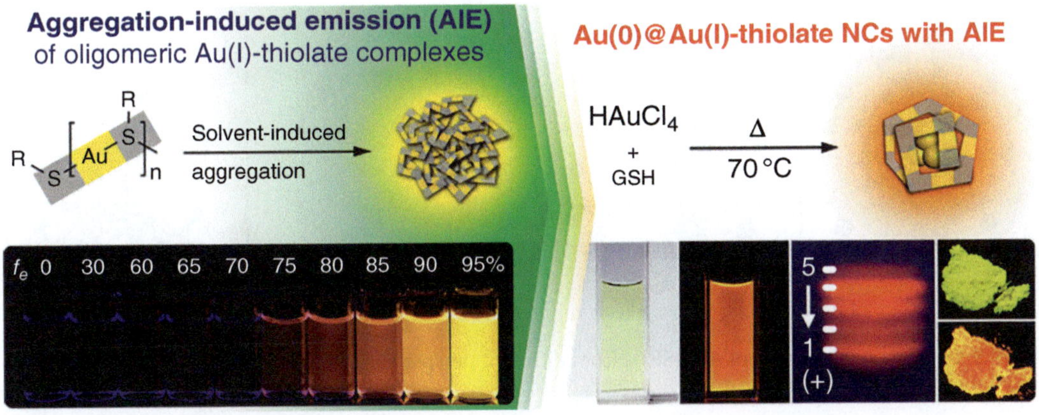

Figure 3.37 (Top left) Solvent-induced AIE properties of oligomeric Au(I)-SR complexes. (bottom left) Photograph of Au(I)-SR complexes in mixed ethanol and water with different volume percentages of ethanol under UV light. (Top right) AIE-guided synthesis of photoluminescent NCs and (bottom right) the corresponding images. *Source:* Reproduced with permission from Ref. [149]. Copyright 2012 American Chemical Society.

clusters [155], that is, the crystalline state shows strong luminescence, whereas the amorphous state or the solution of NCs is weakly (or even non-) emissive. The CIE mechanism can be explained by the AIE mechanism. The structural analysis and DFT calculations confirm that the CIE effect is mainly caused by more restricted vibrations and rotations in the crystalline state. The luminescence of Au NCs can also be enhanced by doping other metal atoms in the clusters. For example, high QY (40–60%) $[Au_{25-x}Ag_x(PPh_3)_{10}(SR)_5Cl_2]^{2+}$ alloy clusters was prepared by site-specific doping of low QY (0.1%) rod-shaped $[Au_{25}(PPh_3)_{10}(SR)_5Cl_2]^{2+}$ cluster with 13 Ag atoms [148]. Moreover, when the number of doped Ag atoms was less than 13, the alloy clusters did not show highly extensive fluorescence. In addition, the PL properties of NCs can be improved by rigidifying the surface of NCs (e.g. coating with tetraoctylammonium (TOA^+) cations) [156], tailoring the surface structure [157], and impregnating NCs into polymers (e.g. chitosan, polypeptide, etc.) [158].

3.4.1.3 Chirality

Chirality is a universal phenomenon in nature and occurs at all length scales. An object becomes chiral when its mirror image cannot be superimposed with the original image. With recent advances in atomically precise nanochemistry, the overall structures of ligand-protected metal NCs have been successfully obtained, and the origin of chiral properties has been found to be related to different parts of the cluster. When achiral ligands are used, there are three sources of intrinsic chirality: one is the chiral kernels (e.g. $Au_{20}(SR)_{16}$ [110], $Au_{44}(SR)_{28}$ [51], $Au_{52}(SR)_{32}$ [52], $Au_{133}(SR)_{52}$ [129]), another is the chiral arrangement of the staple motifs on the surface of the kernel (e.g. $Au_{38}(SR)_{24}$ [56], $Au_{102}(SR)_{44}$ [26], $Au_{130}(SR)_{50}$ [128], $Au_{144}(SR)_{60}$ [9], $Au_{246}(SR)_{80}$ [130]), and the last one is the chiral whirls of carbon tails in the ligand assembly (e.g. $Au_{133}(SR)_{52}$ [129] and $Au_{246}(SR)_{80}$ [130]).

Except for $Au_{36}(SR)_{24}$, all the other clusters in the $Au_{8n+4}(SR)_{4n+8}$ magic series are chiral with quasi-D_2 symmetry for the Au—S framework [51]. In the Au_4-interlinked double-helical view, $Au_{28}(SR)_{20}$, $Au_{44}(SR)_{28}$, and $Au_{52}(SR)_{32}$ have a chiral Au_{14}, Au_{26}, and Au_{32} core, respectively. However, the cubic $Au_{36}(SR)_{24}$ cluster exhibits higher symmetry (quasi-D_{2d} Au—S framework) than the other three clusters, resulting in an achiral core structure. The $Au_{133}(SR)_{52}$ cluster was also found to have a chiral Au_{107} core [129]. The core structure can be divided into three shells, arranged as Au_{13}@Au_{42}@Au_{52}. The first two shells are achiral, but the third Au_{52} shell exhibits chirality.

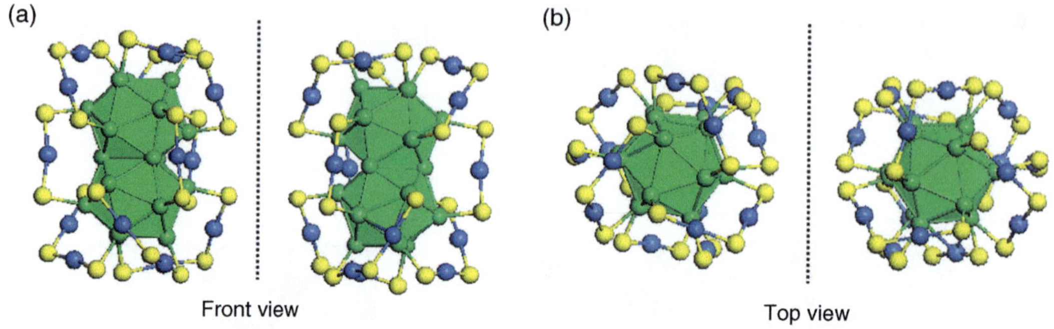

Figure 3.38 Two enantiomers of $Au_{38}(SR)_{24}$ from different perspectives. (a) Front and (b) top views. Color labels: yellow = S, others = Au.

The chirality of most chiral Au NCs results from the asymmetric arrangement of the staple motifs, for example, $Au_{38}(SR)_{24}$ (Figure 3.38) [56]. Dolamic et al. successfully isolated the chiral enantiomers of $Au_{38}(SR)_{24}$ cluster by using high-performance liquid chromatography (HPLC) [159]. The chirality of $Au_{38}(SR)_{24}$ originates from the rotary arrangements of dimeric staples, rather than its achiral Au kernel. In the interpenetrating cuboctahedral view, the fcc Au_{20} core of $Au_{28}(SR)_{20}$ is achiral [67]. However, the arrangements of dimeric staples as well as bridging thiolates result in the chirality of $Au_{28}(SR)_{20}$ cluster. Furthermore, another chiral feature was found in the assembly pattern of the carbon tails in $Au_{133}(SR)_{52}$ [129] and $Au_{246}(SR)_{80}$ [130]. The carbon tails of $Au_{133}(SR)_{52}$ tend to arrange into six swirls, and each swirl consists of four Ph-^{t}Bu tails. These swirls on the surface of $Au_{133}(SR)_{52}$ surface rotate in the same direction; therefore, the chirality of the entire swirl patterns do not counteract each other. In $Au_{246}(SR)_{80}$ cluster, the stripe patterns appear not only in the arrangement of the S-Au-S staples, but also in the arrangement of the carbon tails. The clockwise and counterclockwise rotational arrangements of the stripes induce chirality in the $Au_{246}(SR)_{80}$. Recently, Li et al. summarized the research progress in atomically precise chiral metal NCs, and analyzed the different surface structures of $Au_n(SR)_m$ clusters and the resulting chirality [160].

3.4.1.4 Magnetism

A single Au atom shows paramagnetism due to its unpaired 6s electron, while bulk Au is diamagnetic because the paramagnetism of conduction electrons is counteracted by the diamagnetism of the orbitals and ions. Therefore, it is of great significance to understand the magnetic evolution process from Au atom to Au NP and then to bulk Au. Previous works reported the size-dependent magnetic properties in 1~5 nm Au NPs [161–163]. However, different researchers had different explanations for magnetism, due to the uneven particle size and unclear atomic structure of those Au NPs. The appearance of Au NCs with well-defined compositions and structures make it possible to explore the magnetic mechanism of Au nanomaterials [164].

In 2009, Zhu et al. took $[Au_{25}(SC_2H_4Ph)_{18}]^q$ ($q = -1, 0$) with different charge states as examples to explore the magnetism of Au NCs. The results show that the magnetism of $[Au_{25}(SC_2H_4Ph)_{18}]^q$ ($q = -1, 0$) is closely related to its charge states [165]. As shown in Figure 3.39, the negative $[Au_{25}(SC_2H_4Ph)_{18}]^-$ shows diamagnetism, while the charge-neutral $[Au_{25}(SC_2H_4Ph)_{18}]^0$ shows paramagnetism. In other words, the reversibility of cluster paramagnetism and diamagnetism can be achieved by controlling the charge state of the $[Au_{25}(SC_2H_4Ph)_{18}]^q$ ($q = -1, 0$) clusters. This switching magnetism of the $[Au_{25}(SC_2H_4Ph)_{18}]^q$ can be used as a paramagnetic probe (Figure 3.39a). DFT calculations show that HOMO with triple degeneracy and LUMO with double degeneracy (Figure 3.39b-d). Based on the crystal structure and DFT results, they concluded that the

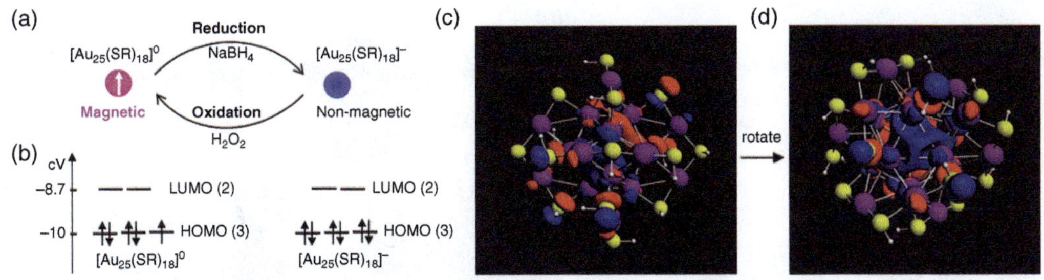

Figure 3.39 Reversible switching of magnetism through control of the charge state of the $[Au_{25}(SC_2H_4Ph)_{18}]^q$. (a) Schematic diagram; (b) quasi-degenerate HOMO and LUMO sets; (c and d) HOMO distributions at different orientations. *Source:* Adapted with permission from Ref. [165]. Copyright 2009 American Chemical Society.

paramagnetism of $[Au_{25}(SC_2H_4Ph)_{18}]^0$ mainly originates from an unpaired single electron in the HOMO orbital. The HOMO orbitals of negatively charged $[Au_{25}(SC_2H_4Ph)_{18}]^-$ are completely filled, and there is no unpaired electron; thus, resulting in diamagnetism of negative $[Au_{25}(SC_2H_4Ph)_{18}]^-$. The strategy of controlling NCs at the single-electron level through oxidation (or reduction) does not apply to $Au_{23}(SR)_{16}^-$ (counterion: TOA^+). The oxidation treatment of the 8-electron $Au_{23}(SR)_{16}^-$ led to size conversion [166], rather than the desired sole change of the charge state as in $Au_{25}(SR)_{18}^-$.

In addition, Negishi et al. investigated the magnetic properties of a series of atomically precise $Au_n(SG)_m$ clusters via X-ray magnetic circular dichroism [167], including $Au_{10}(SG)_{10}$, $Au_{15}(SG)_{13}$, $Au_{22}(SG)_{16}$, $Au_{25}(SG)_{18}$, $Au_{29}(SG)_{20}$, and $Au_{39}(SG)_{24}$. The magnetic moment was found to increase with the core size of cluster. Compared with the quantum size effect of Au NCs, the holes formed by Au—S bonding contribute more to the spin polarization phenomenon. Furthermore, Zeng et al. studied the paramagnetic properties of $Au_{133}(SPh\text{-}^tBu)_{52}$ NCs with 81 nominal "valence electrons" [168]. Through electron paramagnetic resonance (EPR) spectroscopy, it was proved that the $Au_{133}(SPh\text{-}^tBu)_{52}$ hosts one unpaired electron. EPR simulations further suggest that the unpaired spin is mainly delocalized in the inner icosahedral Au_{13} core, similar to the case of paramagnetic $[Au_{25}(SR)_{18}]^0$ (7e). The unpaired spin can be removed by oxidizing $Au_{133}(SPh\text{-}^tBu)_{52}$ to $[Au_{133}(SPh\text{-}^tBu)_{52}]^+$, and the nanocluster transforms from paramagnetism to diamagnetism accordingly.

3.4.2 Applications

Metal NCs have recently attracted a great deal of interest in catalysis, chemical sensors, and biological applications due to their unique properties. In the following section, we highlight some examples of atomically precise metal (mainly Au) NCs in applications of catalysis, sensing, and biology.

3.4.2.1 Sensing

Sensing chemicals or biomolecules using nanomaterials have become an interesting topic in nanomaterials science. The emergence of NCs provides opportunities for understanding the sensing mechanism in detail. In the past two decades, researchers have reported numerous NCs-based luminescent sensors based on luminescence enhancement (or quenching) of clusters, which sufficiently satisfied the diverse sensing needs [24, 169]. For example, Muhammed et al. demonstrated that HSG-protected Au_{23} cluster exhibits quenching of fluorescence selectively in the presence of Cu^{2+} ions, and it can therefore be used as a metal-ion sensor [170]. The hydrophilic NCs are not

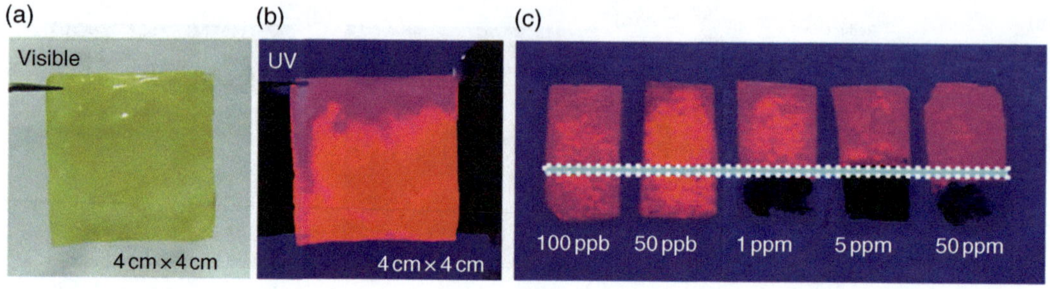

Figure 3.40 Photographs of Au_{15} NC incorporated film under (a) white light; and (b) UV light, respectively. (c) Dependence of Cu^{2+} concentration on luminescence quenching. *Source:* Adapted with permission from Ref. [171]. Copyright 2012 American Chemical Society.

amenable to crystallization, making it difficult to resolve their precise structure by experimental means. However, the skeletal structures of hydrophilic NCs are similar to that of the corresponding hydrophobic ones, regardless of the ligand effect [40]. Consequently, it is possible to elucidate the mechanism of ion detections (or PL changes) by means of the atomically precise structure of the corresponding hydrophobic clusters. Pradeep's group also loaded SG-protected Au_{15} clusters onto chitosan films to make highly photoluminescent composites [171]. Based on the sensitivity of red emission of the clusters to Cu^{2+}, a metal ion sensor similar to the pH paper was prepared (Figure 3.40a,b). The composite film exhibits visual sensitivity to Cu^{2+} up to 1 ppm (Figure 3.40c), which is below the permissible limit (1.3 ppm) in drinking water set by the U.S. environmental protection agency. The specificity of the film for Cu^{2+} sensing may be due to the reduction of Cu^{2+} to Cu^{1+}/Cu^0 by the glutathione ligand or Au_{15} core.

Pradeep also developed a sensor based on a hybrid structure, which is composed of Au mesoflowers (MFs) and bovine serum albumin (BSA)-protected Ag_{15} NCs [20]. Prior to cluster functionalization, a fluorescent isothiocyanate (FITC) was coated on the silica-capped Au MFs, which resulted in bright green luminescence. After further functionalization with Ag_{15} cluster on the Au@SiO_2 FITC MFs, the Au@(SiO_2-FITC)@Ag_{15} MFs showed a red emission. Upon exposure to Hg^{2+} or TNT (trinitrotoluene), the red luminescence was quenched and a green luminescence, characteristic of the underlying FITC, appeared.

Wu et al. found that Ag^+ oxidation of $Au_{25}(SG)_{18}$ could lead to an increase in PL intensity [88]. Based on the PL enhancement, the $Au_{25}(SG)_{18}$ exhibits high sensitivity to Ag^+, with a detection limit of 20 ppb. The luminescence enhancement rarely occurs when the clusters interact with metal ions, which usually cause luminescence quenching. In this study, the unique enhancement induced by Ag^+ was attributed to the change in the oxidation state of Au_{25}, and the interaction of Ag^0 species with Au_{25}, as well as the interaction of Ag^+ with Au_{25}. Aside from Cu^{2+} and Ag^+, PL Au NCs have always been used to detect other ions, e.g. Hg^{2+} [19, 172], As^{3+} [173], S^{2-} [174, 175], NO_2^- [176], CN^- [177]. Due to the harmful effects of Hg^{2+} heavy ions on the environment and human health, efficient detection of Hg^{2+} ions is essential for environmental monitoring of aquatic ecosystems. Xie et al. reported a BSA-stabilized Au_{25} clusters as the fluorometric sensor for Hg^{2+} [172]. The Hg^{2+} can react with the Au(I)-species of BSA-Au_{25} clusters through $5d^{10}$–$5d^{10}$ interactions, thereby changing the electronic structure of Au NCs to quench its strong red emission. In addition to PL sensors, metal NCs also show great potential in electroluminescence (ECL) sensing applications [19, 178]. Tang et al. improved the ECL efficiency of GSH-protected Au NCs by doping Ag to form bimetallic clusters [178]. Based on the enhanced ECL signal of Au NCs, this group constructed an ultrasensitive ECL sensor for detecting dopamine. The proposed ECL sensor exhibits a low detection limit of 2.3 nM, broad linear range (from 10 nM to 1 mM), and good sensitivity and stability.

PL or ECL Au NCs as sensors mainly depend on the interaction between the analytes and Au NCs [179]. The external analytes can interact with the Au atoms in the core as well as with the ligands forming the shell. There are three main interactions between the Au core and the analytes: the analytes cause Au core to decompose; deposition of analytes on the Au surface; and metallophilic interactions. Interaction with the ligand shell of NCs can take place by an enzymatic reaction, incorporation of recognition moieties, or analyte-induced NC aggregation [179].

3.4.2.2 Biological Labeling and Biomedicine

Au NCs have been extensively put in variously biological applications such as biolabeling, cancer treatment, drug screening, and drug delivery. Compared with organic fluorophores or semiconductor quantum dots, PL Au NCs have a number of advantages in biological applications, including ultrasmall size, large Stokes shift, strong photo-stability, good biocompatibility, easy surface modification, and functionalization [180–182]. Below is a discussion of a few biological applications of Au NCs.

To facilitate the biomedical engineering application, the outer ligand shell of clusters can be tailored by employing biological molecules, such as proteins, peptides, and DNA to endow the biocompatibility and enhance the PL intensity, thus establishing Au NCs as a novel ultrasmall biocompatible fluorophores [183]. For example, Gan et al. used 11-mercaptodecanoic acid or BSA as the incoming ligand to perform ligand exchange for hydrophobic $Au_{24}(SCH_2Ph\text{-}^tBu)_{20}$ clusters, in order to improve the PL quantum yield of cluster (7.6% or 9.8%) [68]. The highly photoluminescent Au_{24} clusters after BSA exchange with negligible cytotoxicity can be used for imaging of living macrophages (Figure 3.41). Zhang et al. used GSH-stabilized $Au_{25}(SG)_{18}$ for cancer radiation therapy [184]. The ultrasmall Au_{25} NCs with biocompatible SG surface can preferentially accumulate in tumor via the improved EPR (enhanced permeability and retention) effect, and therefore have a stronger enhancement for cancer radiotherapy than that of large-sized Au NPs.

In addition, Zhang and Xie groups relied on the near infrared (1100–1350 nm) fluorescence properties of Au_{25} clusters protected by peptide ligands [15], and used the clusters for *in vivo* brain vessel imaging and tumor metastasis. Compared with the available NIR dyes, such as nanotubes [185], quantum dots [186], and conjugated polymer [187], the ultrasmall-size Au NCs can cross the glomerular filtration and be excreted fast by urine without any toxicological response at an ultra-high dose. PL Au NCs have also been used for subcellular organelle labelling. Yang et al. developed a mitochondria-targeting Au NCs ($Au_{18}(SG)_{12}(MTPB)_2$) capable of performing strong fluorescence imaging and mass accumulating at mitochondrial sites for precise diagnosis and therapy of cancer cells [188]. This group introduced the MTPB ((4-mercaptobutyl)triphenylphosphonium bromide) into the $Au_{18}(SG)_{14}$ cluster by ligand exchange strategy. The MTPB enables the Au_{18} NCs to have a new ability to target mitochondria, while their initial optical properties are not significantly affected. The $Au_{18}(SG)_{14}$ NCs initially concentrate in lysosomes upon cancer cell internalization, while the ligand-exchanged $Au_{18}(SG)_{12}(MTPB)_2$ NCs are more inclined to accumulate at the mitochondrial site, due to the specific labeling ability of the introduced MTPB.

3.4.2.3 Catalysis

Nanometal-based materials play an important role in catalysis. Compared with conventional Au NPs, Au NCs have better catalytic prospects because of their ultra-small size and higher surface-to-volume ratio. In contrast to the unclear interface between metal core and ligands in Au NPs, the interface environment of Au NCs is clear, with the amount and type of surface ligand also being precisely adjustable, which helps to investigate the surface/interface effect. More importantly, the well-defined structure and monodispersity of NCs will facilitate the analysis of active sites and catalytic mechanisms at the atomic level from the perspective of experiment and theoretical

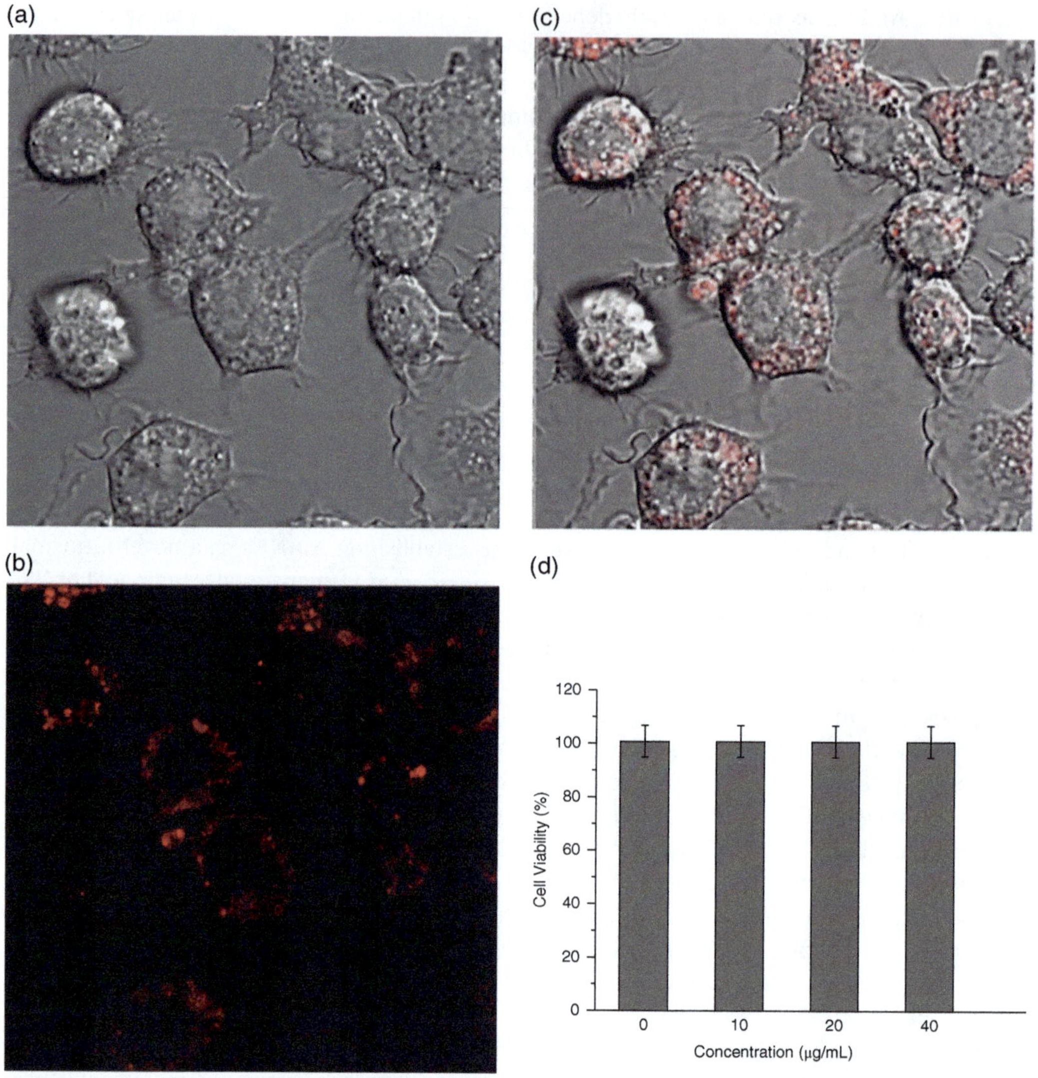

Figure 3.41 Laser confocal images of macrophages after incubation with BSA-exchanged $Au_{24}(SCH_2Ph\text{-}^tBu)_{20}$ for four hours: (a) bright-field image; (b) excitation at 488 nm; (c) merged image; (d) cytotoxicity tests with various concentrations of BSA-exchanged $Au_{24}(SCH_2Ph\text{-}^tBu)_{20}$. *Source:* Reproduced with permission from Ref. [68]. Copyright 2016 Wiley-VCH.

simulation. Therefore, ultra-small Au NCs have the potential to be a new class of model catalysts. Several reviews have already focused on the catalytic properties of Au NCs [189–191]; thus only some representative works are listed here in this section.

In general, the metal NCs catalysts can be divided into two categories: clusters alone acting as catalysts; clusters-supporting materials (such as TiO_2, Fe_2O_3, and CeO_2, etc.) combination acting as catalysts. For example, Tian et al. selected $Au_{38}(SC_2H_4Ph)_{24-Q}$ and $Au_{38}(SC_2H_4Ph)_{24-T}$ NCs with the same composition but different structures as unsupported catalysts, and revealed that the two $Au_{38}(SC_2H_4Ph)_{24}$ isomers exhibit different catalytic activity for nitrophenol reduction [86]. The exploration of catalysis by isomeric Au NCs demonstrated that the catalytic performance is not only affected by size, but also related to the structure of clusters. Researchers can tailor the structure of metal nanoclusters on demand to gain insights into the respective roles of the core and

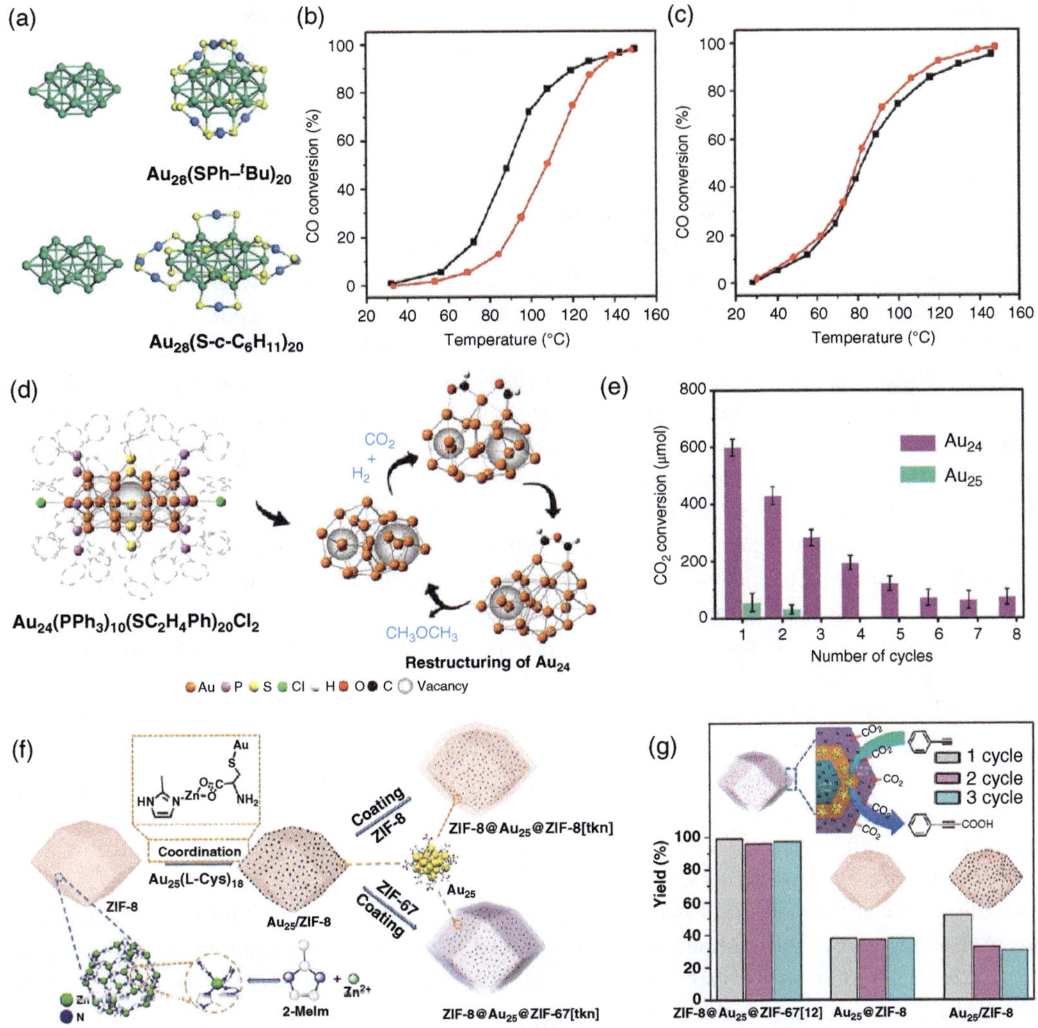

Figure 3.42 (a) Structural comparison of the $Au_{28}(S\text{-}c\text{-}C_6H_{11})_{20}$ and $Au_{28}(SPh\text{-}^tBu)_{20}$. Catalytic activity for $Au_{28}(S\text{-}c\text{-}C_6H_{11})_{20}/CeO_2$ (black profile) and $Au_{28}(SPh\text{-}^tBu)_{20}/CeO_2$ (red); (b) catalysts pretreated with O_2 at 150 °C; (c) pretreated with O_2 at 300 °C to remove ligands. (d) Catalytic performance of $Au_{24}(PPh_3)_{10}(SC_2H_4Ph)_5Cl_2$ on the hydrogenation reaction of CO_2. (e) Recyclability of the $Au_{24}(PPh_3)_{10}(SC_2H_4Ph)_5Cl_2/SiO_2$ and $Au_{25}(PPh_3)_{10}(SC_2H_4Ph)_5Cl_2/SiO_2$ catalysts with each four hours reaction under 2 MPa reaction at 130 °C. (f) Synthetic Route for the sandwich structures of MOFs@Au_{25}@MOFs. (g) Catalytic activity for three different sandwich catalysts. *Source:* (a–c) Adapted with permission from Ref. [47]. Copyright 2016 American Chemical Society. (d, e) Adapted with permission from Ref. [192]. Copyright 2020 American Chemical Society. (f, g) Adapted with permission from Ref. [17]. Copyright 2020 American Chemical Society.

surface. Chen et al. resolved a pair of Au_{28} NCs with identical size and core structure but different staple structure, namely $Au_{28}(S\text{-}c\text{-}C_6H_{11})_{20}$ and $Au_{28}(SPh\text{-}^tBu)_{20}$ [47], as the ideal system to study the effect of surface structure on catalytic performance (Figure 3.42a–c). Before the reaction, two Au_{28} NCs were respectively deposited onto CeO_2 supports, followed by oxygen pretreatment under heating conditions. The results show that the $Au_{28}(S\text{-}c\text{-}C_6H_{11})_{20}/CeO_2$ catalyst exhibits higher catalytic activity toward CO oxidation than the $Au_{28}(SPh\text{-}^tBu)_{20}/CeO_2$, due to the more protruded staple motifs and more accessible surface structure of $Au_{28}(S\text{-}c\text{-}C_6H_{11})_{20}$. DFT shows that the CO adsorption sites of two Au_{28} clusters are both on the staple Au atoms. There was no difference in catalytic activity in both Au_{28} clusters when two Au_{28} clusters were treated at 300 °C (higher than the ligand desorption temperature). Therefore, the catalytic differences of two quasi-isomers are caused by the surface "staple" structure.

Cai et al. selected a pair of Au NCs with only one central Au atom difference, namely $Au_{24}(PPh_3)_{10}(SC_2H_4Ph)_5Cl_2$ and $Au_{25}(PPh_3)_{10}(SC_2H_4Ph)_5Cl_2$ [193], as the ideal system to study the effect of one atom difference in NCs on catalytic performance. One-central-atom loss in the $Au_{24}(PPh_3)_{10}(SC_2H_4Ph)_5Cl_2$ cluster enables better catalytic activity in the methane oxidation toward methanol compared to the $Au_{25}(PPh_3)_{10}(SC_2H_4Ph)_5Cl_2$ cluster. Notably, the activation and deactivation can be reversibly switched by losing the central atom and filling the central vacancy, which effectively avoids the irreversibility of catalytic capability and improves the durability. This work elucidates the contribution of single atom gain and loss to catalytic activity at the atomic level, and provides design rules for how to control catalytic performance of catalysts by single atom removal and addition. Afterward, this group also reported the contribution of the internal vacancy in the $Au_{24}(PPh_3)_{10}(SC_2H_4Ph)_5Cl_2$ to the preformation on the CO_2 hydrogenation (Figure 3.42 d,e) [192]. Compared to the $Au_{25}(PPh_3)_{10}(SC_2H_4Ph)_5Cl_2$ cluster without internal vacancy, the $Au_{24}(PPh_3)_{10}(SC_2H_4Ph)_5Cl_2$ can provide the NCs with much more structural flexibility, inhibit the aggregation and growth of NCs in the catalytic reaction process, further postpone the deactivation of catalyst and thus improve the catalytic stability. The surface ligands of Au NCs also affect the catalytic activity of clusters.

In order to investigate the distinct effects of R groups in mercaptan ligands, Li et al compared the catalytic activity of supported $Au_{25}(SR)_{18}^-/CeO_2$ (R = Nap, Ph, C_2H_4Ph, C_6H_{13}) in the Ullmann heteroreactions [194]. Compared with $Au_{25}(SC_2H_4Ph)_{18}^-$, $Au_{25}(S\text{-}Nap)_{18}^-$ exhibits significantly improved catalytic activity (91% conversion) and selectivity (82% for the heterocoupling product) for this reaction. This study highlight that the aromatic ligands not only lead to a higher conversion in catalytic reaction but also markedly increase the yield of the heterocoupling product. Results demonstrate the effectiveness and promise of ligand engineering for tailoring the electronic and catalytic properties of nanoclusters.

In addition, for ultrasmall-sized Au NCs, their stability is a key parameter for practical applications and is one of the most sought catalytic properties, as its stability guarantees the catalytic recyclability. Therefore, various strategies have been developed for the pursuit Au NCs catalysts with significant activity and stability, such as the combination of Au NCs and MOF. In fact, the introduction of MOF materials can not only endow additional catalytic active sites and coordination effect, but also effectively inhibit the aggregation of Au NCs, thus improving the catalytic activity. As shown in Figure 3.42f,g, Yun et al. sandwiched $Au_{25}(SR)_{18}$ clusters between two MOFs to composite a novel catalyst (MOFs@Au_{25}@MOFs[tkn], tkn = thickness of shell) [17]. Compared to the simple components Au_{25}/MOFs (Au_{25} locates at the outside surface of MOFs) and Au_{25}@MOFs (Au_{25} uniformly disperses in the matrix of MOFs), the sandwich composite shows significantly enhanced catalytic activity and excellent stability in both nitrophenol reduction and terminal alkyne carboxylation with CO_2. The catalytic performance of this "sandwich" composite structure can be optimized by precise adjustment of the thickness of the outer MOF, which opens up a new research approach for the targeted design of Au NCs catalysts with high catalytic activity and stability.

3.5 Conclusion and Future Perspectives

Over the past decade, Au NCs have attracted significant attention in both fundamental research and practical applications owing to their ultra-small size, precise compositions and structures, versatile properties, structure tailorability, and biocompatibility. However, more Au NCs with new functions for applications should be explored in the future. For catalysis, an in-depth understanding of the catalytic site and mechanism remains to be further studied, and the catalytic performance of Au NCs waits to be improved. Current sensor research is mainly based on fluorescence quenching, while sensors based on magnetic response, photothermal effect, and light absorption response are rarely involved. It is thus important to develop multi-type and multifunction sensors of metal NCs for practical applications. Although metal NCs show potential in labeling, diagnosis, antibacteria, and therapy, their bio-related properties such as photoluminescence and phototherapy should be further enhanced for practical applications, based on the elucidation of the structure-property correlations. Overall, atomically precise nanoclusters, as a new class of materials, are expected to stimulate many new directions in future research.

Acknowledgments

This work was supported by National Natural Science Foundation of China (Nos. 21925303, 21701179, 22171268, 91961204, 21771186, 21829501, 21222301, 21528303, and 21171170), Anhui Provincial Natural Science Foundation (Nos. 1708085QB36, 2108085MB56), the Special Foundation of President of HFIPS (No. YZJJ202102), CASHIPS Director's Fund (BJPY2019A02), Key Program of 13th five-year plan, CASHIPS (KP-2017-16), Collaborative Innovation Program of Hefei Science Center, CAS (2020HSC-CIP005), and CAS/SAFEA International Partnership Program for Creative Research Teams.

References

1 Daniel, M.-C. and Astruc, D. (2004). Gold nanoparticles: assembly, supramolecular chemistry, quantum-size-related properties, and applications toward biology, catalysis, and nanotechnology. *Chem. Rev.* 104 (1): 293–346.

2 Saha, K., Agasti, S.S., Kim, C. et al. (2012). Gold nanoparticles in chemical and biological sensing. *Chem. Rev.* 112 (5): 2739–2779.

3 Xia, Y., Gilroy, K.D., Peng, H.-C. et al. (2017). Seed-mediated growth of colloidal metal nanocrystals. *Angew. Chem. Int. Ed.* 56 (1): 60–95.

4 Shi, Y., Lyu, Z., Zhao, M. et al. (2021). Noble-metal nanocrystals with controlled shapes for catalytic and electrocatalytic applications. *Chem. Rev.* 121 (2): 649–735.

5 Kumara, C., Zuo, X., Cullen, D.A. et al. (2014). Faradaurate-940: synthesis, mass spectrometry, Electron microscopy, high-energy X-ray diffraction, and X-ray scattering study of $Au_{\sim 940\pm 20}(SR)_{\sim 160\pm 4}$ nanocrystals. *ACS Nano* 8 (6): 6431–6439.

6 Kumara, C., Zuo, X., Ilavsky, J. et al. (2014). Super-stable, highly monodisperse plasmonic Faradaurate-500 nanocrystals with 500 gold atoms: $Au_{\sim 500}(SR)_{\sim 120}$. *J. Am. Chem. Soc.* 136 (20): 7410–7417.

7 Jin, R. (2010). Quantum sized, thiolate-protected gold nanoclusters. *Nanoscale* 2 (3): 343–362.

8 Jin, R., Zeng, C., Zhou, M. et al. (2016). Atomically precise colloidal metal nanoclusters and nanoparticles: fundamentals and opportunities. *Chem. Rev.* 116 (18): 10346–10413.

9 Yan, N., Xia, N., Liao, L. et al. (2018). Unraveling the long-pursued Au_{144} structure by X-ray crystallography. *Sci. Adv.* 4 (10): eaat7259.

10 Heaven, M.W., Dass, A., White, P.S. et al. (2008). Crystal structure of the gold nanoparticle $[N(C_8H_{17})_4][Au_{25}(SCH_2CH_2Ph)_{18}]$. *J. Am. Chem. Soc.* 130 (12): 3754–3755.

11 Zhu, M., Aikens, C.M., Hollander, F.J. et al. (2008). Correlating the crystal structure of a thiol-protected Au_{25} cluster and optical properties. *J. Am. Chem. Soc.* 130 (18): 5883–5885.

12 Kang, X. and Zhu, M. (2019). Tailoring the photoluminescence of atomically precise nanoclusters. *Chem. Soc. Rev.* 48 (8): 2422–2457.

13 Knoppe, S. and Bürgi, T. (2014). Chirality in thiolate-protected gold clusters. *Acc. Chem. Res.* 47 (4): 1318–1326.

14 Lei, Y., Tang, L., Xie, Y. et al. (2017). Gold nanoclusters-assisted delivery of NGF siRNA for effective treatment of pancreatic cancer. *Nat. Commun.* 8 (1): 15130.

15 Liu, H., Hong, G., Luo, Z. et al. (2019). Atomic-precision gold clusters for NIR-II imaging. *Adv. Mater.* 31 (46): 1901015.

16 Li, Z., Peng, H., Liu, J. et al. (2018). Plant protein-directed synthesis of luminescent gold nanocluster hybrids for tumor imaging. *ACS Appl. Mater. Interfaces* 10 (1): 83–90.

17 Yun, Y., Sheng, H., Bao, K. et al. (2020). Design and remarkable efficiency of the robust Sandwich cluster composite nanocatalysts ZIF-8@Au_{25}@ZIF-67. *J. Am. Chem. Soc.* 142 (9): 4126–4130.

18 Nie, X., Qian, H., Ge, Q. et al. (2012). CO oxidation catalyzed by oxide-supported $Au_{25}(SR)_{18}$ nanoclusters and identification of perimeter sites as active centers. *ACS Nano* 6 (7): 6014–6022.

19 Zhai, Q., Xing, H., Zhang, X. et al. (2017). Enhanced electrochemiluminescence behavior of gold-silver bimetallic nanoclusters and its sensing application for mercury(II). *Anal. Chem.* 89 (14): 7788–7794.

20 Mathew, A., Sajanlal, P.R., and Pradeep, T. (2012). Selective visual detection of TNT at the sub-zeptomole level. *Angew. Chem. Int. Ed.* 51 (38): 9596–9600.

21 Zhou, Y., Liao, L., Zhuang, S. et al. (2021). Traceless removal of two kernel atoms in a gold nanocluster and its impact on photoluminescence. *Angew. Chem. Int. Ed.* 60 (16): 8668–8672.

22 Liu, Y., Chai, X., Cai, X. et al. (2018). Central doping of a foreign atom into the silver cluster for catalytic conversion of CO_2 toward C-C bond formation. *Angew. Chem. Int. Ed.* 57 (31): 9775–9779.

23 Du, Y., Sheng, H., Astruc, D. et al. (2020). Atomically precise Noble metal nanoclusters as efficient catalysts: a bridge between structure and properties. *Chem. Rev.* 120 (2): 526–622.

24 Chakraborty, I. and Pradeep, T. (2017). Atomically precise clusters of noble metals: emerging Link between atoms and nanoparticles. *Chem. Rev.* 117 (12): 8208–8271.

25 Yang, H., Wang, Y., Chen, X. et al. (2016). Plasmonic twinned silver nanoparticles with molecular precision. *Nat. Commun.* 7: 12809.

26 Jadzinsky, P.D., Calero, G., Ackerson, C.J. et al. (2007). Structure of a thiol monolayer-protected gold nanoparticle at 1.1 Å resolution. *Science* 318 (5849): 430–433.

27 Zhou, Q., Kaappa, S., Malola, S. et al. (2018). Real-space imaging with pattern recognition of a ligand-protected Ag_{374} nanocluster at sub-molecular resolution. *Nat. Commun.* 9 (1): 2948.

28 Desireddy, A., Conn, B.E., Guo, J. et al. (2013). Ultrastable silver nanoparticles. *Nature* 501: 399.

29 Nguyen, T.-A.D., Jones, Z.R., Goldsmith, B.R. et al. (2015). A Cu_{25} nanocluster with partial Cu(0) character. *J. Am. Chem. Soc.* 137 (41): 13319–13324.

30 Qu, M., Zhang, F.-Q., Wang, D.-H. et al. (2020). Observation of non-FCC copper in alkynyl-protected Cu_{53} nanoclusters. *Angew. Chem. Int. Ed.* 59 (16): 6507–6512.

31 McPartlin, M., Mason, R., and Malatesta, L. (1969). Novel cluster complexes of Gold(0)-Gold(I). *J. Am. Chem. Soc. D: Chem. Commun.* 7: 334–334.

32 Günter, S., Reinhard, P., Roland, B. et al. (1981). $Au_{55}[P(C_6H_5)_3]_{12}CI_6$-ein Goldcluster ungewöhnlicher Größe. *Chem. Ber.* 114 (11): 3634–3642.

33 Teo, B.K., Shi, X., and Zhang, H. (1992). Pure gold cluster of 1:9:9:1:9:9:1 layered structure: a novel 39-metal-atom cluster $[(Ph_3P)_{14}Au_{39}Cl_6]Cl_2$ with an interstitial gold atom in a hexagonal antiprismatic cage. *J. Am. Chem. Soc.* 114 (7): 2743–2745.

34 Brust, M., Walker, M., Bethell, D. et al. (1994). Synthesis of thiol-derivatised gold nanoparticles in a two-phase liquid-liquid system. *J. Am. Chem. Soc, Chem. Commun.* 7: 801–802.

35 Knoppe, S., Boudon, J., Dolamic, I. et al. (2011). Size exclusion chromatography for semipreparative scale separation of $Au_{38}(SR)_{24}$ and $Au_{40}(SR)_{24}$ and larger clusters. *Anal. Chem.* 83 (13): 5056–5061.

36 Ghosh, A., Hassinen, J., Pulkkinen, P. et al. (2014). Simple and efficient separation of atomically precise noble metal clusters. *Anal. Chem.* 86 (24): 12185–12190.

37 Ma, M.-X., Ma, X.-L., Liang, G.-M. et al. (2021). A nanocluster $[Ag_{307}Cl_{62}(SPh^tBu)_{110}]$: chloride intercalation, specific electronic state, and superstability. *J. Am. Chem. Soc.* 143 (34): 13731–13737.

38 Brust, M., Fink, J., Bethell, D. et al. (1995). Synthesis and reactions of functionalised gold nanoparticles. *J. Am. Chem. Soc, Chem. Commun.* 16: 1655–1656.

39 Negishi, Y., Nobusada, K., and Tsukuda, T. (2005). Glutathione-protected gold clusters revisited: bridging the gap between gold(I)-thiolate complexes and thiolate-protected gold nanocrystals. *J. Am. Chem. Soc.* 127 (14): 5261–5270.

40 Das, A., Li, T., Nobusada, K. et al. (2013). Nonsuperatomic $[Au_{23}(SC_6H_{11})_{16}]^-$ nanocluster featuring bipyramidal Au_{15} kernel and trimeric $Au_3(SR)_4$ motif. *J. Am. Chem. Soc.* 135 (49): 18264–18267.

41 Liao, L., Yao, C., Wang, C. et al. (2016). Quantitatively monitoring the size-focusing of Au nanoclusters and revealing what promotes the size transformation from $Au_{44}(TBBT)_{28}$ to $Au_{36}(TBBT)_{24}$. *Anal. Chem.* 88 (23): 11297–11301.

42 Wu, Z. (2012). Anti-galvanic reduction of thiolate-protected gold and silver nanoparticles. *Angew. Chem. Int. Ed.* 51 (12): 2934–2938.

43 Zhu, M., Lanni, E., Garg, N. et al. (2008). Kinetically controlled, high-yield synthesis of Au_{25} clusters. *J. Am. Chem. Soc.* 130 (4): 1138–1139.

44 Wu, Z., MacDonald, M.A., Chen, J. et al. (2011). Kinetic control and thermodynamic selection in the synthesis of atomically precise gold nanoclusters. *J. Am. Chem. Soc.* 133 (25): 9670–9673.

45 Yuan, X., Zhang, B., Luo, Z. et al. (2014). Balancing the rate of cluster growth and etching for gram-scale synthesis of thiolate-protected Au_{25} nanoclusters with atomic precision. *Angew. Chem. Int. Ed.* 53 (18): 4623–4627.

46 Das, A., Liu, C., Byun, H.Y. et al. (2015). Structure determination of $[Au_{18}(SR)_{14}]$. *Angew. Chem. Int. Ed.* 54 (10): 3140–3144.

47 Chen, Y., Liu, C., Tang, Q. et al. (2016). Isomerism in $Au_{28}(SR)_{20}$ nanocluster and stable structures. *J. Am. Chem. Soc.* 138 (5): 1482–1485.

48 Yang, S., Chen, S., Xiong, L. et al. (2018). Total structure determination of $Au_{16}(S\text{-}Adm)_{12}$ and $Cd_1Au_{14}(StBu)_{12}$ and implications for the structure of $Au_{15}(SR)_{13}$. *J. Am. Chem. Soc.* 140 (35): 10988–10994.

49 Chen, S., Xiong, L., Wang, S. et al. (2016). Total structure determination of $Au_{21}(S\text{-}Adm)_{15}$ and geometrical/electronic structure evolution of thiolated gold nanoclusters. *J. Am. Chem. Soc.* 138 (34): 10754–10757.

50 Higaki, T., Liu, C., Zeng, C. et al. (2016). Controlling the atomic structure of Au_{30} nanoclusters by a ligand-based strategy. *Angew. Chem. Int. Ed.* 55 (23): 6694–6697.

51 Zeng, C., Chen, Y., Iida, K. et al. (2016). Gold quantum boxes: on the periodicities and the quantum confinement in the Au_{28}, Au_{36}, Au_{44}, and Au_{52} magic series. *J. Am. Chem. Soc.* 138 (12): 3950–3953.

52 Zeng, C., Chen, Y., Liu, C. et al. (2015). Gold tetrahedra coil up: Kekulé-like and double helical superstructures. *Sci. Adv.* 1 (9): e1500425.

53 Liao, L., Wang, C., Zhuang, S. et al. (2020). An unprecedented kernel growth mode and layer-number-odevity-dependent properties in gold nanoclusters. *Angew. Chem. Int. Ed.* 59 (2): 731–734.

54 Zeng, C., Liu, C., Chen, Y. et al. (2016). Atomic structure of self-assembled monolayer of thiolates on a tetragonal Au_{92} nanocrystal. *J. Am. Chem. Soc.* 138 (28): 8710–8713.

55 Liao, L., Chen, J., Wang, C. et al. (2016). Transition-sized Au_{92} nanoparticle bridging non-fcc-structured gold nanoclusters and fcc-structured gold nanocrystals. *Chem. Commun.* 52 (81): 12036–12039.

56 Qian, H., Eckenhoff, W.T., Zhu, Y. et al. (2010). Total structure determination of thiolate-protected Au_{38} nanoparticles. *J. Am. Chem. Soc.* 132 (24): 8280–8281.

57 Qian, H. and Jin, R. (2011). Ambient synthesis of $Au_{144}(SR)_{60}$ nanoclusters in methanol. *Chem. Mater.* 23 (8): 2209–2217.

58 Yu, Y., Chen, X., Yao, Q. et al. (2013). Scalable and precise synthesis of thiolated Au_{10-12}, Au_{15}, Au_{18}, and Au_{25} nanoclusters via pH controlled CO reduction. *Chem. Mater.* 25 (6): 946–952.

59 Ghosh, A., Udayabhaskararao, T., and Pradeep, T. (2012). One-step route to luminescent $Au_{18}SG_{14}$ in the condensed phase and its closed shell molecular ions in the gas phase. *J. Phys. Chem. Lett.* 3 (15): 1997–2002.

60 Chen, T., Fung, V., Yao, Q. et al. (2018). Synthesis of water-soluble $[Au_{25}(SR)_{18}]^-$ using a stoichiometric amount of $NaBH_4$. *J. Am. Chem. Soc.* 140 (36): 11370–11377.

61 Zhuang, S., Liao, L., Li, M.-B. et al. (2017). The fcc structure isomerization in gold nanoclusters. *Nanoscale* 9 (39): 14809–14813.

62 Zhuang, S., Liao, L., Zhao, Y. et al. (2018). Is the kernel-staples match a key-lock match?. *Chem. Sci.* 9 (9): 2437–2442.

63 Zhuang, S., Liao, L., Yuan, J. et al. (2018). Kernel homology in gold nanoclusters. *Angew. Chem. Int. Ed.* 57 (47): 15450–15454.

64 Zhuang, S., Liao, L., Yuan, J. et al. (2019). Fcc vs non-fcc structural isomerism of gold nanoparticles with kernel atom packing-dependent photoluminescence. *Angew. Chem. Int. Ed.* 58 (14): 4510–4514.

65 Gan, Z., Chen, J., Wang, J. et al. (2017). The fourth crystallographic closest packing unveiled in the gold nanocluster crystal. *Nat. Commun.* 8: 14739.

66 Zeng, C., Liu, C., Pei, Y. et al. (2013). Thiol ligand-induced transformation of $Au_{38}(SC_2H_4Ph)_{24}$ to $Au_{36}(SPh\text{-}t\text{-}Bu)_{24}$. *ACS Nano* 7 (7): 6138–6145.

67 Zeng, C., Li, T., Das, A. et al. (2013). Chiral structure of thiolate-protected 28-gold-atom nanocluster determined by X-ray crystallography. *J. Am. Chem. Soc.* 135 (27): 10011–10013.

68 Gan, Z., Lin, Y., Luo, L. et al. (2016). Fluorescent gold nanoclusters with interlocked staples and a fully thiolate-bound kernel. *Angew. Chem. Int. Ed.* 55 (38): 11567–11571.

69 Chen, Y., Zeng, C., Kauffman, D.R. et al. (2015). Tuning the magic size of atomically precise gold nanoclusters via isomeric methylbenzenethiols. *Nano Lett.* 15 (5): 3603–3609.

70 Liao, L., Zhuang, S., Yao, C. et al. (2016). Structure of chiral $Au_{44}(2,4\text{-DMBT})_{26}$ nanocluster with an 18-electron shell closure. *J. Am. Chem. Soc.* 138 (33): 10425–10428.

71 Liao, L., Zhuang, S., Wang, P. et al. (2017). Quasi-dual-packed-kerneled $Au_{49}(2,4\text{-DMBT})_{27}$ nanoclusters and the influence of kernel packing on the electrochemical gap. *Angew. Chem. Int. Ed.* 56 (41): 12644–12648.

72 Yao, C., Chen, J., Li, M.-B. et al. (2015). Adding two active silver atoms on Au_{25} nanoparticle. *Nano Lett.* 15 (2): 1281–1287.

73 Yao, C., Lin, Y.-j., Yuan, J. et al. (2015). Mono-cadmium vs mono-mercury doping of Au_{25} nanoclusters. *J. Am. Chem. Soc.* 137 (49): 15350–15353.

74 Zhu, M., Wang, P., Yan, N. et al. (2018). The fourth alloying mode by way of anti-galvanic reaction. *Angew. Chem. Int. Ed.* 57 (17): 4500–4504.

75 Zhuang, S., Chen, D., Liao, L. et al. (2020). Hard-sphere random close-packed $Au_{47}Cd_2(TBBT)_{31}$ nanoclusters with a faradaic efficiency of up to 96% for electrocatalytic CO_2 reduction to CO. *Angew. Chem. Int. Ed.* 59 (8): 3073–3077.

76 Zhang, W., Zhuang, S., Liao, L. et al. (2019). Two-way alloying and dealloying of cadmium in metalloid gold clusters. *Inorg. Chem.* 58 (9): 5388–5392.

77 Wang, S., Song, Y., Jin, S. et al. (2015). Metal exchange method using Au_{25} nanoclusters as templates for alloy nanoclusters with atomic precision. *J. Am. Chem. Soc.* 137 (12): 4018–4021.

78 Li, M.-B., Tian, S.-k., and Wu, Z. (2017). Improving the catalytic activity of Au_{25} nanocluster by peeling and doping. *Chin. J. Chem.* 35 (5): 567–571.

79 Negishi, Y., Takasugi, Y., Sato, S. et al. (2004). Magic-numbered Aun clusters protected by glutathione monolayers (n = 18, 21, 25, 28, 32, 39): isolation and spectroscopic characterization. *J. Am. Chem. Soc.* 126 (21): 6518–6519.

80 Schaaff, T.G., Shafigullin, M.N., Khoury, J.T. et al. (2001). Properties of a ubiquitous 29 kDa Au:SR cluster compound. *J. Phys. Chem. B* 105 (37): 8785–8796.

81 Li, J.-J., Guan, Z.-J., Yuan, S.-F. et al. (2021). Enriching structural diversity of alkynyl-protected gold nanoclusters with chlorides. *Angew. Chem. Int. Ed.* 60 (12): 6699–6703.

82 Nimmala, P.R., Yoon, B., Whetten, R.L. et al. (2013). $Au_{67}(SR)_{35}$ Nanomolecules: characteristic size-specific optical, electrochemical, structural properties and first-principles theoretical analysis. *J. Phys. Chem. A* 117 (2): 504–517.

83 Wu, Z., Chen, J., and Jin, R. (2011). One-pot synthesis of Au_{25} $(SG)_{18}$ 2- and 4-nm gold nanoparticles and comparison of their size-dependent properties. *Adv. Funct. Mater.* 21 (1): 177–183.

84 Schaaff, T.G., Knight, G., Shafigullin, M.N. et al. (1998). Isolation and selected properties of a 10.4 kDa gold:glutathione cluster compound. *J. Phys. Chem. B* 102 (52): 10643–10646.

85 Li, M.-B., Tian, S.-K., and Wu, Z. (2014). Catalyzed formation of α,β-unsaturated ketones or aldehydes from propargylic acetates by a recoverable and recyclable nanocluster catalyst. *Nanoscale* 6 (11): 5714–5717.

86 Tian, S., Li, Y.-Z., Li, M.-B. et al. (2015). Structural isomserism in gold nanoparticles revealed by X-ray crystallography. *Nat. Commun.* 6 (1): 8667.

87 Wu, Z., Gayathri, C., Gil, R.R. et al. (2009). Probing the structure and charge state of glutathione-capped $Au_{25}(SG)_{18}$ clusters by NMR and mass spectrometry. *J. Am. Chem. Soc.* 131 (18): 6535–6542.

88 Wu, Z., Wang, M., Yang, J. et al. (2012). Well-defined nanoclusters as fluorescent nanosensors: a case study on $Au_{25}(SG)_{18}$. *Small* 8 (13): 2028–2035.

89 Zeng, C., Qian, H., Li, T. et al. (2012). Total structure and electronic properties of the gold nanocrystal $Au_{36}(SR)_{24}$. *Angew. Chem. Int. Ed.* 51 (52): 13114–13118.

90 Zhao, Y., Zhuang, S., Liao, L. et al. (2020). A dual purpose strategy to endow gold nanoclusters with both catalysis activity and water solubility. *J. Am. Chem. Soc.* 142 (2): 973–977.

91 Kenzler, S., Schrenk, C., Frojd, A.R. et al. (2018). $Au_{70}S_{20}(PPh_3)_{12}$: an intermediate sized metalloid gold cluster stabilized by the Au_4S_4 ring motif and Au-PPh_3 groups. *Chem. Commun.* 54 (3): 248–251.

92 Wan, X.-K., Tang, Q., Yuan, S.-F. et al. (2015). Au_{19} nanocluster featuring a V-shaped alkynyl-gold motif. *J. Am. Chem. Soc.* 137 (2): 652–655.

93 Lei, Z., Wan, X.-K., Yuan, S.-F. et al. (2018). Alkynyl approach toward the protection of metal nanoclusters. *Acc. Chem. Res.* 51 (10): 2465–2474.

94 Kang, X. and Zhu, M. (2019). Metal nanoclusters stabilized by selenol ligands. *Small* 15 (43): 1902703.

95 Song, Y., Zhong, J., Yang, S. et al. (2014). Crystal structure of $Au_{25}(SePh)_{18}$ nanoclusters and insights into their electronic, optical and catalytic properties. *Nanoscale* 6 (22): 13977–13985.

96 Narouz, M.R., Osten, K.M., Unsworth, P.J. et al. (2019). N-heterocyclic carbene-functionalized magic-number gold nanoclusters. *Nat. Chem.* 11 (5): 419–425.

97 Shen, H., Xu, Z., Hazer, M.S.A. et al. (2021). Surface coordination of multiple ligands endows N-heterocyclic carbene-stabilized gold nanoclusters with high robustness and surface reactivity. *Angew. Chem. Int. Ed.* 60 (7): 3752–3758.

98 Jin, R., Liu, C., Zhao, S. et al. (2015). Tri-icosahedral gold nanocluster $[Au_{37}(PPh_3)_{10}(SC_2H_4Ph)_{10}X_2]^+$: linear assembly of icosahedral building blocks. *ACS Nano* 9 (8): 8530–8536.

99 Zeng, C., Chen, Y., Li, G. et al. (2014). Magic size Au_{64}(S-*c*-$C_6H_{11})_{32}$ nanocluster protected by cyclohexanethiolate. *Chem. Mater.* 26 (8): 2635–2641.

100 Negishi, Y., Sakamoto, C., Ohyama, T. et al. (2012). Synthesis and the origin of the stability of thiolate-protected Au_{130} and Au_{187} clusters. *J. Phys. Chem. Lett.* 3 (12): 1624–1628.

101 Plascencia-Villa, G., Demeler, B., Whetten, R.L. et al. (2016). Analytical characterization of size-dependent properties of larger aqueous gold nanoclusters. *J. Phys. Chem. C* 120 (16): 8950–8958.

102 Azubel, M., Koivisto, J., Malola, S. et al. (2014). Electron microscopy of gold nanoparticles at atomic resolution. *Science* 345 (6199): 909–912.

103 Liu, C., Li, T., Li, G. et al. (2015). Observation of body-centered cubic gold nanocluster. *Angew. Chem. Int. Ed.* 54 (34): 9826–9829.

104 Chen, S., Wang, S., Zhong, J. et al. (2015). The structure and optical properties of the $[Au_{18}(SR)_{14}]$ nanocluster. *Angew. Chem. Int. Ed.* 54 (10): 3145–3149.

105 Sakthivel, N.A., Theivendran, S., Ganeshraj, V. et al. (2017). Crystal structure of Faradaurate-279: $Au_{279}(SPh\text{-}^tBu)_{84}$ plasmonic nanocrystal molecules. *J. Am. Chem. Soc.* 139 (43): 15450–15459.

106 Sakthivel, N.A., Shabaninezhad, M., Sementa, L. et al. (2020). The missing link: $Au_{191}(SPh\text{-}^tBu)_{66}$ Janus nanoparticle with molecular and bulk-metal-like properties. *J. Am. Chem. Soc.* 142 (37): 15799–15814.

107 Dass, A., Theivendran, S., Nimmala, P.R. et al. (2015). $Au_{133}(SPh\text{-}^tBu)_{52}$ nanomolecules: X-ray crystallography, optical, electrochemical, and theoretical analysis. *J. Am. Chem. Soc.* 137 (14): 4610–4613.

108 Li, Q., Huang, B., Yang, S. et al. (2021). Unraveling the nucleation process from a Au(I)-SR complex to transition-size nanoclusters. *J. Am. Chem. Soc.* 143 (37): 15224–15232.

109 Xu, G.-T., Wu, L.-L., Chang, X.-Y. et al. (2019). Solvent-induced cluster-to-cluster transformation of homoleptic gold(I) thiolates between catenane and ring-in-ring structures. *Angew. Chem. Int. Ed.* 58 (45): 16297–16306.

110 Zeng, C., Liu, C., Chen, Y. et al. (2014). Gold-thiolate ring as a protecting motif in the $Au_{20}(SR)_{16}$ nanocluster and implications. *J. Am. Chem. Soc.* 136 (34): 11922–11925.

111 Das, A., Li, T., Li, G. et al. (2014). Crystal structure and electronic properties of a thiolate-protected Au_{24} nanocluster. *Nanoscale* 6 (12): 6458–6462.

112 Xiong, L., Yang, S., Sun, X. et al. (2018). Structure and electronic structure evolution of thiolate-protected gold nanoclusters containing quasi face-centered-cubic kernels. *J. Phys. Chem. C* 122 (26): 14898–14907.

113 Liu, X., Xu, W.W., Huang, X. et al. (2020). De novo design of $Au_{36}(SR)_{24}$ nanoclusters. *Nat. Commun.* 11 (1): 3349.

114 Dong, J., Gan, Z., Gu, W. et al. (2021). Synthesizing photoluminescent $Au_{28}(SCH_2Ph\text{-}^tBu)_{22}$ nanoclusters with structural features by using a combined method. *Angew. Chem. Int. Ed.* 60 (33): 17932–17936.

115 Jiang, D.-e., Overbury, S.H., and Dai, S. (2013). Structure of $Au_{15}(SR)_{13}$ and its implication for the origin of the nucleus in thiolated gold nanoclusters. *J. Am. Chem. Soc.* 135 (24): 8786–8789.

116 Li, Y., Zhou, M., Song, Y. et al. (2021). Double-helical assembly of heterodimeric nanoclusters into supercrystals. *Nature* 594 (7863): 380–384.

117 Xu, W.W., Li, Y., Gao, Y. et al. (2016). Unraveling a generic growth pattern in structure evolution of thiolate-protected gold nanoclusters. *Nanoscale* 8 (14): 7396–7401.

118 Takano, S., Yamazoe, S., Koyasu, K. et al. (2015). Slow-reduction synthesis of a thiolate-protected one-dimensional gold cluster showing an intense near-infrared absorption. *J. Am. Chem. Soc.* 137 (22): 7027–7030.

119 Ma, Z., Wang, P., Zhou, G. et al. (2016). Correlating the structure and optical absorption properties of $Au_{76}(SR)_{44}$ cluster. *J. Phys. Chem. C* 120 (25): 13739–13748.

120 Xu, W.W., Zeng, X.C., and Gao, Y. (2018). The structural isomerism in gold nanoclusters. *Nanoscale* 10 (20): 9476–9483.

121 Pei, Y., Tang, J., Tang, X. et al. (2015). New structure model of $Au_{22}(SR)_{18}$: bitetrahederon golden kernel enclosed by $[Au_6(SR)_6]$ Au(I) complex. *J. Phys. Chem. Lett.* 6 (8): 1390–1395.

122 Dong, H., Liao, L., Zhuang, S. et al. (2017). A novel double-helical-kernel evolution pattern of gold nanoclusters: alternate single-stranded growth at both ends. *Nanoscale* 9 (11): 3742–3746.

123 Dass, A., Jones, T., Rambukwella, M. et al. (2016). Crystal structure and theoretical analysis of green gold $Au_{30}(S\text{-}^tBu)_{18}$ nanomolecules and their relation to $Au_{30}S(S\text{-}^tBu)_{18}$. *J. Phys. Chem. C* 120 (11): 6256–6261.

124 Gu, W., Zhao, Y., Zhuang, S. et al. (2021). Unravelling the structure of a medium-sized metalloid gold nanocluster and its filming property. *Angew. Chem. Int. Ed.* 60 (20): 11184–11189.

125 Briant, C.E., Theobald, B.R.C., White, J.W. et al. (1981). Synthesis and X-ray structural characterization of the centred icosahedral gold cluster compound $[Au_{13}(PMe_2Ph)_{10}Cl_2](PF_6)_3$; the realization of a theoreticael prediction. *J. Am. Chem. Soc, Chem. Commun.* (5): 201–202.

126 Shichibu, Y., Negishi, Y., Watanabe, T. et al. (2007). Biicosahedral gold clusters $[Au_{25}(PPh_3)_{10}(SC_nH_{2n+1})_5Cl_2]^{2+}$ (n = 2–18): a stepping stone to cluster-assembled materials. *J. Phys. Chem. C* 111 (22): 7845–7847.

127 Higaki, T., Liu, C., Zhou, M. et al. (2017). Tailoring the structure of 58-electron gold nanoclusters: $Au_{103}S_2(S\text{-}Nap)_{41}$ and its implications. *J. Am. Chem. Soc.* 139 (29): 9994–10001.

128 Chen, Y., Zeng, C., Liu, C. et al. (2015). Crystal structure of barrel-shaped chiral $Au_{130}(p\text{-}MBT)_{50}$ nanocluster. *J. Am. Chem. Soc.* 137 (32): 10076–10079.

129 Zeng, C., Chen, Y., Kirschbaum, K. et al. (2015). Structural patterns at all scales in a nonmetallic chiral $Au_{133}(SR)_{52}$ nanoparticle. *Sci. Adv.* 1 (2): e1500045.

130 Zeng, C., Chen, Y., Kirschbaum, K. et al. (2016). Emergence of hierarchical structural complexities in nanoparticles and their assembly. *Science* 354 (6319): 1580–1584.

131 Gan, Z., Chen, J., Liao, L. et al. (2018). Surface single-atom tailoring of a gold nanoparticle. *J. Phys. Chem. Lett.* 9 (1): 204–208.

132 Jones, T.C., Sementa, L., Stener, M. et al. (2017). $Au_{21}S(SAdm)_{15}$: crystal structure, mass spectrometry, optical spectroscopy, and first-principles theoretical analysis. *J. Phys. Chem. C* 121 (20): 10865–10869.

133 Crasto, D., Malola, S., Brosofsky, G. et al. (2014). Single crystal XRD structure and theoretical analysis of the chiral $Au_{30}S(S\text{-}t\text{-}Bu)_{18}$ cluster. *J. Am. Chem. Soc.* 136 (13): 5000–5005.

134 Gan, Z., Xia, N., Yan, N. et al. (2021). Compression-driven Internanocluster reaction for synthesis of unconventional gold nanoclusters. *Angew. Chem. Int. Ed.* 60 (22): 12253–12257.

135 Gan, Z., Liu, Y., Wang, L. et al. (2020). Distance makes a difference in crystalline photoluminescence. *Nat. Commun.* 11 (1): 5572.

136 Zhou, M., Higaki, T., Li, Y. et al. (2019). Three-stage evolution from nonscalable to scalable optical properties of thiolate-protected gold nanoclusters. *J. Am. Chem. Soc.* 141 (50): 19754–19764.

137 Yu, Y., Luo, Z., Chevrier, D.M. et al. (2014). Identification of a highly luminescent $Au_{22}(SG)_{18}$ nanocluster. *J. Am. Chem. Soc.* 136 (4): 1246–1249.

138 Li, Q., Zeman, C.J., Schatz, G.C. et al. (2021). Source of bright near-infrared luminescence in gold nanoclusters. *ACS Nano* 15 (10): 16095–16105.

139 Han, X.-S., Luan, X., Su, H.-F. et al. (2020). Structure determination of alkynyl-protected gold nanocluster $Au_{22}(^{t}BuC{\equiv}C)_{18}$ and its thermochromic luminescence. *Angew. Chem. Int. Ed.* 59 (6): 2309–2312.

140 Link, S., Beeby, A., FitzGerald, S. et al. (2002). Visible to infrared luminescence from a 28-atom gold cluster. *J. Phys. Chem. B* 106 (13): 3410–3415.

141 Devadas, M.S., Kim, J., Sinn, E. et al. (2010). Unique ultrafast visible luminescence in monolayer-protected Au_{25} clusters. *J. Phys. Chem. C* 114 (51): 22417–22423.

142 Zhou, M. and Song, Y. (2021). Origins of visible and near-infrared emissions in $[Au_{25}(SR)_{18}]^-$ nanoclusters. *J. Phys. Chem. Lett.* 12 (5): 1514–1519.

143 Lee, D., Donkers, R.L., Wang, G. et al. (2004). Electrochemistry and optical absorbance and luminescence of molecule-like Au_{38} nanoparticles. *J. Am. Chem. Soc.* 126 (19): 6193–6199.

144 Miller, S.A., Womick, J.M., Parker, J.F. et al. (2009). Femtosecond relaxation dynamics of $Au_{25}L_{18}^-$ monolayer-protected clusters. *J. Phys. Chem. C* 113 (22): 9440–9444.

145 Wu, Z. and Jin, R. (2010). On the ligand's role in the fluorescence of gold nanoclusters. *Nano Lett.* 10 (7): 2568–2573.

146 Weerawardene, K.L.D.M. and Aikens, C.M. (2016). Theoretical insights into the origin of photoluminescence of $Au_{25}(SR)_{18}^-$ nanoparticles. *J. Am. Chem. Soc.* 138 (35): 11202–11210.

147 Yao, C., Xu, C.-Q., Park, I.-H. et al. (2020). Giant emission enhancement of solid-state gold nanoclusters by surface engineering. *Angew. Chem. Int. Ed.* 59 (21): 8270–8276.

148 Wang, S., Meng, X., Das, A. et al. (2014). A 200-fold quantum yield boost in the photoluminescence of silver-doped Ag_xAu_{25-x} nanoclusters: the 13th silver atom matters. *Angew. Chem. Int. Ed.* 53 (9): 2376–2380.

149 Luo, Z., Yuan, X., Yu, Y. et al. (2012). From aggregation-induced emission of Au(I)-thiolate complexes to ultrabright Au(0)@Au(I)-thiolate core-shell nanoclusters. *J. Am. Chem. Soc.* 134 (40): 16662–16670.

150 Wu, Z., Du, Y., Liu, J. et al. (2019). Aurophilic interactions in the self-assembly of gold nanoclusters into nanoribbons with enhanced luminescence. *Angew. Chem. Int. Ed.* 58 (24): 8139–8144.

151 Crawford, S.E., Hartmann, M.J., and Millstone, J.E. (2019). Surface chemistry-mediated near-infrared emission of small coinage metal nanoparticles. *Acc. Chem. Res.* 52 (3): 695–703.

152 Ito, S., Takano, S., and Tsukuda, T. (2019). Alkynyl-protected $Au_{22}(C{\equiv}CR)_{18}$ clusters featuring new interfacial motifs and R-dependent photoluminescence. *J. Phys. Chem. Lett.* 10 (21): 6892–6896.

153 Kang, X., Wang, S., Song, Y. et al. (2016). Bimetallic Au_2Cu_6 nanoclusters: strong luminescence induced by the aggregation of copper(I) complexes with gold(0) species. *Angew. Chem. Int. Ed.* 55 (11): 3611–3614.

154 Wu, Z., Yao, Q., Chai, O.J.H. et al. (2020). Unraveling the impact of gold(I)-thiolate motifs on the aggregation-induced emission of gold nanoclusters. *Angew. Chem. Int. Ed.* 59 (25): 9934–9939.

155 Chen, T., Yang, S., Chai, J. et al. (2017). Crystallization-induced emission enhancement: a novel fluorescent Au-Ag bimetallic nanocluster with precise atomic structure. *Sci. Adv.* 3 (8): e1700956.

156 Pyo, K., Thanthirige, V.D., Kwak, K. et al. (2015). Ultrabright luminescence from gold nanoclusters: rigidifying the Au(I)-thiolate shell. *J. Am. Chem. Soc.* 137 (25): 8244–8250.

157 Chen, Y., Zhou, M., Li, Q. et al. (2020). Isomerization-induced enhancement of luminescence in $Au_{28}(SR)_{20}$ nanoclusters. *Chem. Sci.* 11 (31): 8176–8183.

158 Li, Q., Zhou, M., So, W.Y. et al. (2019). A mono-cuboctahedral series of gold nanoclusters: photoluminescence origin, large enhancement, wide tunability, and structure-property correlation. *J. Am. Chem. Soc.* 141 (13): 5314–5325.

159 Dolamic, I., Knoppe, S., Dass, A. et al. (2012). First enantioseparation and circular dichroism spectra of Au_{38} clusters protected by achiral ligands. *Nat. Commun.* 3 (1): 798.

160 Li, Y., Higaki, T., Du, X. et al. (2020). Chirality and surface bonding correlation in atomically precise metal nanoclusters. *Adv. Mater.* 32 (41): 1905488.

161 Crespo, P., Litrán, R., Rojas, T.C. et al. (2004). Permanent magnetism, magnetic anisotropy, and hysteresis of thiol-capped gold nanoparticles. *Phys. Rev. Lett.* 93 (8): 087204.

162 Nealon, G.L., Donnio, B., Greget, R. et al. (2012). Magnetism in gold nanoparticles. *Nanoscale* 4 (17): 5244–5258.

163 Hori, H., Yamamoto, Y., Iwamoto, T. et al. (2004). Diameter dependence of ferromagnetic spin moment in Au nanocrystals. *Phys. Rev. B* 69 (17): 174411.

164 Agrachev, M., Ruzzi, M., Venzo, A. et al. (2019). Nuclear and electron magnetic resonance spectroscopies of atomically precise gold nanoclusters. *Acc. Chem. Res.* 52 (1): 44–52.

165 Zhu, M., Aikens, C.M., Hendrich, M.P. et al. (2009). Reversible switching of magnetism in thiolate-protected Au_{25} superatoms. *J. Am. Chem. Soc.* 131 (7): 2490–2492.

166 Higaki, T., Liu, C., Chen, Y. et al. (2017). Oxidation-induced transformation of eight-Electron gold nanoclusters: $[Au_{23}(SR)_{16}]^-$ to $[Au_{28}(SR)_{20}]^0$. *J. Phys. Chem. Lett.* 8 (4): 866–870.

167 Negishi, Y., Tsunoyama, H., Suzuki, M. et al. (2006). X-ray magnetic circular dichroism of size-selected, thiolated gold clusters. *J. Am. Chem. Soc.* 128 (37): 12034–12035.

168 Zeng, C., Weitz, A., Withers, G. et al. (2019). Controlling magnetism of $Au_{133}(TBBT)_{52}$ nanoclusters at single electron level and implication for nonmetal to metal transition. *Chem. Sci.* 10 (42): 9684–9691.

169 Qian, S., Wang, Z., Zuo, Z. et al. (2022). Engineering luminescent metal nanoclusters for sensing applications. *Coord. Chem. Rev.* 451: 214268.

170 Muhammed, M.A.H., Verma, P.K., Pal, S.K. et al. (2009). Bright, NIR-emitting Au_{23} from Au_{25}: characterization and applications including biolabeling. *Chem. Eur. J.* 15 (39): 10110–10120.

171 George, A., Shibu, E.S., Maliyekkal, S.M. et al. (2012). Luminescent, freestanding composite films of Au_{15} for specific metal ion sensing. *ACS Appl. Mater. Interfaces* 4 (2): 639–644.

172 Xie, J., Zheng, Y., and Ying, J.Y. (2010). Highly selective and ultrasensitive detection of Hg^{2+} based on fluorescence quenching of Au nanoclusters by Hg^{2+}-Au^+ interactions. *Chem. Commun.* 46 (6): 961–963.

173 Roy, S., Palui, G., and Banerjee, A. (2012). The as-prepared gold cluster-based fluorescent sensor for the selective detection of Asiii ions in aqueous solution. *Nanoscale* 4 (8): 2734–2740.

174 Chen, W.-Y., Lan, G.-Y., and Chang, H.-T. (2011). Use of fluorescent DNA-templated gold/silver nanoclusters for the detection of sulfide ions. *Anal. Chem.* 83 (24): 9450–9455.

175 Li, Z., Guo, S., and Lu, C. (2015). A highly selective fluorescent probe for sulfide ions based on aggregation of Cu nanocluster induced emission enhancement. *Analyst* 140 (8): 2719–2725.

176 Zhang, J., Chen, C., Xu, X. et al. (2013). Use of fluorescent gold nanoclusters for the construction of a NAND logic gate for nitrite. *Chem. Commun.* 49 (26): 2691–2693.

177 Liu, Y., Ai, K., Cheng, X. et al. (2010). Gold-nanocluster-based fluorescent sensors for highly sensitive and selective detection of cyanide in water. *Adv. Funct. Mater.* 20 (6): 951–956.

178 Tang, Y., Xu, J., Xiong, C. et al. (2019). Enhanced electrochemiluminescence of gold nanoclusters via silver doping and their application for ultrasensitive detection of dopamine. *Analyst* 144 (8): 2643–2648.

179 Yuan, X., Luo, Z., Yu, Y. et al. (2013). Luminescent noble metal nanoclusters as an emerging optical probe for sensor development. *Chem. Asian J.* 8 (5): 858–871.

180 Yu, H., Rao, B., Jiang, W. et al. (2019). The photoluminescent metal nanoclusters with atomic precision. *Coord. Chem. Rev.* 378: 595–617.

181 Tao, Y., Li, M., Ren, J. et al. (2015). Metal nanoclusters: novel probes for diagnostic and therapeutic applications. *Chem. Soc. Rev.* 44 (23): 8636–8663.

182 Zhang, Y., Zhang, C., Xu, C. et al. (2019). Ultrasmall Au nanoclusters for biomedical and biosensing applications: a mini-review. *Talanta* 200: 432–442.

183 Zuber, G., Weiss, E., and Chiper, M. (2019). Biocompatible gold nanoclusters: synthetic strategies and biomedical prospects. *Nanotechnol.* 30 (35): 352001.

184 Zhang, X., Chen, J., Luo, Z. et al. (2014). Enhanced tumor accumulation of sub-2 nm gold nanoclusters for cancer radiation therapy. *Adv. Healthc. Mater.* 3 (1): 133–141.

185 Welsher, K., Liu, Z., Sherlock, S.P. et al. (2009). A route to brightly fluorescent carbon nanotubes for near-infrared imaging in mice. *Nature Nanotech.* 4 (11): 773–780.

186 Hong, G., Robinson, J.T., Zhang, Y. et al. (2012). In *vivo* fluorescence imaging with Ag_2S quantum dots in the second near-infrared region. *Angew. Chem. Int. Ed.* 51 (39): 9818–9821.

187 Hong, G., Zou, Y., Antaris, A.L. et al. (2014). Ultrafast fluorescence imaging in *vivo* with conjugated polymer fluorophores in the second near-infrared window. *Nat. Commun.* 5 (1): 4206.

188 Yang, Y., Wang, S., Chen, S. et al. (2018). Switching the subcellular organelle targeting of atomically precise gold nanoclusters by modifying the capping ligand. *Chem. Commun.* 54 (66): 9222–9225.

189 Higaki, T., Li, Y., Zhao, S. et al. (2019). Atomically tailored gold nanoclusters for catalytic application. *Angew. Chem. Int. Ed.* 58 (25): 8291–8302.

190 Yan, J., Teo, B.K., and Zheng, N. (2018). Surface chemistry of atomically precise coinage-metal nanoclusters: from structural control to surface reactivity and catalysis. *Acc. Chem. Res.* 51 (12): 3084–3093.

191 Jin, R., Li, G., Sharma, S. et al. (2021). Toward active-site tailoring in heterogeneous catalysis by atomically precise metal nanoclusters with crystallographic structures. *Chem. Rev.* 121 (2): 567–648.

192 Cai, X., Hu, W., Xu, S. et al. (2020). Structural relaxation enabled by internal vacancy available in a 24-atom gold cluster reinforces catalytic reactivity. *J. Am. Chem. Soc.* 142 (9): 4141–4153.

193 Cai, X., Saranya, G., Shen, K. et al. (2019). Reversible switching of catalytic activity by shuttling an atom into and out of gold nanoclusters. *Angew. Chem. Int. Ed.* 58 (29): 9964–9968.

194 Li, G., Abroshan, H., Liu, C. et al. (2016). Tailoring the electronic and catalytic properties of Au_{25} nanoclusters *via* ligand engineering. *ACS Nano* 10 (8): 7998–8005.

4

Structural Design of Thiolate-Protected Gold Nanoclusters

Pengye Liu, Wenhua Han, and Wen Wu Xu

Department of Physics, School of Physical Science and Technology, Ningbo University, Ningbo, 315211, China

4.1 Introduction

In 2007, the breakthrough in the X-ray structure determination of $Au_{102}(p\text{-MBA})_{44}$ (p-MBA = p-mercaptobenzoic acid) nanocluster, which comprises 102 gold atoms and 44 p-MBAs, greatly promoted the theoretical exploration of new methods to understand and design the atomic structures of thiolate-protected gold nanoclusters [1]. In 2008, Häkkinen and coauthors reported the first designed structure of $Au_{25}(SR)_{18}^-$ nanocluster confirmed by two different experimental groups in the same year [2–4]. The same group also reported the atomic structures of $Au_{144}(SR)_{60}$ in 2009 and $Au_{40}(SR)_{24}$ in 2012, with which the simulated optical spectra can well reproduce the experimental measurements [5–6]. The designed structure of $Au_{144}(SR)_{60}$ was not confirmed until 2020 [7]. Pei and Zeng et al. have designed many atomic structures of thiolate-protected gold nanoclusters, including $Au_{20}(SR)_{16}$, $Au_{22}(SR)_{18}$, $Au_{24}(SR)_{20}$, $Au_{28}(SR)_{20}$, $Au_{29}(SR)_{19}$, $Au_{38}(SR)_{24}$, $Au_{44}(SR)_{28}$, $Au_{68}(SR)_{36}$, $Au_{76}(SR)_{44}$, $Au_{145-3n}(SR)_{60-2n}$ (n = 1–8), etc. [8–16]. It is noteworthy that the structures of $Au_{38}(SR)_{24}$, $Au_{44}(SR)_{28}$, and $Au_{29}(SR)_{19}$ were confirmed by the laboratories [17–19]. Recently, Xiong and Pei designed a lot of atomic structures of thiolate-protected cuboid gold nanoclusters with very large sizes such as $Au_{200}(SR)_{80}$, $Au_{232}(SR)_{88}$, $Au_{264}(SR)_{96}$, $Au_{296}(SR)_{104}$, and $Au_{328}(SR)_{112}$ [20]. Jiang et al. predicted several superatom nanoclusters, i.e. $Au_{12}(SR)_9^{2+}$, $Au_{15}(SR)_{13}$, $Au_{44}(SR)_{28}^{2-}$, etc. [21–23]. Xu et al. theoretically presented several structures of gold nanoclusters, i.e. $Au_{22}(SR)_{17}^-$, $Au_{28+4n}(SR)_{20+2n}$ (n = 0–6), $Au_{49}(PR_3)_{10}(SR)_{15}Cl_2$, $Au_{60+8n}(SR)_{36+4n}$ (n = 0–2), $Au_{68}(SR)_{32}$, $Au_{68}(SR)_{34}$, etc. [24–29]. In addition, some thiolate-protected gold nanoclusters such as $Au_{18}(SR)_{14}$, $Au_{25}(SR)_{18}^-$, $Au_{67}(SR)_{35}^{2-}$, $Au_{130}(SR)_{50}$, $Au_{187}(SR)_{68}$, were reported by other research groups [30–34].

Among these theoretically predicted structures, some were designed based on "divide and protect" rule [35, 36], including $Au_{20}(SR)_{16}$, $Au_{22}(SR)_{18}$, $Au_{24}(SR)_{20}$, $Au_{38}(SR)_{24}$, $Au_{68}(SR)_{32}$, $Au_{68}(SR)_{34}$, etc. Some structural predictions hinged on recognition of the common patterns in structural evolution, i.e. $Au_{28}(SR)_{20}$, $Au_{29}(SR)_{19}$, $Au_{44}(SR)_{28}$, $Au_{68}(SR)_{36}$, $Au_{60+8n}(SR)_{36+4n}$ (n = 0–2), etc. Some were designed based on the Grand Unified Model (GUM), i.e. $Au_{28+4n}(SR)_{20+2n}$ (n = 0–6), $Au_{49}(PR_3)_{10}(SR)_{15}Cl_2$, etc. [37, 38]. In addition, a new ligand-binding strategy for predicting the structures of thiolate-protected gold nanoclusters has been developed via redistribution of the Au—S "staple" motifs on the well-determined Au core structures. Employing this strategy, a new

Atomically Precise Nanochemistry, First Edition. Edited by Rongchao Jin and De-en Jiang.

atomic structure of $Au_{22}(SR)_{17}^{-}$ was successfully predicted by redistributing the Au—S "staple" motifs on the Au_{10} core of crystallized $Au_{21}(SR)_{15}$ [29]. In this chapter, four different structural design methods are summarized (see Sections 4.2–4.5). We also provide some future perspectives on the structural design of nanoclusters.

4.2 Structural Design Based on "Divide and Protect" Rule

4.2.1 A Brief Introduction of the Idea

The generic rule states that an thiolate-protected gold nanocluster $Au_m(SR)_n$ can be divided into several groups as illustrated by $[Au]_{a+a'}[Au(SR)_2]_b[Au_2(SR)_3]_c$, where a, a', b, and c are integers [35, 36]. Here, $[Au]_{a+a'}$ represents the gold core, which satisfies the constraint condition that the number of "surface" Au atom (a') in the gold core equals the sum of end points of the exterior motifs ($2b+2c$); i.e. each surface Au atom of the gold core is protected by one end point of the staple motif. Hence, the parameters a, a', b, and c for $Au_m(SR)_n$ must satisfy $a+a'+b+2c = m$, $2b+3c = n$, and $a' = 2b+2c$.

4.2.2 Atomic Structure of $Au_{68}(SH)_{32}$

Kornberg and coworkers reported the atomic structure of 68 Au atoms in Au_{68}(3-MBA)$_{32}$ (3-MBA = 3-mercaptobenzoic acid) via the powerful single-particle transmission electron microscopy (SP-TEM) [39]. SP-TEM can only yield the positions of gold atoms, while it cannot yield the positions of S atoms, suggesting that structural determination of the protection ligands of Au_{68}(3-MBA)$_{32}$ requires further theoretical design. The authors also suggested a theoretical structure of $Au_{68}(SH)_{32}$ containing unusual ring-like $Au_3(SR)_3$ structures based on the atomic positions of Au determined from SP-TEM. Therefore, it is necessary to present a series of new low-energy isomers of $Au_{68}(SH)_{32}$ in which the pattern of thiolate ligands on the gold core is determined according to the generic rule of "divide and protect" (D&P) rule.

According to D&P rule, four predicted formula can be obtained: (i) $[Au]_{24+24}[Au(SR)_2]_4[Au_2(SR)_3]_8$; (ii) $[Au]_{23+26}[Au(SR)_2]_7[Au_2(SR)_3]_6$; (iii) $[Au]_{22+28}[Au(SR)_2]_{10}[Au_2(SR)_3]_4$; and (iv) $[Au]_{21+30}[Au(SR)_2]_{13}[Au_2(SR)_3]_2$. Then, four new isomeric structures, Iso1 to Iso4, of $Au_{68}(SH)_{32}$ were predicted employing the atomic structure of 68 Au atoms and D&P rule (Figure 4.1a). Density functional theory (DFT) calculations show that the relative energy of four new isomers is 3–4 eV lower than that of the isomer reported. The structural analyses on cores of four isomers as well as the experimental core show that the overall structures of inner cores in four isomers are consistent with the experimental one (Figure 4.1b,c). Most importantly, the computed optical absorption spectra of four isomers illustrate that the locations of four prominent absorption peaks (a, b, c, and d) in computed optical absorption spectra of four designed isomers could reproduce the experimental results (Figure 4.1d). The positions of the major absorption peaks of these five isomers are more or less similar to one another as they have similar core structures [24].

4.2.3 Atomic Structure of $Au_{68}(SH)_{34}$

Employing a powerful matrix assisted laser desorption ionization time-of-flight (MALDI-TOF) mass spectrometry (MS) technique, Dass successfully assigned the 14 kDa species with the molecular formula $Au_{68}(SR)_{34}$ containing 34e magic number of valence electrons [40]. However, it was formidable to purify the $Au_{68}(SR)_{34}$ nanocluster experimentally due to the presence of some other

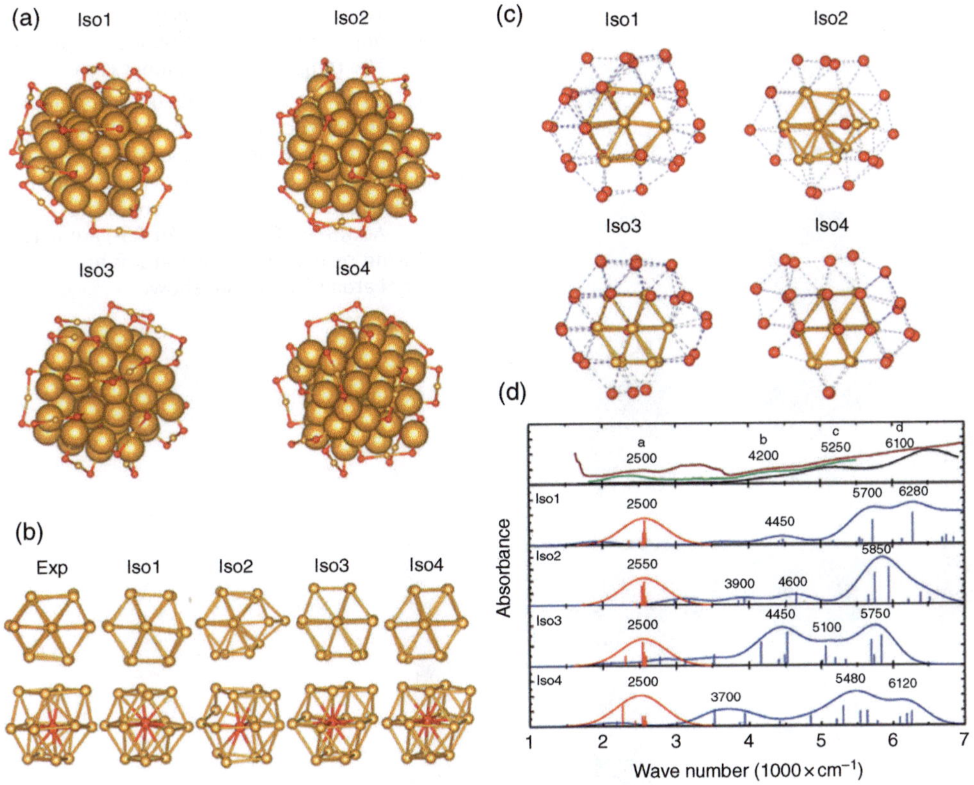

Figure 4.1 (a) Optimized structures of the four isomers Iso1-Iso4 of $Au_{68}(SH)_{32}$. The Au, and S atoms are in golden and red, respectively. The H atoms are not shown. (b) Two orthogonal views of the 15 Au atoms in the core of reported isomer from TEM experiment of Au_{68} and in the cores of Iso1-Iso4. The Au atoms are in golden and red. (c) The face-centered cubic frameworks of Au_{68} in Iso1-Iso4. The 15 Au atoms in the core are in golden, and other Au atoms are in red. (d) Optical absorption spectra of $Au_{68}(SH)_{32}$. *Source:* Reproduced with permission from ref. 24. Copyright 2015 American Association for the Advancement of Science.

lower mass peaks such as $Au_{68}(SR)_{33}$, $Au_{68}(SR)_{32}$, $Au_{68}(SR)_{30}$, and $Au_{64}(SR)_{30}$ in MS, suggesting that the structure of $Au_{68}(SR)_{34}$ nanocluster needs to be theoretically designed via the D&P rule.

Since having the same 34e magic number of valence electrons and similar molecular weight between $Au_{67}(SR)_{35}{}^{2-}$ and $Au_{68}(SR)_{34}$ [32], the joined experimental-theoretical investigations on $Au_{67}(SR)_{35}{}^{2-}$ nanocluster provide us a slice of clues to investigate the structure of $Au_{68}(SR)_{34}$ nanocluster on the theoretical side. Following the D&P concept, the predicted formulations of $Au_{68}(SR)_{34}$ are

i) $[Au]_{17+34}[Au(SR)_2]_{17}[Au_2(SR)_3]_0$
ii) $[Au]_{18+32}[Au(SR)_2]_{14}[Au_2(SR)_3]_2$
iii) $[Au]_{19+30}[Au(SR)_2]_{11}[Au_2(SR)_3]_4$
iv) $[Au]_{20+28}[Au(SR)_2]_8[Au_2(SR)_3]_6$
v) $[Au]_{21+26}[Au(SR)_2]_5[Au_2(SR)_3]_8$
vi) $[Au]_{22+24}[Au(SR)_2]_2[Au_2(SR)_3]_{10}$

According to the predicted formulations and the gold core structure of $Au_{67}(SR)_{35}{}^{2-}$ nanocluster, the four low-lying energy isomers Iso1–Iso4 with high-symmetry inner cores were identified and optimized using the DFT method (Figure 4.2). The energies of Iso1, Iso3, and Iso4 are 1.16, 0.82, and 2.74 eV higher than that of Iso2, respectively, indicating that Iso2 is the most stable among these four isomers. In addition, Iso2 has the largest HOMO-LUMO gap of 0.74 eV, which is consistent with that of $Au_{67}(SR)_{35}{}^{2-}$. These results illustrate that Iso2 is the likely candidate structure for $Au_{68}(SR)_{34}$.

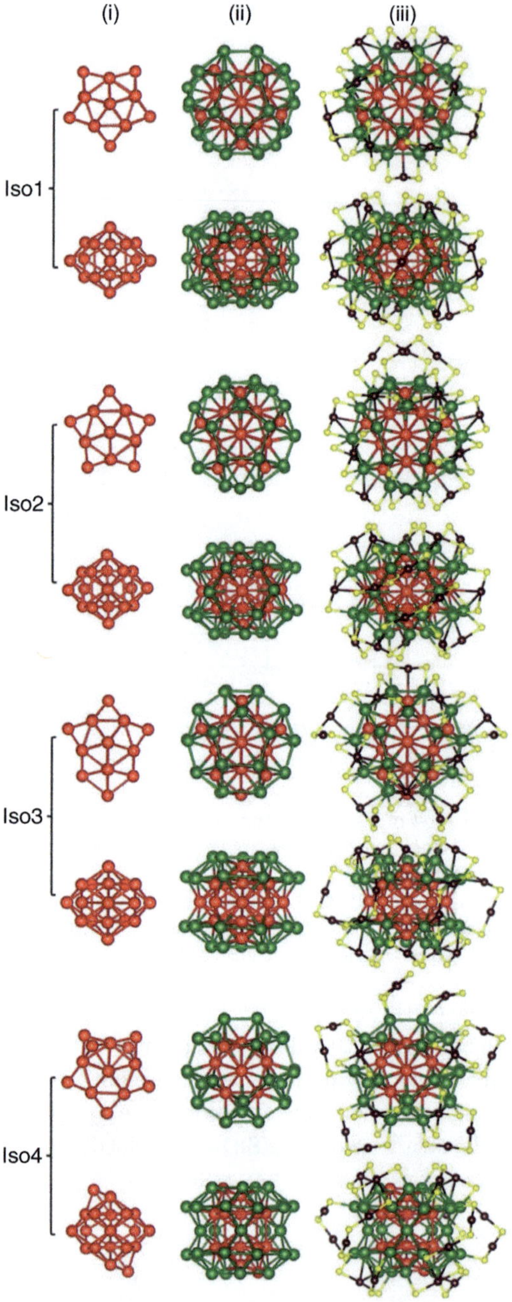

Figure 4.2 Two orthogonal views of the DFT-optimized atomic structure of the neutral $Au_{68}(SR)_{34}$ nanocluster. The $[Au]_a$ (a = 17–20) inner core (column i), $[Au]_{a+a'}$ (a + a′ = 51–48) core (column ii), and the complete structures (column iii) of neutral Iso1-Iso4 are presented. *Source:* Adapted with permission from [32]. © 2015 American Chemical Society. The Au atoms in inner core, shell, and staples (-RS-Au-RS- or -RS-Au-RS-Au-RS-) are in red, olive, and wine, respectively, and S atoms are in yellow. The H atoms (R) are not shown.

4.3 Structural Design via Redistributing the "Staple" Motifs on the Known Au Core Structures

4.3.1 A Brief Introduction of the Idea

Recent experiments have revealed that two thiolate-protected gold nanoclusters $Au_{28}(SR)_{20}$, i.e. $Au_{28}(TBBT)_{20}$ (TBBT = 4-tert-butylbenzenethiolate) and $Au_{28}(S\text{-}c\text{-}C_6H_{11})_{20}$ ($S\text{-}c\text{-}C_6H_{11}$ = cyclohexanethiol) [41, 42], can adopt the same Au_{14} core structure. As presented in Figure 4.3a, the addition of

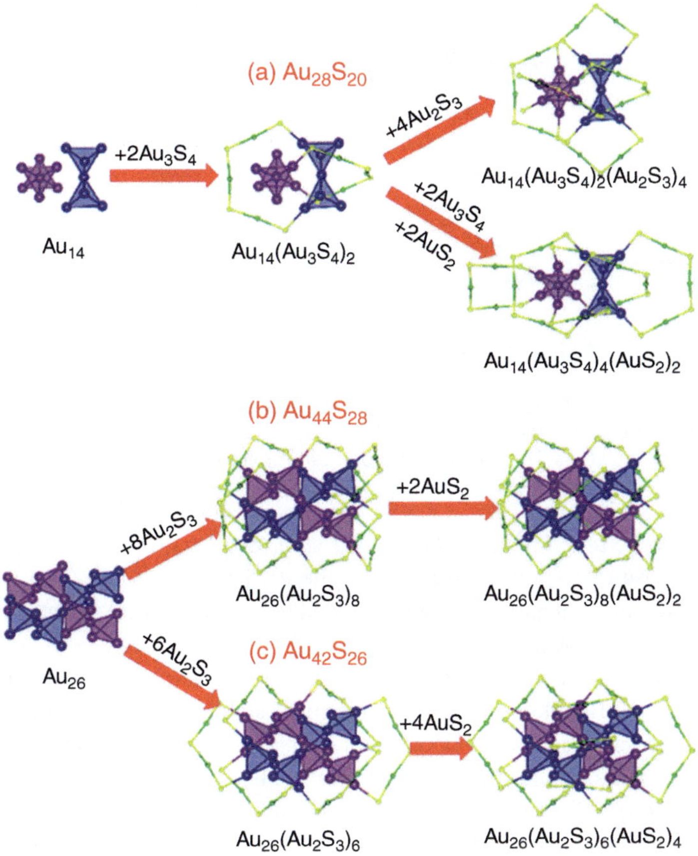

Figure 4.3 Structural formation of two $Au_{28}(SR)_{20}$ (a), $Au_{44}(SR)_{28}$ (b), and $Au_{42}(SR)_{26}$ (c) nanoclusters. *Source:* Reproduced by permission of The Royal Society of Chemistry. Au atoms are presented in wine, blue, and dark green, respectively. S is presented in yellow. The R groups are omitted for clarity.

$[Au_3(SR)_4]_2$ "staple" motif to the Au_{14} core yields the $Au_{14}[Au_3(SR)_4]_2$ structure. Further adding $[Au_2(SR)_3]_4$ and $[Au(SR)_2]_2[Au_3(SR)_4]_2$ "staple" motifs to $Au_{14}[Au_3(SR)_4]_2$, respectively, results in two $Au_{28}(SR)_{20}$ nanoclusters. Hence, both $Au_{28}(SR)_{20}$ have the same Au_{14} core but different distribution of the ligands. Similar behavior can be seen in $Au_{44}(TBBT)_{28}$ and $Au_{42}(TBBT)_{26}$ nanoclusters [18, 43]. Both of them can also adopt the same double-helical Au_{26} core but with different distribution of the ligands, as shown in Figure 4.3b,c. These different ligand binding modes indicate that different thiolate-protected gold nanoclusters can adopt the same core with the same size and morphology but different distribution of "staple" motifs, motiving us to develop a new ligand-binding strategy to design theoretical structure of gold nanoclusters by redistributing the "staple" motifs on the well-determined Au core structures. Such a strategy enables us to construct a highly stable structure of thiolate-protected gold nanoclusters without invoking the impractical global-minimum search.

4.3.2 Atomic Structure of $Au_{22}(SH)_{17}^-$

Employing the new ligand-binding strategy mentioned above, a new atomic structure of chiral thiolate-protected gold nanocluster $Au_{22}(SR)_{17}^-$ is predicted by redistributing "staple" motifs on the well-known Au_{10} core from previously crystallized $Au_{21}(SR)_{15}$ [29, 44]. As shown in Figure 4.4, in the first step, two $[Au_4(SR)_5]$ "staple" motifs were bound to the tetrahedral Au_4 via four Au—S

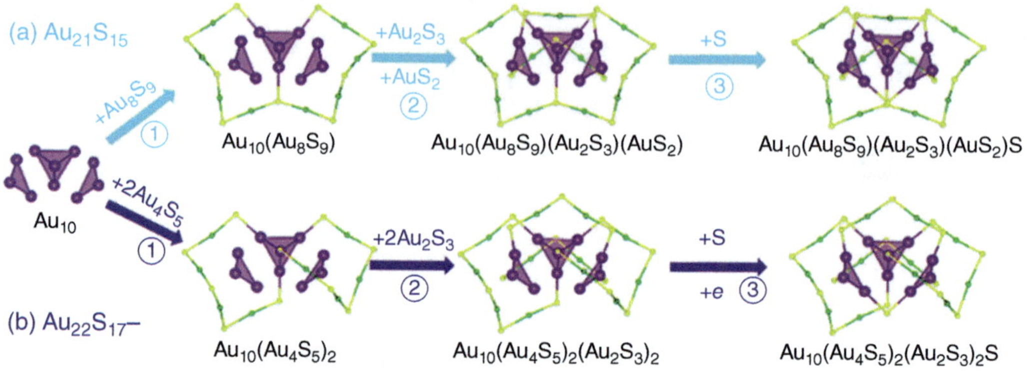

Figure 4.4 Structural formations of $Au_{21}(SR)_{15}$ (a) and $Au_{22}(SR)_{17}^-$ (b). *Source:* Reproduced by permission of The Royal Society of Chemistry. Au atoms are presented in wine and dark green, respectively. S is presented in yellow. ①, ②, and ③ represent the formation steps. The R groups are omitted for clarity.

bonds to form the $Au_4[Au_4(SR)_5]_2$ structure. Then, in the second step, two $[Au_2(SR)_3]$ "staple" motifs are further bound to the $Au_4[Au_4(SR)_5]_2$ structure to form the $Au_4[Au_2(SR)_3]_2[Au_4(SR)_5]_2$ structure. Finally, a bridging SR ligand and a negative charge are added to form the entire structure of $Au_{22}(SR)_{17}^-$.

DFT calculations show that the predicted $Au_{22}(SR)_{17}^-$ has a HOMO-LUMO gap of 1.88 eV, larger than that of $Au_{21}(SH)_{15}$, suggesting the likelihood of higher chemical stability. The computed optical absorption spectrum and fluorescent emission spectrum of predicted $Au_{22}(SR)_{17}^-$ can well reproduce the experimental measurements, suggesting that this structure is very likely the realistic structure for the synthesized $Au_{22}(SR)_{17}^-$. The new structure of $Au_{22}(SR)_{17}^-$ not only offers new chemical insights into the interaction between gold core and thiolate ligands, and validates the new ligand-binding strategy to predict stable thiolate-protected gold nanocluster, but also will motivate future experimental crystallization of the $Au_{22}(SR)_{17}^-$ cluster to confirm our prediction.

4.3.3 Atomic Structures of $Au_{27}(SH)_{20}^-$, $Au_{32}(SR)_{21}^-$, $Au_{34}(SR)_{23}^-$, and $Au_{36}(SR)_{25}^-$

Furthermore, employing the same method, four new atomic structures of thiolate-protected gold nanoclusters $Au_{27}(SR)_{20}^-$, $Au_{32}(SR)_{21}^-$, $Au_{34}(SR)_{23}^-$, and $Au_{36}(SR)_{25}^-$ were predicted via the redistribution of the Au—S "staple" motifs on the known face-centered cubic (FCC) Au_{13} core from experimentally determined $Au_{23}(SR)_{16}^-$ and FCC Au_{20} core from crystallized $Au_{30}(SR)_{18}$, respectively [45–47]. As shown in Figure 4.5, when redistributing the Au—S "staple" motifs, i.e. two $[Au_4(SR)_5]$, two $[Au_2(SR)_3]$, and two $[Au(SR)_2]$, on the FCC Au_{13} core, a new atomic structure of $Au_{27}(SR)_{20}^-$ can be obtained (Figure 4.5c). Then, we focused on another crystallized $Au_{30}(SR)_{18}$ nanocluster with a FCC Au_{20} core (Figure 4.6), whose structure has one more bitetrahedron Au_7 than the FCC Au_{13} core of $Au_{23}(SR)_{16}^-$ nanocluster. By redistributing the Au—S "staple" motifs on the FCC Au_{20} core of the $Au_{30}(SR)_{18}$ nanocluster, three new atomic structures of thiolate-protected gold nanoclusters $Au_{32}(SR)_{21}^-$, $Au_{34}(SR)_{23}^-$, and $Au_{36}(SR)_{25}^-$ with 12*e* valence electrons can be obtained.

It can be seen that $Au_{23}(SR)_{16}^-$, $Au_{25}(SR)_{18}^-$, and $Au_{27}(SR)_{20}^-$ can be written as $Au_{23}(SR)_{16}^- + 2[Au(SR)] \rightarrow Au_{25}(SR)_{18}^- + 2[Au(SR)] \rightarrow Au_{27}(SR)_{20}^-$, namely the growth pattern underlying this sequence of three nanoclusters can be viewed as sequential addition of two [Au(SR)] on the ligands. The redistribution of the Au—S "staple" motifs on the FCC Au_{13} core is simply extending of Au—S "staple" motifs, i.e. the extension of $[Au_2(SR)_3]$ to $[Au_3(SR)_4]$ by adding a [Au(SR)], without changing the ligand-binding modes of Au—S "staple" motifs on the core. Actually the

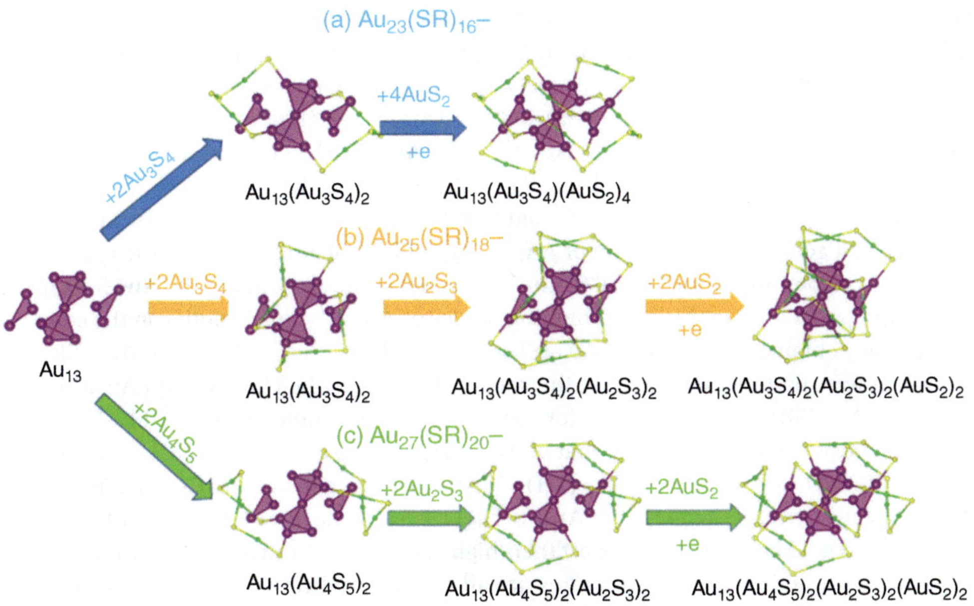

Figure 4.5 Structural formations of $Au_{23}(SR)_{16}^-$ (a), $Au_{25}(SR)_{18}^-$ (b), and $Au_{27}(SR)_{20}^-$ (c). *Source:* Reproduced by permission of The Royal Society of Chemistry. Au atoms are presented in wine and green, respectively. S is presented in yellow. The R groups were omitted for clarity.

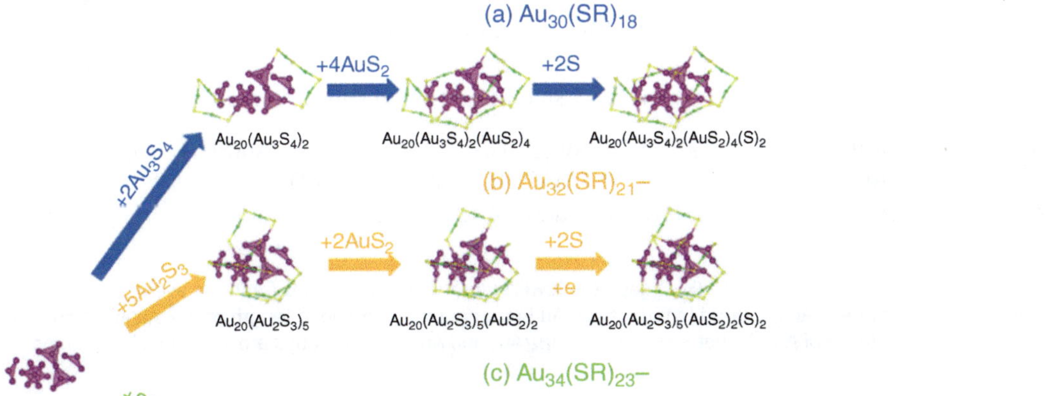

Figure 4.6 Structural formations of $Au_{30}(SR)_{18}$ (a), $Au_{32}(SR)_{21}^-$ (b), $Au_{34}(SR)_{23}^-$ (c), and $Au_{36}(SR)_{25}^-$ (d). *Source:* Reproduced by permission of The Royal Society of Chemistry. Au atoms are presented in wine and green, respectively. S is presented in yellow. The R groups were omitted for clarity.

ligand-binding modes of Au—S "staple" motifs on the same FCC Au_{13} core in three nanoclusters are completely different. For example, two $[Au_3(SR)_4]$ and two $[Au_4(SR)_5]$ were employed to bind the gold atoms in bitetrahedron Au_7, respectively, resulting in two completely different ligand-binding modes of $Au_{13}[Au_3(SR)_4]_2$ (Figure 4.5a) and $Au_{13}[Au_4(SR)_5]_2$ (Figure 4.5c). Two $[Au_3(SR)_4]$ were bound with gold atoms at both ends of Au_7 and two $[Au_4(SR)_5]$ were bound with gold atoms at each end of Au_7. Further binding four $[Au(SR)_2]$ with $Au_{13}[Au_3(SR)_4]_2$ can form the complete structure of $Au_{23}(SR)_{16}^-$ (Figure 4.5a), while binding two $[Au_2(SR)_3]$ and two $[Au(SR)_2]$ with $Au_{13}[Au_4(SR)_5]_2$ can generate the structure of $Au_{27}(SR)_{20}^-$ (Figure 4.5c). Therefore, the redistribution of the Au—S "staple" motifs on the FCC Au_{13} core is not simply extending the Au—S "staple" motifs but completely changing the ligand-binding modes of Au—S "staple" motifs on the core.

With these newly obtained structures of $Au_{27}(SR)_{20}^-$, $Au_{32}(SR)_{21}^-$, $Au_{34}(SR)_{23}^-$, and $Au_{36}(SR)_{25}^-$, DFT calculations were performed to describe their stabilities (Table 4.1). The results of $Au_{23}(SR)_{16}^-$, $Au_{25}(SR)_{18}^-$, and $Au_{30}(SR)_{18}$ were also given for comparison. The computed HOMO-LUMO gap of $Au_{27}(SH)_{20}^-$ is 1.58 eV, slightly lower than that of $Au_{23}(SH)_{16}^-$ (1.85 eV) and $Au_{25}(SH)_{18}^-$ (1.96 eV). The computed HOMO-LUMO gaps of $Au_{32}(SH)_{21}^-$, $Au_{34}(SH)_{23}^-$, and $Au_{36}(SH)_{25}^-$ are 169, 1.41, and 1.44 eV, respectively, larger than that of $Au_{30}(SH)_{18}$ (1.22 eV). The large HOMO-LUMO gaps of these four predicted gold nanoclusters suggest their high chemical stabilities. The minimum vibrational frequencies of these predicted structures are all positive, indicating they are at least local minimums on the potential energy surfaces.

Furthermore, the average Au—Au bond length or distance of the Au_{13} and Au_{20} cores were examined. It can be found that the values of d_1 (average bond length of Au—Au in triangular Au_3), d_2 (average bond length of Au—Au in tetrahedral Au_4), and d_3 (the distance of Au—Au between Au_3 and Au_4/Au_4 and Au_4) of Au_{13} core in predicted $Au_{27}(SH)_{20}^-$ are very close to those of $Au_{23}(SH)_{16}^-$ and $Au_{25}(SH)_{18}^-$, indicating that the structure of Au_{13} core does not change significantly after the redistribution of Au—S "staple" motifs. This behavior can also be observed in the $Au_{32}(SR)_{21}^-$, $Au_{34}(SR)_{23}^-$, and $Au_{36}(SR)_{25}^-$ with the same Au_{20} core. In addition, due to the same gold core structures, it is convenient to evaluate their relative stabilities. The substitution reaction can be written as

$Au_{23}(SH)_{16}^- + 2Au_4(SH)_5^- + 2Au_2(SH)_3^- \rightarrow Au_{27}(SH)_{20}^- + 2Au_3(SH)_4^- + 2Au(SH)_2^-$ (1),
$Au_{30}(SH)_{18} + 5Au_2(SH)_3^- \rightarrow Au_{32}(SH)_{21}^- + 2Au_3(SH)_4^- + 2Au(SH)_2^-$ (2),
$Au_{30}(SH)_{18} + 3Au_2(SH)_3^- \rightarrow Au_{34}(SH)_{23}^- + 2Au(SH)_2^-$ (3), and

Table 4.1 Computed HOMO-LUMO (H-L) gaps, the lowest vibrational frequencies *f*, and the average bond length or distance of Au—Au (d_1: bond length of Au—Au in triangular Au_3, d_2: bond length of Au—Au in tetrahedral Au_4, and d_3: the distance of Au—Au between Au_3 and Au_4/Au_4 and Au_4). The R groups are simplified by H atoms.

				Average bond length or distance of Au—Au		
		H-L/eV	***f*/cm^{-1}**	**d_1/Å**	**d_2/Å**	**d_3/Å**
Au_{13}	$Au_{23}(SH)_{16}^-$	1.85	8.62	2.73	2.78	3.07
	$Au_{25}(SH)_{18}^-$	1.96	7.56	2.71	2.80	3.03
	$Au_{27}(SH)_{20}^-$	1.58	7.28	2.74	2.79	3.07
Au_{20}	$Au_{30}(SH)_{18}$	1.22	8.8	2.75	2.80	3.01
	$Au_{32}(SH)_{21}^-$	1.69	8.08	2.81	2.82	3.01
	$Au_{34}(SH)_{23}^-$	1.41	9.44	2.72	2.82	3.02
	$Au_{36}(SH)_{25}^-$	1.44	5.7	2.74	2.81	3.06

$Au_{30}(SH)_{18} + 3Au_2(SH)_3^- + 2Au_4(SH)_5^- \rightarrow Au_{36}(SH)_{25}^- + 2Au_3(SH)_4^- + 2Au(SH)_2^-$ (4). The substitution energies for four reactions are 0.17, −1.02, −1.77, and − 2.17 eV, respectively, suggesting that the predicted $Au_{27}(SH)_{20}^-$ has close stability with the crystallized $Au_{23}(SH)_{16}^-$, and the predicted $Au_{32}(SH)_{21}^-$, $Au_{34}(SH)_{23}^-$, and $Au_{36}(SH)_{25}^-$ are more stable than the crystallized $Au_{30}(SH)_{18}$. These results further demonstrated the rationality of the predicted nanoclusters.

4.4 Structural Design via Structural Evolution

4.4.1 A Brief Introduction of the Idea

With the crystallized or predicted structures, the structural evolution of some ligand-protected gold nanoclusters can be identified via recognition of the common patterns, as shown in Figure 4.7. The theoretical structure of $Au_{44}(SR)_{28}$ was predicted on the basis of evolution rule $Au_{28}(SR)_{20}$ + $[Au_8(SR)_4] \rightarrow Au_{36}(SR)_{24}$ + $[Au_8(SR)_4] \rightarrow Au_{44}(SR)_{28}$, that is, the sequential addition of a common structural motif $[Au_8(SR)_4]$ [12]. Thanks to the structural determination of $Au_{44}(SR)_{28}$ and $Au_{52}(SR)_{32}$ [43, 48], the above one-direction growth pattern among $Au_{28}(SR)_{20}$, $Au_{36}(SR)_{24}$, $Au_{44}(SR)_{28}$, and $Au_{52}(SR)_{32}$ was experimentally confirmed, while three new nanoclusters $Au_{60}(SR)_{36}$, $Au_{68}(SR)_{40}$, and $Au_{76}(SR)_{44}$ were predicted [25]. The first 10-electron (10*e*) thiolate-protected Au cluster $Au_{29}(SR)_{19}$ was predicted based on the structural evolution from $Au_{28}(SR)_{20}$ to $Au_{29}(SR)_{19}$ and then to the $Au_{30}(SR)_{18}$ cluster via sequential addition of the triangular Au_3 unit in the Au core [16]. An unprecedented ~14 kDa core-mass $Au_{68}(SR)_{36}$ nanocluster is theoretically predicted from the two-dimensional growth based on $Au_{44}(SR)_{28}$ to $Au_{68}(SR)_{36}$ to the $Au_{92}(SR)_{44}$

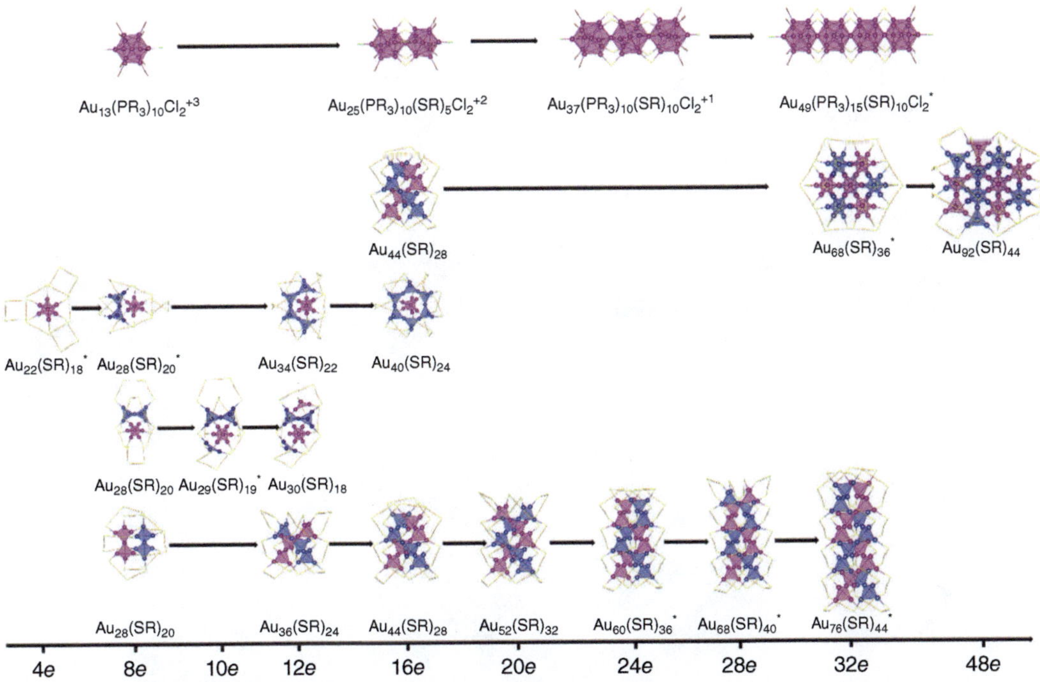

Figure 4.7 Structural evolutions of ligand-protected gold nanoclusters. *Source:* Adapted with permission from [38]. © 2018 American Chemical Society. * denotes the predicted structures.

cluster [14]. A new $Au_{28}(SR)_{20}$ isomer is found to fill the vacant position in the structural evolution among the sequence of $Au_{22}(SR)_{18}$, $Au_{28}(SR)_{20}$, $Au_{34}(SR)_{22}$, and $Au_{40}(SR)_{24}$ [15]. In addition, the structural evolution of $Au_{13}(PR_3)_{10}Cl_2^{+3}$, $Au_{25}(PR_3)_{10}(SR)_5Cl_2^{+2}$, and $Au_{37}(PR_3)_{10}(SR)_{10}Cl_2^{+1}$ [49–51], have been elucidated, and a longer $Au_{49}(PR_3)_{10}(SR)_{15}Cl_2$ cluster with high stability has been predicted based on the structural evolution [28].

4.4.2 Atomic Structures of $Au_{60}(SR)_{36}$, $Au_{68}(SR)_{40}$, and $Au_{76}(SR)_{44}$

Based on previously known (via X-ray crystallography) and/or theoretically predicted structures of a series of thiolate-protected gold nanoclusters, i.e. $Au_{20}(SR)_{16}$, $Au_{28}(SR)_{20}$, $Au_{36}(SR)_{24}$, $Au_{44}(SR)_{28}$, and $Au_{52}(SR)_{32}$ [8, 12, 41, 48, 52], the structural evolution from the starting cluster $Au_{20}(SR)_{16}$ via sequential addition of a $[Au_8(SR)_4]$ motif, i.e. $Au_{20}(SR)_{16} + [Au_8(SR)_4] \rightarrow Au_{28}(SR)_{20} + [Au_8(SR)_4] \rightarrow Au_{36}(SR)_{24} + [Au_8(SR)_4] \rightarrow Au_{44}(SR)_{28} + [Au_8(SR)_4] \rightarrow Au_{52}(SR)_{32}$ can be identified. Figure 4.8 illustrates the structural evolution of the face-centered-cubic (FCC)-type of Au kernels in these clusters via sequential addition of the "boat-like" Au_8 motif.

The unique growth-pattern rule derived among the series of clusters $Au_{20}(SR)_{16}$, $Au_{28}(SR)_{20}$, $Au_{36}(SR)_{24}$, $Au_{44}(SR)_{28}$, and $Au_{52}(SR)_{32}$ suggests possible existence of larger-sized clusters through

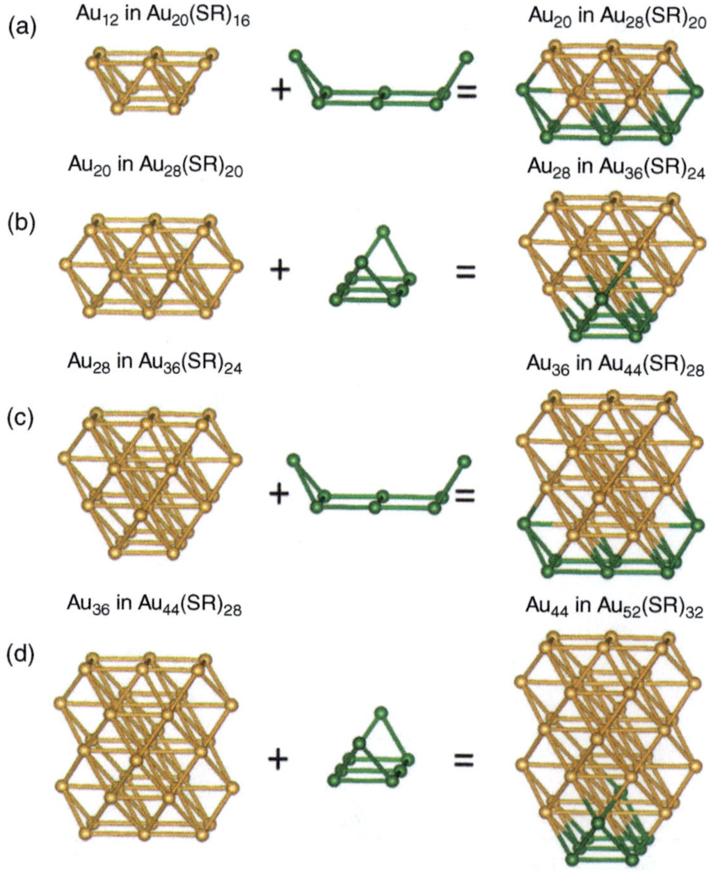

Figure 4.8 Au-kernel growth by adding the "boat-like" Au_8 motif (olive). (a) Au_{12} in $Au_{20}(SR)_{16}$ to Au_{20} in $Au_{28}(SR)_{20}$, (b) Au_{20} in $Au_{28}(SR)_{20}$ to Au_{28} in $Au_{36}(SR)_{24}$, (c) Au_{28} in $Au_{36}(SR)_{24}$ to Au_{36} in $Au_{44}(SR)_{28}$, (d) Au_{36} in $Au_{44}(SR)_{28}$ to Au_{44} in $Au_{52}(SR)_{32}$. *Source:* Reproduced by permission of The Royal Society of Chemistry.

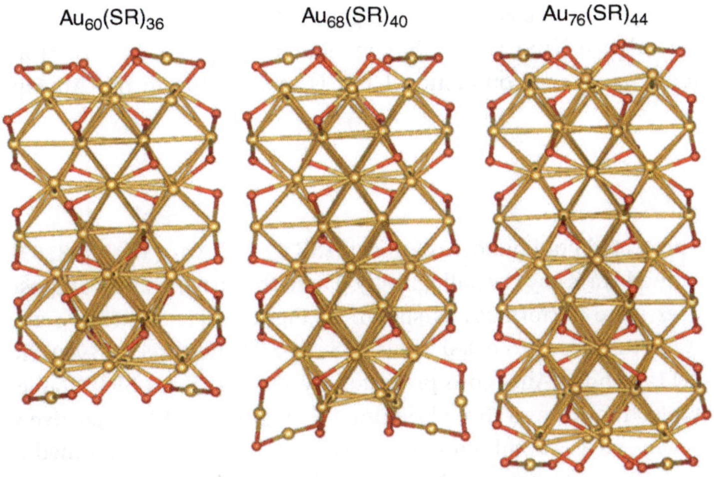

Figure 4.9 The optimized structures of $Au_{60}(SR)_{36}$, $Au_{68}(SR)_{40}$, and $Au_{76}(SR)_{44}$, where the methyl groups are omitted for clarity. *Source:* Reproduced by permission of The Royal Society of Chemistry. Au and S atoms are in gold and red, respectively.

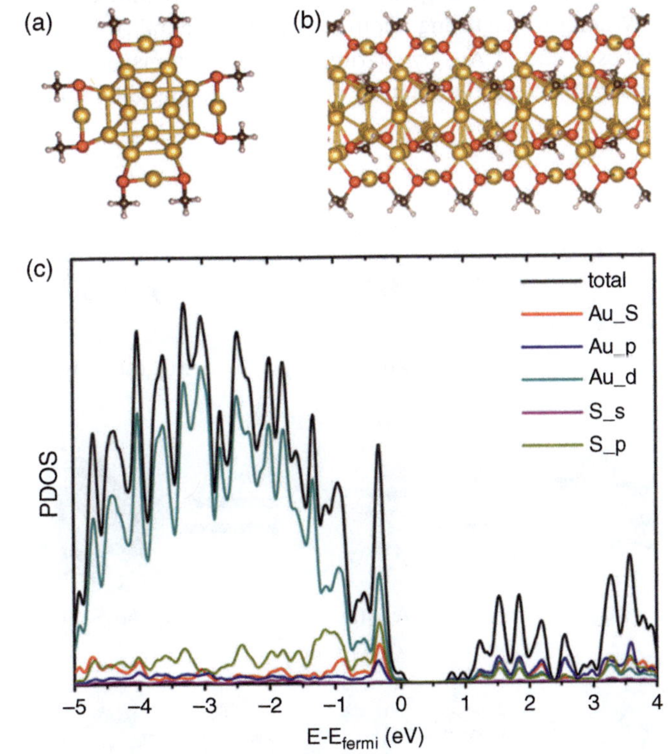

Figure 4.10 The proposed structure of thiolate-protected gold nanowire: (a) viewed along the wire; (b) side view. *Source:* Reproduced by permission of The Royal Society of Chemistry. Au, S, C, and H atoms are in gold, red, dark gray, and white, respectively. (c) Computed projected density of state (PDOS) of thiolate-protected gold nanowire. Au_s, Au_p, and Au_d denote the s, p, and d orbital of Au atoms, respective. S_s and S_p denote the s and p orbital of S atoms, respective.

continuously adding the motif $[Au_8(SR)_4]$, e.g. $Au_{52}(SR)_{32} + [Au_8(SR)_4] \rightarrow Au_{60}(SR)_{36} + [Au_8(SR)_4] \rightarrow Au_{68}(SR)_{40} + [Au_8(SR)_4] \rightarrow Au_{76}(SR)_{44}$, where the newly created $Au_{60}(SR)_{36}$, $Au_{68}(SR)_{40}$, and $Au_{76}(SR)_{44}$ all possess the FCC-type Au-kernels (Figure 4.9).

Lastly, if the growth-pattern rule is extended to the infinitely long nanowire limit by repeatedly adding the $[Au_8(SR)_4]$ units in one direction, the thiolate-protected gold nanowire (RS-AuNW) can be obtained, as shown in Figure 4.10a,b. Figure 4.10c shows the computed projected density of state (PDOS) of the RS-AuNW, which shows an electronic band gap of 0.78 eV, suggesting that the RS-AuNW is semiconducting. The valence band is mainly contributed from the Au(5d), S(3p) Au(6s), and Au(6p) atomic orbitals, while the conduction band is mainly due to the Au(6sp) atomic orbitals.

4.4.3 Atomic Structure of $Au_{58}(SR)_{30}$

First, we focused on two thiolate-protected gold nanoclusters $Au_{40}(SR)_{24}$ and $Au_{49}(SR)_{27}$ [48, 53], whose structures have been determined via X-ray crystallography. Then, the structural decomposition of two experimental structures were performed, as shown in Figure 4.11. $Au_{40}(SR)_{24}$ and $Au_{49}(SR)_{27}$ can be decomposed into a Au_{25} core protected by six [RS-Au-SR] and three [RS-Au-SR-Au-SR] "staple" motifs (Figure 4.11a) and a Au_{34} core protected by three [RS-Au-SR] "staple" motifs, six [RS-Au-SR-Au-SR] "staple" motifs, and three bridging SR (Figure 4.11b), respectively. The Au_{25} core in $Au_{40}(SR)_{24}$ can be viewed as a combination of a Au_7 composed of two tetrahedral Au_4 units fusing together by sharing a gold atom and a Kekulé-like Au_{18} composed of six tetrahedral Au_4 units fusing together by sharing six gold atoms (Figures 4.11a and 4.12). The Au_{34} core in $Au_{49}(SR)_{27}$ can be formed by further adding three tetrahedral Au_4 units on the Au_{25} core in $Au_{40}(SR)_{24}$, namely, fusing three tetrahedral Au_4 units with three unfused gold atoms in one end of Au_7 by sharing three gold atoms (Figures 4.11b and 4.12).

When further fusing another three tetrahedral Au_4 units with three unfused gold atoms of the opposite end of Au_7 by sharing three gold atoms, a new Au_{43} core can be formed (Figures 4.10c and 4.11). Next, the "staple" motifs or bridging SR should be bonded to the Au_{43} core to form a

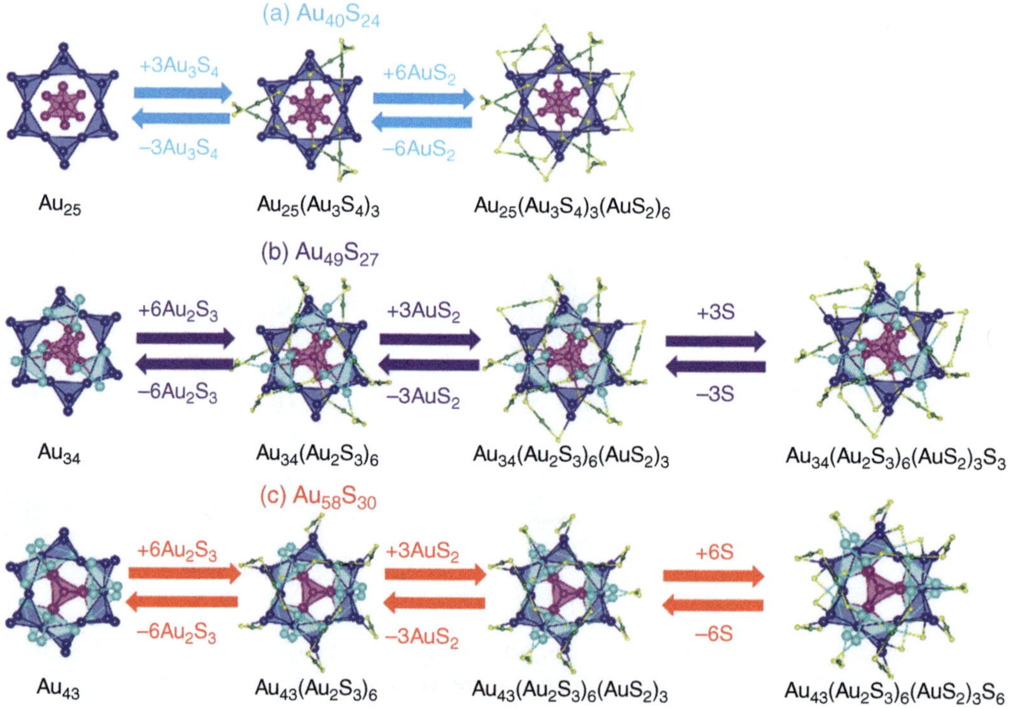

Figure 4.11 The Structural formation or decomposition of $Au_{40}(SR)_{24}$ (a), $Au_{49}(SR)_{27}$ (b), and $Au_{58}(SR)_{30}$ (c). *Source:* Adapted with permission from [54]. © 2020 American Chemical Society. Au atoms are presented in wine, blue, cyan, and dark green, respectively. S is presented in yellow. The R groups are omitted for clarity.

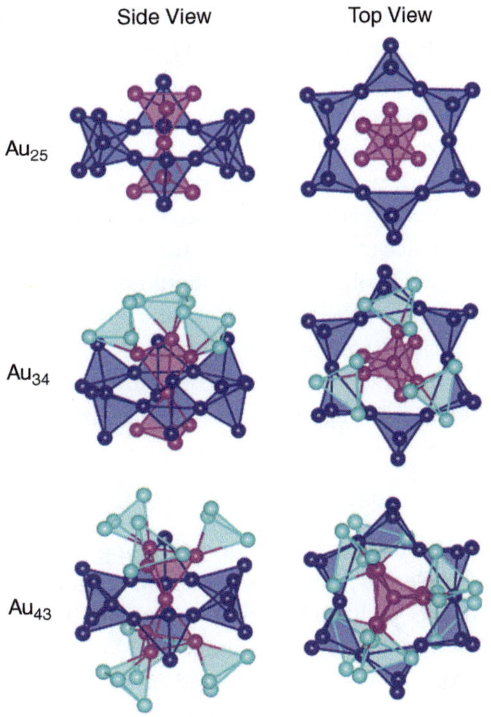

Figure 4.12 The side and top views of Au_{25}, Au_{34}, and Au_{34} cores. *Source:* Adapted with permission from [54]. © 2020 American Chemical Society. Au atoms are presented in wine, blue, and cyan, respectively.

complete structure. Due to the structural similarity between Au_{34} and Au_{43} cores, the protection ligands on the Au_{43} core can be conveniently obtained by following the bonding mode of protection ligand with the Au_{34} core in $Au_{49}(SR)_{27}$. As shown in Figure 4.10c, six $[Au_2(SR)_3]$ "staple" motifs were bonded to Au_{43} core, as the first step. Next, three [SR-Au-SR] "staple" motifs were further bonded to the core, as the second step. In the third step, six bridging SR ligands are added to form the entire structure of $Au_{58}(SR)_{30}$. With the newly predicted structure of $Au_{58}(SR)_{30}$, a new evolution rule, i.e. $Au_{40}(SR)_{24} + [Au_9(SR)_3] \rightarrow Au_{49}(SR)_{27} + [Au_9(SR)_3] \rightarrow Au_{58}(SR)_{30}$, can be unrevealed [54].

With the obtained $Au_{58}(SR)_{30}$ structure, DFT calculations show that the computed HOMO-LUMO gaps of the predicted $Au_{40}(SH)_{24}$, $Au_{49}(SR)_{27}$, and $Au_{58}(SH)_{30}$ are 1.58, 1.03, and 0.84 eV, respectively, decreasing with the increment of size. The all-positive harmonic vibrational frequencies indicate that the predicted $Au_{58}(SH)_{30}$ structure is local minima on the potential energy surface. Furthermore, the conversion in equation $2Au_{49}(SH)_{27} \rightarrow Au_{40}(SH)_{24} + Au_{58}(SH)_{30}$ is exothermic by 0.64 eV, suggesting close stability for the predicted $Au_{58}(SH)_{30}$ compared to $Au_{40}(SH)_{24}$ and $Au_{49}(SR)_{27}$. In addition, the Au_{25} core of $Au_{40}(SR)_{24}$, Au_{34} core of $Au_{49}(SR)_{27}$, and Au_{43} core of $Au_{58}(SR)_{30}$, can be viewed as 8, 11, and 14 tetrahedral $Au_4(2e)$ elementary blocks packing together (Figures 4.12), respectively. This shows that these three nanoclusers can be described by a GUM, suggesting their high stabilities.

4.5 Structural Design via Grand Unified Model

4.5.1 A Brief Introduction of the Idea

A GUM is developed to achieve fundamental understanding of the rich and complex structures of all the 71 ligand-protected gold clusters reported up to date. Inspired by the quark model in particle physics where composite particles such as protons, neutrons, or tetraquark are formed by combing three or four quarks (or flavors), here, gold atoms are assigned three "flavors" (namely, bottom, middle, and top) to represent three possible valence states, i.e. 1*e*, 0.5*e*, and 0*e*. The "composite particles" in the GUM are categorized into two groups: the triangular elementary block Au_3 (analogue to protons and neutrons) and the tetrahedral elementary block Au_4 (analogue to tetraquark). Both groups of elementary blocks are found to satisfy the duet rule of the valence shell, akin to the octet rule widely taught in general chemistry. Under the duet rule, the triangular elementary block exhibits 10 variants of valence states whereas the tetrahedral elementary block exhibits 15 variants of valence states. These variants of elementary blocks, when packed together, form the gold cores of ligated gold clusters. It should be noted that GUM is a predictive heuristic, but not necessarily reflective of the actual electronic structure. With the GUM, a plethora of complex and seemingly unrelated structures of the ligated gold clusters and their structure evolution can be deciphered and understood altogether. Moreover, a series of highly stable ligated gold clusters are predicted, thereby offering the GUM-guided synthesis of ligand-protected gold clusters by design [37, 38].

4.5.2 Atomic Structures of Hollow $Au_{36}(SR)_{12}$ and $Au_{42}(SR)_{14}$

GUM can be utilized for predicting new structures of ligand-protected gold nanoclusters. For example, $Au_{36}(SR)_{12}$ can be constructed by using the C_{12} fullerene as a template. The C_{12} fullerene exhibits eight polygons: four quadrilaterals and four pentagons (Figure 4.13a). Replacing all C atoms of the C_{12} fullerene with 12 fused tetrahedral Au_4 gives rise to the Au_{30} core (Figure 4.13b) of $Au_{36}(SR)_{12}$, followed by adding [-RS-Au-SR-] staple motifs on the unfused Au atoms to build the complete $Au_{36}(SR)_{12}$ cluster (Figure 4.13c). The Au_{30} hollow cage in $Au_{36}(SR)_{12}$ is composed of 12 fused Au_4 elementary blocks. Figure 4.13f shows another example of ligand-protected hollow Au cluster, namely, $Au_{42}(SR)_{14}$, constructed by using the C_{14} fullerene (Figure 4.13d) as a template. The Au_{35} hollow cage (Figure 4.13e) in $Au_{42}(SR)_{14}$ is composed of 14 fused Au_4 elementary blocks. The computed HOMO-LUMO gap of $Au_{42}(SR)_{14}$ is 2.00 eV, suggesting high chemical stability.

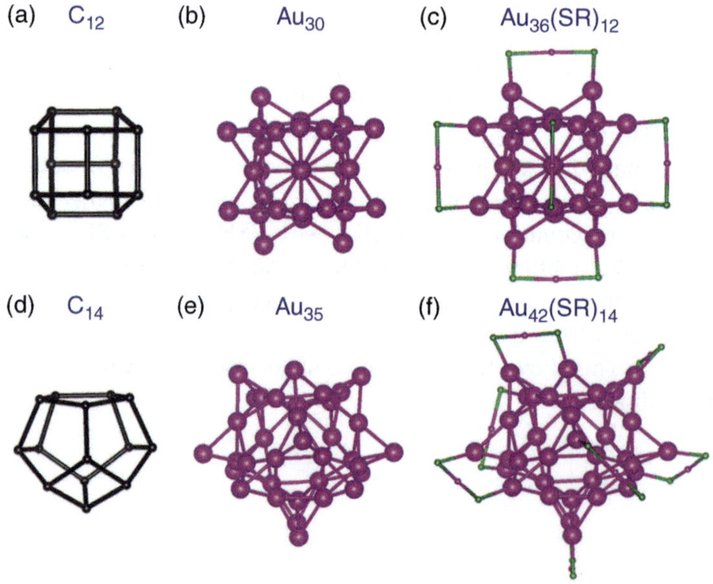

Figure 4.13 Structures of a C_{12} fullerene (a), Au_{30} core (b), and $Au_{36}(SR)_{12}$ cluster (c), as well as C_{14} fullerene (d), Au_{35} core (e), and $Au_{42}(SR)_{14}$ cluster (f). *Source:* Reprinted from ref. 37. Copyright 2016 Springer Nature. Color code: Au – magenta; S – dark green; C – black. The R groups are omitted for clarity.

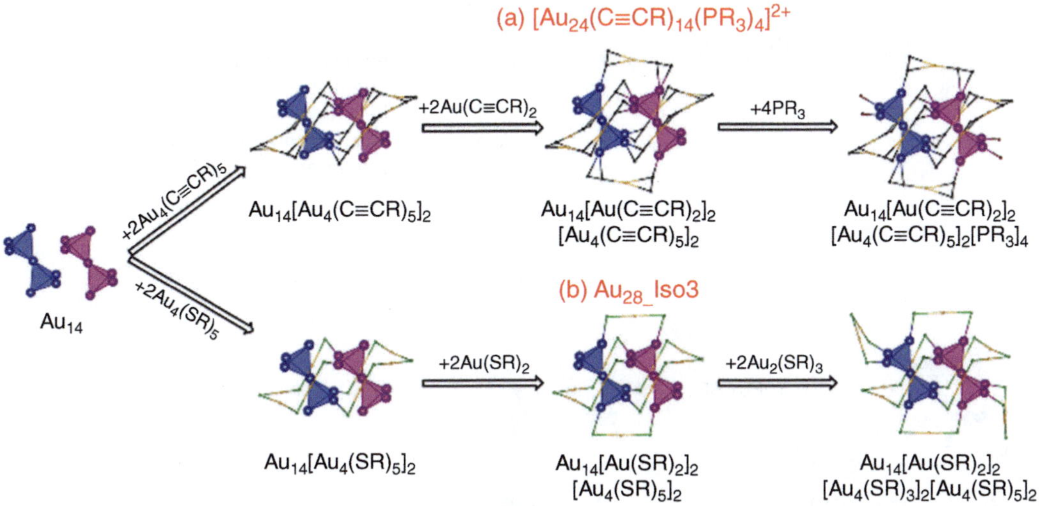

Figure 4.14 Structural decompositions of $[Au_{24}(C{\equiv}CR)_{14}(PR_3)_4]^{2+}$ (a) and Au_{28}_Iso3 (b). *Source:* Reproduced by permission of The Royal Society of Chemistry. The Au atoms are presented in yellow, wine, and blue, respectively. S and C are presented in dark green and black, respectively. The R groups are omitted for clarity.

Interestingly, the $Au_{36}(SR)_{12}$ and $Au_{42}(SR)_{14}$ can be rewritten as $Au_{30}[Au(SR)_2]_6$ and $Au_{35}[Au(SR)_2]_7$, respectively, consistent with the "divide-and-protection" rule [35, 36].

4.5.3 Atomic Structures of $Au_{28}(SR)_{20}$

It is known that both $Au_{28}(SR)_{20}$ isomers (referred as Au_{28}_Iso1 and Au_{28}_Iso2) have the same Au_{14} cores formed by packing two Au_7 units together [41, 42]. In this double helix core, two Au_4 units in each Au_7 pack up face to face with two Au_4 units in another Au_7. Thanks to the structural determinations of the alkynyl-protected $[Au_{24}(C{\equiv}CR)_{14}(PR_3)_4]^{2+}$ nanocluster [55], another Au_{14} core with a totally different packing pattern has been experimentally confirmed. This Au_{14} core is also formed by packing two Au_7 units together. While only one Au_4 unit in each Au_7 packs up face to face with one Au_4 unit in another Au_7. Replacing 14 C≡CR and 4 PR_3 in $[Au_{24}(C{\equiv}CR)_{14}(PR_3)_4]^{2+}$ with 14 SR and 2 $[Au_2(SR)_3]$, respectively, a new $Au_{28}(SR)_{20}$ isomer (Au_{28}_Iso3) can be obtained, as shown in Figure 4.14. Similar with the crystallized Au_{28}_Iso1 and Au_{28}_Iso2, the Au_{28}_Iso3 also has two quasi-octahedral Au_{13} units. Employing the same core as the crystallized Au_{28}_Iso1 and Au_{28}_Iso2, a new $Au_{28}(SR)_{20}$ isomer (Au_{28}_Iso4) can be constructed. As shown in Figure 4.15, the addition of one $[Au_3(SR)_4]$ staple on the Au_{14} core yields the $Au_{14}[Au_3(SR)_4]$ structure. Then, further adding three $[Au_3(SR)_4]$ staples to $Au_{14}[Au_3(SR)_4]$ results in the formation of the $Au_{14}[Au_3(SR)_4]_4$ structure. Finally, the Au_{28}_Iso4 is generated by adding two $[Au(SR)_2]$ staples. It can be seen that the staples of Au_{28}_Iso4 are completely different with those of Au_{28}_Iso1 and Au_{28}_Iso2 [56].

4.6 Conclusion and Perspectives

Four methods for designing the atomic structures of thiolate-protected gold nanoclusters were presented in this chapter. A large number of designed structures were reported. Some structures have been confirmed by experiments. The simulated optical spectra of some structures can well reproduce the experimental measurements. What is left is to wait for experimental confirmation

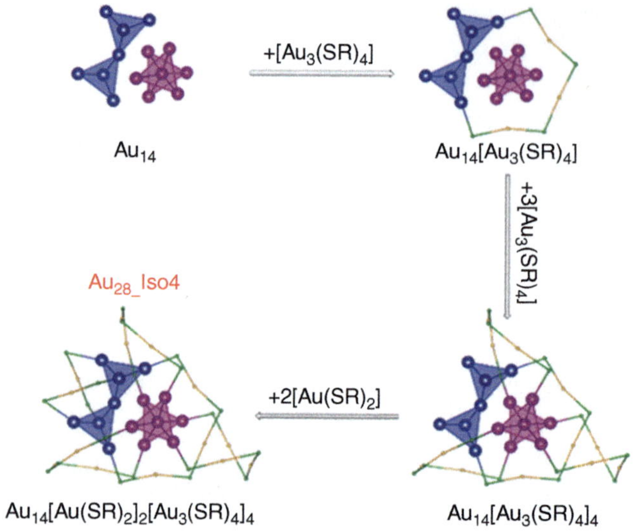

Figure 4.15 Structural decomposition of Au_{28}_Iso4. *Source:* Reproduced by permission of The Royal Society of Chemistry. The Au atoms are presented in yellow, wine, and blue, respectively. S is presented in dark green. The R groups are omitted for clarity.

in the future. Although much progress has been made in the structural design of thiolate-protected gold nanoclusters, it is still challenging to obtain the full structures of thiolate-protected gold nanoclusters without simplifying the organic groups R to CH_3 or H. In future work, it will be necessary to develop accurate molecular forces to describe the complex structures of thiolate-protected gold nanoclusters.

Acknowledgment

The authors acknowledge support by Natural Science Foundation of China (11974195).

References

1 Jadzinsky, P.D., Calero, G., Ackerson, C.J. et al. (2007). Structure of a thiol monolayer-protected gold nanoparticle at 1.1 Å resolution. *Science* 318: 430–433.
2 Akola, J., Walter, M., Whetten, R.L. et al. (2008). On the structure of thiolate-protected Au_{25}. *J. Am. Chem. Soc.* 130: 3756–3757.
3 Zhu, M., Aikens, C.M., Hollander, F.J. et al. (2008). Correlating the crystal structure of a thiol-protected Au_{25} cluster and optical properties. *J. Am. Chem. Soc.* 130: 5883–5885.
4 Heaven, M.W., Dass, A., White, P.S. et al. (2008). Crystal structure of the gold nanoparticle $[N(C_8H_{17})_4][Au_{25}(SCH_2CH_2Ph)_{18}]$. *J. Am. Chem. Soc.* 130: 3754–3755.
5 Lopez-Acevedo, O., Akola, J., Whetten, R.L. et al. (2019). Structure and bonding in the ubiquitous icosahedral metallic gold cluster $Au_{144}(SR)_{60}$. *J. Phys. Chem. C* 113: 5035–5038.
6 Malola, S., Lehtovaara, L., Knoppe, S. et al. (2012). $Au_{40}(SR)_{24}$ cluster as a chiral dimer of 8-electron superatoms: structure and optical properties. *J. Am. Chem. Soc.* 134: 19560–19563.
7 Yan, N., Xia, N., Liao, L. et al. (2018). Unraveling the long-pursued Au_{144} structure by X-ray crystallography. *Sci. Adv.* 4: eaat7259.
8 Pei, Y., Gao, Y., Shao, N., and Zeng, X.C. (2009). Thiolate-protected $Au_{20}(SR)_{16}$ cluster: prolate Au_8 core with new $[Au_3(SR)_4]$ staple motif. *J. Am. Chem. Soc.* 131: 13619–13621.

9 Pei, Y., Tang, J., Tang, X. et al. (2015). New structure model of $Au_{22}(SR)_{18}$: bitetrahederon golden kernel enclosed by $[Au_6(SR)_6]$ Au(I) complex. *J. Phys. Chem. Lett.* 6: 1390–1395.

10 Pei, Y., Pal, R., Liu, C. et al. (2012). Interlocked catenane-like structure predicted in $Au_{24}(SR)_{20}$: implication to structural evolution of thiolated gold clusters from homoleptic gold(I) thiolates to core-stacked nanoparticles. *J. Am. Chem. Soc.* 134: 3015–3024.

11 Pei, Y., Gao, Y., and Zeng, X.C. (2008). Structural prediction of thiolate-protected Au_{38}: a face-fused bi-icosahedral Au Core. *J. Am. Chem. Soc.* 130: 7830–7832.

12 Pei, Y., Lin, S.S., Su, J., and Liu, C. (2013). Structure prediction of $Au_{44}(SR)_{28}$: a chiral superatom cluster. *J. Am. Chem. Soc.* 135: 19060–19063.

13 Ma, Z., Wang, P., and Pei, Y. (2016). Geometric structure, electronic structure and optical absorption properties of one-dimensional thiolate-protected gold clusters containing a quasi-face-centered-cubic (quasi-Fcc) Au-Core: a density-functional theoretical study. *Nanoscale* 8: 17044–17054.

14 Wang, P., Sun, X., Liu, X. et al. (2017). Theoretical predictions of a new ~14 kDa core-mass thiolate-protected gold nanoparticle: $Au_{68}(SR)_{36}$. *J. Phys. Chem. Lett.* 8: 1248–1252.

15 Sun, X., Wang, P., Xiong, L., and Pei, Y. (2018). Theoretical prediction of a new stable structure of $Au_{28}(SR)_{20}$ cluster. *Chem. Phys. Lett.* 704: 68–75.

16 Xiong, L., Peng, B., Ma, Z. et al. (2017). A ten-Electron (10e) thiolate-protected $Au_{29}(SR)_{19}$ cluster: structure prediction and a 'gold-atom insertion, thiolate-group elimination' mechanism. *Nanoscale* 9: 2895–2902.

17 Qian, H., Eckenhoff, W.T., Zhu, Y. et al. (2010). Total structure determination of thiolate-protected Au_{38} nanoparticles. *J. Am. Chem. Soc.* 132: 8280–8281.

18 Zeng, C., Chen, Y., Iida, K. et al. (2016). Gold quantum boxes: on the periodicities and the quantum confinement in the Au_{28}, Au_{36}, Au_{44}, and Au_{52} magic series. *J. Am. Chem. Soc.* 138: 3950–3953.

19 Li, Y., Zhou, M., Song, Y. et al. (2021). Double-helical assembly of heterodimeric nanoclusters into supercrystals. *Nature* 594: 380–384.

20 Xiong, L. and Pei, Y. (2021). Evolution of electronic structure of cuboidthiolate-monolayer-protected gold nanocrystals. *J. Phys. Chem. C* 125: 20670–20675.

21 Jiang, D., Overbury, S.H., and Dai, S. (2013). Structure of $Au_{15}(SR)_{13}$ and its implication for the origin of the nucleus in thiolated gold nanoclusters. *J. Am. Chem. Soc.* 135: 8786–8789.

22 Jiang, D., Whetten, R.L., Luo, W., and Dai, S. (2009). The smallest thiolated gold superatom complexes. *J. Phys. Chem. C* 113: 17291–17295.

23 Jiang, D., Walter, M., and Akola, J. (2010). On the structure of a thiolated gold cluster: $Au_{44}(SR)_{28}{}^{2-}$. *J. Phys. Chem. C* 114: 15883–15889.

24 Xu, W.W., Gao, Y., and Zeng, X.C. (2015). Unraveling structures of protection ligands on gold nanoparticle $Au_{68}(SH)_{32}$. *Sci. Adv.* 1: e1400211.

25 Xu, W.W., Li, Y., Gao, Y., and Zeng, X.C. (2016). Unraveling a generic growth pattern in structure evolution of thiolate-protected gold nanoclusters. *Nanoscale* 8: 7396–7401.

26 Xu, W.W. and Gao, Y. (2015). Unraveling the atomic structures of the $Au_{68}(SR)_{34}$ nanoparticles. *J. Phys. Chem. C* 119: 14224–14229.

27 Liu, P., Han, W., Zheng, M., and Xu, W.W. (2020). Two-dimensional growth mode of thiolate-protected gold nanoclusters $Au_{28+4n}(SR)_{20+2n}$ (n = 0–8): compared with their one-dimensional growth mode. *Nanoscale* 12: 20677–20683.

28 Xu, W.W., Zeng, X.C., and Gao, Y. (2017). Au_{13}(8e): a secondary block for describing a special group of liganded gold clusters containing icosahedral Au_{13} motifs. *Chem. Phys. Lett.* 675: 35–39.

29 Xu, W.W., Lin, D., Fu, J. et al. (2020). Chiral $Au_{22}(SR)_{17}{}^{-}$: a new ligand-binding strategy for structural prediction of thiolate-protected gold nanocluster. *Chem. Commun.* 56: 2995–2998.

30 Tlahuicea, A. and Garzón, I.L. (2012). On the structure of the $Au_{18}(SR)_{14}$ cluster. *Phys. Chem. Chem. Phys.* 14: 3737–3740.

31 Omoda, T., Takano, S., Yamazoe, S. et al. (2018). An $Au_{25}(SR)_{18}$ cluster with a face-centered cubic core. *J. Phys. Chem. C* 122: 13199–13204.

32 Nimmala, P.R., Yoon, B., Whetten, R.L. et al. (2013). $Au_{67}(SR)_{35}$ nanomolecules: characteristic size-specific optical, electrochemical, structural properties and first-principles theoretical analysis. *J. Phys. Chem. A* 117: 504–517.

33 Tlahuice-Flores, A., Santiago, U., Bahena, D. et al. (2013). Structure of the thiolated Au_{130} cluster. *J. Phys. Chem. A* 117: 10470–10476.

34 Tlahuice-Flores, A. (2015). New insight into the structure of thiolated gold clusters: a structural prediction of the $Au_{187}(SR)_{68}$ cluster. *Phys. Chem. Chem. Phys.* 17: 5551–5555.

35 Pei, Y. and Zeng, X.C. (2012). Investigating the structural evolution of thiolate protected gold clusters from first-principles. *Nanoscale* 4: 4054–4072.

36 Häkkinen, H., Walter, M., and Gronbeck, H. (2006). Divide and protect: capping gold nanoclusters with molecular gold-thiolate rings. *J. Phys. Chem. B* 110: 9927–9931.

37 Xu, W.W., Zhu, B., Zeng, X.C., and Gao, Y. (2016). A grand unified model for liganded gold clusters. *Nat. Commun.* 7: 13574.

38 Xu, W.W., Zeng, X.C., and Gao, Y. (2018). Application of electronic counting rules for ligand-protected gold nanoclusters. *Acc. Chem. Res.* 51: 2739–2747.

39 Azubel, M., Koivisto, J., Malola, S. et al. (2014). Electron microscopy of gold nanoparticles at atomic resolution. *Science* 345: 909–912.

40 Dass, A. (2009). Mass spectrometric identification of $Au_{68}(SR)_{34}$ molecular gold nanoclusters with 34-electron shell closing. *J. Am. Chem. Soc.* 131: 11666–11667.

41 Zeng, C., Li, T., Das, A. et al. (2013). Chiral structure of thiolate-protected 28-gold-atom nanocluster determined by X-ray crystallography. *J. Am. Chem. Soc.* 135: 10011–10013.

42 Chen, Y., Liu, C., Tang, Q. et al. (2016). Isomerism in $Au_{28}(SR)_{20}$ nanocluster and stable structures. *J. Am. Chem. Soc.* 138: 1482–1485.

43 Zhuang, S., Liao, L., Zhao, Y. et al. (2018). Is the kernel–staples match a key–lock match? *Chem. Sci.* 9: 2437–2442.

44 Chen, S., Xiong, L., Wang, S. et al. (2016). Total structure determination of $Au_{21}(S\text{-}Adm)_{15}$ and geometrical/electronic structure evolution of thiolated gold nanoclusters. *J. Am. Chem. Soc.* 138: 10754–10757.

45 Lin, D., Zheng, M., and Xu, W.W. (2020). Structural predictions of thiolate-protected gold nanoclusters via redistribution of the Au-S "staple" motifs on the known cores. *Phys. Chem. Chem. Phys.* 22: 16624–16629.

46 Das, A., Li, T., Nobusada, K. et al. (2013). Nonsuperatomic $[Au_{23}(SC_6H_{11})_{16}]^-$ nanocluster featuring bipyramidal Au_{15} kernel and trimeric $Au_3(SR)_4$ motif. *J. Am. Chem. Soc.* 135: 18264–18267.

47 Crasto, D. and Dass, A. (2013). Green gold: $Au_{30}(S\text{-}t\text{-}C_4H_9)_{18}$ molecules. *J. Phys. Chem. C* 117: 22094–22097.

48 Zeng, C., Chen, Y., Liu, C. et al. (2015). Gold tetrahedra coil up: Kekulé-like and double helical superstructures. *Sci. Adv.* 1: e1500425.

49 Briant, C.E., Tobald, B.R.C., White, J.W. et al. (1981). Synthesis and X-ray structural characterization of the centered icosahedral gold cluster compound $[Au_{13}(PMe_2Ph)_{10}Cl_2](PF_6)_3$; the realization of a theoretical prediction. *J. Chem. Soc. Chem. Commun.* 5: 201–202.

50 Shichibu, Y., Negishi, Y., Watanabe, T. et al. (2007). Biicosahedral gold clusters $[Au_{25}(PPh_3)_{10}(SC_nH_{2n+1})_5Cl_2]^{2+}$ (n = 2–18): a stepping stone to cluster-assembled materials. *J. Phys. Chem. C* 111: 7845–7847.

51 Jin, R., Liu, C., Zhao, S. et al. (2015). Tri-icosahedral gold nanocluster $[Au_{37}(PPh_3)_{10}(SC_2H_4Ph)_{10}X_2]^+$: linear assembly of icosahedral building blocks. *ACS NANO* 9: 8530–8536.

52 Zeng, C., Qian, H., Li, T. et al. (2012). Total structure and electronic properties of the gold nanocrystal $Au_{36}(SR)_{24}$. *Angew. Chem. Int. Ed.* 51: 13114–13118.

53 Liao, L., Zhuang, S., Wang, P. et al. (2017). Quasi-dual-packed-kerneled $Au_{49}(2,4\text{-DMBT})_{27}$ nanoclusters and the influence of kernel packing on the electrochemical gap. *Angew. Chem. Int. Ed.* 129: 12818–12822.

54 Zheng, M. and Xu, W.W. (2020). Medium-sized $Au_{58}(SR)_{30}$: a new chiral structure evolving from crystallized $Au_{40}(SR)_{24}$ and $Au_{49}(SR)_{27}$. *J. Phys. Chem. C* 124: 9077–9081.

55 Wan, X., Xu, W.W., Yuan, S. et al. (2015). A near-infrared-emissive alkynyl-protected Au_{24} nanocluster. *Angew. Chem. Int. Ed.* 54: 9683–9686.

56 Xu, W.W., Zeng, X.C., and Gao, Y. (2018). The structural isomerism in gold nanoclusters. *Nanoscale* 10: 9476–9483.

5

Electrocatalysis on Atomically Precise Metal Nanoclusters

Hoeun Seong[1], Woojun Choi[2], Yongsung Jo[1], and Dongil Lee[1]

[1] *Department of Chemistry, Yonsei University, Seoul 03722, Republic of Korea*
[2] *Department of Chemistry and Medical Chemistry, Yonsei University, Wonju, Gangwon 26493, Republic of Korea*

5.1 Introduction

One of the global issues facing humankind today is the development of sustainable and fossil-fuel-free pathways to produce useful fuels and chemicals to mitigate the growing climate crisis [1, 2]. Developing electrochemical conversion processes that can convert molecules in the atmosphere, such as water, nitrogen, and carbon dioxide, into value-added products by coupling with renewable energy is a promising approach [3–5]. Electrocatalysts play a key role in these processes because they can affect the rate, efficiency, and selectivity of chemical conversion reactions [6, 7]. The main focus of electrocatalyst research is the development of highly efficient and sustainable materials to drive stable and high-current conversion at low overpotentials. Despite notable progress recently made in the development of advanced nanocatalysts and solid surfaces [8, 9], an atomic-level understanding of the reaction mechanisms and structure–activity relationships remains elusive owing to the inherent heterogeneity of the catalyst surface. A structurally well-defined model electrocatalyst is necessary to elucidate the details of the reactions and provide rational guidance for the development of efficient and selective electrocatalysts. This chapter discusses the electrocatalytic applications of atomically precise metal nanoclusters (NCs).

5.1.1 Materials Design Strategy for Electrocatalysis

Over the past two decades, considerable progress has been made toward understanding the mechanism and structure–activity relationships in several key electrochemical transformations, such as hydrogen evolution/oxidation reactions (HER/HOR) and oxygen reduction/evolution reactions (ORR/OER) [10, 11]. Theoretical models and computational methods have advanced significantly in recent years and have proven to be powerful in understanding catalytic trends using a descriptor-based approach [12]. Descriptors are key properties of a catalyst surface that can describe and/or predict catalytic performance. The ability to predict catalytic trends using descriptors, including electronic descriptors represented by *d*-band theory and structural descriptors, has enabled high-throughput screening of potential catalysts and rational design of improved catalysts [13]. For example, HER activity is predominantly dictated by the hydrogen adsorption energy (ΔG_{H}) [14].

Atomically Precise Nanochemistry, First Edition. Edited by Rongchao Jin and De-en Jiang.

The HER is a two-electron-transfer reaction with one catalytic intermediate (*H), where * denotes the active site [15].

$$\text{Volmer step: } * + H^+ + e^- \rightarrow *H \tag{5.1}$$

$$\text{Heyrovsky step: } *H + H^+ + e^- \rightarrow * + H_2 \tag{5.2}$$

$$\text{Tafel step: } 2*H \rightarrow 2* + H_2 \tag{5.3}$$

The HER may occur through either the Volmer–Heyrovsky [Eqs. (5.1) and (5.2)] or Volmer–Tafel [Eqs. (5.1) and (5.3)] mechanism. If hydrogen binds weakly to the catalyst surface, the adsorption step will limit the HER rate. By contrast, the desorption step limits the rate if the binding is too strong. Thus, an active HER catalyst should have a thermoneutral adsorption energy, that is, $\Delta G_H \approx 0$ [14, 15].

Figure 5.1a shows the exchange current densities (j_0) for a wide range of catalysts as a function of ΔG_H calculated from density functional theory (DFT), showing a clear volcano relationship.

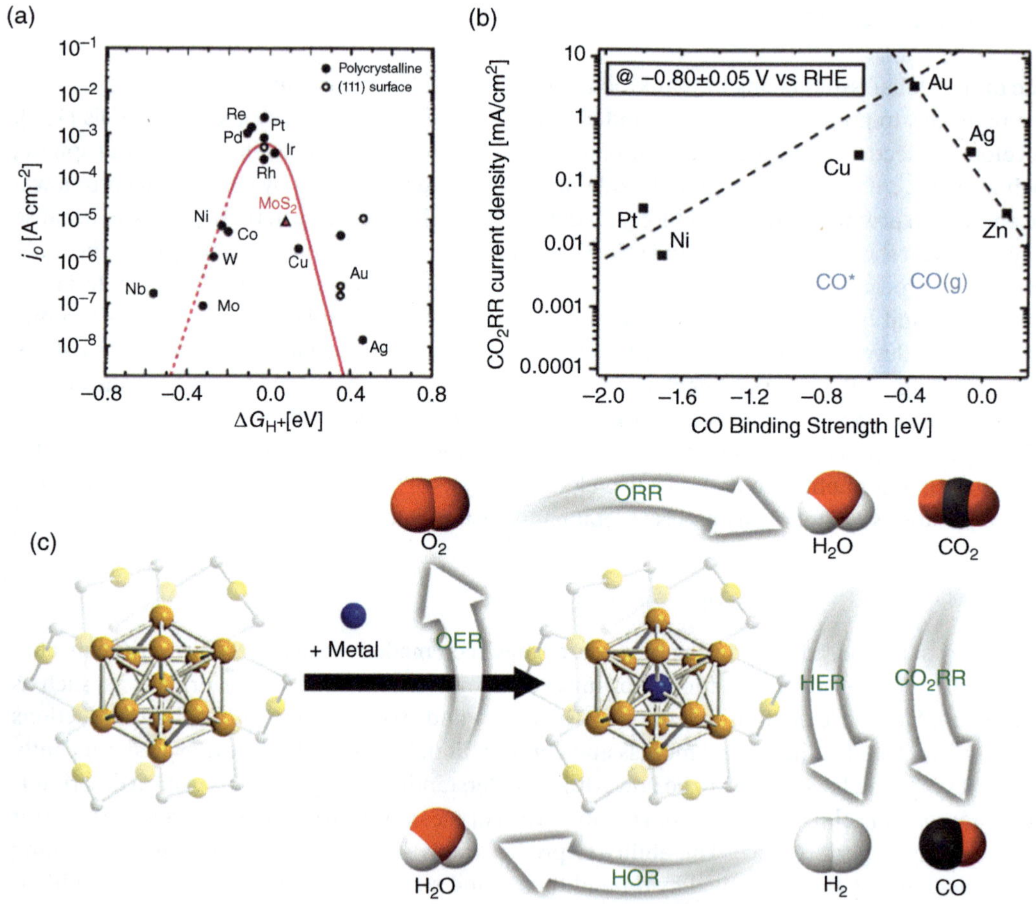

Figure 5.1 (a) Volcano plot for HER on metals. *Source:* Reproduced with permission from Ref. [16]. © 2017 American Association for the Advancement of Science. (b) Volcano plot for CO_2-to-CO electroreduction on metals. *Source:* Reproduced with permission from Ref. [17]. © 2014 American Chemical Society. (c) Schematic of tailored electrocatalysis by atomic-level engineering of metal NCs. *Source:* Reproduced with permission from [18]. © 2019 American Chemical Society.

That is, platinum resides near the top of the hydrogen volcano, and the current densities decrease as ΔG_H moves in either direction along the ΔG_H axis. In fact, it is well known that Pt is the best-performing electrocatalyst for the HER, having negligible overpotentials in acidic media [19]. The volcano plot clearly shows that ΔG_H is the key descriptor of the HER, providing the design principle for active HER electrocatalysts. Another important electrocatalytic reaction includes the electroreduction of CO_2 into value-added products [5]. This is a multielectron reduction reaction involving several reaction intermediates [20]. The CO_2 reduction reaction (CO_2RR) may produce a variety of products, including CO, $HCOO^-$, CH_4, CH_3OH, and C_{2+} hydrocarbons and oxygenates, probably produced through different intermediates [20]. In addition, most of the CO_2RR occurs near 0 V vs. the reversible hydrogen electrode (RHE), making the HER an additional competing reaction [21, 22]. Therefore, controlling the reaction selectivity among the possible products has been a major challenge in optimizing the CO_2RR.

Figure 5.1b shows a volcano plot of CO_2RR current density against CO binding strength. In this plot, Au sits near the top of the volcano with optimal CO binding strength and is known to produce CO as the major product from the CO_2RR. C_{2+} products are generally produced on Cu that binds CO* rather strongly through hydrogenation or dimerization of the adsorbed CO intermediate. Hydrogen is predominantly produced on Pt or Ni that binds CO* too tightly. The combination of theory and experiment has proven fruitful in electrocatalysis research and has provided design principles for the development of improved electrocatalysts.

5.1.2 Atomically Precise Metal Nanoclusters as Electrocatalysts

Over the past few decades, metal nanoparticles have attracted much attention as electrocatalysts for water splitting and CO_2RR [20, 23]. Combining theory and experiment, researchers have revealed that the relatively high catalytic profiles of the metal nanoparticles (compared to the bulk surface) result from the exposed low-coordinated metal surface and quantum size effects [16]. Extraordinary catalytic activities have been reported for various electrocatalytic reactions by reducing the particle size [24, 25]. However, metal nanoparticles are typically polydisperse in their sizes and morphologies and thus contain multiple active sites exhibiting different catalytic activities [24, 26]. This heterogeneity results in a significant discrepancy between the theoretical calculations on an ideal surface and experiments performed on a real surface. Developing atomically precise metal nanocatalysts with well-defined surface structures is crucial for bridging this discrepancy.

Recent synthetic advances have created new opportunities to produce ultrasmall metal NCs comprising a few to few hundred metal atoms [27]. These metal NCs are molecularly pure and exhibit well-defined atomic structures, which have been confirmed using X-ray crystallography [28]. They are commonly described by their molecular formulas (e.g. $Au_{25}L_{18}$, $Au_{38}L_{24}$, $Au_{102}L_{44}$, and $Ag_{44}L_{30}$), where L is a protecting ligand. The metal NCs with the formula M_NL_X, where N and X denote the number of metal atoms and ligands, respectively, are abbreviated as M_N in the following sections. Significant progress has been made in their synthesis and structural determination, which has enabled elucidation of their structure–property relationships at the atomic level. Numerous size-dependent optical and electrochemical properties of metal NCs have been reported [29]. However, much less progress has been made with respect to their electrocatalytic applications.

The atom-precise metal NCs have special stability at certain compositions and can thus be isolated as a single species. Molecularly pure metal NCs with well-defined atomic structures can serve as effective model catalysts for electrocatalytic reactions. The catalytic properties of metal NCs can be finely tuned by atomic-level engineering, such as metal doping [30, 31], morphology control [32], and ligand modification [33–35], for a specific reaction (Figure 5.1c). By comparing the catalytic properties with the atomic and electronic structures of the model NC catalysts, the descriptor at the atomic level can be determined. Furthermore, atomically precise metal NCs can act as an ideal platform for precisely identifying the active sites for specific electrocatalytic reactions. The atomic-level identification of the active sites can be achieved by combining theoretical computations and experimental analyses of the well-known atomic structure of NCs.

This chapter does not discuss the synthesis and characterization of metal NCs in detail; readers are referred to the recommended extensive reviews [36, 37]. Instead, we focus on the redox electrochemistry and electrocatalytic properties related to electrocatalytic water splitting and CO_2 reduction reactions.

5.2 Electrochemistry of Atomically Precise Metal Nanoclusters

Atomically precise metal NCs cover the transition region between the bulk and molecule, where unique optical, electrochemical, and catalytic features are observed [18, 38]. These NCs have been proven to have rich electrochemistry and electron-transfer chemistry, which are directly related to their electrocatalytic properties. This section discusses how their redox properties are controlled by their size, structure, and composition.

5.2.1 Size-Dependent Voltammetry

The voltammetry of NC solutions was extensively studied by Murray et al., who distinguished three voltammetric regimes based on the range of core sizes: continuum density of states (DOS), quantized double-layer charging, and molecule-like (Figure 5.2a) [39]. Whereas only unresolved current–potential responses were observed for particles larger than 3–4 nm, well-resolved peaks related to successive single-electron transfers (SET) were observed for $Au_{\sim 140}$ NCs (Figure 5.2b). The observed SET properties were analogous to the Coulomb staircase behavior, which is typically observed in scanning tunneling spectroscopy of single nanoparticles at reduced temperatures [41]. Later, it was recognized that the double-layer property of small NCs was the more fundamental aspect of the phenomenon, and thus it was dubbed quantized double-layer charging (QDL) [39]. A simple relation that distinguishes quantum capacitors exhibiting QDL behavior from bulk-continuum nanoparticles is as follows:

$$\Delta V = e / C_{NC} \tag{5.4}$$

where ΔV is the average potential change of a NC with double-layer capacitance C_{NC}, which is incurred upon transfer of an electron to/from the NC. This Eq. (5.4) indicates that the potential interval between single-electron changes in the NC core can be observable when C_{NC} becomes very small. In other words, when the C_{NC} is reduced to the subattofarad (aF) regime, the potential interval drastically exceeds the Boltzmann thermal energy distribution factor k_BT (25.7 meV at 25 °C). Successive electron-transfer processes to/from the NC results in a stepwise change in the NC's potential, even at room temperature. Figure 5.2b shows a well-resolved QDL-charging voltammogram of the dissolved hexanethiolate-protected $Au_{\sim 140}$ NCs. The C_{NC} of the NC was estimated to be 0.60 aF from the average potential interval (263 mV) between successive single-electron changes in the electronic charge on the NC core [40]. Amplitude differential voltammetric techniques, such as differential pulse voltammetry and square-wave voltammetry, exhibit more prominent current peaks compared to linear voltammetry because the differentiation method can magnify small current features. QDL charging of NCs in electrolyte solutions was successfully modeled using a

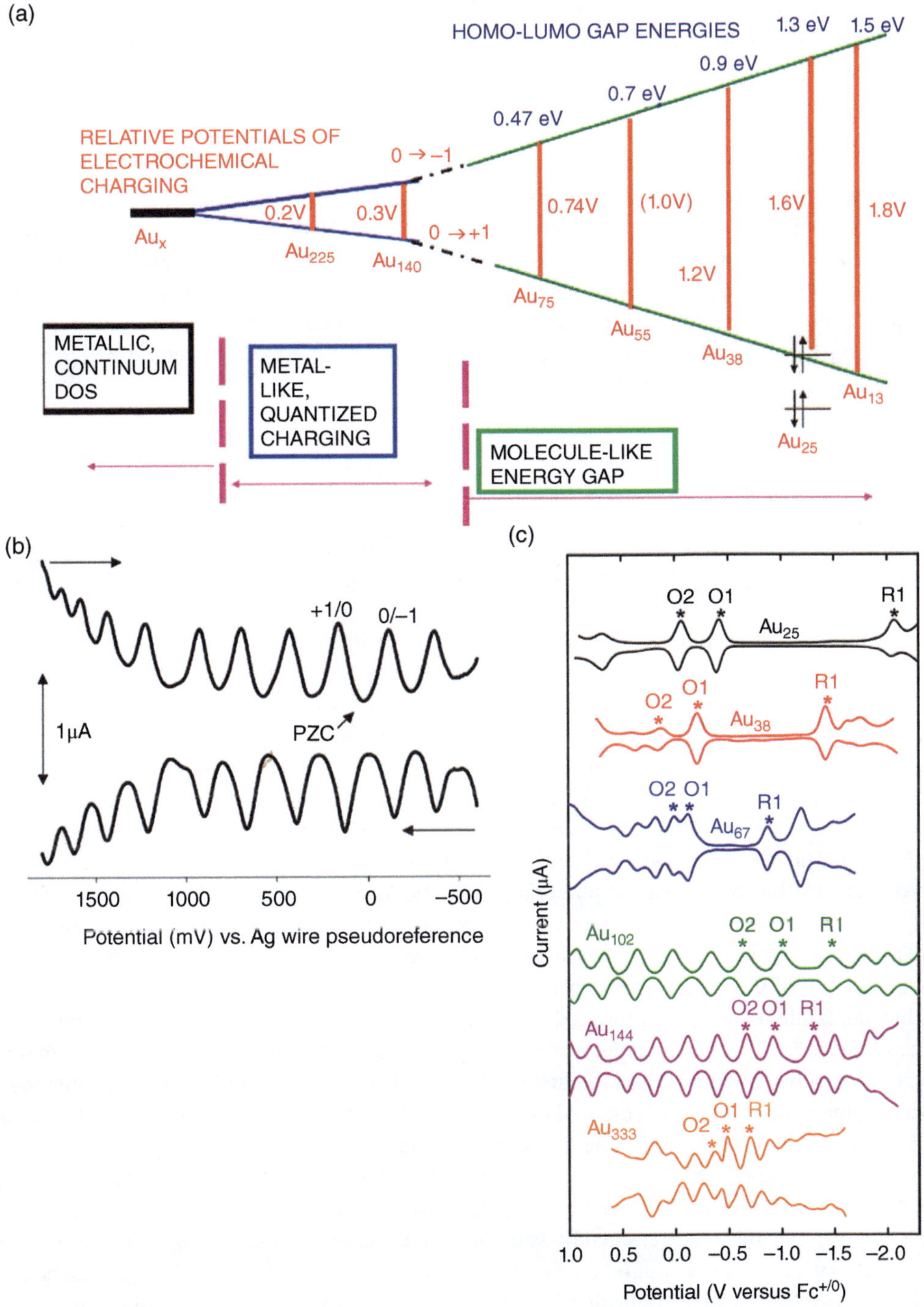

Figure 5.2 (a) Size-dependent voltammetric regimes and energy gaps. *Source:* Reproduced with permission from Ref. [39]. © 2008 American Chemical Society. (b) Differential pulse voltammogram of hexanethiolate-protected $Au_{\sim 140}$ NCs in CH_2Cl_2. Arrows indicate the direction of the potential scan. *Source:* Reproduced with permission from Ref. [40]. © 2002 American Chemical Society. (c) SWVs of Au_{25}, Au_{38}, Au_{67}, Au_{102}, Au_{144}, and Au_{333} NCs in CH_2Cl_2. *Source:* Reproduced with permission from Ref. [29]. © 2019 American Chemical Society.

concentric sphere capacitor [42]. In this model, the double-layer capacitance of an NC is related to the core radius (r) and thiolate monolayer thickness (d) with an effective dielectric constant (ε):

$$C_{\mathrm{NC}} = A_{\mathrm{NC}} \frac{\varepsilon\varepsilon_0}{r} \frac{r+d}{d} = 4\pi\varepsilon\varepsilon_0 \frac{r}{d}(r+d) \tag{5.5}$$

where ε_0 is the permittivity of free space, and A_{NC} is the Au-core surface area.

Recent advances in the synthesis and characterization of metal NCs have enabled the atomic-level determination of their structures and compositions [36, 43]. Mass spectrometry, single-crystal X-ray diffraction (SC-XRD) measurements, and DFT calculations have been employed to determine their atomic and electronic structures [44–46]. The square-wave voltammograms (SWVs) of hexanethiolate-protected Au_{102}, Au_{144}, and Au_{333} NCs in Figure 5.2c show the size-dependent QDL-charging behavior. The C_{NC} values of Au_{102}, Au_{144}, and Au_{333} NCs were determined to be 0.49, 0.57, and 0.88 aF by the following Eq. (5.6):

$$E^{o}_{z,z-1} = \frac{\left(z - \frac{1}{2}\right)e}{C_{\mathrm{NC}}} + E_{\mathrm{PZC}} \tag{5.6}$$

where z, $E^{o}_{z,z-1}$, and E_{PZC} are the charge state, the formal potential of the z/z−1 change, and the potential of zero charge of a NC. It was demonstrated that C_{NC} increased with increasing NC core size, as predicted from Eq. (5.5).

As the NC size decreases further, an energy gap emerges, indicating the onset of molecularity of the Au NCs. The molecule-like behavior of the Au NCs was detected by optical and/or electrochemical methods [18]. The optical energy gap for an NC is the absorbance spectrum onset for transitions from the highest-occupied molecular orbital (HOMO) to the lowest-unoccupied molecular orbital (LUMO). The electrochemical energy gap is the difference between the first oxidation and first reduction potentials of the parent NC. However, in this gap, there is a charging-energy term associated with the additional energy involved in producing positive or negative species by charging a neutral species or further charging already charged species. The charging energy can be approximated by the spacing of the potentials between the QDL peaks. Subtracting the charging energy from the measured electrochemical energy gap leads to the electrochemically determined HOMO–LUMO gaps of the NCs. Uncertainties in the values of the HOMO–LUMO gap are related to the estimation of the charging-energy correction.

Within the family of Au NCs protected by hexanethiolate, size-dependent voltammograms were obtained (Figure 5.2c). The electrochemical energy gap of Au_{25} determined by the potential difference between the first oxidation (O1) and the first reduction (R1) potentials is 1.65 V. The charging-energy correction (approximated by the potential difference between the O1 and O2 peaks) for the electrochemical gap results in a corrected energy gap that constitutes an electrochemical determination of the HOMO–LUMO gap (1.32 eV) for Au_{25}. The HOMO–LUMO gap energies obtained for Au_{38}, Au_{67}, Au_{102}, and Au_{144} NCs in a similar manner were 0.99, 0.61, 0.18, and 0.15 V, respectively. The HOMO–LUMO gap energies matched reasonably well with those determined by the absorbance spectrum onset at 1.3, 0.9, 0.7, 0.5, 0.4, and 0.2 eV for Au_{25}, Au_{38}, Au_{67}, Au_{102}, Au_{144}, and Au_{333}, respectively, showing the metal-to-molecule transition for Au NCs. In other words, whereas the larger Au_{102}, Au_{144}, and Au_{333} NCs are metallic and display QDL-charging voltammograms, the smaller Au_{25}, Au_{25}, and Au_{67} NCs exhibit molecule-like voltammograms with distinct HOMO–LUMO gap energies.

5.2.2 Metal-Doped Gold Nanoclusters

It has been found that the redox potentials of Au NCs are rather insensitive to changes in the ligand shell but are critically sensitive to alteration of the core, especially when a change in the electronic structure is incurred [47, 48]. Thus, heterometallic doping of stable gold NCs is a powerful means to tune the electrochemical properties of the NCs [49]. Anionic $[Au_{25}(SR)_{18}]^-$ composed of an Au_{13} core and six protecting $Au_2(SR)_3$ semi-rings has been extensively studied as a stable host for metal doping [43]. The extraordinary stability of $[Au_{25}(SR)_{18}]^-$ was explained by the shell-closing model of the superatoms [50], in which the number of superatomic electrons (n) for a spherical metal NC protected by thiolate ligands, $[M_N(SR)_X]^z$, is given by

$$n = NV_A - X - z \tag{5.7}$$

where V_A is the valence of the metal atoms. For the $[Au_{25}(SR)_{18}]^-$ NC, $n = (25 \times 1) - 18 - (-1) = 8$. The eight superatomic electrons of $[Au_{25}(SR)_{18}]^-$ fully occupy the 1S and 1P levels of the superatomic electron shell ($1S^2 1P^6$).

Many examples of heteroatom doping of Au_{25} NCs exist [43]. Singly doped NCs, forming $MAu_{24}(SR)_{18}$ (M = Hg, Cd, Pt, and Pd), have been prepared using various synthetic procedures [51–53]. When the dopant is within the same group (e.g. Ag and Cu), multiply doped, anionic $[Ag_xAu_{25-x}(SR)_{18}]^-$ (x = 1–11) and $[Cu_xAu_{25-x}(SR)_{18}]^-$ (x = 1–5) NCs are produced [54–56]. Unsurprisingly, the voltammograms of these NCs with eight superatomic electrons were very similar to those of the undoped NC [55]. On the other hand, neutral $[HgAu_{24}(SR)_{18}]^0$ NCs were produced upon doping with group 12 metals, and their voltammetry was similar to that of the $[Au_{25}(SR)_{18}]^-$ NC [52].

Doped NCs with significantly different electronic structures were produced when the dopant was a group 10 metal such as Pd or Pt [51]. SC-XRD analysis revealed that the produced NCs were singly doped $PdAu_{24}(SR)_{18}$ and $PtAu_{24}(SR)_{18}$ NCs, where the dopant was located in the central position of the core [57]. Furthermore, the produced NCs were neutrally charged, that is, $[PdAu_{24}(SR)_{18}]^0$ and $[PtAu_{24}(SR)_{18}]^0$, which have six superatomic electrons. Voltammetry experiments have been proven to provide very useful information on their electronic structures. Figure 5.3a displays the dramatically different SWVs from those of the undoped Au_{25} NC. The electrochemical gap energies observed for $PdAu_{24}$ and $PtAu_{24}$ NCs from their voltammograms were 0.75 and 0.73 V, respectively. The electrochemical HOMO–LUMO gaps determined for $PdAu_{24}$ and $PtAu_{24}$ NCs, after charging-energy correction, were 0.32 and 0.33 V, respectively, which are substantially smaller than that of Au_{25} NC (1.32 V). These values agree well with those determined by the DFT calculations [51].

The drastic changes in the electronic structures of $PdAu_{24}$ and $PtAu_{24}$ NCs were ascribed to the Jahn–Teller effect, which induced geometrical distortion of the doped NCs. The $8e^-$ superatom $[Au_{25}(SR)_{18}]^-$ adopts a noble-gas-like electronic configuration ($1S^2 1P^6$) with triply degenerate 1P orbitals and doubly degenerate 1D orbitals (Figure 5.3b). The $[PdAu_{24}(SR)_{18}]^0$ and $[PtAu_{24}(SR)_{18}]^0$ NCs adopt the superatomic orbitals of the $6e^-$ system, in which the triply degenerate 1P orbitals split into a doubly degenerate HOMO and LUMO. The drastic reduction in the HOMO–LUMO gap energies observed for the $PdAu_{24}$ and $PtAu_{24}$ NCs were the consequence of the Jahn–Teller effect, which was accompanied by icosahedral core distortion (Figure 5.3b).

Jahn–Teller distortion was also observed for the $[Pd_2Au_{36}(SR)_{24}]^0$ NC [59]. In a SC-XRD study, Qian et al. revealed that $Au_{38}(SC_2H_4Ph)_{24}$ consisted of a face-fused bi-icosahedral Au_{23} core and three monomeric and six dimeric protecting staples [60]. Interestingly, Pt doping of Au_{38} resulted in dianionic $[Pt_2Au_{36}(SR)_{24}]^{2-}$ NCs, while neutral $[Pd_2Au_{36}(SR)_{24}]^0$ NCs were produced upon Pd doping [59]. A SWV investigation revealed that the HOMO–LUMO gap energy of the former was 0.95 V, which is similar to that of the undoped Au_{38} NC (0.86 V), whereas that of the latter was drastically reduced to 0.26 V. DFT calculations revealed that the $[Au_{38}(SCH_3)_{24}]^0$ and $[Pt_2Au_{36}(SCH_3)_{24}]^{2-}$ model NCs adopt the $14e^-$ superatom electronic configuration ($1S^2 1P^6 1D^6$), while $[Pd_2Au_{36}(SCH_3)_{24}]^0$ adopts a $12e^-$ system in which the doubly degenerate HOMO splits into

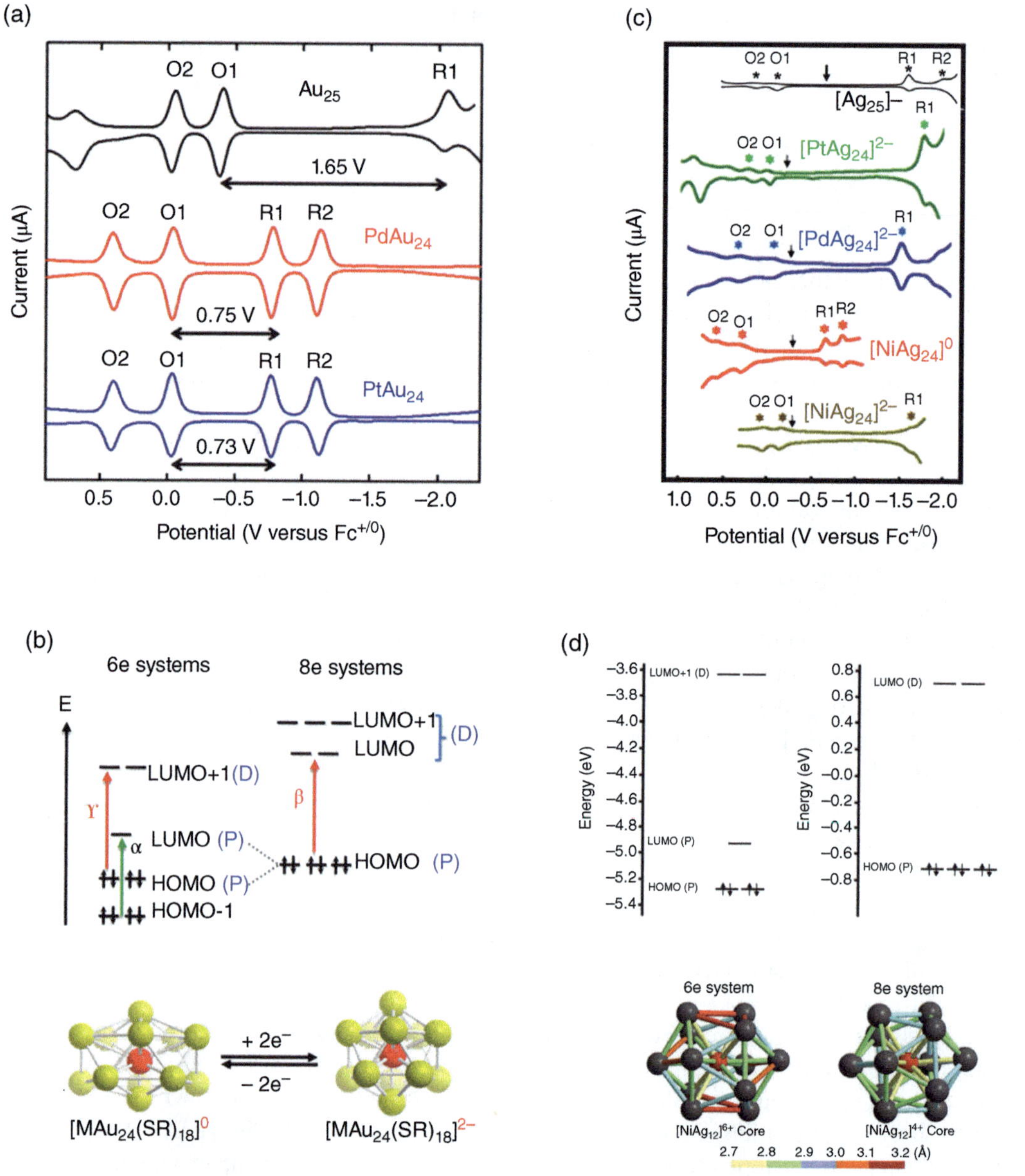

Figure 5.3 (a) SWVs of $[Au_{25}]^-$, $[PdAu_{24}]^0$, and $[PtAu_{24}]^0$ NCs in CH_2Cl_2 containing 0.1 M Bu_4NPF_6. (b) Electronic structures of the superatomic $6e^-$ and $8e^-$ systems. Optical transitions occurring in the superatom systems are denoted by α, β, and γ. The MAu_{12} core of the Jahn–Teller-distorted $6e^-$ $[MAu_{24}(SR)_{18}]^0$ (left) undergoes structural conversion to approximately spherical $8e^-$ $[MAu_{24}(SR)_{18}]^{2-}$ (M = Pd and Pt) upon reduction (right). *Source:* Reproduced with permission from Ref. [51]. © 2015 American Chemical Society. (c) SWVs of $[PtAg_{24}]^{2-}$, $[PdAg_{24}]^{2-}$, $[NiAg_{24}]^0$, and $[NiAg_{24}]^{2-}$ in CH_2Cl_2 containing 0.1 M Bu_4NPF_6. The downward-pointing arrows indicate the solution open-circuit potential. (d) DFT-predicted electronic energy levels of $6e^-$ $[NiAg_{24}(SH)_{18}]^0$ and $8e^-$ $[NiAg_{24}(SH)_{18}]^{2-}$ NCs and bond-length comparison of the icosahedral core predicted for the $6e^-$ $[NiAg_{12}]^{6+}$ (left) and $8e^-$ $[NiAg_{12}]^{4+}$ (right) cores. *Source:* Reproduced with permission from Ref. [58]. © 2020 American Chemical Society.

a HOMO and LUMO as a result of the Jahn–Teller effect. The origin of the Jahn–Teller distortion observed only for $[Pd_2Au_{36}(SR)_{24}]^0$ (not for $[Pt_2Au_{36}(SR)_{24}]^{2-}$) was ascribed to the extent of size mismatch between the dopant and host.

5.2.3 Metal-Doped Silver Nanoclusters

Unlike the aforementioned Au NCs, numerous metal-doped Ag NCs can be produced in high yields by galvanic metal-exchange synthesis. Mechanistic investigations revealed that Pt(Pd)-doped $MAg_{24}(SPhMe_2)_{18}$ NCs (M = Pd and Pt) were produced almost exclusively via the galvanic metal exchange of the initially formed Ag_{25} NC with the metal dopant [58]. The galvanic metal-exchange reaction can be considered as a redox process involving the reduction of noble-metal cations (Pt^{2+} and Pd^{2+}) by the less noble Ag host with a favorable driving force. It should be noted here that small Au and Ag NCs have distinct charge states, and thus the charge-state-dependent reducing power of the Ag_{25} NC host should be considered instead of that of the bulk metal in order to understand the driving force of the galvanic reaction. A two-step metal-exchange reaction between the Ag_{25}^- host and dopant ions was proposed. The metal-exchange reaction is initiated by reducing metal ions on the surface to form an adduct (Pt^0Ag_{25}), which subsequently undergoes an alloying process to produce $PtAg_{24}$ following Ag^+ dissolution (Eq. (5.8)). The first step is driven by the relative potential difference between the host and the metal ion, while the second step is driven by the stability of the doped NC. Furthermore, in the metal-exchange synthesis of $NiAg_{24}^0$ NCs, a yield of >70% was realized, in which the less noble Ni atom substitutes the central Ag atom of the Ag_{25} NC in the presence of a co-reductant (Eq. (5.9)).

$$Ag_{25}^- + Pt^{2+} + 2e^- \rightarrow \left[Pt^0Ag_{25}\right]^{1-} \rightarrow PtAg_{24}^{2-} + Ag^+ \tag{5.8}$$

$$Ag_{25}^- + Ni^{2+} + 2e^- \rightarrow \left[Ni^0Ag_{25}\right]^{1-} \rightarrow NiAg_{24}^0 + Ag^+ + 2e^- \tag{5.9}$$

The SWVs of the Ag_{25} and doped MAg_{24} (M = Pt, Pd, and Ni) NCs show well-resolved current peaks at the formal potentials of the NC charge-state couples (Figure 5.3c). The HOMO–LUMO gap energies determined for $[Ag_{25}]^-$, $[PtAg_{24}]^{2-}$, $[PdAg_{24}]^{2-}$, and $[NiAg_{24}]^0$ were 1.61, 1.76, 1.29, 0.76 V, respectively. DFT calculations predicted that the $[Ag_{25}]^-$, $[PtAg_{24}]^{2-}$, and $[PdAg_{24}]^{2-}$ NCs are 8e⁻ superatoms with $1S^21P^6$ electron configuration, accounting for the relatively large HOMO–LUMO gap energies observed for these NCs. By contrast, the $[NiAg_{24}]^0$ NC is a 6e⁻ superatom with $1S^21P^4$ configuration (Figure 5.3d). Again, the substantially reduced HOMO–LUMO gap energy observed for $[NiAg_{24}]^0$ can be explained by the Jahn–Teller effect. In addition, 8e⁻ $[NiAg_{24}]^{2-}$ NCs were obtained by the chemical reduction of the $[NiAg_{24}]^0$ NCs. The SC-XRD investigation revealed that $[NiAg_{24}]^{2-}$ consists of a centered icosahedral $NiAg_{12}$ core protected by six $Ag_2(SR)_3$ motifs, with an overall framework structure similar to that of $[Ag_{25}]^-$. Combined structural analysis of the $[NiAg_{24}]^0$ and $[NiAg_{24}]^{2-}$ NCs using SC-XRD and DFT calculations revealed that the $NiAg_{12}$ core of $[NiAg_{24}]^0$ was significantly distorted to an oblate spheroid shape (Figure 5.3d) compared to the nearly spherical core of the dianionic $[NiAg_{24}]^{2-}$, demonstrating the Jahn–Teller effect for the 6e⁻ $[NiAg_{24}]^0$. Figure 5.3c shows that the SWVs of $[NiAg_{24}]^0$ and $[NiAg_{24}]^{2-}$ NCs are significantly different, suggesting that they adopt distinctly different electronic structures. The former displays a reduced HOMO–LUMO gap owing to the Jahn–Teller effect, while the latter displays a large HOMO–LUMO gap energy (1.37 V), which is comparable to that of 8e⁻ superatom NCs [58].

5.3 Electrocatalytic Water Splitting on Atomically Precise Metal Nanoclusters

Hydrogen is considered an attractive alternative energy carrier and artificial catalytic systems that can efficiently produce H_2 from water have generated intense interest [15, 23, 61]. A variety of synthetic molecular HER catalysts, including nickel, iron, and molybdenum complexes, have been developed to mimic hydrogenase enzymes, natural biological catalysts [62]. However, the turnover frequency (TOF) values reported for these complexes remain quite low (4.1 mol_{H2} mol_{cat}^{-1} s^{-1}), and they often show limited stability in aqueous media [63]. Recent progress in the computational design of solid catalysts has revealed the importance of engineering the structure and adsorption energies for catalysis at surfaces [64, 65]. However, the inherent heterogeneity in the surface structure and composition often found in these solid catalysts hampers the fine control of these properties at the atomic level.

OER is a crucial process in energy conversion and storage, especially in water electrolysis, electrochemical CO_2 conversion, and in the functioning of metal–air rechargeable batteries and regenerative fuel cells [66]. State-of-the-art OER catalysts include precious metals and their oxides (Ru, Ir, RuO_2, and IrO_2) used to reduce energy consumption and enhance energy conversion efficiency. However, these precious catalysts also introduce significant economic barriers to large-scale applications owing to their limited supply and high cost. Finding alternative precious-metal-free catalysts to boost the kinetically sluggish OER is critical for practical applications in energy conversion and storage. This section discusses the results of electrocatalytic water splitting based on atomically precise metal NCs.

5.3.1 Hydrogen Evolution Reaction: Core Engineering

Many metal NCs consist of a metallic core and metal–ligand shells. The well-defined core–shell structure of NCs offers special advantages in the development of efficient electrocatalysts because the core and shell can be independently engineered to exhibit suitable binding energies and electron-transfer properties for specific electrocatalysis.

We examined the electrocatalytic activity of a core-doped $PtAu_{24}$ NC as a homogeneous HER electrocatalyst in tetrahydrofuran (THF) solution by increasing the concentration of trifluoroacetic acid (TFA) at a glassy carbon electrode (GCE) [30]. As shown in the linear sweep voltammograms (LSVs) in Figure 5.4a, the cathodic current for proton reduction was significantly increased in the presence of $PtAu_{24}$ NC. The HER catalytic activity of the $PtAu_{24}$ NCs can be quantified by calculating the pseudo-first-order rate constant (k_{obs}) for H_2 evolution catalyzed by freely diffusing catalysts. At a sufficiently high acid concentration relative to the catalyst, the following equation can be used to calculate the k_{obs} values [68]:

$$\frac{I_c}{I_p} = \frac{2}{0.446}\sqrt{\frac{RTk_{obs}}{Fv}} \tag{5.10}$$

where I_c is the catalytic current, I_p is the peak current in the absence of acid (taken in this case from the wave of $[PtAu_{24}]^{0/1-}$), 2 is the number of electrons involved in the catalytic reaction, R is the ideal gas constant, T is the temperature in Kelvin, F is Faraday's constant, and v is the scan rate.

Figure 5.4b shows the calculated k_{obs} for $PtAu_{24}$ as a function of potential. The k_{obs} value found at −0.89 V is 1300 s^{-1} and sharply increases to 62 000 s^{-1} at −1.1 V, followed by a gradual increase to 186 000 s^{-1} at −2.0 V vs. $Fc^{+/0}$. The k_{obs} values observed for $PtAu_{24}$ are much higher than those for Au_{25} under the same potentials. The HER mechanisms studied by correlating the catalytic current

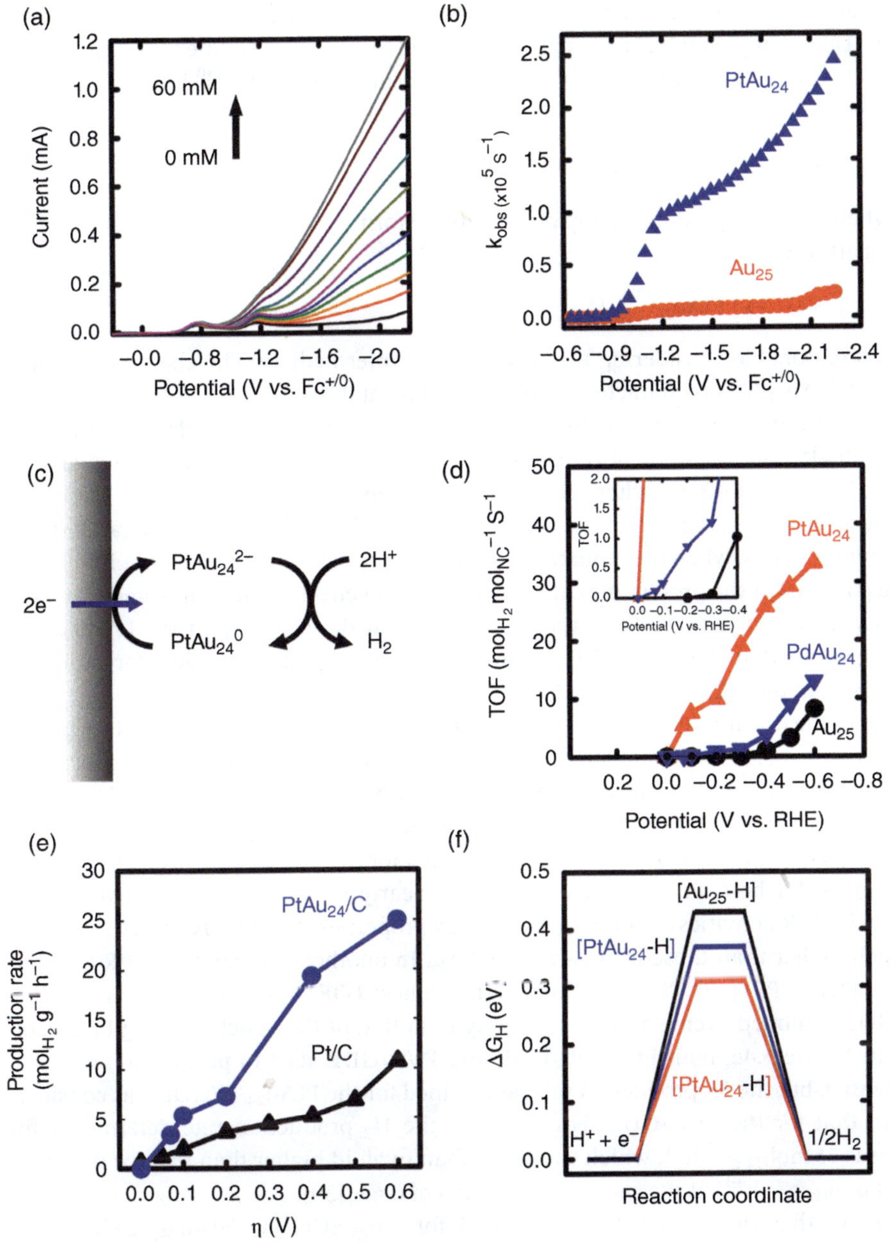

Figure 5.4 (a) LSVs of $PtAu_{24}$ (1 mM) in THF (0.1 M Bu_4NPF_6) in the presence of 0–60 mM TFA. (b) k_{obs}-potential plots for Au_{25} and $PtAu_{24}$ NCs in the presence of 1.0 M TFA. *Source:* Reproduced with permission from Ref. [30]. © 2017 Nature Publishing Group. (c) Schematic illustration of mediated catalytic hydrogen production by $PtAu_{24}$. *Source:* Reproduced with permission from Ref. [33]. © 2018 Royal Society of Chemistry. (d) Plots of TOF vs. potential obtained from CPE experiments on MAu_{24}/C/GDE, where MAu_{24} = Au_{25}, $PtAu_{24}$, and $PdAu_{24}$, in 1.0 M Britton–Robinson buffer solution and 2.0 M KCl (pH 3). The inset shows the enlarged graph in the low-overpotential region. *Source:* Reproduced with permission from Ref. [67]. © 2018 American Chemical Society. (e) HER mass activities (mol H_2 g^{-1} h^{-1}) of $PtAu_{24}$/C/GDE and Pt/C/GDE (1 cm^2) at various overpotentials. *Source:* Reproduced with permission from Ref. [30]. © 2017 Nature Publishing Group. (f) Calculated Gibbs free-energy diagram for ΔG_H on Au_{25}, $PdAu_{24}$, and $PtAu_{24}$. *Source:* Reproduced with permission from Ref. [67]. © 2018 American Chemical Society.

with the concentrations of TFA and catalyst revealed the charge-state-dependent catalytic activity of the $PtAu_{24}$ NC. Depending on the charge state of NC, the HER occurs via different pathways. In other words, at −1.0 V, $[PtAu_{24}]^-$ reacts with a proton to form an $[H–PtAu_{24}]^0$ intermediate that favors the homolytic HER pathway:

$$\left[H-PtAu_{24}\right]^0+\left[H-PtAu_{24}\right]^0 \rightarrow 2\left[PtAu_{24}\right]^0+H_2 \tag{5.11}$$

By contrast, when $PtAu_{24}$ is predominantly present in the form of $[PtAu_{24}]^{2-}$, the HER proceeds via a heterolytic pathway:

$$\left[H-PtAu_{24}\right]^-+H^+ \rightarrow \left[PtAu_{24}\right]^0+H_2 \tag{5.12}$$

Equation (5.12) is reasonable considering the charge of the intermediate. The charge-dependent catalytic activity of $PtAu_{24}$ NCs is characteristic of molecular catalysts.

The charge state is not the only factor that affects the HER activity of the NCs. Comparing $[PtAu_{24}]^{2-}$ with $[Au_{25}]^{2-}$, the k_{obs} value of $[PtAu_{24}]^{2-}$ (229 000 s^{-1}) is more than 10-fold higher than that of $[Au_{25}]^{2-}$ (22 000 s^{-1}) at −2.2 V, where both NCs are present in the dianionic form. The vastly different catalytic activity was ascribed to the reduction potential match between the catalyst and the substrate (proton), that is, when the reduction potential of the catalyst matches closely with the thermodynamic potential of the proton, the catalyst can act as an effective electron-transfer mediator to enhance its activity. The increase in the HER current at the potential of the $[PtAu_{24}]^{-/2-}$ couple strongly suggests that the $PtAu_{24}$ NC acts as an electron-transfer mediator for the HER, shuttling electrons from the GCE to the proton in the solution (Figure 5.4c).

Figure 5.4d shows the dopant-dependent HER activities evaluated by TOF from constant potential electrolysis (CPE) experiments conducted on a composite electrode fabricated by dropcasting a NC suspension containing carbon black (C) and Nafion on a gas diffusion electrode (NC/C/GDE). The HER onset potentials (E_{onset}) for both $PtAu_{24}$ and $PdAu_{24}$ were found to be −0.07 V vs. RHE (Figure 5.4d inset), significantly lower compared to that for Au_{25} [67]. The drastically reduced E_{onset} values observed for both $PtAu_{24}$ and $PdAu_{24}$ NCs clearly indicate that the Pt(Pd) doping directly affects their HER activities by altering their redox properties. The observed TOF values of $PtAu_{24}$ were much higher than those of Au_{25} and $PdAu_{24}$. In another comparative HER study of M_2Au_{36} (M = Au, Pt, and Pd) [67], Pt_2Au_{36} exhibited the highest TOF values among the three NCs. The $PtAu_{24}$/C/GDE exhibited even better mass activity than that of the benchmarking Pt/C catalyst. As shown in Figure 5.4e, both $PtAu_{24}$/C/GDE and Pt/C/GDE start to produce H_2 near the equilibrium potential, but the H_2 production rate determined for the $PtAu_{24}$/C/GDE was considerably higher than that for the Pt/C/GDE. For instance, the H_2 production rate determined for $PtAu_{24}$/C/GDE was 25 $mol_{H2}\ g^{-1}\ h^{-1}$, which was more than twofold higher than that for the Pt/C/GDE catalyst (11 $mol_{H2}\ g^{-1}\ h^{-1}$) at the same overpotential of 0.6 V [30].

Figure 5.4f shows that the ΔG_H values computed for $Au_{25}(SCH_3)_{18}$, $PdAu_{24}(SCH_3)_{18}$, and $PtAu_{24}(SCH_3)_{18}$ model NCs are 0.43, 0.36, and 0.31 eV, respectively. The significantly reduced ΔG_H calculated for $PtAu_{24}(SCH_3)_{18}$ clearly explains the higher HER activity observed in Figure 5.4d. The catalytic HER was greatly facilitated by metal doping of the core, which induced a substantial change in the ΔG_H value, highlighting another key descriptor for HER on metal NCs.

5.3.2 Hydrogen Evolution Reaction: Shell Engineering

Hydrogenases are key enzymes in the energy metabolism of many microorganisms [69]. They catalyze the HER from water with TOFs of 100 to 10 000 $mol_{H2}\ mol_{cat}^{-1}\ s^{-1}$ near their thermodynamic potential [70]. Structural analyses of the hydrogenase enzymes (e.g. [FeFe] hydrogenase) revealed that the active sites are located at the intersections of electron transport and proton transport chains [71]. The active site has a hydride-binding site adjacent to a pendant base that acts as a proton relay. Structural studies of hydrogenase enzymes have triggered numerous research efforts to synthesize structural analogs and investigate their catalytic activities. Rationally tailored NCs can be developed to possess an active core with suitable hydrogen adsorption energy and to be effectively surrounded by proton relays, as illustrated in Figure 5.5a.

In addition to the core engineering effect discussed in the previous section, the ligand effect on the HER was investigated using Au_{25} NCs protected by three different ligands: 1-hexanethiolate (C_6S), 3-mercaptopropionic acid (MPA), and 3-mercapto-1-propanesulfonic acid (MPS) [33]. The LSVs presented in Figure 5.5b show a remarkable ligand effect on the HER; for example, the HER currents are 0.01, 0.13, and 1.0 mA for C_6S-Au_{25}, MPA-Au_{25}, and MPS-Au_{25} NCs, respectively, at −0.7 V vs. RHE. The significantly higher catalytic current observed for MPA-Au_{25} and MPS-Au_{25} suggests that the functional group in the ligand exerts a major influence on the HER activity of the NC catalyst. The origin of this effect was ascribed to the ligand's ability to release protons; whereas the hexanethiolate ligand is impotent, the sulfonic acid ($pK_a < 1$) and carboxylic acid ($pK_a = 3.7$) groups are capable of releasing protons. A ligand-bearing acidic group releases more protons when its pK_a is lower than the local pH, which is subsequently transferred to the active site on the NC (Figure 5.5a). In other words, the proton relays effectively lower the energy barriers associated with the inter- and intramolecular proton-transfer steps, leading to an efficient HER. The proton relay effect has been observed in a number of hangman porphyrins bearing sulfonic acids, carboxylic acids, and amines [72, 73].

In Figure 5.5c, the TOF values obtained from homogeneous electrocatalysis in the presence of MPA-Au_{25}, MPS-Au_{25}, and MPS-$PtAu_{24}$ in water clearly reflect the core- and shell-engineering effects. The TOF value determined for MPS-Au_{25} was 28 mol_{H2} mol_{NC}^{-1} s^{-1}, which is approximately threefold higher than that for MPA-Au_{25} (9.5 mol_{H2} mol_{NC}^{-1} s^{-1}) at $\eta = 700$ mV. This result clearly demonstrates the effect of the proton-relaying ligand on the HER. The MPS-$PtAu_{24}$ NC exhibited further improvement in HER activity with a near-zero overpotential ($\eta = 100$ mV). The TOF value obtained from the MPS-$PtAu_{24}$ solution was 127 mol_{H2} mol_{NC}^{-1} s^{-1} at $\eta = 700$ mV, which is more than fourfold higher than that obtained for MPS-Au_{25}. This result clearly shows the extraordinary electrocatalytic activity of the core-engineered $PtAu_{24}$ protected with proton-relaying ligands.

As discussed earlier, Pt doping has a significant impact on the hydrogen binding energy of the host NC [30, 67]. The hydrogen adsorption energy calculated for MPS-$PtAu_{24}$ (Figure 5.5d) was thermodynamically neutral (−0.09 eV), whereas that for MPS-Au_{25} was considerably endothermic (0.37 eV). This result clearly shows that the thermoneutral catalyst for the hydrogen adsorption step could be prepared by Pt doping of the MPS-Au_{25} NCs. In addition, during geometry optimization, it was found that the adsorbed H atom spontaneously penetrated the Au_{12} subsurface to directly interact with the central Pt. The enhanced interaction appears to be the main factor that determines the favorable hydrogen adsorption energy of MPS-$PtAu_{24}$.

5.3.3 Hydrogen Evolution Reaction on Ag Nanoclusters

Similar metal-doping effects on the HER were observed for Ni-doped silver NCs [74]. Although HER is most effectively catalyzed by Pt-group metals in acidic media, the slow HER kinetics of Pt in alkaline media hinder the development of alkaline electrolyzers [75, 76]. Ni has often been the metal of choice for alloying and doping of a host metal [77], and thus Ni-doped $NiAg_{24}$ NCs were examined for the alkaline HER. As can be seen in Figure 5.6a, the Ni dopant is located at the center of the $NiAg_{12}$ core protected by six $Ag_2(SR)_3$ staples. The $NiAg_{24}$ NCs exhibited unique electronic structures and redox properties distinctly different from those of the undoped $Ag_{25}(SR)_{18}$ NCs. In alkaline media, the HER begins with the adsorption and dissociation of H_2O, producing *H on the

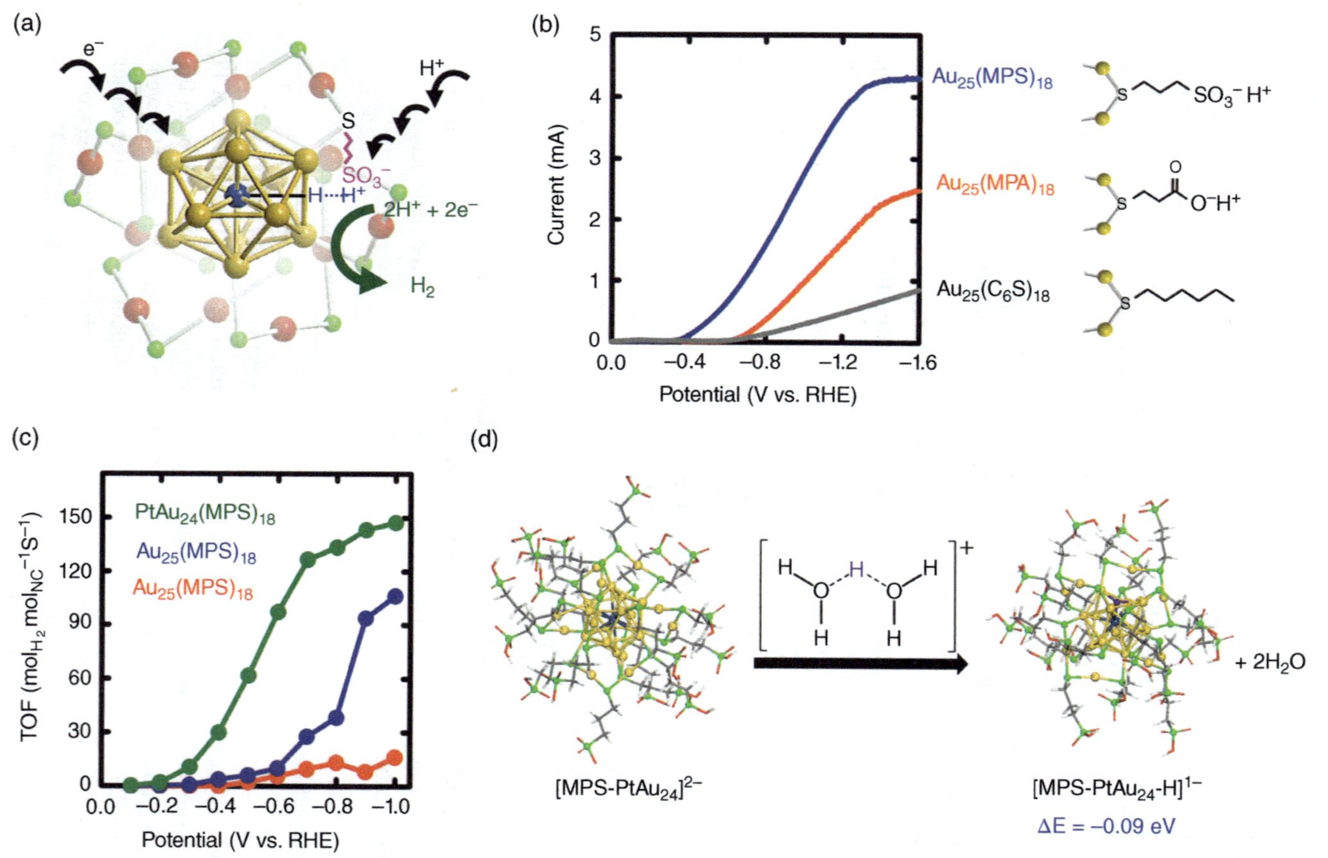

Figure 5.5 (a) Schematic illustration of the HER on a core–shell $PtAu_{24}(SR)_{18}$ NC. Only one ligand, SR = $S(CH_2)_3SO_3^-$, is shown for clarity. Color codes: blue = Pt; golden = core Au; red = shell Au; and green = S. The positioned proton relay adjacent to the bound hydrogen facilitates the efficient HER. (b) LSVs of 1 mM C_6S-Au_{25} in THF solution (0.1 M Bu_4NPF_6) containing 1.0 M TFA; 1 mM MPA-Au_{25} and 1 mM MPS-Au_{25} in aqueous solution (0.1 M KCl) containing 180 mM acetic acid. (c) TOFs obtained at various potentials after five minutes CPE experiments in water (3.0 M KCl) containing 180 mM acetic acid with a glassy carbon plate (1 cm^2) in the presence of 2.5 mM MPA-Au_{25}, 2.5 mM MPS-Au_{25}, and 2.5 mM MPS-$PtAu_{24}$. (d) Calculated reaction energy for H adsorption from the solvated proton to MPS-$PtAu_{24}$. Color codes: blue = Pt; golden = core Au and shell Au; green = S; gray = C, red = O; and white = H. A solvated proton is transferred from water molecules to [MPS-$PtAu_{24}]^{2-}$ to form [MPS-$PtAu_{24}$-H]$^{1-}$. *Source:* Reproduced with permission from Ref. [33]. © 2018 Royal Society of Chemistry.

catalyst surface (Eq. (5.13), Volmer step) [78]. The *H species then reacts with H_2O (Heyrovsky step) or recombines with neighboring *H (Tafel step) to produce H_2 (Eqs. (5.14) and (5.15)):

$$* + H_2O + e^- \rightarrow *H + OH^- \tag{5.13}$$

$$*H + H_2O + e^- \rightarrow H_2 + OH^- \tag{5.14}$$

$$*H + *H \rightarrow H_2 \tag{5.15}$$

The alkaline HER mechanism is described by the Volmer–Tafel or Volmer–Heyrovsky steps, and the mechanism can be investigated using Tafel analysis. In this analysis, the slopes are expected to be 120, 40, and 30 mV/dec when the Volmer, Heyrovsky, and Tafel steps, respectively, are associated with the rate-determining step (RDS). In the alkaline HER, water dissociation is usually the RDS because additional energy is required to dissociate the water molecule [78, 79]. Hence, facilitating the Volmer step is one of the strategies for developing an efficient HER catalyst.

Figure 5.6b shows that the Tafel slope of the Ag_{25} NC was 125 mV/dec, indicating that the HER was determined by the Volmer step. In contrast, the Tafel slope observed for the $NiAg_{24}$ NC was considerably reduced to 58 mV/dec, indicating that the Volmer step was significantly accelerated. The enhanced HER activities observed for the $NiAg_{24}$ NC in Figure 5.6b can be attributed to the enhanced water dissociation upon Ni doping. Furthermore, the TOF value obtained for $NiAg_{24}$ was

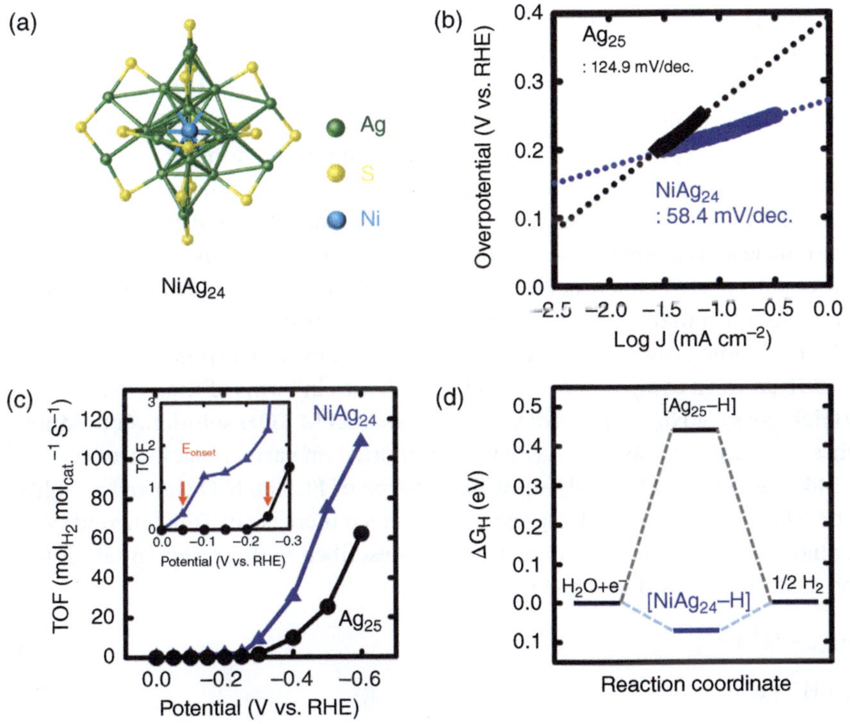

Figure 5.6 (a) Crystal structure of $NiAg_{24}(SPhMe_2)_{18}$ (green, Ag; blue, Ni; and yellow, S). Only the sulfur atoms of thiolate ligands are shown for clarity. (b) Tafel plots constructed for the electroreduction current in LSVs recorded at 20 mV/s on Ag_{25}/C/GCE and $NiAg_{24}$/C/GCE in 1.0 M KOH. (c) Average TOFs of H_2 obtained from CPE experiments in 1.0 M KOH for one hour on Ag_{25}/C/GDE and $NiAg_{24}$/C/GDE at different potentials. Inset shows the enlarged graph in the low-potential region. (d) Calculated Gibbs free-energy diagram for ΔG_H on $Ag_{25}(SCH_3)_{17}$ and $NiAg_{24}(SCH_3)_{17}$. *Source:* Reproduced with permission from Ref. [74]. © 2021 Wiley-VCH.

substantially higher; e.g. 108.5 mol_{H2} mol_{NC}^{-1} s^{-1} than that of Ag_{25} (62.2 mol_{H2} mol_{NC}^{-1} s^{-1}) at −0.6 V vs. RHE (Figure 5.6c). In addition, the Ni doping effect is clearly demonstrated by the positive shift of the onset potential for H_2 production (Figure 5.6c inset). The E_{onset} for Ag_{25} was determined to be −0.25 V, while that for $NiAg_{24}$ was positively shifted to −0.05 V, which is quite close to the equilibrium potential of H^+/H_2. The positive shift in E_{onset} can be the consequence of the reduction potential matching.

Dethiolation has also been reported to occur under HER conditions [80]. Hence, ΔG_H was computed on partially dethiolated NCs, namely, $Au_{25}(SCH_3)_{17}$ and $NiAu_{24}(SCH_3)_{17}$. The ΔG_H of Ag_{25} and $NiAg_{24}$ NCs calculated using DFT (Figure 5.6d) were 0.43 and −0.07 eV, respectively (in the latter case, the limiting step is the hydrogen-releasing step). Therefore, it can be concluded that the hydrogen adsorption property of NCs is substantially altered by Ni doping, which promotes HER activity in alkaline media.

5.3.4 Oxygen Evolution Reaction

The OER represents the anodic half-reaction in both the electrochemical water splitting and electroreduction of CO_2 electrolyzer systems. In these systems, the OER has limited the overall efficiency because it is a kinetically sluggish four-electron-transfer reaction [66]. The substantial overpotential involved in the OER process requires more energy consumption in addition to the thermodynamic potential E^0 (O_2/H_2O) = 1.23 V (vs. RHE). Thus, many research efforts have been devoted to the development of efficient OER catalysts with lower overpotentials [7, 11, 16]. In addition to the precious catalysts, nonprecious metal oxides, including nickel oxides, cobalt oxides, manganese oxides, and multication perovskites, have been extensively studied recently [66, 81]. Although these materials are promising OER candidates, large variations in the OER rates, onset potentials, and RDSs have been reported for these materials owing to the difficulty in accurately determining the reactive surface structure and number of active sites in these catalysts [81]. Structurally well-defined catalysts are indispensable for understanding the structure–property relationships at the atomic level that are needed to advance OER catalyst design.

In 2016, Kauffman et al. reported experimental and computational OER studies with a nickel NC, $Ni_6(PET)_{12}$ (PET = 2-phenylethanethiol) [82]. The general structure of $Ni_6(SR)_{12}$ NCs was first reported in 1965 [83]. The double-crown structure, in which all Ni sites were exposed, was resolved (Figure 5.7a). Figure 5.7b shows a comparison of the OER activities in terms of the TOF (left axis) and current density (right axis) of $Ni_6(PET)_{12}$, NiO, Ir, and Pt in a 0.1 M KOH solution. Ir produced the smallest OER onset potential of 1.493 V. $Ni_6(PET)_{12}$ (supported on carbon black) produced an onset potential of 1.544 V, which is statistically equivalent to that of Pt. The NiO showed a slightly higher OER onset at 1.575 V vs. RHE. $Ni_6(PET)_{12}$ exhibited a steep increase in TOF (current) with increasing overpotential. The OER mechanism is generally described by four sequential proton-coupled one-electron oxidation steps [Eqs. (5.16)–(5.19)].

$$H_2O + * \rightarrow OH_{ads} + H^+ + e^- \quad (5.16)$$

$$OH_{ads} \rightarrow O_{ads} + H^+ + e^- \quad (5.17)$$

$$O_{ads} + H_2O \rightarrow OOH_{ads} + H^+ + e^- \quad (5.18)$$

$$OOH_{ads} \rightarrow O_2(g) + H^+ + e^- \quad (5.19)$$

The initial OER step involves OH_{ads} formation/adsorption at a free metal site (denoted as *). Subsequent oxidation of OH_{ads} into O_{ads} (step 2, Eq. (5.17)) is followed by the combination of O_{ads} with a second water molecule to produce OOH_{ads} (step 3, Eq. (5.18)). The final oxidation step

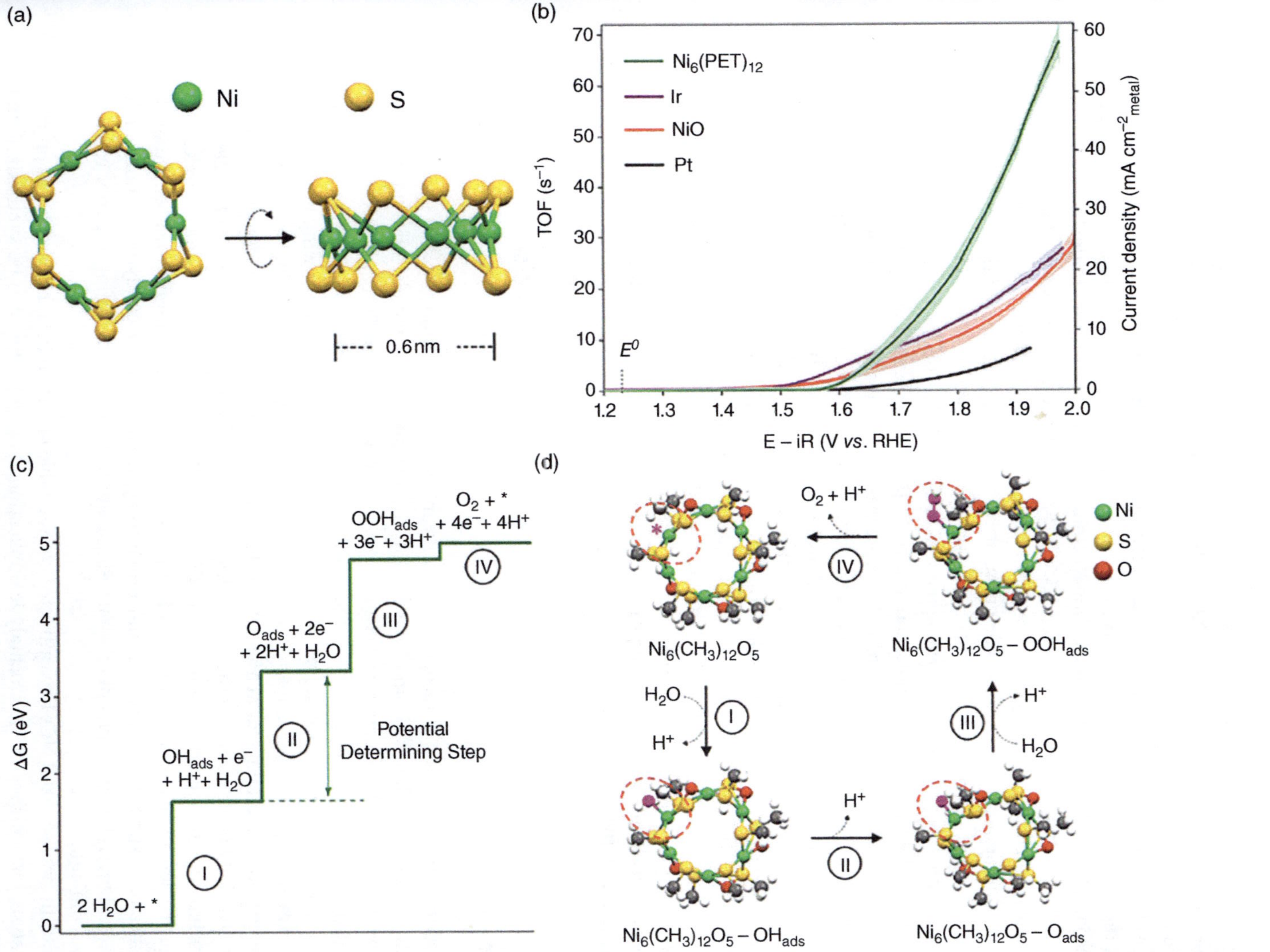

Figure 5.7 (a) Crystal structure of the $Ni_6(PET)_{12}$ NC. The organic component of the PET ligand is not shown. (b) Average OER voltammograms of $Ni_6(PET)_{12}$, NiO, Ir, and Pt in N_2 purged 0.1 M KOH. (c) Free-energy diagram for OER on O-covered $Ni_6(SCH_3)_{12}O_5$ at T = 298 K, pH = 0, and zero applied potential. The starting point consists of a free metal site (denoted as *) at $\Delta G = 0$, and each step corresponds to a one-electron oxidation process. (d) Predicted structures of OER intermediates on O-covered $Ni_6(SCH_3)_{12}O_5$. The organic ligands are included in this representation, and the vacant Ni site is marked with a pink asterisk. *Source:* Reproduced with permission from Ref. [82]. © 2016 American Chemical Society.

releases O_2 gas and leaves a free metal site available for additional catalytic cycles (step 4, Eq. (5.19)). Tafel analysis can help identify the RDS within the OER mechanism. For example, Tafel slopes of 120, 60, and 40 mV/dec indicate that the formation of OH_{ads} (step 1), O_{ads} (step 2), and OOH_{ads} (step 3), respectively, are the RDSs. An average Tafel slope of 69 mV/dec suggests that step 2 is the RDS. In another study, Joya et al. reported that both $Ni_4(PET)_8$ and $Ni_6(PET)_{12}$ NCs are highly active OER catalysts [84]. The $Ni_4(PET)_8$ NCs exhibited slightly higher activity than that of $Ni_6(PET)_{12}$, with a peak OER current density of ~150 mA cm^{-2} at 2.0 V and a Tafel slope of 38.69 mV/dec. The OER activities of the $Ni_4(PET)_8$ NCs were comparable to those of the benchmarking RuO_2 electrocatalysts [85].

The well-defined Ni NCs allowed detailed mechanistic studies and helped identify the active sites at the atomic level. Figure 5.7c presents a free-energy diagram for the OER using an oxygen-adsorbed model catalyst, $Ni_6(SCH_3)_{12}O_5$. One Ni atom is left vacant to represent the OER active site based on the mechanism presented in Eqs. (5.16)–(5.19). A vacant surface site is typically chosen as the starting point because O_2 evolution from the catalyst surface leaves a bare metal atom to continue additional catalytic cycles. The cumulative free-energy change for the overall reaction was $\Delta G = 4.96$ eV, and the resulting theoretical potential of 1.24 V (4.96 eV/4e$^-$) was in excellent agreement with the expected OER formal potential of $E^0 = 1.23$ V. The ΔG for each reaction step is denoted in the figure. The onset potential and corresponding potential-determining step can be extracted by identifying ΔG_{max}. Accordingly, step 2 was identified as a potential-determining step that agreed with the Tafel slope analysis. All electrochemical reaction steps become downhill in free energy with an applied potential larger than 1.68 V. The theoretical onset potential of 1.68 V was within 0.14 V of the experimentally determined value for $Ni_6(PET)_{12}$ in Figure 5.7b. Figure 5.7d shows the predicted structures of the adsorbed intermediates. $Ni_6(SCH_3)_{12}$ retained its structure throughout the OER process. Analysis of the optimized structure further indicated that step 2 ($OH_{ads} \rightarrow O_{ads}$) would be the RDS if the O_{ads} intermediate binds in a onefold coordination with Ni, as opposed to the twofold coordination with a Ni–S pair. The joint experimental–computational work on atomically precise catalysts appears to be a powerful combination for identifying atomic-level structure–activity relationships.

5.4 Electrocatalytic Conversion of CO_2 on Atomically Precise Metal Nanoclusters

The electrochemical CO_2RR is recognized as a promising method for sustainably transforming CO_2 into value-added chemicals and fuels [5, 16]. Among the various metals, Au- and Ag-based electrodes were found to exhibit high selectivity toward CO production. Cu-based electrodes were shown to produce C_{2+} products owing to their unique CO binding properties [17]. Nanostructured catalysts exhibit higher electrocatalytic activity than that of bulk electrodes owing to their large surface area and high number of low-coordinated sites. However, these nanocatalysts are typically polydisperse in shape and size at the atomic level [24–26], making the understanding of the structure–activity relationships at the atomic level difficult. In recent years, the use of single-atom catalysts has attracted significant attention in the CO_2RR [86, 87]. They exhibit superior selectivity and catalytic activity due to their high atomic uniformity and atom utilization. However, the single metal sites on a support are unstable and prone to aggregation under catalytic reaction conditions [88]. Hence, the development of robust single-atom catalysts that exhibit superior catalytic activity is difficult.

Atomically precise metal NCs have attracted considerable attention from researchers as promising catalysts for a variety of electrocatalytic applications [18, 89]. They can be synthesized and isolated with molecular purity and have well-defined structures that allow an atomic-level understanding of structure–activity relationships [36]. Furthermore, their electrocatalytic activity can be considerably enhanced via atomic-level modifications of their core and ligand shells. This section discusses how atomically precise NCs have been used in the CO_2RR to elucidate catalytic reaction mechanisms and identify active sites, which are crucial for developing highly efficient and selective electrocatalysts for the CO_2RR.

5.4.1 Mechanistic Investigation of CO_2RR on Au Nanoclusters

In 2012, Kauffman et al. first reported that a Au_{25} NC could efficiently catalyze the electroreduction of CO_2 to CO in an aqueous solution [90]. Using DFT calculations, they identified a CO_2-binding site comprising three sulfur atoms on the NC surface. However, the current density for CO production (j_{CO}) was quite low; for example, j_{CO} was less than 30 mA cm^{-2} at $\eta = 1.1$ V in a H-type cell. The CO_2RR is considered more cost-effective compared to that of conventional fossil fuel-based processes when a high current density for CO production ($j_{CO} > 300$ mA cm^{-2}) and high CO selectivity are achieved at a low overpotential [91]. Recently, GDE-based flow electrolyzers have been extensively studied for the CO_2RR [92, 93]. They provide a solution to the high-current CO_2RR by removing the solubility limit of CO_2 in water. Kim et al. first examined the CO_2RR activity of Au_{25} supported on a GDE in a flow cell using an alkaline electrolyte [94]. The Au_{25} catalysts supported on a GDE were covered by a thin electrolyte layer (Figure 5.8a), and the CO_2RR occurred at the catalyst/electrolyte interface. Under a continuous CO_2 stream, CO_2 molecules were continuously dissolved in the electrolyte and supplied to the reaction sites via short diffusion paths, overcoming the CO_2 mass transport limitation in aqueous solutions. In the flow cell, the Au_{25} electrode produced an industrially relevant current density of 540 mA cm^{-2} at $\eta = 0.7$V with a selectivity higher than 90% (Figure 5.8b).

To understand the origin of the extraordinary activity of the Au_{25} NC, an electrokinetic study was performed by plotting j_{CO} as a function of the applied potential. Figure 5.8c shows the resulting Tafel plot of log(j_{CO}) vs. η for the Au_{25} and Au/C (average Au diameter = 25.2 ± 3.9 nm) catalysts for comparison. The Tafel slope obtained for the Au_{25} NC was very small (43 mV/dec) in the kinetically controlled potential region, while that of Au/C was much larger (112 mV/dec). This result suggests that CO_2-to-CO electroreduction occurs via distinct pathways on these catalysts. The CO_2RR to CO on the catalyst surface in neutral–alkaline media can be described by the following elementary steps [94]:

$$* + CO_2 + e^- \rightarrow *CO_2^- \tag{5.20}$$

$$*CO_2^- + H_2O \rightarrow *COOH + OH^- \tag{5.21}$$

$$*COOH + e^- \rightarrow *COOH^- \tag{5.22}$$

$$*COOH^- \rightarrow *CO + OH^- \tag{5.23}$$

$$*CO \rightarrow * + CO \tag{5.24}$$

The CO_2RR activity was independent of the electrolyte pH in the pH range of 8–13, indicating that the proton-transfer step was not involved in the RDS. This result indicates that the RDS was associated with the first (Eq. (5.20)) and second (Eq. (5.22)) electron-transfer steps, for which the theoretical Tafel slopes of 120 and 40 mV/dec are expected. The experimentally determined Tafel slope of 43 mV/dec for the Au_{25} NC indicates that the second electron-transfer step (Eq. (5.22)) is the RDS for the CO_2RR, which is uniquely different from other gold surfaces.

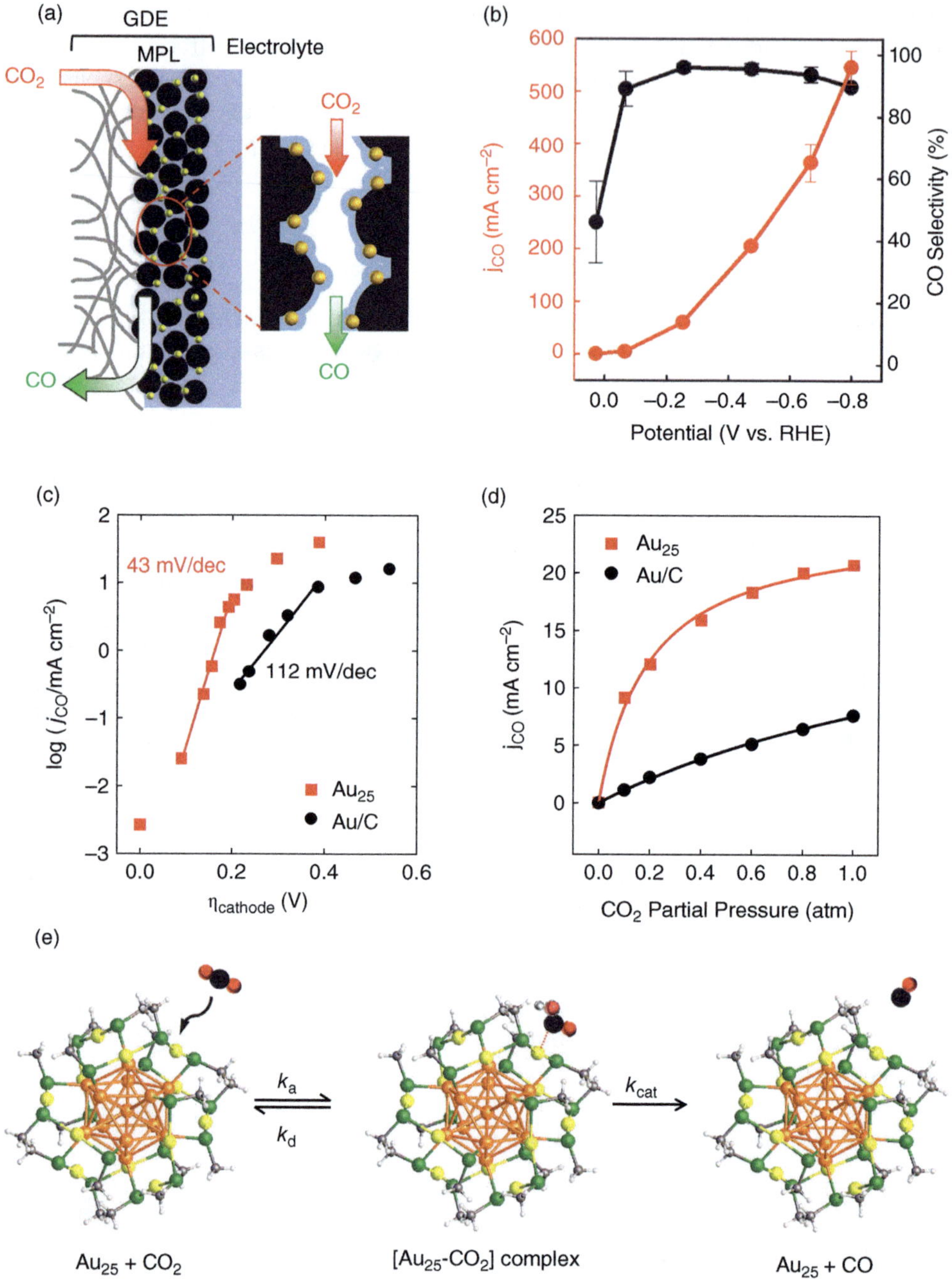

Figure 5.8 (a) Schematic of the CO_2RR on the Au_{25}-immobilized GDE (Au_{25}/GDE) in a flow cell. Au_{25} NCs were directly anchored on the microporous layer of the GDE. (b) j_{CO} and CO selectivity obtained via CPE experiments on the Au_{25}/GDE in a 5.0 M KOH solution (a 5 M KOH electrolyte was used to reduce the solution resistance). Error bars represent the standard deviation of three separate measurements. (c) Tafel plots for the electroreduction of CO_2 to CO on Au_{25} (red) and Au/C (black) in a 1.0 M KOH solution. (d) j_{CO} on Au_{25} and Au/C at −0.36 V vs. RHE as a function of the CO_2 partial pressure (P_{CO2}). (e) Two-step electroreduction process of CO_2 to CO on the Au_{25} NC. k_a, k_d, and k_{cat} denote the rate constants for CO_2 adsorption and desorption and the catalytic reaction, respectively. *Source:* Reproduced with permission from Ref. [94]. © 2020 American Chemical Society.

The investigation of the reaction order with respect to CO_2 concentration provides further insight into the mechanistic origin of the CO_2RR on the Au_{25} NC. Figure 5.8d shows the j_{CO} vs. CO_2 partial pressure (P_{CO2}) plot measured at −0.36 V vs. RHE on the Au_{25} and Au/C supported on a GDE in a flow electrolyzer. The experiments were conducted at a very low catalyst loading (16 μg/cm^2 of Au_{25}) to rapidly develop a steady state. Interestingly, a highly concave curve was observed for the Au_{25} NC, while Au/C exhibited the expected near-linear dependence of j_{CO} on P_{CO2}.

The well-defined Au_{25} NC directly interacts with the CO_2 molecule in a stoichiometric manner, resulting in a (quasi)equilibrium with a precisely defined intermediate. Therefore, we built a kinetic model based on Michaelis–Menten kinetics [95], which is widely used to describe enzyme kinetics. In this model, five elementary steps [Eqs. (5.20)–(5.24)] for the CO_2RR are reduced to a two-step process, as illustrated in Figure 5.8e. CO_2 binds to Au_{25} to generate a [Au_{25}–CO_2] complex intermediate (first step), which subsequently releases the CO product and regenerates the Au_{25} NC for the catalytic cycle (second step). This kinetic model works well for the CO_2RR on the Au_{25} NC, where the second step is the RDS, in adequate agreement with the Tafel analysis. The CO production rate is then expressed as follows:

$$v_{CO} = k_{cat}\left[\mathrm{Au}_{25} - \mathrm{CO}_2\right] \tag{5.25}$$

Under steady-state conditions, j_{CO} is expressed as:

$$j_{CO} = \frac{nFk_{cat}^{0}K\left[\mathrm{Au}_{25}\right]_0 P_{\mathrm{CO}_2}}{1 + KP_{\mathrm{CO}_2}} \exp\left(\beta\eta F / RT\right), \tag{5.26}$$

where $K = k_a/(k_d + k_{cat})$ is the CO_2 binding affinity of the catalyst, $k_{cat}{}^0$ is the standard rate constant, β is the symmetry factor, and the other symbols are as commonly known. By fitting the j_{CO}–P_{CO2} plot with Eq. (5.26), the K value of the catalyst was determined. While the K value of Au/C was only 0.5 atm^{-1}, a high K value of 5.2 atm^{-1} was obtained for the Au_{25} NC, explaining the highly concave shape and the high CO_2 binding affinity. The first electron-transfer step was significantly facilitated by the strong Au_{25}–CO_2 interaction, resulting in the rapid conversion of CO_2 to CO at modest overpotentials.

5.4.2 Identification of CO_2RR Active Sites

In the recent DFT studies on the CO_2RR on the thiolate-protected Au_{25} NC, Alfonso et al. and Austin et al. independently reported the promotion of the CO_2RR by the ligand removal from $Au_{25}(SR)_{18}$ under electrochemical CO_2RR conditions [96, 97]. The former predicted the removal of a thiolate ligand and the dethiolated Au was considered the active site [96]. The latter identified the dealkylated S site in $Au_{25}S(SR)_{17}$ as the active CO_2RR site based on the limiting potential comparison between the CO_2RR and the HER [97]. To experimentally determine the active sites, three structurally well-defined Au NCs, $Au_{25}(SR)_{18}$, $Au_{38}(SR)_{24}$, and $Au_{144}(SR)_{60}$ (Figure 5.9a), were used as the model catalysts for the electrochemical CO_2RR. Figure 5.9b displays the CO_2RR performances of the Au NCs supported on GDEs compared by conducting CPE experiments as functions of applied potentials in a 3.0 M KOH-fed flow cell. As shown in the figure, clear size-dependent CO_2RR activities are observed in these model catalysts, where CO_2RR activity improves with increasing size [80].

To investigate the structural changes in the thiolate-protected Au NCs during the CO_2RR, the chemical compositions of the Au NCs were analyzed using X-ray photoelectron spectroscopy (XPS) after electrochemical activation. Figure 5.9c shows the changes in the CO_2RR activity and the chemical composition of the Au_{25} NCs as functions of the electrochemical activation time.

As shown in Figure 5.9c, j_{CO} gradually increased with increasing activation time and leveled off after 60 minutes. Interestingly, the S/Au ratio determined using XPS gradually decreased with activation time, indicating that dethiolation occurred during the activation process. The chemical composition of the Au_{25} NCs determined after 60 minutes of activation was $Au_{25}(SR)_{12}$, indicating that approximately six thiolate ligands were removed from the Au_{25} NCs during the activation process. Considering the symmetrical structure of Au_{25}, which comprises six staple motifs, it can be inferred that a ligand was detached from each staple of this NC. The Au_{38} and Au_{144} NCs were

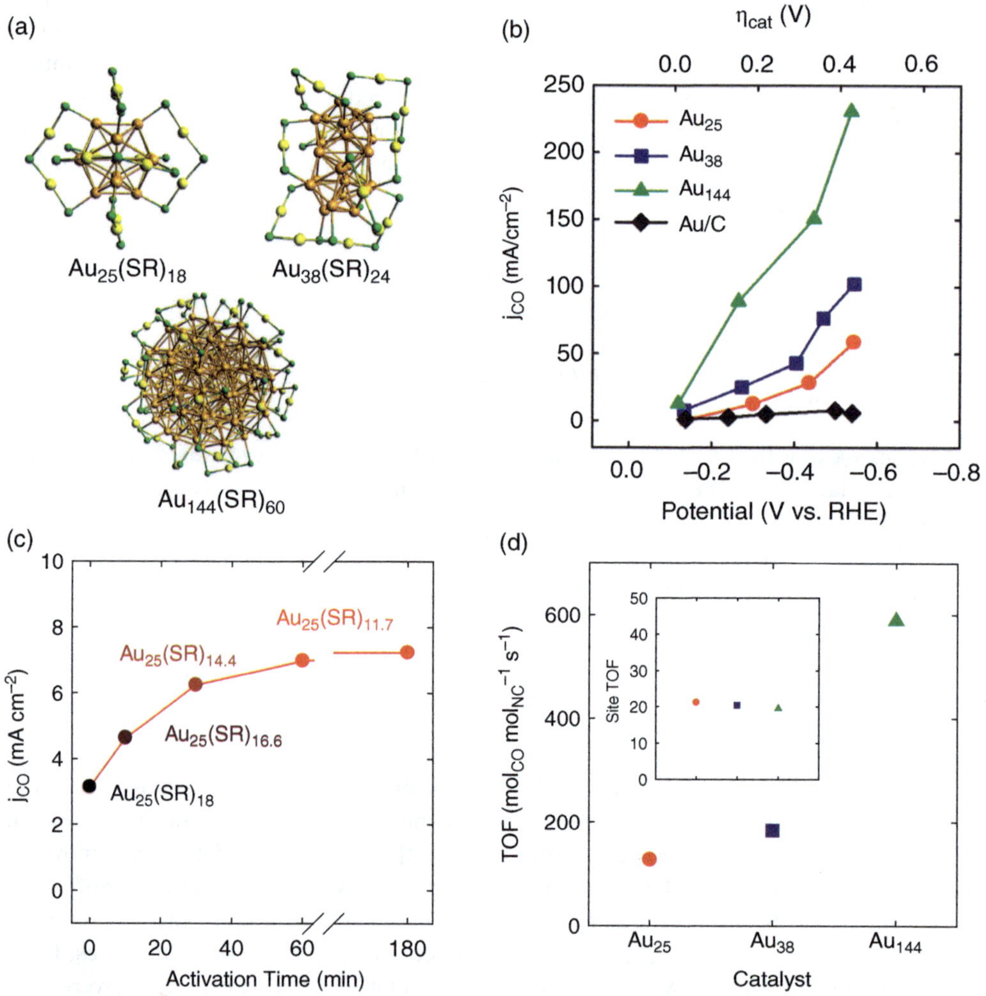

Figure 5.9 (a) Crystal structures of the $Au_{25}(SR)_{18}$, $Au_{38}(SR)_{24}$, and $Au_{144}(SR)_{60}$ NCs, where SR is a thiolate ligand (*Source:* redrawn from Refs. [98, 60, 99], respectively). Color codes: orange, core Au; yellow, shell Au; green, S (only the sulfur atoms of thiolate ligands are shown for clarity). (b) j_{CO} values measured for different Au NC/GDEs in the 3.0 M KOH electrolyte solution at various applied potentials. Note: the loading of the Au NCs on GDE was 2.1 nmol/cm^2, while the Au/C loading on GDE was 400 mg/cm^2. (c) j_{CO} values measured on Au_{25}/GDE at −0.16 V vs. RHE in the 1.0 M KOH electrolyte solution after electrochemical activation at −0.46 V vs. RHE for 0, 10, 30, and 60 minutes. The compositions of the dethiolated Au_{25} NCs are shown above the graph. (d) TOFs obtained for the Au_{25} (red circle), Au_{38} (blue square), and Au_{144} (green triangle) NC/GDEs in the 1.0 M KOH solution at −0.56 V vs. RHE. The inset shows the site TOFs ($mol_{CO}\ mol_{site}^{-1}\ s^{-1}$) determined for each active site of the Au NCs. *Source:* Reproduced with permission from Ref. [80]. © 2021 Wiley-VCH.

similarly activated, resulting in electrochemically activated NCs with formulars of $Au_{38}(SR)_{15}$ and $Au_{144}(SR)_{33}$, respectively. This result strongly suggests that dethiolation is universally observed for thiolate-protected Au NCs, and that one thiolate ligand is removed from each staple regardless of the staple type (monomeric $Au(SR)_2$ or dimeric $Au_2(SR)_3$). As shown in Figure 5.9d, the TOFs for CO production measured at −0.56 V vs. RHE for Au_{25}, Au_{38}, and Au_{144} NCs were significantly high (128, 186, and 590 mol_{CO} mol_{NC}^{-1} s^{-1}, respectively). Because each NC contained a different number of active sites, the site TOFs normalized to the number of active sites were calculated, which were equal to 21, 21, and 20 mol_{CO} mol_{site}^{-1} s^{-1}, respectively. The fact that close-site TOFs were obtained for all Au NCs suggests that site activity is independent of the staple type, and the CO_2RR activity was mainly determined by the total number of dethiolated sites.

The free energy diagrams presented in Figure 5.10a corroborate the experimental observations. The formation of the *COOH intermediate on the $Au_{25}(SCH_3)_{18}^-$ NC model represents the most uphill process with a reaction energy of 1.57 eV. However, when a single ligand is removed from the Au_{25} NC, its value sharply decreases to 0.09 eV, indicating that the CO_2RR process is significantly facilitated upon dethiolation. After the removal of five thiolate groups from the NC, the reaction energy remained low (0.14 eV). This result indicates that the thermodynamic barrier significantly decreases with dethiolation, and that the active sites of $Au_{25}(SCH3)_{17}^-$ and $Au_{25}(SCH_3)_{13}^-$ have similar binding properties. Figure 5.10b shows the free energy profiles of the HER, which is a competitor of the CO_2RR in aqueous media. It can be observed from the figure that dethiolation also significantly accelerates the formation of the *H intermediate by reducing the limiting potential from 1.68 to 0.32 eV for both dethiolated NCs. The experimentally observed high selectivity toward CO_2 reduction over H_2 evolution can be qualitatively explained by the higher barrier of the HER compared to that of the CO_2RR (0.32 vs. 0.14 eV). In these calculations, we assumed that the charge states of all NCs were −1. Surprisingly, selectivity was reversed for the neutral $Au_{25}(SCH_3)_{13}^0$ NC, which favors the thermoneutral HER (0.01 eV) over the CO_2RR (0.46 eV).

For both intact $Au_{25}(SCH_3)_{18}^-$ and dethiolated $Au_{25}(SCH_3)_{13}^-$ NCs, the sulfur-bound shell Au sites were theoretically predicted as the active sites for the CO_2RR, as shown in Figure 5.10c. The projected density of states (PDOS) in Figure 5.10d shows that the *d*-states of the shell Au in the $Au_{25}(SCH_3)_{13}^-$ NC are considerably higher than those of the shell Au in the $Au_{25}(SCH_3)_{18}^-$ NC, while the *sp*-states are similar. The energy level of the *d*-states (vs. the Fermi level, E_F) is an adequate indicator of the bond strength of a metal catalyst with an adsorbate [100, 101]. The metal-adsorbate bond strength typically increases with the energy level of the *d*-states. However, if the *d*-state energy is excessively high, the strong *CO intermediate produces additional reduced products, such as CH_4 and C_{2+} molecules [102]. Consequently, the metal surfaces with optimal *d*-state energies lead to high CO_2-to-CO conversion activities [102]. As shown in Figure 5.10d, the *d*-state energy of the dethiolated shell Au of the $Au_{25}(SCH_3)_{13}^-$ NC increases from −3.2 to −2.6 eV vs. E_F. The exceptional CO_2-to-CO conversion activity observed in the dethiolated NCs can be ascribed to the moderate upshift of the *d*-states, which endowed the dethiolated Au NCs with an appropriate binding strength for the CO_2 intermediates. The atomic-level investigations using structurally well-defined NCs clearly revealed the atomic origin of their high CO_2RR activities.

5.4.3 CO_2RR on Cu Nanoclusters

Among the various electrocatalysts, Cu catalysts have attracted significant attention due to their ability to convert CO_2 to CO and even to C_{2+} products through the strongly bound *CO intermediate [20]. Despite numerous research efforts, the key descriptor controlling the selectivity of CO_2RR on the Cu surface remains unaddressed at the atomic level. The atomically well-defined Cu-hydride NC, $Cu_{32}H_{20}L_{12}$ [L = $S_2P(O^iPr)_2$], shows that the CO_2RR selectivity can be significantly

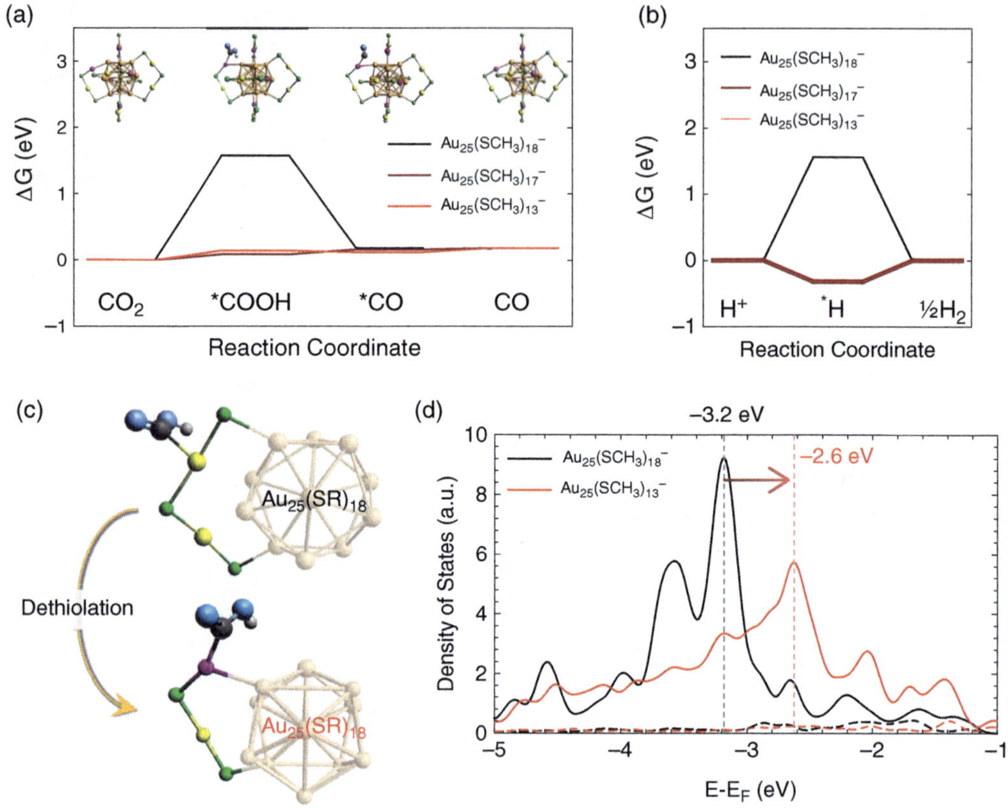

Figure 5.10 Free energy (ΔG) diagrams describing (a) CO_2 reduction to CO and (b) H^+ reduction to H_2 on the $Au_{25}(SCH_3)_{18}^-$, $Au_{25}(SCH_3)_{17}^-$, and $Au_{25}(SCH_3)_{13}^-$ NCs at zero applied potential. The inset illustrates the CO_2RR occurring at the active site of the dethiolated $Au_{25}(SCH_3)_{13}^-$ NC. (c) Schematic of the CO_2 adsorption on the intact and dethiolated Au_{25} NCs. Color codes: gray, C; blue, O; white, H; orange, core Au; yellow, intact shell Au; magenta, dethiolated shell Au; green, S (the alkyl chains of the thiolate ligands and other five staple motifs are omitted for clarity). (d) Projected density of states (PDOS) of the sp- (dashed lines) and d-states (solid lines) of the staple Au in the $Au_{25}(SCH_3)_{18}^-$ (black) and $Au_{25}(SCH_3)_{13}^-$ (red) NCs. *Source:* Reproduced with permission from Ref. [80]. © 2021 Wiley-VCH.

altered by the surface chemistry of the NC catalysts [103]. Figure 5.11a shows the results of the CO_2RR on the Cu_{32} NC. As shown in the figure, HCOOH (as $HCOO^-$ at pH 6.8) is predominantly produced at low overpotentials (89% at 0.3 V and 83% at 0.4 V vs. RHE) with minor amounts of CO and H_2. The product selectivity significantly changes when the overpotential is higher than 0.5 V, where H_2 is predominantly produced. It has been observed that HCOOH and CO formations are competitive at the early stage of the CO_2RR on the Cu surface, and CO is the major product [104]. In contrast, in the Cu_{32} NC, the presence of the negatively charged hydride facilitated the formation of the HCOO* intermediate, favoring the formation of HCOOH instead of CO. The Cu_{32} NC, which contains both capping and interstitial hydrides, offers distinctly different pathways and product selectivity.

The free-energy diagrams for HCOOH and CO formations presented in Figure 5.11b corroborate the experimentally observed selectivity. It can be observed that COOH formation is energetically favored by the HCOO* intermediate (activation energy: 0.89 eV) over CO formation, which is limited by the formation of the COOH* intermediate (activation energy: 2.12 eV). It was further observed that HCOOH formation proceeds through two steps of lattice-hydride reduction: CO_2

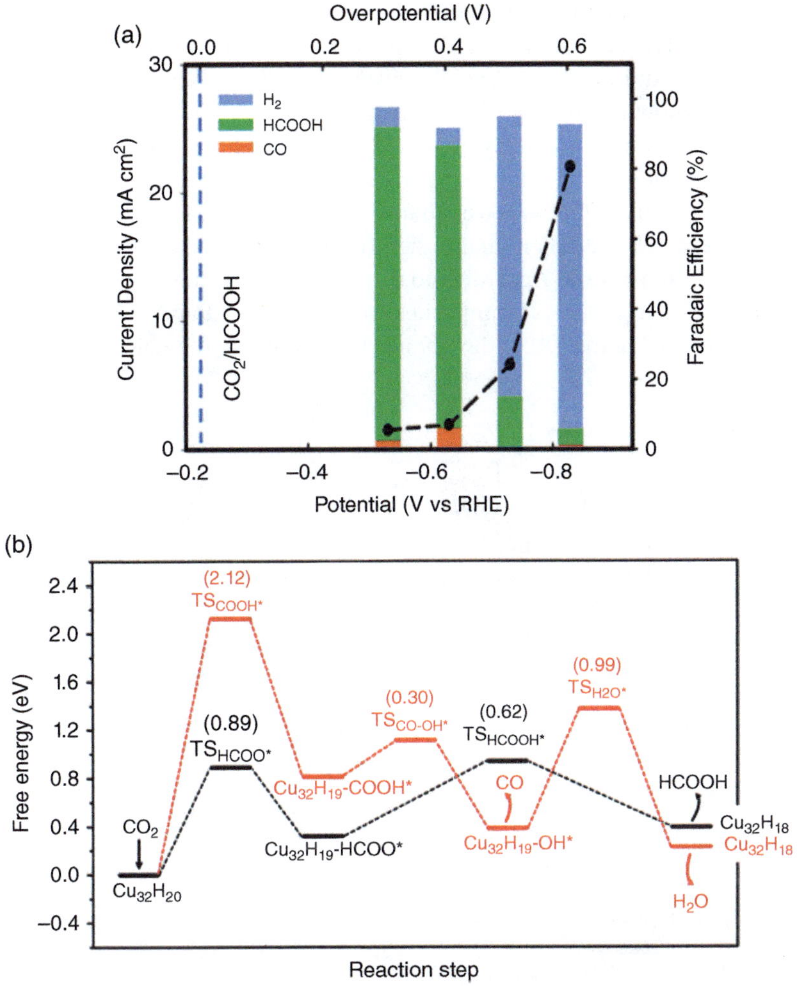

Figure 5.11 (a) Total current densities (black circles) and Faradaic efficiencies for H_2, HCOOH, and CO at various applied potentials. CPEs were conducted for 90 minutes in the 0.1 M $KHCO_3$ and 0.4 M KCl (pH 6.8) electrolyte solutions on a Cu_{32}/C supported on the GDE. (b) Free energy profiles for the intermediates produced during the HCOOH and CO formation on the $Cu_{32}H_{20}L_{12}$ NC via the lattice-hydride mechanism. TS is transition state, and the L is omitted for clarity. The reaction energies are written in parentheses. *Source:* Reproduced with permission from Ref. [103]. © 2017 American Chemical Society.

directly reacts with capping hydride to form HCOO*, which subsequently reacts with another interstitial hydride to form HCOOH. The lost hydrides are replenished via proton reduction processes from the solution.

5.4.4 Syngas Production on Formulated Metal Nanoclusters

In the previous sections, we discussed the significant effects of the core composition on the electrocatalytic activity and selectivity. Although the $PtAu_{24}$ NC exhibits remarkable HER activity [30, 67], the Au_{25} NC selectively produces CO during the CO_2RR in aqueous media [90, 94]. The core-dependent selectivity of metal NCs can be exploited to produçe synthetic gas (syngas), a gaseous mixture of CO and H_2. Syngas is a fundamental industrial feedstock that can readily be converted into various advanced products via the alteration of the H_2/CO ratio of the syngas; for example, a H_2-to-CO ratio of 3 for producing CH_4, 2 for CH_3OH, and 1 for hydroformylation [24, 105].

Figure 5.12a,b show that the Au_{25} and $PtAu_{24}$ NCs exhibit distinctly different selectivities in the CO_2RR in aqueous solutions. As shown in the figure, the Au_{25} NCs produced CO almost exclusively at low overpotentials. The TOF values gradually increased with increasing overpotential and stabilized at approximately 7 mol_{CO} mol_{NC}^{-1} s^{-1} due to the mass transport limitation (in a H-type cell). The CO production on the $PtAu_{24}$ NC was lower than that on the Au_{25} NC; however, H_2 production was dominant even at a low overpotential and significantly increased with overpotential, reaching 19 mol_{H2} mol_{NC}^{-1} s^{-1} at $\eta = 1.0$ V. According to their TOFs for CO and H_2 production at -1.0 V vs. RHE, four Au_{25} and $PtAu_{24}$ catalyst composites were formulated and subsequently examined for syngas production. In Figure 5.12c, the comparison between the calculated and

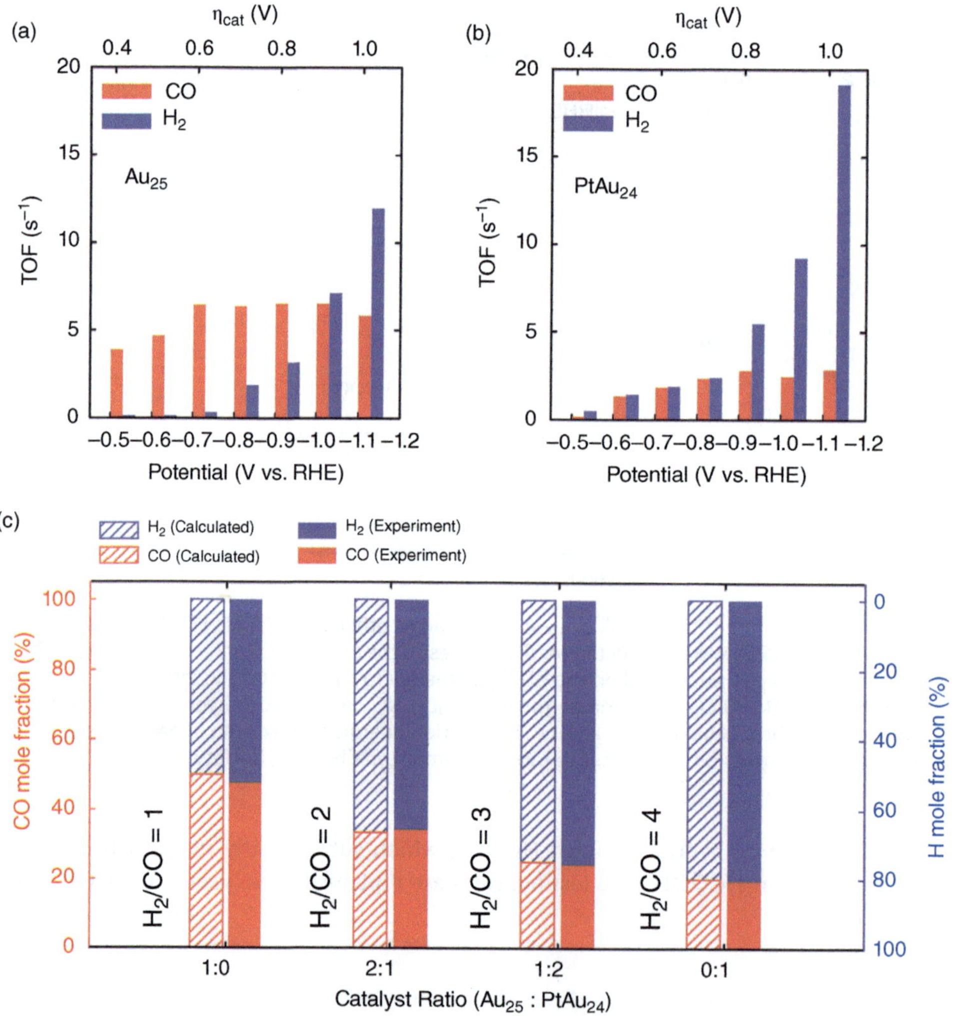

Figure 5.12 TOFs ($mol_{product}$ mol_{NC}^{-1} s^{-1}) of CO and H_2 obtained from CPE experiments on the (a) Au_{25}/C/GDE and (b) $PtAu_{24}$/C/GDE at various potentials. The mass loadings of both Au_{25} and $PtAu_{24}$ NCs were 80 μg/cm^2. (c) Calculated (shaded) and experimentally determined (filled) H_2/CO ratio on the formulated Au_{25} and $PtAu_{24}$ catalysts (total mass loading was 80 μg/cm^2 for all experiments). The CPE experiments were conducted at −1.0 V vs. RHE for one hour in a CO_2-saturated solution of 0.1 M $KHCO_3$ and 0.4 M KCl. *Source:* Reproduced with permission from Ref. [106]. © 2021 AIP Publishing.

experimental syngas H_2/CO production on the four formulated composites shows that the H_2/CO syngas ratios experimentally obtained were 1.1, 1.9, 3.1, and 4.1, which matched well with the calculated ratios of 1.0, 2.0, 3.0, and 4.0, respectively. The substantially different selectivities of the Au_{25} and $PtAu_{24}$ NCs and their undisturbed catalytic activities in the formulated electrodes appear to be the enabling factors for the precise control of syngas production [106].

5.5 Conclusions and Outlook

The electrochemistry of atomically precise metal NCs has revealed significant information about their redox properties, which are very responsive to the changes in their size, structure, and composition. Voltammetry is particularly informative regarding their electronic structures, especially near the HOMO and LUMO levels. Understanding the structural parameters that control the redox properties is essential for their practical applications. Crucial to achieving this goal is the synthesis of a variety of well-defined metal NCs via size-focusing syntheses, metal doping, and ligand modifications. Advancing the synthetic routes to tunable atom-precise metal NCs with molecular uniformity will further enable electrochemical methods to elucidate the kinetics of electron transfer, which is crucial for their electrocatalytic utilization. These concerted efforts will enable a profound understanding of the structure–electrochemical property relationship.

Combined theoretical and experimental approaches based on atom-precise metal NCs have been helpful in the development of electrocatalysts for water splitting and the CO_2RR. Metal NCs with well-defined core-shell structures offer special advantages in that the core and shell can be independently engineered to exhibit adequate properties for electrocatalysis. The doped metal in the core, although buried in the central position, has a significant impact on the binding properties of NC catalysts. Highly selective chemical transformations have been demonstrated with the $PtAu_{24}$ (HER), Au_{25} ($CO_2 \rightarrow CO$), Cu_{32} ($CO_2 \rightarrow HCOOH$), and $Au_{25} + PtAu_{24}$ (syngas) NCs that possess suitable binding properties for specific reactions. Redox potentials are another key descriptor for electrocatalysis, which, depending on their size, structure, and composition, can make NC an efficient electron-transfer mediator for targeted electrocatalysis. Future studies are required to elucidate the details of binding property tuning with atomically well-defined model catalysts.

We have shown how atom-precise model catalysts help in identifying the active sites for the HER, OER, and CO_2RR. The combined efforts of theoretical calculations and electrochemical methods enabled the identification of the key intermediates involved, their bonding to the catalyst, and the energetics of each step. An in-depth understanding of the reaction mechanisms will lead to successful design strategies for the development of improved electrocatalysts. In electrocatalytic reactions that involve multiple intermediates in particular, the intrinsic overpotential is largely set by scaling relations that exist between the energies of different adsorbed intermediates. For two-dimensional solid surfaces with monotonous binding sites, it is difficult to decouple the binding energies of different intermediates. Highly tunable metal NCs will provide a powerful platform to break the scaling relation by stabilizing one intermediate relative to another. The free energy profile in Figure 5.10a shows that this is indeed possible with the dethiolated Au_{25} NC, i.e. the *COOH intermediate could be selectively stabilized without significantly affecting the energy of the *CO intermediate.

The presence of a protecting ligand shell in metal NCs plays a crucial role in electrocatalysis. Unlike large nanoparticles, where the ligand shell inhibits access of the reactant to the catalytic sites, the ligand shell of the metal NC is quite open, allowing access of small reactants such as H^+/H_2O, O_2, and CO_2. In addition, they can be further tailored to increase the catalytic activity, selectivity, and durability of the active site. Further efforts are needed to elucidate the details of the active site-ligand-electrolyte interface. A molecular-level understanding of the intermediates, solvents, cations, and anions near the interface will be the focus of future research. The use of appropriate in situ and *operando* techniques will considerably accelerate these studies. We envision that a systematic investigation of active sites and corresponding catalytic environments using atom-precise model catalysts will uncover key descriptors for electrocatalysis that can further guide the community to develop improved electrocatalysts with optimal efficiency and selectivity.

Acknowledgments

This research was supported by the grants from the National Research Foundation of Korea (NRF; Grant No. NRF-2022R1A2C3003610) and the Carbon-to-X Project (Project No. 2020M3H7A1096344) through the NRF funded by the Ministry of Science and ICT, Republic of Korea.

References

1 Turner, J.A. (1999). A realizable renewable energy future. *Science* 285: 687–689.

2 Lewis, N.S. and Nocera, D.G. (2006). Powering the planet: chemical challenges in solar energy utilization. *Proc. Natl. Acad. Sci. U. S. A.* 103: 15729–15735.

3 Giddey, S., Badwal, S.P.S., and Kulkarni, A. (2013). Review of electrochemical ammonia production technologies and materials. *Int. J. Hydrogen Energy* 38: 14576–14594.

4 Luna, P.D., Hahn, C., Higgins, D. et al. (2019). What would it take for renewably powered electrosynthesis to displace petrochemical processes? *Science* 364: eaav3506.

5 Qiao, J., Liu, Y., Hong, F., and Zhang, J. (2014). A review of catalysts for the electroreduction of carbon dioxide to produce low-carbon fuels. *Chem. Soc. Rev.* 43: 631–675.

6 Mistry, H., Varela, A.S., Kühl, S. et al. (2016). Nanostructured electrocatalysts with tunable activity and selectivity. *Nat. Rev. Mater.* 1: 16009.

7 McCrory, C.C.L., Jung, S., Peters, J.C., and Jaramillo, T.F. (2013). Benchmarking heterogeneous electrocatalysts for the oxygen evolution reaction. *J. Am. Chem. Soc.* 135: 16977–16987.

8 Zhang, L., Zhao, Z.-J., and Gong, J. (2017). Nanostructured materials for heterogeneous electrocatalytic CO_2 reduction and their related reaction mechanisms. *Angew. Chem. Int. Ed.* 56: 11326–11353.

9 Li, X., Hao, X., Abudula, A., and Guan, G. (2016). Nanostructured catalysts for electrochemical water splitting: current state and prospects. *J. Mater. Chem. A* 4: 11973–12000.

10 Tian, X., Zhao, P., and Sheng, W. (2019). Hydrogen evolution and oxidation: mechanistic studies and material advances. *Adv. Mater.* 31: 1808066.

11 Wang, X.-R., Liu, J.-Y., Liu, Z.-W. et al. (2018). Identifying the key role of pyridinic-N–Co bonding in synergistic electrocatalysis for reversible ORR/OER. *Adv. Mater.* 30: 1800005.

12 Ramaswamy, N., Tylus, U., Jia, Q., and Mukerjee, S. (2013). Activity descriptor identification for oxygen reduction on nonprecious electrocatalysts: linking surface science to coordination chemistry. *J. Am. Chem. Soc.* 135: 15443–15449.

13 Rebollar, L., Intikhab, S., Oliveira, N.J. et al. (2020). Beyond adsorption. Descriptors in Hydrogen Electrocatalysis. *ACS Catal.* 10: 14747–14762.

14 Parsons, R. (1958). The rate of electrolytic hydrogen evolution and the heat of adsorption of hydrogen. *Trans. Faraday Soc.* 54: 1053–1063.

15 Nørskov, J.K., Bligaard, T., Logadottir, A. et al. (2005). Trends in the exchange current for hydrogen evolution. *J. Electrochem. Soc.* 152: J23.

16 Seh, Z.W., Kibsgaard, J., Dickens, C.F. et al. (2017). Combining theory and experiment in electrocatalysis: insights into materials design. *Science* 355: eaad4998.

17 Kuhl, K.P., Hatsukade, T., Cave, E.R. et al. (2014). Electrocatalytic conversion of carbon dioxide to methane and methanol on transition metal surfaces. *J. Am. Chem. Soc.* 136: 14107–14113.

18 Kwak, K. and Lee, D. (2019). Electrochemistry of atomically precise metal nanoclusters. *Acc. Chem. Res.* 52: 12–22.

19 Benck, J.D., Hellstern, T.R., Kibsgaard, J. et al. (2014). Catalyzing the hydrogen evolution reaction (HER) with molybdenum sulfide nanomaterials. *ACS Catal.* 4: 3957–3971.

20 Nitopi, S., Bertheussen, E., Scott, S.B. et al. (2019). Progress and perspectives of electrochemical CO_2 reduction on copper in aqueous electrolyte. *Chem. Rev.* 119: 7610–7672.

21 Zhang, Y.-J., Sethuraman, V., Michalsky, R., and Peterson, A.A. (2014). Competition between CO_2 reduction and H_2 evolution on transition-metal electrocatalysts. *ACS Catal.* 4: 3742–3748.

22 Goyal, A., Marcandalli, G., Mints, V.A., and Koper, M.T.M. (2020). Competition between CO_2 reduction and hydrogen evolution on a gold electrode under well-defined mass transport conditions. *J. Am. Chem. Soc.* 142: 4154–4161.

23 Popczun, E.J., McKone, J.R., Read, C.G. et al. (2013). Nanostructured nickel phosphide as an electrocatalyst for the hydrogen evolution reaction. *J. Am. Chem. Soc.* 135: 9267–9270.

24 Mistry, H., Reske, R., Zeng, Z. et al. (2014). Exceptional size-dependent activity enhancement in the electroreduction of CO_2 over Au nanoparticles. *J. Am. Chem. Soc.* 136: 16473–16476.

25 Gao, D., Zhou, H., Wang, J. et al. (2015). Size-dependent electrocatalytic reduction of CO_2 over Pd nanoparticles. *J. Am. Chem. Soc.* 137: 4288–4291.

26 Zhu, W., Michalsky, R., Metin, Ö. et al. (2013). Monodisperse Au nanoparticles for selective electrocatalytic reduction of CO_2 to CO. *J. Am. Chem. Soc.* 135: 16833–16836.

27 Jin, R., Qian, H., Wu, Z. et al. (2010). Size focusing: a methodology for synthesizing atomically precise gold nanoclusters. *J. Phys. Chem. Lett.* 1: 2903–2910.

28 Jin, R. (2010). Quantum sized, thiolate-protected gold nanoclusters. *Nanoscale* 2: 343–362.

29 Kwak, K., Thanthirige, V.D., Pyo, K. et al. (2017). Energy gap law for exciton dynamics in gold cluster molecules. *J. Phys. Chem. Lett.* 8: 4898–4905.

30 Kwak, K., Choi, W., Tang, Q. et al. (2017). A molecule-like $PtAu_{24}(SC_6H_{13})_{18}$ nanocluster as an electrocatalyst for hydrogen production. *Nat. Commun.* 8: 14723.

31 Li, S., Alfonso, D., Nagarajan, A.V. et al. (2020). Monopalladium substitution in gold nanoclusters enhances CO_2 electroreduction activity and selectivity. *ACS Catal.* 10: 12011–12016.

32 Zhao, S., Austin, N., Li, M. et al. (2018). Influence of atomic-level morphology on catalysis: the case of sphere and rod-like gold nanoclusters for CO_2 electroreduction. *ACS Catal.* 8: 4996–5001.

33 Kwak, K., Choi, W., Tang, Q. et al. (2018). Rationally designed metal nanocluster for electrocatalytic hydrogen production from water. *J. Mater. Chem. A* 6: 19495–19501.

34 Narouz, M.R., Osten, K.M., Unsworth, P.J. et al. (2019). N-heterocyclic carbene-functionalized magic-number gold nanoclusters. *Nat. Chem.* 11: 419–425.

35 Li, S., Nagarajan, A.V., Li, Y. et al. (2021). The role of ligands in atomically precise nanocluster-catalyzed CO_2 electrochemical reduction. *Nanoscale* 13: 2333–2337.

36 Jin, R., Zeng, C., Zhou, M., and Chen, Y. (2016). Atomically precise colloidal metal nanoclusters and nanoparticles: fundamentals and opportunities. *Chem. Rev.* 116: 10346–10413.

37 Du, Y., Sheng, H., Astruc, D., and Zhu, M. (2020). Atomically precise noble metal nanoclusters as efficient catalysts: a bridge between structure and properties. *Chem. Rev.* 120: 526–622.

38 Jin, R. (2015). Atomically precise metal nanoclusters: stable sizes and optical properties. *Nanoscale* 7: 1549–1565.

39 Murray, R.W. (2008). Nanoelectrochemistry: metal nanoparticles, nanoelectrodes, and nanopores. *Chem. Rev.* 108: 2688–2720.

40 Hicks, J.F., Miles, D.T., and Murray, R.W. (2002). Quantized double-layer charging of highly monodisperse metal nanoparticles. *J. Am. Chem. Soc.* 124: 13322–13328.

41 Ingram, R.S., Hostetler, M.J., Murray, R.W. et al. (1997). 28 kDa alkanethiolate-protected Au clusters give analogous solution electrochemistry and STM Coulomb staircases. *J. Am. Chem. Soc.* 119: 9279–9280.

42 Green, S.J., Stokes, J.J., Hostetler, M.J. et al. (1997). Three-dimensional monolayers: nanometer-sized electrodes of alkanethiolate-stabilized gold cluster molecules. *J. Phys. Chem. B* 101: 2663–2668.

43 Kang, X., Chong, H., and Zhu, M. (2018). $Au_{25}(SR)_{18}$: the captain of the great nanocluster ship. *Nanoscale* 10: 10758–10834.

44 Dharmaratne, A.C., Krick, T., and Dass, A. (2009). Nanocluster size evolution studied by mass spectrometry in room temperature $Au_{25}(SR)_{18}$ synthesis. *J. Am. Chem. Soc.* 131: 13604–13605.

45 Jadzinsky, P.D., Calero, G., Ackerson, C.J. et al. (2007). Structure of a thiol monolayer-protected gold nanoparticle at 1.1 Å resolution. *Science* 318: 430–433.

46 Zhu, M., Aikens, C.M., Hollander, F.J. et al. (2008). Correlating the crystal structure of a thiol-protected Au_{25} cluster and optical properties. *J. Am. Chem. Soc.* 130: 5883–5885.

47 Negishi, Y., Nobusada, K., and Tsukuda, T. (2005). Glutathione-protected gold clusters revisited: bridging the gap between gold(I)–thiolate complexes and thiolate-protected gold nanocrystals. *J. Am. Chem. Soc.* 127: 5261–5270.

48 Park, S. and Lee, D. (2012). Synthesis and electrochemical and spectroscopic characterization of biicosahedral Au_{25} clusters. *Langmuir* 28: 7049–7054.

49 Ghosh, A., Mohammed, O.F., and Bakr, O.M. (2018). Atomic-level doping of metal clusters. *Acc. Chem. Res.* 51: 3094–3103.

50 Walter, M., Akola, J., Lopez-Acevedo, O. et al. (2008). A unified view of ligand-protected gold clusters as superatom complexes. *Proc. Natl. Acad. Sci. U. S. A.* 105: 9157–9162.

51 Kwak, K., Tang, Q., Kim, M. et al. (2015). Interconversion between superatomic 6-electron and 8-electron configurations of $M@Au_{24}(SR)_{18}$ clusters (M = Pd, Pt). *J. Am. Chem. Soc.* 137: 10833–10840.

52 Thanthirige, V.D., Kim, M., Choi, W. et al. (2016). Temperature-dependent absorption and ultrafast exciton relaxation dynamics in $MAu_{24}(SR)_{18}$ clusters (M = Pt, Hg): role of the central metal atom. *J. Phys. Chem. C* 120: 23180–23188.

53 Zhou, M., Yao, C., Sfeir, M.Y. et al. (2018). Excited-state behaviors of $M_1Au_{24}(SR)_{18}$ nanoclusters: the number of valence electrons matters. *J. Phys. Chem. C* 122: 13435–13442.

54 Negishi, Y., Iwai, T., and Ide, M. (2010). Continuous modulation of electronic structure of stable thiolate-protected Au_{25} cluster by Ag doping. *Chem. Commun.* 46: 4713–4715.

55 Negishi, Y., Munakata, K., Ohgake, W., and Nobusada, K. (2012). Effect of copper doping on electronic structure, geometric structure, and stability of thiolate-protected Au_{25} nanoclusters. *J. Phys. Chem. Lett.* 3: 2209–2214.

56 Wang, S., Song, Y., Jin, S. et al. (2015). Metal exchange method using Au_{25} nanoclusters as templates for alloy nanoclusters with atomic precision. *J. Am. Chem. Soc.* 137: 4018–4021.

57 Tofanelli, M.A., Salorinne, K., Ni, T.W. et al. (2016). Jahn–Teller effects in $Au_{25}(SR)_{18}$. *Chem. Sci.* 7: 1882–1890.

58 Kim, M., Weerawardene, K.L.D.M., Choi, W. et al. (2020). Insights into the metal-exchange synthesis of $MAg_{24}(SR)_{18}$ (M = Ni, Pd, Pt) nanoclusters. *Chem. Mater.* 32: 10216–10226.

59 Kim, M., Tang, Q., Narendra Kumar, A.V. et al. (2018). Dopant-dependent electronic structures observed for $M_2Au_{36}(SC_6H_{13})_{24}$ clusters (M = Pt, Pd). *J. Phys. Chem. Lett.* 9: 982–989.

60 Qian, H., Eckenhoff, W.T., Zhu, Y. et al. (2010). Total structure determination of thiolate-protected Au_{38} nanoparticles. *J. Am. Chem. Soc.* 132: 8280–8281.

61 Luo, Y., Tang, L., Khan, U. et al. (2019). Morphology and surface chemistry engineering toward pH-universal catalysts for hydrogen evolution at high current density. *Nat. Commun.* 10: 269.

62 Simmons, T.R., Berggren, G., Bacchi, M. et al. (2014). Mimicking hydrogenases: from biomimetics to artificial enzymes. *Coord. Chem. Rev.* 270–271: 127–150.

63 Zhang, P., Wang, M., Yang, Y. et al. (2014). A molecular copper catalyst for electrochemical water reduction with a large hydrogen-generation rate constant in aqueous solution. *Angew. Chem. Int. Ed.* 53: 13803–13807.

64 Zeng, M. and Li, Y. (2015). Recent advances in heterogeneous electrocatalysts for the hydrogen evolution reaction. *J. Mater. Chem. A* 3: 14942–14962.

65 Wodrich, M.D., Busch, M., and Corminboeuf, C. (2016). Accessing and predicting the kinetic profiles of homogeneous catalysts from volcano plots. *Chem. Sci.* 7: 5723–5735.

66 Tahir, M., Pan, L., Idrees, F. et al. (2017). Electrocatalytic oxygen evolution reaction for energy conversion and storage: a comprehensive review. *Nano Energy* 37: 136–157.

67 Choi, W., Hu, G., Kwak, K. et al. (2018). Effects of metal-doping on hydrogen evolution reaction catalyzed by MAu_{24} and M_2Au_{36} nanoclusters (M = Pt, Pd). *ACS Appl. Mater. Interfaces* 10: 44645–44653.

68 Helm, M.L., Stewart, M.P., Bullock, R.M. et al. (2011). A synthetic nickel electrocatalyst with a turnover frequency above 100,000 s^{-1} for H_2 production. *Science* 333: 863–866.

69 Lubitz, W., Ogata, H., Rüdiger, O., and Reijerse, E. (2014). Hydrogenases. *Chem. Rev.* 114: 4081–4148.

70 Frey, M. (2002). Hydrogenases: hydrogen-activating enzymes. *ChemBioChem* 3: 153–160.

71 Schilter, D., Camara, J.M., Huynh, M.T. et al. (2016). Hydrogenase enzymes and their synthetic models: the role of metal hydrides. *Chem. Rev.* 116: 8693–8749.

72 Graham, D.J. and Nocera, D.G. (2014). Electrocatalytic H_2 evolution by proton-gated hangman iron porphyrins. *Organometallics* 33: 4994–5001.

73 Solis, B.H., Maher, A.G., Dogutan, D.K. et al. (2016). Nickel phlorin intermediate formed by proton-coupled electron transfer in hydrogen evolution mechanism. *Proc. Natl. Acad. Sci. U. S. A.* 113: 485–492.

74 Jo, Y., Choi, M., Kim, M. et al. (2021). Promotion of alkaline hydrogen production via Ni-doping of atomically precise Ag nanoclusters. *Bull. Korean Chem. Soc.* 42: 1672–1677.

75 Zheng, Y., Jiao, Y., Vasileff, A., and Qiao, S.-Z. (2018). The hydrogen evolution reaction in alkaline solution: from theory, single crystal models, to practical electrocatalysts. *Angew. Chem. Int. Ed.* 57: 7568–7579.

76 Safizadeh, F., Ghali, E., and Houlachi, G. (2015). Electrocatalysis developments for hydrogen evolution reaction in alkaline solutions – a review. *Int. J. Hydrogen Energy* 40: 256–274.

77 Tang, M.H., Hahn, C., Klobuchar, A.J. et al. (2014). Nickel–silver alloy electrocatalysts for hydrogen evolution and oxidation in an alkaline electrolyte. *Phys. Chem. Chem. Phys.* 16: 19250–19257.

78 Mahmood, N., Yao, Y.D., Zhang, J.W. et al. (2018). Electrocatalysts for hydrogen evolution in alkaline electrolytes: mechanisms, challenges, and prospective solutions. *Adv. Sci.* 5: 1700464.

79 Tang, L., Yu, J.J., Zhang, Y. et al. (2021). Boosting the hydrogen evolution reaction activity of Ru in alkaline and neutral media by accelerating water dissociation. *RSC Adv.* 11: 6107–6113.

80 Seong, H., Efremov, V., Park, G. et al. (2021). Atomically precise gold nanoclusters as model catalysts for identifying active sites for electroreduction of CO_2. *Angew. Chem. Int. Ed.* 60: 14563–14570.

81 Wu, Z.-P., Lu, X.F., Zang, S.-Q., and Lou, X.W. (2020). Non-noble-metal-based electrocatalysts toward the oxygen evolution reaction. *Adv. Funct. Mater.* 30: 1910274.

82 Kauffman, D.R., Alfonso, D., Tafen, D.N. et al. (2016). Electrocatalytic oxygen evolution with an atomically precise nickel catalyst. *ACS Catal.* 6: 1225–1234.

83 Woodward, P., Dahl, L.F., Abel, E.W., and Crosse, B.C. (1965). A new type of cyclic transition metal complex, $[Ni(SC_2H_5)_2]_6$. *J. Am. Chem. Soc.* 87: 5251–5253.

84 Joya, K.S., Sinatra, L., AbdulHalim, L.G. et al. (2016). Atomically monodisperse nickel nanoclusters as highly active electrocatalysts for water oxidation. *Nanoscale* 8: 9695–9703.

85 de Respinis, M., Joya, K.S., De Groot, H.J.M. et al. (2015). Solar water splitting combining a $BiVO_4$ light absorber with a Ru-based molecular cocatalyst. *J. Phys. Chem. C* 119: 7275–7281.

86 Li, X., Bi, W., Chen, M. et al. (2017). Exclusive $Ni–N_4$ sites realize near-unity CO selectivity for electrochemical CO_2 reduction. *J. Am. Chem. Soc.* 139: 14889–14892.

87 Yang, H.B., Hung, S.-F., Liu, S. et al. (2018). Atomically dispersed Ni(I) as the active site for electrochemical CO_2 reduction. *Nat. Energy* 3: 140–147.

88 Ding, S., Guo, Y., Hülsey, M.J. et al. (2019). Electrostatic stabilization of single-atom catalysts by ionic liquids. *Chem* 5: 3207–3219.

89 Zhao, S., Jin, R., and Jin, R. (2018). Opportunities and challenges in CO_2 reduction by gold- and silver-based electrocatalysts: from bulk metals to nanoparticles and atomically precise nanoclusters. *ACS Energy Lett.* 3: 452–462.

90 Kauffman, D.R., Alfonso, D., Matranga, C. et al. (2012). Experimental and computational investigation of Au_{25} clusters and CO_2: a unique interaction and enhanced electrocatalytic activity. *J. Am. Chem. Soc.* 134: 10237–10243.

91 Weekes, D.M., Salvatore, D.A., Reyes, A. et al. (2018). Electrolytic CO_2 reduction in a flow cell. *Acc. Chem. Res.* 51: 910–918.

92 Kim, B., Ma, S., Molly Jhong, H.-R., and Kenis, P.J.A. (2015). Influence of dilute feed and pH on electrochemical reduction of CO_2 to CO on Ag in a continuous flow electrolyzer. *Electrochim. Acta* 166: 271–276.

93 Verma, S., Hamasaki, Y., Kim, C. et al. (2018). Insights into the low overpotential electroreduction of CO_2 to CO on a supported gold catalyst in an alkaline flow electrolyzer. *ACS Energy Lett.* 3: 193–198.

94 Kim, B., Seong, H., Song, J.T. et al. (2020). Over a 15.9% solar-to-CO conversion from dilute CO_2 streams catalyzed by gold nanoclusters exhibiting a high CO_2 binding affinity. *ACS Energy Lett.* 5: 749–757.

95 Johnson, K.A. and Goody, R.S. (2011). The original Michaelis constant: translation of the 1913 Michaelis–Menten paper. *Biochemistry* 50: 8264–8269.

96 Alfonso, D.R., Kauffman, D., and Matranga, C. (2016). Active sites of ligand-protected Au_{25} nanoparticle catalysts for CO_2 electroreduction to CO. *J. Chem. Phys.* 144: 184705.

97 Austin, N., Zhao, S., McKone, J.R. et al. (2018). Elucidating the active sites for CO_2 electroreduction on ligand-protected Au_{25} nanoclusters. *Catal. Sci. Technol.* 8: 3795–3805.

98 Heaven, M.W., Dass, A., White, P.S. et al. (2008). Crystal structure of the gold nanoparticle $[N(C_8H_{17})_4][Au_{25}(SCH_2CH_2Ph)_{18}]$. *J. Am. Chem. Soc.* 130: 3754–3755.

99 Yan, N., Xia, N., Liao, L. et al. (2018). Unraveling the long-pursued Au_{144} structure by X-ray crystallography. *Sci. Adv.* 4: eaat7259.

100 Nilsson, A., Pettersson, L.G.M., Hammer, B. et al. (2005). The electronic structure effect in heterogeneous catalysis. *Catal. Lett.* 100: 111–114.

101 Nørskov, J.K., Abild–Pedersen, F., Studt, F., and Bligaard, T. (2011). Density functional theory in surface chemistry and catalysis. *Proc. Natl. Acad. Sci. U. S. A.* 108: 937–943.

102 Kim, D., Resasco, J., Yu, Y. et al. (2014). Synergistic geometric and electronic effects for electrochemical reduction of carbon dioxide using gold–copper bimetallic nanoparticles. *Nat. Commun.* 5: 4948.

103 Tang, Q., Lee, Y., Li, D.-Y. et al. (2017). Lattice-hydride mechanism in electrocatalytic CO_2 reduction by structurally precise copper-hydride nanoclusters. *J. Am. Chem. Soc.* 139: 9728–9736.

104 Hori, Y., Murata, A., Takahashi, R., and Suzuki, S. (1987). Electroreduction of carbon monoxide to methane and ethylene at a copper electrode in aqueous solutions at ambient temperature and pressure. *J. Am. Chem. Soc.* 109: 5022–5023.

105 Foit, S.R., Vinke, I.C., de Haart, L.G.J., and Eichel, R.-A. (2017). Power-to-syngas: an enabling technology for the transition of the energy system? *Angew. Chem. Int. Ed.* 56: 5402–5411.

106 Choi, W., Seong, H., Efremov, V. et al. (2021). Controlled syngas production by electrocatalytic CO_2 reduction on formulated $Au_{25}(SR)_{18}$ and $PtAu_{24}(SR)_{18}$ nanoclusters. *J. Chem. Phys.* 155: 014305.

6

Atomically Precise Metal Nanoclusters as Electrocatalysts: From Experiment to Computational Insights

Fang Sun[1], *Qing Tang*[1], *and De-en Jiang*[2]

[1] *School of Chemistry and Chemical Engineering, Chongqing Key Laboratory of Theoretical and Computational Chemistry, Chongqing University, Chongqing 401331, China*
[2] *Department of Chemical and Biomolecular Engineering, Vanderbilt University, Nashville, TN 37235, USA*

6.1 Introduction

The goal of this chapter is to briefly introduce factors affecting the activity and selectivity of atomically precise metal nanoclusters (NCs) for electrocatalytic behaviors and then discuss the recent advances in electrocatalytic applications. As compared to the larger nanoparticles or nanocrystals counterparts, metal NCs occupy the gap between discrete atoms and plasmonic nanomaterials, and are an emerging class of atomically precise nanomaterials with diameter typically below 2 nm, thus, the well-defined NCs with their total structures available constitute a new class of model catalysts and hold great promise in fundamental catalysis studies [1]. In particular, the ultrasmall-size imparted strong quantum confinement effects render the NCs with significantly different physical and chemical properties, including absorption, photoluminescence, surface chemistry, and catalytic behaviors [2]. Moreover, the ultra-small size also makes molecular purity possible, in which the atomic structure can be crystallographically determined. The last decade has witnessed the great development of NCs research since the Kornberg group in 2007 pioneered the crystal structure report of the $Au_{102}(p\text{-}MBA)_{44}$ (MBA = mercaptobenzoic acid) cluster [3]. Thus far, more than 200 coinage metal (Au, Ag, Cu) NCs protected by different ligands (including phosphine, thiolate, alkynyl molecule, halogen, selenide, and mixed capping agents) have been successfully synthesized and characterized [4]. Meanwhile, recent experiments and theoretical studies have also indicated that these metal NCS show great promise in many electrocatalytic reactions, such as hydrogen evolution reaction (HER), oxygen evolution reaction (OER), oxygen reduction reaction (ORR), and CO_2 reduction reaction (CO_2RR), which dramatically advance the relationships understanding between unique structures and potential properties in the relevant field.

With rising global population and expanding industrialization, major concerns have been raised over the sustainability and security of our energy future [5, 6]. Therefore, creating global-scale green and sustainable energy technologies for the future, while maintaining our benign environment, is imperative [7–11]. Electrocatalytic energy conversion has been deemed as one of the most effective and cleanest pathways to convert and store electrical energy [12–14]. To realize the above clean energy technologies, a major goal is to pursue highly selective and efficient catalytic processes under mild conditions, thus, the fundamental aspects of catalysis, such as the size dependence, structure-reactivity relationships, active sites, and catalytic mechanisms becomes very

Atomically Precise Nanochemistry, First Edition. Edited by Rongchao Jin and De-en Jiang.

critical. Certainly, the success of atomically precise nanochemistry in ultrasmall metal NPs have offered a new type of model catalysts and exciting opportunities for fundamental electrocatalysis research, including the atomically precise size dependent catalytic activity, control of selectivity by structure and ligand engineering, structure–property relationships, atomic-level insight into molecular activation and catalytic mechanisms, as well as the identification of active-sites. In this chapter, we focus our discussions on the electrocatalytic studies enabled by atomically precise NCs that are structurally characterized by X-ray crystallography. There is no doubt that atomically precise NCs mark a major step forward in designing well-defined catalysts, which provide new opportunities for basic research. With the available library of atomically precise metal NCs, complex effects of size, shape, structure, ligand, metal–ligand interface and other factors can be isolated using related clusters, such as the same size but different atomic structures [15, 16]. Such correlation studies have previously been difficult to achieve with other catalysts, so the newly developed atomic precision metal NCs constitute a rare opportunity. In the second part of this chapter, we mainly discuss the potential factors that have been reported to affect the electrocatalytic performance and selectivity of ligand-protected NCs, either positively or negatively. These factors include two categories: (i) size, shape, and doping – which are also important in traditional nanocatalysts, but NCs provide unprecedented atomic-level information; (ii) cluster isomerization, charge state effects, and ligand effects – which are unique in terms of ligand-protected NCs, as these effects would not have been found without finely controlled NCs at single-atom, single-electron level. The third section outlines electro-catalysis by NCs based on some paradigms including HER, OER, ORR, and CO_2RR. The last part offers our summary and perspectives on future efforts. Overall, model catalysts based on the atomically precise NCs will prompt fundamental understanding of the catalytic mechanisms in both experimental and computational studies, tailoring of active-sites at the atomic level, and design of new catalysts with high selectivity and activity as well as durability for electrocatalysis.

6.2 Factors Affecting the Activity and Selectivity of NCs Electrocatalysis

6.2.1 Size Effect

In general, smaller particles usually lead to higher catalytic activity (i.e. monotonically falling [17]) due to larger specific surface area and potential electronic contribution, [18] but other trends may also be seen, such as volcanic type [19], monotonically rising, and independent of size. Obviously, the same series of nanoparticle sizes may have different activity trends in different catalytic reactions, and different metal nanocatalysts may have different behaviors in the same reaction. The availability of $Au_n(SR)_m$ NCs of different sizes provides an excellent opportunity to study their size dependence at the previously inaccessible atomic level.

As mentioned before, the smaller NCs show higher activity in most cases, ascribed to the larger surface-to-volume ratio and more active sites such as corner atoms and edge atoms. Low-coordinated surface atoms tend to act as active sites, resulting in enhanced molecular adsorption, which is related to the coordination number of surface atoms and the upward shift of the d-band center of transition metal with decreasing size [20]. Vallejo et al. advocated a more accurate descriptor of size effects than the usual coordination number and d-band center methods, namely the generalized coordination number of surface sites, thereby linking geometry and adsorption energies [21]. In addition, the electronic effect is also an aspect: as the size of the metal particles decreases, the hybridization of electron orbitals becomes weaker, and the decrease in the number of electrons also leads to energy quantization. Taken together, the geometric effect and the electronic effect fundamentally affect the binding strength of the adsorbate in the whole reaction, thus showing a size-dependent monotonically decreasing activity trend.

Of course, there may be some complex scenarios. A recent experimental survey vividly illustrates this trend. Negishi's group [22] measured the HER, OER, and ORR activities of $Au_n(PET)_m$ (PET = 2-phenylethanethiolate) clusters containing different numbers of constituent atoms under identical experimental conditions, and it was found that the activity of these clusters decreases as the number of constituent atoms increases. A monotonic descending trend was observed, showing the order of activity from the highest to the lowest: $Au_{25} > Au_{38} > Au_{130} > Au_{144}$ (Figure 6.1a).

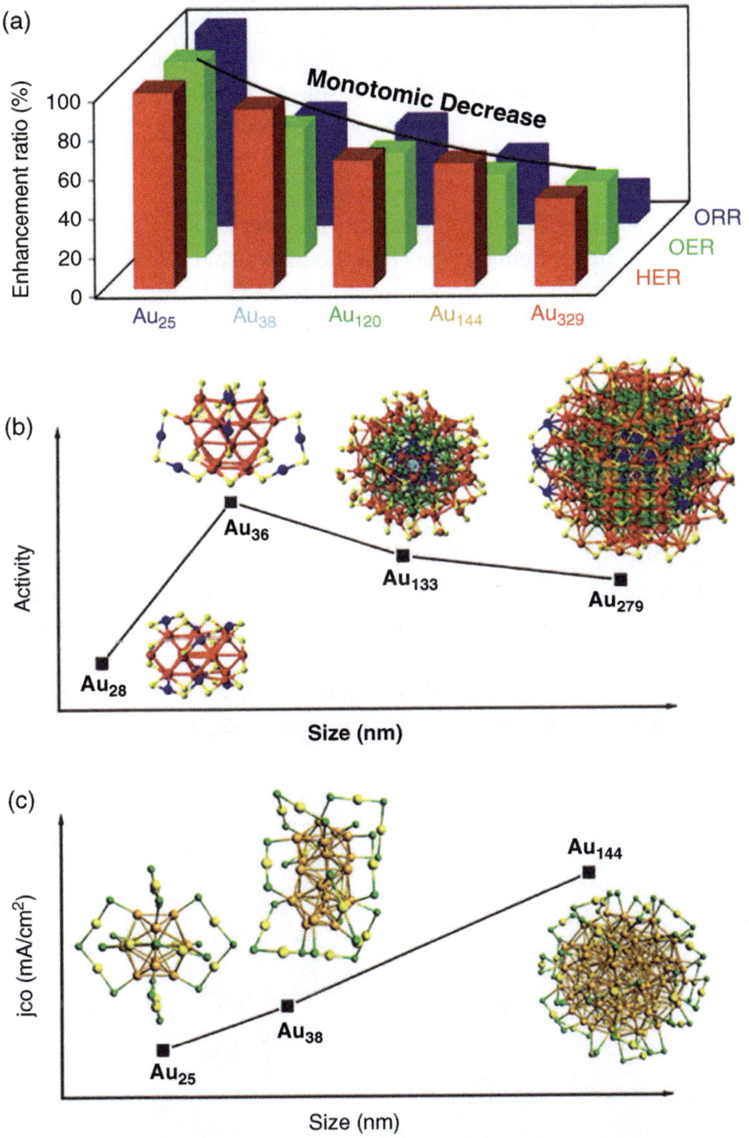

Figure 6.1 (a) Enhancement ratio of HER, OER, and ORR activities for gold clusters with different size. *Source:* Reprinted with permission from [22]. © 2020 Royal Society of Chemistry. Scenarios of electrocatalytic activity as a function of Au NC size in the (b) ORR and (c) CO_2RR. *Source:* Data referred to Ref. [23, 24].

A recent study from Sumner et al. [23] argues for a trend similar to volcanic activity. They have investigated the ORR kinetics of thiolate-protected gold nanoclusters of sizes ranging from 1 to 2.2 nm in diameter consisting of 28, 36, 133, and 279 atoms of Au, among which $Au_{36}(SR)_{24}$ (SR = 4-tert-butylbenzenethiol) is the most active NC for ORR demonstrating >threefold lower overpotential of 160 mV than the smallest $Au_{28}(SR)_{20}$ and quantitatively yielding the $4e^-$ reduced product OH^- (Figure 6.1b). In addition, they found that the activity trend $Au_{36} > Au_{133} > Au_{279} > Au_{28}$ using overpotential and product selectivity as matrices can be qualitatively correlated with the thermochemical stability of the NCs, suggesting that other contributing factors such as stability can also dictate the activity outcomes as opposed to the core size alone. Another trend is that the bigger nanocluster gives the higher catalytic activity. For example, Lee's group [24] conducted an atomic-level CO_2RR investigation of three atomically well-defined Au NCs protected by thiolate ligands: Au_{25}, Au_{38}, and Au_{144}. They found that these NCs exhibited very high selectivity toward CO formation (greater than 90%) and extraordinary size-dependent CO_2RR activity, where the current densities for CO production (j_{CO}) value significantly increases with increasing Au NC size (Figure 6.1c).

6.2.2 Shape Effect

Shape effects, also commonly known as morphological effects, have long been a topic of interest in the catalysis community. The success of nanochemistry has provided a feasible route for various shape-controllable nanoparticles, which has greatly promoted the research on this topic [25]. Some specific crystal planes such as (111) and (100) planes were found to be active or inactive in some specific reactions. However, atomic-level details are still lacking because the surface atomic structures in which surface stabilizers exist are often illusory. In this subsection, we are mainly concerned with atomically precise NC with different shapes.

The development of atom-precise nanochemistry offers even more exciting opportunities for this type of research, allowing researchers to control the shape of nanoclusters with the same number of metal atoms. A good example is the 25-atom Au NC, including spherical $Au_{25}(SR)_{18}$ and rod-like $[Au_{25}(PPh)_{10}(SR)_5Cl_2]^{2+}$ (Figure 6.2). Zhao et al. [26] investigated two Au_{25} clusters with diameters of approximately 1 nm for electrocatalytic CO_2RR. In terms of catalytic activity, Au_{25} spheres show higher Faraday efficiency with the increase of CO yield. To explain the shape effect at the atomic level, they performed first-principles calculations, which showed that the negatively charged $[Au_{25}(SR)_{18}]^-$ and the energetically more favorable removal of a ligand on the sphere to create an active site can better stabilize the important *COOH intermediate, therefore, exhibit

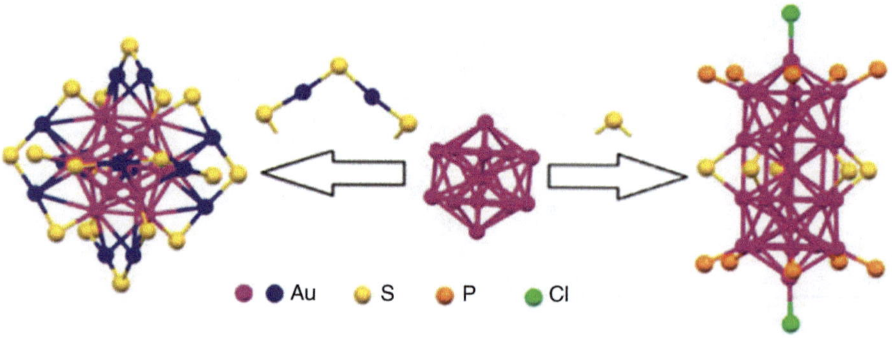

Figure 6.2 Atom packing structures of $Au_{25}(SR)_{18}$ nanosphere (left; R=C_2H_4Ph, omitted for clarity) and $Au_{25}(PPh_3)_{10}(SR)_5Cl_2$ nanorod (right; carbon tails omitted for clarity). *Source:* Reprinted with permission from [26]. © 2018 American Chemical Society.

higher activity and selectivity for CO formation than rod Au_{25}. To be clear, shape effects tend to be related to the response type. In contrast to the aforementioned CO_2 electroreduction, Au_{25} spheres and rods do not exhibit shape effects in the H_2 semi-hydrogenated alkyne reaction [27]. In the field of heterogeneous catalysis, the study of structural sensitivity has always received much attention [25]. All surface atoms are regarded as active centers if they have equal catalytic activity; therefore, the catalytic system is structurally insensitive, and its TOF appears to be flat relative to size and shape when normalized by the number of surface atoms. But more commonly, not all surface atoms are equally catalytic, for example, surface atoms on edges and corners (or vertices) are often more active than atoms within the facets; such systems become structure-sensitive, and different activities can be exhibited depending on the size and shape of the catalyst, both monotonic and volcanic trends are possible.

6.2.3 Ligands Effect

6.2.3.1 Different -R Groups in Thiolate Ligands

Ligands can largely alter the chemical properties, activity, and selectivity of nanocatalysts, and sometimes the local polarity of the metal surface, thereby significantly affecting the catalytic performance [28]. To investigate the potential effect of the -R group of the thiolate, Kumar et al. [22] used three types of hydrophobic -SR (2-Phenylethanethiolate (PET), hexanethiolate (C6T), and dodecanethiolate (C12T)) as Au_{25} protecting ligands to measure their HER, OER, and ORR electrocatalytic activities under the identical experimental conditions. They claim that since C12T has a longer hydrocarbon moiety than PET (Figure 6.3a), it can be inferred that when $[Au_{25}(C12T)_{18}]^0$ is used as a catalyst, the insulating layer formed between the electrode and the metal core is thicker than $[Au_{25}(PET)_{18}]^0$, resulting in more difficult electron transfer, and consequently, its activity is lower than that of $[Au_{25}(PET)_{18}]^0$ (Figure 6.3b). On the other hand, $[Au_{25}(C6T)_{18}]^0$ showed higher activity than that of $[Au_{25}(PET)_{18}]^0$ (Figure 6.3b). C6T has a slightly longer hydrocarbon moiety than PET (Figure 6.2a), however, since the ligand in $[Au_{25}(PET)_{18}]^0$ includes the phenyl group thereby π–π interaction should occur between the ligands in $[Au_{25}(PET)_{18}]^0$. Thus, it is assumed that the ligand layer in $[Au_{25}(C6T)_{18}]^0$ is more liquid-like and thereby thinner on the electrode than that in $[Au_{25}(PET)_{18}]^0$. They believe that this may be the main reason why $[Au_{25}(C6T)_{18}]^0$ showed higher HER activity than $[Au_{25}(PET)_{18}]^0$. Such an experimental insight highlights that ligands of $Au_n(SR)_m$ clusters may play an important role in electrocatalytic reactions as insulating layers between electrodes and metal cores.

6.2.3.2 Different Types of Ligands

The protecting ligands, as the outmost surface structures, play a critical role in determining the structure–property of the whole system. Their presence can not only prevent aggregation or facilitate the isolation of target nanoclusters, but also functionalize the nanostructures and generate unique and chemically active interfaces for many important fields such as electronic, magnetic, optical, catalytic, biological, and sensing [29]. Without a doubt, different metal–ligand interface (e.g. Au—S, Au—Se, and Au—C bonding) can affect the electronic properties of surface metal atoms, thereby impacting their catalytic reactivity. Zhao et al. [30] constructed two novel Au_{25}/MoS_2 nanocomposites with the same metal core structure but protected by thiolate ($-SC_2H_4Ph$) or selenolate (–SePh) ligand for catalyzing electrochemical water reduction to H_2, in which MoS_2 catalyzes the HER and Au_{25} acts as a co-catalyst. The $Au_{25}(SR)_{18}$ and $Au_{25}(SePh)_{18}$ NCs possess the same atomic packing structure, i.e. an Au_{13} icosahedral core protected by six –Se(R)–Au–Se(R)–Au–Se(R)– dimeric staple motifs. Compared to plain MoS_2 nanosheets, these two nanocomposites can enhance HER activity, exhibiting smaller onset overpotentials (−0.20 and−0.22 V) and higher

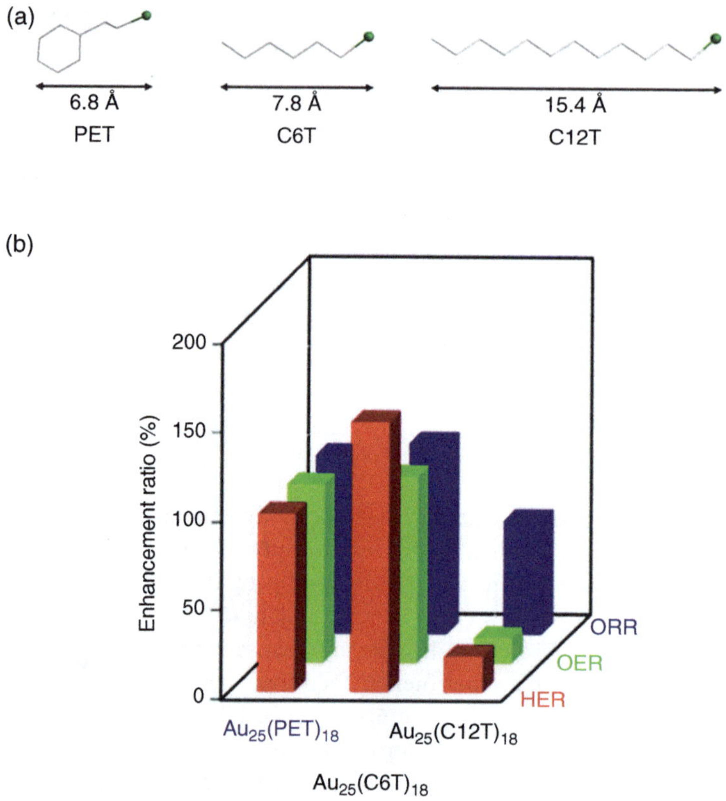

Figure 6.3 (a) Geometrical structures of PET, C6T, and C12T. The length of ligand was estimated from the geometrical structures. (b) Enhancement ratio of HER, OER, and ORR activities for Au_{25} clusters with different –R group of the thiolate ligand. *Source:* Reprinted with permission from [22]. © 2020 Royal Society of Chemistry.

current densities (59.3 and 39.0 mA cm^{-2}) at −0.4 V vs. RHE, but thiolate is better than selenolate. The rationalization of the interface between the Au_{25} core and the surface ligands (–SC_2H_4Ph vs. –SePh) is considered to be the main factor affecting the catalytic performance. This report undoubtedly highlights the interfacial electronic interactions between metal cores and surface ligands, which lead to different charge densities on the surface metal atoms and even different surface motifs (e.g. staple motif vs. bridging motif on Au—S or Au–C≡CR interface). These factors can be used to design ligand-protected NC catalysts in the future.

6.2.3.3 Ligand-on and -off Effect

Jin et al. [31] successfully synthesized the heavily Ag-doped nanoclusters $[Ag_xAu_{25-x}(SC_6H_{11})_{18}]^-$ in 2013. They found that the ligand-off NCs supported by carbon black were obtained by thermal removal of the $[Ag_xAu_{25-x}(SC_6H_{11})_{18}]^-$ surface ligands at 300°C for two hours in N_2 atmosphere. They then experimentally assessed the electrocatalytic performance for ORR in alkaline solutions. Compared to ligand-on $[Ag_xAu_{25-x}(SC_6H_{11})_{18}]^-$, ligand-off NCs show distinctly improved ORR features with an onset potential of −0.09 V (Figure 6.4a) and a much higher mass activity (217.4 A g^{-1} $_{meta}$) at the potential of −0.3 V (vs. Ag/AgCl) (Figure 6.4b), which can be attributed to the increased exposure of active metal sites on NC surfaces after the removal of ligands; thus, the interactions between oxygen molecules and nanoclusters are enhanced effectively.

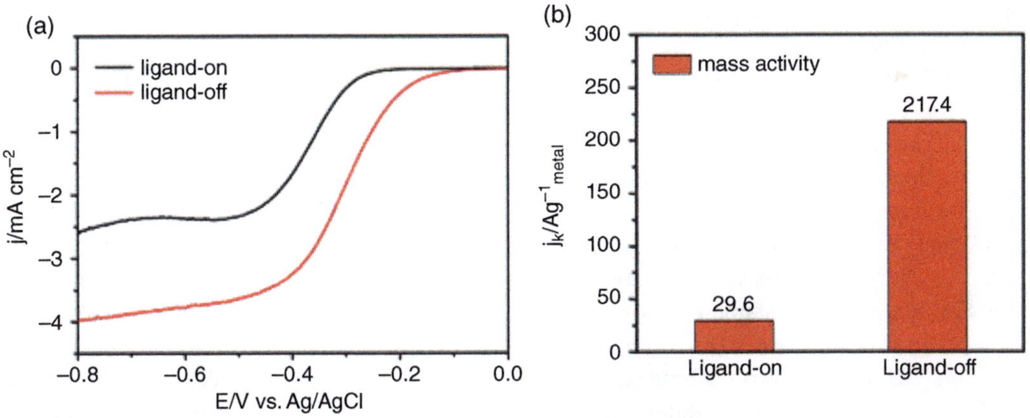

Figure 6.4 Electrochemical ORR activities of ligand-on, ligand-off $[Ag_xAu_{25-x}(SC_6H_{11})_{18}]^-$ nanoclusters: (a) polarization curves for catalysts in O_2-saturated 0.1 M KOH solutions at 1600 rpm. (b) Mass activities for catalysts at −0.3 V (vs. Ag/AgCl). *Source:* Reprinted with permission from [31]. © 2017 Royal Society of Chemistry.

In addition, some studies have proposed that the partial −SR ligand removal or cleaving the organic −R moiety of thiolate-protected Au NCs to create exposed Au or S atom sites may act as key steps to form electrocatalytic reaction centers, which may be promising to boost the catalytic performance [32–36]. Based on first-principles density functional theory (DFT) [32], it was revealed that the fully ligand protected $Au_{25}(SR)_{18}$ cluster is not an active CO_2 reduction catalyst because the formation of the crucial carboxyl intermediate required very high electrochemical potentials. Theoretical simulations showed that the surface S atoms of the thiolate ligand are identified as the active sites (with the Au_{13} core as the electron reservoir) for selective CO_2RR, whereas under-coordinated Au atom active sites are predicted to prefer H_2 evolution. At present, many works have indicated that the removal of ligands can significantly increase the CO_2RR performance [24, 33, 36, 37]. Recent work by Sun et al. [38] systematically examined the ORR activities of prototype $[Au_{25}(SR)_{18}]^q$ cluster, mono-atom-doped alloy clusters, and their singly deligated M-exposure and S-exposure systems as electrocatalysts toward ORR at the acidic medium by means of large-scale DFT computations. Their simulations reveal that the fully ligand-protected clusters prefer H_2O_2 formation through $2e^-$ mechanism, whereas −SR ligand removal can change ORR selectivity significantly because the exposed, uncoordinated sites favor $4e^-$ pathway for generating H_2O, and the partially ligated $[HgAu_{24}(SCH_3)_{17}]^0$ with an exterior $-SCH_3$ removal is revealed as the most effective $4e^-$ ORR catalyst with the lowest overpotential of 0.43 V among the studied NCs. Overall, the ligand-protected nanoclusters offer new opportunities for electrocatalysis research. Although ligands may be undesirable, they may play some key positive roles in certain reactions, such as selective access to desired products by the ligand-on and -off control.

6.2.4 Charge State Effect

The charge state here refers to the total charge of the metal core. Atomically precise NCs usually possess non-zero ground state charges and exhibit unprecedented catalytic effect that would not be possible to discover without NCs with precisely controlled charge states [39–41]. The ultrasmall $[Au_{25}(SR)_{18}]^q$ cluster of about 1 nm is one such example with a tunable ground state charge ($q = -1$, 0, +1). Although they all have the same core structure, there will be some contraction or expansion of localized bonds at different charge states. In short, the special size regime and electronic properties of differently charged Au_{25}^{q} clusters bring exciting opportunities to probe the chemistry of charged active sites.

In 2014 Kauffman's group [42] investigated the CO_2 electrocatalytic reduction to CO of carbon black (CB)-supported $Au_{25}(SR)_{18}$ with different ground state charges (q = 0, −1, and +1), and found that the charge state does play a critical role in the catalytic performance. The experimental and DFT calculation results confirm that the charge state affects the stability of the adsorbed reactant or product, among which the anionic $[Au_{25}(SR)_{18}]^-$ showed the best activity, followed by $[Au_{25}(SR)_{18}]^0$ and then $[Au_{25}(SR)_{18}]^+$ (Figure 6.5a,b). They assumed that this apparent difference is due to the charge-state dependent stabilization of the co-adsorption reactants of CO_2 and H^+, where H is bound to one staple Au atom, and CO_2 coordinated with three S atoms of −SR in the pocket position of the cluster (Figure 6.5c). Computationally, it is found that the binding of CO_2 and H^+ on $[Au_{25}(SR)_{18}]^-$ is the strongest compared with other charge states, although the adsorption of CO_2 is relatively weak. Besides, the characteristic optical spectra of these clusters after CO_2 reduction did not change significantly, from which they concluded that their catalytic results are intrinsic to the charge state rather than originating from any degraded catalyst. Meanwhile, they also investigated a similar dependency on charge-states for ORR in alkaline media, and found that they had a weak interaction with oxygen (Figure 6.5d), whose activity trend was $Au_{25}^- > Au_{25}^0 > Au_{25}^+$ (Figure 6.5e,f). The positively charged Au_{25}^+ exhibited poor O_2 reduction kinetics because strongly bound OH^- reaction products blocked the Au_{25}^+ surface-active sites (Figure 6.5g). In another report, Chen and co-workers [43] examined the $[Au_{25}(SC_{12}H_{25})_{18}]^q$ (q = 0, −1, and +1) for ORR, which displays electron transfer numbers of less than 2.6 in all cases. Based on this result, they identified the charge dependences for the H_2O_2 production from the $2e^-$ reduction of dioxygen in aqueous media and the activity followed an order of $-1 > 0 > +1$. The most efficient electron transfer involved in this mechanism arises from the $Au_{25}(SR)_{18}^-$ to the LUMO (π^*) of O_2.

Taken together, these reports demonstrate the effect of charge state on an electrocatalytic reaction, the success of which is largely due to the exquisite control of NCs via atomically precise nanochemistry. Since the charge is distributed in the Au_{13} core of $[Au_{25}(SC_{12}H_{25})_{18}]^q$, the above results suggest that the core participates in the active site in addition to the outer moiety trapping reactants, which is reasonable because the Au_{13} nucleus is redox active. We believe that the charge state adjustment strategy of electrocatalytic properties can be extended to other reactions involving H^+ or OH^-.

6.2.5 Doping and Alloying Effect

Doping plays an important role in the functional adjustment of materials. Since the 1960s, a variety of bimetallic nanocatalysts have been reported to improve catalytic activity, selectivity and stability. Single-atom doping into gold NCs has been achieved, which brings a precious opportunity to study the catalytic effect at the level of "one atom at a time." The tunable properties of bimetallic NCs enable improved catalytic performance, selectivity, and persistence of a specific reaction. Kwak et al. [44] reported a single Pt-doped bimetallic cluster ($PtAu_{24}[SC_6H_{13}]_{18}$), where Pt can selectively replace the central gold atom in the Au_{25} cluster, and found that its electronic structure and HER catalytic activity can be finely controlled by the doping Pt atom. The bimetallic clusters are molecular-like structures with a high HER catalytic activity and can generate H_2 with TOFs of 4.8 in tetrahydrofuran (THF) and 34 mol H_2 (mol cat)$^{-1}$ s^{-1} in water at a moderate overpotential of 0.6 V. Mechanistic studies disclosed that the H* bonding on this bimetallic cluster is thermodynamically neutral, and the central Pt atom can form a Pt—H chemical bond, indicating a key role of the dopant. Recent work by Jin's group [32] reported the single-atom doping effect on CO_2RR by comparing monopalladium-doped Pd_1Au_{24} and homogold Au_{25} NCs. Experimental and

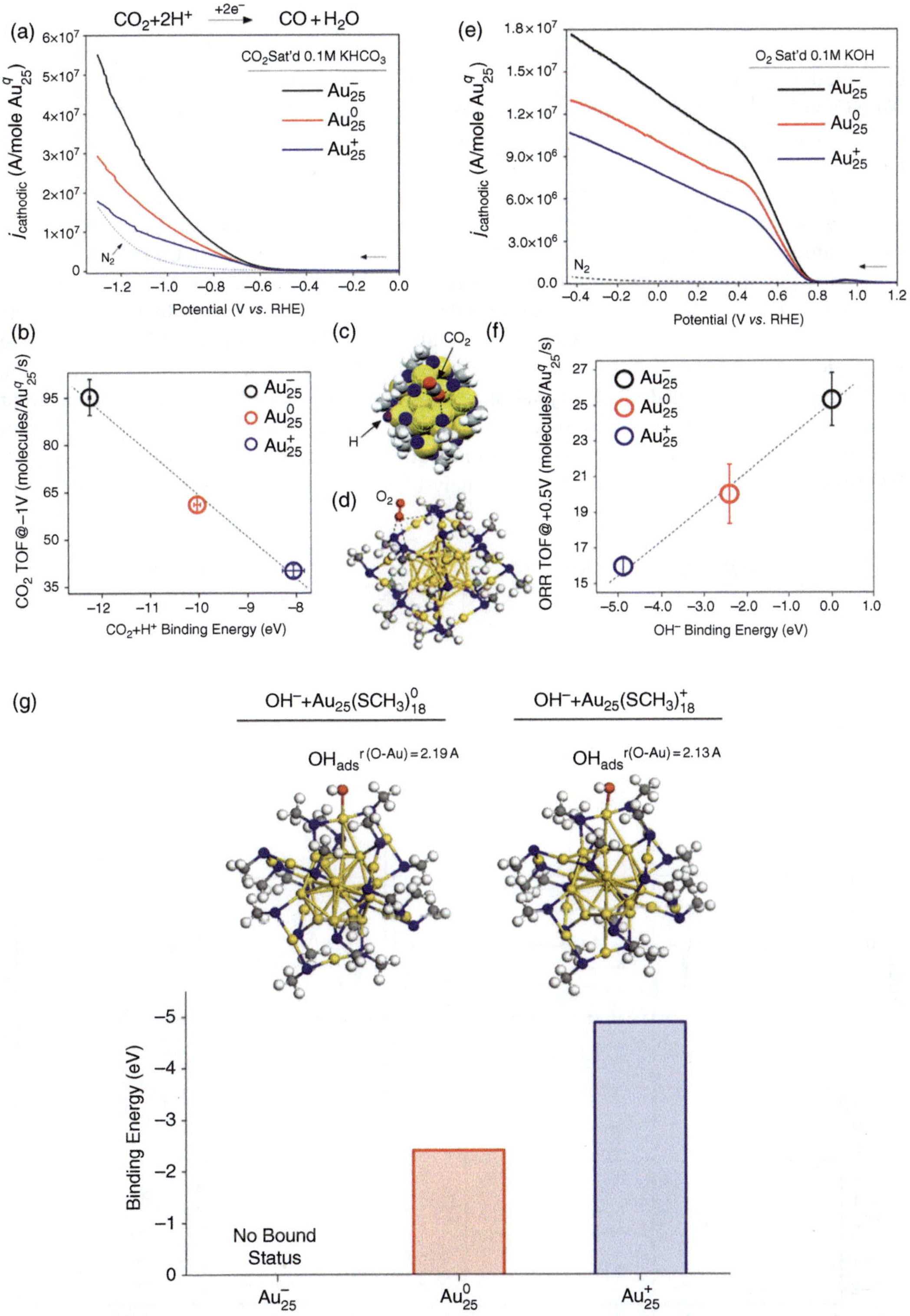

Figure 6.5 Rotating disk electrode polarization curves for CB-supported $Au_{25}{}^q$ with different charge states in (a) CO_2 saturated 0.1 M $KHCO_3$ and (e) O_2 saturated 0.1 M KOH (ω = 2500 rpm). Current densities were normalized to the moles of Au_{25} on the electrode and equivalent Au_{25} loadings were used to compare electrocatalytic activity. Correlation between reactant binding energy and reaction TOF for (b) CO_2RR at −1 V and (f) OH^- at +0.50 V. (c), (d) Representative Au_{25}-coadsorbate models. (g) Calculated binding energies and OH^- adsorption models at $Au_{25}{}^q$. Color code: Au, yellow; S, blue; C, gray; H_{ligand}, white; H_{ads}, purple; O, red. *Source:* Reprinted with permission from [42]. © 2014 Royal Society of Chemistry.

computational results show that single Pd-substitution can substantially improve the selectivity of Au NCs at large potentials due to the retained more S atom active sites compared with Au_{25}, showing a faraday efficiency of about 100% for CO_2 conversion to CO from −0.6 (onset) to −1.2 V (vs RHE).

In recent years, considerable efforts have been devoted to investigating the doping of foreign atoms, which further diversified the chemical space that can be built on $Au_{25}(SR)_{18}$. So far, the foreign atoms such as Ag [31, 45–47], Cu [48, 49], Pt [44, 50, 51], Pd [51, 52], Cd [53, 54], Hg [53, 55], and Ir [56] have been successfully incorporated into the Au_{25} structure. These foreign atom doping would modify the electronic structure and physical properties of Au_{25}, which in turn affects its catalytic performance. In this context, Sun et al. [38] insight into the ORR behaviors of the mono-atom-doped $MAu_{24}{}^{q}$ clusters (M = Pt, Pd, Ag, Cu, Hg, or Cd; q = −1, or 0) based on the experimentally determined single-crystal structures. Discovery results revealed that, different from the strong d-electron effect in $Au_{25}{}^{q}$ and other doped $MAu_{24}{}^{q}$ clusters, the main contribution from the 6s state of the staple-doped Hg atom near the Fermi level promotes $[HgAu_{24}(SCH_3)_{18}]^0$/$[HgAu_{24}(SCH_3)_{17}]^0$ as a distinguished electrocatalyst for O_2 reduction (Figure 6.6). In addition to

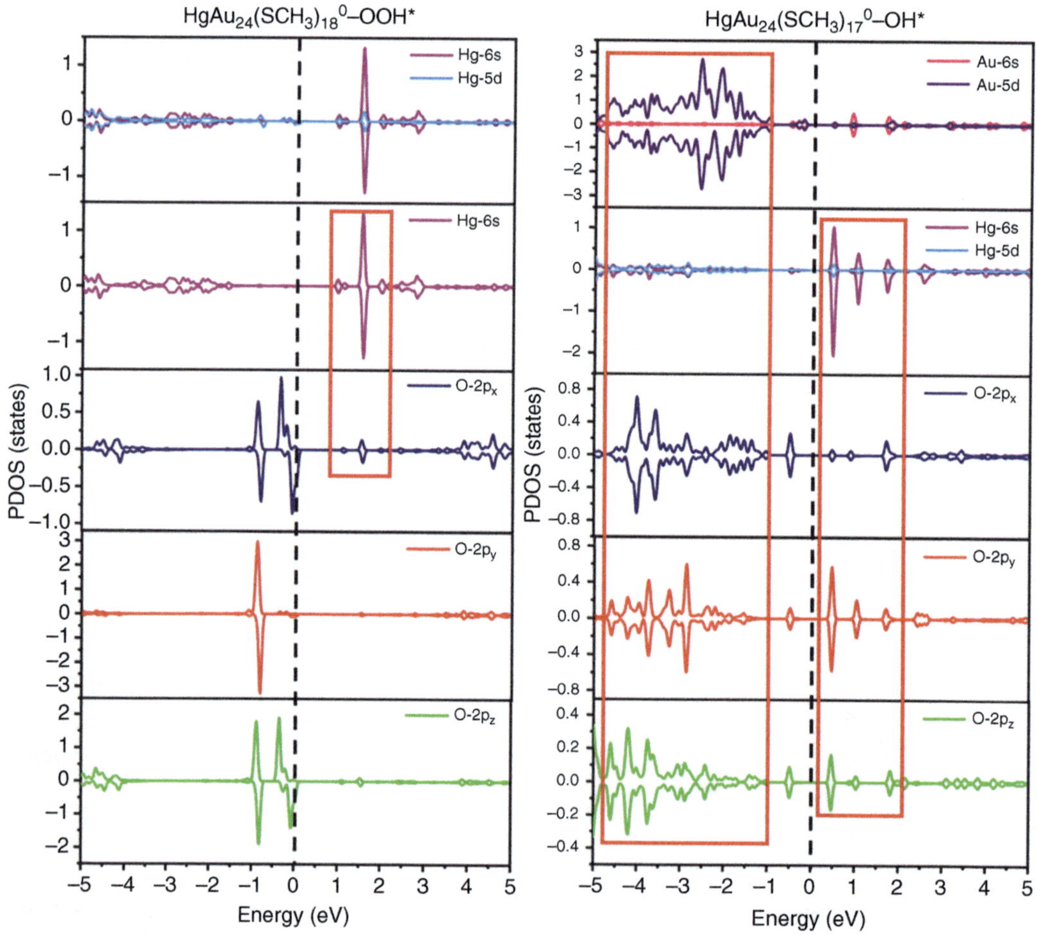

Figure 6.6 Projected density of states for *OOH/*OH adsorbed on the $HgAu_{24}$. The Fermi energy levels have been normalized to zero (black dotted line). *Source:* Reprinted with permission from [38]. © 2021 American Chemical Society.

the s-electron effect, doping p-block metals (e.g. Sn, Bi) with strong p-electron effects may also be a good choice for tuning the metal NCs activity.

In another work, Jin's group [31] investigated the heavy Ag-doping alloy NC of $[Ag_xAu_{25-x}(SC_6H_{11})_{18}]^-$ for ORR in alkaline solutions. Experimental characterization indicated that Ag dopants have a distribution in both the icosahedral M_{13} core and the 6 S-Ag-S-Ag(Au)-S staple motifs, with the average composition being x = 21 and the maximum number of Ag dopants being up to 24. Compared to the low-doped Ag_xAu_{25-x} NCs synthesized by the one-phase method, Jin's work revealed that the Ag(I)-SC_6H_{11} complex plays a critical role in achieving the heavily Ag-doped NCs even with a low Ag: Au precursor ratio of 2:1. The ORR catalytic activity of $Au_{20.5}Ag_{4.5}(PET)_{18}$, $Au_{23.7}Cu_{1.3}(PET)_{18}$, and $Au_{24}Pd(PET)_{18}$ has been evaluated by Negishi's experiments [22]. They claimed that the introduction of Pd is beneficial to the ORR process, whereas Ag and Cu substitution degrade the ORR activity.

Trimetallic and even tetrametallic NCs have also been reported in the literature [57, 58], which may be attractive for some catalytic applications. The atomic-level regulation of catalytic activity and selectivity has considerable scope for future research. It is worth noting that, in some cases, the doping may also have adverse effects [1]. Understanding the negative and positive effects of doping/alloying in NCs has great implications for advancing future work. Such catalysts are certainly more complex than homometal ones, and therefore both experimental detection and theoretical simulation are needed in order to understand the doping effect [59].

6.3 Important Electrocatalytic Applications

6.3.1 Electrocatalytic Water Splitting

6.3.1.1 Water Electrolysis Process

Electrocatalytic splitting of water has shown particular promise in terms of its high efficiency [60–64]. To minimize charge transport losses during electrochemical processes, conventional water electrolysis is usually performed with proton exchange membranes in acidic electrolytes or with membranes in alkaline environments (Figure 6.7a) [65]. The electrocatalytic water splitting consists of two half-reactions, i.e. cathodic water reduction–HER and anodic water

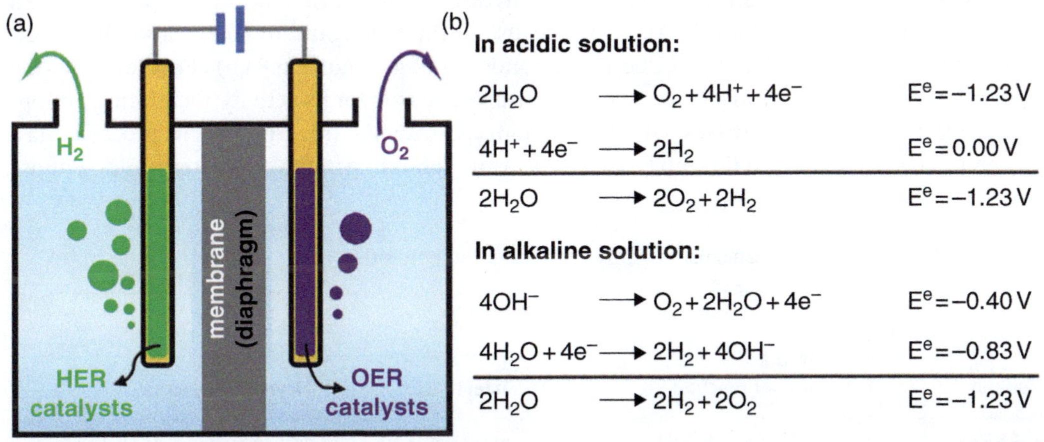

Figure 6.7 (a) Scheme and (b) reactions of water splitting under acidic and alkaline conditions. *Source:* Reprinted with permission from [65]. © 2018 American Chemical Society.

oxidation–OER, which can be expressed in different ways according to the reaction conditions, as shown in Figure 6.7b. Under standard conditions, the thermodynamic potential for water splitting is 1.23 V, but the actual potential is always higher.

6.3.1.2 Cathodic Water Reduction–HER

As mentioned earlier, HER is the cathodic half-reaction of water electrolysis, which follows a two-step reduction: (i) the initial step–Volmer reaction, followed by (ii) Tafel or Heyrovsky process (Table 6.1). Consequently, the HER mechanism in aqueous electrolytes can be described as following either a Volmer-Tafel or Volmer-Heyrovsky pathway.

The bimetallic $[PtAu_{24}(SR)_{18}]$ NC (1.1 nm) was the first to be used for HER electrocatalysis by Kwak et al. [44] The combined experimental and theoretical results revealed that this NC can exhibit an exceptionally higher catalytic activity than state-of-the-art Pt/C catalysts, and the Volmer-Heyrovsky route is the thermodynamically preferred pathway: the reduced $[PtAu_{24}]^{2-}$ species firstly reacts with one H^+ to form an $[H\text{-}PtAu_{24}]^-$ intermediate with a ΔE_1 of −0.059 eV, and then the $[H\text{-}PtAu_{24}]^-$ is coupled with another H^+ from the solution to generate H_2 with a ΔE_{2b} of −0.155 eV. On the other hand, the geometric optimization reveals that H prefers to occupy the surface hollow site of the $PtAu_{12}$ core to form a relatively strong H—Pt bond, which is a crucial factor in the improved HER activity observed in experiment. Similarly, the team also examined the HER electrocatalytic activity of $PdAu_{24}$, Pd_2Au_{36}, and Pt_2Au_{36} [51]. The combination of experiment and theory explains the structure-performance correlation of these atomically precise metal NCs, and provides in-depth understanding of the HER mechanism. Computationally, Jiang's group [66] studied how H interacts with $[Au_{25}(SR)_{18}]^q$ (q = −1, 0, 1) and bimetallic $[M_1Au_{24}(SR)_{18}]^q$ NCs (M = Pt, Pd, Ag, Cu, Hg or Cd; q = −2, −1, 0, +1) from DFT. Particularly, they found that H behaves as a metal and contributes its 1s electron to the superatomic free-electron count of the Au_{25} NCs. Different from Cu- and Ag hydride clusters, the small hydrogen can be doped at the interstice of the Au clusters and tune further their superatomic electronic structure. Calculations of the Gibbs free energy of H* adsorption (ΔG_{H^*}) indicated that $PtAu_{24}$, $PdAu_{24}$, and center-doped $CuAu_{24}$ are potential alternative catalysts for HER, among which $PtAu_{24}$ shows the most promising activity. Choi et al.'s prior experimental study [51] also confirmed this conclusion, and the same trend was observed in HER with a larger Au_{36} NCs.

Recently, Li et al. [67] investigated the HER activities of a trimeric $Au_{36}Ag_2(SR)_{18}$ NC with low ligand coverage, in which three icosahedral (I_h) units are face-fused together in a cyclic manner. They found that this NC can show higher HER catalytic performance and lower overpotential than the monomeric Au_{25}^- and dimeric Au_{38} NC because of the low ligand-to-metal ratio, the lower-coordinated Au atoms and unfilled 20e superatomic orbitals (Figure 6.8a,b). Furthermore, the greater negative electron affinity (−2.65 eV) may also be a reason for its activity, facilitating proton-coupled electron transfer processes. DFT calculations indicate that this trimeric NC proceeds via a Volmer-Tafel mechanism and thermodynamically favors H binding in the rate-determining Volmer

Table 6.1 Aqueous HER mechanisms (* refers to an active catalytic site).

Medium	Step 1 Volmer reaction	Step 2 Tafel reaction	 Heyrovsky reaction
Acidic solution	$H^+ + 2e^- + * = H^*$	$2H^* = H_2 + 2*$	$H^+ + e^- + H^* = H_2 + *$
Alkaline solution	$H_2O + e^- + * = H^* + OH^-$	$2H^* = H_2 + 2*$	$H_2O + e^- + H^* = H_2 + * + OH^-$

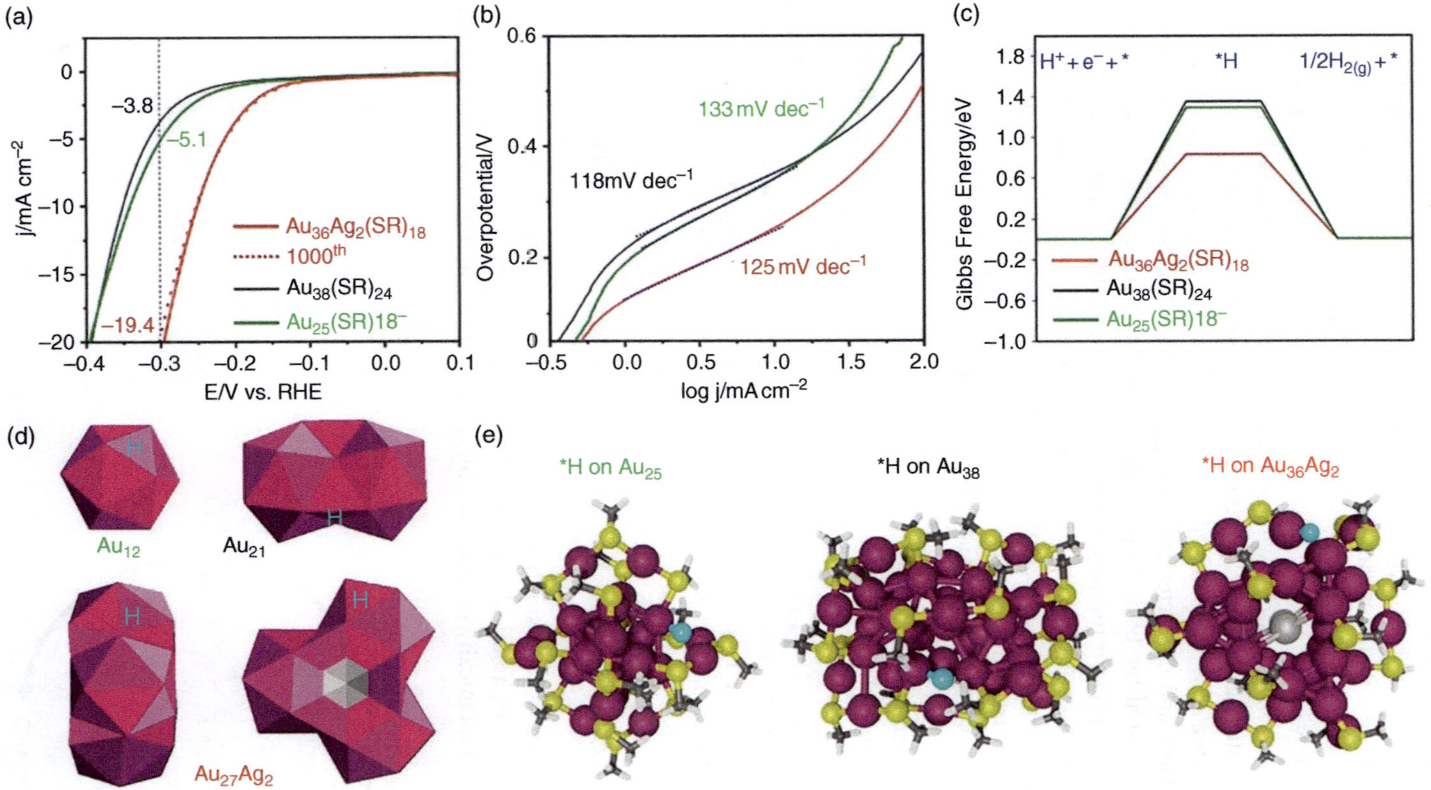

Figure 6.8 (a) HER voltammograms, (b) Tafel plots, (c) calculated ΔG_{H^*}, (d) illustrated kernel surfaces for H* adsorption, and (e) H* adsorption configuration on the NCs studied here. *Source:* Reprinted with permission from [67]. © 2021 American Chemical Society.

step (Figure 6.8c), where H* is inclined to occupy the exposed Au atom on each I_h unit of the fused $Au_{27}Ag_2$ shell (Figure 6.8d,e). In another case, the diphosphine-covered Au_{22} is structurally unique, which exposes eight coordinately unsaturated (cus) Au sites at the interface of two Au_{11} units, i.e. in situ uncoordinated site, thus partially protecting the cluster. Jiang and co-workers [68] found that these unsaturated sites can adsorb up to six H atoms with favorable energetics for HER based on the DFT results. Among them, the first four H atoms tend to occupy the long bridge sites of cus Au atoms, showing moderate H_2 activation and dissociation barriers and favorable H adsorption strength (ΔG_{H*}~−0.40~0.10 eV), while the latter two H atoms would be bonded to bridging Au atoms at the opposite ends of the NC protected by phosphine ligands. Charge analysis showed that the adsorbed H in the Au_{22} NCs behaved as negatively charged hydrides. Later, Wang's group [69] successfully isolated and characterized a $[Au_{22}H_4(dppo)_6]^{2+}$ nanohydride cluster and observed four H atoms located at the bridging position, confirming Jiang et al.'s prediction. These works suggest a design prospect for tuning the HER activity of metallic NCs by tailoring the geometric and electronic structures. Atomically precise metal NCs are arguably very unique model catalysts, and the promising examples we present here are just the beginning of their application in electrochemical HER. The large number of existing NCs offer many opportunities for the establishment of structure–activity correlations and active site design.

6.3.1.3 Anodic Water Oxidation–OER

OER is a complex multi-electron charge transfer process in acid and alkaline media. In general, the electrochemical reaction that occurs at the anode (OER) electrolytes are: $2H_2O_{(aq)} \rightarrow 4H^+_{(aq)} + 4e^- + O_{2(g)}$(acid medium) and $4OH^-_{(aq)} \rightarrow 2H_2O_{(aq)} + 4e^- + O_{2(g)}$(alkaline medium), shown in Figure 6.7b. Possible OER mechanisms have also been proposed by different research groups and are presented in Figure 6.9 [70]. Most of the proposed mechanisms generally involve the same intermediate such as M-OH and M-O. Obviously, there are two different routes involved here, one is a direct combination of 2MO to produce $O_{2(g)}$ (green pathway), and the other is the decomposition of the formed MOOH into $O_{2(g)}$ (black pathway). For the heterogeneous OER process, the bonding interactions between each intermediate and MO is the key to determine the overall electrocatalytic performance.

Precise control of the composition is of great interest to gain deep insight into the OER mechanism, in such a scenario, atomically precise NCs provide a better platform for manipulation. Zhao et al. [71] explored atomically precise $Au_{25}(SR)_{18}$, $Au_{144}(SR)_{60}$ and $Au_{333}(SR)_{79}$ NCs to construct

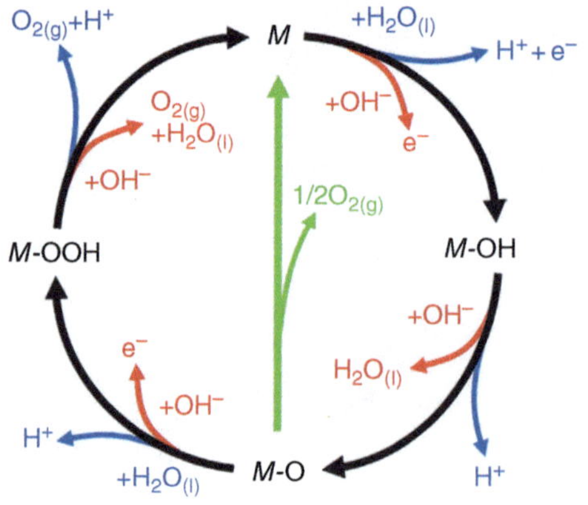

Figure 6.9 OER mechanism in acid (blue line) and alkaline (red line) conditions. *Source:* Reprinted with permission from [70]. © 2017 Royal Society of Chemistry.

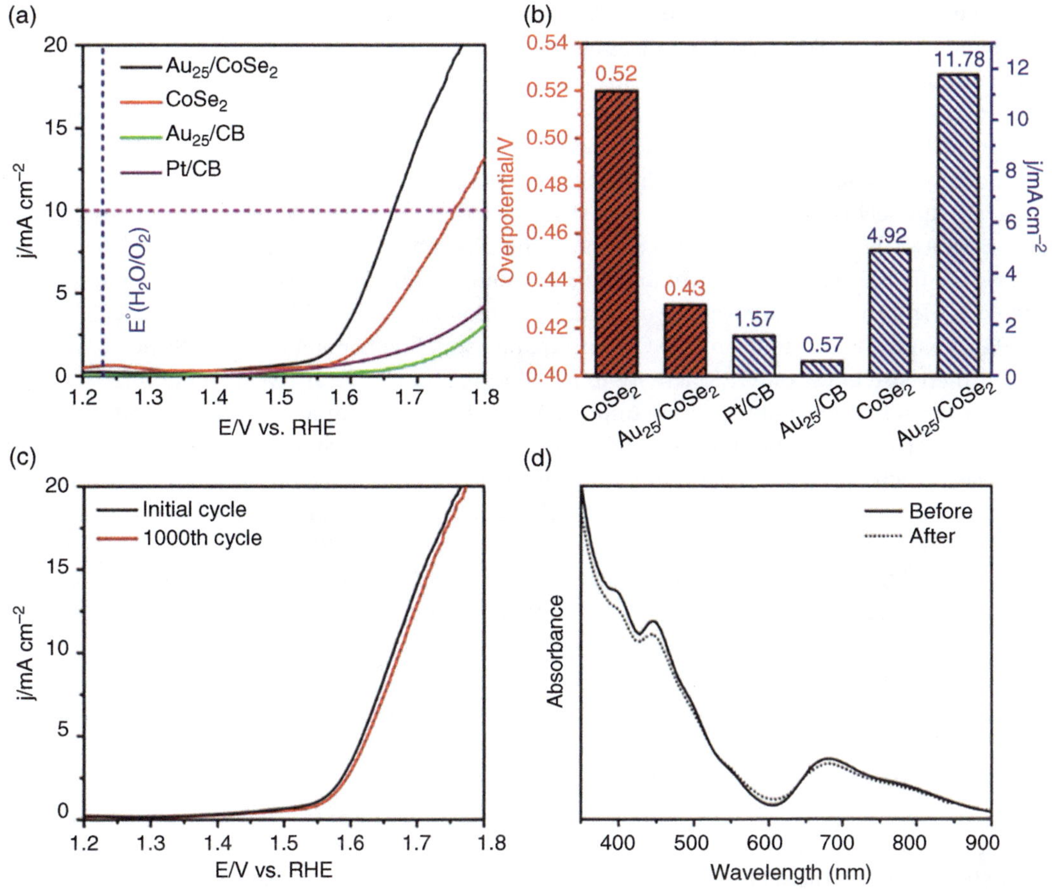

Figure 6.10 Electrocatalytic performance: (a) OER polarization curves; (b) comparison of the overpotential required for achieving the current density of 10 mA cm^{-2}, and the current density at the overpotential of 0.45 V. Stability test: (c) OER polarization curves for Au_{25}-$CoSe_2$ before and after 1000th cycle, and (d) UV–vis spectra of $Au_{25}(SR)_{18}$ before and after 1000th cycle. Catalyst loading: ~0.2 mg cm^{-2}. Sweep rate: 5 mV s^{-1}. All data were reported without iR compensation. *Source:* Reprinted with permission from [71]. © 2017 American Chemical Society.

$Au_n/CoSe_2$ nanocomposites for electrocatalytic OER under 0.1 M KOH solution. Electrochemical tests observed that $Au_{25}/CoSe_2$ exhibits the highest OER activity with a low overpotential of 0.43 V, a high current density of 11.78 mA/cm^{-2}, and a longer durability (Figure 6.10). After the removal of ligands, the OER activity was slightly improved. For the size dependencies of $Au_{10}(SR)_{10}$, $Au_{25}(SR)_{18}$, $Au_{144}(SR)_{60}$ and $Au_{333}(SR)_{79}$ loaded on $CoSe_2$ (~2wt.%), OER polarization curves showed a moderate increase of OER activity with increasing size. XPS and Raman measurements of $Au_{25}/CoSe_2$ found that the binding energy of its Co 2p was about 1 eV lower than that of $CoSe_2$, indicating an electronic interaction between Au_{25} and $CoSe_2$ that can stabilize the critical OOH* intermediate and optimize $CoSe_2$/oxygen interactions to enhance OER performance. DFT calculations also show that the Co—Au interface created by the complex facilitates the formation of the intermediate O* with the help of the base (OH^-), which further accelerates the key intermediate OOH* production, thereby improving the OER behavior of $CoSe_2$-Au. As mentioned earlier, Negishi's recent work [22] systematically clarified the dependences of each parameter (size, ligands, heteroatom doping, and charge effects) on the HER, OER, and ORR activities of $Au_n(SR)_m$

under identical experimental conditions. Taking both activity and stability into consideration, it was found that $[Au_{24}Pd(PET)_{18}]^0$ can be considered as a metal cluster with high potential in all reactions due to its favorable electronic structure.

Gao and Chen [72] supported the $Pd_6(SC_{12}H_{26})_{12}$ on activated carbon (AC) to obtain Pd_6/AC and calcined Pd_6/AC-V composites with ligands desorbed, and described their OER activity. Interestingly, Pd_6/AC was found to exhibit the highest mass and specific current densities, which was approximately 10–12 times higher than that of Pd_6/AC-V and Pt/C at the same potentials. The higher electron density was rationalized to favor the desorption of O_2 molecule (the rate determining step), thus Pd_6/AC is more favorable for OER than Pd_6/AC-V, but Pd_6/AC-V was found to be more suitable for HER. Moreover, Kwon et al. [73] obtained a series of bare Pd_n clusters (Pd_4, Pd_6, and Pd_{17}) that were soft landed on UNCD (ultrananocrystalline diamond)-coated Si electrode, and studied their size effect on OER activity in alkaline medium. Linear sweep voltammetry (LSV) after 500 seconds of cleaning by chronamperometery displayed a large increase of anodic current for Pd_6 and Pd_{17} and a sudden increase after the subtraction of current of blank UNCD support, yet there was no activity for Pd_4, suggesting the enhancement of OER activity with size. The drop-in activity in Pd_4 is suggested to arise from site blocking or interactions that alter the electronegativity of the UNCD surface atoms. The turnover rate results confirmed that even though Pd_6 and Pd_{17} were comparable, only a fraction of the surface sites of Pd_{17} were occupied, and additional theoretical calculations indicated that the bridging Pd—Pd sites in the Pd_6 cluster were OER active. Nickel-based materials are also regarded as promising OER catalysts. Currently, some atomically and structurally precise Ni NCs for OER have been reported [74, 75]. For instance, Bakr and co-workers [75] found that $Ni_4(SR)_8$ behaves as ultraefficient OER electrocatalysts, exhibiting a lower overpotential ~1.51 V and Tafel slope~38 mV dec^{-1}, which can rival the state-of-the-art RuO_2 OER catalyst. In another work, experiments and calculations from Kauffman et al. [74] confirmed that $Ni_6(SR)_{12}$ exhibits highly efficient catalytic activity, achieving the TOF ~70 s^{-1} at 2 V and a steady current voltage ~1.7 V at 10 s^{-1} in N_2 purged 0.1 M KOH. Overall, the above findings are expected to lead to the development of highly active OER catalysts, which in turn will lead to design guidelines to realize ultimate energy conversion systems.

6.3.2 Oxygen Reduction Reaction (ORR)

The ORR is the cathodic electrode reaction and is regarded as the rate-determining step of proton exchange membrane fuel cells (PEMFCs) due to its sluggish kinetics, which can proceed by a direct $4e^-$ charge transfer process to form H_2O (acid medium) or OH^-(alkaline medium), or generate HO_2^- and H_2O_2 with the consumption of $2e^-$, as shown in Table 6.2 below [76].

The Pt-based materials hold great promise as practical ORR catalysts because of the 4e reduction of O_2 at low overpotentials. Until now, a series of Pt clusters have been extensively studied to understand their catalytic behavior at an atomic level, revealing that their catalytic activity is not linearly dependent on the size while other factors, such as the number of surface atoms, specific edge structure, metal-oxygen binding energy, may play a key role [1, 77]. Anderson's group [78] investigates the size selected Pt_n NCs for ORR in 0.1 M $HClO_4$, and found their onset potentials are size dependent, varying from ~0.66 V (vs. NHE) to ~0.78 V, among which the maximum current/g_{Pt} of Pt_{10} is an order of magnitude higher than Pt nanoparticles of ~5 nm. In addition, they observed that the resulting products (H_2O or H_2O_2) are strongly dependent on size, with H_2O_2 generation increasing with increasing size, suggesting that such Pt NCs can selectively synthesize H_2O_2. Unfortunately, the structures of these Pt clusters are unknown.

Compared with Pt, Au is much less used as ORR electrocatalysts due to poor activity and $2e^-$ mechanism of bulk gold. Thanks to the great successes in determining the structures by nine-ray

Table 6.2 The ORR electrochemical reactions and corresponding thermodynamic potentials.

Electrolyte	ORR reaction	Thermodynamic electrode potential (V)
Acid medium	$O_2 + 4H^+ + 4e^- \rightarrow 2H_2O$	1.23
	$O_2 + 2H^+ + 2e^- \rightarrow H_2O_2$	0.70
	$H_2O_2 + 2H^+ + 2e^- \rightarrow 2H_2O$	1.76
Alkaline medium	$O_2 + 2H_2O + 4e^- \rightarrow 4OH^-$	0.40
	$O_2 + H_2O + 2e^- \rightarrow OH^- + (HO_2)^-$	−0.07
	$(HO_2)^- + H_2O + 2e^- \rightarrow 3OH^-$	0.87

crystallography, structurally precise Au NCs with ultra-small size are promoted as ideal systems to explore the ORR behavior at the atomic level. For example, Chen's group [79] and Sumner's group [23] investigated the core size effects of $Au_n(SR)_m$ clusters as ORR electrocatalysts. Chen et al. [79] found the Au_{11} without removal of ligands exhibited the highest ORR activity in 0.1 M KOH aqueous solutions, and the activity followed an order of $Au_{11} > Au_{25} > Au_{55} > Au_{140}$. In 2007, Negishi's team [100] first reported that Au_{25} cluster has tunable ground state charge ($q = -1, 0, +1$), allowing to probe the chemistry of charged active sites. In 2014, Lu's group [43] and Kaufman et al. [42] then independently examined $Au_{25}(SR)_{18}{}^q$ with different charge states for ORR evolution in 0.1 M KOH. Based on the electrochemical test results, Lu et al. [43] proved that the $2e^-$ reduction of O_2 is dominant, where the $Au_{25}{}^-$ can achieve high activity and selectivity toward H_2O_2 production with a yield of 86% and their activity follow an order of $-1 > 0 > +1$. The authors suggest that this high activity is possibly due to the electron transfer from the anionic $Au_{25}{}^-$ core into the LUMO (π^*) of O_2 to activate the O_2 molecule, resulting in the formation of peroxo-like species. A similar trend of activity was observed by Kaufman et al. [42], namely, $Au_{25}{}^- > Au_{25}{}^0 > Au_{25}{}^+$; however, a combination of $4e^-$ and $2e^-$ ORR mechanism is thought to favor the generation of OH^- and OOH^- on $Au_{25}{}^q$. In addition, some excellent work has been done on Au_{25} composite NCs for ORR applications, such as the composite films of Au_{25} clusters on reduced graphene oxide (rGO) [80]. Moreover, Sumner and co-workers [23] observed that $Au_{36}(SR)_{24}$ is the most active ORR catalyst with the lowest overpotential requirement (0.16 V vs. RHE) in alkaline condition, and the activity trend is $Au_{36}(SR)_{24} > Au_{133}(SR)_{52} > Au_{279}(SR)_{84} > Au_{28}(SR)_{20}$, which can be qualitatively related to their trends in thermochemical stability. Obviously, the significant difference in the catalytic activity of ORR is not only determined by the size of the metal core, but other factors such as stability can also determine the activity result. Negishi and co-workers [22] recently examined the ORR behaviors of $Au_n(SR)_m$ in acidic conditions. They asserted that the $4e^-$ reduction pathway is dominant, and their activity is remarkably dependent on the number of constituent atoms, increasing as the number of constituent atoms decreased.

At the computational side, two recent works from Tang's group have provided some insight. By means of DFT simulations, Deng et al. [81] explored the potential of the experimentally well-defined $Au_{22}(L^8)_6$ (where L^8 = 1,8-bis[diphenylphosphino]) as an ORR electrocatalyst. They found that $Au_{22}(L^8)_6$ exhibits distinguished activity for H_2O formation through the $4e^-$ reduction, with a small predicted overpotential of 0.49 V. The coordination unsaturated Au atoms are the active sites for O_2 activation, and can bind well to all ORR intermediates. Further extended studies of ORR activity on gold surfaces showed that the order of activity is $Au_{22}(L^8)_6$ (0.49 V) > Au(211) (0.70 V) > Au(110) (0.82 V) > Au(100) (0.92 V) > Au(111) (1.22 V), which is closely related to the bonding of *OH intermediate and the generalized coordination number of Au atoms. In another contribution, Tang's team [38] presented a comprehensive understanding of the oxygen electrocatalysis by $[Au_{25}(SR)_{18}]$. DFT results showed that fully ligand-protected NCs do prefer to produce

H_2O_2 via the 2e⁻ pathway, while removal of a $-SCH_3$ ligand significantly enhances the adsorption of *OOH and O*, resulting in catalytic selectivity toward the four-electron route (Figure 6.11a). For two-electron ORR, the adsorption of OOH* intermediate is weak, but relatively ideal for H_2O_2*(Figure 6.11b); for four-electron ORR, the affinity to *OOH is significantly optimized, but over strong for O* and OH* because of the weakened ligand assembly (Figure 6.11c). Bader charge

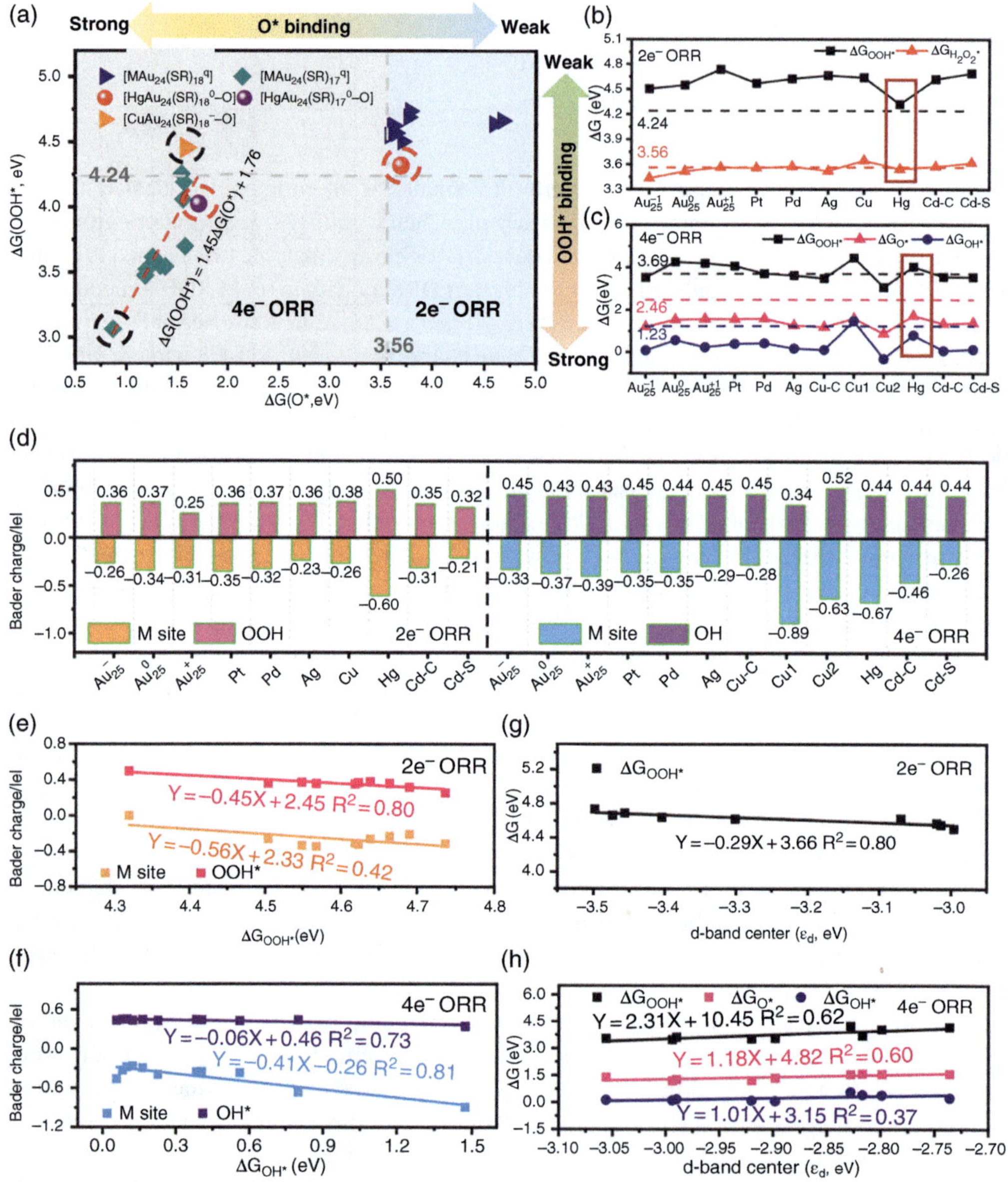

Figure 6.11 (a) The variations of computed ΔG_{*OOH} vs. ΔG_{O*} and (d) Bader charge analysis of the studied NCs. (b and c) The comparison between the calculated ΔG of each intermediate and their corresponding optimal ΔG level. Correlations of the ΔG of oxygenated intermediates with Bader charge (e–f) and d-band center of metal active site (g–h). 2e⁻ ORR: $[MAu_{24}(SCH_3)_{18}]^q$; 4e⁻ ORR: dethiolated $[MAu_{24}(SCH_3)_{17}]^q$; Cu1 and Cu2 represent $[CuAu_{24}(SCH_3)_{18}]^-$-O and $[CuAu_{24}(SCH_3)_{17}]^-$-O, respectively. *Source:* Reprinted with permission from [38]. © 2021 American Chemical Society.

analysis (Figure 6.11d) intuitively shows that electrons depleted from the metal atoms are transferred to the *OOH ($2e^-$ ORR) and *OH ($4e^-$ ORR). Then, the correlation of adsorption free energy of oxygenated intermediates with the Bader charge and d-band center (ε_d) shed light on the underlying origin of selectivity and activity: for $2e^-$ ORR catalysts, the charge transfer (Figure 6.11e) and band hybridization (Figure 6.11g) play an essential role in *OOH adsorption; in $4e^-$ ORR cases, the charge transfer between OH* and NCs exerts an enormous function (Figure 6.11f), while the band hybridization presents a poor correlation (Figure 6.11h). In particular, benefited from the s-electron effect from the staple-doped Hg atom, the fully ligated $[HgAu_{24}(SCH_3)_{18}]^0$ and dethiolated $[HgAu_{24}(SCH_3)_{17}]^0$ hold great promise in high-efficiency $2e^-$ and $4e^-$ ORR. Experimentally, Jin and co-workers [31] obtained Ag-doped $[Ag_xAu_{25-x}(SC_6H_{11})_{18}]^-$, which shows a more positive onset potential and higher mass activity for ORR in alkaline solutions after removing some ligands at high temperature. The calculated electron transfer numbers indicate that the ligand removal can tune the ORR kinetic process. Based on these findings, they concluded that ligand engineering on the surface may be the key to the ORR features of such NCs rather than alloying inside the metal core.

Ag NCs have also functioned as potential electrocatalysts for O_2 reduction. For instance, Yang et al. [82] demonstrated that the 2-mercaptobenzothiazole protected Ag NCs (~2–5nm) display a more positive onset potential and higher currents than commercial Pt/C and Ag NPs toward $4e^-$ reduction of O_2 to H_2O. Chen's group [83] prepared the Ag NCs supported on carbon nanodots as the ORR catalyst that displays excellent electrocatalytic activity with 4e transfer process. Jin et al. [84] synthesized the Ag nanocomposite (<2 nm) using single-stranded oligonucleotides (ss-DNA) as a template on nitrogen-doped graphene (NG) by the clean electrochemical reduction method, which exhibits competitive onset potential, half-wave potential, and mass activity for ORR compared to commercial Pt/C catalyst. These advances further confirm that the power of subnanoclusters in the field of electrocatalysis should not be underestimated.

6.3.3 Electrochemical CO_2 Reduction Reaction (CO_2RR)

In fundamental research, the electrochemical CO_2RR has been gaining extensive attentions, mainly owing to its mild conditions under ambient temperature and pressure, and controllable reaction rates and product selectivity using renewable energy sources [2]. Such reaction is generally involved in multiple proton and electron transfer process, and can generate multiple products, e.g. CO ($2e^-$, −0.10 V vs. RHE), HCOOH ($2e^-$, −0.12 V vs. RHE), HCOH ($4e^-$, −0.07 V vs. RHE), CH_3OH ($6e^-$, 0.03 V vs. RHE), CH_4 ($8e^-$, 0.17 V vs. RHE), etc., and can be accompanied by competing HER, as summarized in Table 6.3 [85]. The atomically precise NCs have shown extraordinary catalytic properties for CO_2RR [2]. Below we highlight some representative examples in several different types of nanocatalysts, including Au NCs, Au-based alloy NCs, Ag NCs, Cu NCs.

For Au NCs, Au_{25}^q are widely studied for CO_2RR. As early as in 2012, pioneering work by Kauffman et al. reported the negatively charged Au_{25}^- molecule as electrocatalyst for CO_2 reduction [86]. Based on the change in Au_{25} valence electrons after the CO_2 adsorption, they confirmed that the electron depletion of the S atom may affect the reaction performance. In the CO_2RR test, they observed that the Au_{25}^- could boost the CO_2RR to CO within 90 mV of the formal potential, and the product efficiency could reach 100%. With the aid of DFT simulations, the authors discovered that Au_{25}^- has the strongest adsorption capability hence it possessed higher cathodic current density, TOF value, and faraday efficiency (FE). Two years later, the experimental investigation by Kauffman et al. [42] confirmed the charge-dependent electrocatalytic activity of Au_{25}^q (q = 0, −1, and +1) for CO_2RR, and their activity followed an order of $-1 > 0 > +1$.

A following theoretical study conducted by Alfonso et al. [33] uncovered that the fully ligand covered $[Au_{25}(SCH_3)_{17}]^-$ clusters are not active due to the high potential of forming *COOH

Table 6.3 The elementary reactions along with corresponding reduction potentials (E_0 vs. RHE) and products for CO_2RR.

Reaction	E_0 (V vs. RHE)	Product
$CO_2 + 2H^+ + 2e^- \rightarrow CO_{(q)} + 2H_2O$	−0.10	Carbon monoxide
$CO_2 + 2H^+ + 2e^- \rightarrow HCOOH_{(aq)}$	−0.12	Formic acid
$2CO_2 + 2H^+ + 2e^- \rightarrow (CO_2H)_{2(s)}$	−0.47	Oxalic acid
$CO_2 + 4H^+ + 4e^- \rightarrow HCOH_{(aq)}$	−0.07	Formaldehyde
$CO_2 + 4H^+ + 4e^- \rightarrow C_{(s)} + 2H_2O$	0.21	Graphite
$CO_2 + 6H^+ + 6e^- \rightarrow CH_3OH_{(aq)}$	0.03	Methanol
$CO_2 + 8H^+ + 8e^- \rightarrow CH_{4(g)} + 2H_2O$	0.17	Methane
$2CO_2 + 8H^+ + 8e^- \rightarrow CH_3CO_2H_{(aq)} + 2H_2O$	0.11	Acetic acid
$2CO_2 + 10H^+ + 10e^- \rightarrow CH_3CHO_{(aq)} + 3H_2O$	0.06	Acetaldehyde
$2CO_2 + 12H^+ + 12e^- \rightarrow CH_3CH_2OH_{(aq)} + 3H_2O$	0.09	Ethanol
$2CO_2 + 12H^+ + 12e^- \rightarrow C_2H_{4(g)} + 4H_2O$	0.08	Ethylene
$2H^+ + 2e^- \rightarrow H_2$	0.00	Hydrogen

species, while the partially ligated NC would be energetically preferred to promote the CO_2 reduction since the de-thiolated Au sites can greatly stabilize the formation of key *COOH intermediate. A similar conclusion was also given in another ab-initio electronic structure calculation study on CO_2RR over fully ligand-capped $Au_{25}{}^{q}$ and their partially ligand-removed (removing –SR and –R) NCs [87]. Regardless of the charge state, fully ligand-protected NC is chemically inert. However, the partially ligand removed NCs are feasible for CO_2RR owing to the exposure of Au and S sites, manifesting that the generation of exposed surface active sites are critical for catalytic activity.

Recently, Kim et al. [88] showcased that the 1-hexanethiolate protected Au_{25} NCs were able to initiate CO_2RR to CO, which can achieve a large current density about 540 mA cm^{-2} in a gas-phase reactor. Interestingly, Jin's group [26] synthesized two Au_{25} NCs with different morphologies (Figure 6.2: nanosphere and nanorod) for CO_2RR. They discovered that such apparently different configurations significantly affect the reaction. As illustrated in Figure 6.12a,b, the total current density and the FE for CO over the spherical Au_{25} are larger than that of rodlike Au_{25} at all voltages. Selectivity comparison (Figure 6.12c) showed that the FE for CO is much higher than that for H_2 at $U = -1.07$V, where the Au_{25} nanosphere exhibits much higher selectivity with smaller FE toward H_2 (24.9%) compared to the Au_{25} nanorod (41.2%). Besides the high FE, the CO production rate of Au_{25} nanosphere is also faster (Figure 6.12d), reaching 33.3 μl/min at −1.17 V, which is 2.8 times that of the nanorod. According to their DFT calculations (Figure 6.12e), the removal of PH_3 and—Cl on nanorod is more favorable, but the removal of -SCH_3 is less endergonic from the nanosphere (red line, $\Delta G = 0.49$ eV), suggesting that more active sites can be released from the surface of nanosphere NCs by ligand removal. From Figure 6.12f, it can be seen that regardless of which ligand is removed, the key *COOH intermediate could be more stable on the -SCH_3 removed Au_{25} nanosphere, which can be responsible for the improved CO_2RR reactivity. In addition to the widely studied Au_{25}, some other relatively larger NC such as $Au_{137}(SR)_{56}$ also exhibits considerable CO_2RR activity, with a sixfold increase in current in CO_2-saturated solutions compared to N_2-saturated solutions [89].

Moreover, surface ligand engineering is regarded as an effective strategy to regulate the catalytic performance of NCs [15, 90, 91]. Narouz et al. [92] reported the novel Au_{11} NCs protected by

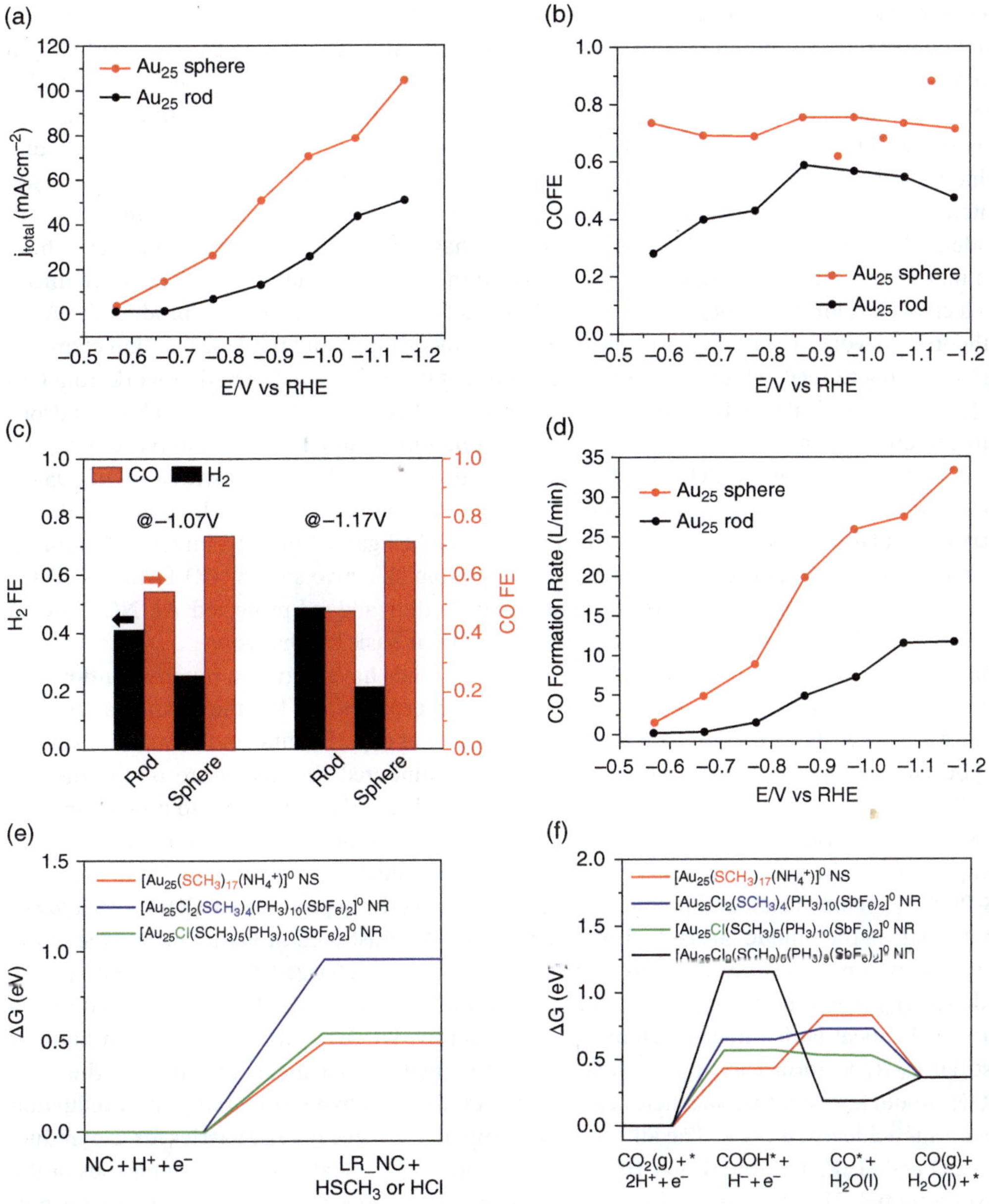

Figure 6.12 Electrocatalytic CO_2 reduction performance of the Au_{25} nanosphere and nanorod: (a) total current density; (b) FE for CO production; (c) FEs for CO and H_2 at the potential of −1.07 and −1.17 V, and (d) CO formation rates. DFT calculation results: ΔG values for (e) ligand removal and for (f) CO_2 reduction to CO on the ligand-removed Au_{25} nanosphere and nanorod. *Source:* Reprinted with permission from [26]. © 2018 American Chemical Society.

N-heterocyclic carbenes (NHCs), which possess strong metal-carbon single bonds and endow the corresponding NC with high stability. When examined for CO_2RR, $Au_{11}(PPh_3)_7$ $(NHC^{Me})Cl_3$ has the highest FE, the largest current density and mass activity duo to its ability to maintain intact structure during the reaction process. For Au-based alloy NCs for CO_2RR, Jin's group [32] and Wu's team [93] made some interesting discoveries; the work of Jin et al. has been described in detail in the discussion of doping and alloying effect in Section 6.2.5 and will not be repeated here. Wu and co-workers [93] found that the Cd-doped $Au_{47}Cd_2(SR)_{31}$ could efficiently convert CO_2 to CO with ~96% FE at a low potential (−0.57 vs. RHE), because the introduction of Cd was conducive to the formation of Cd-O-C(OH)-Au and promoted the stability of COOH* intermediate.

Ag is also widely considered to have high activity and selectivity for CO, mainly due to its unique CO binding affinity [2]. Theoretically, Zhang et al. [94] executed a DFT calculation to investigate the interaction between CO_2 and Ag NCs, and study the active site and the mechanism. The time-dependent DFT calculations indicated that Ag_{18} NC has a strong orbital coupling effect, which would make the electrons on Ag easily transferred to the LUMO orbital of CO_2. The mechanistic and selectivity exploration determined that CH_3OH is the most probable final product for Ag_{18}. This theoretical work provides an example to study the mechanism and possible reaction channels of Ag NC driving CO_2RR, which need further experimental verification. In another work from Qin et al. [37], they fabricated a homoleptic alkynyl-protected $Ag_{15}(C{\equiv}C^t\text{-Bu})_{12}{}^+$, which can adsorb CO_2 in the air to spontaneously self-assemble into one-dimensional linear material during the crystal growth process. In the CO_2RR test, such NC can convert CO_2 into CO with a FE of ~95.0% at −0.6 V and reach a maximal TOF of $6.37\,s^{-1}$ at −1.1 V along with excellent long-term stability. The DFT calculations disclosed that releasing one −C≡CR ligand from the intact NC, namely, $Ag_{15}(C{\equiv}C^t\text{-Bu})_{11}{}^+$, can expose the uncoordinated Ag atom as active site for CO formation. This study undoubtedly opens an avenue for synthesizing other alkynyl-protected Ag NCs toward CO_2RR, and will further advance the understanding of their basic mechanisms.

Among various CO_2RR electrocatalysts, Cu-based materials have attracted much attention due to their ability to reduce CO_2 into hydrocarbons at high overpotentials, but the product selectivity is unsatisfactory [95, 96]. To this end, albeit a large number of experimental and simulation studies have poured, the elusive surface structure has greatly hindered the disclosure of the mechanism [77]. Excitingly, the atomically precise Cu NCs provide a unique platform to understand the CO_2RR mechanisms at the atomic level. In 2017, Tang et al. [97] conducted a combined theoretical and experimental investigation regarding the electrochemical CO_2RR by $Cu_{32}H_{20}L_{12}$ (where, $L = S_2P[O^iPr]_2$) nanocluster, and predicted that it could selectively produce HCCOH with a lower overpotential via the unique lattice-hydride mechanism. As illustrated in Figure 6.13a, due to the attraction between negatively charged hydride and positively charged C of CO_2, one surface hydride (μ_3-H_1) is first added to CO_2 forming a five membered ring HCOO* intermediate via a nonelectrochemical lattice hydride channel. Subsequently, HCOO* further reacts with another interstitial μ_4-H_1 to form $Cu_{32}H_{18}L_{12}$ cluster with two hydride vacancies and release the final HCOOH product, where two vacancies can regenerate through two sequential proton reduction process, thus achieving a cycle. The authors also compared the free energy barrier for CO_2 reduction to CO and HER, and found that HCOOH formation is kinetically more advantageous at the low overpotential. The experimental results (Figure 6.13b,c) also verified that the HCOOH is predominantly generated at low overpotential (89% at 0.3 V, 83% at 0.4 V), while the H_2 is the major product at higher overpotential (>0.5 V). Another copper-hydride NC, $[Cu_{25}H_{22}(PH_3)_{12}]Cl$, was examined by Li et al. to study its activity and selectivity for electrochemical CO_2RR [98]. One major structure difference with the Cu_{32} nanocluster (containing both surface and interstitial hydrides) lies in that all the hydride ligands in Cu_{25} are located at the surface sites.

In contrast to the previously described lattice-hydride pathway, the simulation results show that this Cu_{25} NC mainly generates HCOOH through the hybrid mechanism (Figure 6.14), i.e. the proton-hydride, or the hydride-proton channel. For the hydride-proton pathway (blue line), a hydride first attracts CO_2 to form a HCOO* species and then adds to another proton to generate the final HCOOH, where the formed HCOOH is easily desorbed from this NC surface ($\Delta G = 0.33\,eV$), and vacant hydride is readily regenerated by adsorbing one proton in solution ($\Delta G = -0.87\,eV$). For

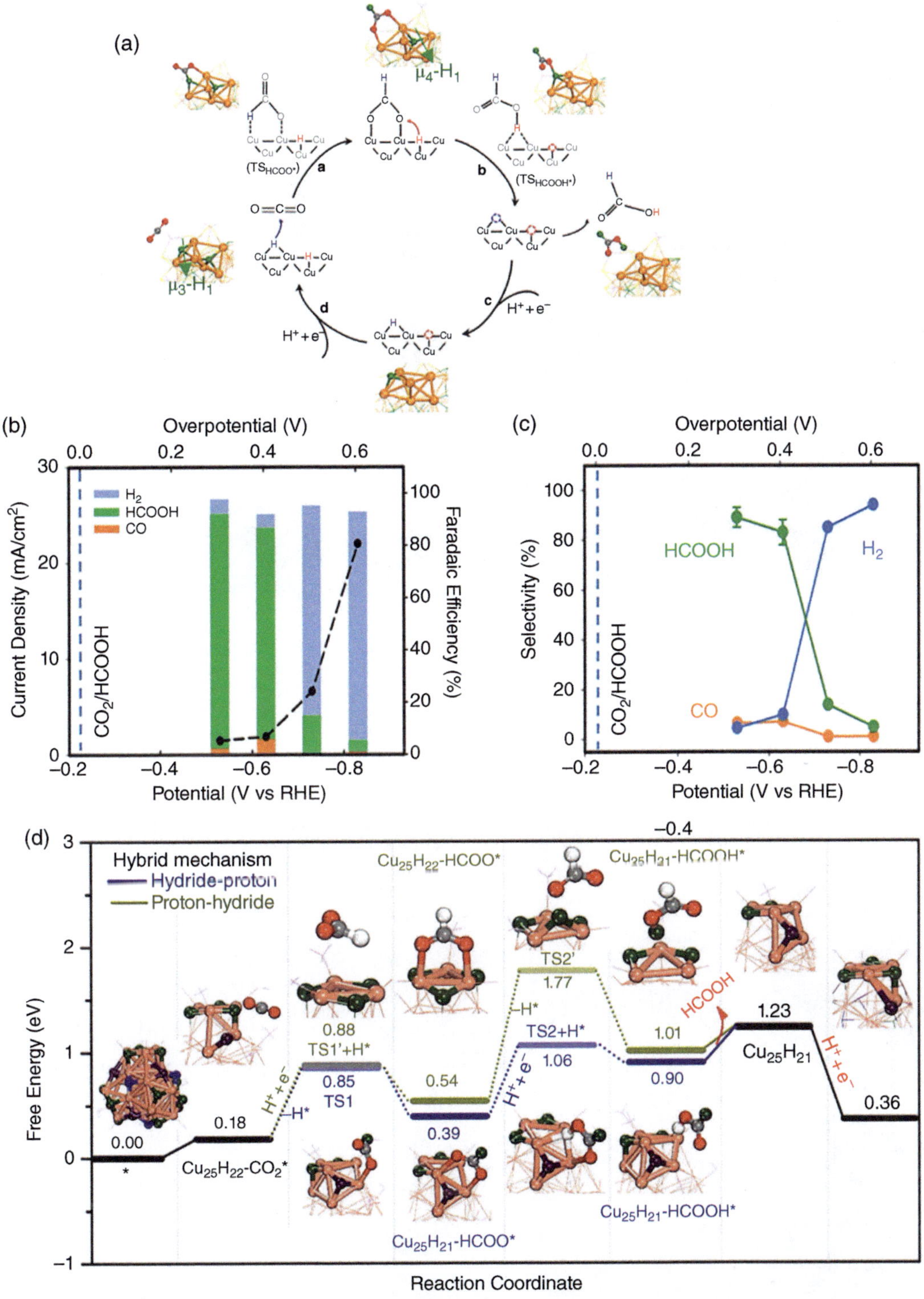

Figure 6.13 (a) The mechanism of HCOOH formation, (b) average current densities, and (c) selectivity comparison of three products from CO_2 reduction on $Cu_{32}H_{20}L_{12}$. Reprinted with permission from [97]. © 2017 American Chemical Society. (d) The scheme of hybrid mechanism for electrocatalytic CO_2 reduction on $[Cu_{25}H_{22}(PH_3)_{12}]Cl$ to form HCOOH. *Source:* Adapted with permission from [98]. © 2020 Elsevier Inc.

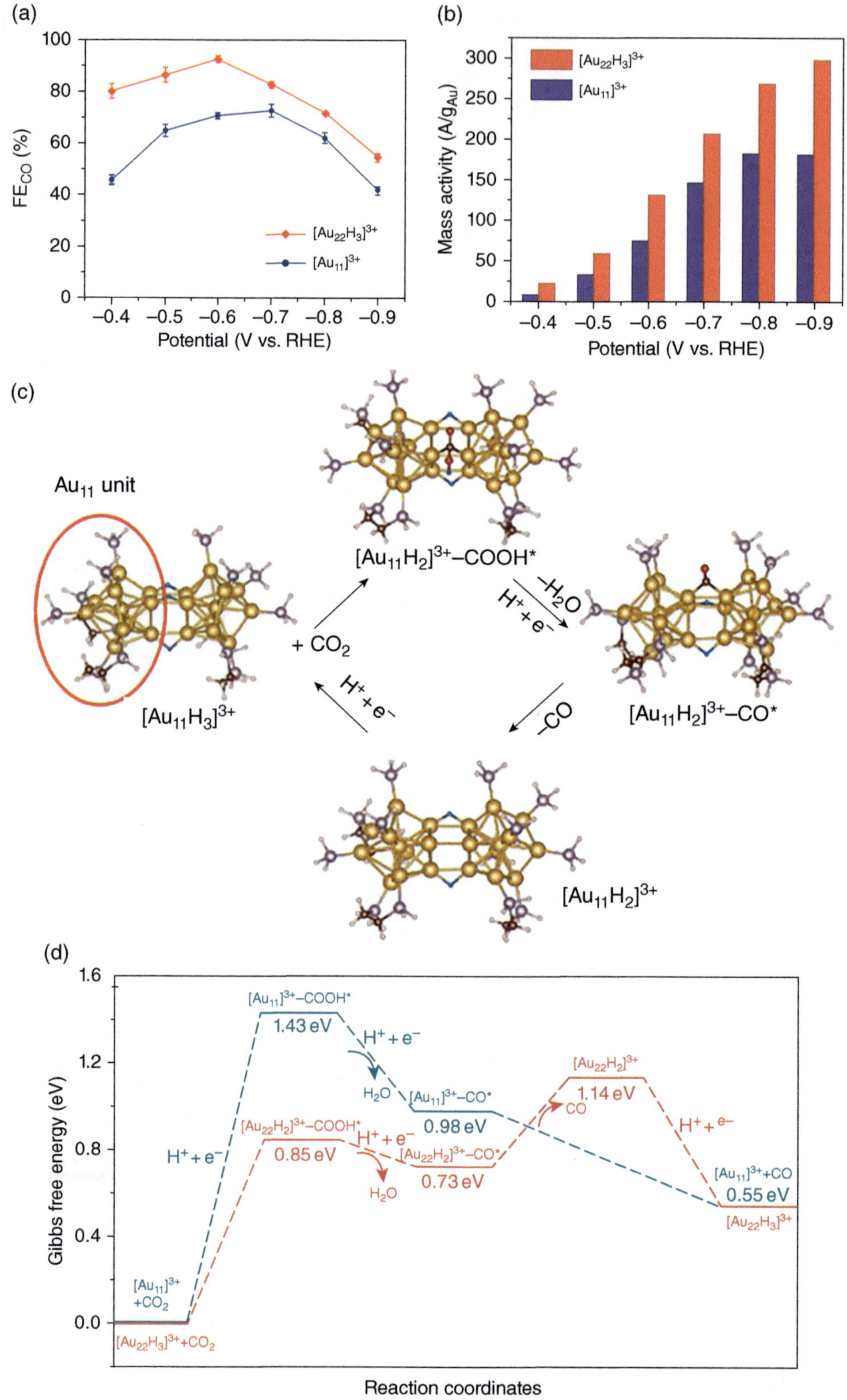

Figure 6.14 (a) FE$_{CO}$ and (b) potential-dependent mass activity for electrocatalytic reduction of CO_2 to CO. (c) DFT-optimized structures of the CO_2 reduction to CO cycle and (d) corresponding free energy profiles on $[Au_{22}H_3]^{3+}$ NC. *Source:* Reprinted with permission from [99]. © 2022 American Chemical Society.

the proton-hydride pathway (brown line), a proton in the solution first reduces CO_2 to HCOO* intermediate and is then attacked by the neighboring hydride to generate the $Cu_{25}H_{21}$-HCOOH*. Finally, HCOOH* is released from the surface of the catalyst to get the defective $Cu_{25}H_{21}$ and HCOOH product ($\Delta G = 0.21$ eV) and simultaneously replenishes the consumed hydride to ensure a complete electrochemical cycle ($\Delta G = -0.87$ eV). These unique mechanisms may provide new guidance for designing well-defined Cu-hydride NCs to achieve desired CO_2RR selectivity and greatly promote their development in other field of catalysis.

Note that the hydride ligand, H^-, is considered the smallest closed-shell spherical anion ever known. At present, polyhydrido Cu NCs are the most widely studied, while Au hydride clusters are relatively lacking. Unlike H in Cu NCs, the calculated Bader charges from Jiang and Hu et al. [66] showed that H in ligand-protected Au cluster (ranging from +0.034 to −0.004 |e|) behaves like a metal. In other words, the H atoms do not withdraw electrons from the metal core but donate their electrons to the cluster. This can be understood by the fact that H's electronegativity (2.2) is smaller than that of Au (2.54) but larger than that of Cu (1.90). Interestingly, Wang's team [99] recently synthesized a unique heteroleptic Au trihydride NC, $[Au_{22}H_3(dppe)_3(PPh_3)_8]^{3+}$ [dppe = 1,2-bis(di phenylphosphino)ethane, PPh_3 = triphenylphosphine], and simultaneously investigated its activity for CO_2RR. Structural characterization confirmed that this cluster consists of two Au_{11} units (circled in red in Figure 6.14c) that are bonded together via two triangular faces with three H atoms bridging six uncoordinated Au atoms. Electrochemical tests found that $[Au_{22}H_3]^{3+}$ exhibits remarkable performance and selectivity to CO with a 92.7% FE at −0.6 V vs. RHE (Figure 6.14a), 134 A/g_{Au} mass activity (Figure 6.14b), and long-term stability. The presence of hydride in $[Au_{22}H_3]^{3+}$ greatly improves the catalytic efficiency as compared to the $[Au_{11}]^{3+}$ nanocluster. DFT calculations uncovered a lattice hydride mechanism similar to CO_2RR on Cu hydride NCs (Figure 6.14c): first, the bridge H atom could directly hydrogenate CO_2 to facilitate the key COOH* intermediate formation ($\Delta G = 0.85$ eV, Figure 6.14d), thereby favoring the generation of adsorbed CO* and H_2O through an electrochemical proton reduction process ($\Delta G = -0.12$ eV, Figure 6.14d). Moreover, the H vacancy in $[Au_{22}H_2]^{3+}$ can be readily recovered by proton reduction to complete the catalytic cycle (Figure 6.14c). Obviously, the Au—H active sites of the lattice hydrides in $[Au_{22}H_3]^{3+}$ behave very differently from those in the Cu—H systems, especially in the product selectivity. This work suggests a possibility of designing catalytically active centers using Au nanohydrides for potential applications.

6.4 Conclusion and Perspectives

In recent years, the combination of experiment and theory has greatly promoted the development of atomically precise metal NCs in catalysis. An important goal is to link its catalytic properties to the corresponding atomic-scale structure, and ultimately to map the factors that determine catalytic activity and selectivity. While multiple factors are often puzzling, well-defined nanocatalysts can differ with only one factor but the others remain the same, which can provide an ideal platform for studying various effects to realize the fundamental understanding of nanometal catalysis. Therefore, the synthesis of metal NCs and the determination of their overall structure is undoubtedly an important step to be overcome, which requires a variety of advanced characterization techniques and feasible theoretical methods to complement each other. In this chapter, we particularly focus on the potential factors affecting electrocatalytic performance and selectivity (e.g. size, shape, charge state, ligand, doping, and alloying effect), either positively or negatively. Furthermore, we outline some experimental and theoretical advances of these clusters for some important electrocatalysis, including HER, OER, ORR, and CO_2RR.

The thiolate- or alkynyl-protected metal NCs usually have a core-shell structure, i.e. a polyhedral metal core and a metal-containing staple-motif shell, which can create special sites, such as pocket site on $Au_{25}(SR)_{18}$, for reactant adsorption and activation. Interesting electronic interactions between an electron-rich core and electron-poor shell may arise, which have been shown to play an important role in catalysis. In this regard, the kernel-shell interface deserves further exploration in future studies. Alloy NC also offer many new opportunities to understand the synergy in bimetallic/multimetallic nanocatalysis. It remains an important goal to unravel the relationship between geometrical/electronic properties (e.g. heteroelement's atomic distribution, oxidation state) and reactivity of doped NCs. Although creating bare uncoordinated metal sites as much as possible may improve the electrocatalytic process, the ligand-protected NCs are much more robust in structure. At present, it is still unclear whether the structure will undergo certain dynamic changes at high temperature, and whether the ligands on its surface will be displaced and recovered after the reaction. The development of in situ/operando tools (including imaging and spectroscopy) is expected to precisely analyze some complex problems such as partial ligand removal effects. Another important aspect worthy of special attention is that many theoretical revelations based on DFT also urgently need experimental verification, such as the ORR and CO_2RR mechanism predicted on $Au_{22}(L^8)_6$ and $[Cu_{25}H_{22}(PH_3)_{12}]Cl$, respectively.

In the field of cluster research, the theoretical community also faces rigorous challenges. Computationally, to reduce the cost, experimentally larger and more complex groups are always simplified. However, this treatment may not be applicable to bulky ligands, since vdW interactions or steric effects of large ligands are critical in determining the relative stability of the isomers. Even with such assumptions, it is still time-consuming and expensive using DFT in terms of structure prediction and mechanism understanding; thus, it is necessary to develop more time-efficient methods, of which the force-field based approach may be a direction for future efforts.

Acknowledgments

F.S. and Q.T. were supported by the National Natural Science Foundation of China (No.21903008), the Chongqing Science and Technology Commission (cstc2020jcyj-msxmX0382), and the Fundamental Research Funds for the Central Universities (2020CDJQY-A031, 2020CDJ-LHZZ-063). D.J. was supported by the U.S. Department of Energy, Office of Science, Office of Basic Energy Sciences, Chemical Sciences, Geosciences, and Biosciences Division, Catalysis Science Program.

References

1 Jin, R., Li, G., Sharma, S. et al. (2021). Toward active-site tailoring in heterogeneous catalysis by atomically precise metal nanoclusters with crystallographic structures. *Chem. Rev.* 121: 567–648.

2 Qin, L., Ma, G., Wang, L., and Tang, Z. (2021). Atomically precise metal nanoclusters for (photo) electroreduction of CO_2: recent advances, challenges and opportunities. *J. Energy Chem.* 57: 359–370.

3 Jadzinsky, P.D., Calero, G., Ackerson, C.J. et al. (2007). Structure of a thiol monolayer-protected gold nanoparticle at 1.1 Å resolution. *Science* 318: 430–432.

4 Du, Y., Sheng, H., Astruc, D., and Zhu, M. (2020). Atomically precise Noble metal nanoclusters as efficient catalysts: a bridge between structure and properties. *Chem. Rev.* 120: 526–622.

5 Fu, G. and Lee, J.-M. (2019). Ternary metal sulfides for electrocatalytic energy conversion. *J. Mater. Chem. A* 7: 9386–9405.

6 Ang, B.W., Choong, W.L., and Ng, T.S. (2015). Energy security: definitions, dimensions and indexes. *Renewable Sustainable Energy Rev.* 42: 1077–1093.

7 Dresselhaus, M.S. and Thomas, I.L. (2001). Alternative energy technologies. *Nature* 414: 332–337.

8 Chu, S. and Majumdar, A. (2012). Opportunities and challenges for a sustainable energy future. *Nature* 488: 294–303.

9 Seh, Z.W., Kibsgaard, J., Dickens, C.F. et al. (2017). Combining theory and experiment in electrocatalysis: insights into materials design. *Science* 355: eaad4998.

10 Chu, S., Cui, Y., and Liu, N. (2016). The path towards sustainable energy. *Nat. Mater.* 16: 16–22.

11 Gür, T.M. (2018). Review of electrical energy storage technologies, materials and systems: challenges and prospects for large-scale grid storage. *Energy Environ. Sci.* 11: 2696–2767.

12 Hong, W.T., Risch, M., Stoerzinger, K.A. et al. (2015). Toward the rational design of non-precious transition metal oxides for oxygen electrocatalysis. *Energy Environ. Sci.* 8: 1404–1427.

13 Jiao, Y., Zheng, Y., Jaroniec, M., and Qiao, S.Z. (2015). Design of electrocatalysts for oxygen- and hydrogen-involving energy conversion reactions. *Chem. Soc. Rev.* 44: 2060–2086.

14 Stamenkovic, V.R., Strmcnik, D., Lopes, P.P., and Markovic, N.M. (2016). Energy and fuels from electrochemical interfaces. *Nat. Mater.* 16: 57–69.

15 Jin, R., Zeng, C., Zhou, M., and Chen, Y. (2016). Atomically precise colloidal metal nanoclusters and nanoparticles: fundamentals and opportunities. *Chem. Rev.* 116: 10346–10413.

16 Higaki, T., Li, Y., Zhao, S. et al. (2019). Atomically tailored gold nanoclusters for catalytic application. *Angew. Chem. Int. Ed.* 58: 8291–8302.

17 Watanabe, Y. (2014). Atomically precise cluster catalysis towards quantum controlled catalysts. *Sci. Technol. Adv. Mater.* 15: 063501.

18 Cao, S., Tao, F., Tang, Y. et al. (2016). Size- and shape-dependent catalytic performances of oxidation and reduction reactions on nanocatalysts. *Chem. Soc. Rev.* 45: 4747–4765.

19 Takahashi, M., Imaoka, T., and Yamamoto, K. (2015). Reactivities of platinum subnanocluster catalysts for the oxidation reaction of alcohols. *RSC Adv.* 5: 100693–100696.

20 Norskov, J.K., Abild-Pedersen, F., Studt, F., and Bligaard, T. (2011). Density functional theory in surface chemistry and catalysis. *Proc. Natl. Acad. Sci.* 108: 937–943.

21 Calle-Vallejo, F., Martinez, J.I., Garcia-Lastra, J.M. et al. (2014). Fast prediction of adsorption properties for platinum Nanocatalysts with generalized coordination numbers. *Angew. Chem. Int. Ed.* 53: 8316–8319.

22 Kumar, B., Kawawaki, T., Shimizu, N. et al. (2020). Gold nanoclusters as electrocatalysts: size, ligands, heteroatom doping, and charge dependences. *Nanoscale* 12: 9969–9979.

23 Sumner, L., Sakthivel, N.A., Schrock, H. et al. (2018). Electrocatalytic oxygen reduction activities of thiol protected Nanomolecules ranging in size from $Au_{28}(SR)_{20}$ to $Au_{279}(SR)_{84}$. *J. Phys. Chem. C* 122: 24809–24817.

24 Seong, H., Efremov, V., Park, G. et al. (2021). Atomically precise gold nanoclusters as model catalysts for identifying active sites for electroreduction of CO_2. *Angew. Chem. Int. Ed.* 60: 14563–14570.

25 Semagina, N. and Kiwi-Minsker, L. (2009). Recent advances in the liquid-phase synthesis of metal nanostructures with controlled shape and size for catalysis. *Catal. Rev.* 51: 147–217.

26 Zhao, S., Austin, N., Li, M. et al. (2018). Influence of atomic-level morphology on catalysis: the case of sphere and rod-like gold nanoclusters for CO_2 electroreduction. *ACS Catal.* 8: 4996–5001.

27 Li, G. and Jin, R. (2014). Gold nanocluster-catalyzed semihydrogenation: a unique activation pathway for terminal alkynes. *J. Am. Chem. Soc.* 136: 11347–11354.

28 Rossi, L.M., Fiorio, J.L., Garcia, M.A.S., and Ferraz, C.P. (2018). The role and fate of capping ligands in colloidally prepared metal nanoparticle catalysts. *Dalton Trans.* 47: 5889–5915.

29 Kobayashi, R., Nonoguchi, Y., Sasaki, A., and Yao, H. (2014). Chiral monolayer-protected bimetallic Au-Ag nanoclusters: alloying effect on their electronic structure and chiroptical activity. *J. Phys. Chem. C* 118: 15506–15515.

30 Zhao, S., Jin, R., Song, Y. et al. (2017). Atomically precise gold nanoclusters accelerate hydrogen evolution over MoS_2 nanosheets: the dual interfacial effect. *Small* 13: 1701519.

31 Jin, R., Jin, R., Zhao, S. et al. (2017). Controlling Ag-doping in $[Ag_xAu_{25-x}(SC_6H_{11})_{18}]$ nanoclusters: cryogenic optical, electronic and electrocatalytic properties. *Nanoscale* 9: 19183–19190.

32 Li, S., Alfonso, D., Nagarajan, A.V. et al. (2020). Monopalladium substitution in gold nanoclusters enhances CO_2 electroreduction activity and selectivity. *ACS Catal.* 10: 12011–12016.

33 Alfonso, D.R., Kauffman, D., and Matranga, C. (2016). Active sites of ligand-protected Au_{25} nanoparticle catalysts for CO_2 electroreduction to CO. *J. Chem. Phys.* 144: 1598–10243.

34 Akola, J., Kacprzak, K.A., Lopez-Acevedo, O. et al. (2010). Thiolate-protected Au_{25} superatoms as building blocks: dimers and crystals. *J. Phys. Chem. C* 114: 15986–15994.

35 Kauffman, D.R., Alfonso, D., Matranga, C. et al. (2013). Photomediated oxidation of atomically precise $Au_{25}(SC_2H_4Ph)_{18}^-$ nanoclusters. *J. Phys. Chem. Lett.* 4: 195–202.

36 Li, S., Nagarajan, A.V., Alfonso, D.R. et al. (2021). Boosting CO_2 electrochemical reduction with atomically precise surface modification on gold nanoclusters. *Angew. Chem. Int. Ed.* 60: 6351–6356.

37 Qin, L., Sun, F., Ma, X. et al. (2021). Homoleptic alkynyl-protected Ag_{15} nanocluster with atomic precision: structural analysis and electrocatalytic performance toward CO_2 reduction. *Angew. Chem. Int. Ed.* 60: 26136–26141.

38 Sun, F., Deng, C., Tian, S., and Tang, Q. (2021). Oxygen electrocatalysis by $[Au_{25}(SR)_{18}]$: charge, doping, and ligand removal effect. *ACS Catal.* 11: 7957–7969.

39 Venzo, A., Antonello, S., Gascón, J. et al. (2011). Effect of the charge state ($z = -1, 0, +1$) on the nuclear magnetic resonance of monodisperse $Au_{25}[S(CH_2)_2Ph]_{18}^z$ clusters. *Anal. Chem.* 83: 6355–6362.

40 Liu, Z., Zhu, M., Meng, X. et al. (2011). Electron transfer between $[Au_{25}(SC_2H_4Ph)_{18}]^-$ TOA^+ and oxoammonium cations. *J. Phys. Chem. Lett.* 2: 2104–2109.

41 Zhu, M., Eckenhoff, W.T., Pintauer, T., and Jin, R. (2008). Conversion of anionic $[Au_{25}(SCH_2CH_2Ph)_{18}]^-$ cluster to charge neutral cluster via air oxidation. *J. Phys. Chem. C* 112: 14221–14224.

42 Kauffman, D.R., Alfonso, D., Matranga, C. et al. (2014). Probing active site chemistry with differently charged Au_{25}^q nanoclusters ($q = -1, 0, +1$). *Chem. Sci.* 5: 3151–3157.

43 Lu, Y., Jiang, Y., Gao, X., and Chen, W. (2014). Charge state-dependent catalytic activity of $Au_{25}(SC_{12}H_{25})_{18}$ nanoclusters for the two-electron reduction of dioxygen to hydrogen peroxide. *Chem. Commun.* 50: 8464–8467.

44 Kwak, K., Choi, W., Tang, Q. et al. (2017). A molecule-like $PtAu_{24}(SC_6H_{13})_{18}$ nanocluster as an electrocatalyst for hydrogen production. *Nat. Commun.* 8: 14723.

45 Negishi, Y., Iwai, T., and Mao, I. (2010). Continuous modulation of electronic structure of stable thiolate-protected Au_{25} cluster by Ag doping. *Chem. Commun.* 46: 4713–4715.

46 Krishnadas, K.R., Baksi, A., Ghosh, A. et al. (2016). Structure-conserving spontaneous transformations between nanoparticles. *Nat. Commun.* 7: 13447.

47 Li, Y., Biswas, S., Luo, T.Y. et al. (2020). Doping effect on the magnetism of thiolate-capped 25-atom alloy nanoclusters. *Chem. Mater.* 32: 9238–9244.

48 Yamazoe, S., Kurashige, W., Nobusada, K. et al. (2014). Preferential location of coinage metal dopants (M = Ag or Cu) in $[Au_{25-x}M_x(SC_2H_4Ph)_{18}]^-$ ($x \sim 1$) as determined by extended 9-ray absorption fine structure and density functional theory calculations. *J. Phys. Chem. C* 118: 25284–25290.

49 Negishi, Y., Munakata, K., Ohgake, W., and Nobusada, K. (2012). Effect of copper doping on electronic structure, geometric structure, and stability of thiolate-protected Au_{25} nanoclusters. *J. Phys. Chem. Lett.* 3: 2209–2214.

50 Qian, H., Jiang, D.-E., Li, G. et al. (2012). Monoplatinum doping of gold nanoclusters and catalytic application. *J. Am. Chem. Soc.* 134: 16159–16162.

51 Choi, W., Hu, G., Kwak, K. et al. (2018). Effects of metal-doping on hydrogen evolution reaction catalyzed by MAu_{24} and M_2Au_{36} nanoclusters (M = Pt, Pd). *ACS Appl. Mater. Interfaces* 10: 44645–44653.

52 Nair, L.V., Sakiat, H., Shunjiro, T. et al. (2018). Hetero-biicosahedral $[Au_{24}Pd(PPh_3)_{10}(SC_2H_4Ph)_5Cl_2]^+$ nanocluster: selective synthesis and optical and electrochemical properties. *Nanoscale* 10: 18969–18979.

53 Yao, C., Lin, Y.-J., Yuan, J. et al. (2015). Mono-cadmium vs. mono-mercury doping of Au_{25} nanoclusters. *J. Am. Chem. Soc.* 137: 15350–15353.

54 Wang, S., Song, Y., Jin, S. et al. (2015). Metal exchange method using Au_{25} nanoclusters as templates for alloy nanoclusters with atomic precision. *J. Am. Chem. Soc.* 137: 4018–4021.

55 Liao, L., Zhou, S., Dai, Y. et al. (2015). Mono-mercury doping of Au_{25} and the HOMO/LUMO energies evaluation employing differential pulse voltammetry. *J. Am. Chem. Soc.* 137: 9511–9515.

56 Bhat, S., Baksi, A., Mudedla, S.K. et al. (2017). $Au_{22}Ir_3(PET)_{18}$: an unusual alloy cluster through Intercluster reaction. *J. Phys. Chem. Lett.* 8: 2787–2793.

57 Hossain, S., Ono, T., Yoshioka, M. et al. (2018). Thiolate-protected trimetallic $Au{\sim}_{20}Ag{\sim}_4Pd$ and $Au{\sim}_{20}Ag{\sim}_4Pt$ alloy clusters with controlled chemical composition and metal positions. *J. Phys. Chem. Lett.* 9: 2590–2594.

58 Kang, X., Wei, X., Jin, S. et al. (2019). Rational construction of a library of M_{29} nanoclusters from monometallic to tetrametallic. *Proc. Natl. Acad. Sci.* 116: 18834–18840.

59 Panapitiya, G., Wang, H., Chen, Y. et al. (2018). Structural and catalytic properties of the $Au_{25-x}Ag_x(SCH_3)_{18}$ (x = 6, 7, 8) nanocluster. *Phys. Chem. Chem. Phys.* 20: 13747–13756.

60 Sultan, S., Tiwari, J.N., Singh, A.N. et al. (2019). Single atoms and clusters based nanomaterials for hydrogen evolution, oxygen evolution reactions, and full water splitting. *Adv. Energy Mater.* 9: 1900624.

61 Dubouis, N. and Grimaud, A. (2019). The hydrogen evolution reaction: from material to interfacial descriptors. *Chem. Sci.* 10: 9165–9181.

62 Zou, X. and Zhang, Y. (2015). Noble metal-free hydrogen evolution catalysts for water splitting. *Chem. Soc. Rev.* 44: 5148–5180.

63 Xu, Y., Kraft, M., and Xu, R. (2016). Metal-free carbonaceous electrocatalysts and photocatalysts for water splitting. *Chem. Soc. Rev.* 45: 3039–3052.

64 Anantharaj, S., Ede, S.R., Karthick, K. et al. (2018). Precision and correctness in the evaluation of electrocatalytic water splitting: revisiting activity parameters with a critical assessment. *Energy Environ. Sci.* 11: 744–771.

65 You, B. and Sun, Y. (2018). Innovative strategies for electrocatalytic water splitting. *Acc. Chem. Res.* 51: 1571–1580.

66 Hu, G., Tang, Q., Lee, D. et al. (2017). Metallic hydrogen in atomically precise gold nanoclusters. *Chem. Mater.* 29: 4840–4847.

67 Li, Y., Li, S., Nagarajan, A.V. et al. (2021). Hydrogen evolution electrocatalyst design: turning inert gold into active catalyst by atomically precise nanochemistry. *J. Am. Chem. Soc.* 143: 11102–11108.

68 Hu, G., Wu, Z., and Jiang, D.-E. (2018). Stronger-than-Pt hydrogen adsorption in a Au_{22} nanocluster for the hydrogen evolution reaction. *J. Mater. Chem. A* 6: 7532–7537.

69 Dong, J., Gao, Z.H., Zhang, Q.F., and Wang, L.S. (2021). The synthesis, bonding, and transformation of a ligand-protected gold nanohydride cluster. *Angew. Chem. Int. Ed.* 60: 2424–2430.

70 Suen, N.T., Hung, S.F., Quan, Q. et al. (2017). Electrocatalysis for the oxygen evolution reaction: recent development and future perspectives. *Chem. Soc. Rev.* 46: 337–365.

71 Zhao, S., Jin, R., Abroshan, H. et al. (2017). Gold nanoclusters promote electrocatalytic water oxidation at the nanocluster/$CoSe_2$ interface. *J. Am. Chem. Soc.* 139: 1077–1080.

72 Gao, X. and Chen, W. (2017). Highly stable and efficient $Pd_6(SR)_{12}$ cluster catalysts for the hydrogen and oxygen evolution reactions. *Chem. Commun.* 53: 9733–9736.

73 Kwon, G., Ferguson, G.A., Heard, C.J. et al. (2013). Size-dependent subnanometer Pd luster (Pd_4, Pd_6, and Pd_{17}) water oxidation electrocatalysis. *ACS Nano* 7: 5808–5817.

74 Kauffman, D.R., Alfonso, D., Tafen, D.N. et al. (2016). Electrocatalytic oxygen evolution with an atomically precise nickel catalyst. *ACS Catal.* 6: 1225–1234.

75 Joya, K.S., Sinatra, L., AbdulHalim, L.G. et al. (2016). Atomically monodisperse nickel nanoclusters as highly active electrocatalysts for water oxidation. *Nanoscale* 8: 9695–9703.

76 Zhang, J. (2008). *PEM Fuel Cell Electrocatalysts and Catalyst Layers: Fundamentals and Applications*. London: Springer-Verlag London Limited.

77 Sun, F. and Tang, Q. (2021). The ligand effect on the interface structures and electrocatalytic applications of atomically precise metal nanoclusters. *Nanotechnology* 32.

78 von Weber, A., Baxter, E.T., White, H.S., and Anderson, S.L. (2015). Cluster size controls branching between water and hydrogen peroxide production in electrochemical oxygen reduction at Pt n/ ITO. *J. Phys. Chem. C* 119: 11160–11170.

79 Chen, W. and Chen, S. (2010). Oxygen electroreduction catalyzed by gold nanoclusters: strong core size effects. *Angew. Chem. Int. Ed.* 121: 4450–4453.

80 Kwak, K., Azad, U.P., Choi, W. et al. (2016). Efficient oxygen reduction electrocatalysts based on gold nanocluster-graphene composites. *Chemelectrochem* 3: 1253–1260.

81 Deng, C., Li, F., and Tang, Q. (2019). Electrocatalytic oxygen reduction reaction over the $Au_{22}(L^8)_6$ nanocluster with promising activity: a DFT study. *J. Phys. Chem. C* 123: 27116–27123.

82 Yang, X., Gan, L., Zhu, C. et al. (2014). A dramatic platform for oxygen reduction reaction based on silver nanoclusters. *Chem. Commun.* 50: 234–236.

83 Liu, M. and Chen, W. (2013). Green synthesis of silver nanoclusters supported on carbon nanodots: enhanced photoluminescence and high catalytic activity for oxygen reduction reaction. *Nanoscale* 5: 12558–12564.

84 Jin, S., Chen, M., Dong, H. et al. (2015). Stable silver nanoclusters electrochemically deposited on nitrogen-doped graphene as efficient electrocatalyst for oxygen reduction reaction. *J. Power Sources* 274: 1173–1179.

85 Nitopi, S., Bertheussen, E., Scott, S.B. et al. (2019). Progress and perspectives of electrochemical CO_2 reduction on copper in aqueous electrolyte. *Chem. Rev.* 119: 7610–7672.

86 Kauffinan, D.R., Alfonso, D., Matranga, C. et al. (2012). Experimental and computational investigation of Au_{25} clusters and CO_2: a unique interaction and enhanced electrocatalytic activity. *J. Am. Chem. Soc.* 134: 10237–10243.

87 Austin, N., Zhao, S., McKone, J.R. et al. (2018). Elucidating the active sites for CO_2 electroreduction on ligand-protected Au25 nanoclusters. *Catal. Sci. Technol.* 8: 3795–3805.

88 Kim, B., Seong, H., Song, J.T. et al. (2020). Over a 15.9% solar-to-CO conversion from dilute CO_2 streams catalyzed by gold nanoclusters exhibiting a high CO_2 binding affinity. *ACS Energy Lett.* 5: 749–757.

89 Jupally, V.R., Dharmaratne, A.C., Crasto, D. et al. (2014). $Au_{137}(SR)_{56}$ nanomolecules: composition, optical spectroscopy, electrochemistry and electrocatalytic reduction of CO_2. *Chem. Commun.* 50: 9895–9898.

90 Krishnadas, K.R., Natarajan, G., Baksi, A. et al. (2019). Metal–ligand Interface in the chemical reactions of ligand-protected noble metal clusters. *Langmuir* 35: 11243–11254.

91 Chakraborty, I. and Pradeep, T. (2017). Atomically precise clusters of Noble metals: emerging link between atoms and nanoparticles. *Chem. Rev.* 117: 8208–8271.

92 Narouz, M.R., Osten, K.M., Unsworth, P.J. et al. (2019). N-heterocyclic carbene-functionalized magic-number gold nanoclusters. *Nat. Chem.* 11: 419–425.

93 Zhuang, S., Chen, D., Liao, L. et al. (2020). Hard-sphere random close-packed $Au_{47}Cd_2(TBBT)_{31}$ nanoclusters with a faradaic efficiency of up to 96% for electrocatalytic CO_2 reduction to CO. *Angew. Chem. Int. Ed.* 132: 397–3101.

94 Zhang, X.-G., Liu, Y., Zhan, C. et al. (2019). Reaction selectivity for plasmon-driven carbon dioxide reduction on silver clusters: a theoretical prediction. *J. Phys. Chem. C* 123: 11101–11108.

95 Li, J., Wang, Z., McCallum, C. et al. (2019). Constraining CO coverage on copper promotes high-efficiency ethylene electroproduction. *Nat. Catal.* 2: 1124–1131.

96 Wang, Y., Wang, Z., Dinh, C.-T. et al. (2020). Catalyst synthesis under CO_2 electroreduction favours faceting and promotes renewable fuels electrosynthesis. *Nat. Catal.* 3: 98–106.

97 Tang, Q., Lee, Y., Li, D.-Y. et al. (2017). Lattice-hydride mechanism in electrocatalytic CO_2 reduction by structurally precise copper-hydride nanoclusters. *J. Am. Chem. Soc.* 139: 9728–9736.

98 Li, F. and Tang, Q. (2020). The critical role of hydride (H^-) ligands in electrocatalytic CO_2 reduction to HCOOH by $[Cu_{25}H_{22}(PH_3)_{12}]Cl$ nanocluster. *J. Catal.* 387: 95–101.

99 Gao, Z.-H., Wei, K., Wu, T. et al. (2022). A heteroleptic gold hydride nanocluster for efficient and selective electrocatalytic reduction of CO_2 to CO. *J. Am. Chem. Soc.* 144: 5258–5262.

100 Negishi, Y., Chaki, N.K., Shichibu, Y. et al. (2007). Origin of magic stability of thiolated gold clusters: a case study on $Au_{25}(SC_6H_{13})_{18}$. *J. Am. Chem. Soc.* 129: 11322–11323.

7

Ag Nanoclusters: Synthesis, Structure, and Properties

Manman Zhou and Manzhou Zhu

Department of Chemistry and Centre for Atomic Engineering of Advanced Materials, Key Laboratory of Structure and Functional Regulation of Hybrid Materials of Ministry of Education, and Anhui Province Key Laboratory of Chemistry for Inorganic/Organic Hybrid Functionalized Materials, Anhui University, Hefei, Anhui, 230601, P. R. China

7.1 Introduction

In recent years, atomically precise metal nanoclusters (NCs) have attracted wide interest because of their novel molecule-like properties and promising applications, such as catalysis, sensors, chirality, and electrochemistry [1–9]. These diverse applications of NCs have led to rapid progress in pursuing the functional NCs [5, 7, 10–13]. Structurally, the NCs contain a metal core and an organic ligand shell, which are of ultrasmall size (typically 1–3 nm) and homogeneity of chemical components [14–16]. Such NCs exhibit discrete energy levels, leading to the absence of plasmonic features, and thus molecular-like properties [17–19] come into play. In addition, even a single-atom alternation in NCs may cause significant changes in their chemical and physical properties. Besides, a great diversity of peripheral ligands allows for structure tailoring, which renders NCs a good platform to study their electronic structures and structure–property relationships at the atomic level [20–23]. The metallophilic interactions are also commonly studied based on NCs [20, 24–29]. Most of these NCs can be characterized by single-crystal X-ray diffraction (SC-XRD), conducive to the study of their structure–property relationships.

The peripheral protecting ligands of NCs are such as thiolate, phosphine, alkyne, protein, and DNA, among which the thiolated NCs have been studied intensively [19, 21, 29–32]. The organic ligands covering the surface of the NCs prevent the cores from agglomerating during the synthesis, and also play an important role in determining the size, architecture, and ultimate properties [26, 27, 33–35]. Several studies have also shown that the surface ligands can induce spatial and electronic effects of NCs, thereby changing their catalytic and fluorescence performances [36–43]. Among them, Au- and Ag-based NCs are intensively investigated [37, 38, 40–43]. The sulfur atom that binds to gold can be inorganic sulfur (i.e., S^{2-}) or organic sulfur (i.e. SR; R = carbon tails) [1, 19, 44–46]. Compared to Au NCs that are at the frontier in current research, Ag-based NCs are less in research [47–50] perhaps due to being less stable than Au NCs and the resultant difficulties in synthesis. Nevertheless, with the improvement of experimental methodology, research on Ag NCs is gaining momentum. To date, a number of Ag NCs with definite formulas have been synthesized, such as Ag_{25} [51], Ag_{29} [52, 53], Ag_{44} [38], Ag_{50} [54], Ag_{67} [55], and Ag_{374} [56]. The research on the synthesis and characterization of Ag NCs is of important theoretical significance for

Atomically Precise Nanochemistry, First Edition. Edited by Rongchao Jin and De-en Jiang.

understanding the relationship between their structures and properties, and such NCs hold potential in many fields such as life sciences and materials science.

This chapter summarizes the literature work on ligand-protected Ag NCs. The sections include the synthesis, structure, and properties. While Au NCs have been studied widely in the last decade, Ag NCs are relatively less explored. The electronic and geometrical structures of $Ag_n(SR)_m$ NCs will obviously change via the metal and ligand control, among which processes their physicochemical properties will also be tailored. The combination of single crystal X-ray diffraction (SC-XRD) and electrospray ionization mass spectrometry (ESI-MS) measurements can determine their precise atomic structures and molecular formulas. In terms of applications, silver NCs have been applied in biosensing, fluorescent imaging, temperature sensing, and ratiometric pH sensing due to their fluorescence, biocompatibility, chirality, and other properties.

7.2 Synthetic Methods

The synthesis of Ag nanoclusters has been developed over the past years. The main difficulty in the synthesis is that the zero-valent state of silver atoms is less stable than gold, that is, Ag(0) tends to be oxidized. Similar to the synthesis method of Au NCs, Ag clusters can be synthesized by the bottom-up methods or top-down routes [24, 57, 58]. Typically, the synthetic methods are one-pot synthesis, two-phase ligand exchange method, chemical etching, solid-phase synthesis, seeded growth, and electroplating method. In addition, recent works have reported the synthesis of new structures induced by different solvents. Based on these synthesis strategies, the research progress in silver nanoclusters is advanced by combining various testing methods, such as UV–vis absorption spectroscopy, ESI-MS, and SC-XRD, to determine the formula and atomic structure of Ag NCs [59–61].

7.2.1 One-Pot Synthesis

Benefiting from the Au NC synthetic method – reduction of Au^{3+} in the presence of thiol ligands – *one-pot synthesis* has been developed as the easiest way to synthesize silver NCs [1, 62]. The one-pot synthesis can directly reduce the Ag^+ salt in both organic and aqueous media in the presence of reducing agents. Sodium borohydride, sodium cyanoborohydride, and borane tert-butylamine complex, for example, are commonly used as reducing agents [63, 64]. In this context, the method is also regarded as the reduction growth method, which is commonly used in the synthesis of Ag-based alloy nanoclusters. The prepared crude product is usually separated by solvent extraction, electrophoresis, or thin layer chromatography in order to obtain the final pure product.

7.2.2 Ligand Exchange

The ligand-induced structural change is a good strategy to prepare NCs. This method can enrich the properties of nanoclusters by tailoring the structures of NCs at the atomic level via surface ligand engineering [65–68]. Besides, the rich chemistry of peripheral ligands will have a significant impact on the structures or sizes of NCs. Through ligand exchange, the original structure may occur in two main types of conversion: (i) maintaining the overall structure via changing the peripheral ligands only (such as the conversion from $Au_{25}(S\text{-}C_2H_4Ph)_{18}$ to $Au_{25}(SePh)_{18}$) [65]; and (ii) completely altering the structure of the starting NCs (such as the conversion from Ag_{44} (4-FTP)$_{30}$ with a hollow kernel to Ag_{25}(2,4-DMBT)$_{18}$ with a nonhollow kernel) [69].

7.2.3 Chemical Etching

Some Ag NCs can be synthesized via a chemical etching process, i.e. etching larger Ag nanoparticles to small Ag NCs via adding metals or ligands [70–72]. This method is more time-consuming and the yield is usually relatively low, so the reaction time and temperature and other conditions are more stringently required. For example, the Xie group proposed a reduction–decomposition–reduction cycle to etch Ag NC intermediates in water to obtain the nonluminescent Ag NC. Then, these modified clusters were etched by thiol ligands and subjected to a size/structure focusing environment. Finally, the intermediates were transformed to highly luminescent Ag NCs, which included $Ag_{16}(SG)_9$ and $Ag_9(SG)_6$ [73]. Ag_{16} and Ag_9 have excellent stability in water, therefore facilitating their potential in biomedical applications.

7.2.4 Seeded Growth Method

The seeded growth method involves smaller NCs as seeds, which gradually grow into a new structure of larger size. This method is similar to the reduction growth method [54], that is, the regrowth of nanoclusters on the basis of existing structures by adding reagents such as ligands, metals, and reducing agents. For example, using Ag_{44} as seeds, one can obtain a larger size (Ag_{50} NC), and choosing Au_{25} as seeds, a larger size (Au_{44}) was obtained [54, 74].

7.3 Structure of Ag NCs

Analysis of the geometric structures of ligands-protected Ag nanoclusters has revealed several growth modes. The structure of the NCs reported so far exhibits a broad range of structural forms, consisting of face-centered cubic (fcc), hexagonally close-packed (hcp), icosahedral, and decahedral structures [75, 76]. For example, icosahedral Ag_{13} and its assembly were found in Ag_{20} [66], Ag_{21} [77], Ag_{25} [51], Ag_{29} [52], Ag_{32} [78], Ag_{33} [79], and Ag_{45} [78]. Furthermore, fcc structures are also very common, such as Ag_{14} [80], Ag_{15} [81], Ag_{23} [82], Ag_{38} [83], Ag_{46} [84, 85], Ag_{62} [86], Ag_{63} [83], and Ag_{67} [55]. Based on these small building blocks, the kernel fusion, vertex sharing, and interpenetration, etc. further construct 1D, 2D, and 3D structures. In addition to these regular assembly methods to form metal clusters, there are also some special irregular structures [17, 87].

Jin et al. synthesized a well-defined monodisperse $Ag_7(DMSA)_4$ nanocluster determined by mass spectrometry in 2009 [57]. This cluster is the first reported exact composition of silver cluster and fully analyzed by MS (Figure 7.1a, b). Subsequently, some silver nanoclusters with certain compositions were reported successfully. Unfortunately, none of these clusters could grow into single crystals with high quality, which limited further study of their composition and structure. But with the improvement of experimental methods and the development of technology, some measurements have been used to characterize the Ag NCs, such as UV–vis absorption spectroscopy, ESI-MS, photoluminescence spectroscopy, and X-ray photoelectron spectroscopy. Therefore, a large number of silver clusters have been reported, which has facilitated the development of their applications. Recently, Bakr and coworkers reported the smallest, crystallized silver nanocluster of $[Ag_9(1,2\text{-BDT})_6]^{3-}$. Using various testing methods to clarify its precise structure and applying density functional theory, they calculated the experimental optical absorption and molecular orbitals [88]. The structure of Ag_9 can be seen as two Ag_5 square pyramids sharing a single Ag vertex, and each Ag_5 is capped with one bidentate thiolate ligand in the vertex position and two in the equatorial plane, which leads to a body-centered cage (Figure 7.1c, d). This structure provides a reference for the subsequent synthesis and research of ultra-small silver nanoclusters; meanwhile, it can be a template for studying single-atom catalysis.

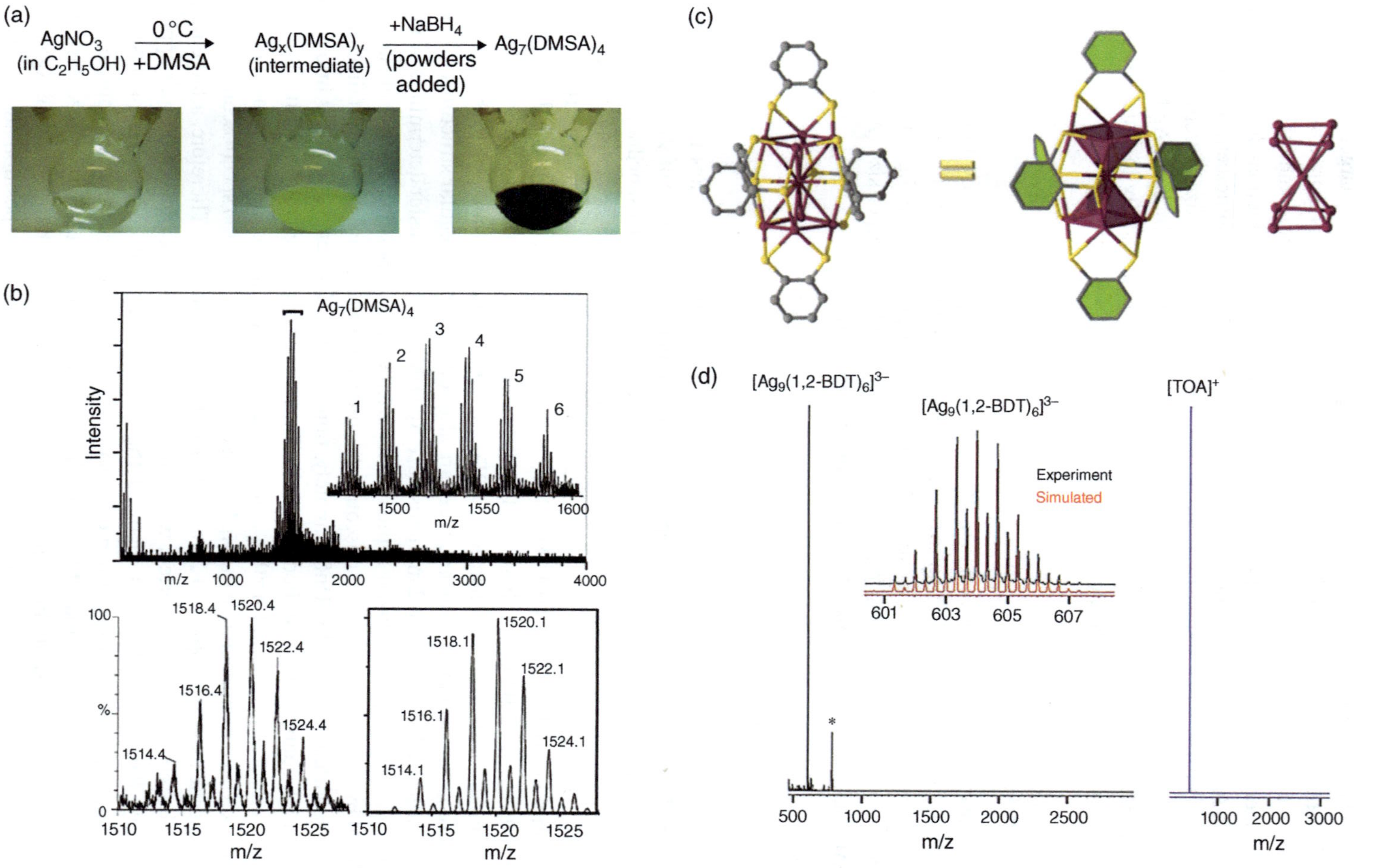

Figure 7.1 (a) Synthetic method of $Ag_7(DMSA)_4$; (b) ESI spectra of Ag_7; (c) crystal structure of the Ag_9 cluster; (d) ESI-MS spectra of $[Ag_9(1,2\text{-}BDT)_6]^{3-}$. *Source:* © 2009 and 2021 American Chemical Society.

7.3.1 Based on Icosahedral Units' Assembly

Icosahedral M_{13} units have a Ne-type superatomic electronic structure that commonly exists in Au and Ag NCs; thus, they can be seen as a building block, which can self-assemble to construct extremely stable metal clusters with different structures [87]. Changes in the number or assembly method of the M_{13} units will directly affect the overall structure and properties of the clusters. Therefore, exploring other assembly methods of icosahedral M_{13} will not only help to understand the relationship between the assembly method and structure of the basic unit of M_{13} but also provide an important reference method for effectively regulating the metal core, and ultimately for the new assembly of nanomaterials, which provides a certain guiding significance.

The first mono-icosahedral $Au_{13}(PPhMe_2)_{10}Cl_2$ was reported in 1981 by the Mingos group [89]. Subsequently, M_{13}-based structures that are common in Au NCs, such as Au_{25} [90], $Au_{38}(SR)_{24}$ [91], and Au_2Ag_{42} [92], etc., are also seen in Ag-based NCs. Most of their geometric/electronic structures are with the characteristics of the superatomic structure.

$[Ag_{25}(SR)_{18}]^-$ can be seen as a silver analog of $[Au_{25}(SR)_{18}]^-$, not only the composition, superatom electronic configuration, but also the charge and the crystal structure [9]. This cluster provides an ideal model for investigating the fundamental electronic structure and physicochemical properties of Ag NCs. $[Ag_{25}(SR)_{18}]^-$ is protected by 2,4-dimethylbenzenethiolate, being negatively charged and associated with a PPh_4^+ counterion. Characterization by single-crystal X-ray diffraction (SC-XRD) revealed that the metal kernel is an icosahedral Ag_{13} structure, which is protected by six $Ag_2(SR)_3$ motifs (Figure 7.2a). The number of free electrons in Ag_{25} is eight, which is in accordance with the superatom electronic configuration. Based on this structure, the Zhu group reported doping of Au or Pt into Ag_{25} and synthesized $Au_1Ag_{24}(SR)_{18}$, $Au_xAg_{25\text{-}x}(SR)_{18}$ (x = 4–7), $Pt_1Ag_{24}(SR)_{18}$, and $Pt_1Au_xAg_{24\text{-}x}(SR)_{18}$ (x = 3–7) NCs to explore the alloying effect on the fluorescence properties of NCs [93]. When doping a metal of strong electron affinity (e.g. Au) into the center of the icosahedral Ag_{13}, the free electrons tend to approach the core. Then, the fluorescence intensity is enhanced. But after doping more Au atoms to the kernel surface, the free valence electrons expand outward and the fluorescence intensity instead declines. In addition, Pradeep et al. reported the first example of a covalently bound dimer of $[Ag_{25}(SR)_{18}]^-$, linked by the $[Ru(bpy)_2bpy(CH_2SH)_2]^{2+}$ linker [94]. The formation was determined by optical absorption spectroscopy, IR, and ESI-MS. DFT calculations predicted the most stable binding site, and it was confirmed by the spicule positional notation and diagrams. Especially, the UV–vis of Ag_{25} and its dimer are slightly different, which indicates that the cluster core remains complete even in the formation of a dimer. In this way, the properties of clusters and linkers can be combined into one substance, which is conducive to the follow-up exploration of new applications.

Different from Ag_{25}, the $Ag_{29}(BDT)_{12}(TPP)_4$ nanocluster is co-protected by thiol and bidentate phosphine ligands, so the structure and properties are somewhat different from the single-ligand protected Ag_{25} [52]. The core of Ag_{29} also is an icosahedral Ag_{13} unit, capped by four Ag_3S_6 and four $Ag_1S_3P_1$ motifs (Figure 7.2b). Later, Pradeep and coworkers reported that, if the synthesized Ag_{29} was without TPP ligands, then the resulting product $[Ag_{29}(BDT)_{12}]^{3-}$ was unstable (stable only for a few hours) [95]. But adding the functional fullerenes C_{60} linkers gave rise to $[Ag_{29}(BDT)_{12}(C_{60})_n]^{3-}$ ($n = 1-4$) complexes, which can be stable for more than a week. These linkers can form tetrahedral concave cavities at the cluster surface, compatible with the convex π-surface of fullerenes. The detailed structure of $[Ag_{29}(BDT)_{12}(C_{60})_n]^{3-}$ ($n = 1-4$) was determined by UV–vis, ESI, collision-induced dissociation, and nuclear magnetic resonance spectroscopy (Figure 7.2c). This study can be extended to other nanoclusters and linkers. It can not only stabilize the clusters, but also add functional groups so they have optical and mechanical properties and can be applied in sensors, photovoltaics, medical science, etc.

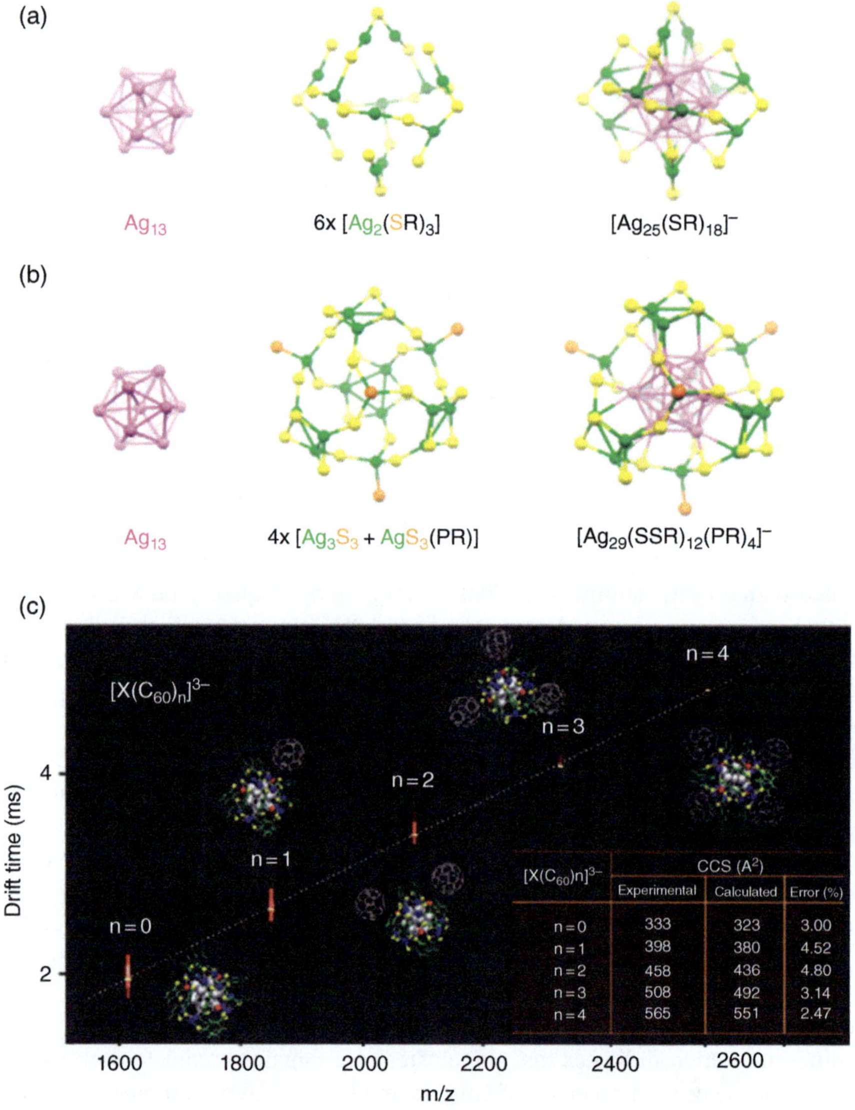

Figure 7.2 The core, shell and framework of (a) $Ag_{25}(SPhMe_2)_{18}$; and (b) $Ag_{29}(BDT)_{12}(TPP)_4$ nanoclusters; (c) Plot of drift time of the complexes with $[Ag_{29}(BDT)_{12}(C_{60})_n]^{3-}$ m/z values. *Source:* © 2016 and 2018 American Chemical Society.

The Chen group reported a novel $Ag_{33}(SCH_2CH_2Ph)_{24}(PPh_3)_4$ NC (abbrev. Ag_{33}), which was formed by a keplerate Ag_{13} icosahedron core and encapsulated with an $Ag_{20}S_{24}P_4$ unit [79]. Interestingly, the presence of some Pd reagent as a catalyst to form the $Pd(PPh_3)_4$ is essential in the synthesis of the Ag_{33} NC, because without this modifier, the product became the previously reported Ag_{23}. Crystallographic structure analysis found that Ag_{33} has a keplerate Ag_{13} icosahedral core, and the other 20 Ag atoms form a chiral $Ag_{20}S_{24}P_4$ shell, which adopts T symmetry constituted by -SR-Ag-SR- motifs and -Ag-P units (Figure 7.3). Furthermore, no counterions were observed in the unit cell, so the Ag_{33} NC should be charge neutral. Nuclear magnetic resonance (NMR), electronic circular dichroism (ECD), and time-dependent density functional theory (TD-DFT) analyses were further performed to verify the geometry, atomic composition, and chirality. This work has some valuable implications for the intrinsic chirality of metal cores.

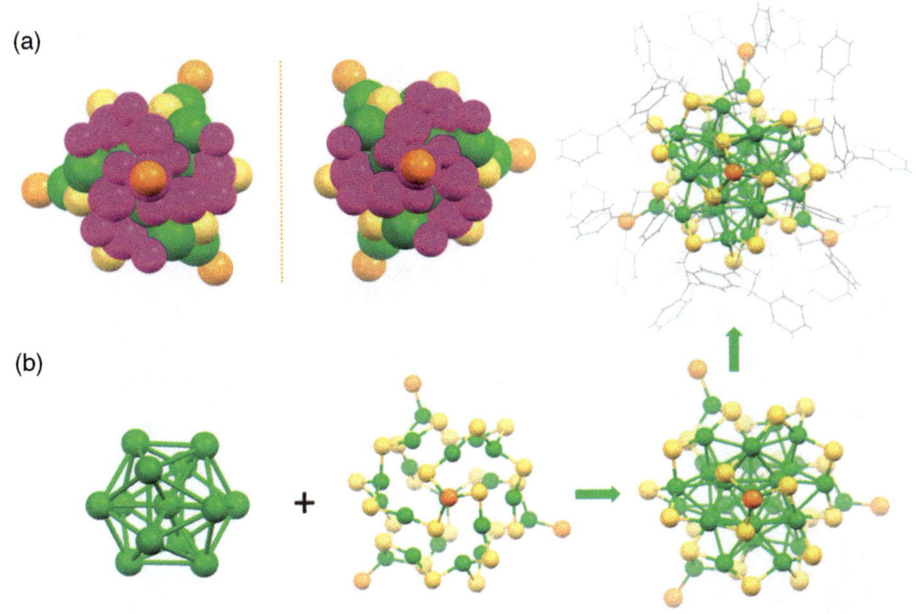

Figure 7.3 (a) Atomic packing of Ag_{33} enantiomers; and (b) the framework of the Ag_{33} nanocluster.
Source: © 2019 American Chemical Society.

In 2013, Bigioni and coworkers reported the first all thiolate-protected silver nanocluster with high yield and good stability, determined to be $[Ag_{44}(p\text{-}MBA)_{30}]^{4-}$ (Ag_{44}-1) [53]. The same year, Zheng's group also reported the crystal structures and theoretical analysis of three $Ag_{44}(SR)_{30}$ (SR = SPhF, $SPhF_2$, or $SPhCF_3$) (Ag_{44}-2) [38]. The skeletons of these clusters are all similar and have a two-shell Ag_{32} core, which includes an icosahedral hollow Ag_{12} and a dodecahedral Ag_{20}, capped by six $Ag_2(SR)_5$ motifs (Figure 7.4a). The difference between Ag_{44}-1 and Ag_{44}-2 is the Ag atoms in the $Ag_2(SR)_5$ of the latter can be replaced by Au atoms, forming the $Au_{12}Ag_{32}(SR)_{30}$ cluster. The structure is distinctly different from those thiolate-capped Au NCs (Figure 7.4c). The thermal stability of the doped alloy NCs is better than the all-silver one. Furthermore, using Ag_{44} as seeds and adding $AgNO_3$, dppm, TBBM, and $NaBH_3CN$ to the solution of Ag_{44}, one can obtain the controlled growth of a larger size $Ag_{50}(TBBM)_{30}(dppm)_6$ (abbrev. Ag_{50}) confirmed by ESI-MS and XPS (Figure 7.4, b,d,f) [54]. The structure of Ag_{50} retained some features of the parent NC such as the two-shell $Ag_{12}@Ag_{20}$, but the composition of the peripheral ligand shell becomes two symmetrical $Ag_9(SR)_{15}P_6$ ring motifs. In addition, the Ag_{50} can also be alloyed by templated/galvanic metal exchange to the $Au_xAg_{50-x}(SR)_{30}(dppm)_6$ NCs, and the stability of the alloy cluster is also better than the Ag_{50}. The peripheral ligands may have a great influence on the overall structure, as well as the optical and photophysical properties of the NCs. Recently, Hyeon and Zheng et al. reported a mercaptan/phosphine co-protected Ag_{44} cluster, charge-neutral and the structure identified by X-ray diffraction as $Ag_{44}(EBT)_{26}(TPP)_4$ (Figure 7.4e) [96]. The core is still $Ag_{12}@Ag_{20}$, similar to the all-thiolate-protected $[Ag_{44}(SR)_{30}]^{4-}$. Because the four negative thiolates are replaced by four neutral phosphines, the 4- charge switches to zero, meanwhile, the surface structures and properties are also changed. The motifs in $Ag_{44}(EBT)_{26}(TPP)_4$ are three-dimensional $Ag_2(EBT)_5$, long V-shaped $Ag_4(EBT)_7(TPP)_2$, and bridging thiolates. Compared with the all-thiolate-stabilized anionic $[Ag_{44}(SR)_{30}]^{4-}$ NCs, the changes in the peripheral ligands resulted in an enhancement of the NIR-II photo-luminescence QY by about 25-fold. Although they all have the metal core of Ag_{32}, the Ag—Ag bonds of the Ag_{12} inner core in $Ag_{44}(EBT)_{26}(TPP)_4$ with circumjacent atoms are

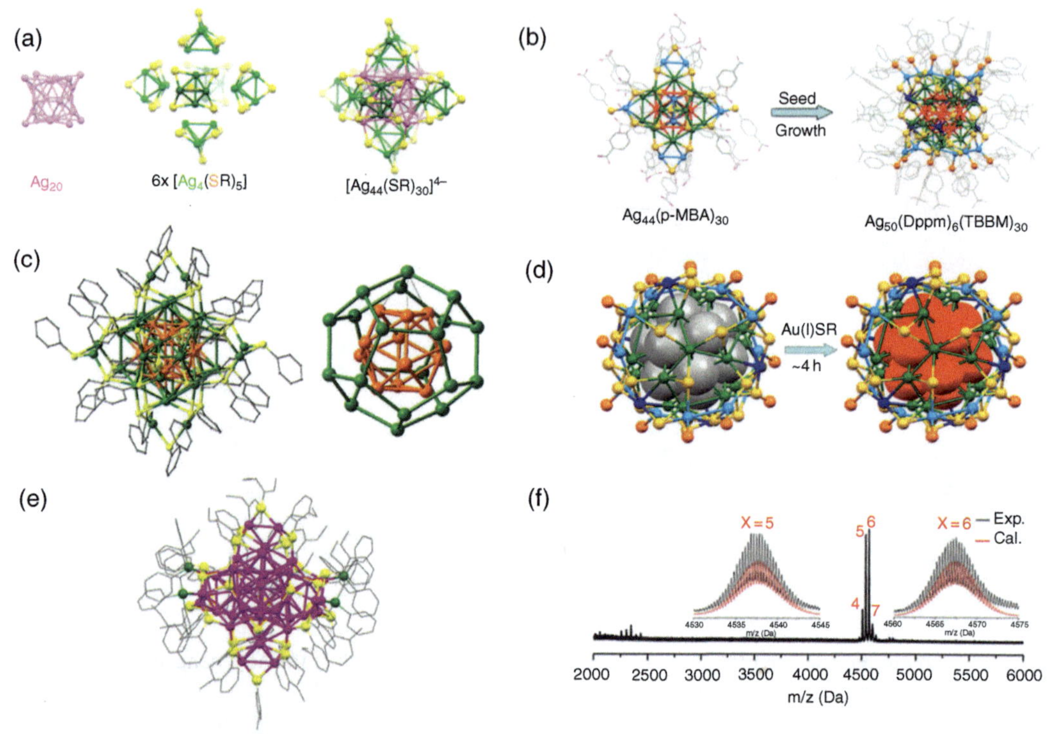

Figure 7.4 (a) The core, shell, and framework of $Ag_{44}(SR)_{30}$; (b) Total structure transformation from the Ag_{44}(p-MBA)$_{30}$ to the $Ag_{50}(Dppm)_6(SR)_{30}$ nanocluster; (c) The overall structure and two-shell $M_{12}@Ag_{20}$ core of the $[M_{12}Ag_{32}(SPhF_2)_{30}]$; (d) Metal exchange from Ag_{50} to Au_xAg_{50-x} by a "templated/galvanic metal exchange" method; (e) Total structure of $Ag_{44}(EBT)_{26}(TPP)_4$ NC; (f) ESI mass spectrum of Au_xAg_{50-x}. *Source:* © 2016 and 2018 American Chemical Society, 2013 Macmillan Publishers Limited, 2021 Wiley-VCH GmbH, respectively.

significantly elongated, indicating the critical role of ligands in distorting the metal core of $Ag_{44}(EBT)_{26}(TPP)_4$. These works combine theory with an experimental study on the effects of ligands and alloying on the structure and properties of NCs in different aspects, promoting the understanding of structure–property relationships.

The 0D, 1D, 2D, and 3D structures assembled by M_{13} icosahedral units exhibit fascinating optical properties and have broad application potential in solar cells, energy storage, and catalysis. Among the 1D structures, rod-shaped NCs have emerged as an ideal model to explore the structure–property relationships of metal nanorods and nanowires because of their definite structure and atomic monodispersity. For the reported Au NCs, Au_{13}, rod-shaped Au_{25} and Au_{37} consist of one, two, and three vertex-sharing Au_{13} units, respectively [89, 90, 97].

Recently, Wang and co-workers prepared a linear supercluster of $Ag_{61}(dpa)_{27}(SbF_6)_4$ (abbrev. Ag_{61}, where dpa = dipyridylamine), which has 30 free electrons (Figure 7.5) [98]. The metal core Ag_{41} structure in the Ag_{61} can be seen as four icosahedral Ag_{13} blocks arranged via sharing vertexes to form a 1D linear structure. Every dpa ligand has one amido and two pyridyl N donors, so the dpa can bridge two neighboring Ag_{13} blocks and form various dpa-Ag motifs. The symmetry in Ag_{49} is D_{5h}, but due to the core, each end is capped by six silver atoms and arranged in an irregular form, so the whole symmetry of Ag_{61} is lowered to C_2. Among the reported atomically precise metal superclusters, Ag_{61} is currently the longest with a resolved structure.

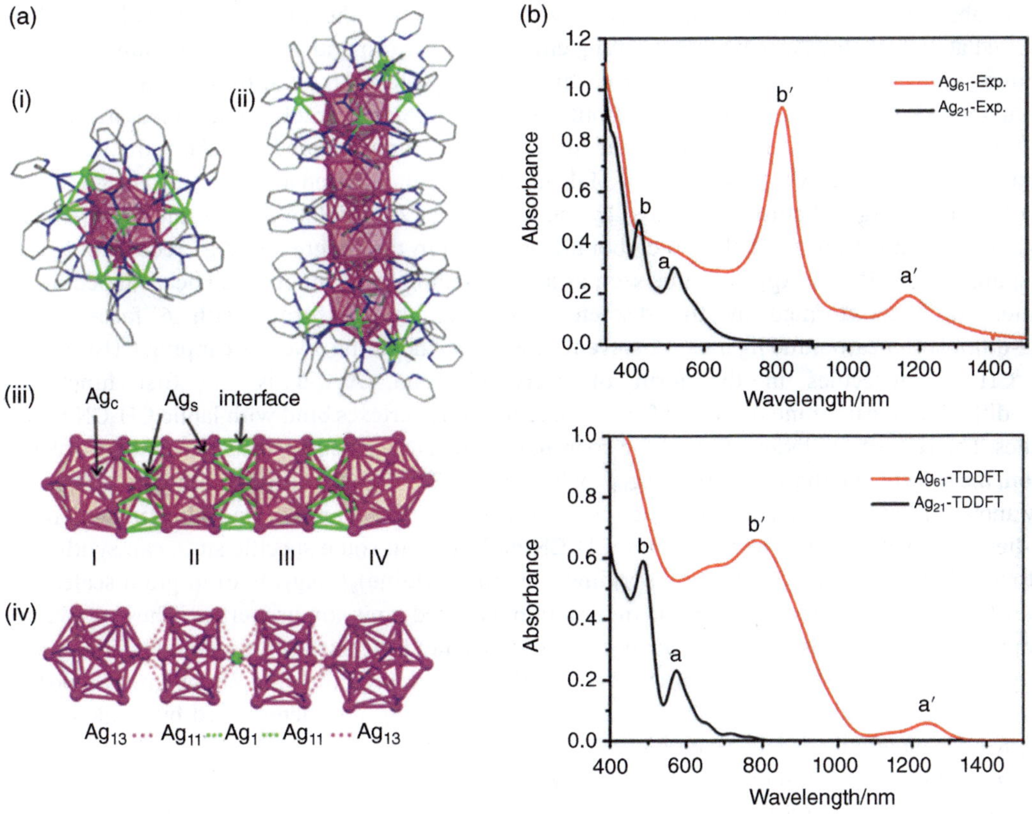

Figure 7.5 (a) Structures of (i) $[Ag_{21}(dpa)_{12}]^+$ and (ii) $[Ag_{61}(dpa)_{27}]^{4+}$ showing the Ag_{13} polyhedra. (iii) Four Ag_{13} icosahedral units of the Ag_{49} core are labeled I–IV along the chain. (iv) Alternative anatomy of the Ag_{49} core structure. Middle two units can be viewed as a Ag_{23} cylinder with two drum-like Ag_{11} connected by one central Ag atom (green); (b) Experimental and simulated optical absorption spectra of $[Ag_{21}(dpa)_{12}]^+$ and $[Ag_{61}(dpa)_{27}]^{4+}$. *Source:* © 2021 American Chemical Society.

Ag_{61} is extremely stable in CH_2Cl_2 solution, which is attributed to three reasons: the 8e electron configuration of the Ag_{13} unit, the interaction between the adjacent Ag_{13} units, and the ligand effect in the dpa-Ag motifs. The experimental and TD-DFT calculated absorption spectra of the mono-icosahedral Ag_{21} are almost the same, so is the case tetra-icosahedral Ag_{61}, confirming that the enhanced absorption of Ag_{61} results from strong longitudinal absorption due to the increase in aspect ratio.

This work provides new insight into the assembly of Ag_{13} icosahedral units and the 1D packing patterns of the supercluster, further promoting the application of amido ligands in the synthesis of NCs.

7.3.2 Based on Ag_{14} Units' Assembly

Except for Ag_{13}-based NCs, the fcc structures also have attracted much attention because of their importance in understanding the origin of fcc structures in nanomaterials. The fcc unit contains eight vertices and six face centers. Successively, fcc-based structures have been reported and the related property research has also gradually deepened. $Ag_{14}(SC_6H_3F_2)_{12}(PPh_3)_8$ (Ag_{14}) superatom NC has an octahedral Ag_6^{4+} metal core covered by eight $[Ag^+(SC_6H_3F_2^-)_2PPh_3]$ tetrahedra arranged in a cubical configuration and shares one corner with the tetrahedra (Figure 7.6a) [80]. When excited at 360 nm, the Ag_{14} exhibits a strong emission at 536 nm and a minor emission at 420 nm shows yellow luminescent. Different from the thiolate-protect Au NCs, the thiolate ligands all bond to three Ag atoms, thus, no staple motifs in Ag_{14}. The Ag_6^{4+} kernel can be seen as a fragment cut from fcc metals. This reminds us that Ag structures cannot be predicted from the structural characteristics of Au NCs. Subsequently, Mak and Zang et al. reported a $[Ag_{14}(C_2B_{10}H_{10}S_2)_6(CH_3CN)_8]\cdot4CH_3CN$ (Ag_{14}-a) cluster protected by the bidentate 1,2-dithiolate-o-carborane ($C_2B_{10}H_{10}S_2$) ligands, which is from the 1-thiol-o-carborane oxidation in the progress of Ag^+ reduction to Ag^0 (Figure 7.6b) [75]. The Ag_{14}-a can be seen as an octahedral Ag_6^{4+} inner core encapsulated by an outer Ag_8^{8+}cube formed by the discrete cubic Ag_{14} metal kernel with 6 face-capping 1,2-dithiolate-o-carborane ligands, 12 solvent molecules, including 8 vertex-capping CH_3CN, and 4 CH_3CN molecules in the form of cocrystallization. Ag_{14}-a is the first functional 1,2-dithiolate-o-carborane protected fcc-Ag_{14} and its cubic vertexes bind with labile CH_3CN molecules. The NCs suffer dissociation of the coordinated solvent molecules, thus they will be isolated from the solvent and the structure of Ag_{14}-a will collapse. So Ag_{14}-a has the flexibility to be replaced by another monodentate ligand at some specific vertex sites. For example, via adding monodentate N-heteroaromatic ligands to replace the CH_3CN molecules in some specific sites, can synthesize ultrastable NCs $[Ag_{14}(C_2B_{10}H_{10}S_2)_6$ (pyridine/p-methylpyridine)$_8]$ (Ag_{14}-b, c) in gram scale, and they have thermostability over 150 °C in air and modulated emission properties. These modifications in surface ligands provide a perspective to understand the surface functionalization–property relationship. Based on the $Ag_{14}(SC_6H_3F_2)_{12}(PPh_3)_8$ building units, the Zheng group reported the comprised $2\times2 = 4$ and $2\times2\times2 = 8$ metal cubes of Ag_{14} and determined by single-crystal X-ray crystallography as $[Ag_{38}(SPhF_2)_{26}(PR'^3)_8]$ (Ag_{38}, 2^2) and $[Ag_{63}(SPhF_2)_{36}(PR'^3)_8]+$($Ag_{63}$, 2^3), where $HSPhF_2$ = 3,4-difluorothiophenol and R′ = alkyl/aryl, and the frameworks can be likened to Nichol's half (2^2) and full (2^3) cubes (Figure 7.6c). These also illustrate that ligands play an important role in the shape-controlled synthesis of metal NCs. The cubes in Ag_{38} and Ag_{63} are different. In the former, eight fcc units are arranged in a cube of frequency two, whereas in the Ag_{63}, simply half of 2^3 – i.e. two sets of four fcc units – are arranged in a square pattern (Figure 7.6d). For the split of the structure, one can view it as an octacapped octahedron of $Ag_6@Ag_8$ unit bonded to the bridging thiolate ligands situated at the midpoints of the 12 edges of the cube (or half cube). The eight phosphines are ligated to the Ag atoms at the corners of the cube (or half cubes). Based on the cube patterns in Ag_{38} and Ag_{63}, we can predict the next member is frequency three, named 3^3, and would have $3\times3\times3 = 27$ basic fcc units. This work provides well-defined structure models for specific surface modification of faceted fcc metal nanocrystals.

The synthesis and properties of individual NC are charming and have been studied widely. However, there are few reports on the formation of co-crystallization between different metal nanoclusters by intermolecular interactions. Pradeep et al. reported a co-crystal structure protected by phosphine and thiol ligands in 2018, and used single-crystal X-ray diffraction to show that the cluster consists of $[Ag_{46}(SPhMe_2)_{24}(PPh_3)_8](NO_3)_2$ (Ag_{46}) and $[Ag_{40}(SPhMe_2)_{24}(PPh_3)_8](NO_3)_2$ (Ag_{40}) NCs and each cluster occupied 50% [84]. Both Ag_{46} and Ag_{40} can be viewed as superatoms with 20 and 14 free electrons, respectively. Interestingly, Ag_{46} and Ag_{40} were synthesized by a ligand-exchange-induced structure transformation process, in which the reported $[Ag_{18}H_{16}(PPh_3)_{10}]^{2+}$ served as the precursor and 2,4-dimethylbenzenethiol ($SPhMe_2$) as the exchange ligands. According to the SCXRD analysis, there are two kinds of molecules in the lattice which are nearly isostructural but differ by six silver atoms. The crystal structure also shows nitrate as the counter ion, and the packing structure indicates two counter ions per cluster. Although they are protected by the same ligands and synthesized simultaneously, they have almost the same shell but have different cores. The same outer shell is $Ag_{32}S_{24}P_8$ and then encapsulates the well-defined

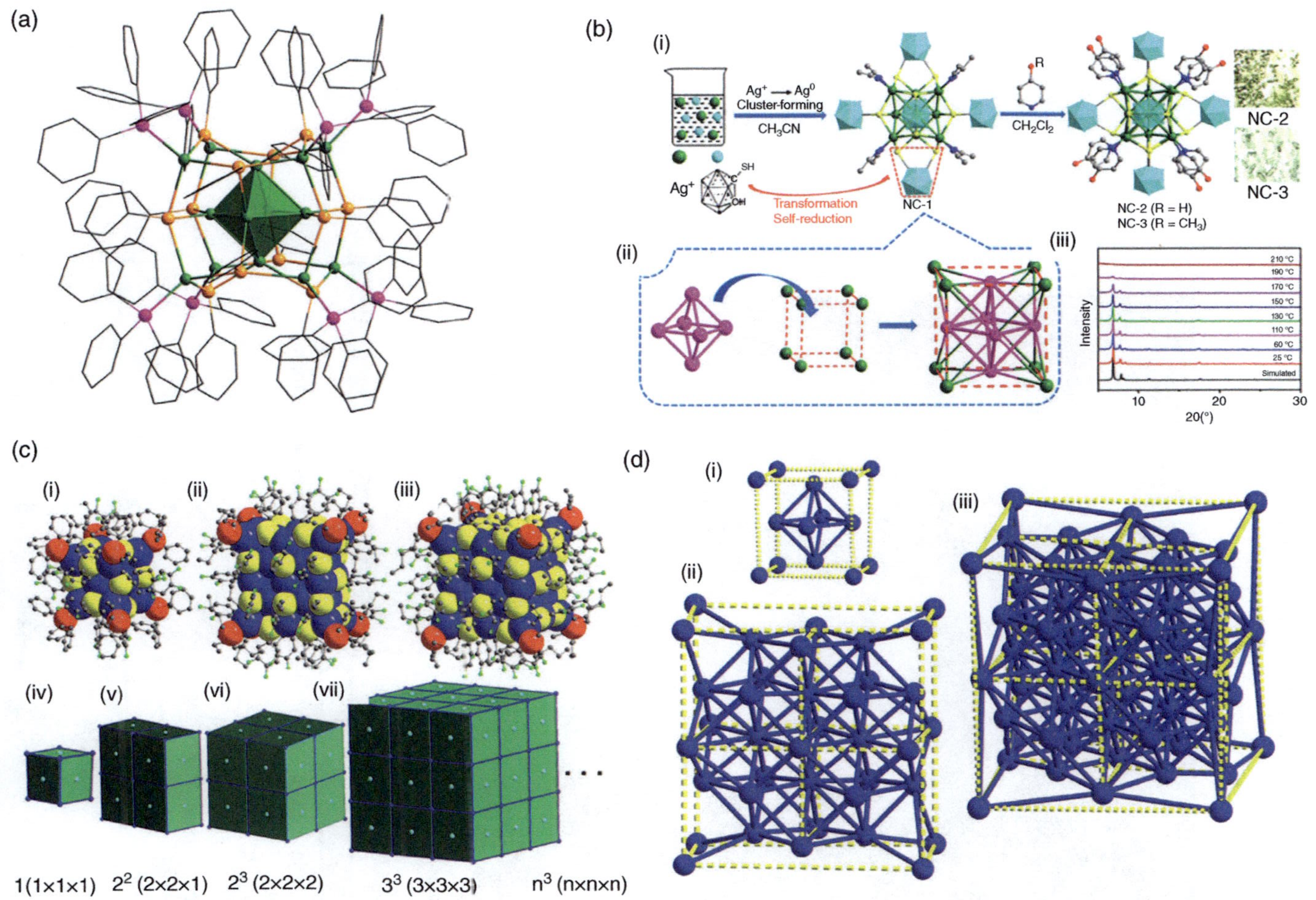

Figure 7.6 (a) Overall structure of $Ag_{14}(SC_6H_3F_2)_{12}(PPh_3)_8$; (b) (i) synthesis of desired Ag_{14}-a NCs, (ii) structural dissection of NC-1, (iii) variable-temperature PXRD patterns of NC-2; (c) (i) $Ag_{14}(SC_6H_3F_2)_{12}(PPh_3)_8$ (1), (ii) $[Ag_{38}(SPhF_2)_{26}(PR'^3)_8]$ (2^2), (iii) $[Ag_{63}(SPhF_2)_{36}(PR'^3)_8]^+$ (2^3) clusters, (iv–vii): the idealized fcc close-packing growth sequence of corresponding cubes; (d) metal frameworks of (i) Ag_{14}, (ii) Ag_{38}, (iii) Ag_{63}. *Source:* © 2013 Royal Society of Chemistry, 2016 and 2018 American Chemical Society, respectively.

fcc Ag_{14} core and a novel simple cube Ag_8 unit in Ag_{46} and Ag_{40}, respectively (Figure 7.7a). The common $Ag_{32}S_{24}P_8$ shell can be regarded as an Ag_{24} core capped by eight AgS_3P motifs and these motifs are located in the faces of eight hexagons of Ag_{24} unit to form the framework of the $Ag_{32}S_{24}P_8$ outer shell. For Ag_{46}, the Ag_{24} outer core combined with the fcc Ag_{14} inner core formed the Ag_{38}

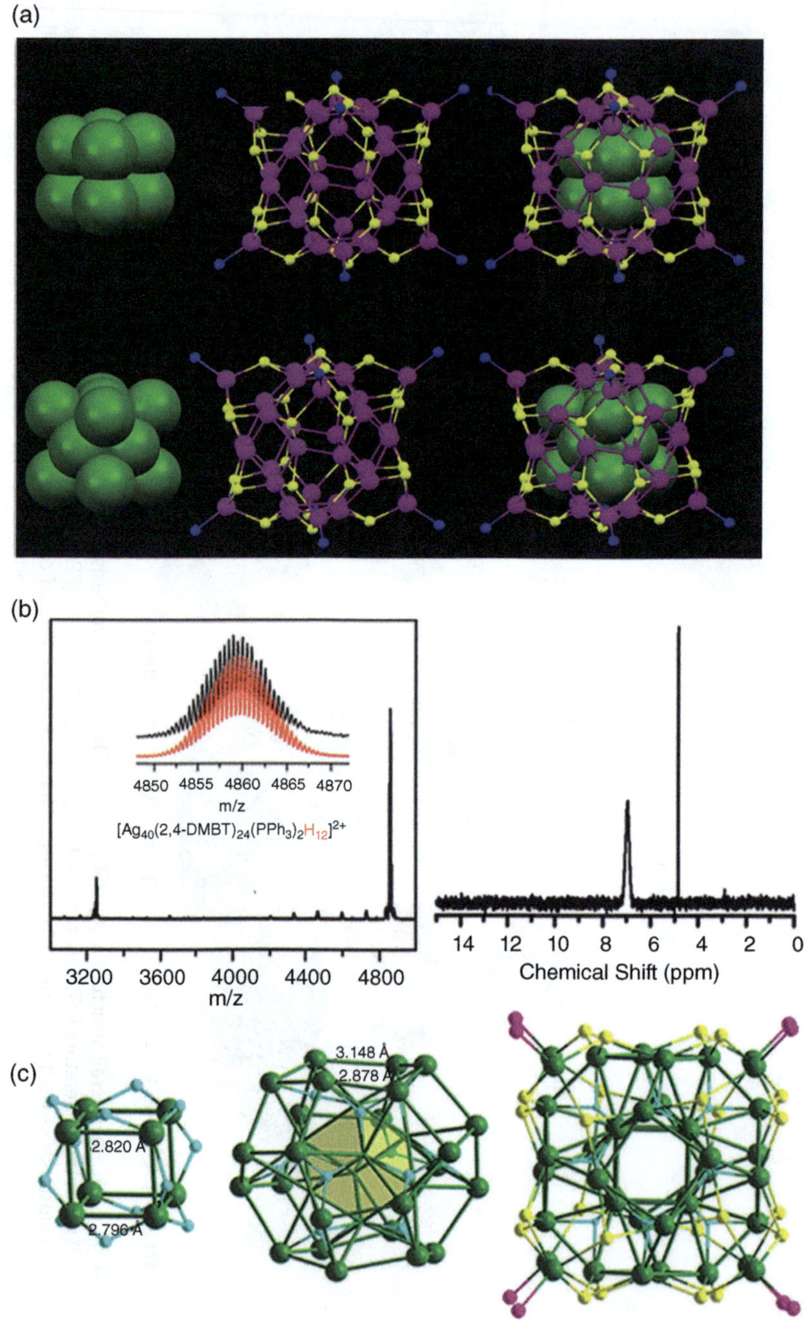

Figure 7.7 (a) The different cores and the same shell of co-crystallization Ag_{40} and Ag_{46}; (b) the determination of the hydrides in Ag_{40}; and (c) its structural anatomy. *Source:* © 2019 Wiley-VCH Verlag GmbH & Co. KGaA, Weinheim and 2019 American Chemical Society.

core, while Ag_{24} encapsulated Ag_8 inner core formed the Ag_{32} core in Ag_{40}. In the structure of Ag_{40} exists a cavity but this is not found in Ag_{46}, and for the kernels, the stackings of silver atoms in the two NCs are different. The optical properties also showed special features and were consistent with the results from the TD-DFT. They tried to use thin-layer chromatography (TLC) and high-performance liquid chromatography (HPLC)-based procedures to separate the Ag_{46} and Ag_{40} but these methods didn't work.

The Zheng group reported the presence of hydrides in the ligated Ag_{40} nanocluster with a non-compact Ag_8 kernel [85]. Structurally, the metal framework of Ag_{40} consisted of three concentric shells of Ag_8@Ag_{24}@Ag_8, which could be described as (ν1-cube)@(truncatedν3-octahedron)@(ν2-cube), respectively. Significantly, the presence of the 12 hydrides in each cluster framework was systematically identified by various techniques [85]. Based on a detailed analysis of the structural features and ^1H and ^2H NMR spectra, the positions of the 12 hydrides were determined to be residing on the 12 edges of the cubic core. As a result, the electron count of the Ag_{40} cluster was a two-electron superatomic system instead of a 14-electron system. Moreover, based on the DFT calculations and experimental probing, it has been demonstrated that the 12 hydrides played a crucial role in stabilizing both the electronic and geometric structure of the $Ag_{40}H_{12}$ NC (Figure 7.7b).

High-nuclearity property is one of the most important parts of NCs. In 2010, the Wang group reported a core-shell NC protected by thiol and halogen ligands. This structure was determined by SC-XRD, with a precise chemical formula of $[Ag_{62}S_{13}(SBu^t)_{32}](BF_4)_4$ (Ag_{62}-1) [99]. The sulfur atoms in this cluster come from the tert-butyl thiolates in the presence of reductive hydrazine. In both liquid and solid-state, the Ag_{61}-1 was highly red-emissive under UV light. The PL originated from a ligand to metal charge transfer (LMCT, $S^{2-} \rightarrow M_n$). Such an NC was the first report that revealed the mechanism of photoluminescence with atomically precise Ag_2S-based NCs. SC-XRD revealed that the Ag_{62}-1 NC is centrosymmetric and the S atom is located in the innermost center of the NC. Ag_{62}-1 has a fcc $Ag_{14}S_3$ core and each edge of this core was bridged by an S^{2-} ligand. The obtained $Ag_{14}S_{13}$ structure was encapsulated by a $Ag_{48}(SBu^t)_{32}$ shell (Figure 7.8a). Subsequently, the Zhu group reported a similar Ag_{62} nanocluster with a chemical formula of $[Ag_{62}S_{12}(SBu^t)_{32}]^{2+}$ (Ag_{62}-2) [86]. The Ag_{62}-2 has a complete fcc Ag_{14} kernel, which is encapsulated by a $Ag_{48}(SBu^t)_{32}$ shell and 12 sole S ligands (Figure 7.8b). Alternatively, the Ag_{62}-2 NC also can be divided into an octahedral Ag_6 core without the central S atom relative to the Ag_{62}-1 NC. In this context, the innermost S atom can significantly affect the electronic structure of such Ag_{62} NCs. The main differences between the two Ag_{62} NCs are their electronic properties. First, although both NCs exhibited fcc units, the Ag_{62}-2 NC contained an Ag_{14} core that was more complete than that in Ag_{62}-1. Besides, the Ag_{62}-2 NC has four free valence electrons while Ag_{62}-1 has an extra sulfur atom and contained no free valence electrons. Such differences in electronic structures led to the fluorescence quenching of Ag_{62}-2, resulting from the changes of the ligand to metal charge transfer (LMCT, S 3p → Ag 5s) mechanism in these NCs. The optical properties between the two clusters also showed some differences. The UV–vis absorption spectrum of Ag_{62}-1 NCs showed three peaks at 330, 370, and 543 nm, while that of Ag_{62}-2 showed two peaks at 420 and 520 nm. Ag_{62}-1 emitted red photoluminescence whereas Ag_{62}-2 was almost nonemissive (Figure 7.8c). The DFT calculations were carried out to study the differences in the electronic structures of Ag_{62}-1 and Ag_{62}-2 NCs. The comparison between these two similar structures provides a new insight to probe the effect of a single atom on the overall structure of NCs, and also elucidates the structure–property relationship of the silver NCs at the atomic level.

In addition to these conventional FCC structures, $Ag_{67}(2,4\text{-DMBT})_{32}(PPh_3)_8$ (Ag_{67}) can be divided into the infrequent box-shaped type [55]. The Ag_{67} was reported by Bakr et al. in 2016, and the crystal structure comprises an Ag_{23} core and a capping layer of $Ag_{44}S_{32}P_8$, which is arranged in the

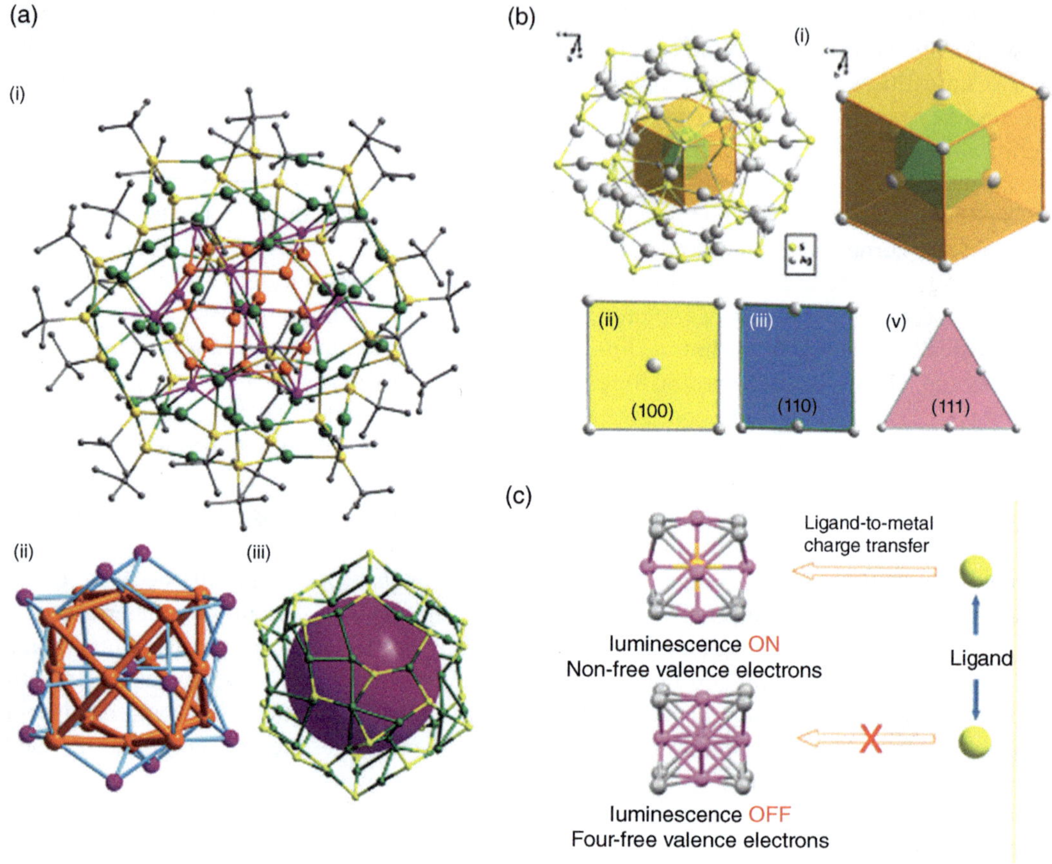

Figure 7.8 (a) (i) Centrosymmetric structure of $[Ag_{62}S_{13}(SBu^t)_{32}](BF_4)_4$ (Ag_{62}-1), (ii) the $Ag_{14}S_{13}$ core in Ag_{62}-1, (iii) the $Ag_{48}(SBu^t)_{32}$ shell structure in Ag_{62}-1; (b) (i) the total structure of $[Ag_{62}S_{12}(SBu^t)_{32}]^{2+}$ (Ag_{62}-2), (ii) Ag_{14} core in Ag_{62}-2 as a compete for FCC unit, (iii–v) three complete crystal faces in the Ag_{14} cube; (c) the difference of PL properties between Ag_{62}-1 and Ag_{62}-2. *Source:* © 2010, 2014 American Chemical Society.

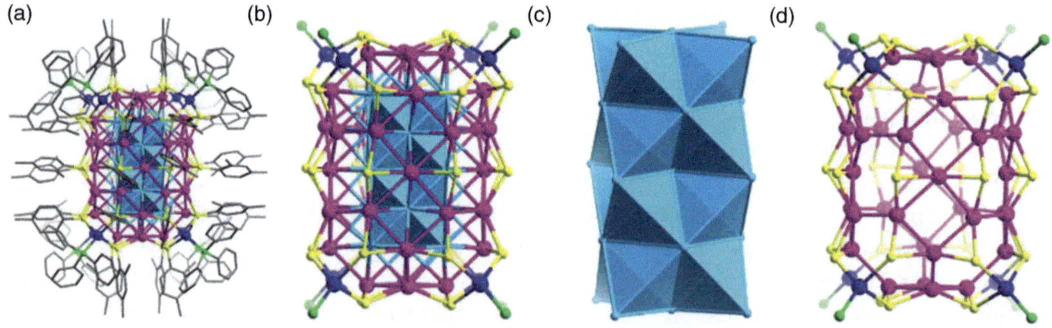

Figure 7.9 (a) Overall structure of $Ag_{67}(2,4\text{-DMBT})_{32}(PPh_3)_8$ cluster; (b) the structure of $Ag_{67}S_{32}P_8$ obtained by disconnecting carbon atoms in (a); (c) the Ag_{23} metal core; (d) the structure of the Ag_{67} without the Ag_{23} metal core, i.e. $Ag_{44}(SPhMe_2)_{32}(PPh_3)_8$. *Source:* © 2017, American Chemical Society.

shape of a box (Figure 7.9a,b). In particular, the central Ag_{23} unit is unlike the traditional icosahedron Ag_{13} unit, whereas it was built from an unprecedented centered Ag_{13} cuboctahedron sharing opposite square faces with two Ag_8 crowns to form the Ag_{21} unit and then capped by two silver atoms at the open crown positions further constitute the Ag_{23} core (Figure 7.9c). This Ag_{13} cuboctahedron is why the Ag_{67} clusters grow into a box-shape configuration. The protective group on the periphery of the $Ag_{44}(SPhMe_2)_{32}(PPh_3)_8$ shell can be dissected into two Ag_{16} units, which arrange as two bowls and are then joined by four 1D S—Ag—S motifs to form the $Ag_{36}S_8$ unit. Finally, the $Ag_{36}S_8$ covers eight 3D AgS_3P motifs to produce the whole $Ag_{44}S_{32}P_8$ shell. It is interesting to note that if we remove the $Ag_{21}(SR)_8$ unit from the Ag_{67} we can predict another box-like structure of $Ag_{46}(SR)_{24}(PPh_3)_8$, which recently has been synthesized and crystallized by the Zhu group and Pradeep group and determined as Ag_{46}(2,5-DMBT)$_{24}(PPh_3)_8$. Different from the Ag_{67}, the Ag_{46} cluster has an FCC structure of Ag_{14} core and is capped by the $Ag_{32}S_{24}P_8$ shell (Figure 7.9d).

The largest fcc-based silver nanocluster reported so far is $[Ag_{100}(SC_6H_3{}^{3,4}F_2)_{48}(PPh_3)_8]^-$ (abbrev. Ag_{100}), synthesized by the Zhu group in 2020 [100]. Ag_{100} is co-protected by 3,4-difluoro-benzenethiol and triphenylphosphine, using the conventional one-step synthetic method in an ice bath. The crystal structure was determined by SCXRD. With the ESI-MS and SCXRD together, the cluster was confirmed to carry a 1- charge and formulated as $[Ag_{100}(SC_6H_3{}^{3,4}F_2)_{48}(PPh_3)_8](PPh_4)$. The structure can be viewed as four parts: a Ag_6 octahedron unit, an octahedron Ag_{38} shell, a further $Ag_{48}S_{24}$ shell, and finally the outmost layer formed by eight tetrahedral AgS_3P motifs that constitute $Ag_8S_{24}P_8$ (Figure 7.10). To sum up, Ag_{100} exhibits a rhombicuboctahedron structure and the multishell can be written as $Ag_6@Ag_{38}@Ag_{48}S_{24}@Ag_8S_{24}P_8$. Time-dependent UV–vis tests showed that Ag_{100} is not particularly stable in dichloromethane compared to in 1,2-dichloroethane. DFT calculations on a simplified model demonstrate that the steric effect of the thiolate ligands plays an important role in regulating the framework of the Ag_{100} cluster other than adopting a high symmetry. Overall, this work also provides a reference for future studies on the synthesis and exploration of the growth of silver NCs with fcc-based structures.

7.3.3 Other Special Ag NCs

In addition to the silver NCs based on icosahedral and fcc units, there are also NCs with Ag_4, Ag_7, Ag_{19}, and Ag_{23} building blocks. The Sun group reported two Ag NCs with tetrahedral Ag_4 units, including $[Ag_{10}@(MoO_4)_7@Ag_{60}]$ (SD/Ag70a) and $[Ag_4S_4(MoO_4)_5@Ag_{66}]$ (SD/Ag70b); the Ag_{10} core in the former can be seen as five tetrahedra sharing faces and edges, then capped by the Ag_{60} shell and surround by seven $MoO_4{}^{2-}$ (Figure 7.11a), while the SD/Ag70b NC consists of just a Ag_4 core and a Ag_{66} shell as well as five $MoO_4{}^{2-}$ [101]. The SD/Ag70b cluster can be converted to the SD/Ag70a cluster by adding DMF.

The $Ag_{146}Br_2(SPh^iPr)_{80}$ cluster was reported by the Jin group, in which a gram-scale synthesis with high yield was achieved by one-pot reaction [102]. The structure contains an incomplete decahedron Ag_{51} core, which can be further divided into three layers of decahedron, i.e. Ag_7, Ag_{32},

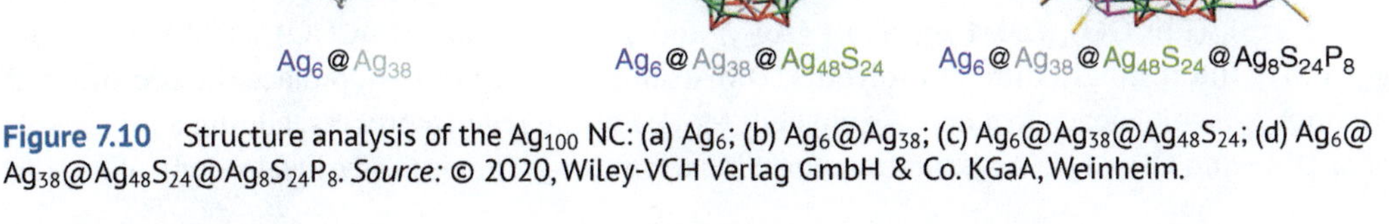

Figure 7.10 Structure analysis of the Ag_{100} NC: (a) Ag_6; (b) $Ag_6@Ag_{38}$; (c) $Ag_6@Ag_{38}@Ag_{48}S_{24}$; (d) $Ag_6@Ag_{38}@Ag_{48}S_{24}@Ag_8S_{24}P_8$. *Source:* © 2020, Wiley-VCH Verlag GmbH & Co. KGaA, Weinheim.

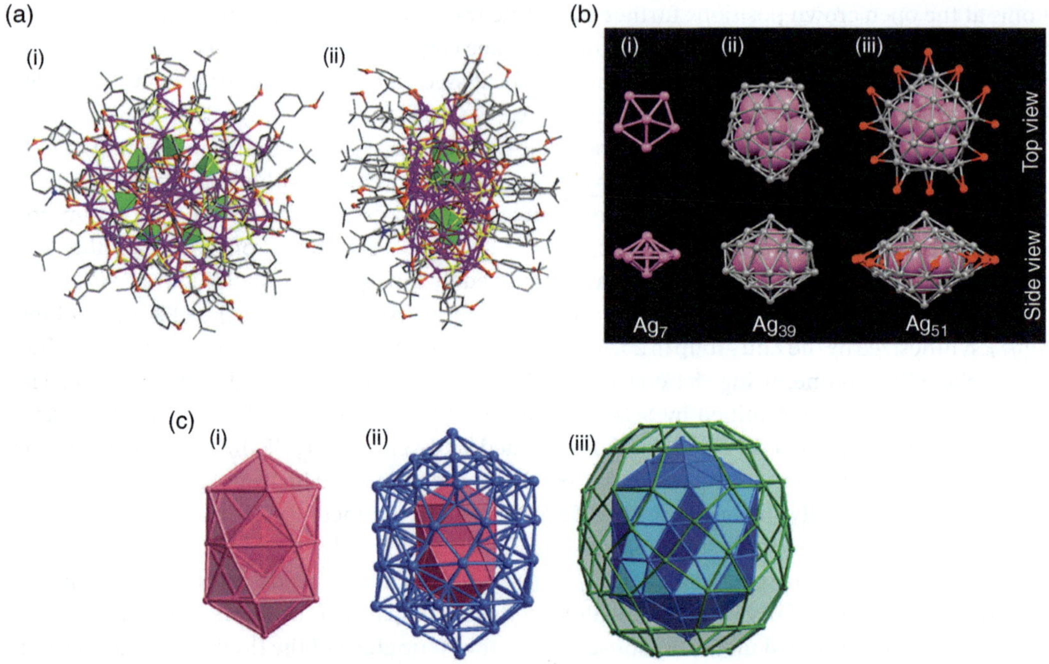

Figure 7.11 (a) (i) top view and (ii) side view of the structure of SD/Ag70a; (b) (i) the decahedral seven-atom shell (magenta), (ii) the 32-atom shell (gray), (iii) the ring-shaped 12-atom (red) in the $Ag_{146}Br_2(SPh^iPr)_{80}$ cluster; (c) (i) the inner shell of Ag_{19}, (ii) the two-shelled $Ag_{19}@Ag_{52}$ of the nanocluster, (iii) the three-shelled $Ag_{19}@Ag_{52}@Ag_{70}$ in the $[Ag_{141}X_{12}(S\text{-}Adm)_{40}]^{3+}$ cluster. *Source:* © 2019, Wiley-VCH Verlag GmbH & Co. KGaA, Weinheim, 2017 and 2018 American Chemical Society, respectively.

and Ag_{12}, respectively, then capped by a shell of $Ag_{95}Br_2S_{80}$ (Figure 7.11b). The outer $Ag_{95}Br_2S_{80}$ shell has four hierarchical layers: $Ag_{30}S_{10}$, $Ag_{30}S_{20}$, $Ag_{25}S_{30}$, and $Ag_{10}S_{20}$. Especially, Ag_{146} shows molecular-like properties, such as an optical bandgap and power-independent electron dynamics.

Compared to Ag_{146}, the $[Ag_{141}X_{12}(S\text{-}Adm)_{40}]^{3+}$ (X = Cl, Br, I) nanocluster has a multishell structure with an $Ag_{19}@Ag_{52}$ core and $Ag_{70}(SAdm)_{68}F_2Cl_2$ shell (Figure 7.11c). The Ag_{71} core shows an elongated shape rather than a regular icosahedron. In addition, the thiolate ligands have a low fraction of surface coverage, perhaps because of the multiple-twinned metal core, which may suggest the important role of surface ligands in controlling the shape of metal NCs. Surprisingly, if using phenylacetylene (PA) ligand as the exchange reagent, two $[PA]^-$ would replace two $[S\text{-}Adm]^-$ ligands, and ligand-exchange with water-soluble mercaptosuccinic acid (MSA) can also occur, resulting in water-soluble NCs. These phenomena are related to the low thiolate ligand coverage on the cluster surface.

The direct synthesis methods have led to some new structures. One popular strategy for enriching the structures is using external stimuli to transform NCs. The external stimuli are not only ligands or metal salts but also some solvents that can promote the transformation reactions. Recently, the Lang group reported a reversible transformation between thiolate-protected Ag_{25} and Ag_{26} clusters driven by solvent [103]. They choose Tab (abbrev. for 4-(trimethylammonio)-benzenethiolate) as the ligand and functional solvent to synthesize two novel NCs: $[Ag_{25}Cl_2(Tab)_{14}(PhCOO)_{11}(DMF)_4](PF_6)_{12}$ (Ag_{25}) and $[Ag_{26}Cl_2(Tab)_{14}(PhCOO)_{13}(DMAc)_4](PF_6)_{11}$ (Ag_{26}) [103]. Interestingly, different solvents (DMAc and CH_3CH_2OH) can replace the coordinated DMF on Ag_{25}, which leads to a new Ag cluster (Ag_{26}). For the reverse process, adding a mixed solvent of DMF and CH_2Cl_2 to Ag_{26} can transform it back to Ag_{25}. That is to say, just by controlling the

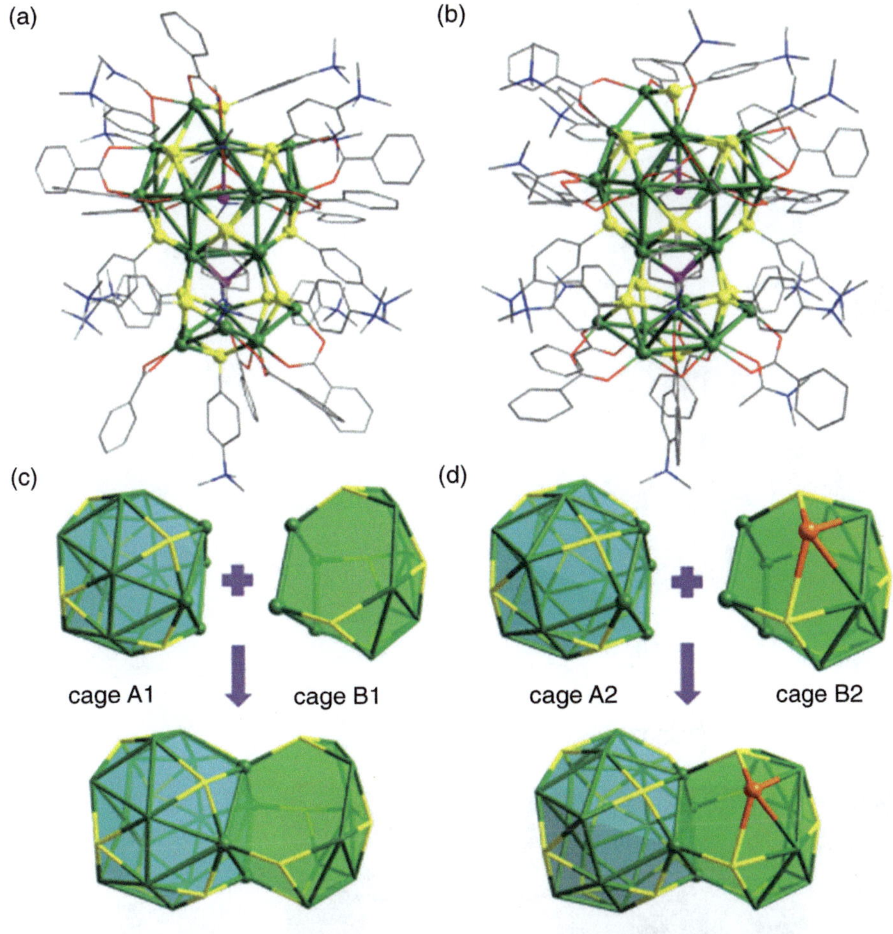

Figure 7.12 Structures of (a) Ag_{25} and (b) Ag_{26}. Structural dissection of the core Ag—S skeleton of (c) Ag_{25} and (d) Ag_{26}; other parts are all omitted for clarity. *Source:* © 2021, Science China Chemistry.

solvent, reversible conversion between Ag_{25} and Ag_{26} can be achieved (Figure 7.12). Furthermore, the two clusters have inherent chirality because of the peripheral arrangement of Ag, S, and Cl atoms. But the enantiomers have not been isolated. Anyway, this work provides a new strategy to synthesize novel Ag NCs and will promote the development of Ag clusters.

$[Ag_{307}Cl_{62}(SPh^tBu)_{110}]$ (abbreviated as Ag_{307}) is the second-largest Ag NC to date and is also the first Ag NC that contains so many chlorides [104]. Ag_{307} has a Ag_{167} core, which is capped by a $[Ag_{140}Cl_2S_{110}]$ shell which is linked by a Cl_{60} intermediate layer (Figure 7.13a). The synthesis of Ag_{307} is by a $Ag_m(SPh^tBu)_n$ precursor crystallized in DMF/$CHCl_3$ at ~4 °C and the chlorine atoms are from the $CHCl_3$ solvent. Changing the solvent from $CHCl_3$ to CH_2Cl_2 can also lead to the NC but the yield would be lower. The Ag_{307} NC can exist stably under natural light at room temperature or 55 °C for more than a week, and the solid state can be stable and stored for a longer time. DPV measurements show continuous charging/discharging behaviors and has a surface plasmon resonance; these phenomena indicate that Ag_{307} is of metallic behavior. The EPR signals show electron spin magnetic behavior in Ag_{307}.

The Zheng group reported two giant thiolated Ag nanoclusters in 2016 and determined their crystal structures, including $[Ag_{136}(SR)_{64}Cl_3Ag_{0.45}]^-$ (denoted as Ag_{136}) and $[Ag_{374}(SR)_{113}Br_2Cl_2]$

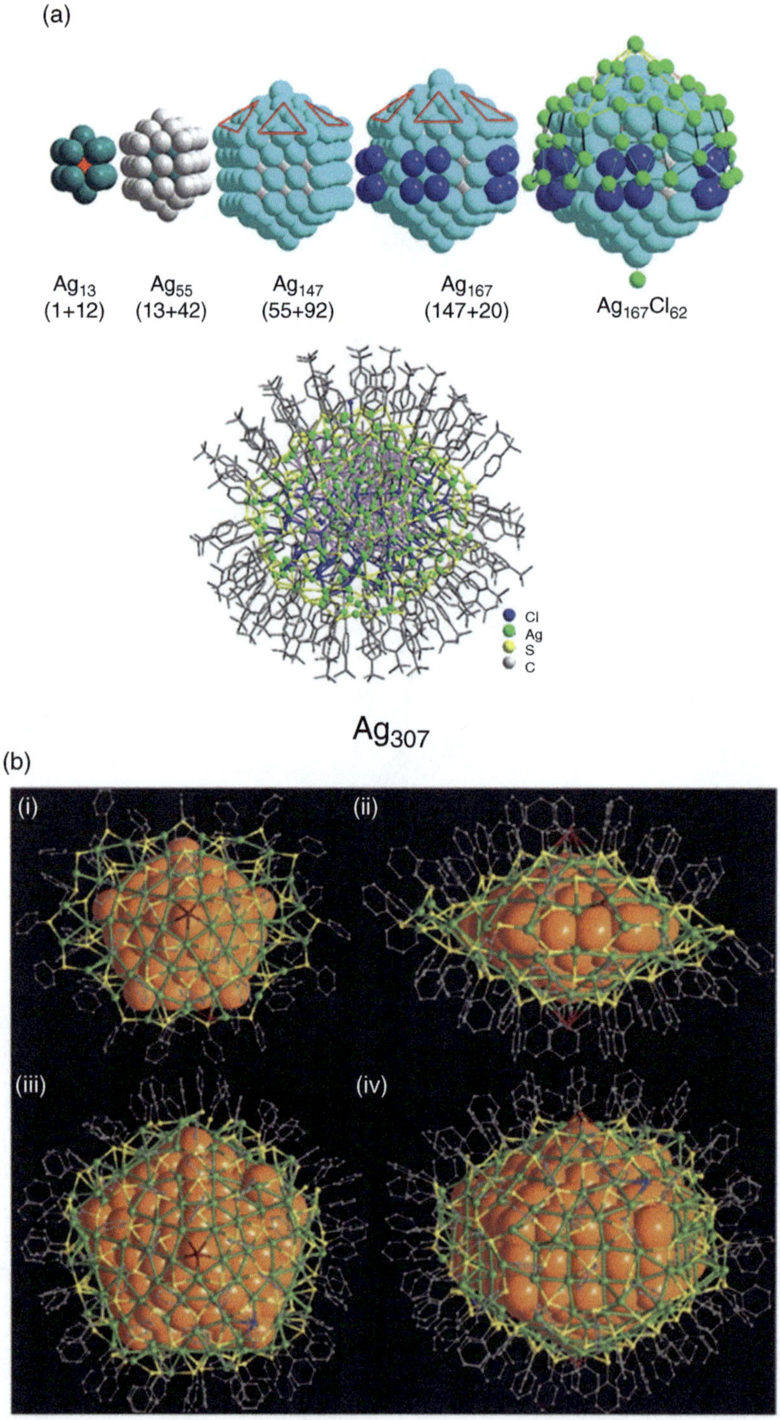

Figure 7.13 (a) Structural anatomy of the $[Ag_{167}Cl_{62}]$ part of the Ag_{307} total structure; (b) (i, ii) the overall structure of the Ag_{136} NC via top and side views, (iii, iv) the overall structure of the Ag_{374} NC via top and side views. *Source:* © 2021 American Chemical Society, 2016 Springer Nature.

(denoted as Ag_{374}) [56]. These two NCs can be divided into a fivefold core and Ag-SR complex shells. The core in Ag_{136} is two pentagonal pyramids (Ag_{54}) and each is surmounted by a bowl-like $[Ag_{30}(SR)_{15}Cl]$ unit (Figure 7.13). In Ag_{374}, the core is Ag_{207} and similarly, each pentagonal pyramid is capped by a bowl-like $[Ag_{30}(SR)_{15}Br]$. The UV–vis absorption spectruma of Ag_{136} exhibits a peak at 450 nm and a shoulder at 772 nm, while Ag_{374} shows just one strong peak at 465 nm. DFT calculations indicate that the 465 nm peak of Ag_{374} origins from the collective oscillation, and the redshift compared with Ag_{136} is attributed to the much larger size. The two structures show the phenomenon from the molecular to the metallic regime.

7.4 Properties of Ag NCs

Compared with nanomaterials of large size such as nanoparticles and nanorods, metal NCs show the most advantageous feature, that is, the atomically precise structures. One can control the overall metal core size and structure at the atomic level and further explore the mechanisms/origins of the intriguing properties. The correlations between the structures and properties can be used as a guideline for other nanomaterials. Meanwhile, studies on the properties of NCs provide chemical possibilities for potentially modifying the structures and electronic properties. The ultrasmall size (<3 nm) of NCs brings many molecular-like properties that have found applications in biomarkers, biomedicine, heavy metal identification, and so on.

In recent years, a number of novel NCs structures have been reported, which also provide models for studying properties and the surface coordination chemistry of metal nanomaterials. In this part, we discuss three aspects of properties: chirality, fluorescence, and catalysis.

7.4.1 Chirality of Ag NCs

Chirality is ubiquitous and plays a key role in nature. Chiral molecules can be used in enantioselective catalysis, pharmaceutical sensors, optoelectronics, and optical devices [105–107]. The different binding modes of M—N (M = metal, and N = the binding atom of ligand) lead to different surface structures on NCs, which in turn give rise to various characteristic features of chirality to the NCs. The first obtained optical activity in the gold-thiolate NC was $Au_{28}(SG)_{16}$ reported in 1998 [108] (*Note:* Later corrected as $Au_{25}(SG)_{18}$). Different from the Au NCs, the origin of chirality for Ag NCs mostly depends on the distortions of the outmost Ag-SR shell. Nonetheless, most chiral NCs from the direct synthesis involve achiral ligands, resulting in racemates in the products, thus, the enantioselective synthesis is still a challenge. Commonly used methods of chiral separation are HPLC, chiral ion-pairing, or other methodologies [105]. For the structure of NCs, the origins of intrinsic chirality may come from four sources: (i) intrinsically chiral metal kernel; (ii) chiral arrangement of the achiral staple motifs on the surface of the kernel; (iii) chiral arrangement of carbon tails; and (iv) the chiral ligands induction to achiral metal core [105, 107].

So far, many chiral structures in the field of NCs have been discovered, and their atomically precise features are good models for understanding the origin of chirality. Generally speaking, the surface motifs of Ag-SR NCs are very different from those of Au-SR NCs, but in addition to the case of $M_{25}(SR)_{18}$ (M = Au, Ag), the case of chirality in Ag NCs and Au NCs also varies. The chiral Au NCs are common, such as Au_{20}, Au_{38}, Au_{102}, Au_{133}, and Au_{246} [91, 109–112], but chiral Ag NCs are relatively less. Benefiting from the advances in the synthesis methods and characterization techniques for chiral structures, more and more chiral Ag NCs have been obtained, such as Ag_{23} [82], Ag_{30} [113], Ag_{32} [78], Ag_{33} [79], Ag_{45} [78], Ag_{47} [114], and Ag_{78} [115].

Jin and co-workers synthesized and characterized a chiral silver nanocluster with 23 silver atoms protected by phosphine and phenylethanethiolate ligands, $Ag_{23}(PPh_3)_8(SC_2H_4Ph)_{18}$, which crystallizes in the chiral monoclinic group Cc [82]. The source of chirality is the chiral arrangement of the metal core, not the configuration of peripheral ligands. Regrettably, Ag_{23} enantiomers have not been separated, but the simulated ECD spectra reproduce the experimental spectral curves. The total structure is composed of two twisted fcc units and possesses an open-shell electronic structure (23(Ag $5s^1$)–18(SR) = 5e); thus, this cluster has an unpaired electron and shows an EPR signal. The 18 thiolates have two binding modes with the surface Ag atoms: tri- and tetra-podal types, respectively. It is worth noting that the tetra-podal mode of thiolate had not been found in earlier reported Ag nanoclusters.

Mak et al. reported a racemic anisotropic nanocluster, $Ag_{30}(C_2B_{10}H_9S_3)_8Dppm_6$ (Ag_{30}-rac) coprotected by achiral carboranetrithiolate and phosphine ligands in a one-pot method, and realized a spontaneous self-resolution upon recrystallization in DMAc solution [113]. The chirality of Ag_{30}-race comes from the chiral arrangement of ligands and is directed by the unusual B—H···π and C—H···π hydrogen-bonding interactions between the carboranyl groups and the phenyl units of phosphine ligands. The overall framework of the chiral Ag_{30} is built up by triangles via vertex- or edge-sharing with a threefold axis passing through the center of the regular triangle (Figure 7.14a). In addition to hydrogen-bonding interactions between the carboranyl and phosphine ligands, the adjacent phenyl groups from the adjacent Dppm ligands also exhibit C—H···π interactions, which results in a reverse spiral arrangement of the two sets of benzene rings. The chiral arrangement of the peripheral ligands is then connected to the Ag_{30} unit, which makes the cluster chiral. In the enantiomorphous crystal structures, the homochiral NCs adopt enantiomeric helical assemblies, and along with the [001] direction, the superlattice contains a parallel array of helical tubes built with the cluster in the double-helical arrangement. When the Ag_{30} is dissolved in DMAc, the solvent molecules located in the reverse helices bridge the nanoclusters together through the reinforced C—H···O bonds and H···H interactions, and these noncovalent weak interactions support the superstructure. The crystals of R/L-Ag_{30} show mirror-symmetric, circularly polarized luminescence (CPL) signals (Figure 7.14b). The Zhu group synthesized two different sizes of NCs via engineering the surface ligands, $[Ag_{32}(Dppm)_5(SAdm)_{13}Cl_8]^{3+}$ (Ag_{32}) and $[Ag_{45}(Dppm)_4(S\text{-}Bu^t)_{16}Br_{12}]^{3+}$ (Ag_{45}), and the total structures were determined [78]. The Ag_{32} and Ag_{45} are two chiral silver NCs, and their chirality comes from the asymmetric distributions of peripheral ligands' shell, not the achiral metal core. Ag_{32} has an achiral Ag_{13} kernel, but the $Ag_{19}S_{13}Cl_8P_{10}$ shell with irregular arrangement leads to the chiral Ag_{32} (Figure 7.14c). The Ag_{45} is protected by three types of ligands: Dppm, HS-Bu^t, and Br. The core of Ag_{45} is two icosahedral Ag_{13} units self-assembled through the three faces sharing Ag atoms, then capped by the $Ag_{22}S_{16}P_8Br_{12}$ units, forming the Ag_{45} structure. In other words, this work reported the synthesis and determination of two clusters with achiral metal cores but chiral shells via engineering the ligands, and provided insights into the synthesis of chiral Ag NCs.

The Zheng group reported an intrinsically chiral nanocluster $[Ag_{78}(DPPP)_6(SR)_{42}]$ (Ag_{78}) in 2017, which crystallizes as racemates in a centric space group but can be separated via chiral diphosphines [115]. The overall structures of Ag_{78} can be seen as an Ag@Ag_{21} kernel capped by a Ag_{44} unit, then the Ag_{66} core encapsulated in a shell containing three $[Ag_4(DPPP)_2(SR)_8]^{4-}$ units and 18 SR^- (Figure 7.15). The Ag@Ag_{21} kernel can be described as three icosahedra mutually integrated, which contains 38 triangular Ag_3 faces and the different faces are capped by a 44-atom Ag shell of the Ag_{66} core. Then the Ag_{66} unit is encapsulated by three twisted trigonal prism-shaped $Ag_4(DPPP)_2(SR)_8$ complex shells to form the complete structure of Ag_{78}. The achiral DPPP ligand-protected Ag_{78} is an intrinsically chiral nanocluster. The single-crystal structure analysis shows that the chirality of this cluster is dictated by the C—C—C bond angles of diphosphines, which

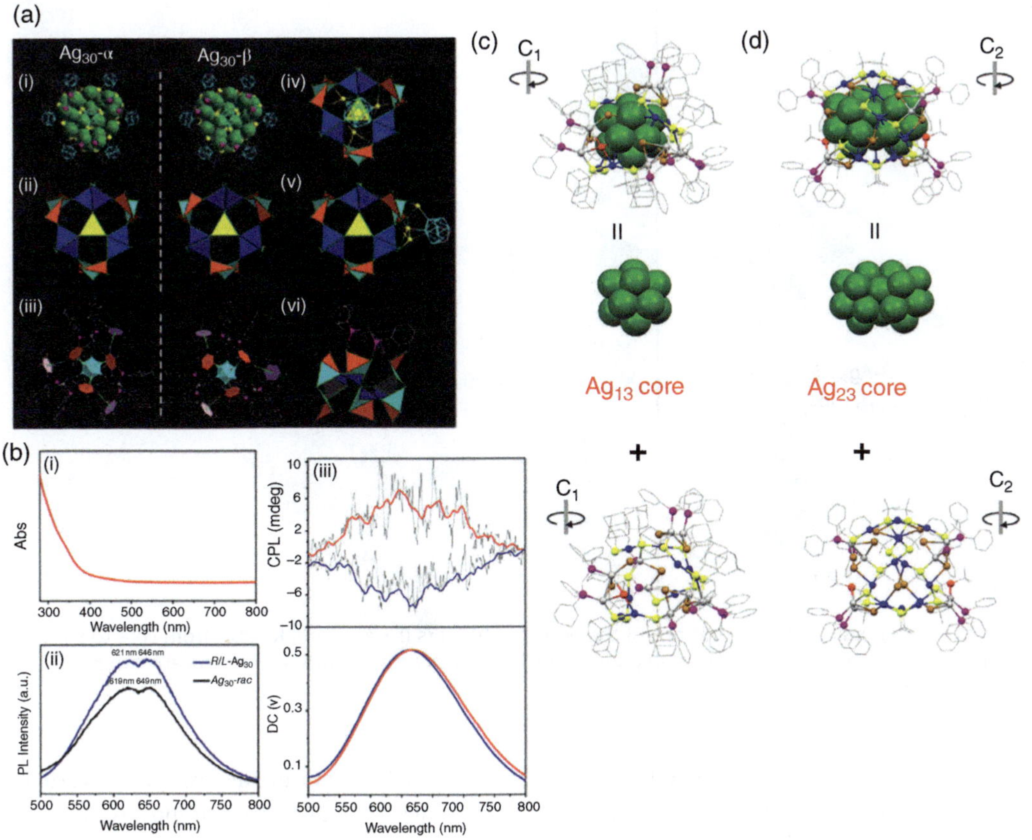

Figure 7.14 (a) Atomic structure of the cluster in $Ag_{30\text{-rac}}$. (i) the total structure of the enantiomers, (ii) chiral triangle-based metal architectures in the enantiomers, (iii) illustration of the intracluster noncovalent interactions in the enantiomers, (iv–vi) binding modes of carboranetrithiolate and phosphine in Ag_{30}-α; (b) (i) UV–vis absorption spectrum of $Ag_{30\text{-rac}}$ in DMAc, (ii) solid-state emission spectra of $Ag_{30\text{-rac}}$ and the racemic conglomerates R/L-Ag_{30}, (iii) CPL spectra of R-Ag_{30} and L-Ag_{30} excited at 367 nm; (c) the structure of Ag_{32} cluster; (d) the structures of Ag_{45} cluster. *Source:* © 2020, American Chemical Society and 2017, The Royal Society of Chemistry.

restricts the relative orientation of the tetrahedral $R_3PAg(SR)_3$ groups. This work, instead of using the traditional method of chiral separation, chose chiral diphosphines as the substitute for achiral DPPP ligands, which led to enantioselective synthesis and 100% optical purity of R/S-Ag_{78}. This strategy provides an effective synthesis idea for the preparation of chiral metal NCs.

7.4.2 Photoluminescence of Ag NCs

Compared with traditional fluorescent materials, such as organic dyes, quantum dots, and nanoparticles, metal NCs have the characteristics of high photostability, adjustable fluorescence wavelength, good biocompatibility, and simple preparation. Photoluminescence is one of the advantages of Ag NCs, which have precise sizes, brightness, photostability, and low toxicity, and most of the Ag clusters have potential applications in the field of biology, medicine, and diagnosis [9, 112]. The fluorescence originates from the metallic kernel or outer complexes. Nevertheless, the quantum yields (QY) of most Ag NCs are still low and the luminescence wavelength is uncontrollable. The luminescence property of Ag NCs is related to size, peripheral ligands, ligand coordination modes,

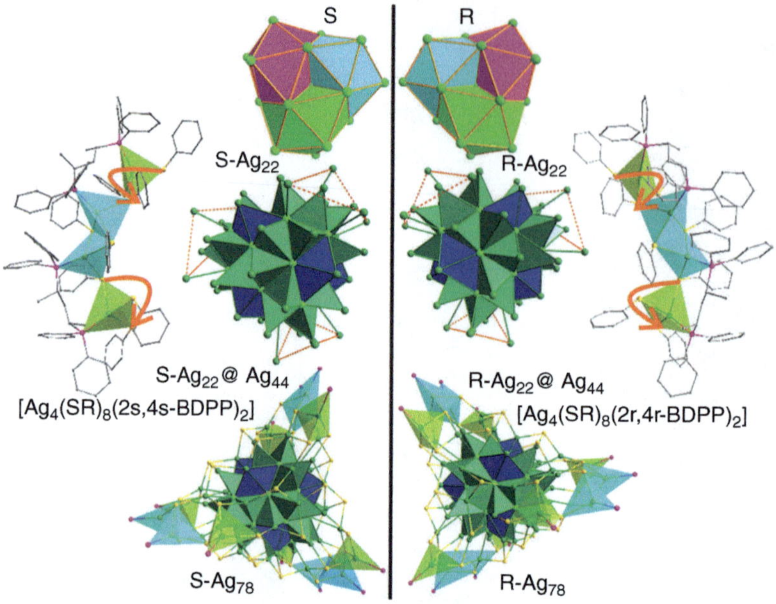

Figure 7.15 Anatomical representations of the two enantiomers (S—Ag_{78} and R-Ag_{78}) from core to surface. The clusters display idealized D_3 symmetry and perfect mirror symmetry with respect to each other. *Source:* © 2017, American Chemical Society.

and doped metal atoms. In addition, for the aqueous clusters, temperature and pH are also crucial and influence the PL intensity. At present, the photoluminescence properties of most Ag NCs are still unsatisfactory, the PL quantum yield is low (generally less than 20%), and the emission lifetime is generally in the nanosecond range. However, the luminescence mechanism is still unclear, which greatly hinders the design and synthesis of highly fluorescent NCs [116].

The Zhu group chose Ag_{25} NCs as the parent cluster and Au and Pt atoms as the dopants for the synthesis of $Au_1Ag_{24}(SR)_{18}$, $Au_xAg_{25-x}(SR)_{18}$ (x = 4–7), $Pt_1Ag_{24}(SR)_{18}$, and $Pt_1Au_xAg_{24-x}(SR)_{18}$ (x = 3–7), and then explored the metal interaction effect on the FL of Ag NCs [68]. The Ag_{13} core of $Ag_{25}(SR)_{18}$ can be viewed as a Ne-like superatom structure, and the different electron affinities of the innermost metal atom contribute to different free valence electrons, which shrink to the center of the superatom. When doping one Au or Pt atom into the M_{13} core, the free electrons will shrink to the innermost site, leading to a remarkable enhancement of PL. Unexpectedly, if adding more heteroatoms, quenching of the PL was instead observed. This work contributed some new perspectives on the influence of metal core on the fluorescence of NCs.

Besides the metal core, the peripheral ligands also have a great influence on the PL properties of Ag clusters. The Ag_{29} cluster is one of the most studied Ag NCs, which has fluorescent properties. For example, Mark et al. synthesized $[Ag_{29}(BDT)_{12}]^{3-}$ and $[Ag_{29}(BDT)_{12}(X)_4]^{3-}$ (X = phosphine ligands of different chain lengths, including PPh_3, DPPM, DPPE, DPPP, TTP, TFPP, and TCPP), and these NCs were characterized by optical absorption and ESI MS [52]. Among these NCs, the weakest fluorescence corresponds to $Ag_{29}(BDT)_{12}(PPh_3)_4]^{3-}$ with monophosphine ligand of the shortest chain length. DFT and TD-DFT calculations were used to understand their structure and electronic properties. As the chain length of the bisphosphine ligand increases, the rigidity of the cluster increases via the more noncovalent interactions, which then promotes the radiative transitions by increasing RIR. In short, as the chain length of diphosphines increases, the PLQY of the Ag_{29} cluster is enhanced by about 30-fold. This phenomenon also appeared when introducing heavy ligands (such as TTP, TFPP, and TCPP). In addition, silver-based alloy clusters also show this nature. $Pt_1Ag_{28}(S\text{-}Adm)_{18}(PPh_3)_4$ with binary ligands was converted from $Pt_1Ag_{24}(SPhMe_2)_{18}$ [117], and the PL quantum yield (QY) was increased from 0.1% to 4.9% (i.e., enhanced by about 50-fold). This phenomenon was explained by the mechanism that the tetrahedral structure in Pt_1Ag_{28} suppressed phonon emission and other nonradiative pathways. Except for the ligand study on the Ag_{29}-based NCs, the PL properties of Ag NCs of other structures have also been reported. For example, Liu et al. reported the two homo-silver $[Ag_{20}\{S_2P(OiPr)_2\}_{12}]$ (abbreviated Ag_{20}—S) and $[Ag_{21}\{S_2P(OiPr)_2\}_{12}](PF_6)$ (abbreviated Ag_{21}—S) [66]. Their emissions are in the near-IR region at 921 nm and 950 nm, respectively. After ligand exchange with the Se-donor ligands, $[Ag_{20}\{Se_2P(OiPr)_2\}_{12}]$ (abbreviated Ag_{20}—Se) and $[Ag_{21}\{Se_2P(OEt)_2\}_{12}](PF_6)]$ (abbreviated Ag_{21}—Se) were obtained, which can only be made via ligand displacement reactions from their sulfur counterparts. They have similar geometric structures, but the emission wavelength was blue-shifted to 720 and 754 nm, respectively. The different ligands have an effect on not only the fluorescence intensity but also the emission wavelength.

Water-soluble metal nanoclusters, which are protected by carboxyl groups (such as mercaptosuccinic acid, glutathione, dihydrolipoic acid, etc.), are common and have higher PLQY. Based on the $Ag_{29}(BDT)_{12}$ cluster, Chakraborty's group studied the effect of ligands on the structure and optical properties of $Ag_{29}(DHLA)_{12}$ by ligand exchange with DHLA ligands [118]. They synthesized a series of $Ag_{29}(BDT)_{12-x}(DHLA)_x$ NCs (where $x = 1–6$). The PLQY of these ligand-exchanged products was remarkably enhanced (~44-fold), and DFT showed that the $Ag_{29}(BDT)_{11}(DHLA)^{3-}$ species is the most stable one, which suggests that the mechanism of PLQY enhancement is due to LMCT. The carboxyl group in DHLA can promote the charge transfer and reduce the nonradiative relaxation in $Ag_{29}(BDT)_{12-x}(DHLA)_x^{3-}$ ($x = 1–6$), which enhances the structural rigidity and then generates the enhancement of PLQY. That is to say, the origin and mechanism of PL properties are related to every metal atom and the ligands as well, which is different from semiconductor QDs and regular nanoparticles.

In addition to the metal core and ligands of NCs that play an important role in the PL properties, aggregation-induced emission, self-assembly, or external environments also influence the PL. Aggregation-induced emission (AIE) is a recently developed strategy to design highly luminescent metal NCs, which means that nonluminescent or weakly luminescent materials become highly luminescent upon aggregation, either in poor solvents or in the solid state [119, 120]. NCs can produce strong PL upon aggregation, which is regulated by their surface M(I)-SR motifs. Most of the AIE enhancement of PL in NCs is induced by the restriction of intramolecular motion (RIM) [121]. Crystallization-induced emission enhancement (CIEE) is a special kind of AIE, in which the enhancement of PL is caused by their transition from the amorphous (solution, solid, or other aggregated states) to the crystal state. This strategy was first introduced by Tang et al. by using organic molecules such as hexaphenylsiloles and tetraphenylethylene, and progress has been reported in many areas, such as materials development, mechanism analyses, and AIE gens applications [122]. The Xie group introduced the AIE idea to enhance the PL properties of thiolate-protected metal NCs [119]. Except for the mentioned NCs' internal factors that would influence the PL properties, the external environment (gas, solvent, pressure, and temperature, etc.) also affects the PL properties. The AIE phenomenon has been found in many NCs, such as $Au_{22}(SR)_{18}$ [123], $[Cu_{14}(R/S\text{-}DPM)_8](PF_6)_6$ [124], $Au_4Ag_5(DPPM)_2(S\text{-}Adm)_6$ [125], $Au_2Cu_6(S\text{-}Adm)_6(PPh_3)_2$ [126], and $(Au_4Ag_{13}(DPPM)_3(SR)_9)$ crystallization-induced emission enhancement [127], but is rarely reported in Ag NCs. Zhu's group reported a novel type of AIE involving the restriction of the "dissociation–aggregation pattern" of ligands, and by using $Ag_{29}(BDT)_{12}(TPP)_4$ as a model to illustrate this phenomenon [120]. First, they dissolved $Ag_{29}(BDT)_{12}(TPP)_4$ clusters in N, N-dimethylformamide solution and then gradually added TPP ligand. With the progress of promoting the aggregation of TPP onto the easy-to-dissociate NC surface, the PL intensity and quantum yield were significantly enhanced (about 13-fold and 11-fold, respectively). Furthermore, with a decrease in temperature, the $Ag_{29}(BDT)_{12}(TPP)_4$ fluorescence intensity was gradually increased by 25-fold.

7.4.3 Catalytic Properties of Ag NCs

Understanding the chemical bonding and activity at the surface of NCs can benefit the design of their properties and applications, especially the design of catalysts. The relative instability of Ag NCs compared to Au NCs is detrimental. The ligands containing amine groups can enhance the stability of Ag NCs in the air. Compared with thiolates and phosphines, alkynyl is easier to be released from the NCs by thermochemical methods, so they are often used in studying catalytic reactions. The above-mentioned Ag_{141} structure has bulky and very rigid adamantyl ligands, resulting in low coverage on the surface of Ag_{141} and making interspace for the halogen atoms [34]. In addition, ligand exchange with water-soluble thiols leads to water-soluble NCs, which can be applied to the biological and catalytic reactions. The Suzuki group reported a mixed-valence (Ag^+/Ag^0) and POM-stabilized Ag cluster in 2021, which is composed to a $\{Ag_{27}\}^{17+}$ core surrounded by three C-shaped $\{Si_2W_{18}O_{66}\}$ units [128]. Similar to most Ag NCs, the low stability and redox reactions hinder the in-depth study of the intrinsic reactivity and catalytic properties. They found the special structure between Ag_{27} and POMs can convert H_2 molecules (under 1 atm) into electrons and protons efficiently, and these products are stored on the Ag NCs and POM frameworks, respectively. In this process, POMs not only help stabilize the Ag NCs structures but also provide an active site to promote the cleavage of H_2 and store the generated protons. This work is the first report of the H_2 reactivity of Ag NC supported by POMs, and is conducive to the future investigation of the catalytic activity of hybrid materials with silver NCs. Overall, the research on the catalytic performance of silver NCs is still quite limited compared to the rich work done on alloys and Au nanocluster materials. Due to their geometric and electronic structure characteristics, silver NCs are expected to gain more exciting progress in future research.

7.5 Conclusion and Perspectives

In this chapter, we have summarized the recent development of Ag NCs in three aspects: the synthesis, structures, and properties. In terms of the synthesis, many effective synthesis strategies have been reported, which can produce Ag NCs in high yields, such as the one-pot reduction, ligand exchange, and chemical etching. In addition, by combining some high-resolution analytical techniques such as UV–vis, ESI, X-ray, XPS, NMR, etc., the structures of Ag NCs have been analyzed at the atomic level. On the basis of these measurements, the precise structures and properties of silver NCs have been studied. Typically, most of the structures can be dissected into several basic building blocks, including the tetrahedron Ag_4, the decahedron Ag_7, the icosahedron Ag_{13}, and the fcc Ag_{14}. The PL, chirality, and catalytic properties are affected by the electronic structure, heteroatom doping, the protective group of the Ag NCs. Besides, the precise structures are of great help in studying the structure–property correlations. In the field of Ag NCs, although more and more structures have been reported, the research on their water-solubility is still limited and the application in catalysis calls for more efforts. One of the immediate needs in this field is the in-depth understanding of their structure- and size-dependent properties.

There are still many challenges for the development of silver nanoclusters as functional nanomaterials, such as the crystallization and characterization of Ag NCs. The current research is still relatively limited, and in-depth analysis of the relationship between their sizes/structures and properties at the atomic level calls for more efforts in future research. We believe that future research on Ag NCs will lead to more significant progress.

Acknowledgment

We acknowledge the financial support of the NSFC (21631001, 21871001), the Ministry of Education, and the University Synergy Innovation Program of Anhui Province (GXXT-2020-053).

References

1 Jin, R.C. (2010). Quantum sized, thiolate-protected gold nanoclusters. *Nanoscale* 2: 343–362.
2 Fernando, A., Weerawardene, K.L., Karimova, N.V., and Aikens, C.M. (2015). Quantum mechanical studies of large metal, metal oxide, and metal chalcogenide nanoparticles and clusters. *Chem. Rev.* 115: 6112–6216.
3 Yamazoe, S., Koyasu, K., and Tsukuda, T. (2014). Nonscalable oxidation catalysis of gold clusters. *Acc. Chem. Res.* 47: 816–824.
4 Zeng, S., Yong, K.-T., Roy, I. et al. (2011). A review on functionalized gold nanoparticles for biosensing applications. *Plasmonics* 6: 491–506.
5 Rosi, N.L. and Mirkin, C.A. (2005). Nanostructures in biodiagnostics. *Chem. Rev.* 105: 1547–1562.
6 Pei, Y. and Zeng, X.C. (2012). Investigating the structural evolution of thiolate protected gold clusters from first-principles. *Nanoscale* 4: 4054–4072.
7 Dolamic, I., Varnholt, B., and Bürgi, T. (2015). Chirality transfer from gold nanocluster to adsorbate evidenced by vibrational circular dichroism. *Nat. Commun.* 6: 7117.
8 Xu, Q., Kumar, S., Jin, S.S. et al. (2014). Chiral 38-gold-atom nanoclusters: synthesis and chiroptical properties. *Small* 10: 1008–1014.
9 Kang, X. and Zhu, M.Z. (2019). Tailoring the photoluminescence of atomically precise nanoclusters. *Chem. Soc. Rev.* 48: 2422.
10 Higaki, T., Li, Q., Zhou, M. et al. (2018). Toward the tailoring chemistry of metal nanoclusters for enhancing functionalities. *Acc. Chem. Res.* 51: 2764–2773.
11 Tian, S.B., Li, Y.-Z., Li, M.-B. et al. (2015). Structural isomerism in gold nanoparticles revealed by X-ray crystallography. *Nat. Commun.* 6: 8667.
12 Liu, C., Li, T., Li, G. et al. (2015). Observation of body-centered cubic gold nanocluster. *Angew. Chem., Int. Ed.* 54: 9826–9829.
13 Zhuang, S.L., Liao, L.W., Li, M.-B. et al. (2017). The Fcc structure isomerization in gold nanoclusters. *Nanoscale* 9: 14809–14813.
14 Zhou, M., Zeng, C.J., Li, Q. et al. (2019). Gold nanoclusters: bridging gold complexes and plasmonic nanoparticles in photophysical properties. *Nanomaterials* 9: 933.
15 Yang, J. and Jin, R.C. (2019). New advances in atomically precise silver nanoclusters. *ACS Materials Lett.* 1: 482–489.
16 Du, X.S., Chai, J.S., Yang, S. et al. (2019). Fusion growth patterns in atomically precise metal nanoclusters. *Nanoscale* 11: 19158–19165.
17 Bonacchi, S., Antonello, S., Dainese, T., and Maran, F. (2021). Atomically precise metal nanoclusters: novel building blocks for hierarchical structures. *Chem. Eur. J.* 27: 30–38.

18 Gan, Z.B., Lin, Y.J., Luo, L. et al. (2016). Fluorescent gold nanoclusters with interlocked staples and a fully thiolate-bound kernel. *Angew. Chem., Int. Ed.* 55: 11567–11571.

19 Chakraborty, I. and Pradeep, T. (2017). Atomically precise clusters of noble metals: emerging link between atoms and nanoparticles. *Chem. Rev.* 117: 8208–8271.

20 Aikens, C.M. (2018). Electronic and geometric structure, optical properties, and excited state behavior in atomically precise thiolate-stabilized noble metal nanoclusters. *Acc. Chem. Res.* 51: 3065–3073.

21 Jin, R.C., Zeng, C., Zhou, M.Z., and Chen, Y. (2016). Atomically precise colloidal metal nanoclusters and nanoparticles: fundamentals and opportunities. *Chem. Rev.* 116: 10346–10413.

22 Niihori, Y., Hossain, S., Kumar, B. et al. (2017). Perspective: exchange reactions in thiolate-protected metal clusters. *APL Mater.* 5: 053201.

23 Higaki, T., Zhou, M.Z., Lambright, K. et al. (2018). Sharp transition from nonmetallic Au_{246} to metallic Au_{279} with nascent surface plasmon resonance. *J. Am. Chem. Soc.* 140: 5691–5695.

24 Udayabhaskararao, T. and Pradeep, T. (2013). New protocols for the synthesis of stable Ag and Au nanocluster molecules. *J. Phys. Chem. Lett.* 4: 1553–1564.

25 Chen, T.K., Lin, H.B., Cao, Y.T. et al. (2022). Interactions of metal nanoclusters with light: fundamentals and applications. *Adv.Mater* 34: 2103918.

26 Qin, R.X., Liu, P.X., Fu, G., and Zheng, N.F. (2018). Strategies for stabilizing atomically dispersed metal catalysts. *Small Methods* 2: 1700286.

27 Liu, P.X., Qin, R.X., Fu, G., and Zheng, N.F. (2017). Surface coordination chemistry of metal nanomaterials. *J. Am. Chem. Soc.* 139: 2122–2131.

28 Campisi, S., Schiavoni, M., Chan-Thaw, C.E., and Villa, A. (2016). Untangling the role of the capping agent in nanocatalysis: recent advances and perspectives. *Catalysts* 6: 185.

29 Konishi, K. (2014). Phosphine-coordinated pure-gold clusters: diverse geometrical structures and unique optical properties/responses. *Struct Bond* 161: 49–86.

30 Negishi, Y., Iwai, T., and Ide, M. (2010). Continuous modulation of electronic structure of stable thiolate-protected Au_{25} cluster by Ag doping. *Chem. Commun.* 46: 4713–4715.

31 Mohanty, J.S., Xavier, P.L., Chaudhari, K. et al. (2012). Luminescent, bimetallic AuAg alloy quantum clusters in protein templates. *Nanoscale* 4: 4255–4262.

32 Liu, G.L., Feng, D.-Q., Zheng, W.J. et al. (2013). An anti-galvanic replacement reaction of DNA templated silver nanoclusters monitored by the light-scattering technique. *Chem. Commun.* 49: 7941–7943.

33 Hornstein, B.J. and Finke, R.G. (2004). Transition-metal nanocluster kinetic and mechanistic studies emphasizing nanocluster agglomeration: demonstration of a kinetic method that allows monitoring of all three phases of nanocluster formation and aging. *Chem. Mater.* 16: 139–150.

34 Yan, J.B., Teo, K., and Zheng, N.F. (2018). Surface chemistry of atomically precise coinage-metal nanoclusters: from structural control to surface reactivity and catalysis. *Acc. Chem. Res.* 51: 3084–3093.

35 Zheng, K.Y., Yuan, X., and Xie, J.P. (2017). Effect of ligand structure on the size control of mono- and bi-thiolate-protected silver nanoclusters. *Chem. Commun.* 53: 9697–9700.

36 Zeng, C.J., Chen, Y.X., Iida, K. et al. (2016). Gold quantum boxes: on the periodicities and the quantum confinement in the Au_{28}, Au_{36}, Au_{44}, and Au_{52} magic series. *J. Am. Chem. Soc.* 138: 3950–3953.

37 Bootharaju, M.S., Joshi, C.P., Parida, M.R. et al. (2016). Templated atom-precise galvanic synthesis and structure elucidation of a $(Ag_{24}Au[SR]_{18})-$ nanocluster. *Angew. Chem. Int. Ed.* 55: 922–926.

38 Yang, H.Y., Wang, Y., Huang, H.Q. et al. (2013). All-thiol-stabilized Ag_{44} and $Au_{12}Ag_{32}$ nanoparticles with single-crystal structures. *Nat. Commun.* 4: 2422.

39 Udayabhaskararao, T., Sun, Y., Goswami, N. et al. (2012). Ag_7Au_6: a 13-atom alloy quantum cluster. *Angew. Chem. Int. Ed.* 51: 2155–2159.

40 Yao, C.H., Lin, Y.-J., Yuan, J.Y. et al. (2015). Mono-cadmium vs mono-mercury doping of Au_{25} nanoclusters. *J. Am. Chem. Soc.* 137: 15350–15353.

41 Chen, Y.X., Zeng, C.J., Kauffman, D.R., and Jin, R.C. (2015). Tuning the magic size of atomically precise gold nanoclusters via isomeric Methylbenzenethiols. *Nano Lett.* 15: 3603–3609.

42 Krommenhoek, P.J., Wang, J.Q., Hentz, N. et al. (2012). Bulky adamantanethiolate and cyclohexanethiolate ligands favor smaller gold nanoparticles with altered discrete sizes. *ACS Nano.* 6: 4903–4911.

43 Yao, Q.F., Yuan, X., Yu, Y. et al. (2015). Introducing amphiphilicity to noble metal nanoclusters via phase-transfer driven ion-pairing reaction. *J. Am. Chem. Soc.* 137: 2128–2136.

44 Deng, C.L., Sun, C.F., Wang, Z. et al. (2020). A sodalite-type silver orthophosphate cluster in a globular silver nanocluster. *Angew. Chem. Int. Ed.* 59: 12659–12663.

45 Abbas, M.A., Kamat, P.V., and Bang, J.H. (2018). Thiolated gold nanoclusters for light energy conversion. *ACS Energy Lett.* 3: 840–854.

46 Wang, E.D., Xu, W.W., Zhu, B.E., and Gao, Y. (2021). Understanding the chemical insights of staple motifs of thiolate-protected gold nanoclusters. *Small* 17: 2001836.

47 Corrigan, B.J.F., Fuhr, O., and Fenske, D. (2009). Metal chalcogenide clusters on the border between molecules and materials. *Adv. Mater.* 21: 1867–1871.

48 Wang, Q.-M., Lin, Y.-M., and Liu, K.-G. (2015). Role of anions associated with the formation and properties of silver clusters. *Acc. Chem. Res.* 48: 1570–1579.

49 Sabin, F., Ryu, C.-K., Ford, P.-C., and Vogler, A. (1992). Photophysical properties of hexanuclear copper [I] and silver[I] clusters. *Inorg. Chem.* 31: 1941–1945.

50 Wang, Q.-M. and Mak, T.C.W. (2000). Novel honeycomb-like layered structure: the first isomorphous triple salts of silver acetylide. *J. Am. Chem. Soc.* 122: 7608–7609.

51 Joshi, C.P., Bootharaju, M.S., Alhilaly, M.J., and Bakr, O.M. (2015). $[Ag_{25}(SR)_{18}]^-$: the "golden" silver nanoparticle. *J. Am. Chem. Soc.* 137: 11578–11581.

52 AbdulHalim, L.G., Bootharaju, M.S., Tang, Q. et al. (2015). $Ag_{29}[BDT]_{12}[TPP]_4$: a tetravalent nanocluster. *J. Am. Chem. Soc.* 137: 11970–11975.

53 Desireddy, A., Conn, B.E., Guo, J. et al. (2013). Ultrastable silver nanoparticles. *Nature* 501: 399–402.

54 Du, W.J., Jin, S., Xiong, L. et al. (2017). $Ag_{50}(Dppm)_6(SR)_{30}$ and its homologue Au_xAg_{50-x} $(Dppm)_6(SR)_{30}$ alloy nanocluster: seeded growth, structure determination, and differences in properties. *J. Am. Chem. Soc.* 139: 1618–1624.

55 Alhilaly, M.J., Bootharaju, M.S., Joshi, C.P. et al. (2016). $[Ag_{67}(SPhMe_2)_{32}(PPh_3)_8]^{3+}$: synthesis, total structure, and optical properties of a large box-shaped silver nanocluster. *J. Am. Chem. Soc.* 138: 14727–14732.

56 Yang, H.Y., Wang, Y., Chen, L. et al. (2016). Plasmonic twinned silver nanoparticles with molecular precision. *Nat. Commun.* 7: 12809.

57 Wu, Z.K., Lanni, E., Chen, W.Q. et al. (2009). High yield, large scale synthesis of thiolate-protected Ag_7 clusters. *J. Am. Chem. Soc.* 131: 16672–16674.

58 Dhanalakshmi, L., Udayabhaskararao, T., and Pradeep, T. (2012). Conversion of double layer charge-stabilized Ag@citrate colloids to thiol passivated luminescent quantum clusters. *Chem. Commun.* 48: 859–861.

59 Yang, J., Xia, N., Wang, X.N. et al. (2015). One-pot one-cluster synthesis of fluorescent and bio-compatible Ag_{14} nanoclusters for cancer cell imaging. *Nanoscale* 7: 18464–18470.

60 Liu, X., Chen, J.S., Yuan, J.Y. et al. (2018). A silver nanocluster containing interstitial sulfur and unprecedented chemical bonds. *Angew. Chem., Int. Ed.* 57: 11273–11277.

61 Cao, Y.T., Guo, J.G., Shi, R. et al. (2018). Evolution of thiolate-stabilized Ag nanoclusters from Ag-thiolate cluster intermediates. *Nat. Commun.* 9: 2379.

62 Goswami, N., Yao, Q.F., Chen, T.K., and Xie, J.P. (2016). Mechanistic exploration and controlled synthesis of precise thiolate-gold nanoclusters. *Coord. Chem. Rev.* 329: 1–15.

63 Muhammed, M.A., Aldeek, F., Palui, G. et al. (2012). Growth of in situ functionalized luminescent silver nanoclusters by direct reduction and size focusing. *ACS Nano* 6: 8950–8961.

64 Adhikari, B. and Banerjee, A. (2010). Facile synthesis of water-soluble fluorescent silver nanoclusters and Hg^{II} sensing. *Chem. Mater.* 22: 4364–4371.

65 Meng, X.M., Xu, Q., Wang, S.X., and Zhu, M.Z. (2012). Ligand-exchange synthesis of selenophenolate-capped Au_{25} nanoclusters. *Nanoscale* 4: 4161–4165.

66 Chang, W.-T., Lee, P.-Y., Liao, J.-H. et al. (2017). Eight-electron silver and mixed gold/silver nanoclusters stabilized by selenium donor ligands. *Angew. Chem., Int. Ed.* 56: 10178–10182.

67 Kang, X., Huang, L., Liu, W. et al. (2019). Reversible nanocluster structure transformation between face-centered cubic and icosahedral isomers. *Chem. Sci.* 10: 8685–8693.

68 Zhou, M.M., Jin, S., Wei, X. et al. (2020). Reversible Cu—S motif transformation and Au_4 distortion via thiol ligand exchange engineering. *J. Phys. Chem. C.* 124: 7531–7538.

69 Bootharaju, M.S., Joshi, C.P., Alhilaly, M.J., and Bakr, O.M. (2016). Switching a nanocluster core from hollow to nonhollow. *Chem. Mater.* 28: 3292–3297.

70 Le Guével, X., Spies, C., Daum, N. et al. (2012). Highly fluorescent silver nanoclusters stabilized by glutathione: a promising fluorescent label for bioimaging. *Nano Res.* 5: 379–387.

71 Yuan, X., Setyawati, M.I., Tan, A.S. et al. (2013). Highly luminescent silver nanoclusters with tunable emissions: cyclic reduction-decomposition synthesis and antimicrobial properties. *NPG Asia Mater.* 5: e39.

72 Mrudula, K.V., Udayabhaskararao, T., and Pradeep, T. (2009). Interfacial synthesis of luminescent 7 kDa silver clusters. *J. Mater. Chem.* 19: 4335–4342.

73 Feng, Y., Yao, Q.F., Li, J.G. et al. (2016). Converting ultrafine silver nanoclusters to monodisperse silver sulfide nanoparticles via a reversible phase transfer protocol. *Nano Res.* 9: 942–950.

74 Yao, Q.F., Yuan, X., Fung, V. et al. (2017). Understanding seed-mediated growth of gold nanoclusters at molecular level. *Nat. Commun.* 8: 927.

75 Wang, Z.-Y., Wang, M.-Q., Li, Y.-L. et al. (2018). Atomically precise site-specific tailoring and directional assembly of superatomic silver nanoclusters. *J. Am. Chem. Soc.* 140: 1069–1076.

76 Piotrowski, M.J., Ungureanu, C.G., Tereshchuk, P. et al. (2016). Theoretical study of the structural, energetic, and electronic properties of 55-atom metal nanoclusters: a DFT investigation within van der Waals corrections, spin–orbit coupling, and PBE+U of 42 metal systems. *J. Phys. Chem. C.* 120: 28844–28856.

77 Dhayal, R.S., Liao, J.-H., Liu, Y.-C. et al. (2015). $[Ag_{21}\{S_2P(OiPr)_2\}_{12}]^+$: an eight-electron Superatom. *Angew. Chem. Int. Ed.* 54: 3702–3706.

78 Zou, X.J., Jin, S., Du, W.J. et al. (2017). Multi-ligand-directed synthesis of chiral silver nanoclusters. *Nanoscale* 9: 16800–16805.

79 Tian, F. and Chen, R. (2019). Pd-mediated synthesis of Ag_{33} chiral nanocluster with Core–Shell structure in T point group. *J. Am. Chem. Soc.* 141: 7107–7114.

80 Yang, H.Y., Lei, J., Wu, B.H. et al. (2013). Crystal structure of a luminescent thiolated Ag nanocluster with an octahedral Ag_6^{4+} core. *Chem. Commun.* 49: 300–302.

81 Shen, X.-T., Ma, X.-L., Ni, Q.-L. et al. (2018). $[Ag_{15}(N\text{-triphos})_4(Cl_4)](NO_3)_3$: a stable Ag—P superatom with eight electrons (N-triphos = tris((diphenylphosphino)methyl)amine). *Nanoscale* 10: 515–519.

82 Liu, C., Li, T., Abroshan, H. et al. (2018). Chiral Ag_{23} nanocluster with open shell electronic structure and helical face-centered cubic framework. *Nat. Commun.* 9: 744–749.

83 Yang, H.Y., Yan, J.Z., Wang, Y. et al. (2017). Embryonic growth of face-Center-cubic silver nanoclusters shaped in nearly perfect half-cubes and cubes. *J. Am. Chem. Soc.* 139: 31–34.

84 Bodiuzzaman, M., Ghosh, A., Sugi, K.S. et al. (2019). Camouflaging structural diversity: co-crystallization of two different nanoparticles having different cores but the same shell. *Angew. Chem. Int. Ed.* 58: 189–194.

85 Yuan, X.T., Sun, C.F., Li, X.H. et al. (2019). Combinatorial identification of hydrides in a ligated Ag_{40} nanocluster with noncompact metal Core. *J. Am. Chem. Soc.* 141: 11905–11911.

86 Jin, S., Wang, S.X., Song, Y.B. et al. (2014). Crystal structure and optical properties of the $[Ag_{62}S_{12}(SBut)_{32}]^{2+}$ nanocluster with a complete face-Centered cubic kernel. *J. Am. Chem. Soc.* 136: 15559–15565.

87 Kang, X. and Zhu, M.Z. (2019). Intra-cluster growth meets inter-cluster assembly: the molecular and supramolecular chemistry of atomically precise nanoclusters. *Coordination Chemistry Reviews* 394: 1–38.

88 Alamer, B.J., Bootharaju, M.S., Kozlov, S.M. et al. (2021). $[Ag_9(1,2\text{-}BDT)_6]^{3-}$: how square-pyramidal building blocks self-assemble into the smallest silver nanocluster. *Inorg. Chem.* 60: 4306–4312.

89 Mingos, D. (2015). Structural and bonding patterns in gold cluster. *Dalton Trans.* 44: 6680–6695.

90 Zhu, M.Z., Aikens, C.M., Hollander, F.J. et al. (2008). Correlating the crystal structure of a thiol-protected Au_{25} cluster and optical properties. *J. Am. Chem. Soc.* 130: 5883–5885.

91 Rambukwella, M., Burrage, S., Neubrander, M. et al. (2017). $Au_{38}(SPh)_{24}$: Au_{38} protected with aromatic thiolate ligands. *J. Phys. Chem. Lett.* 8: 1530–1537.

92 Jin, S., Zou, X.J., Xiong, L. et al. (2018). Bonding of two 8-electron Superatom clusters. *Angew. Chem. Int. Ed.* 57: 16768–16772.

93 Yuan, Q.Q., Kang, X., Hu, D.Q. et al. (2019). Metal synergistic effect on cluster optical properties: based on Ag_{25} series nanocluster. *Dalton Trans.* 48: 13190–13196.

94 Bodiuzzaman, M., Nag, A., Narayanan, P.R. et al. (2019). A covalently linked dimer of $[Ag_{25}(DMBT)_{18}]^-$. *Chem. Commun.* 55: 5025–5028.

95 Chakraborty, P., Nag, A., Paramasivam, G. et al. (2018). Fullerene-functionalized monolayer-protected silver clusters: $[Ag_{29}(BDT)_{12}(C_{60})_n]^{3-}$ (n = 1–9). *ACS Nano* 12: 2415–2425.

96 Bootharaju, M.S., Lee, S., Deng, G.H. et al. (2021). $Ag_{44}(EBT)_{26}(TPP)_4$ nanoclusters with tailored molecular and electronic structure. *Angew. Chem. Int. Ed.* 60: 9038–9044.

97 Jin, R.X., Liu, C., Zhao, S. et al. (2015). Tri-icosahedral gold nanocluster $[Au_{37}(PPh_3)_{10}(SC_2H_4Ph)_{10}X_2]^+$: linear assembly of icosahedral building blocks. *ACS Nano* 9: 8530–8536.

98 Yuan, S.-F., Xu, C.-Q., Liu, W.-D. et al. (2021). Rod-shaped silver supercluster unveiling strong electron coupling between substituent icosahedral units. *J. Am. Chem. Soc.* 143: 12261–12267.

99 Li, G., Lei, Z., and Wang, Q.-M. (2010). Luminescent molecular Ag-S nanocluster $[Ag_{62}S_{13}(SBut)_{32}](BF_4)_4$. *J. Am. Chem. Soc.* 132: 17678–17679.

100 Ma, X.Y., Bai, Y.Y., Song, Y.B. et al. (2020). A rhombicuboctahedral Ag_{100}: four-layered octahedral silver nanocluster adopting the Russian nesting doll model. *Angew. Chem. Int. Ed.* 59: 17234–17238.

101 Liu, J.-W., Wang, Z., Chai, Y.-M. et al. (2019). Core modulation of 70-nuclei core-shell silver nanoclusters. *Angew. Chem. Int. Ed.* 58: 6276–6279.

102 Song, Y.B., Lambright, K., Zhou, M. et al. (2018). Large-scale synthesis, crystal structure, and optical properties of the $Ag_{146}Br_2(SR)_{80}$ nanocluster. *ACS Nano* 12: 9318–9325.

103 Tan, Y.-L., Yang, L., Yu, T.-C. et al. (2021). Solvent-driven reversible transformation between electrically neutral thiolate protected Ag_{25} and Ag_{26} clusters. *Sci. China Chem.* 64: 948–952.

104 Ma, M.-X., Ma, X.-L., Liang, G.-M. et al. (2021). A nanocluster $[Ag_{307}Cl_{62}(SPh^tBu)_{110}]$: chloride intercalation, specific electronic state, and superstability. *J. Am. Chem. Soc.* 143: 13731–13737.

105 Zhu, Y.F., Guo, J., Qiu, X.Y. et al. (2021). Optical activity of chiral metal nanoclusters. *Acc. Mater. Res.* 2: 21–35.

106 Cao, Z.L., Gao, H., Qiu, M. et al. (2020). Chirality transfer from sub-nanometer biochemical molecules to sub-micrometer plasmonic metastructures: physiochemical mechanisms, biosensing, and bioimaging opportunities. *Adv. Mater.* 32: 1907151.

107 Li, Y.W., Higaki, T., Du, X.S., and Jin, R.C. (2020). Chirality and surface bonding correlation in atomically precise metal nanoclusters. *Adv. Mater.* 32: 1905488.

108 Schaaff, T.G., Knight, G., Shafigullin, M.N. et al. (1998). Isolation and selcted properties of a 10.4 kDa gold:glutathione cluster compound. *J. Phys. Chem. B* 102: 10643–10646.

109 Zhu, Y.F., Wang, H., Wan, K.W. et al. (2018). Enantioseparation of $Au_{20}(PP_3)_4Cl_4$ clusters with intrinsically chiral cores. *Angew. Chem. Int. Ed.* 57: 9059–9063.

110 Sakthivel, N.A., Sementa, L., Yoon, B. et al. (2020). Isomeric thiolate monolayer protected Au_{92} and Au_{102} Nanomolecules. *J. Phys. Chem. C.* 124: 1655–1666.

111 Dass, A., Theivendran, S., Nimmada, P.R. et al. (2015). $Au_{133}(SPh\text{-}tBu)_{52}$ nanomolecules: X-ray crystallography, optical, electrochemical, and theoretical analysis. *J. Am. Chem. Soc.* 137 (14): 4610–4613.

112 Zhou, M., Zeng, C.J., Song, Y.B. et al. (2017). On the non-metallicity of 2.2 nm $Au_{246}(SR)_{80}$ nanoclusters. *Angew. Chem. Int. Ed.* 56: 16257–16261.

113 Huang, J.-H., Wang, Z.-Y., Zang, S.-Q., and Mak, T.C.W. (2020). Spontaneous resolution of chiral multi-thiolate-protected Ag_{30} nanoclusters. *ACS Cent. Sci.* 6: 1971–1976.

114 Liu, W.-D., Wang, J.-Q., Yuan, S.-F. et al. (2021). Chiral superatomic nanoclusters Ag_{47} induced by the ligation of amino acids. *Angew. Chem. Int. Ed.* 60: 11430–11435.

115 Yang, H.Y., Yan, J.Z., Wang, Y. et al. (2017). From racemic metal nanoparticles to optically pure enantiomers in one pot. *J. Am. Chem. Soc.* 139: 16113–16116.

116 Yang, J. and Jin, R.C. (2021). Advances in enhancing luminescence of atomically precise ag nanoclusters. *J. Phys. Chem. C* 125: 2619–2625.

117 Kang, X., Zhou, M., Wang, S.X. et al. (2017). The tetrahedral structure and luminescence properties of Bi-metallic $Pt_1Ag_{28}(SR)_{18}(PPh_3)_4$ nanocluster. *Chem. Sci.* 8: 2581–2587.

118 Zeng, Y., Havenridge, H., Gharib, M. et al. (2021). Impact of ligands on structural and optical properties of Ag_{29} nanoclusters. *J. Am. Chem. Soc.* 143: 9405–9414.

119 Wu, Z.N., Yao, Q.F., Zang, S.-Q., and Xie, J.P. (2021). Aggregation-induced emission in luminescent metal nanoclusters. *Natl. Sci. Rev.* 8: 208.

120 Kang, X., Wang, S.X., and Zhu, M.Z. (2018). Observation of a new type of aggregation-induced emission in nanoclusters. *Chem. Sci.* 9: 3062–3068.

121 Bain, D., Maity, S., and Patra, A. (2020). Surface motifs regulated aggregation induced emission in gold–silver nanoclusters. *Chem. Commun.* 56: 9292–9295.

122 Ding, D., Li, K., Liu, B., and Tang, B.Z. (2013). Bioprobes based on AIE fluorogens. *Acc. Chem. Res.* 46: 2441–2453.

123 Pyo, K., Ly, N.H., Yoon, S.Y. et al. (2017). Highly luminescent folate-functionalized Au22 nanoclusters for bioimaging. *Advanced Healthcare Materials* 6: 1700203.

124 Li, Y.-L., Wang, J., Luo, P. et al. (2019). Cu_{14} cluster with partial cu(0) character: difference in electronic structure from isostructural silver Analog. *Adv. Sci.* 6: 1900833.

125 Jin, S., Liu, W., Hu, D.Q. et al. (2018). Aggregation-induced emission (AIE) in Ag–Au Bimetallic nanocluster. *Chem. Eur.J.* 24: 3712–3715.

126 Kang, X., Li, X.W., Yu, H.Z. et al. (2017). Modulating photo-luminescence of Au_2Cu_6 nanoclusters via ligand-engineering. *RSC Adv.* 7: 28606–28609.

127 Chen, T., Yang, S., Chai, J.S. et al. (2017). Crystallization-induced emission enhancement: a novel fluorescent Au-Ag bimetallic nanocluster with precise atomic structure. *Sci. Adv.* 3: e1700956.

128 Yonesato, K., Yamazoe, S., Yokogawa, D. et al. (2021). A molecular hybrid of an atomically precise silver nanocluster and polyoxometalates for H_2 cleavage into protons and electrons. *Angew. Chem. Int. Ed.* 60: 16994–16998.

8

Atomically Precise Copper Nanoclusters: Syntheses, Structures, and Properties

Chunwei Dong, Saidkhodzha Nematulloev, Peng Yuan, and Osman M. Bakr

KAUST Catalysis Center (KCC), Division of Physical Sciences and Engineering, King Abdullah University of Science and Technology (KAUST), Thuwal 23955-6900, Saudi Arabia

8.1 Introduction

The past decades have witnessed rapid progress in the development of ligand-protected coinage metal (copper, silver, gold) nanoclusters (NCs), which exhibit molecule-like properties, well-defined dimensions, and tunable compositions. Thanks to the atomically precise structures resolved by X-ray crystallography, they provide an ideal model for studying surfaces and building a correlation between structures and fundamental properties [1, 2]. According to the jellium model, an electronic shell model describing the delocalized valence electrons in a metal nanocluster, some metal NCs can be regarded as superatoms, and their electronic structures can be qualitatively described [3]. In this regard, each coinage metal atom with a valence electronic configurations of $nd^{10}(n+1)s^1$ can contribute one valence electron. A simple electron-counting scheme has been proposed that considers the electron transfer between the ligands and metal atoms. Generally, metal NCs with a "magic number" of electrons ($n^* = 2, 8, 18, 20, 34, 40, \ldots$) have high stability because of the closure of the electronic shells [4]. Therefore, the free electron count of a metal nanocluster plays a critical role in determining its stability and architecture.

Compared with gold and silver NCs, copper NCs often exhibit different but more complex geometrical motifs, thereby generating a potentially wider variety of nanocluster species [5]. However, the M^I/M^0 half-cell potential of copper (0.52 V) is lower than that of silver (0.80 V) and gold (1.68 V) [6], making it more difficult to synthesize and isolate copper NCs. Nevertheless, ligand-protected NCs based on copper have long been studied. For instance, tetranuclear cubane NCs of copper(I) halide $[Cu_4X_4L_4]$ (X=Cl, Br, I; L= N- or P-donor ligands), which display remarkably rich photoluminescence (PL) properties, have been extensively investigated since the 1960s [7, 8]. In addition, copper(I) chalcogenide NCs with a protective organic ligand shell have attracted much interest because of their rich structural variety [9, 10]. Over the past decade, copper hydride NCs have gained increasing attention because of their promising applications in various fields, such as catalysis, sensing, self-assembly, and bio-related chemistry [11]. In particular, copper NCs with copper(0) character, referred to as copper superatoms, have emerged as a new frontier in nanocluster chemistry and have become a highly pursued target. In this chapter, we focus on superatom-like copper NCs and copper(I) hydride NCs, including their synthesis, structural diversity, and properties.

Atomically Precise Nanochemistry, First Edition. Edited by Rongchao Jin and De-en Jiang.

8.2 Syntheses of Copper NCs

In this section, we focus on the synthesis of copper NCs with atomically precise structure. Such copper NCs are readily prepared by two main approaches (Figure 8.1): direct synthesis via the chemical reaction of cationic metal precursors (Cu^{2+} or Cu^{+}) and surface ligands (thiol, alkyne, or phosphine) and indirect synthesis, or nanocluster-to-nanocluster transformation, which is used to obtain novel copper-based NCs with different surface ligands, metal compositions, and dimensions. Although copper NCs have attracted increasing attention for many applications, scalable preparation remains a major problem hindering further development. Recently, pioneering studies into improving stability [12], enhancing crystallization [13], and developing high-yield reductants [14] have paved the way for large-scale preparation of copper NCs.

8.2.1 Direct Synthesis

In a typical direct synthesis of metal NCs, metal precursors are dissolved in a solvent and then chemically reduced by certain reducing agents in the presence of surface stabilizing agents. Discussed below are several key parameters for controlling the direct synthesis of metal NCs.

$$\text{Method 1: copper}\left(\text{II}\right) + \text{copper powder} + \text{ligands} - \text{copper}\left(\text{I}\right)\text{NCs}$$
$$\left(\text{a.k.a. the comproportionation reaction}\right)$$

The most common approach to generate copper(I) alkynyl complexes is the comproportionation reaction using a copper(II) salt and copper metallic powder. Inspired by copper-catalyzed alkyne–azide cycloaddition (CuAAC) reactions, where copper(I) alkynyl complexes/NCs have been proven to be key intermediates (e.g. $Cu_{14}(^{t}BuC{\equiv}C)_{14}$) [15, 16], Mak's group introduced the concept of the comproportionation reaction to the field of copper(I) alkynyl NCs [17, 18]. The authors found that the comproportionation reaction of copper(II) salt and copper powder in the presence of a terminal alkyne yielded two novel copper(I) NCs in high yield, $[Cu_{17}(^{t}BuC{\equiv}C)_{16}(CH_3OH)]^{+}$ and $[Cu_{18}(^{t}BuC{\equiv}C)_{16}(H_2O)_2]^{2+}$. Importantly, the coexisting anion played an important role in the progression of the total synthesis. Based on this interesting discovery, the authors added polyoxomolybdate to the reaction system and produced two high-nuclearity core–shell copper(I) NCs, $[Cu_{33}(^{t}BuC{\equiv}C)_{24}(Mo_4O_{16})]^{+}$ and $[Cu_{62}(^{t}BuC{\equiv}C)_{34}(Mo_5O_{19})_2(MoO_4)_2(CF_3COO)_2(OH)_4]^{2+}$ [17]. Subsequently, they succeeded in preparing three ethynediide/isopropylethynide-protected copper(I) NCs through

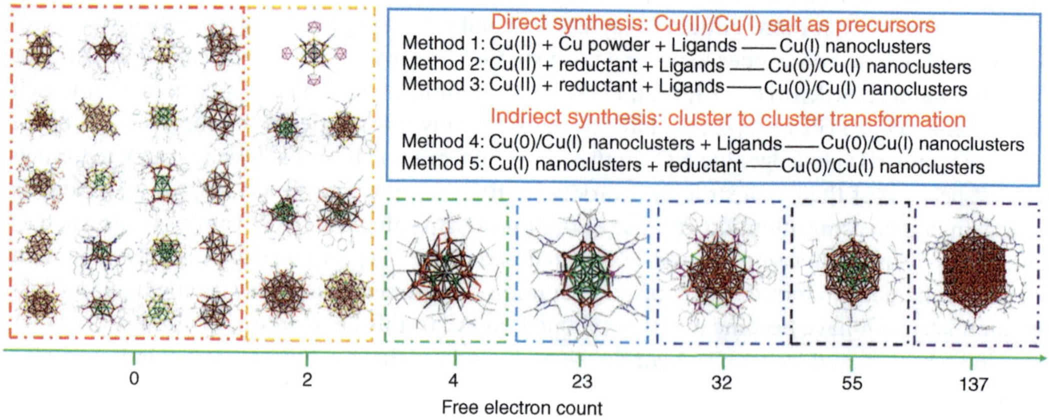

Figure 8.1 A summary of published copper NCs and the synthetic methods used to produce them.

temperature-mediated template release and the comproportionation reaction [18]. In addition, Sun's group used the same comproportionation-based synthetic route to obtain two air-stable copper(I) NCs, $[Cu^{I}_{15}(^{t}BuC{\equiv}C)_{10}(CF_3COO)_5]$ and $[Cu^{I}_{16}(^{t}BuC{\equiv}C)_{12}(CF_3COO)_4(CH_3OH)_2]$ [19]. These studies provide insight into the structural aspects of copper(I) alkynyl chemistry while creating a new synthetic paradigm. However, the reported comproportionation-based synthetic route can yield only copper(I) alkynyl NCs, which limits access to superatom-like copper NCs.

$$\text{Method 2}: \text{copper}(\text{II}) + \text{reductant} + \text{ligands} - \text{Cu}^0/\text{Cu}^+ \text{ NCs}$$

With the rapid development of organic ligand-protected metal NCs, increasing attention has been directed toward the synthesis of Cu^0/Cu^+ NCs. Hayton's group deeply explored the reaction of copper(II) salts with thiols by a comparative synthetic method and spectroscopic techniques. Results suggested that copper(0) is unlikely to be formed by the reaction of copper(II) salts with thiols [20]. To realize the goal of achieving Cu^0/Cu^+ NCs, strong reducing agents were introduced into the synthesis based on copper(II) precursors. Recently, Burgi's group reported a thiolate-protected copper sulfide nanocluster with the tentative composition $Cu_{74}S_{15}(PET)_{45}$ (PET = 2-phenylethanethiolate) by self-reducing a complex solution of $Cu(NO_3)_2$ and thiolate ligands [21]. Zheng's group used aqueous sodium borohydride ($NaBH_4$) to reduce the copper(II) precursor of $Cu(CH_3COO)_2$, resulting in the formation of $[Cu_{25}H_{10}(SPhCl_2)_{18}]^{3-}$ [22]. Unfortunately, the charge states of all copper atoms in $[Cu_{25}H_{10}(SPhCl_2)_{18}]^{3-}$ and $Cu_{74}S_{15}(PET)_{45}$ are +1. As researchers learned more about copper NCs, alkyne ligands were introduced into the reaction with copper(II) salts and reducing agents, because the interaction between alkyne ligands and copper(II) precursors may be different from that of thiols. Consequently, Zhang's group reported an alkynyl-protected copper nanocluster with copper(0) character, $Cu_{53}(C{\equiv}CPhPh)_9(dppp)_6Cl_3(NO_3)_9$ (dppp = 1,3-bis(diphenylphosphino)propane), via the reduction of a mixed solution of $Cu(NO_3)_2{\cdot}3H_2O$, alkyne, and dppp. By tracking the formation process, it was demonstrated that copper(0) was reduced from partial copper(I) species, rather than copper(II) precursors. The analysis of the valence variation from copper(II) to copper(I) and then to a mixture of copper(I) and copper(0) provided new insight into the formation of copper NCs [23].

$$\text{Method 3}: \text{copper}(\text{I}) + \text{reductant} + \text{ligands} - \text{copper}(\text{I})\text{NCs or Cu}^0/\text{Cu}^+ \text{ NCs}$$

Compared with copper(II), copper(I) precursors have been demonstrated to be more attractive during synthesis. Liu's group reported a series of chalcogenolate-protected copper hydride NCs by the reaction of a mixture of bidentate chalcogenolate ligands and reductants with $[Cu(CH_3CN)_4]PF_6$, $[Cu_{20}H_{11}(S_2P(O^iPr)_2)_9]$ [24], $[Cu_{20}H_{11}\{Se_2P(O^iPr)_2\}_9]$ [25], $[Cu_{28}H_{15}\{S_2CN^nPr_2\}_{12}]^+$ [26], and $[Cu_{32}H_{20}\{S_2P(O^iPr)_2\}_{12}]$ [27]. In addition, Bakr's group has developed a synthetic strategy for various copper NCs. In brief, $[Cu(CH_3CN)_4]BF_4$ is dissolved in a mixture of acetonitrile and chloroform, followed by the addition of different thiolate ligands (such as 2-phenylethanethiol, PhSeH, and PhSH) and triphenylphosphine (PPh_3). Reductant such as $NaBH_4$ is added quickly to achieve a dark-red crude solution, yielding copper(I) NCs, $[Cu_{15}(PPh_3)_6(PET)_{13}]^{2+}$ [28], $[Cu_{23}(PhSe)_{16}(PPh_3)_8H_6]^+$ [29], $[Cu_{36}H_{10}(PET)_{24}(PPh_3)_6Cl_2]$ [13], and $[Cu_{81}(PhS)_{46}(^tBuNH_2)_{10}H_{32}]^{3+}$ [30].

Generally, crystallization is the simplest and most effective strategy to purify the product of metal NCs. Therefore, 2-phenylethanethiol and PPh_3 have been used as protective ligands, because their relatively rigid and bulky structure can be conducive to the crystallization of NCs. The scalable synthesis of $[Cu_{36}H_{10}(PET)_{24}(PPh_3)_6Cl_2]$ (~1.5 g of pure crystals was obtained through a one-pot synthesis) with a moderate yield (~47% based on copper) paves the way for studying the basic physicochemical properties of copper NCs in detail and exploring their potential applications [13].

Subsequently, Hyeon's group developed a diamine-assisted synthetic strategy to synthesize copper nanocluster, $[Cu_{32}(PET)_{24}H_8Cl_2]^{2+}$. The copper(I) source and tetramethylethylenediamine are crucial for the formation of the nanocluster [31]. Zhu's group also explored the ligand effect (RSeH vs. RSH) in forming copper(I) NCs. Using a copper(I) precursor, two copper(I) NCs, $[Cu_{13}(SePh)_{13}(PPh_3)_4]$ and $[Cu_8(SPh)_8(PPh_3)_4]$, were synthesized and structurally characterized [32]. The ligand effect can also greatly influence the stability of copper NCs, which may be very closely related to the product yield. Besides, the same group achieved gram-level production of stable NCs, $Cu_{25}H_{22}((p\text{-}FPh)_3P)_{12}$, under ambient conditions using electron-withdrawing ligands. These simple synthetic strategies lay the foundation for large-scale synthesis and the future exploration of copper NCs [12].

Amid increasing reports of copper(I) NCs, the burgeoning of copper(0) NCs truly began in 2015. Based on previous work, Hayton's group believed that the failure to produce a copper(0) nanocluster could be rationalized based on the higher stability of copper(I) hydrides with excess ligands. The authors hypothesized that reducing a copper(I) salt in a ligand-deficient environment would result in the formation of unstable Cu_xH_x oligomers that would be more amenable to nanocluster growth [33, 34]. Ultimately, they created two copper NCs with partial copper(0) character, $[Cu_{20}(PhC{\equiv}C)_{12}(CH_3COO)_6]$ and $[Cu_{25}H_{22}(PPh_3)_{12}]^+$, both of which were two-electron copper superatoms. Another representative example of a superatom-like copper nanocluster is $[Cu_{61}(S^tBu)_{26}S_6Cl_6H_{14}]^+$ [35]. During the synthesis, a chloroform solution of Cu(I)–S^tBu was reduced by a mild reducing agent, borane *tert*-butylamine complex. They also performed a control experiment with a stronger reducing agent to elucidate the crucial role of the mild reducing agent in this synthesis. It was claimed that NCs synthesized with borane *tert*-butylamine complex were more stable than ones synthesized via $NaBH_4$.

8.2.2 Indirect Synthesis: Nanocluster-to-Nanocluster Transformation

The synthesis of metal NCs based on transformation chemistry has been widely used to prepare gold and silver NCs. Using structurally characterized metal NCs as the precursors would lead to unanticipated NCs with transformed morphology or structure and size. The original structures can be transformed into novel NCs by ligand exchange, using a stronger reductant, and/or foreign metal doping. The transformation chemistry of metal NCs is rapidly maturing in nanocluster science, providing us with rich opportunities to explore novel structures and potential applications.

$$\text{Method 4}: Cu^0/Cu^+\ \text{NCs} + \text{ligands} - Cu^0/Cu^+\ \text{NCs}$$

$[(PPh_3)CuH]_6$ is the most famous and earliest copper hydride nanocluster to be widely applied in catalysis and nanocluster-to-nanocluster transformation [36]. The addition of 26 equiv. of 1,10-phenanthroline (phen) to $[(PPh_3)CuH]_6$ in dichloroethane resulted in the immediate formation of a dark-red solution yielding another novel copper nanocluster $[Cu_{14}H_{12}(phen)_6(PPh_3)_4]^{2+}$ [37]. Hayton's group cleanly isolated the novel two-electron superatom-like copper nanocluster $[Cu_{29}Cl_4H_{22}(Ph_2phen)_{12}]^+$ (Ph_2phen = 4,7-diphenyl-1,10-phenanthroline) from the smaller two-electron superatom system $[Cu_{25}H_{22}(PPh_3)_{12}]^+$. Surprisingly, the yield of Cu_{29} reached 84%. The formation of nanocluster from Cu_{25} demonstrates that atomically precise copper NCs can be used as precursors to generate larger NCs that retain the fundamental electronic and bonding properties of the original NCs [38].

Liu's group also reported the reaction of terminal alkynes with $[Cu_{28}H_{15}(S_2CN^nBu_2)_{12}]^+$ to yield superatom-like NCs. A 10-fold excess of methyl propiolate was added to a suspension of $[Cu_{28}H_{15}(S_2CN^nBu_2)_{12}]^+$, and the reaction mixture was stirred at 30 °C for 24 hours. The color of the suspension changed from orange to deep red, and a dark-green solid deposited from the solution. Workup of this mixture led to the isolation of $[Cu_{13}\{S_2CN^nBu_2\}_6\{C{\equiv}CC(O)OMe\}_4]^+$ as a dark-red solid in 73% yield [39]. The replacement of methyl propiolate with phenylacetylene led to the isolation of $[Cu_{15}H_2\{S_2CN^nBu_2\}_6\{PhC{\equiv}C\}_6]^+$ as an orange-yellow solid in 20% yield [40]. Both of the resulting NCs have two free electrons and can be regarded as two-electron superatom systems.

$$\text{Method 5: copper}(\text{I})\text{NCs} + \text{reductant} - \text{Cu}^0/\text{Cu}^+\ \text{NCs}$$

Yuan et al. reported an in situ continuous reduction synthetic method for producing a high-nuclearity Cu^0/Cu^+ nanocluster, $[Cu_{53}(CF_3COO)_{10}(^tBuC{\equiv}C)_{20}Cl_2H_{18}]^+$. The high-nuclearity Cu_{53} nanocluster was controlled by the combination of different reducing agents, copper powder, and Ph_2SiH_2. Copper powder reduced the copper(II) complexes through comproportionation reaction to yield intermediate copper(I) NCs, $Cu_{15}(^tBuC{\equiv}C)_{10}(CF_3COO)_5$ and $Cu_{16}(^tBuC{\equiv}C)_{12}(CF_3COO)_4(CH_3OH)_2$, which were subsequently reduced by Ph_2SiH_2 to produce high-nuclearity Cu_{53} [41]. Sun's group later used the same strategy to form the air- and moisture-stable nanocluster $[Cu_{23}(^tBuC{\equiv}C)_{13}(CF_3COO)_6]$ [42]. The continuous reduction of copper(II) by copper powder and Ph_2SiH_2 makes it more convenient to achieve high-nuclearity Cu^0/Cu^+ NCs.

8.3 Structures of Copper NCs

In contrast to gold and silver NCs, the presence of hydrides, the smallest closed-shell anion, is prevalent in copper NCs [43], which appears to be related to the electronegativity of copper. Indeed, hydrides are considered to play an indispensable role in constructing and stabilizing the structure of copper NCs. However, the accommodation of hydrides within the copper skeleton would result in distinct structural diversity due to multiple coordination modes, high reactivity, and fluxionality. Therefore, in this section, copper NCs are discussed based on the electron count and hydrides in copper NCs. Superatom-like copper NCs are referred to as copper NCs with mixed Cu^0/Cu^+ character, whereas copper(I) hydride NCs involve only copper(I).

8.3.1 Superatom-like Copper NCs without Hydrides

To date, in most reported crystal structures, the copper atom is predominantly in the valence state of +1, and copper NCs with metallic character are significantly scarcer due to their much higher susceptibility to oxidation. Only a handful of superatom-like copper NCs without hydrides have been reported.

In 2016, Liu's group reported a two-electron superatom without hydrides, $[Cu_{13}\{S_2CN(^nBu)_2\}_6(\text{acetylide})_4]^+$ [39]. This was the first structurally characterized copper nanocluster with a centered cuboctahedral Cu_{13} core, a model of the bulk copper *fcc* structure. Based on the total composition, there are two free electrons; thus, the nanocluster could be considered an $N^* = 2$ superatom with a $1S^2$ closed-shell configuration. The Cu_{13} skeleton is capped by acetylide groups on four of the eight triangular faces, and each of the six square faces of the cuboctahedron is bridged by a dithiocarbamate ligand. Four of the eight triangular faces of the cuboctahedron are capped by acetylide groups. Each of the six square faces is bridged by a dithiolate ligand in a μ_2, μ_2 fashion, which leads to a truncated tetrahedron of 12 sulfur atoms (Figure 8.2a–c). Later, the same group reported a series of Cu_{13} and Cu_{12} NCs with similar compositions and core structures. The center copper atom in the cuboctahedral Cu_{13} core could be replaced with S, Cl, or Br, leading to the formation of inverse-coordination NCs [47]. Subsequently, silver and gold atoms were introduced into Cu_{13} host NCs as well. The silver and gold dopants were demonstrated to preferentially occupy the center of the cuboctahedral core [48].

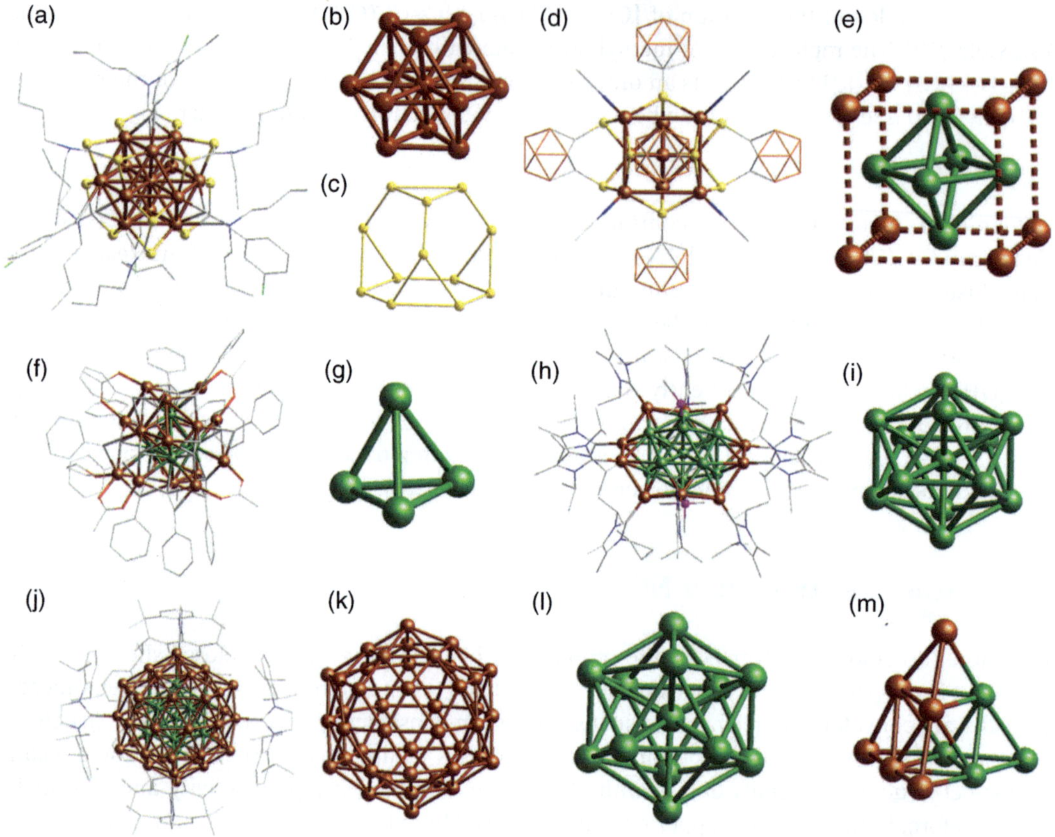

Figure 8.2 Structures of copper superatoms without hydrides. (a–c) Structure of $[Cu_{13}\{S_2CN(^nBu)_2\}_6(acetylide)_4]^+$ (a: the total structure; b: Cu_{13} cuboctahedron; c: S_{12} truncated tetrahedron) [39]. (d, e) Structure of $Cu_{14}(C_2B_{10}H_{10}S_2)_6(CH_3CN)_8$ (d: the total structure; e: Cu_{14} core) [44]. (f, g) Structure of $[Cu_{20}(PhC{\equiv}C)_{12}(OAc)_6]$ (f: the total structure; g: Cu_4 tetrahedron) [34]. (h, i) Structure of $[(Me_2I^iPr)_{10}Cu_{23}(PMe_3)_2]$ (h: the total structure; i: Cu_{13} icosahedron) [45]. (j–m) Structure of $[(IDipp)_6Cu_{55}]$ (j: the total structure; k: Cu_{42} shell; l: Cu_{13} cuboctahedron; m: Cu_{10} subunit of the copper skeleton) [46]. Brown (shell)/green (core): Cu; yellow: S; blue: N; pink: P; bright green: F; red: O; orange: B; gray: C.

Very recently, Zang's group reported another copper nanocluster with partial copper(0) character, $Cu_{14}(C_2B_{10}H_{10}S_2)_6(CH_3CN)_8$ (Figure 8.2d) [44], which is a two-electron superatom without hydrides. This nanocluster has an exact silver analog in terms of composition and structure of the metal framework, where the 14 copper atoms form a *fcc* Cu_{14} core with a cubic geometry. The six faces of the core are capped by bidentate 1,2-dithiolate-o-carborane ligands, whereas the eight vertices are connected to acetonitrile ligands (Figure 8.2e). This work also provides a model platform for understanding the fundamental differences between copper NCs and other coinage metal (silver and gold) NCs.

By using phenylacetylene ligands, Hayton's group reported the synthesis of a mixed-valent nanocluster $[Cu_{20}(PhC{\equiv}C)_{12}(OAc)_6]$ [34]. According to the electron count, the nanocluster can be considered a two-electron superatom system. A unique tetrahedral Cu_4 core in this nanocluster was first observed for copper-based superatoms. The tetrahedral Cu_4 core is further trapped inside the $[Cu_{16}(PhC{\equiv}C)_{12}(OAc)_6]^{2-}$ shell, which has an overall tetrahedral arrangement (Figure 8.2f, g).

Another example of a superatom-like copper nanocluster without hydrides is $[Cu_{23}(^tBuC{\equiv}C)_{13}(CF_3COO)_6]$, composing a Cu_4 tetrahedral core encircled by a Cu_{19} shell [42]. The nanocluster can be described as a four-electron superatom with a $1S^2 1P^2$ electronic shell. In addition, Zhang's group reported a high-nuclearity copper nanocluster with 32 free electrons, $Cu_{53}(C{\equiv}CPhPh)_9(dppp)_6Cl_3(NO_3)_9$. Notably, 41 copper atoms in the nanocluster adopt a close-packed configuration with an ABABC stacking sequence, which can be regarded as having a mixed *fcc* and *hcp* structure [23].

Recently, the reductive decomposition of copper(I) boryl complexes, the central intermediates in borylation reactions by copper catalysts with diboranes, has been reported to be an efficient strategy for the synthesis of copper NCs. Due to their poor stability, the complexes are easily decomposed to form elemental copper in solution. In this context, low-valence copper NCs without hydrides as the decomposition products can be obtained. In 2018, Kleeberg and Borner isolated a new nanocluster with a fascinating structure, $[(Me_2I^iPr)_{10}Cu_{23}(PMe_3)_2]$ (Me_2I^iPr = 1,3-bis(isopropyl)-imidazol-4,5-dimethyl-2-ylidene), during the decomposition of an N-heterocyclic carbene (NHC) copper(I) boryl complex [45]. The nanocluster comprises 23 copper atoms and 12 neutral ligands (10 NHCs and 2 phosphines); hence, all the copper atoms should be zero. As shown in Figure 8.2h, i, the nanocluster bears a centered icosahedral Cu_{13} core, where two apical copper atoms are coordinated by two phosphine ligands and 10 equatorial trigonal faces are capped by another 10 copper atoms, each of which is further connected to one NHC ligand. Very recently, the same group reported another two unprecedent copper NCs protected by neutral NHC ligands [46], $[(IDipp)_6Cu_{55}]$ and $[(IDipp)_{12}Cu_{179}]$ (IDipp = 1,3-bis(2,6-diisopropylphenyl)imidazol-2-ylidene). Similarly to the reported heterometallic nanocluster $[Cu_{43}Al_{12}](Cp^*)_{12}$ ($Cp^* = \eta^5\text{-}C_5Me_5$) [49], $[(IDipp)_6Cu_{55}]$ features a Mackay-type icosahedral structure. The nanocluster could be dissected into a two-shell icosahedron: an inner-centered icosahedral Cu_{13} core and an outer Cu_{42} shell. The outer shell comprises a Cu_{12} icosahedron with 30 additional copper atoms in the middle of all icosahedron edges (Figure 8.2j–m). In addition to $[(IDipp)_6Cu_{55}]$, $[(IDipp)_{12}Cu_{179}]$ was isolated via the decomposition of the same complex. $[(IDipp)_{12}Cu_{179}]$ is reported to be the largest copper nanocluster formed to date. The core of $[(IDipp)_{12}Cu_{179}]$ has the same icosahedral motif found in $[(IDipp)_6Cu_{55}]$. Hence, $[(IDipp)_{12}Cu_{179}]$ could be described as an extension of $[(IDipp)_6Cu_{55}]$. Despite the fascinating structures, these copper NCs have not been fully characterized, and the presence of additional hydrides in the NCs cannot be excluded. In addition, the method used to synthesize these structures suffers from poor reproducibility and extremely low yield, and the ambient stability of the NCs has not been studied.

8.3.2 Superatom-like Copper NCs with Hydrides

The first known copper nanocluster with mixed-valent superatom-like character, $[Cu_{25}H_{22}(PPh_3)_{12}]^+$, was reported by Hayton's group in 2015 [33]. The nanocluster features a distorted icosahedral Cu_{13} core, which is connected to four triangular $[Cu(PPh_3)]_3$ motifs forming a tetrahedral arrangement (Figure 8.3a). The Cu_{13} core is virtually isostructural to the known M_{13} core in known and well-characterized monolayer-protected gold and silver NCs, such as in the $Au_{25}(SR)_{18}$ nanocluster. Based on the composition charge, the electron count for the nanocluster gives a value of 2; thus, the nanocluster could be described as a two-electron superatom with a closed-shell $1S^2$ configuration. One year later, the same group reported that this nanocluster could be transformed into another nanocluster, $[Cu_{29}Cl_4H_{22}(Ph_2phen)_{12}]^+$ [38]. Like its precursor nanocluster ($[Cu_{25}H_{22}(PPh_3)_{12}]^+$), the new nanocluster features an icosahedral Cu_{13} core with two-electron superatom character. The Cu_{13} core is connected to four triangular $[Cu_4(Ph_2phen)_3Cl]$ motifs in a tetrahedral arrangement (Figure 8.3b). Therefore, the 29 metal atoms in $[Cu_{29}Cl_4H_{22}(Ph_2phen)_{12}]^+$ adopt spatial arrangements nearly identical to those in the reported $[Ag_{29}(BDT)_{12}(PPh_3)_4]^{3-}$, despite the great difference in ligands.

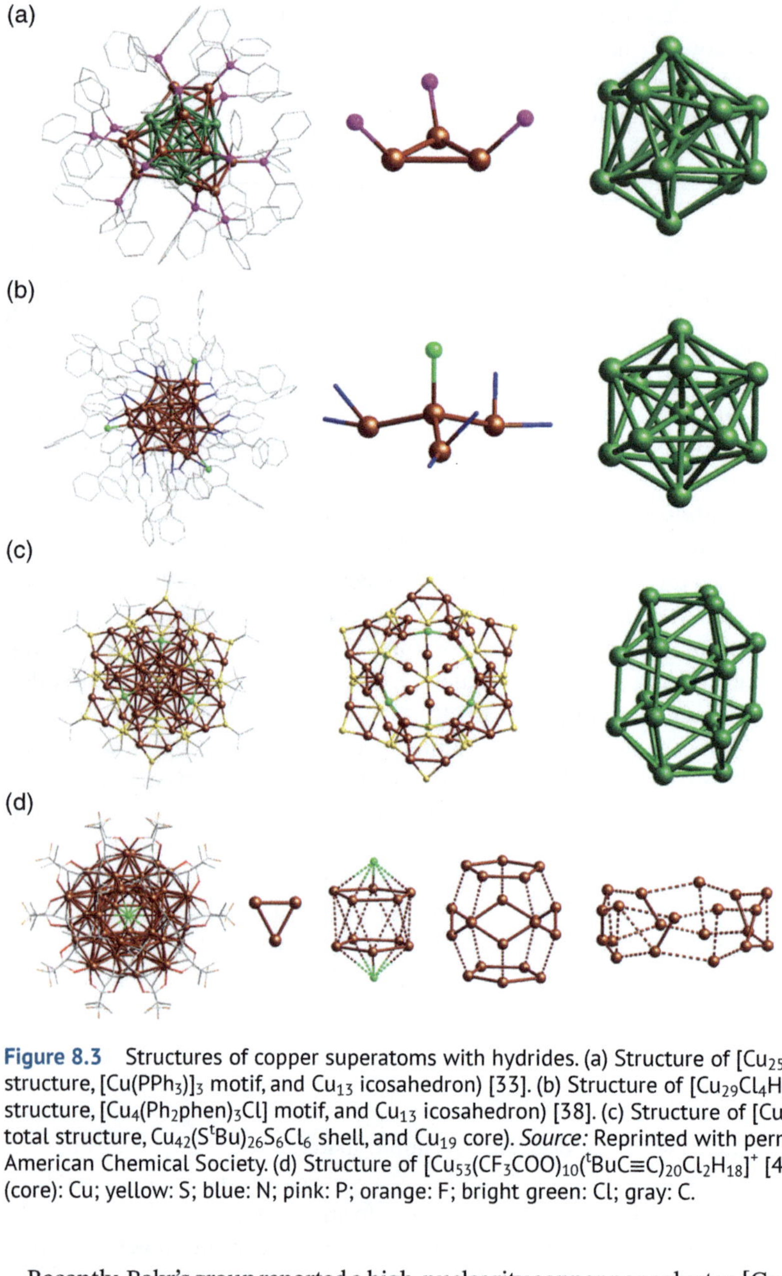

Figure 8.3 Structures of copper superatoms with hydrides. (a) Structure of $[Cu_{25}H_{22}(PPh_3)_{12}]^+$ (the total structure, $[Cu(PPh_3)]_3$ motif, and Cu_{13} icosahedron) [33]. (b) Structure of $[Cu_{29}Cl_4H_{22}(Ph_2phen)_{12}]^+$ (the total structure, $[Cu_4(Ph_2phen)_3Cl]$ motif, and Cu_{13} icosahedron) [38]. (c) Structure of $[Cu_{61}(S^tBu)_{26}S_6Cl_6H_{14}]^+$ (the total structure, $Cu_{42}(S^tBu)_{26}S_6Cl_6$ shell, and Cu_{19} core). *Source:* Reprinted with permission from [35]. © 2019 American Chemical Society. (d) Structure of $[Cu_{53}(CF_3COO)_{10}(^tBuC{\equiv}C)_{20}Cl_2H_{18}]^+$ [41]. Brown (shell)/green (core): Cu; yellow: S; blue: N; pink: P; orange: F; bright green: Cl; gray: C.

Recently, Bakr's group reported a high-nuclearity copper nanocluster, $[Cu_{61}(S^tBu)_{26}S_6Cl_6H_{14}]^+$ [35]. The nanocluster features an unprecedented quasi-elongated triangular gyrobicupola (quasi-J_{36}, J_{36} = Johnson solid) Cu_{19} core, which is trapped in a giant $Cu_{42}(S^tBu)_{26}S_6Cl_6$ shell (Figure 8.3c). Because of the strong Cu—S bond in the shell and good stability, the nanocluster yields a single molecular ion peak in its mass spectrum. The electron count of the nanocluster is 2, and the nanocluster can be considered a two-electron superatom system.

In addition to thiolates and phosphines, alkynes have been demonstrated to be efficient ligands for copper superatoms. In addition, Yuan et al. reported an ether-soluble copper nanocluster, $[Cu_{53}(CF_3COO)_{10}(^tBuC{\equiv}C)_{20}Cl_2H_{18}]^+$, which is another two-electron superatom protected by alkyne ligands [41]. The nanocluster could be depicted as four concentric shells, $Cu_3@Cu_{10}Cl_2@Cu_{20}@Cu_{20}$

(Figure 8.3d). Around the triangular Cu_3 core, 10 copper atoms and 2 Cl atoms form a $Cu_{10}Cl_2$ icosahedron. The second shell of the Cu_{20} dodecahedron is further trapped inside a Cu_{20} nanowheel.

A series of copper NCs, $[Cu_{15}H_2\{S_2CNR_2\}_6\{PhC{\equiv}C\}_6]^+$ (R=^{n}Bu, ^{n}Pr, ^{i}Bu), were recently isolated by Liu's group [40]. Regardless of the ligand, the NCs share a common structure, where a linear CuH_2 unit is encapsulated in a bicapped icosahedral Cu_{14} cage. According to the composition, there are two electrons in each nanocluster in addition to two hydrides. Because the nanocluster could be spontaneously converted to a two-electron superatom $[Cu_{13}(S_2CNR_2)_6(PhC{\equiv}C)_4]^+$ in solution, they are considered to be active intermediates in the formation of latter nanocluster. Notably, the hydride positions in the nanocluster were determined by neutron diffraction, which indicated that each hydride is located within a trigonal-bipyramidal cavity with the μ_5 coordination mode.

8.3.3 Copper(I) Hydride NCs

8.3.3.1 Determination of Hydrides

Determining the location of hydrides could aid in understanding the structure of copper NCs and developing potential applications. In the early days, the hydrides in certain copper complexes with low nuclearity were identified via X-ray crystallography, where μ_2-H and μ_3-H were the most common coordination modes [11]. However, it is notoriously difficult to identify hydrides in high-nuclearity copper NCs by X-ray crystallography due to the weak scattering of X-rays by hydrides. In this context, neutron diffraction experiments have been employed to unequivocally determine the assignment and location of hydrides in copper NCs, revealing more coordination modes for hydrides (μ_3-H, μ_4-H, μ_5-H, and μ_6-H). However, because neutron diffraction has stringent requirements for the quality and size of single crystals, only a few of the reported copper NCs have been characterized by this technique.

Through the continuous efforts of scientists, other advanced technologies, including electrospray ionization mass spectroscopy (ESI-MS), nuclear magnetic resonance (NMR) spectroscopy, and density functional theory (DFT) calculations, have been applied in solving the puzzle of hydrides. For high-nuclearity copper NCs, a systematic strategy for determining the positions of hydrides has been widely accepted. Generally, ESI-MS and ^{1}H/^{2}H NMR experiments are used to determine the precise composition of copper NCs, including the number of hydrides and different coordination environments [22]. The single crystals of copper NCs have subsequently been characterized by low-temperature X-ray crystallography, providing possible positions of hydrides. Based on the simplified model, theoretical optimization can be carried out to verify the rationality of each structure.

However, the initial position of hydrides determined by low-temperature X-ray crystallography is only available for some of the reported copper NCs, which is related to the symmetry of the copper NCs, the quality of the single crystals, and the instruments used. Today, predicting the location of hydrides in copper NCs by machine learning represents a promising research direction. A recently developed deep-learning method showed great potential in accelerating the determination of hydride sites in copper NCs [50, 51].

8.3.3.2 Copper(I) Hydride NCs Determined by Single-Crystal Neutron Diffraction

In 1989, to reveal the exact positions of the elusive hydride ligands, Stevens et al. performed single-crystal neutron diffraction experiments on a hexameric $[HCuP(p\text{-tolyl})_3]_6$ copper nanocluster [36]. The result was consistent with previous low-temperature X-ray diffraction studies and revealed that the six μ_3-bridging hydrides in the nanocluster are located along the six small faces in the distorted octahedron (Figure 8.4a). Over the past decade, Liu's group has expanded this research to a

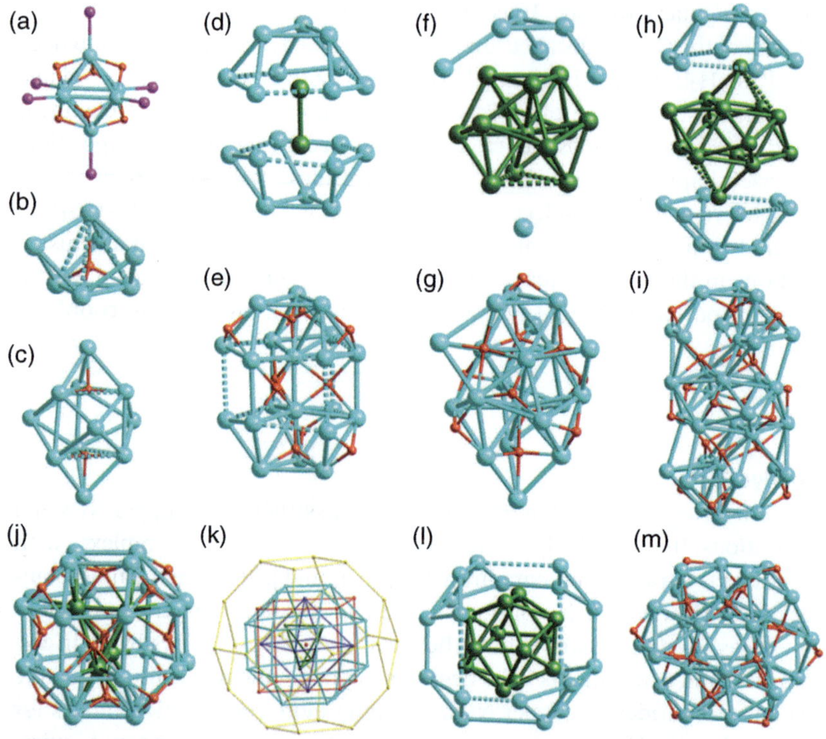

Figure 8.4 Structures of copper(I) hydride NCs determined by single-crystal neutron diffraction. (a) Structure of $[HCuP(p\text{-tolyl})_3]_6$ [36]. (b) Structure of $[Cu_7H\{S_2CR\}_6]$ [52]. (c) Structure of $[Cu_{11}H_2\{S_2P(O^iPr)_2\}_6(PhC{\equiv}C)_3]$ [53]. (d, e) Structure of $[Cu_{20}H_{11}(S_2P(O^iPr)_2)_9]$ (d: Cu_{20} skeleton; e: $Cu_{20}H_{11}$ skeleton) [54]. (f, g) Structure of $[Cu_{20}H_{11}\{Se_2P(O^iPr)_2\}_9]$ (f: Cu_{20} skeleton; g: $Cu_{20}H_{11}$ skeleton) [25]. (h, i) Structure of $[Cu_{32}H_{20}\{S_2P(O^iPr)_2\}_{12}]$ (h: Cu_{32} skeleton; i: $Cu_{32}H_{20}$ skeleton) [27]. (j, k) Structure of $[Cu_{28}H_{15}\{S_2CN^nPr_2\}_{12}]^+$ (j: Cu_{28} skeleton; k: $Cu_{28}H_{15}S_{24}$ skeleton) [26]. (l, m) Structure of $[Cu_{30}H_{18}\{S_2P(O^nPr)_2\}_{12}]$ (l: Cu_{30} skeleton; m: $Cu_{30}H_{18}$ skeleton) [55]. Turquoise (shell)/green (core): Cu; blue/red: hydride; yellow: S; pink: P.

diversity of copper NCs protected by bidentate chalcogenolate (S, Se) ligands, which proved highly efficient in the stabilization of copper NCs. Significant progress has been made in determining the hydrides in the copper NCs through neutron diffraction studies. In the following section, we focus on neutron diffraction studies of copper NCs recently performed by Liu's group.

In 2009, a series of octanuclear $[Cu_8HL_6]^+$ NCs were reported for the first time [56, 57]. The framework of the NCs features a tetracapped tetrahedral Cu_8 cage, which is further inscribed in a S_{12} icosahedron constituted by bidentate ligands. A hydride with a μ_4 coordination mode was determined by X-ray diffraction to be located at the center of a tetrahedron. Later, the tetracapped tetrahedral nanocluster $[Cu_8H\{S_2CR\}_6]^+$ ($R{=}N^nPr_2$, NEt_2, aza-15-crown-5) was demonstrated to be converted to a neutral tricapped tetrahedral $[Cu_7H\{S_2CR\}_6]$ nanocluster. Single-crystal neutron diffraction analysis indicated that the central μ_4-coordinated hydride occupies a central tetrahedral site in the heptanuclear nanocluster (Figure 8.4b) [52].

In 2013, an air- and moisture-stable copper nanocluster protected by dithiophosphate ligands, $[Cu_{20}H_{11}(S_2P(O^iPr)_2)_9]$, was successfully isolated [24]. The copper skeleton in the nanocluster displays an elongated triangular orthobicupola (Cu_{18}) array, which encapsulates an inner Cu_2 unit. Later, in 2014, neutron diffraction experiments unequivocally revealed the location of the hydrides in this nanocluster [54]. As shown in Figure 8.4d,e, the 11 hydrides can be grouped into six capping μ_3-H, two tetrahedral μ_4-H, and three μ_4-H in a near-square-planar geometry.

A ligand-exchange reaction based on this nanocluster was then demonstrated to be an effective strategy to synthesize other copper NCs. For example, the achiral core in the original nanocluster could be transformed into a chiral metal skeleton in $[Cu_{20}H_{11}\{Se_2P(O^iPr)_2\}_9]$.

The nanocluster can be described as a distorted cuboctahedral Cu_{13} core that is further capped by a Cu_6 cupola and a single copper atom (Figure 8.4f,g) [25]. The 11 hydrides were determined to be seven (1 + 3 + 3) capping μ_3-H, three μ_5-H in a square-pyramidal cavity, and one μ_5-H in trigonal-bipyramidal cavity. In addition, $[Cu_{11}H_2\{S_2P(O^iPr)_2\}_6(PhC{\equiv}C)_3]$ could be isolated during the ligand-exchange process of $[Cu_{20}H_{11}(S_2P(O^iPr)_2)_9]$ with excess terminal alkynes [53]. The 11 copper atoms in the new nanocluster adopt a 3,3,4,4,4-pentacapped trigonal prismatic geometry, and two hydrides with a trigonal pyramidal coordination mode are accommodated inside the Cu_{11} cage, as determined by neutron diffraction (Figure 8.4c).

As an extended form of $[Cu_{20}H_{11}(S_2P(O^iPr)_2)_9]$, $[Cu_{32}H_{20}\{S_2P(O^iPr)_2\}_{12}]$ with an elongated triangular gyrobicupola geometry was synthesized and structurally characterized [27]. The structural correlation between the Cu_{20} and Cu_{32} metal framework was established. The structure of $[Cu_{32}H_{20}\{S_2P(O^iPr)_2\}_{12}]$ can be described as a distorted hexa-capped rhombohedral Cu_{14} core sandwiched between two nest-like triangular Cu_9 cupola fragments (Figure 8.4h,i), as observed in $[Cu_{20}H_{11}(S_2P(O^iPr)_2)_9]$. Twenty hydrides were demonstrated by neutron diffraction to reside inside the metal skeleton and display different coordination modes: twelve (6 + 6) capping μ_3-H, six (2 + 4) μ_4-H in a tetrahedral cavity, and two μ_5-H in an approximately square pyramidal configuration.

In 2014, $[Cu_{28}H_{15}\{S_2CN^nPr_2\}_{12}]^+$ was synthesized as the first rhombicuboctahedral copper nanocluster [26]. The metal framework of the nanocluster is composed of a Cu_4 tetrahedron, which is enclosed by a concentric Cu_{24} rhombicuboctahedron. The metal framework is further capped by 12 dithiocarbamate ligands with a S_{24} truncated octahedron (Figure 8.4j, k). Neutron diffraction clearly defined the positions of the 15 hydrides in the nanocluster. At the center of the nanocluster, one interstitial hydride (μ_4-H) is coordinated to four copper atoms in the Cu_4 tetrahedron. Six face-truncating hydrides (four μ_5-H and two μ_6-H) bridge the inner Cu_4 tetrahedron and Cu_{24} rhombicuboctahedron. Each of the eight triangular faces of the Cu_{24} rhombicuboctahedron is capped by one hydride (μ_3-H). Thus, the overall structure of the nanocluster can be described as $H@Cu_4@H_6@Cu_{24}@H_8@S_{24}$, corresponding to the configuration of a center, tetrahedron, octahedron, rhombicuboctahedron, cube, and truncated octahedron, respectively.

A hollow M_{12} icosahedral core framework has been observed in gold and silver NCs. However, such a hollow icosahedral core has been rarely observed in copper NCs. Recently, a dithiophosphate-protected copper nanocluster with a central Cu_{12} hollow icosahedron, $[Cu_{30}H_{18}\{S_2P(O^nPr)_2\}_{12}]$, was isolated and fully characterized [55]. Six faces of the icosahedron are capped by Cu_3 triangles, forming an incomplete rhombicuboctahedral Cu_{18} framework (Figure 8.4l,m). The 18 hydrides were found to possess three coordination modes by neutron diffraction: capping μ_3-H, interstitial seesaw μ_4-H, and square pyramidal μ_5-H. Notably, the hydride with an unprecedent seesaw coordination mode is rarely observed, whereas tetrahedral and near-square-planar coordination modes are most common for four-coordinated hydrides.

Collectively, in addition to hexameric $[HCuP(p\text{-tolyl})_3]_6$, several bidentate dithiolate ligand-protected copper NCs with nuclearities of Cu_7, Cu_{11}, Cu_{15} (two-electron superatom), Cu_{20} (achiral), Cu_{20} (chiral), Cu_{28}, Cu_{30}, and Cu_{32} have been characterized by neutron diffraction. The environments of the hydrides (capping and interstitial hydrides) in these NCs have been unambiguously confirmed, including μ_3-H (pyramid), μ_4-H (tetrahedron or trigonal pyramid, seesaw, and near-square plane), μ_5-H (square pyramid and trigonal bipyramid), and μ_6-H (trigonal prism).

8.3.3.3 Copper(I) Hydride NCs Determined by Single-Crystal X-ray Diffraction

In 2014, Huertos et al. reported the isolation of an air- and moisture-stable copper nanocluster, $[Cu_{18}H_7L_{10}I]$ ($L={}^{-}S(C_6H_4)PPh_2$), whose structure features a central Cu_8 core composed of two Cu_4 butterflies [58]. The remaining 10 copper atoms around the Cu_8 core are connected to S and P atoms (Figure 8.5a, e). Later, Hayton's group reported a novel copper nanocluster protected with neutral donor ligands, $[Cu_{14}H_{12}(phen)_6(PPh_3)_4]^{2+}$ [37]. The nanocluster is composed of a tetrahedral Cu_4 core encapsulated in a diamondoid shell of 10 copper atoms, each of which is coordinated to a PPh_3 or phen ligand (Figure 8.5b,f). The 12 hydrides are suggested to be μ_3-H. In 2017, the octanuclear nanocluster $[Cu_8H_6dppm_5]^{2+}$ (dppm = bis(diphenylphosphino)methane) was isolated by Tanase's group [59]. The resulting nanocluster features a *trans*-bicapped octahedral Cu_8 framework (Figure 8.5c,g). DFT optimization indicates that each of the six hydrides caps the Cu_3 triangular face and adopts the μ_3 coordination mode. Both the organic ligands and hydrides exhibit interesting fluxional behaviors in solution. In 2018, O'Hair's group reported the isolation of a copper nanocluster with a distorted Cu_6 octahedral core, $[Cu_{18}H_{16}(dppe)_6]^{2+}$ (dppe = bis (diphenylphosphino)ethane) [60]. The remaining 12 copper atoms bond to the P atoms of dppe and encircle the central Cu_6 core (Figure 8.5d, h).

In general, a compact core is rarely observed in copper(I) hydride NCs. Recently, Zheng's group reported the ambient-stable copper nanocluster $[Cu_{25}H_{10}(SPhCl_2)_{18}]^{3-}$, whose core can be described as a Cu_{13}-centered twinned cuboctahedron (ctco) with an *hcp* arrangement [22]. Note that the Cu_{13} ctco core is different from the reported cuboctahedral Cu_{13} core with an *fcc* arrangement (Figure 8.6a–c), although both of them are close-packed. The remaining copper atoms form a Cu_{12} truncated tetrahedral shell, and the edges are further bridged with 18 thiolates. Despite the close-packed character of the Cu_{13} ctco core, all the copper atoms in the nanocluster are determined to be copper(I), and 10 hydrides exist in each nanocluster. A full experimental and theoretical characterization of the nanocluster confirmed the positions of 10 hydrides (6 μ_6-H and 4 μ_3-H) inside the metallic core.

Recently, a high-nuclearity copper nanocluster with a large number of hydrides and copper atoms, $[Cu_{81}(PhS)_{46}({}^{t}BuNH_2)_{10}H_{32}]^{3+}$, was reported by Bakr's group [30]. Full structural analysis revealed that the nanocluster is composed of an unusual Cu_{17} planar core, a hemispherical shell, and different surface protective motifs (Figure 8.6d–f). Both the large planar core and hemispherical shell are unprecedented in coinage metal NCs. The formation of such a framework is considered to be related to the synergistic effects between the ligands and the interstitial hydrides.

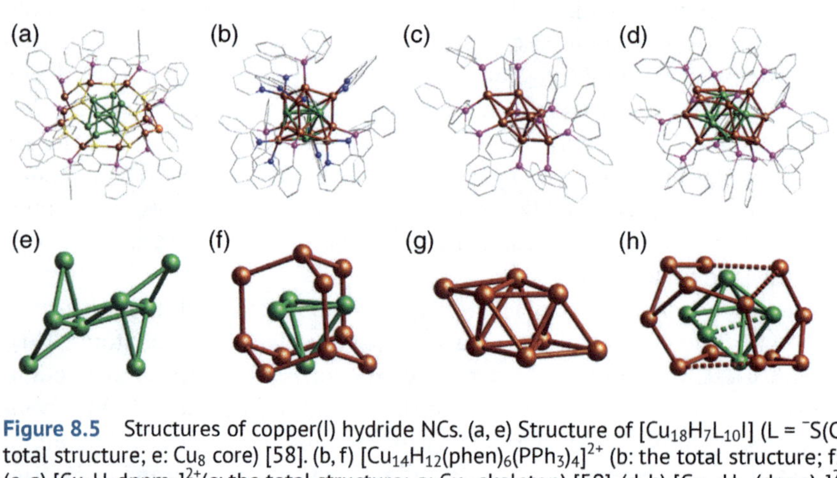

Figure 8.5 Structures of copper(I) hydride NCs. (a, e) Structure of $[Cu_{18}H_7L_{10}I]$ ($L = {}^{-}S(C_6H_4)PPh_2$) (a: the total structure; e: Cu_8 core) [58]. (b, f) $[Cu_{14}H_{12}(phen)_6(PPh_3)_4]^{2+}$ (b: the total structure; f: Cu_{14} skeleton) [37]. (c, g) $[Cu_8H_6dppm_5]^{2+}$(c: the total structure; g: Cu_8 skeleton) [59]. (d, h) $[Cu_{18}H_{16}(dppe)_6]^{2+}$(d: the total structure; h: Cu_{18} skeleton) [60]. Brown (shell)/green (core): Cu; yellow: S; blue: N; pink: P; orange: I; gray: C.

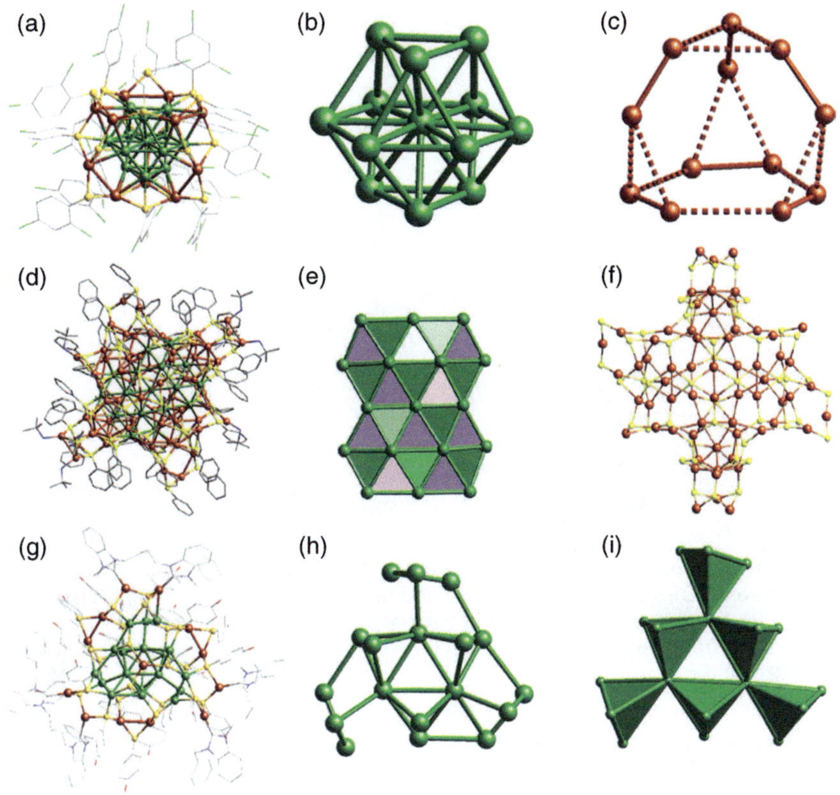

Figure 8.6 Structures of copper(I) hydride NCs. (a–c) $[Cu_{25}H_{10}(SPhCl_2)_{18}]^{3-}$(a: the total structure; b: Cu_{13} ctco; c: Cu_{12} truncated tetrahedron) [22]. (d–f) Structure of $[Cu_{81}(PhS)_{46}(^{t}BuNH_2)_{10}H_{32}]^{3+}$ (d: the total structure; e: Cu_{17} core; f: shell). *Source:* Reprinted with permission from [30]. © 2020 American Chemical Society. (g–i) Structure of $[Cu_{31}(SPhF)_{25}(NHC)_3H_6]$ (g: the total structure; h, i: Cu_{18} core) [61]. Brown (shell)/ green (core): Cu; yellow: S; blue: N; pink: P; red: F; bright green: Cl; gray: C.

In general, copper NCs suffer from poor stability, which hinders their storage, operation, and utilization. To achieve highly stable copper NCs, Shen et al. introduced N-heterocyclic carbene as a stabilizing ligand, in addition to thiolate, during synthesis, and the novel nanocluster $[Cu_{31}(SPhF)_{25}(NHC)_3H_6]$ (NHC = 1,4-bis(1-benzyl-1H-benzimidazol-1-ium-3-yl) butane) was successfully isolated [61]. Structurally, the inner core of the nanocluster can be regarded as a superstructure assembled from six Cu_4 tetrahedrons via vertex sharing (Figure 8.6g–i). Each of the Cu_4 units accommodates one hydride with the μ_4 coordination mode, according to X-ray diffraction and DFT calculations. Thanks to the unusually strong metal-carbene bonds, the nanocluster possesses ultrahigh thermal and air stability.

Morphologically, most of the coinage metal NCs reported to date are spherical, while the study of box- or cube-like NCs is also of great importance due to their unusual self-assembly behavior. In this context, some cube-like copper NCs have been reported. For example, Lee et al. reported the synthesis and characterization of the box-like copper nanocluster $[Cu_{32}(PET)_{24}H_8Cl_2]^{2-}$, which possesses a bisquare antiprismatic Cu_{14} core assembled from two Cu_8 square antiprisms by sharing one common edge (Figure 8.7a, d) [31]. The rod-shaped Cu_{14} core is further clamped by two $Cu_7(PET)_{11}Cl$ and two Cu_2PET units. The eight hydrides were ascertained to lie on the faces of the Cu_{14} core based on experimental and theoretical analyzes. Another cuboidal copper nanocluster, $[Cu_{23}(PhSe)_{16}(PPh_3)_8H_6]^+$, with a distorted cuboctahedral Cu_{13} core sandwiched by two square

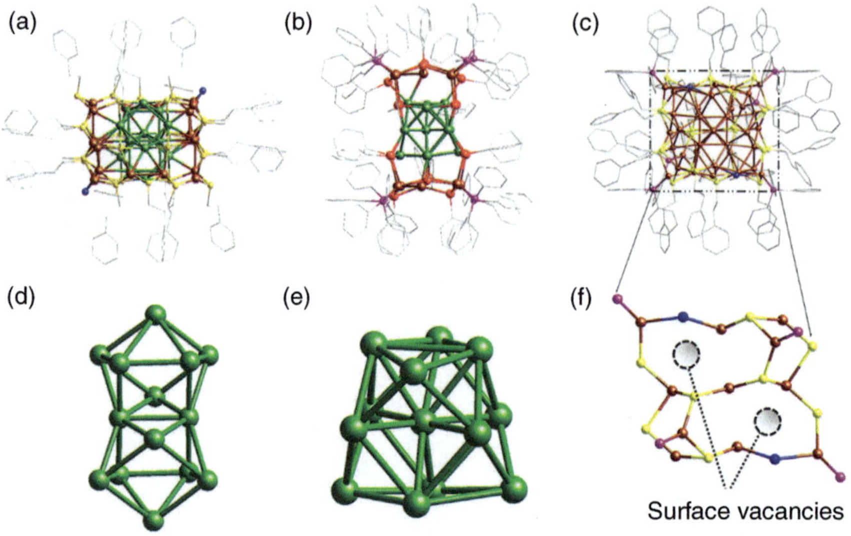

Figure 8.7 Structures of copper(I) hydride NCs. (a, d) Structure $[Cu_{32}(PET)_{24}H_8Cl_2]^{2-}$ (a: the total structure; d: Cu_{14} core) [31]; (b, e) Structure of $[Cu_{23}(PhSe)_{16}(PPh_3)_8H_6]^+$ (b: the total structure; e: Cu_{13} core) [29]; (c, f) Structure of $[Cu_{36}H_{10}(PET)_{24}(PPh_3)_6Cl_2]$ (c: the total structure; f: surface skeleton with vacancies). *Source:* Reprinted with permission from [13]. © 2021 American Chemical Society. Brown (shell)/green (core): Cu; yellow: S; blue: Cl; pink: P; red: Se; gray: C.

protecting motifs (Figure 8.7b, e), was isolated by Bakr's group [29]. All six hydrides of three different types adopt the μ_4 coordination mode, as determined by DFT analysis. A quasi-simple cubic packed pattern was observed via the face-to-face assembly of this cuboidal nanocluster. In addition, the same group reported the synthesis and full characterization of a defective copper nanocluster with a distorted half-cubic shape, $[Cu_{36}H_{10}(PET)_{24}(PPh_3)_6Cl_2]$ [13]. The nanocluster can be regarded as a defective derivative of a perfect Nichol's half cubic metal nanocluster, missing two surface copper atoms (Figure 8.7c,f). DFT calculations indicate that the hydrides link the inner Cu_4 core to the outer shell.

8.4 Properties

Due to their small size (1–3 nm) and enhanced quantum confinement effect, coinage metal NCs exhibit unusual optical and electronic properties. With the rapid development of copper NCs, an increasing number of exciting properties and applications have been explored, such as PL and catalysis. The exact atomic structure of NCs allow us to dive deeper into understanding the PL origin of NCs as well as their structure–property relationships, which cannot be achieved for large nanoparticles.

8.4.1 Photoluminescence of Copper NCs

PL is one of the most fascinating properties of nanoparticles. However, the origin of PL in nanoparticles has not yet been fully studied. A deep understanding of the origin of their PL can open new horizons for the controlled synthesis of functional materials in biomedicine, imaging, light-emitting devices, catalysis, and so on. In general, metal NCs exhibit outstanding optical properties, including a large Stokes shift, good photostability, biocompatibility, and tunable PL by controlling their size and peripheral ligands, which do not occur in conventional organic dyes. However, the PL quantum yield (PLQY) of metal NCs is much lower than that of organic dyes and semiconductor quantum dots (QDs), which has limited their practical application. The primary reason is the lack of an in-depth understanding of the origin of their PL at the atomic level. Herein, we focus mainly on the PL of copper NCs and the strategies for tailoring their optical properties.

Previously, a simple and straightforward spherical jellium model was used to describe the relationship between the emission energy and the number of metal atoms in NCs. Several follow-up publications have claimed that indeed copper NCs follow this simple spherical jellium model. It appears that "naked," ligand-free copper NCs or those stabilized by weakly bound ligands, such as tetrabutylammonium and polyethylenimine, tend to follow the simple spherical jellium model. These NCs generally have an emission peak in the UV-blue region and possess a short PL lifetime on the scale of nanoseconds, which can be attributed to the singlet excited state of the copper core. Vilar-Vidal et al. prepared a series of copper NCs with varying core sizes using tetrabutylammonium salt as a ligand. Cu_5, Cu_{13}, and Cu_{20} NCs displayed a redshift in the first absorption maximum and a concomitant redshift in the PL peak, which agree well with the jellium model [62].

However, the jellium model cannot be applied to all types of copper NCs [28, 63–65]. The simple jellium model cannot explain the differences in the PL peak positions of copper NCs with a metal core of the same size. Clearly, the number of metal atoms cannot be the only factor that determines the PL properties of NCs. The nature of the stabilizing ligands, such as whether they are aromatic or aliphatic; the functional groups in the ligands and their surface density; and the oxidation state of copper (0, +1, +2) in the nanocluster play tremendously important roles in determining the PL properties of NCs. Copper NCs stabilized by strongly bounded ligands, such as thiolates, amines, and carboxyl groups, usually display an orange-red PL peak with a much longer lifetime on the scale of microseconds. Such emission can be attributed to a ligand-to-metal-metal charge transfer (LMMCT) mechanism [66]. Li et al. synthesized and structurally characterized a new copper(I) nanocluster $[Cu_{11}(TBBT)_9(PPh_3)_6]^{2+}$ (where TBBT = 4-*tert*-butylbenzenethiol) that displays bright-red emission both in solution (685 nm) and in the solid state (675 nm) with a large Stokes shift (~280 nm). Temperature-dependent emission characteristics and theoretical calculations suggest that the PL derives mainly from a combination of metal–ligand charge transfer and cluster-centered triplet excited states [63].

Recently, Zang's group successfully synthesized and characterized a new fluorescent copper nanocluster formulated as $[Cu_{13}(SR)_{12}]NO_3$ (SR: 2-mercaptobenzimidazole). The nanocluster exhibits good stability under ambient conditions and bright-red emission in the solid state (627 nm). The long PL lifetime on the order of microseconds and large Stokes shift (~280 nm) indicate spin-forbidden triplet phosphorescence. The nanocluster shows a prompt response and great selectivity in the electrochemical detection of H_2O_2, making them a low-cost, advanced H_2O_2-sensing material. This work provides an opportunity to investigate the structure–optical and electrochemical property relationships of copper NCs [67].

8.4.1.1 Aggregation-Induced Emission

Aggregation-induced emission (AIE) refers to a photophysical phenomenon in which a nonemissive or weakly emissive material becomes highly luminescent in poor solvents or in the solid state as an aggregate [68]. In 2001, Tang's group was the first to observe AIE phenomenon in small organic molecules [69]. The AIE effect provides researchers a large platform for the in-depth study of light-emitting processes and offers insights into structure–property relationships in the aggregated state.

In the field of metal NCs, Xie's group was the first to observe the AIE phenomenon in thiolate-capped gold NCs in 2012 [70]. Nonluminescent oligomeric Au(I)-thiolate complexes were found to exhibit strong emission after aggregation induced by a solvent or cation. Regarding the origin of the AIE phenomenon, it has been suggested that the ordered assembly of staple motifs restricts vibrational motion compared with their disordered counterparts. However, because the precise structure of this nanocluster has not been determined, the authors failed to correlate the nanocluster's structure with the AIE phenomenon. Numerous studies have since been published regarding atomically precise gold and silver NCs exhibiting AIE. The AIE effect in metal NCs can be conveniently triggered by using solvents of varying polarity, pH, temperature, or the addition of certain ions that induce self-assembly and aggregation [71].

The AIE phenomenon has been observed and studied in copper NCs as well. AIE has provided a powerful approach for enhancing the PLQY of copper NCs [72]. The restriction of intramolecular motion (RIM) is the widely accepted reason for the AIE phenomenon. RIM can be affected by such factors as hydrophilicity, electrostatic charges, and the length of the peripheral ligands of NCs, which makes the peripheral ligands of copper NCs critical to their AIE behavior [73]. Several physical, chemical, and technical strategies have been proposed for controlling the AIE behavior of copper NCs, such as taking advantage of the viscosity and polarity of solvent mixtures, modulating surface charge, and incorporating NCs into templates [74]. Understanding and correlating structure with AIE phenomenon in copper NCs holds tremendous potential for developing smart nanomaterials for sensing, catalysis, bioimaging, etc. Numerous studies have been published on the AIE of copper NCs without a defined crystallographic structure [75]. However, to more deeply explore the relationship between structure and AIE, we must know the exact structure of NCs, particularly the structure of the shell. Several studies have been published on AIE phenomenon of copper NCs with a precisely defined structure.

Bakr's group successfully synthesized a novel copper nanocluster co-protected by thiolate and phosphine ligands, formulated as $[Cu_{15}(PPh_3)_6(PET)_{13}]^{2+}$, which exhibits bright emission in the near-infrared (NIR) region (~720 nm) and crystallization-induced emission enhancement (CIEE) [28]. An in-depth structural investigation of the ligand shell revealed that the extended C—H···π and π-π intermolecular ligand interactions significantly restrict intramolecular rotation and vibration and thus are a major reason for the CIEE phenomenon (Figure 8.8).

Recently, Fang et al. synthesized and characterized two nanocluster-based coordination chain polymers, $[Cu_{16}(^tBuC{\equiv}C)_{12}(PhOPO_3)_2]_n$ ($PhOPO_3$ = phenyl phosphate) and $[Cu_{16}(^tBuC{\equiv}C)_{12}(1\text{-}NaphOPO_3)_2]_n$ ($1\text{-}NaphOPO_3$ = 1-naphthyl phosphate), by self-assembly during crystallization. By introducing VO_4^{3-} and PO_4^{3-} anions, the lantern-like $[VO_4@Cu_{25}(^tBuC{\equiv}C)_{19}(1\text{-}NaphOPO_3)]^+$ and

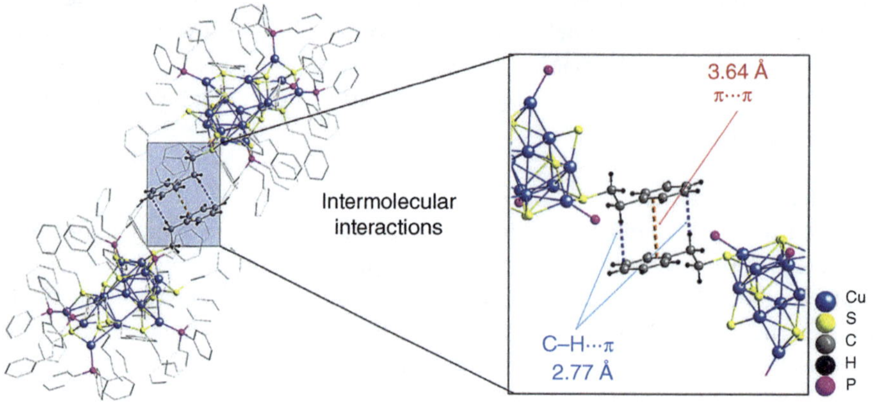

Figure 8.8 Intercluster stacking between the benzene rings of PET ligands enhanced by C—H···π interactions. *Source:* Reprinted with permission from [28]. © 2021 Wiley-VCH GmbH.

$[PO_4@Cu_{25}(^tBuC{\equiv}C)_{19}(1\text{-NaphOPO}_3)]^+$ were synthesized, which possess excellent photocurrent-generating properties. All of the abovementioned NCs exhibited the CIEE phenomenon [76].

8.4.1.2 Circularly Polarized Luminescence (CPL)

Chirality in atomically precise metal NCs has become an intensively investigated field in nanoscience and nanotechnology due to the potential applications of such NCs in asymmetric catalysis, medicine design, chiral recognition, and separation [77]. Recently, the CPL-active enantiomeric copper nanocluster $[D/L\text{-valinol}(18\text{-crown-}6)]^+[Cu_5(S^tBu)_6]^-$ was reported by Zang's group [78]. First, the novel chiral, luminescent, and anionic $[Cu_5(S^tBu)_6]^-$ was prepared. By tuning the cationic units, racemic crystals of $[K\text{-}(CH_3OH)_2(18\text{-crown-}6)]^+[Cu_5(S^tBu)_6]^-$ or a mesomeric crystal of $[K(18\text{-crown-}6)]^+[Cu_5(S^tBu)_6]^-$ could be obtained. Furthermore, cationic chiral amino alcohol was used to induce the chiral assembly of enantiomers, leading to an optically pure nanocluster with impressive CPL. Hydrogen-bonding interactions between the methanol molecules and sulfur atoms of the anionic units play a critical role in the chiral arrangement of the nanocluster.

The same group also reported the synthesis and characterization of a pair of optically pure, atomically precise copper NCs, formulated as $[Cu_{14}(R/S\text{-DPM})_8](PF_6)_6$ (R/S-Cu_{14}; R/S-DPM = (R/S)-2-diphenyl-2-hydroxylmethylpyrrolidine-1-propyne). Two designed R/S-DPM alkynyl ligands were used as chiral ligands to synthesize intrinsically chiral nanocluster. The resulting nanocluster exhibits bright-red luminescence and CPL with a high luminescence anisotropy factor. AIE contributes to triggering the CPL of R/S-Cu_{14} in the crystalline and aggregated states. With an increasing volume fraction of *n*-hexane, the main peak in the CD spectra shows a gradual hyperchromic shift, and the CPL response increases gradually due to emission changes at different degrees of aggregation (Figure 8.9). Both single-crystal structure analysis and theoretical calculations revealed the origin of the nanocluster's chiroptical activity at the atomic level. Because of its good biocompatibility, R/S-Cu_{14} has been used to image HeLa and NG108-15 cells [65].

8.4.2 Catalytic Properties of Copper NCs

Over the past decades, numerous nanoparticles have been extensively studied in the field of catalysis due to their small size (5–100 nm) – and thus high surface area – and their extraordinary activity. However, nanoparticles are not well-defined and uniform at the atomic level, which hinders fundamental aspects of catalysis, such as size dependence, structure–activity relationship, active sites, and catalytic mechanism. In this regard, well-defined, atomically precise metal NCs represent a new type of model catalyst and open up exciting opportunities in the field of catalysis [79–81].

Although copper complexes are well-known and good catalysts for various reactions [82–84], the catalytic properties of copper NCs with a crystallographic structure have not been widely studied to date. Nevertheless, the literature is saturated with examples of copper species with sub-nanometer dimensions, mostly deposited on a substrate, which are highly active and selective catalysts in certain catalytic reactions [85]. For these copper species, structural information is missing regarding their geometry and coordination mode. Copper NCs with the same number of atoms may have different geometric configurations, which can influence the catalytic process in different ways. However, the influence of geometric structure on catalytic properties has rarely been studied. The strong interest in copper NCs is also due to their size-dependent and molecular-like properties. In the following section, we summarize studies on the catalytic applications of copper NCs with a crystallographic structure.

8.4.2.1 Reduction of CO_2

Recently, copper-based nanomaterials have been widely investigated for their ability to regulate the electrocatalytic activity and product selectivity of CO_2 electroreduction [86]. The morphology

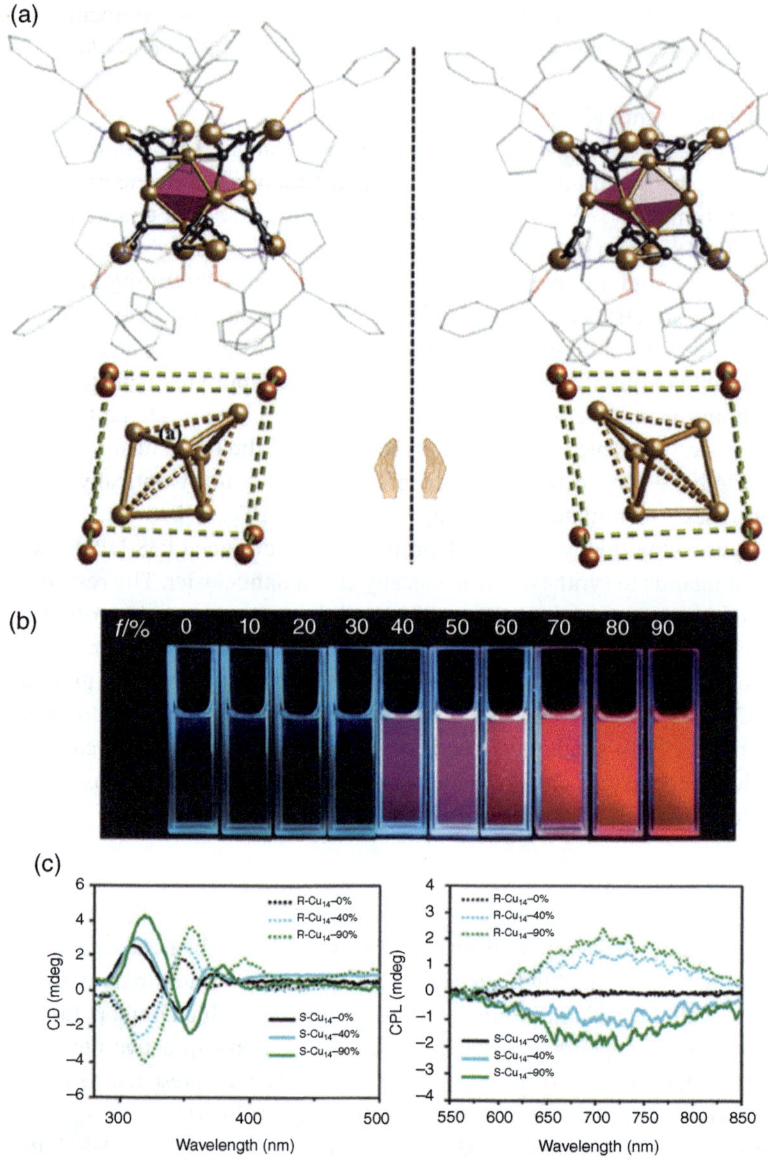

Figure 8.9 (a) The structures of R/S-Cu_{14} and the metal frameworks in R/S-Cu_{14} with mirror symmetry. (b) Photographs of R-Cu_{14} in solutions with different fractions of n-hexane under UV light. (c) CD spectra and CPL spectra of R-Cu_{14} in dichloromethane/n-hexane mixtures with different fractions of n-hexane. Brown/orange: Cu; red: O; blue: N; gray: C. *Source:* Reprinted with permission from [65]. © 2019 Wiley-VCH Verlag GmbH & Co. KGaA, Weinheim.

and particle size of copper-based catalysts have a tremendous effect on their catalytic activity. Therefore, atomically precise copper NCs with well-defined structures could serve as a model for understanding the mechanism of the electrocatalytic reduction of CO_2.

In fact, copper hydride NCs have been reported to exhibit high activity toward CO_2 since 1981. Beguin et al. reported the first example of a CO_2 reduction reaction with $[(PPh_3)CuH]_6$. They observed that CO_2 reacts with $[(PPh_3)CuH]_6$ in benzene solution at room temperature to give a complex mixture of products containing metallic copper, the formate $(PPh_3)_2CuOCOH$, and a noncrystalline material [87]. Similarly, $[Cu_{14}H_{12}(phen)_6(PPh_3)_4]^{2+}$ was found to react with CO_2 in the presence of excess PPh_3. Monitoring the reaction by NMR spectroscopy revealed the rapid consumption of $[Cu_{14}H_{12}(phen)_6(PPh_3)_4]^{2+}$ and the formation of the formate complex $[(PPh_3)_2Cu(\kappa^2\text{-}O_2CH)]$ in 37% yield, along with $[(phen)(PPh_3)CuCl]$ [37].

In 2017, Nakamae et al. reported the asymmetrical octanuclear complex $[Cu_8H_6dppm_5]^{2+}$, which showed catalytic activity in the hydrogenation of CO_2 under mild conditions and yielded linear tetranuclear copper complexes with formate bridges. Results showed that new motifs of copper hydride NCs could be established by the tetraphosphine ligands and that the structural features affect the NCs' reactivity. The reaction of CO_2 with copper hydride NCs can provide useful information in developing base metallic hydrogenation catalyst alternatives to noble metals [88].

Tang et al. studied thiolate-protected copper hydride NCs with a precise structure, formulated as $Cu_{32}H_{20}L_{12}$ (where L is a dithiophosphate ligand), which showed unique selectivity toward the electrocatalytic reduction of CO_2 at low overpotentials [89]. DFT calculations predict that the presence of the negatively charged hydrides in the copper NCs plays a critical role in determining the selectivity of the reaction, yielding HCOOH over CO with a lower overpotential. Briefly, HCOOH formation proceeds through two steps of lattice-hydride reduction: CO_2 reacts directly with the capping hydride to form HCOO*, which then reacts with another interstitial hydride to form HCOOH (Figure 8.10). The reacted hydrides can be regenerated easily by proton reduction. The predictions were confirmed by electrochemical testing of CO_2 reduction on the synthesized $Cu_{32}H_{20}L_{12}$: $HCOO^-$ was found to be the main product (>80% selectivity). Moreover, CO_2

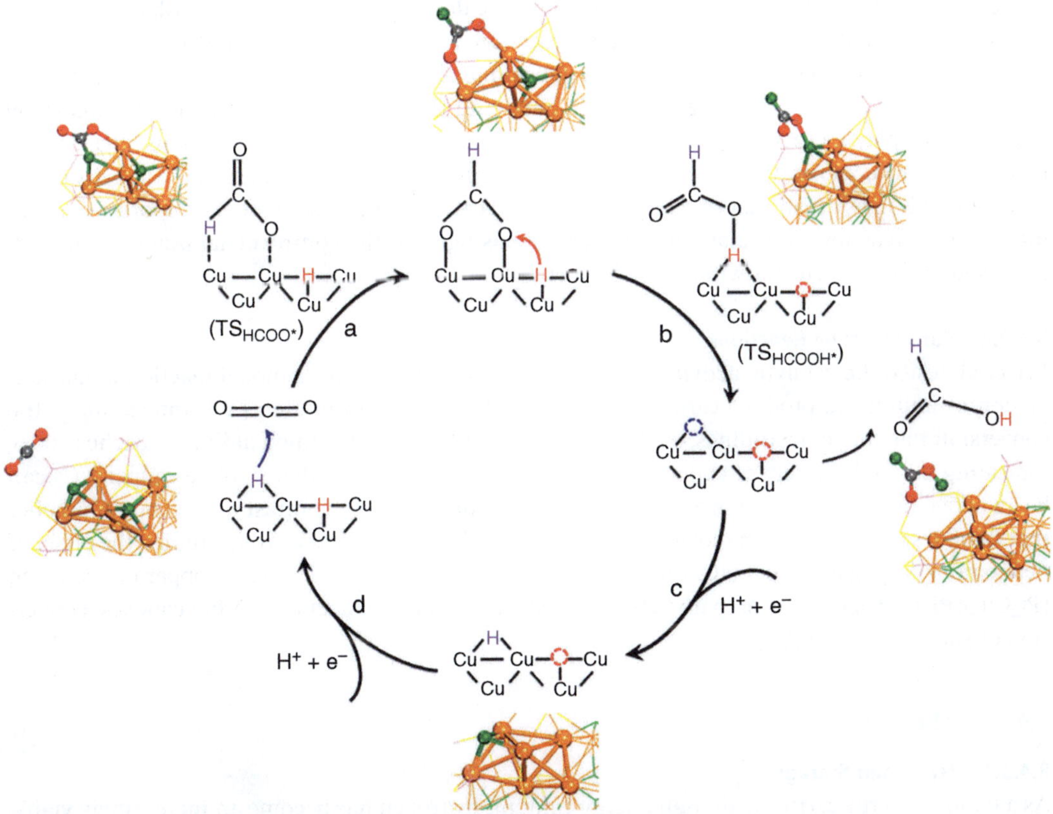

Figure 8.10 The mechanism of HCOOH formation from CO_2 reduction of copper hydride NCs. *Source:* Reprinted with permission from [89]. © 2017 American Chemical Society.

reduction was also predicted to be kinetically preferred over hydrogen evolution. The lattice-hydride mechanism can be a general path on hydride-containing transition-metal NCs and hence may offer more unique product selectivity than traditional electrocatalysts for CO_2 reduction.

8.4.2.2 "Click" Reaction

Click chemistry originally described reactions that generate high yields and selectivity through carbon-heterobonding reactions. The word *click* referred to the ease with which molecular building blocks could be joined, like the two ends of a seatbelt buckle. In 2013, Lee et al. developed a new catalyst, $[Cu_8H\{S_2P(OEt)_2\}_6]^+$, for the 1,3-dipolar cycloaddition of organic azides and alkynes for preparing substituted triazoles. With a required catalyst loading of only 0.4 mol%, the reactions of terminal alkynes with BnN_3 all proceeded smoothly under ambient conditions to form exclusively 1,4-triazoles in good yields. This study shed light into the effect of sulfur-based ligands on copper ion activity [90]. In addition, the two-electron superatom of $[Cu_{20}(PhC{\equiv}C)_{12}(OAc)_6]$, which could be immobilized on partially dehydroxylated silica, has shown the capacity to catalyze Huisgen [3+2] cycloadditions [34]. Cu K-edge EXAFS confirmed that the nanocluster adopts similar structures before and after deposition on silica. It has also been confirmed that the immobilized nanocluster, both pre- and post-catalysis, does not undergo any major structural transformations.

8.4.2.3 Hydrogenation

Between 1987 and 1990, Stryker's group discovered that $[(PPh_3)CuH]_6$ is effective for the selective conjugate reduction of unsaturated carbonyl and ketone compounds, as well as the selective reduction of alkynes to cis-alkenes. This mild hydride donor is chemically compatible with added chlorotrimethylsilane, affording an efficient procedure for reductive silylation [91–93].

Recently, the catalytic hydrogenation of 2-hexanone and 3-hexanone by $[Cu_{25}H_{10}(SPhCl_2)_{18}]^{3-}$ was reported by Zheng's group [22]. Experimental and theoretical characterization results show that hydrides play a crucial role in the hydrogenation of ketones. DFT calculations suggest that hydrogenation occurs only around a single site among the 10 hydrides. Although the activity of the nanocluster is lower than that of a typical industrial catalyst, this system offers important insight into catalytic reactions at the atomic level and sheds light on the controversial issue of a ligand-protected metal nanocluster with catalytic function.

8.4.2.4 Carbonylation Reactions

Lee et al. tested the catalytic activity of $[Cu_{32}(PET)_{24}H_8Cl_2]^{2-}$ toward a model reaction of carbonylation of anilines to produce carbamates through C—N bond formation [31]. Interestingly, the conversion rate for some anilines, such as methyl- and halide-substituted anilines, reaches 100%, indicating the high activity of copper NCs. This study demonstrates that thiolate copper NCs can be an active homogeneous catalyst even under mild conditions, in contrast to the common belief that thiolates poison copper nanocluster catalysts. In addition, the same group studied the effect of heteroatom doping in copper NCs on catalytic performance. A novel Pt-doped copper nanocluster, $[Pt_2Cu_{34}(PET)_{22}Cl_4]^{2-}$, was prepared and showed a 300-fold enhancement in the catalytic conversion of silane to silanol [94].

8.4.3 Other Properties

8.4.3.1 Hydrogen Storage

As a famous energy carrier with high energy content, hydrogen has become an increasingly viable clean and green option for transportation and energy storage. Therefore, hydrogen storage has attracted tremendous attention among researchers, and a wide range of materials have been tested for hydrogen storage, including metal hydrides, complex hydrides, chemical hydrides, carbon materials, and sorbents such as metal–organic frameworks (MOFs) [95, 96]. Among them, metal hydrides are one of the most important groups that can be used for chemical sorption due to the low-pressure requirements for this type of storage and its safety [97].

Coinage metal hydride NCs can release hydrogen under certain levels of solar irradiation, which is not achieved by other hydride systems that mainly rely on thermal activation. A hydrogen evolution study was performed with high-nuclearity copper hydride NCs under different physicochemical conditions, including solar irradiation, thermolysis, and acidification [24–26, 55]. It was observed that the hydrogen evolution rate is slower under sunlight irradiation than under thermolysis. $[Cu_{32}H_{20}\{S_2P(O^iPr)_2\}_{12}]$ is capable of releasing ~7, ~4, ~11, and ~19 equiv. of hydrogen per molecule when exposed to solar energy or reflux or in the presence of weak or strong acids [98]. When $[Cu_{28}H_{15}\{S_2CN^nPr_2\}_{12}]^+$ is exposed to solar energy, to reflux, or a weak or strong acid, the amount of hydrogen released is 2.6, 4.0, 2.0, and 12.2 equiv., respectively. These copper hydride NCs are excellent models for hydrogen storage as they are stable and release hydrogen under mild conditions.

8.4.3.2 Electronic Devices

Benefiting from the bright and tunable PL of copper NCs, Wang et al. developed a method for preparing down-conversion white LEDs. The synthetic method involves the formation of PVP-stabilized copper NCs and further treatment with various electron-rich ligands, such as GSH, cysteine, and cysteamine. The group showed that the initial QY of copper NCs, reported to be 8%, can be greatly enhanced (several fold) with the treatment of the abovementioned ligands [99]. Unfortunately, no copper NCs with a crystallographic structure have been reported in this field. In addition, Yuan et al. reported the application of $[Cu_{53}(CF_3COO)_{10}(^tBuC{\equiv}C)_{20}Cl_2H_{18}]^+$ as a precursor of CuI films for perovskite solar cells. Briefly, the iodination of alkynyl-protected nanocluster at room temperature resulted in the formation of CuI films that were used as hole transport layers (HTLs) for solar cells. As a result, a high PCE of 14.3% with good stability was achieved [41].

8.5 Summary Comparison with Gold and Silver NCs

Compared with gold and silver NCs [1, 2], the reported copper NCs with superatom character are scarce and suffer from low stability. Furthermore, most copper NCs with copper(0) character are two-electron superatoms – which are the smallest type of superatoms – whereas the electron count of reported superatoms based on gold and silver ranges from two to hundreds. Both metallic and nonmetallic (molecular) states, which are mainly dependent on the size of NCs, have been observed in gold and silver NCs. In this regard, the transition from molecular to metallic behavior among copper NCs has not been reported due to the lack of high-nuclearity copper superatoms. In addition, gold and silver NCs generally show multiple absorption peaks due to their discrete energy levels, which is the nonmetallic behavior exhibited by molecules. However, no distinct absorption peak has been observed for copper NCs, which is likely related to the electronic structure of copper or the small free electron count. Further study is expected to solve this puzzle.

With respect to structure, superatom-like copper NCs often exhibit some similarities with gold and silver NCs. Even for copper superatoms with hydrides, the core is likely to adopt a compact arrangement, such as an icosahedron, a cuboctahedron, or an *fcc* packing arrangement. The structural variety of the surface is quite different from that of gold NCs. Generally, staple $[Au_n(SR)_{n+1}]$ ($n = 1, 2$) units have been observed as common surface structure motifs of thiolate-protected gold NCs [1]. In contrast, the surface thiolate ligands in copper NCs generally adopt the μ_3 or μ_4

coordination mode with copper atoms, which have been widely observed in silver NCs [2]. Another feature of copper NCs is the prevalent hydrides in the structure. Single-crystal neutron diffraction reveals the multiple coordination modes of hydrides in copper hydride NCs, including μ_3-H (pyramid), μ_4-H (tetrahedron or trigonal pyramid, seesaw, and near-square plane), μ_5-H (square pyramid and trigonal bipyramid), and μ_6-H (trigonal prism). The presence of hydrides gives rise to more diverse core and surface structures. A noncompact or highly distorted core with hydrides is usually found in copper(I) hydride NCs. Some unprecedented core structures, such as a planar core, which have not been encountered in gold and silver NCs, have been discovered. Therefore, a compact or noncompact core could be a rough rule for estimating the presence of copper(0) in copper NCs. One exception is $[Cu_{25}H_{10}(SPhCl_2)_{18}]^{3-}$, in which all the copper atoms are copper(I) but the core is an *hcp* Cu_{13} ctco [22]. The surface structure of copper(I) hydride NCs is more diverse and complex due to the multiple coordination modes of organic ligands and hydrides. Therefore, the overall shape and surface units in copper(I) hydride NCs are distinct from those of their gold and silver counterparts.

In addition, copper NCs are more active during catalytic reactions due to the presence of hydrides. Indeed, hydrides have been demonstrated to play a crucial role in certain catalytic reactions. For the electrocatalytic CO_2 reduction, DFT calculations and experiments reveal that the surface hydrides participate in the reaction with CO_2 and mediate the product species [89]. Similar results have been observed for the hydrogenation of ketones [22]. Moreover, copper hydride NCs have been studied as promising materials for hydrogen storage, as they can release hydrogen under mild conditions.

8.6 Conclusion and Perspectives

With significant advances in recent years, numerous copper NCs with diverse crystal structures have been added to the nanocluster library. Multiple approaches have been established for synthesizing copper NCs, such as direct and indirect syntheses. Various precursors, ligands, and reductants have been employed for synthesis. Nevertheless, the isolation of superatom-like copper NCs with high nuclearity remains challenging due to the susceptibility to oxidation. In this regard, the reductive decomposition of copper(I) boryl complexes with NHC ligands provides a promising strategy for obtaining high-nuclearity copper NCs with copper(0) character [46]. Several high-nuclearity copper NCs with a large electron count, such as $[(IDipp)_{12}Cu_{179}]$ and $[(IDipp)_6Cu_{55}]$, have been obtained based on this method. However, these copper NCs suffer from poor stability and have not been fully characterized. This novel method also provides some inspiration for the synthesis of superatom-like copper NCs, such as the use of NHC ligands and nonhydride reductant.

Similar to gold and silver NCs, copper NCs have been shown to exhibit AIE, which offers insights into structure–property relationships in the aggregated state. What's more, by combining chirality with the AIE effect, CPL has been achieved in copper NCs, which can enable widespread potential applications. However, to date, copper NCs with CPL properties are still very rare. Therefore, many challenges remain, and more intensive research efforts should be devoted to this area. The surface modification of metal clusters with a combination of chiral ligands and luminescent ligands may support efforts to synthesize desirable CPL-active copper NCs.

Heteroatom doping has been shown to lead to large improvements in the physicochemical properties of gold and silver NCs and thus benefit their corresponding applications. However, doping based on copper NCs has not been exploited well yet. Some pioneering studies indicate that copper(I) hydride NCs can be transformed into copper superatoms after heteroatom doping [48, 94].

The NCs' optical properties and electronic structure may be altered by doping, further enhancing the stability and performance of copper NCs.

Considering the widespread applications of copper-based catalysts (both copper complexes and nanoparticles), copper NCs have the potential to be active and selective catalysts in certain reactions. Atomically precise copper NCs protected by thiolate and phosphine ligands have been demonstrated to be efficient catalysts under mild conditions. These structures provide a model platform for studying the mechanism of catalytic reactions. However, their activities are still lower than that of a typical popular catalyst, thus, there is room for significant improvement. Overall, we believe that further research into copper NCs can lead to novel, smart nanomaterials that offer solutions to problems related to the environment, catalysis, and human health.

References

1 Jin, R., Zeng, C., Zhou, M., and Chen, Y. (2016). Atomically precise colloidal metal nanoclusters and nanoparticles: fundamentals and opportunities. *Chem. Rev.* 116: 10346–10413.

2 Chakraborty, I. and Pradeep, T. (2017). Atomically precise clusters of noble metals: emerging link between atoms and nanoparticles. *Chem. Rev.* 117: 8208–8271.

3 Takano, S. and Tsukuda, T. (2021). Chemically modified gold/silver superatoms as artificial elements at nanoscale: design principles and synthesis challenges. *J. Am. Chem. Soc.* 143: 1683–1698.

4 Walter, M., Akola, J., Lopez-Acevedo, O. et al. (2008). A unified view of ligand-protected gold clusters as superatom complexes. *Proc. Natl. Acad. Sci. U. S. A.* 105: 9157–9162.

5 Baghdasaryan, A. and Bürgi, T. (2021). Copper nanoclusters: designed synthesis, structural diversity, and multiplatform applications. *Nanoscale* 13: 6283–6340.

6 Bratsch, S.G. (1989). Standard electrode potentials and temperature coefficients in water at 298.15 K. *J. Phys. Chern. Ref. Data* 18: 1–21.

7 Raston, C.L. and White, A.H. (1976). Crystal structure of the copper (I) iodide–pyridine (1/1) tetramer. *J. Chem. Soc. Dalton Trans.* 2153–2156.

8 Ford, P.C., Cariati, E., and Bourassa, J. (1999). Photoluminescence properties of multinuclear copper (I) compounds. *Chem. Rev.* 99: 3625–3647.

9 Fenske, D. and Krautscheid, H. (1990). New copper clusters containing Se and PEt_3 as ligands: $[Cu_{70}Se_{35}(PEt_3)_{22}]$ and $[Cu_{20}Se_{13}(PEt_3)_{12}]$. *Angew. Chem. Int. Ed. Engl.* 29: 1452–1454.

10 Fuhr, O., Dehnen, S., and Fenske, D. (2013). Chalcogenide clusters of copper and silver from silylated chalcogenide sources. *Chem. Soc. Rev.* 42: 1871–1906.

11 Dhayal, R.S., van Zyl, W.E., and Liu, C.W. (2016). Polyhydrido copper clusters: synthetic advances, structural diversity, and nanocluster-to-nanoparticle conversion. *Acc. Chem. Res.* 49: 86–95.

12 Chen, A., Kang, X., Jin, S. et al. (2019). Gram-scale preparation of stable hydride M@ Cu_{24}(M = Au/Cu) nanoclusters. *J. Phys. Chem. Lett.* 10: 6124–6128.

13 Dong, C., Huang, R.-W., Chen, C. et al. (2021). $[Cu_{36}H_{10}(PET)_{24}(PPh_3)_6Cl_2]$ reveals surface vacancy defects in ligand-stabilized metal nanoclusters. *J. Am. Chem. Soc.* 143: 11026–11035.

14 Shen, H., Han, Y.-Z., Wu, Q. et al. (2020). Simple and selective synthesis of copper-containing metal nanoclusters using $(PPh_3)_2CuBH_4$ as reducing agent. *Small Methods* 5: 2000603.

15 Hein, J.E. and Fokin, V.V. (2010). Copper-catalyzed azide-alkyne cycloaddition (CuAAC) and beyond: new reactivity of copper(I) acetylides. *Chem. Soc. Rev.* 39: 1302–1315.

16 Kuang, G.-C., Guha, P.M., Brotherton, W.S. et al. (2011). Experimental investigation on the mechanism of chelation-assisted, copper(II) acetate-accelerated azide-alkyne cycloaddition. *J. Am. Chem. Soc.* 133: 13984–14001.

17 Zhang, L.-M. and Mak, T.C. (2016). Comproportionation synthesis of copper(I) alkynyl complexes encapsulating polyoxomolybdate templates: bowl-shaped Cu_{33} and peanut-shaped Cu_{62} nanoclusters. *J. Am. Chem. Soc.* 138: 2909–2912.

18 Zhang, L.L.-M. and Mak, T.C.W. (2017). Temperature-mediated template release: facile growth of Copper(I) mixed ethynediide/isopropylethynide nanoclusters. *Angew. Chem. Int. Ed.* 56: 16228–16232.

19 Zhuo, H.-Y., Su, H.-F., Cao, Z.-Z. et al. (2016). High-nuclear organometallic Copper(I)-Alkynide clusters: thermochromic near-infrared luminescence and solution stability. *Chem. Eur. J.* 22: 17619–17626.

20 Cook, A.W., Jones, Z.R., Wu, G. et al. (2019). Synthesis and characterization of "Atlas-sphere" copper nanoclusters: new insights into the reaction of Cu^{2+} with thiols. *Inorg. Chem.* 58: 8739–8749.

21 Baghdasaryan, A., Besnard, C., Lawson Daku, L.M. et al. (2020). Thiolato protected copper sulfide cluster with the tentative composition $Cu_{74}S_{15}$(2-PET)$_{45}$. *Inorg. Chem.* 59: 2200–2208.

22 Sun, C., Mammen, N., Kaappa, S. et al. (2019). Atomically precise, thiolated copper-hydride nanoclusters as single-site hydrogenation catalysts for ketones in mild conditions. *ACS Nano* 13: 5975–5986.

23 Qu, M., Zhang, F.-Q., Wang, D.-H. et al. (2020). Observation of non-FCC copper in alkynyl-protected Cu_{53} nanoclusters. *Angew. Chem. Int. Ed.* 59: 6507–6512.

24 Dhayal, R.S., Liao, J.-H., Lin, Y.-R. et al. (2013). A nanospheric polyhydrido copper cluster of elongated triangular orthobicupola array: liberation of H_2 from solar energy. *J. Am. Chem. Soc.* 135: 4704–4707.

25 Dhayal, R.S., Liao, J.-H., Wang, X. et al. (2015). Diselenophosphate-induced conversion of an Achiral $[Cu_{20}H_{11}\{S_2P(OiPr)_2\}_9]$ into a chiral $[Cu_{20}H_{11}\{Se_2P(OiPr)_2\}_9]$ polyhydrido nanocluster. *Angew. Chem. Int. Ed.* 54: 13604–13608.

26 Edwards, A.J., Dhayal, R.S., Liao, P.-K. et al. (2014). Chinese puzzle molecule: a 15 hydride, 28 copper atom nanoball. *Angew. Chem. Int. Ed.* 53: 7214–7218.

27 Dhayal, R.S., Liao, J.-H., Kahlal, S. et al. (2015). $[Cu_{32}(H)_{20}\{S_2P(OiPr)_2\}_{12}]$: the largest number of hydrides recorded in a molecular nanocluster by Neutron diffraction. *Chem. Eur. J.* 21: 8369–8374.

28 Nematulloev, S., Huang, R.-W., Yin, J. et al. (2021, 17). $[Cu_{15}(PPh_3)_6(PET)_{13}]^{2+}$: a copper nanocluster with crystallization enhanced photoluminescence. *Small* 2006839.

29 Huang, R.-W., Yin, J., Dong, C. et al. (2020). $[Cu_{23}(PhSe)_{16}(Ph_3P)_8(H)_6]\cdot BF_4$: atomic-level insights into cuboidal polyhydrido copper nanoclusters and their quasi-simple cubic self-assembly. *ACS Materials Lett.* 3: 90–99.

30 Huang, R.-W., Yin, J., Dong, C. et al. (2020). $[Cu_{81}(PhS)_{46}(tBuNH_2)_{10}(H)_{32}]^{3+}$ reveals the coexistence of large planar cores and hemispherical shells in high-nuclearity copper nanoclusters. *J. Am. Chem. Soc.* 142: 8696–8705.

31 Lee, S., Bootharaju, M.S., Deng, G. et al. (2020). $[Cu_{32}(PET)_{24}H_8Cl_2](PPh_4)_2$: a copper hydride nanocluster with a bisquare antiprismatic core. *J. Am. Chem. Soc.* 142: 13974–13981.

32 Ke, F., Song, Y., Li, H. et al. (2019). Sub-nanometer Cu(I) clusters: coordination-modulated (Se vs. S) atom-packing mode and emission. *Dalton Trans.* 48: 13921–13924.

33 Nguyen, T.-A., Jones, Z.R., Goldsmith, B.R. et al. (2015). A Cu_{25} nanocluster with partial Cu(0) character. *J. Am. Chem. Soc.* 137: 13319–13324.

34 Cook, A.W., Jones, Z.R., Wu, G. et al. (2018). An organometallic Cu_{20} nanocluster: synthesis, characterization, immobilization on Silica, and "Click" chemistry. *J. Am. Chem. Soc.* 140: 394–400.

35 Ghosh, A., Huang, R.-W., Alamer, B. et al. (2019). $[Cu_{61}(StBu)_{26}S_6Cl_6H_{14}]^+$: a Core–Shell superatom nanocluster with a Quasi-J_{36} Cu_{19} core and an "18-Crown-6" Metal-Sulfide-like stabilizing belt. *ACS Materials Lett.* 1: 297–302.

36 Stevens, R.C., McLean, M.R., Bau, R., and Koetzle, T.F. (1989). Neutron diffraction structure analysis of a hexanuclear copper hydrido complex, $H_6Cu_6[P(p\text{-tolyl})_3]_6$: an unexpected finding. *J. Am. Chem. Soc.* 111: 3472–3473.

37 Nguyen, T.A., Goldsmith, B.R., Zaman, H.T. et al. (2015). Synthesis and characterization of a Cu_{14} hydride cluster supported by neutral donor ligands. *Chem. Eur. J.* 21: 5341–5344.

38 Nguyen, T.-A.D., Jones, Z.R., Leto, D.F. et al. (2016). Ligand-exchange-induced growth of an atomically precise Cu_{29} nanocluster from a smaller cluster. *Chem. Mater.* 28: 8385–8390.

39 Chakrahari, K.K., Liao, J.-H., Kahlal, S. et al. (2016). $[Cu_{13}\{S_2CN^nBu_2\}_6(\text{acetylide})_4]^+$: a two-electron superatom. *Angew. Chem. Int. Ed.* 55: 14704–14708.

40 Chakrahari, K.K., Liao, J., Silalahi, R.P.B. et al. (2021). Isolation and structural elucidation of 15-Nuclear copper dihydride clusters: an intermediate in the formation of a two-electron copper superatom. *Small* 17: e2002544.

41 Yuan, P., Chen, R., Zhang, X. et al. (2019). Ether-soluble Cu_{53} nanoclusters as an effective precursor of high-qality CuI films for optoelectronic applications. *Angew. Chem. Int. Ed.* 58: 835–839.

42 Han, B.-L., Liu, Z., Feng, L. et al. (2020). Polymorphism in atomically precise Cu_{23} nanocluster incorporating tetrahedral $[Cu_4]^0$ kernel. *J. Am. Chem. Soc.* 142: 5834–5841.

43 Jordan, A.J., Lalic, G., and Sadighi, J.P. (2016). Coinage metal hydrides: snthesis, characterization, and reactivity. *Chem. Rev.* 116: 8318–8372.

44 Li, Y.-L., Wang, J., Luo, P. et al. (2019). Cu_{14} cluster with partial Cu(0) character: difference in electronic structure from isostructural silver analog. *Adv. Sci.* 6: 1900833.

45 Kleeberg, C. and Borner, C. (2018). Syntheses, structures, and reactivity of NHC copper(I) boryl complexes: a systematic study. *Organometallics* 37: 4136–4146.

46 Drescher, W., Borner, C., and Kleeberg, C. (2021). Stability and decomposition of copper (I) boryl complexes: [(IDipp)Cu–Bneop], [(IDipp*)Cu–Bneop] and copper clusters. *New J. Chem.* 45: 14957–14964.

47 Chakrahari, K.K., Silalahi, R.P.B., Liao, J.-H. et al. (2018). Synthesis and structural characterization of inverse-coordination clusters from a two-electron superatomic copper nanocluster. *Chem. Sci.* 9: 6785–6795.

48 Silalahi, R.P.B., Chakrahari, K.K., Liao, J.-H. et al. (2018). Synthesis of two-electron bimetallic Cu-Ag and Cu-Au clusters by using $[Cu_{13}(S_2CN^nBu_2)_6(C{\equiv}CPh)_4]^+$ as a template. *Chem. Asian J.* 13: 500–504.

49 Wessing, J., Ganesamoorthy, C., Kahlal, S. et al. (2018). The mackay-type cluster $[Cu_{43}Al_{12}](Cp^*)_{12}$: open-shell 67-electron superatom with emerging metal-like electronic structure. *Angew. Chem. Int. Ed.* 57: 14630–14634.

50 Wang, S., Wu, Z., Dai, S., and Jiang, D.-e. (2021). Deep learning accelerated determination of hydride locations in metal nanoclusters. *Angew. Chem. Int. Ed.* 60: 12289–12292.

51 Wang, S., Liu, T., and Jiang, D.-e. (2021). Locating hydrides in ligand-protected copper nanoclusters by deep learning. *ACS Appl. Mater. Interfaces* 13: 53468–53474.

52 Liao, P.-K., Fang, C.-S., Edwards, A.J. et al. (2012). Hydrido copper clusters supported by dithiocarbamates: oxidative hydride removal and neutron diffraction analysis of $[Cu_7(H)\{S_2C(\text{aza-15-crown-5})\}_6]$. *Inorg. Chem.* 51: 6577–6591.

53 Brocha Silalahi, R.P., Huang, G.-R., Liao, J.-H. et al. (2020). Copper clusters containing hydrides in trigonal pyramidal geometry. *Inorg. Chem.* 59: 2536–2547.

54 Liao, J.-H., Dhayal, R.S., Wang, X. et al. (2014). Neutron diffraction studies of a four-coordinated hydride in near square-planar geometry. *Inorg. Chem.* 53: 11140–11145.

55 Barik, S.K., Huo, S.-C., Wu, C.-Y. et al. (2020). Polyhydrido copper nanoclusters with a hollow icosahedral core: $[Cu_{30}H_{18}\{E_2P(OR)_2\}_{12}]$ (E=S or Se; R=^{n}Pr, ^{i}Pr or ^{i}Bu). *Chem. Eur. J.* 26: 10471–10479.

56 Liao, P.-K., Sarkar, B., Chang, H.-W. et al. (2009). Facile entrapment of a hydride inside the tetracapped tetrahedral Cu^I_8 cage inscribed in a S_{12} icosahedral framework. *Inorg. Chem.* 48: 4089–4097.

57 Liu, C.W., Sarkar, B., Huang, Y.-J. et al. (2009). Octanuclear copper (I) clusters inscribed in a Se_{12} icosahedron: Anion-induced modulation of the core size and symmetry. *J. Am. Chem. Soc.* 131: 11222–11233.

58 Huertos, M.A., Cano, I., Bandeira, N.A. et al. (2014). Phosphinothiolates as ligands for polyhydrido copper nanoclusters. *Chem. Eur. J.* 20: 16121–16127.

59 Nakamae, K., Tanaka, M., Kure, B. et al. (2017). A fluxional Cu_8H_6 cluster supported by bis(diphenylphosphino)methane and its facile reaction with CO_2. *Chem. Eur. J.* 23: 9457–9461.

60 Li, J., Ma, H.Z., Reid, G.E. et al. (2018). Synthesis and X-ray crystallographic characterisation of frustum-shaped ligated $[Cu_{18}H_{16}(DPPE)_6]^{2+}$ and $[Cu_{16}H_{14}(DPPA)_6]^{2+}$ nanoclusters and studies on their H_2 evolution reactions. *Chem. Eur. J.* 24: 2070–2074.

61 Shen, H., Wang, L., López-Estrada, O. et al. (2021). Copper-hydride nanoclusters with enhanced stability by N-heterocyclic carbenes. *Nano Res.* 14: 3303–3308.

62 Vilar-Vidal, N., Rivas, J., and López-Quintela, M.A. (2012). Size dependent catalytic activity of reusable subnanometer copper (0) clusters. *ACS Catal.* 2: 1693–1697.

63 Li, H., Zhai, H., Zhou, C. et al. (2020). Atomically precise copper cluster with intensely near-infrared luminescence and its mechanism. *J. Phys. Chem. Lett.* 11: 4891–4896.

64 Wu, Z., Liu, H., Li, T. et al. (2017). Contribution of metal defects in the assembly induced emission of Cu nanoclusters. *J. Am. Chem. Soc.* 139: 4318–4321.

65 Zhang, M.-M., Dong, X.-Y., Wang, Z.-Y. et al. (2020). AIE triggers the circularly polarized luminescence of atomically precise enantiomeric copper (I) alkynyl clusters. *Angew. Chem. Int. Ed.* 59: 10052–10058.

66 Qing, T., Zhang, K., Qing, Z. et al. (2019). Recent progress in copper nanocluster-based fluorescent probing: a review. *Microchim. Acta* 186: 1–20.

67 Qin, H.-N., Wang, Z.-Y., and Zang, S.-Q. (2021). Photoluminescence and electrochemical sensing of atomically precise Cu_{13} cluster. *Acta Chim. Sin.* 79: 1037–1041.

68 Mei, J., Hong, Y., Lam, J.W. et al. (2014). Aggregation-induced emission: the whole is more brilliant than the parts. *Adv. Mater.* 26: 5429–5479.

69 Luo, J., Xie, Z., Lam, J.W. et al. (2001). Aggregation-induced emission of 1-methyl-1,2,3,4,5-pentaphenylsilole. *Chem. Commun.* 1740–1741.

70 Luo, Z., Yuan, X., Yu, Y. et al. (2012). From aggregation-induced emission of Au(I)–thiolate complexes to ultrabright Au(0)@Au(I)–thiolate core–shell nanoclusters. *J. Am. Chem. Soc.* 134: 16662–16670.

71 Zhao, Z., Zhang, H., Lam, J.W., and Tang, B.Z. (2020). Aggregation-induced emission: new vistas at the aggregate level. *Angew. Chem. Int. Ed.* 59: 9888–9907.

72 Mei, J., Leung, N.L., Kwok, R.T. et al. (2015). Aggregation-induced emission: together we shine, united we soar! *Chem. Rev.* 115: 11718–11940.

73 Wang, H., Zhao, E., Lam, J.W., and Tang, B.Z. (2015). AIE luminogens: emission brightened by aggregation. *Mater. Today* 18: 365–377.

74 Liu, Y., Yao, D., and Zhang, H. (2017). Self-assembly driven aggregation-induced emission of copper nanoclusters: a novel technology for lighting. *ACS Appl. Mater. Interfaces* 10: 12071–12080.

75 Shi, Y.-e., Ma, J., Feng, A. et al. (2021). Aggregation-induced emission of copper nanoclusters. *Aggregate* 2: e112.

76 Fang, J.-J., Shen, Y.-L., Liu, Z. et al. (2021). Copper(I) alkynyl clusters with crystallization-induced emission enhancement. *Inorg. Chem.* 60: 13493–13499.

77 Zhang, M.-M., Li, K., and Zang, S.-Q. (2020). Progress in atomically precise coinage metal clusters with aggregation-induced emission and circularly polarized luminescence. *Adv. Optical Mater.* 8: 1902152.

78 Jin, Y., Li, S., Han, Z. et al. (2019). Cations controlling the chiral assembly of luminescent atomically precise copper (I) clusters. *Angew. Chem. Int. Ed.* 58: 12143–12148.

79 Jin, R., Li, G., Sharma, S. et al. (2021). Toward active-site tailoring in heterogeneous catalysis by atomically precise metal nanoclusters with crystallographic structures. *Chem. Rev.* 121: 567–648.

80 Lei, Y., Mehmood, F., Lee, S. et al. (2010). Increased silver activity for direct propylene epoxidation via subnanometer size effects. *Science* 328: 224–228.

81 Qin, R., Liu, K., Wu, Q., and Zheng, N. (2020). Surface coordination chemistry of atomically dispersed metal catalysts. *Chem. Rev.* 120: 11810–11899.

82 Munro-Leighton, C., Delp, S.A., Blue, E.D., and Gunnoe, T.B. (2007). Addition of N—H and O—H bonds of amines and alcohols to electron-deficient olefins catalyzed by monomeric Copper (I) systems: reaction scope, mechanistic details, and comparison of catalyst efficiency. *Organometallics* 26: 1483–1493.

83 Gong, J., Yue, H., Zhao, Y. et al. (2012). Synthesis of ethanol via syngas on Cu/SiO_2 catalysts with balanced Cu^0–Cu^+ sites. *J. Am. Chem. Soc.* 134: 13922–13925.

84 Cheng, L.-J. and Mankad, N.P. (2020). C–C and C–X coupling reactions of unactivated alkyl electrophiles using copper catalysis. *Chem. Soc. Rev.* 49: 8036–8064.

85 Halder, A., Curtiss, L.A., Fortunelli, A., and Vajda, S. (2018). Perspective: size selected clusters for catalysis and electrochemistry. *J. Chem. Phys.* 148: 110901.

86 Jin, H., Guo, C., Liu, X. et al. (2018). Emerging two-dimensional nanomaterials for electrocatalysis. *Chem. Rev.* 118: 6337–6408.

87 Beguin, B., Denise, B., and Sneeden, R. (1981). Hydrocondensation of CO_2: II. Reaction of carbon dioxide and carbon monoxide with $[HCuPPh_3]_6$. *J. Organomet. Chem.* 208: C18–C20.

88 Nakamae, K., Nakajima, T., Ura, Y. et al. (2020). Facially dispersed Polyhydride Cu_9 and Cu_{16} clusters comprising apex-truncated supertetrahedral and square-face-capped cuboctahedral copper frameworks. *Angew. Chem. Int. Ed.* 59: 2262–2267.

89 Tang, Q., Lee, Y., Li, D.-Y. et al. (2017). Lattice-hydride mechanism in electrocatalytic CO_2 reduction by structurally precise copper-hydride nanoclusters. *J. Am. Chem. Soc.* 139: 9728–9736.

90 Lee, B.-H., Wu, C.-C., Fang, X. et al. (2013). $[Cu_8(\mu_4\text{-}H)\{S_2P(OEt)_2\}_6](PF_6)$: a novel catalytic hydride-centered copper cluster for azide-alkyne cycloaddtion. *Catal. Lett.* 143: 572–577.

91 Mahoney, W.S., Brestensky, D.M., and Stryker, J.M. (1988). Selective hydride-mediated conjugate reduction of. alpha.,.beta.-unsaturated carbonyl compounds using $[(Ph_3P)CuH]_6$. *J. Am. Chem. Soc.* 110: 291–293.

92 Mahoney, W.S. and Stryker, J.M. (1989). Hydride-mediated homogeneous catalysis. Catalytic reduction of α,β-unsaturated ketones using $[(Ph_3P)CuH]_6$ and H_2. *J. Am. Chem. Soc.* 111: 8818–8823.

93 Daeuble, J.F., McGettigan, C., and Stryker, J.M. (1990). Selective reduction of alkynes to cis-alkenes by hydrometallation using $[(Ph_3P)CuH]_6$. *Tetrahedron Lett.* 31: 2397–2400.

94 Lee, S., Bootharaju, M.S., Deng, G. et al. (2021). $[Pt_2Cu_{34}(PET)_{22}Cl_4]^{2-}$: an atomically precise, 10-electron PtCu bimetal nanocluster with a direct Pt-Pt bond. *J. Am. Chem. Soc.* 143: 12100–12107.

95 Moradi, R. and Groth, K.M. (2019). Hydrogen storage and delivery: Review of the state of the art technologies and risk and reliability analysis. *Int. J. Hydrog. Energy* 44: 12254–12269.

96 Chanchetti, L.F., Leiva, D.R., de Faria, L.I.L., and Ishikawa, T.T. (2020). A scientometric review of research in hydrogen storage materials. *Int. J. Hydrog. Energy* 45: 5356–5366.

97 Nguyen, H.Q. and Shabani, B. (2021). Review of metal hydride hydrogen storage thermal management for use in the fuel cell systems. *Int. J. Hydrog. Energy* 46: 31699–31726.

98 Dhayal, R.S., Chen, H.-P., Liao, J.-H. et al. (2018). Synthesis, structural characterization, and H_2 evolution study of a spheroid-shape hydride-rich copper nanocluster. *ChemistrySelect* 3: 3603–3610.

99 Wang, Z., Susha, A.S., Chen, B. et al. (2016). Poly(vinylpyrrolidone) supported copper nanoclusters: glutathione enhanced blue photoluminescence for application in phosphor converted light emitting devices. *Nanoscale* 8: 7197–7202.

9

Atomically Precise Nanoclusters of Iron, Cobalt, and Nickel: Why Are They So Rare?

Trevor W. Hayton

Department of Chemistry and Biochemistry, University of California Santa Barbara, Santa Barbara, CA, 93106, USA

9.1 Introduction

Atomically precise nanoclusters (APNCs) are an emerging class of materials that feature properties common to both traditional metal complexes and bulk metals [1]. Their unique mix of properties make them attractive for a variety of applications, [2] including solid-state memory, [3] catalysis, [4–6] sensing, [7] and imaging [7, 8]. Nanoclusters can also feature single molecule magnet (SMM) behavior, [9] and have been proposed to function as novel spin qubits [10–12].

While a wide variety of APNC definitions are found in the literature, there are three requirements that are commonly agreed upon, namely, that APNCs must be monodisperse, nano-sized, and feature a well-defined arrangement of their capping ligands [13, 14]. For the purposes of this review, we will include a fourth requirement. Specifically, an APNCs must feature at least one metal atom that is bound to at least eight other metal atoms *and* no nonmetal atoms. This criterion is met by those APNCs that feature an $[M_{13}]$ kernel, such as the iconic Au_{25} APNC, $[NOct_4][Au_{25}(SCH_2CH_2Ph)_{18}]$, which features a central gold atom that is bound to 12 other gold atoms in an icosahedral geometry, but no ligand atoms (Figure 9.1) [15, 18]. However, other kernels also meet this criterion, including the $[M_9]$ cubic kernel and the $[M_{11}]$ sphenocorona kernel [16, 17]. Importantly, this requirement restricts our discussion to APNCs that feature considerable amounts of metal–metal bonding and low average metal oxidation states (approaching zero in many cases). Put differently, these APNCs feature "metal-like" character. APNCs made of transition metal oxides, chalcogenides, or pnictinides are not included within this definition and so will not be discussed. APNCs with interstitial carbide and hydride ligands are also not included. Additionally, for convenience, we will generally restrict our discussion to homometallic APNCs.

Finally, we will restrict our discussion to APNCs containing nickel, cobalt, and iron. This restriction was implemented because APNCs of these metals have not been widely reviewed. APNC reviews tend to focus on the group 11 metals [2, 19–21] and do not focus much attention on the mid-transition metals. This emphasis makes sense, given that most APNCs contain one of the coinage metals and given that APNCs of nickel, cobalt, and iron are exceptionally rare (for reasons that should become apparent in following sections). However, the emphasis placed on the group 11 APNCs means that the interesting properties of nickel, cobalt, and iron APNCs are often underappreciated. In particular, APNCs of these metals are reasonably expected to be open shell, in contrast to group 11 APNCs. Open-shell APNCs are generally rare, and their magnetic properties are

Atomically Precise Nanochemistry, First Edition. Edited by Rongchao Jin and De-en Jiang.

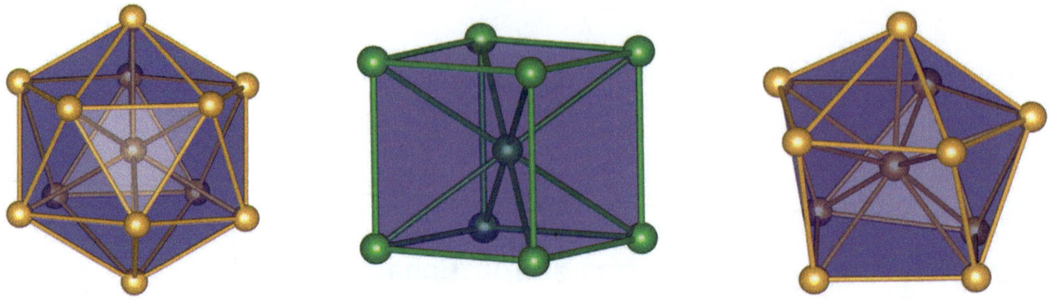

Figure 9.1 The $[M_{13}]$, $[M_9]$, and $[M_{11}]$ kernels found in $[NOct_4][Au_{25}(SCH_2CH_2Ph)_{18}]$, $[Ni_9Te_6(PEt_3)_8]$, and $[Au_{11}Cl_3(PR_3)_7]$ (R = 3-$CF_3C_6H_4$), respectively. *Source:* X-ray data taken from Refs. [15–17].

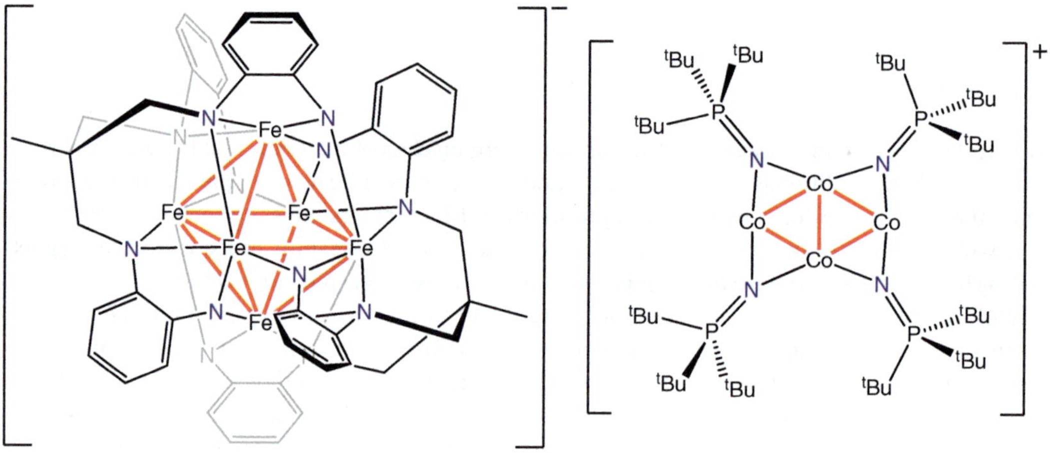

Figure 9.2 Molecular structures of $[(^{H}L)_2Fe_6(py)_2]^-$ and $[Co_4(N{=}P^tBu_3)_4]^+$. Direct exchange interactions are shown in red.

not well understood. However, preliminary evidence suggests that low-valent, open-shell APNCs of Fe, Co, and Ni should be good single molecule magnets (SMMs), [22] as evidenced by the SMM behavior of the small cluster complexes, $[NBu_4][(^{H}L)_2Fe_6(py)_2]$ ($^{H}LH_6 = MeC(CH_2NHPh$-*o*-$NH_2)_3$) [23] and $[Co_4(N{=}P^tBu_3)_4][B(C_6F_5)_4]$ (Figure 9.2) [24, 25]. Persistent molecular magnetism requires the presence of a barrier (U) to magnetic reversal. The larger the barrier, in principle, the better the SMM performance. In integer spin systems, U is directly correlated to spin state (S) and the zero field splitting (D) according to $U = |D|S^2$ [26]. Thus, the ability of $[(^{H}L)_2Fe_6(py)_2]^-$ ($S = 11$) and $[Co_4(N{=}P^tBu_3)_4]^+$ ($S = 9/2$) to function as SMMs relates partly to their large spin ground states, which are a consequence of the direct metal–metal bonding interactions that are found in their structures. This metal–metal bonding promotes ferromagnetic coupling of the unpaired spins via the "direct exchange" mechanism [24]. Importantly, this relationship suggests that higher nuclearity metal–metal bonded APNCs with larger spin ground states should possess even better SMM performance.

In this chapter, we will attempt to identify the reasons for the paucity of Fe, Co, and Ni APNCs and highlight potential synthetic strategies for accessing APNCs of these metals. In addition, we will discuss the unique properties of these open-shell APNCs. We hope that a discussion of these topics will accelerate the discovery of these uncommon materials.

9.2 General Considerations

As already mentioned, there are very few examples of homometallic Fe, Co, and Ni clusters that meet our definition of an APNC. Indeed, a search of the Cambridge Structural Database (CSD) reveals that there are zero, three, and nine structurally characterized homometallic APNCs of these metals, respectively (Table 9.1). In contrast, APNCs of silver and gold are quite common. Indeed, there are now 194 structurally characterized homometallic Au APNCs, many of which have been made in the past five years, which reflects the excitement surrounding this field.

To understand this disparity, it is useful to examine the reduction potentials as well as the diatomic M—M bond dissociation enthalpies (BDEs) for the relevant transition metal ions (Table 9.1). For instance, the BDE for gas phase Au_2 is 226 kJ/mol, whereas the BDEs for gas phase Fe_2 and Co_2 are 118 and < 127 kJ/mol, respectively. Thus, it appears that Au APNCs are so abundant, in part, because the strong Au—Au bonds can successfully drive APNC self-assembly [19]. A comparison of the metal reduction potentials is also informative. The standard $Au^{+1/0}$ reduction potential is +1.69 V, whereas the standard $M^{+2/0}$ reduction potentials for Fe and Co are −0.44 and −0.28 V, respectively (Table 9.1). Thus, it is much easier to reduce an Au(I) starting material to Au(0) than to reduce a Fe(II) or Co(II) starting material to the zero-valent state. Indeed, many attempted APNC syntheses are unsuccessful because the reducing agents employed are not strong enough to reduce the metal salts to M(0) [29, 30]. Additionally, for elements with more negative potentials, it can be a challenge to keep an APNC in a low-valent state once it is formed. These APNCs readily react with O_2, necessitating the use of inert atmosphere techniques during their isolation and

Table 9.1 Number of structurally characterized APNCs, M—M bond strengths, and standard redox potentials for selected transition metal ions.

Group	M	M—M BDE (kJ/mol)[a]	E° (V, vs. SHE)[b]	n	Number of examples[c]
8	Fe	118	−0.44	2	0
	Ru	190	+0.8	2	0
	Os	420	+0.9	3	0
9	Co	<127	−0.28	2	3
	Rh	234	+0.76	3	11
	Ir	360	+1.0	3	0
10	Ni	204	−0.24	2	9
	Pd	>136	+0.915	2	24
	Pt	306	+1.18	2	13
11	Cu	201	+0.52	1	10
	Ag	162	+0.80	1	47
	Au	226	+1.69	1	194

[a] Values correspond to experimentally measured gas phase BDEs for the neutral diatomic species. Data taken from Ref. [27]. While the M—M bond strengths in an APNCs will be different, values for the diatomic species were chosen for comparison because they are experimentally known for a wide variety of metals.
[b] Standard redox potentials for the $M^{n+}(aq) + ne^- \rightarrow M(s)$ couple. Data taken from Ref. [28].
[c] Data taken from a search of the Cambridge Structural Database (CSD), Version 5.42 (September 2021 update). Defined here as a cluster that contains at least one transition metal atom coordinated to at least eight other transition metal atoms and no nonmetal atoms. The search was restricted to homometallic APNCs.

characterization. These strongly reducing metal atoms can also reduce many common stabilizing ligands, such as thiolates and phosphines, and several examples of this phenomenon will be discussed in the following sections. Accordingly, their isolation necessitates the use of carefully designed, redox-stable ligands.

The analysis just described has its limitations, however. For example, no homometallic APNCs are known for Os or Ir, despite their apparently favorable M—M BDEs and reduction potentials. In these two examples, it may be that the paucity of examples is due to the relatively high cost of the elements, which has perhaps limited the amount of exploratory synthesis that has been performed on these metals. The case of Ru is more challenging to explain. It has a favorable M—M BDE (190 kJ/mol), a favorable reduction potential (+0.8 V), and is relatively cheap, but no APNCs have been reported for this metal, suggesting that the factors affecting APNC formation are more complex than suggested by this analysis. Nickel also has an apparently favorable M—M BDE of 204 kJ/mol, which is similar to that of Au. However, only a handful of Ni APNCs have ever been isolated. In this particular case, though, Ni displays an unfavorable $M^{+2/0}$ reduction potential (−0.24 V), in line with those of Fe and Co.

9.3 Synthesis of Ni APNCs

The earliest attempts to form large metal clusters focused on the use of carbon monoxide as the supporting ligand [31, 32]. Indeed, within group 10, several large homoleptic carbonyl-supported APNCs have been reported over the years, including $[Pd_{66}(CO)_{45}(PEt_3)_{16}]$ and $[Pt_{38}(CO)_{44}]^{2-}$ [33, 34], but these very large clusters are restricted to Pd and Pt. Homoleptic Ni carbonyl clusters are known, including $[Ni_6(CO)_{12}]^{2-}$, $[Ni_9(CO)_{18}]^{2-}$, and $[Ni_{12}(CO)_{21}]^{4-}$ [35, 36], but they tend to be much smaller, at least relative to known homoleptic Pd and Pt carbonyl clusters [34, 37–39]. In the case of Ni, cluster growth is typically achieved by addition of $[Ni(CO)_4]$ to a low nuclearity Ni carbonyl precursor. For example, $[Ni_9(CO)_{18}]^{2-}$ can be prepared by addition of three equiv. of $Ni(CO)_4$ to $[Ni_6(CO)_{12}]^{2-}$ [32]. The largest homoleptic Ni carbonyl cluster appears to be the aforementioned $[Ni_{12}(CO)_{21}]^{4-}$, along with its protonated analogs, $[Ni_{12}(CO)_{21}H_{4-n}]^{n-}$ (n = 2, 3) [40, 41]. The structure of D_{3h}-symmetric $[Me_4N]_3[Ni_{12}(CO)_{21}H]$ consists of a planar $[Ni_6(CO)_3(\mu\text{-}CO)_6]$ fragment sandwiched by two planar $[Ni_3(CO)_3(\mu\text{-}CO)_3]$ fragments (Figure 9.3) [42]. As a result of this arrangement, every Ni atom is bound by at least two CO ligands, and thus it does not qualify

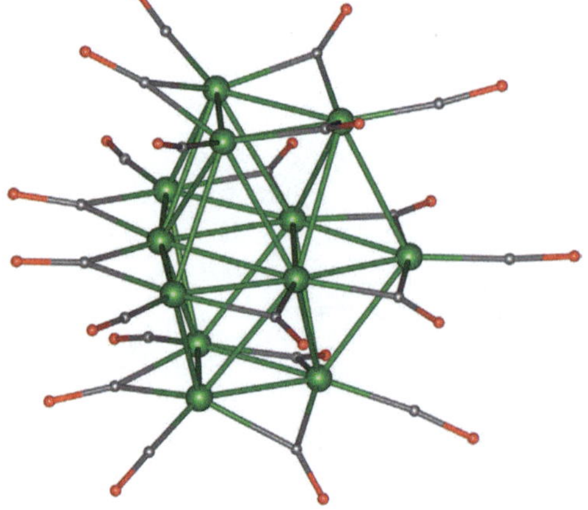

Figure 9.3 Solid-state structure of $[Me_4N]_3[Ni_{12}(CO)_{21}H]$. Ni, C, and O atoms are shown in green, gray, and red, respectively. Hydrogen atom and $[Me_4N]^+$ counterions are omitted for clarity. *Source:* X-ray data taken from Ref. [42].

as an APNC. Larger clusters are possible, but they feature interstitial main group atoms, such as carbon, in their structures. Examples include $[HNi_{38}(CO)_6(\mu\text{-}CO)_{36}C_6]^{5-}$ and $[Ni_{32}C_6(CO)_{36}]^{6-}$ [43, 44]. Because of the presence of interstitial main group atoms, these clusters also do not obey our definition of an APNC and will not be discussed in detail. Presumably, larger homoleptic carbonyl clusters cannot be made because any structure with $n > 12$ would require an interstitial Ni atom (assuming formation of a cuboctahedral kernel), which is evidently unfavorable because Ni(0) is so reducing.

A number of large carbonyl clusters have been reported that contain mixtures of Ni with other metals, including Pd, Pt, and Au. Their structures are also informative. For example, $[Pt_6Ni_{38}(CO)_{48}H_{6-n}]^{n-}$ (n = 3, 4, 5, 6) features an octahedral $[Pt_6]$ kernel surrounded by a shell of 38 Ni atoms [45–47]. Each Ni atom is further bound by a mix of terminal and bridging CO ligands. The partitioning of the Pt atoms within the core is consistent with the greater redox stability of Pt(0) vs. Ni(0) (Table 9.1). Stated differently, the more strongly reducing Ni(0) prefers to be bound by the strongly π-accepting CO ligands, which ameliorates its high reducing power. A similar picture emerges for other mixed-metal group 10 clusters [48–50], including $[Pd_{33}Ni_9(CO)_{41}(PPh_3)_6]^{4-}$, [51] $[Au_6Ni_{32}(CO)_{44}]^{6-}$, [52] $[Ni_{29}Pd_6(CO)_{42}]^{6-}$, [53], and $[HNi_{24}Pt_{17}(CO)_{25}(CO)_{21}]^{5-}$ [54]. In all of these examples, and despite the presence of substitutional disorder in some cases, every Ni atom resides on the cluster surface.

While the synthesis of Ni APNCs via $Ni(CO)_4$ condensation has been unsuccessful thus far, a handful of Ni APNCs have been made via different routes, [55] including $[Ni_{23}Se_{12}Cl_3(PEt_3)_{10}]$, $[Ni_{21}Se_{14}(PEt_2Ph)_{12}]$, [56], and $[Ni_9Te_6(PEt_3)_8]$ [16]. For example, we recently prepared $[Ni_{23}Se_{12}Cl_3(PEt_3)_{10}]$ via reaction of $[Ni(1,5\text{-}COD)_2]$ (1.0 equiv.), PEt_3 (0.04 equiv.), $SePEt_3$ (0.52 equiv.), and $[NiCl_2(PEt_3)_2]$ (0.07 equiv.), in a mixture of toluene and THF [57]. It can be isolated in 35% yield when prepared via this route. Interestingly, this cluster was first synthesized in 1992 by Steigerwald and co-workers [58]; however, it was originally formulated as $[Ni_{23}Se_{12}(PEt_3)_{13}]$ on the basis of an incomplete X-ray crystallographic analysis. In the solid-state, $[Ni_{23}Se_{12}Cl_3(PEt_3)_{10}]$ features a central $[Ni_{13}]$ anti-cuboctahedral kernel, making it only one of two Ni APNCs to contain this structural feature (Figure 9.4). One hemisphere of the $[Ni_{13}]$ kernel is encapsulated by a $[Ni_{10}(\mu\text{-}Se)_9Cl_3(PEt_3)_7]^-$ shell, whereas the other hemisphere is ligated by three $(\mu_4\text{-}Se)^{2-}$ ligands and three PEt_3 ligands. Within the $[Ni_{13}]$ cuboctahedron, the average $Ni_{central}$–Ni_{shell} distance is 2.54 Å (range = 2.516(4) – 2.578(6) Å), which is somewhat longer than the Ni—Ni distance in bulk fcc Ni (2.48 Å) [59]. Evans method magnetism measurements suggest that $[Ni_{23}Se_{12}Cl_3(PEt_3)_{10}]$ has three unpaired electrons.

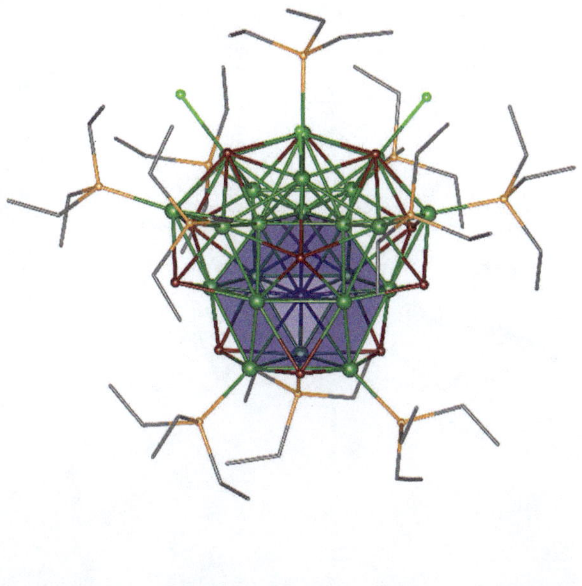

Figure 9.4 Solid-state structure of $[Ni_{23}Se_{12}Cl_3(PEt_3)_{10}]$. Ni atoms are shown in green, except for the central Ni atom, which is shown in blue. The blue polyhedron represents the coordination sphere of the central Ni atom. P, Se, and Cl atoms are shown in saffron, maroon, and bright green, respectively. C atoms are depicted in gray wireframe. Hydrogen atoms and THF solvate atoms are omitted for clarity. *Source:* X-ray data taken from Ref. [57].

In 1992, Fenske and co-workers reported the synthesis of $[Ni_{21}Se_{14}(PEt_2Ph)_{12}]$, which can be prepared by reaction of $[NiCl_2(PEt_2Ph)_2]$ with $Se(SiMe_3)_2$ over the course of two weeks at room temperature, followed by crystallization from heptane [56]. A yield was not reported for this cluster. The average Ni oxidation state is +1.33. The reducing agent required to convert the Ni(II) starting material to the reduced APNC product is most likely PEt_2Ph, as $SePEt_2Ph$ is formed as a byproduct in the reaction. Me_3SiCl is also a byproduct of the transformation. The use of an Ni(II) starting material here is notable, as other Ni APNC syntheses that started from Ni(II) precursors fail to achieve any metal reduction (see below for further discussion). $[Ni_{21}Se_{14}(PEt_2Ph)_{12}]$ features a central $[Ni_{13}]$ cuboctahedral kernel, which is capped by two $[Ni_4(\mu_4\text{-}Se)_5(PEt_2Ph)_4]$ terraces that are found at the north and south poles (Figure 9.5). Additionally, there are four $(\mu_4\text{-}Se)^{2-}$ ligands and four PEt_2Ph ligands located along the equatorial belt of the $[Ni_{13}]$ cuboctahedron. Within the $[Ni_{13}]$ cuboctahedron, the $Ni_{central}—Ni_{shell}$ distances range from 2.576 to 2.800 Å, whereas the $Ni_{shell}—Ni_{shell}$ distances range from 2.606 to 2.724 Å. Similar values were observed for $[Ni_{23}Se_{12}Cl_3(PEt_3)_{10}]$. The magnetic properties of $[Ni_{21}Se_{14}(PEt_2Ph)_{12}]$ were not mentioned in the original report; however, they would be of interest for comparison to those of $[Ni_{23}Se_{12}Cl_3(PEt_3)_{10}]$ and $[Ni_9Te_6(PEt_3)_8]$. Because of these unknowns, coupled with the overall rarity of Ni APNCs, this cluster is a candidate for further study.

Reaction of $[Ni(1,5\text{-}COD)_2]$ with Et_3PTe and Et_3P in a 2 : 1 : 22 ratio leads to formation of $[Ni_9Te_6(PEt_3)_8]$ in 45% yield [16, 60]. This cluster features an $[Ni_9]$ nickel-centered cubic kernel that is encapsulated by six $(\mu_4\text{-}Se)^{2-}$ ligands arranged on the square faces of the cube (Figure 9.6). Additionally, the eight outer Ni atoms are each bound by a PEt_3 ligand. The average $Ni_{central}—Ni_{shell}$ bond length is 2.47 Å, whereas average $Ni_{shell}—Ni_{shell}$ distance is 2.85 Å. The average $Ni_{central}—Te$ bond length is 2.98 Å, suggesting that there is no meaningful $Ni_{central}—Te$ interaction. As a result, $[Ni_9Te_6(PEt_3)_8]$ qualifies as an APNC by our definition. $[Ni_9Te_6(PEt_3)_8]$ exhibits a room temperature magnetic moment of 4.9 μ_B, which corresponds to an $S = 2$ spin state [61, 62]. Electronic structure calculations performed by Saillard and co-workers suggest a ground state configuration of $(e_g)^4(t_{1g})^4(t_{2g})^2$, which is consistent with the magnetism measurements [63, 64]. Additionally, they found that the Ni—Ni bonds within the Ni_8 cage are derived primarily from 4s to 4p atomic orbitals. Similarly, the central Ni atom interacts with the cage via its 4s and 4p orbitals. The 3d subshell of the central Ni atom is fully filled. Earlier calculations by Wheeler uncovered a similar bonding picture [65, 66]. Calculations have also been performed on the CO-substituted analogue, $[Ni_9Te_6(CO)_8]$ [67]. Magnetic measurements on $[Ni_9Te_6(PEt_3)_8][BF_4]$ and $[Ni_9Te_6(PEt_3)_8][BF_4]_2$ have also been reported; [61, 62] although no synthetic or structural details were provided for these clusters. $[Ni_9Te_6(PEt_3)_8][BF_4]$ and $[Ni_9Te_6(PEt_3)_8][BF_4]_2$ feature

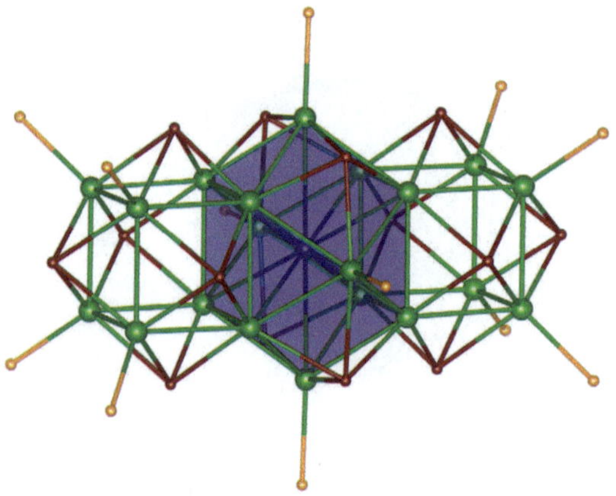

Figure 9.5 Solid-state structure of $[Ni_{21}Se_{14}(PEt_2Ph)_{12}]$. Ni atoms are shown in green, except for the central Ni atom, which is shown in blue. The blue polyhedron represents the coordination sphere of the central Ni atom. P and Se atoms are shown in saffron and maroon, respectively. Carbon and hydrogen atoms are omitted for clarity. *Source:* X-ray data taken from Ref. [56].

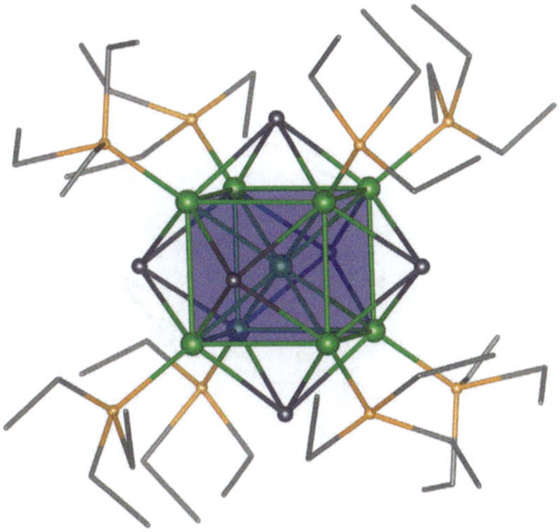

Figure 9.6 Solid-state structure of $[Ni_9Te_6(PEt_3)_8]$. Ni, P, and Te atoms are shown in green, orange, and silver, respectively. The blue polyhedron defines the coordination sphere of the central Ni atom. Carbon atoms are shown in wireframe and hydrogen atoms are omitted for clarity. *Source:* X-ray data taken from Ref. [16].

moments of 3.7 and 1.9 μ_B at room temperature, respectively, which are both lower than the moment observed for the neutral cluster, as expected.

$[Ni_9Te_6(PEt_3)_8]$ reacts with C_{60} to form $[Ni_9Te_6(PEt_3)_8][C_{60}]$, a material with a NaCl-type lattice [68]. This material exhibits ferromagnetic ordering below 4 K in the solid-state. X-ray crystallography reveals that the Ni—Ni and Ni—Te distances within the $[Ni_9Te_6(PEt_3)_8]^+$ cation are essentially identical to those of the neutral cluster. The closely related analogues, $[Ni_9Te_6(PMe_3)_8][C_{60}]$ and $[Ni_9Te_6(PEt_3)_8][C_{70}]$ have also been reported [69].

Several clusters with closely related $[Ni_9]$ kernels are also known [70], including $[Ni_9(GeEt)_6(CO)_8]$, [71] $[Ni_9(As)_6(PPh_3)_5Cl_3]$, [72] $[Ni_9(As)_6(PPh_3)_6Cl_2]$ [72], and $[Ni_9(\mu_4\text{-}P)_6(PCy_3)_6Cl_2]$ [73]. Generally, these clusters are synthesized by reaction of $[NiCl_2(PR_3)_2]$ with $E(SiMe_3)_3$ or $PhE(SiMe_3)_2$ (E = main group element) [72, 73]. However, $[Ni_9(GeEt)_6(CO)_8]$ is formed by reaction of $[Ni_6(CO)_{12}]^{2-}$ with $EtGeCl_3$ [71]. In $[Ni_9(As)_6(PPh_3)_5Cl_3]$, the average Ni_{center}—As distance is 2.61 Å, [72] whereas in $[Ni_9(\mu_4\text{-}P)_6(PCy_3)_6Cl_2]$, the average Ni_{center}—P distance is 2.44 Å [73]. These distances lie just outside of their respective single bond covalent radii, [74] suggesting that these clusters also fit our definition of an APNC. Nonetheless, there may still be weak Ni_{center}—E interactions in these materials.

No other Ni clusters with $[M_{13}]$ or $[M_9]$ kernels are known; however, a number Ni clusters with mixed-element $[Ni_{11}E_2]$ icosahedral kernels have been reported. While these clusters do not fall within the definition of APNCs used in this chapter, as defined in Section 9.1, their synthesis, structure, and bonding are highly informative. For example, in 1992, Dahl and co-workers reported the synthesis of $[PPh_3Me]_2[Ni_{11}Se_2(CO)_{18}]$ via reaction of PhSeCl with $[PPh_3Me]_2[Ni_6(CO)_{12}]$ [75]. Also formed in the reaction is $NiCl_2$ and $Ni(CO)_4$. The selenium source for this transformation, PhSeCl, presumably undergoes a C—Se oxidative addition during the reaction, losing its phenyl group as biphenyl. A similar sequence of reaction steps has been reported for the synthesis of $[Ni_8(S)_5(PPh_3)_7]$ from PhSSPh and $[Ni(1,5\text{-}COD)_2]$ [76]. In the solid state, $[PPh_3Me]_2[Ni_{11}Se_2(CO)_{18}]$ features a $[Ni_{11}Se_2]$ Ni-centered icosahedron, wherein the two $(\mu_5\text{-}Se)^{2-}$ ligands are found at the north and south poles (Figure 9.7). Additionally, each outer Ni atom is bound by a terminal CO ligand, and there are eight μ-CO ligands arranged along the equatorial belt. The $Ni_{central}$—Se distance is 2.19 Å, the average $Ni_{central}$—Ni_{shell} bond length is 2.55 Å, and the average intra- and inter-pentagonal Ni_{shell}—Ni_{shell} distances are 2.74 and 2.52 Å, respectively. The central Ni atom is thought to feature a d^{10} electronic configuration. A bonding analysis of the closely related cluster, $[Ni_{10}Sb_2(\mu_{12}\text{-}Ni)\{Ni(CO)_3\}_2(CO)_{18}]^{2-}$ suggests that the 3d orbitals of the interstitial Ni(0) atom do

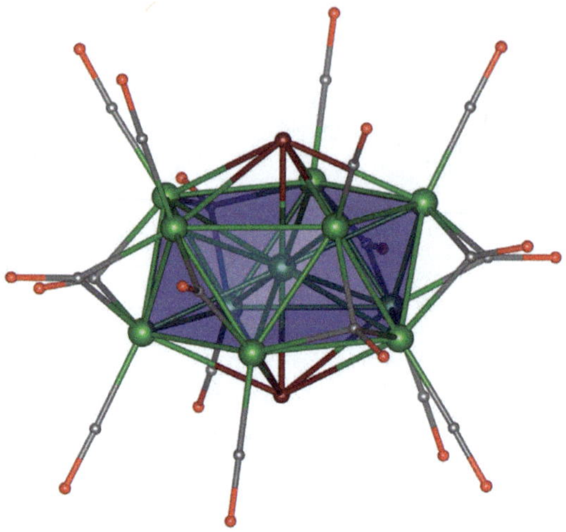

Figure 9.7 Solid-state structure of $[PPh_3Me]_2[Ni_{11}Se_2(CO)_{18}]$. Ni, Se, C, and O atoms are shown in green, maroon, gray, and red, respectively. The blue polyhedron defines the 10 Ni—Ni bonds from the central Ni atom. Counterions are omitted for clarity. *Source:* X-ray data taken from Ref. [75].

not interact very strongly with the cage atoms [77, 78]. Moreover, reducing this cluster by one electron to form $[Ni_{10}Sb_2(\mu_{12}\text{-}Ni)\{Ni(CO)_3\}_2(CO)_{18}]^{3-}$ does not change the M—M bond lengths, suggesting that the extra electron occupies a nonbonding or slightly antibonding orbital within $[Ni_{11}Sb_2]$ core [77]. Magnetism measurements of $[EtV]_8[Ni_{10}Sb_2(\mu_{12}\text{-}Ni)\{Ni(CO)_3\}_2(CO)_{18}]_3$ (EtV = 1,1′-diethyl-4,4′-bipyridinium cation), which features two open-shell $[Ni_{10}Sb_2(\mu_{12}\text{-}Ni)\{Ni(CO)_3\}_2(CO)_{18}]^{3-}$ trianions and one closed-shell $[Ni_{10}Sb_2(\mu_{12}\text{-}Ni)\{Ni(CO)_3\}_2(CO)_{18}]^{2-}$ dianion, are consistent with an $S = 1$ ground state for the trianion [79]. Other clusters with the same $[Ni_{11}E_2]$ skeleton include $[Ni_{11}(SnR)_2(CO)_{18}]^{2-}$ (R = ^{n}Bu, Me) [70] and $[Ni_{11}Bi_2(CO)_{18}]^{n-}$ (n = 2 or 3) [80].

There have been many other attempts to prepare low-valent Ni APNCs, but all have been unsuccessful. For example, the reaction of $NiCl_2$ with $PhCH_2CH_2SH$ and excess $NaBH_4$, in the presence of $[NOct_4][Br]$, resulted in isolation of $[Ni_6(SCH_2CH_2Ph)_{12}]$, which adopts a "tiara-like" structure in the solid-state (Scheme 9.1) [81–83]. Importantly, every Ni atom in this cluster is in the 2+ oxidation state, demonstrating that, despite the use of Brust-Schiffrin reaction conditions, no metal ion reduction has occurred [29]. The contrast between this synthetic outcome and those involving the coinage metal triad, which readily undergo reduction using the Brust-Schiffrin protocol, can be explained by the much more negative reduction potential found for Ni^{2+} vs. the group 11 metals (Table 9.1).

In an effort to circumvent the challenge of reducing Ni^{2+} in situ during APNC formation, several research groups have employed Ni(0) precursors to synthesize Ni APNCs. For example, reaction of bis(trityl)nickel, $Ni(\eta^3\text{-}CPh_3)_2$, which, on account of the high stability of the trityl radical, functions as a Ni(0) synthon due to the facile homolysis of the Ni-trityl bonds, with 2 equiv. of PPh_3 at

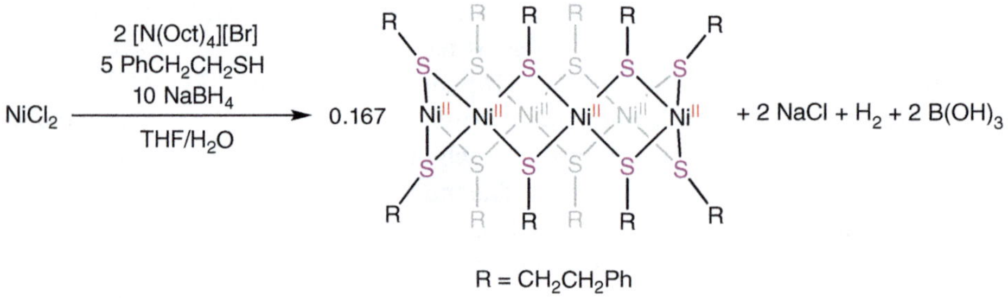

Scheme 9.1 Synthesis of $[Ni_6(SCH_2CH_2Ph)_{12}]$.

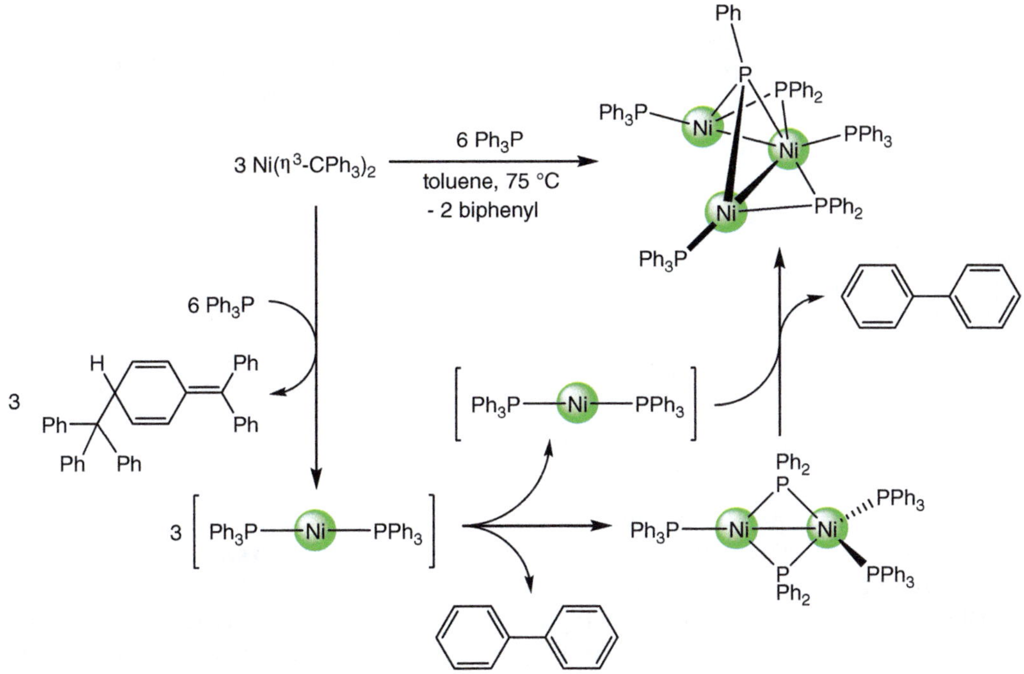

Scheme 9.2 Proposed mechanism for formation of $[Ni_3(\mu_3\text{-PPh})(\mu\text{-PPh}_2)_2(PPh_3)_3]$.

75 °C in toluene results in formation of the phosphinidene cluster, $[Ni_3(\mu_3\text{-PPh})(\mu\text{-PPh}_2)_2(PPh_3)_3]$ in moderate yield [84]. We hypothesize that the reaction proceeds through a $[Ni_2(\mu\text{-PPh}_2)_2(PPh_3)_3]$ intermediate. (Scheme 9.2). Its phosphide ligands are formed by P—C oxidative addition, which is followed by C—C reductive elimination to form biphenyl. The phosphinidene ligand in $[Ni_3(\mu_3\text{-PPh})(\mu\text{-PPh}_2)_2(PPh_3)_3]$ is formed similarly. This unwanted PPh_3 reduction thwarted attempts to make larger Ni APNCs via this route and nicely demonstrates how strongly reducing and reactive Ni(0) can be, when it is unencumbered by good π accepting ligands, such as CO.

A similar phosphine activation is observed during the synthesis of Ni NPs [85]. Thermolysis of $Ni(acac)_2$ and $P(Oct)_3$ in a solution of di(octyl)ether/oleylamine at 230 °C for 1 hour results in production of relatively monodisperse 22 nm Ni(0) nanoparticles. Powder X-ray diffraction shows that the Ni NPs possess an fcc Ni structure. However, ICP-OES and EXAFS experiments reveal 5% incorporation of atomic P into the lattice. The structural data suggest that the P atoms are located at Ni sites and not at interstitial sites. Presumably, phosphorus incorporation occurs via a similar sequence of reaction steps observed in the formation of $[Ni_3(\mu_3\text{-PPh})(\mu\text{-PPh}_2)_2(PPh_3)_3]$, namely, sequential P—C oxidative additions and C—C reductive eliminations, via $[PR_2]^-$ and $[PR]^{2-}$ intermediates. The phosphorus doping affects the magnetic properties of the NPs, but the structure is unchanged from the fcc lattice of Ni. Thus, if only a limited suite of characterization techniques (e.g. powder X-ray diffraction) is performed on the isolated material, the P doping could be missed. These results also suggest that element doping into transition metal nanoparticles may be common, especially for the highly reducing ferrous metals. Interestingly, tri-*n*-octylphosphine oxide (TOPO) can also be used purposefully as a P atom source, as in the case of the Co_2P nanocrystals reported by Robinson and co-workers [86].

Further evidence for the high reducing ability of Ni(0) comes from the isolation of $[Ni_8(CN^tBu)_{12}]$ [Cl], which could be prepared in low yield by reaction of $[Ni(1,5\text{-COD})_2]$ with CN^tBu, followed by oxidation with adventitious $CHCl_3$ [87]. Its formula suggests that the average Ni oxidation

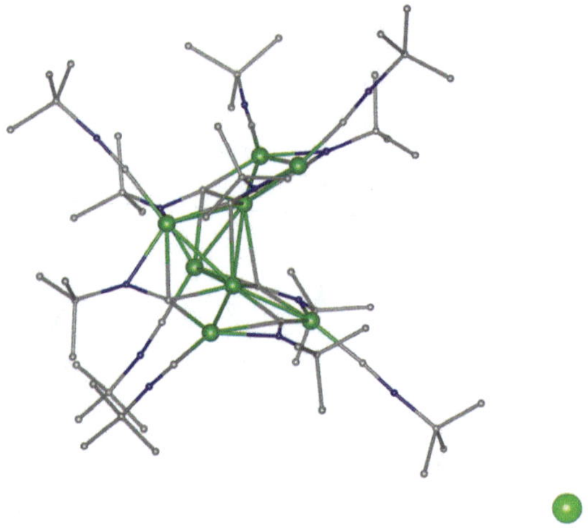

Figure 9.8 Solid-state structure of $[Ni_8(CN^tBu)_{12}]$ [Cl]. Ni, N, Cl, and C atoms are shown in green, blue, gray, and bright green, respectively. Hydrogen atoms are omitted for clarity. *Source:* X-ray data taken from Ref. [87].

state is +0.125; however, the structural analysis suggests that this value is a gross simplification. In particular, this cluster features a folded nanosheet structure containing two symmetry-related planar Ni_4 arrays hinged about an Ni—Ni bond (Figure 9.8). Six of the *tert*-butyl isocyanide ligands are terminally-bound, whereas six feature bridging μ-η^2:η^1 binding modes. The metrical parameters of the bridging isocyanide ligands suggest that they are doubly reduced, i.e. they should be considered to be $[^tBuN{=}C]^{2-}$, which implies that Ni atoms along the outer edges of the nanosheet adopt the 2+ oxidation state. In contrast, the Ni atoms that form the hinge likely have an oxidation state of +0.5 each. This result demonstrates that, for isocyanide-stabilized Ni nanoclusters, a noncompact sheet structure is preferred over a compact spherical structure, despite the fact that isocyanide is a relatively good π-acceptor that should be able to support the formation of a spherical Ni(0) APNC.

Another example that highlights the challenges inherent in making Ni(0)-containing APNCs comes from the study of nanocluster growth within the $[Ni(1,5\text{-}COD)_2]/PhSSPh/PEt_3$ system [76]. Heating a combination of $[Ni(1,5\text{-}COD)_2]$, PhSSPh, and PEt_3 results in formation of $[Ni_8(S)_5(PEt_3)_7]$ via series of intermediate clusters, including $[Ni_3(SPh)_4(PEt_3)_3]$, $[Ni_4(S)(Ph)(SPh)_3(PEt_3)_3]$, and $[Ni_5(S)_2(SPh)_2(PEt_3)_5]$ (Scheme 9.3). Importantly, the isolation of the organometallic cluster $[Ni_4(S)(Ph)(SPh)_3(PEt_3)_3]$, coupled with the observation of biphenyl in the reaction mixture, demonstrates that S^{2-} incorporation into these clusters occurs via sequential C—S oxidative addition and C—C reductive elimination steps. A similar series of steps is likely responsible for Se^{2-} incorporation into $[Ni_{11}Se_2(CO)_{18}]^{2-}$, [75] as well as P incorporation into Ni NPs [85]. These results also help explain why no thiolate-stabilized Ni APNCs have been isolated thus far, despite their ubiquity for group 11 [2, 19–21]. Because Ni(0) is so reducing (Table 9.1), C—S oxidative addition at low-valent Ni is facile, which results in net oxidation of the Ni atom and destruction of the thiolate ligand.

9.4 Synthesis of Co APNCs

As was seen for group 10, the earliest attempts to make large cobalt clusters focused on the use of carbon monoxide as the supporting ligand. For example, Chini reported the synthesis of $[Co_6(CO)_{15}]^{2-}$ in 1967, [32, 88, 89] and despite its synthesis over five decades ago, it remains the largest reported homoleptic cobalt carbonyl cluster known. It was synthesized by reduction of $[Co_4(CO)_{12}]$ with lithium or sodium metal in THF, [88] a route that has also been employed to make Ni and Fe carbonyl clusters [32]. As was observed for Ni, larger cobalt carbonyl clusters

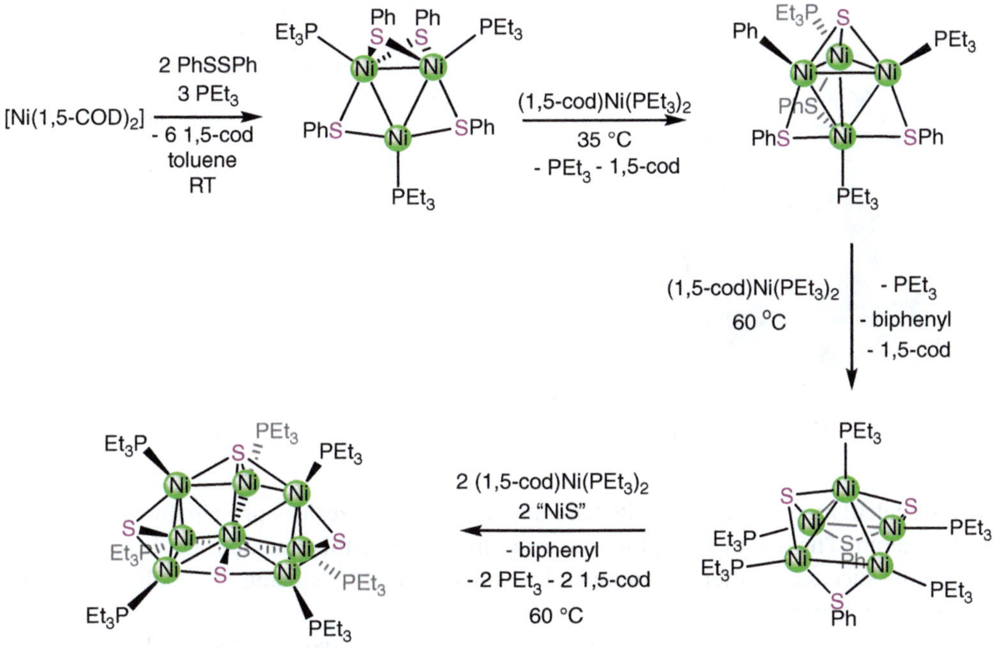

Scheme 9.3 Synthesis of $[Ni_8(S)_5(PEt_3)_7]$ from $[Ni(1,5\text{-}COD)_2]$.

typically feature interstitial main group atoms, most commonly carbon. As a result, these larger clusters do not fit our requirements for an APNC. That said, there are at least two Co-containing clusters that meets our strict definition, namely, $[Co_9Te_6(CO)_8]^{n-}$ ($n = 1, 2$) [90–93]. These clusters were prepared by reaction of $[Cp'_2Nb(\eta^2\text{-}Te_2)H]$ ($Cp' = {}^tBuC_5H_4$) with MeLi, followed by addition of $[Co_2(CO)_8]$, followed by cation exchange with [PPN][Cl]. Mechanistic studies suggest that the first step of the reaction results in formation of a mixture of $[Cp'_2Nb(Te)Me]$, $[Cp'_2Nb(H)_2(TeMe)]$, and $[Cp'_2Nb(TeMe)_2]$ [94]. These species function as good Te-atom transfer reagents, presumably because of the mismatch between the hard acid (Nb) and soft base (Te) results in weak Nb—Te bonds. Additionally, the Cp'_2Nb fragment function as a CO trap during the reaction, forming the Nb(III) cation, $[Cp'_2Nb(CO)_2]^+$. Its formation apparently allows Co cluster growth by decreasing the pool of available CO in the reaction mixture. The mixture of telluride-supported niobium metallocenes oxidize $[Co_2(CO)_8]$ to form a mixture of $[Co_9Te_6(CO)_8]^{n-}$ ($n = 1, 2$), which are then converted into the $[PPN]^+$ salts by cation exchange. The two clusters could be separated by selective crystallization and could be isolated in 28% total yield. In the solid state, both the −1 and −2 clusters feature a $[Co_9]$ cobalt-centered cubic kernel (Figure 9.9). The metrical parameters of the two derivatives are nearly identical. The average Co_{center}—Co_{shell} distance is 2.40 Å, which is considerable shorter that the Co—Co distance in the bulk metal (2.49 Å), [95] whereas the average Co_{center}—Te distance is 2.98 Å, demonstrating that there is little interaction between these atoms. Thus, the interstitial cobalt is only bound to the eight outer cobalt atoms and no other atoms. Calculations suggest that the frontier orbitals are closely-spaced and delocalized over the $[Co_9Te_6(CO)_8]^{n-}$ cage [93]. The −1 cluster is predicted to have a $(e_g)^2(t_{1g})^0$ open shell ground state. However, magnetism measurements reveal a room temperature moment of 4.65 μ_B, which is the value expected for an $S = 2$ state. The authors suggest that mixing of a low-lying excited state with the ground state may account for the observed magnetic properties. The neutral cluster $[Co_9Te_6(CO)_4(PPh_3)_4]$ is also known, but has not been well studied [93].

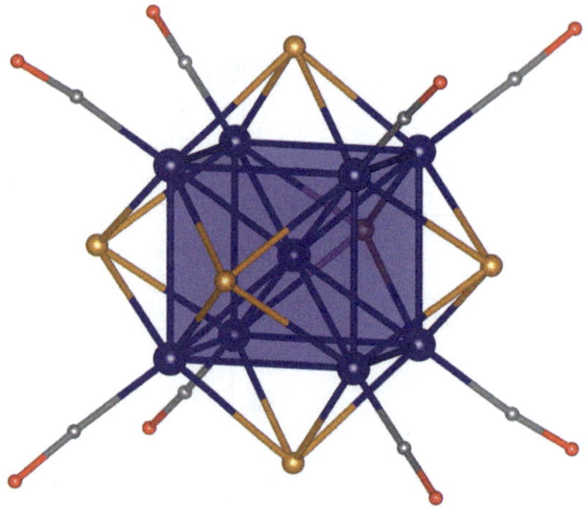

Figure 9.9 Solid-state structure of [PPN][$Co_9Te_6(CO)_8$]. Co, Te, C, and O atoms are shown in blue, goldenrod, gray, and red, respectively. The blue polyhedron represents the coordination sphere of the central Co atom. $[PPN]^+$ counterion omitted for clarity. *Source:* X-ray data taken from Ref. [93].

Nonetheless, calculations suggest that its lone unpaired electron resides principally on the interstitial cobalt atom. The use of a Co(0)-containing precursor in this reaction, i.e. [$Co_2(CO)_8$], is likely key to its success, as attempts to make low-valent Co APNCs using Co^{2+} salts can fail to achieve metal reduction [96].

The reaction of [$Cp^*{}_2Nb(\eta^2$-$Te_2H)$] with [$Co_2(CO)_8$] in refluxing toluene also produces $[Co_9Te_6(CO)_8]^{n-}$ ($n = 1, 2$), in addition to the larger clusters, [$Cp^*{}_2Nb(CO)_2$][$Co_{11}Te_5(CO)_{15}$] and [$Cp^*{}_2Nb(CO)_2$][$Co_{11}Te_7(CO)_{10}$] [97]. While neither of these latter clusters qualifies as an APNC, their formation demonstrates the complex energy landscape that is present during the reaction, and highlights how hard it is to control speciation during the synthesis of APNCs. An analysis of their structures and bonding is also informative. In this solid state, [$Cp^*{}_2Nb(CO)_2$][$Co_{11}Te_5(CO)_{15}$] features a [$Co_{11}Te_2$] cobalt-centered bi-capped pentagonal prism kernel. The two Te^{2-} ligands occupy the capping sites and feature μ_5 binding modes. In addition, three of the five square faces of the pentagonal prism are bridged by μ_4-Te ligands, while the remaining two square faces are bridged by CO ligands. Each outer Co center also features one terminal CO ligand. [$Cp^*{}_2Nb(CO)_2$][$Co_{11}Te_7(CO)_{10}$] also features a [$Co_{11}Te_2$] cobalt-centered bi-capped pentagonal prism kernel [90, 98]. As seen with $[Co_{11}Te_5(CO)_{15}]^-$, the two Te^{2-} ligands occupy the capping sites and feature μ_5 binding modes. However, each square face of the pentagonal prism is now bridged by a μ_4-Te ligand. Each outer Co center also features one terminal CO ligand. The closely related cluster, $[PPh_4]_2[Co_{11}Te_7(CO)_{10}]$, can be prepared by reaction of [$Co_2(CO)_8$] with Na_2Te_2 and [Ph_4P]Cl in methanol under solvothermal conditions [99]. The authors propose that the oxidation state of central Co is 2+, whereas the outer Co atoms are in the 1+ oxidation state. In addition, several Co-containing clusters built around [Co_9Bi_4] anti-cuboctahedral kernels are known, including $[Bi_4Co_9(CO)_{16}]^{2-}$ and $[Bi_8Co_{14}(CO)_{20}]^{2-}$ [100]. These clusters were formed by oxidation of a [PPN][$Bi_2Co_4(CO)_{11}$]/[$Mo(CO)_3(\eta^6$-toluene)] mixture with air. While the reaction did not generate an APNC, owing to the inclusion of Bi into the anti-cuboctahedral kernel, it provides another example of the use of a CO trap (in this case [$Mo(CO)_3(\eta^6$-toluene)]) to increase cluster nuclearity.

The challenges inherent in Co APNC synthesis are also exemplified by the reactivity of $CoCl_2$ with $PhCH_2CH_2SH$ under modified Brust/Schiffrin conditions [96]. In particular, reaction of $CoCl_2$ with excess $PhCH_2CH_2SH$ and $NaBH_4$ in THF at room temperature resulted in formation of [$Co_{10}(SR)_{16}Cl_4$] in 37% yield (Scheme 9.4). In the solid state, this cluster features a T3 supertetrahedral structure with 12 doubly bridged thiolate ligands and four triply bridged thiolate ligands. The four Cl^- ligands occupy the corners of the supertetrahedron. Importantly, every Co center in this cluster is Co(II), demonstrating that, despite the use of Brust/Schiffrin conditions, no metal

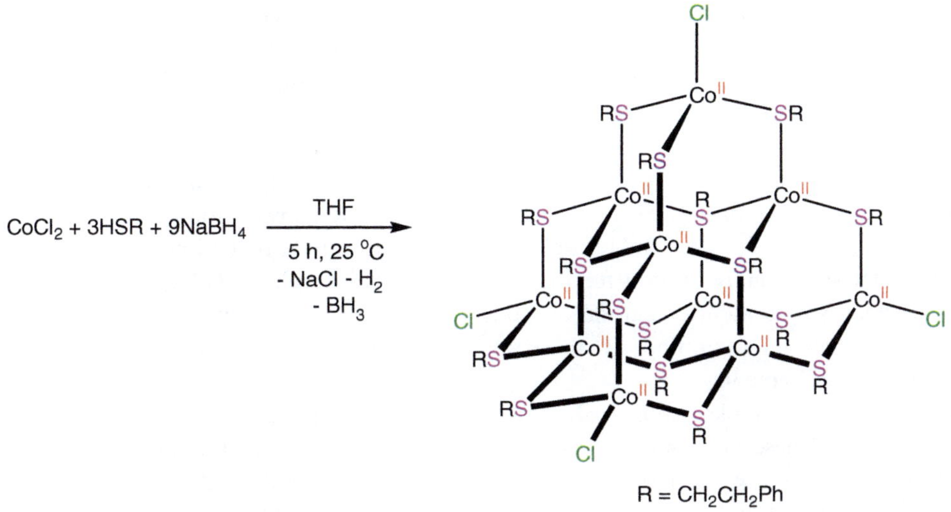

Scheme 9.4 Synthesis of $[Co_{10}(SR)_{16}Cl_4]$ using modified Brust/Schiffrin conditions.

reduction is occurring. This observation can be rationalized by the unfavorable $2e^-$ reduction potential of Co^{2+} (Table 9.1).

9.5 Attempted Synthesis of Fe APNCs

Thus far, no Fe APNCs are known, at least according to the definition of APNCs adopted for this chapter, despite the long and rich history of iron cluster chemistry [31, 101, 102]. As with Sections 9.3 and 9.4, we will start the discussion with metal carbonyl clusters. In this regard, several iron carbonyl clusters have been synthesized via the oligomerization of $[Fe(CO)_5]$; however, the highest nuclearity homoleptic iron carbonyl cluster made via this route is $[Fe_4(CO)_{13}]^{2-}$ [32]. Larger iron carbonyl clusters have been isolated, but they feature other co-ligands as well, most notably carbide (C^{4-}), as observed in the case of $[Fe_5(\mu_5\text{-}C)(CO)_{15}]$ and $[Fe_6(\mu_6\text{-}C)(CO)_{16}]^{2-}$ [32, 103–105]. Intriguingly, the interstitial carbide ligand in these examples is formed via disproportionation of carbon monoxide (Eq. (1), Scheme 9.5), [32] which forms CO_2 as a byproduct. For example, reduction of $[Fe(CO)_5]$ with $[Mn(CO)_5]^-$ in refluxing diglyme results in formation of $[Fe_6(\mu_6\text{-}C)(CO)_{16}]^{2-}$ and CO_2, in low yields (Eq. (2), Scheme 9.5) [104, 105]. This transformation, which proceeds via an $[MnFe_2(CO)_{12}]^-$ intermediate, is likely promoted by the highly reducing nature of Fe(0), and can be rationalized by the Fe : CO ratio, which increases as the cluster nuclearity increases. This increase is significant because fewer CO ligands per Fe atoms means that each Fe atom is more electron rich and therefore more likely to reduce CO. As a result of the CO disproportionation, the

$$2\,CO + 4\,e^- \longrightarrow C^{4-} + CO_2 \qquad (1)$$

$$[Fe(CO)_5] + [Mn(CO)_5]^- \xrightarrow[\text{reflux, 1 h}]{\text{diglyme}} [Fe_6C(CO)_{16}]^{2-} + CO_2 + CO \qquad (2)$$

Scheme 9.5 CO disproportionation reactions.

resulting carbide cluster features an average Fe oxidation state that is higher than the zero oxidation state expected for a homoleptic Fe carbonyl cluster. In particular, the average Fe oxidation state of $[Fe_6(\mu_6\text{-}C)(CO)_{16}]^{2-}$ is +0.33. Similar CO disproportionation chemistry has been reported for Ru [106].

The observation of spontaneous CO disproportionation during the oligomerization of $[Fe(CO)_5]$ implies that the Fe(0) core in a putative Fe APNC will be highly reducing. Accordingly, it is clear that the passivating ligands must be resistant to reduction. However, many commonly used ligands in APNC synthesis, such as thiolates and phosphines, are not resistant to reduction, as was shown in Section 9.3. Not surprisingly, several reports of Fe NP (and amorphous Fe) syntheses using $[Fe(CO)_5]$ as a precursor do mention carbon contamination in the isolated material [107–110]. It seems likely that carbon incorporation into these materials occurs via the CO disproportionation reaction described in Scheme 9.5.

Many chalcogenide- and thiolate-supported Fe clusters are known, [102, 111] but none obey our definition of an APNC. These clusters are typically made by reacting $FeCl_2$ with $E(SiMe_3)_2$ (E = S, Se) [112, 113], often in the presence of a phosphine. For example, reaction of $FeCl_2$, PEt_3, and $S(SiMe_3)_2$ in THF results in formation of $[Fe_7S_6(PEt_3)_4Cl_3]$, which converts to $[Fe_6S_6(PEt_3)_4Cl_2]$ on standing (Eqs. (1) and (2), Scheme 9.6) [114–116]. They feature average Fe oxidation states of +2.14 and +2.33, respectively. To account for the +2.14 oxidation state in $[Fe_7S_6(PEt_3)_4Cl_3]$, and the increase in oxidation state on converting to $[Fe_6S_6(PEt_3)_4Cl_2]$, it is likely that some Fe^{2+} ions are undergoing disproportionation during the reaction, forming unidentified Fe(0) or Fe(I) by-products. In fact, Holm and co-workers note the formation of dark insoluble material during the conversion of $[Fe_7S_6(PEt_3)_4Cl_3]$ to $[Fe_6S_6(PEt_3)_4Cl_2]$ [116], which could correspond to these reduced products. For comparison, reaction of $[NiCl_2(PEt_2Ph)_2]$ with $Se(SiMe_3)_2$ result is successful generation of the Ni APNC, $[Ni_{21}Se_{14}(PEt_2Ph)_{12}]$ (Section 9.3) [56], which features an average Ni oxidation state of +1.33. In this example, the average oxidation state of Ni actually decreases over the course of the reaction. Another relevant Fe cluster, $[NBu_4]_2[Fe_4S_4Br_4]$, was prepared by reaction of $FeBr_3$ with Na_2S and $[NBu_4][Br]$ in DMF [117]. In this case, Na_2S is the reducing agent and S_8 is formed as a byproduct. Its average Fe oxidation state is +2.5. The chloride and iodide congeners are also known [118]. Similarly, reaction of $[Fe(H_2O)_6][BF_4]_2$ with PEt_3 and H_2Se in the presence of $[NBu_4][PF_6]$ in EtOH/acetone results in formation of $[Fe_6(\mu_3-Se)_8(PEt_3)_6][PF_6]$ [119]. The average Fe oxidation state for this cluster is +2.83.

Several Fe chalcogenide clusters have been prepared from Fe(0) starting materials, as well, but they typically display Fe oxidation states ≥2, in contrast to Co and Ni clusters made by similar routes. For example, $[NEt_4]_2[Fe_6S_6I_6]$ was prepared by reaction of Fe metal with S_8, I_2, and $[NEt_4][I]$ (Eq. (3), Scheme 9.6), [120] while $[NBu_4]_2[Fe_6E_6(NO)_6]$ (E = S, Se), was prepared by the solvothermal reactions of $[NBu_4][Fe(CO)_3NO]$ with Se or S_8 in methanol [121]. $[Fe_6S_6I_6]^{2-}$ features an average Fe oxidation state of +2.67. Also of note, reaction of the Fe(0) synthon $[(IEt_2Me_2)_2Fe(dvtms)]$

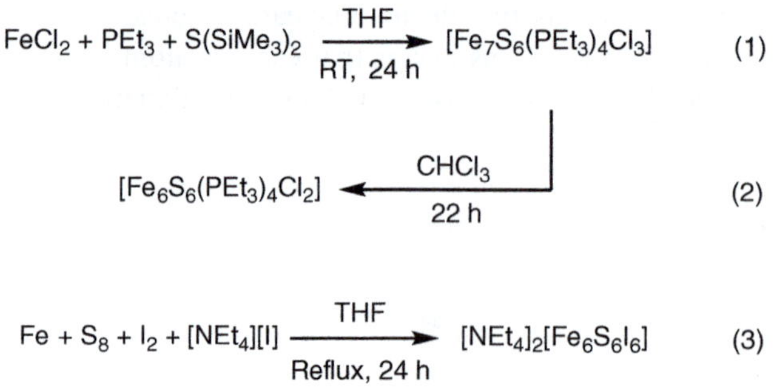

Scheme 9.6 Synthesis of Fe chalcogenide clusters.

(IEt_2Me_2 = 1,3-diethyl-4,5-dimethylimidazol-2-ylidene; dvtms = divinyltetramethyldisiloxane) with 1 equiv. of Se or 0.125 equiv. of S_8 results in the formation of $[(IEt_2Me_2)_4Fe_4E_4]$ (E = S, Se), which represents an unusually reduced iron chalcogenide cluster [113, 122]. Its average Fe oxidation state is +2. Cyclic voltammetry of the closely related cluster, $[NBu_4]_4[Fe_4S_4(CN)_4]$, confirms that the $[Fe_4E_4]^0$ core is highly reducing [123]. In particular, this cluster features a reversible $1e^-$ oxidation potential of −1.42 V in acetonitrile (vs. saturated calomel electrode), despite the presence of the electron withdrawing cyanide ligands. It is not surprising, then, that more highly reduced iron chalcogenide clusters have eluded isolation.

9.6 Conclusions and Outlook

As is hopefully apparent from the sections above, there are a number of interrelated reasons why APNCs of Fe, Co, and Ni are so exceptionally rare. Some of these reasons are inherent to their material properties and some are practical. One practical reason why these materials are so rare is due to their enhanced air sensitivity, especially relative to the group 11 APNCs [28]. This aspect of their chemistry is nicely illustrated by $[Ni_{23}Se_{12}Cl_3(PEt_3)_{10}]$, which is not only air-sensitive, but also rapidly react with CH_2Cl_2 [57]. Both observations demonstrate the care required to work with this cluster. $[Co_9Te_6(CO)_4(PPh_3)_4]$ is also highly air sensitive [93], as is $[Co_{10}(SR)_{16}Cl_4]$ [96], and many of the Fe chalcogenide clusters described in Section 9.5 [102, 111]. While daunting, this challenge can actually be overcome by the implementation of air-free synthetic procedures and rigorous purification of starting materials and solvents. In his regard, it would be helpful to adopt the experimental protocols of organometallic chemists, who have been synthesizing and characterizing highly air-sensitive metal complexes for decades. These procedures and protocols have been described in detail [124, 125].

The high reduction potentials often displayed by these clusters also restricts the choice of available co-ligands. This fact is nicely illustrated by the attempted preparation of phenylthiolate-protected Ni APNCs from $[Ni(1,5\text{-}COD)_2]$ and PhSSPh [76]. In this example, the phenylthiolate is reduced to biphenyl and sulfide, providing $[Ni_8(S)_5(PEt_3)_7]$ among other products. Similar reduction chemistry is observed during the reaction of $Ni(\eta^3\text{-}CPh_3)_2$ with PPh_3 [84], which results in formation of biphenyl, along with $[Ni_3(\mu_3\text{-}PPh)(\mu\text{-}PPh_2)_2(PPh_3)_3]$. The synthesis of $[Ni_8(CN^tBu)_{12}]$ [Cl] illustrates a similar effect [87]. In this example, the ligand does not undergo fragmentation upon reduction. Instead, it undergoes a change in binding mode – from η^1 to $\mu\text{–}\eta^2{:}\eta^1$ – which promotes the formation of the nanosheet structure and disfavors the formation of the highly compact metal core required to form an APNC. Given these examples, it is not surprising then that all authentic Co- and Ni-containing APNCs isolated thus far are stabilized by dianionic chalcogenide ligands, i.e. E^{2-}, which are unable to undergo further reduction chemistry. Consequently, it is clear that future work should focus on the use of reduction-resistant ligands to stabilize APNCs of Fe, Co, and Ni, such as Cl^-, Br^-, I^-, S^{2-}, Te^{2-}, and P^{3-}. These ligands could be used in combination with neutral Lewis bases that are also reduction-resistant, such as PEt_3, which appears to be well-suited for this purpose, although there are certainly other reduction-resistant Lewis bases that could also be employed. That said, it is obvious that this list is somewhat restrictive. Halide, chalcogenide, and pnictide ligands do not possess R groups, which eliminates an important synthetic tool for modifying structure and function. This modifiability has been used to great effect in thiolate-protected group 11 APNC chemistry. However, until thiolate ligands can be designed that do not undergo C—S oxidative addition, this popular ligand class will likely not be useful for stabilizing APNCs of Fe, Co, and Ni. The design and synthesis of reduction-resistant thiolate ligands will likely be an important research direction in the coming years.

Uncontrolled redox chemistry has also played an important role in the speciation of chalcogenide-supported Fe clusters, as exemplified by [$Fe_7S_6(PEt_3)_4Cl_3$], [$Fe_6S_6(PEt_3)_4Cl_2$], and related clusters [114–116]. In these examples, unwanted disproportionation of a lower-valent intermediate cluster leads to the isolation of the high oxidation state products. Thus, access to an Fe(0)-containing APNC will require suppression of this unwanted disproportionation. Alternatively, it could potentially be harnessed to access low-valent materials, provided the desired low-valent product can be separated from the reaction by-products. However, it is not readily apparent how these disproportionation reactions can be controlled, making this task much easier said than done.

Looking to the future, there are several avenues of investigation that promise to expand our knowledge of APNC synthesis, structure, and bonding. For example, it is clear that APNCs of Fe, Co, and Ni usually feature open-shell electronic structures, which contrasts with the closed shell configurations typically observed for group 11 APNCs [2, 19–21]. Open-shell APNCs with large spin states and a high degree of metal–metal bonding, which leads to manifold of closely spaced electronic states near the HOMO-LUMO gap, could feature interesting magnetic properties, such as SMM and metamagnetism. These properties could eventually be harnessed for a variety of potential applications, including magnetic data storage and quantum computing [3, 10–12]. Another area of interest is the synthesis of heterometallic APNCs of Fe, Co, and Ni [126]. Previously work on nanoparticles has shown that alloying can greatly modify their magnetic properties. For example, face-centered tetragonal FePt NPs exhibit higher coercivities than comparably sized Fe NPs, [127, 128] due to the magnetic coupling of the Fe and Pt *n*d states [126]. Heterometallic FeCo APNCs are also an intriguing target, given the high magnetic moment that is expected for this combination [129]. However, the challenges inherent in the synthesis of homometallic APNCs are even more acute for the synthesis of heterometallic APNCs. Their successful synthesis will require the discovery of an appropriate combination of capping ligands and metal atom precursors, coupled with the need to reproducibly incorporate an exact number of each metal atom into the APNC. This challenging synthetic problem will likely remain unsolved for some time.

Acknowledgments

I thank the National Science Foundation (CHE 2055063) for financial support of this work. I also thank Alexander J. Touchton for assistance with the preparation of the VESTA image files.

References

1 Walter, M., Akola, J., Lopez-Acevedo, O. et al. (2008). A unified view of ligand-protected gold clusters as superatom complexes. *Proc. Natl. Acad. Sci.* 105: 9157–9162.

2 Chakraborty, I. and Pradeep, T. (2017). Atomically precise clusters of noble metals: emerging link between atoms and nanoparticles. *Chem. Rev.* 117: 8208–8271.

3 Busche, C., Vilà-Nadal, L., Yan, J. et al. (2014). Design and fabrication of memory devices based on nanoscale polyoxometalate clusters. *Nature* 515: 545.

4 Li, G. and Jin, R. (2013). Atomically precise gold nanoclusters as new model catalysts. *Acc. Chem. Res.* 46: 1749–1758.

5 Liu, L. and Corma, A. (2018). Metal catalysts for heterogeneous catalysis: from single atoms to nanoclusters and nanoparticles. *Chem. Rev.* 118: 4981–5079.

6 Cook, A.W., Jones, Z.R., Wu, G. et al. (2018). An organometallic Cu_{20} nanocluster: synthesis, characterization, immobilization on silica, and "click" chemistry. *J. Am. Chem. Soc.* 140: 394–400.

7 Chen, L.-Y., Wang, C.-W., Yuan, Z., and Chang, H.-T. (2015). Fluorescent gold nanoclusters: recent advances in sensing and imaging. *Anal. Chem.* 87: 216–229.

8 Colombo, M., Carregal-Romero, S., Casula, M.F. et al. (2012). Biological applications of magnetic nanoparticles. *Chem. Soc. Rev.* 41: 4306–4334.

9 Lisnard, L., Tuna, F., Candini, A. et al. (2008). Supertetrahedral and bi-supertetrahedral cages: synthesis, structures, and magnetic properties of Deca- and Enneadecametallic Cobalt(II) clusters. *Angew. Chem. Int. Ed.* 47: 9695–9699.

10 Long, D.-L., Tsunashima, R., and Cronin, L. (2010). Polyoxometalates: building blocks for functional nanoscale systems. *Angew. Chem. Int. Ed.* 49: 1736–1758.

11 Gaita-Ariño, A., Luis, F., Hill, S., and Coronado, E. (2019). Molecular spins for quantum computation. *Nat. Chem.* 11: 301–309.

12 Tejada, J., Chudnovsky, E.M., del Barco, E. et al. (2001). Magnetic qubits as hardware for quantum computers. *Nanotechnology* 12: 181.

13 Jin, R. (2010). Quantum sized, thiolate-protected gold nanoclusters. *Nanoscale* 2: 343–362.

14 Kang, X., Chong, H., and Zhu, M. (2018). $Au_{25}(SR)_{18}$: the captain of the great nanocluster ship. *Nanoscale* 10: 10758–10834.

15 Heaven, M.W., Dass, A., White, P.S. et al. (2008). Crystal structure of the gold nanoparticle $[N(C_8H_{17})_4][Au_{25}(SCH_2CH_2Ph)_{18}]$. *J. Am. Chem. Soc.* 130: 3754–3755.

16 Brennan, J.G., Siegrist, T., Stuczynski, S.M., and Steigerwald, M.L. (1989). The transition from molecules to solids: molecular syntheses of $Ni_9Te_6(PEt_3)_8$, $Ni_{20}Te_{18}(PEt_3)_{12}$ and NiTe. *J. Am. Chem. Soc.* 111: 9240–9241.

17 Keiko, N., Satoru, O., Tadashi, Y. et al. (2003). Synthesis and characterization of the Au_{11} cluster with sterically demanding phosphine ligands by single crystal X-ray diffraction and XPS spectroscopy. *Bull. Chem. Soc. Jpn.* 76: 1601–1602.

18 The $[M_{13}]$ kernel can also adopt cuboctahedral and anti-cuboctahedral arrangements: Jin, R., Zeng, C., Zhou, M., and Chen, Y. (2016). Atomically precise colloidal metal nanoclusters and nanoparticles: fundamentals and opportunities. *Chem. Rev.* 116: 10346–10413.

19 Jin, R., Zeng, C., Zhou, M., and Chen, Y. (2016). Atomically precise colloidal metal nanoclusters and nanoparticles: fundamentals and opportunities. *Chem. Rev.* 116: 10346–10413.

20 Du, X. and Jin, R. (2020). Atomic-precision engineering of metal nanoclusters. *Dalton Trans.* 49: 10701–10707.

21 Kang, X., Li, Y., Zhu, M., and Jin, R. (2020). Atomically precise alloy nanoclusters: syntheses, structures, and properties. *Chem. Soc. Rev.* 49: 6443–6514.

22 Zabala-Lekuona, A., Seco, J.M., and Colacio, E. (2021). Single-molecule magnets: from Mn_{12}-ac to dysprosium metallocenes, a travel in time. *Coord. Chem. Rev.* 441: 213984.

23 Hernández Sánchez, R. and Betley, T.A. (2015). Meta-atom behavior in clusters revealing large spin ground states. *J. Am. Chem. Soc.* 137: 13949–13956.

24 Chakarawet, K., Bunting, P.C., and Long, J.R. (2018). Large anisotropy barrier in a tetranuclear single-molecule magnet featuring low-coordinate cobalt centers. *J. Am. Chem. Soc.* 140: 2058–2061.

25 Chakarawet, K., Atanasov, M., Marbey, J. et al. (2020). Strong electronic and magnetic coupling in M_4 (M = Ni, Cu) clusters via direct orbital interactions between low-coordinate metal centers. *J. Am. Chem. Soc.* 142: 19161–19169.

26 Frost, J.M., Harriman, K.L.M., and Murugesu, M. (2016). The rise of 3-d single-ion magnets in molecular magnetism: towards materials from molecules? *Chem. Sci.* 7: 2470–2491.

27 Luo, Y.R. (2007). *Comprehensive Handbook of Chemical Bond Energies*. Boca Raton, FL: CRC Press.

28 Bratsch, S.G. (1989). Standard electrode potentials and temperature coefficients in water at 298.15 K. *J. Phys. Chem. Ref. Data* 18: 1–21.

29 Cook, A.W. and Hayton, T.W. (2018). Case studies in nanocluster synthesis and characterization: challenges and opportunities. *Acc. Chem. Res.* 51: 2456–2464.

30 Cook, A.W., Jones, Z.R., Wu, G. et al. (2019). Synthesis and characterization of "atlas-sphere" copper nanoclusters: new insights into the reaction of Cu^{2+} with thiols. *Inorg. Chem.* 58: 8739–8749.

31 Muetterties, E.L., Rhodin, T.N., Band, E. et al. (1979). Clusters and surfaces. *Chem. Rev.* 79: 91–137.

32 Cesari, C., Shon, J.-H., Zacchini, S., and Berben, L.A. (2021). Metal carbonyl clusters of groups 8–10: synthesis and catalysis. *Chem. Soc. Rev.* 50: 9503–9539.

33 Mednikov, E.G. and Dahl, L.F. (2010). Syntheses, structures and properties of primarily nanosized homo/heterometallic palladium CO/PR_3-ligated clusters. *Philos. Trans. R. Soc. A* 368: 1301–1332.

34 Ciabatti, I., Femoni, C., Iapalucci, M.C. et al. (2014). Platinum carbonyl clusters chemistry: four decades of challenging nanoscience. *J. Clust. Sci.* 25: 115–146.

35 Beattie, J.K., Masters, A.F., and Meyer, J.T. (1995). Nickel carbonyl cluster complexes. *Polyhedron* 14: 829–868.

36 Masters, A.F. and Meyer, J.T. (1995). Polyhedron report number 51 structural systematics in nickel carbonyl cluster anions. *Polyhedron* 14: 339–365.

37 Stromnova, T.A. and Moiseev, I.I. (1998). Palladium carbonyl complexes. *Russ. Chem. Rev.* 67: 485–514.

38 Femoni, C., Iapalucci, M.C., Kaswalder, F. et al. (2006). The possible role of metal carbonyl clusters in nanoscience and nanotechnologies. *Coord. Chem. Rev.* 250: 1580–1604.

39 Berti, B., Femoni, C., Iapalucci, M.C. et al. (2018). Functionalization, modification, and transformation of platinum Chini clusters. *Eur. J. Inorg. Chem.* 2018: 3285–3296.

40 Ceriotti, A., Chini, P., Della Pergola, R., and Longoni, G. (1983). Carbonylnickelates. 3. Synthesis and chemical characterization of the $[Ni_{12}(CO)_{21}H_{4-n}]^{n-}$ ($n = 2, 3, 4$) clusters. *Inorg. Chem.* 22: 1595–1598.

41 Broach, R.W., Dahl, L.F., Longoni, G. et al. (1978). *Transition Metal Hydrides*, ACS Symposium Series, vol. 167, 93–110. American Chemical Society.

42 Barkley, J.V., Eguchi, T., Harding, R.A. et al. (1999). Solid state studies (X-ray and ^1H-, ^{13}C-NMR) on $(NMe_4)_{4-x}[H_xNi_{12}(CO)_{21}]\cdot S$ ($x = 1$, $S = Me_2CO$; $x = 2$, $S = 2THF$). *J. Organomet. Chem.* 573: 254–260.

43 Ceriotti, A., Fait, A., Longoni, G. et al. (1986). Synthesis and x-ray structure of the $[HNi_{38}(CO)_{42}C_6]^{5-}$ cluster: an extended fragment of the $Cr_{23}C_6$ lattice stabilized in a molecular carbonyl nickel cluster. *J. Am. Chem. Soc.* 108: 8091–8092.

44 Calderoni, F., Demartin, F., Iapalucci, M.C., and Longoni, G. (1996). Synthesis and crystal structure of the $[Ni_{32}C_6(CO)_{36}]^{6-}$ Hexaanion: an extended fragment of the $M_{23}C_6$ lattice stabilized in a Shell of edge-bridging carbonyl ligands. *Angew. Chem. Int. Ed.* 35: 2225–2226.

45 Ceriott, A., Demartin, F., Longoni, G. et al. (1985). Synthesis and structure of the $[Ni_{38}Pt_6(CO)_{48}H_{6-n}]^{n-}$ ($n = 5, 4$) ions: Ni-Pt clusters as models for "cherry" crystallites. *Angew. Chem. Int. Ed.* 24: 697–698.

46 Fabrizi de Biani, F., Femoni, C., Iapalucci, M.C. et al. (1999). Redox behavior of $[H_{6-n}Ni_{38}Pt_6(CO)_{48}]^{n-}$ ($n = 4$–6) anions: a series of metal carbonyl clusters displaying Electron-sink features. *Inorg. Chem.* 38: 3721–3724.

47 de Silva, N. and Dahl, L.F. (2006). Aprotic synthesis and structural determination of the nanosized nonprotonated ν_3-octahedral $[Pt_6Ni_{38}(CO)_{48}]^{6-}$ Hexaanion stabilized as a cubic solvated $[NMe_4]^+$ salt. *Inorg. Chem.* 45: 8814–8816.

48 Femoni, C., Iapalucci, M.C., Longoni, G. et al. (2004). Synthesis and characterisation of ν_3-octahedral $[Ni_{36}Pd_8(CO)_{48}]^{6-}$ and $[Ni_{35}Pt_9(CO)_{48}]^{6-}$ clusters displaying unexpected surface segregation of Pt atoms and molecular and/or crystal substitutional Ni/Pd and Ni/Pt disorder. *Chem. Eur. J.* 10: 2318–2326.

49 Tran, N.T., Kawano, M., Powell, D.R., and Dahl, L.F. (2000). High-nuclearity $[Pd_{13}Ni_{13}(CO)_{34}]^{4-}$ containing a 26-atom $Pd_{13}Ni_{13}$ core with an unprecedented five-layer close-packed triangular stacking geometry: possible substitutional Pd/Ni crystal disorder at specific intralayer nickel sites. *J. Chem. Soc. Dalton Trans.* 4138–4144.

50 Femoni, C., Iapalucci, M.C., Longoni, G. et al. (2000). Homoleptic carbonyl Ni–Pd clusters: synthesis of $[Ni_{16}Pd_{16}(CO)_{40}]^{4-}$ and $[Ni_{26}Pd_{20}(CO)_{54}]^{6-}$ and structural characterization of $[N^nBu_4]_4[Ni_{16}Pd_{16}(CO)_{40}]$. *Angew. Chem. Int. Ed.* 39: 1635–1637.

51 Kawano, M., Bacon, J.W., Campana, C.F. et al. (2001). High-nuclearity close-packed palladium-nickel carbonyl phosphine clusters: Heteropalladium $[Pd_{16}Ni_4(CO)_{22}(PPh_3)_4]^{2-}$ and $[Pd_{33}Ni_9(CO)_{41}(PPh_3)_6]^{4-}$ containing Pseudo-T_d ccp $Pd_{16}Ni_4$ and Pseudo-D_{3h} hcp $Pd_{33}Ni_9$ cores. *Inorg. Chem.* 40: 2554–2569.

52 Tran, N.T., Kawano, M., Powell, D.R. et al. (1999). Isostructural $[Au_6Pd_6(Pd_{6-x}Ni_x)Ni_{20}(CO)_{44}]^{6-}$ and $[Au_6Ni_{32}(CO)_{44}]^{6-}$ clusters containing corresponding nonstoichiometric $Au_6Pd_6(Pd_{6-x}Ni_x)Ni_{20}$ and stoichiometric Au_6Ni_{32} nanosized cores: substitutional Pd/Ni crystal disorder (coloring problem) at only six specific nonadjacent Pseudoequivalent metal sites in the 38-atom trimetallic close-packed framework. *J. Am. Chem. Soc.* 121: 5945–5952.

53 Berti, B., Cesari, C., Femoni, C. et al. (2020). Redox active Ni–Pd carbonyl alloy nanoclusters: syntheses, molecular structures and electrochemistry of $[Ni_{22-x}Pd_{20+x}(CO)_{48}]^{6-}$ ($x = 0.62$), $[Ni_{29-x}Pd_{6+x}(CO)_{42}]^{6-}$ ($x = 0.09$) and $[Ni_{29+x}Pd_{6-x}(CO)_{42}]^{6-}$ ($x = 0.27$). *Dalton Trans.* 49: 5513–5522.

54 Femoni, C., Iapalucci, M.C., Longoni, G., and Svensson, P.H. (2004). New high-nuclearity Ni–Pt carbonyl clusters: synthesis and X-ray structure of the ordered $[HNi_{24}Pt_{17}(CO)_{46}]^{5-}$ and the substitutionally Ni/Pt disordered $[Ni_{32}Pt_{24}(CO)_{56}]^{6-}$ cluster anions. *Chem. Commun.* 2274–2275.

55 Wix, P., Kostakis, G.E., Blatov, V.A. et al. (2013). A database of topological representations of polynuclear nickel compounds. *Eur. J. Inorg. Chem.* 2013: 520–526.

56 Fenske, D., Krautscheid, H., and Müller, M. (1992). New intermediate steps in the synthesis of larger nickel clusters. *Angew. Chem. Int. Ed.* 31: 321–323.

57 Touchton, A.J., Wu, G., and Hayton, T.W. (2021). $[Ni_{23}Se_{12}(PEt_3)_{13}]$ revisited: isolation and characterization of $[Ni_{23}Se_{12}Cl_3(PEt_3)_{10}]$. *Inorg. Chem.* 60: 17586–17592.

58 Brennan, J.G., Siegrist, T., Kwon, Y.U. et al. (1992). Nickel-selenium-triethylphosphine $(Ni_{23}Se_{12}(PEt_3)_{13})$, an intramolecular intergrowth of nickel selenide (NiSe) and nickel. *J. Am. Chem. Soc.* 114: 10334–10338.

59 Jette, E.R. and Foote, F. (1935). Precision determination of lattice constants. *J. Chem. Phys.* 3: 605–616.

60 Nomikou, Z., Schubert, B., Hoffmann, R., and Steigerwald, M.L. (1992). Relationships between extended structures and molecular clusters of nickel and tellurium. *Inorg. Chem.* 31: 2201–2209.

61 Palstra, T.T.M., Steigerwald, M.L., Ramirez, A.P. et al. (1993). Electron correlations on a mesoscopic scale: magnetic properties of transition metal telluride cluster compounds. *Phys. Rev. Lett.* 71: 1768–1771.

62 Palstra, T.T.M., Steigerwald, M.L., Ramirez, A.P., and Zaanen, J. (1994). Electron correlations in transition metal-telluride cluster compounds. *Phys. B Condens. Matter* 199–200: 619–621.

63 Furet, E., Le Beuze, A., Halet, J.-F., and Saillard, J.-Y. (1995). Bonding analysis in inorganic transition-metal cubic clusters. 2. Metal-centered hexacapped $M_9(\mu_4\text{-}E)_6L_8$ species. *J. Am. Chem. Soc.* 117: 4936–4944.

64 Halet, J.-F. and Saillard, J.-Y. (1997). *Structural and Electronic Paradigms in Cluster Chemistry* (ed. D.M.P. Mingos), 81–109. Heidelberg: Springer Berlin Heidelberg.

65 Wheeler, R.A. (1990). Bonding to interstitial main-group or transition-metal atoms in cubic clusters related to nickel tellurium triethylprophine, $Ni_9(\mu_4\text{-}Te)_6(PEt_3)_8$. *J. Am. Chem. Soc.* 112: 8737–8741.

66 Wheeler, R.A. (1991). Bonding to interstitial main-group or transition-metal atoms in cubic clusters related to nickel tellurium triethylphosphine, $Ni_9(\mu_4\text{-}Te)_6(PEt_3)_8$ [Erratum]. *J. Am. Chem. Soc.* 113: 4046–4046.

67 Reber, A.C., Chauhan, V., and Khanna, S.N. (2017). Symmetry and magnetism in Ni_9Te_6 clusters ligated by CO or phosphine ligands. *J. Chem. Phys.* 146: 024302.

68 Roy, X., Lee, C.-H., Crowther, A.C. et al. (2013). Nanoscale atoms in solid-state chemistry. *Science* 341: 157–160.

69 Lee, C.-H., Liu, L., Bejger, C. et al. (2014). Ferromagnetic ordering in superatomic solids. *J. Am. Chem. Soc.* 136: 16926–16931.

70 Zebrowski, J.P., Hayashi, R.K., and Dahl, L.F. (1993). A new family of icosahedral cages with transition metal and main group IV (14) atoms: synthesis and structural-bonding analysis of the $[Ni_{11}(SnR)_2(CO)_{18}]^{2-}$ dianions (R = Bu, Me) containing nickel-centered icosahedral $Ni_{10}Sn_2$ cages and of their unusual $[Ni(SnRCl_2)_4(CO)]^{2-}$ precursors containing a trigonal-bipyramidal d^8 nickel(II) configuration. *J. Am. Chem. Soc.* 115: 1142–1144.

71 Zebrowski, J.P., Hayashi, R.K., Bjarnason, A., and Dahl, L.F. (1992). A new family of 14-vertex hexacapped metal cubes with main group IV (14) atoms: synthesis and structural-bonding analysis of $Ni_9(\mu_4\text{-}GeEt)_6(CO)_8$ containing a nickel-centered $Ni_8(\mu_4\text{-}Ge)_6$ cubic cage with an unusual electron count. *J. Am. Chem. Soc.* 114: 3121–3123.

72 Fenske, D., Merzweiler, K., and Ohmer, J. (1988). $[Ni_9(\mu_4\text{-}As)_6(PPh_3)_5Cl_3]$, a cubic body-centered Ni_9-cluster. *Angew. Chem. Int. Ed.* 27: 1512–1513.

73 Fenske, D. and Persau, C. (1991). Synthesen und Kristallstrukturen von $[Pd_9As_8(PPh_3)_8]$ und $[Pd_9Sb_6(PPh_3)_8]$. *Z. Anorg. Allg. Chem.* 593: 61–68.

74 Pyykkö, P. (2015). Additive covalent radii for single-, double-, and triple-bonded molecules and tetrahedrally bonded crystals: a summary. *J. Phys. Chem. A* 119: 2326–2337.

75 Kahaian, A.J., Thoden, J.B., and Dahl, L.F. (1992). A new family of metal-centred and noncentred icosahedral cages with transition metal and main group VI atoms: synthesis and structural-bonding analysis of the nickel-centred $[Ni_{11}Se_2(CO)_{18}]^{2-}$, nickel-centred $[Ni_{10}Te_3(CO)_{15}]^{2-}$ and noncentred $[Ni_8Te_4(CO)_{12}]^{2-}$ icosahedral dianions. *J. Chem. Soc. Chem. Commun.* 353–355.

76 Touchton, A.J., Wu, G., and Hayton, T.W. (2021). Understanding the early stages of nickel sulfide nanocluster growth: isolation of Ni_3, Ni_4, Ni_5, and Ni_8 intermediates. *Small* 17: 2003133.

77 Albano, V.G., Demartin, F., Iapalucci, M.C. et al. (1991). Icosahedral carbonyl clusters $[Ni_{10}Sb_2(\mu_{12}\text{-}Ni)\{Ni(CO)_3\}_2(CO)_{18}]^{n-}$ ($n = 2$, 3 or 4); synthesis, spectroscopic, electrochemical and bonding analysis. Crystal structures of $[Ni_{10}Sb_2(\mu_{12}\text{-}Ni)\{Ni(CO)_3\}_2(CO)_{18}]^{n-}$ ($n = 2$ or 3). *J. Chem. Soc. Dalton Trans.* 739–748.

78 Albano, V.G., Demartin, F., Iapalucci, M.C. et al. (1990). Synthesis and X-ray structure of the $[Ni_{13}Sb_2(CO)_{24}]^{n-}$ ($n = 2, 3, 4$) cluster anions containing an icosahedral $[Ni_{11}Sb_2(CO)_{18}]^{n-}$ moiety behaving as a distibine ligand and displaying an unexpected electron count. *J. Chem. Soc. Chem. Commun.* 547–548.

79 Femoni, C., Iapalucci, M.C., Longoni, G. et al. (2014). The chemistry of Ni–Sb carbonyl clusters – synthesis and characterization of the $[Ni_{19}Sb_4(CO)_{26}]^{4-}$ Tetraanion and the viologen salts of $[Ni_{13}Sb_2(CO)_{24}]^{n-}$ carbonyl clusters. *Eur. J. Inorg. Chem.* 2014: 4151–4158.

80 Albano, V.G., Demartin, F., Iapalucci, M.C. et al. (1992). New icosahedral heterometallic nickel carbonyl clusters containing bismuth. *J. Chem. Soc. Dalton Trans.* 497–502.

81 Kagalwala, H.N., Gottlieb, E., Li, G. et al. (2013). Photocatalytic hydrogen generation system using a nickel-thiolate Hexameric cluster. *Inorg. Chem.* 52: 9094–9101.

82 Zhu, M., Zhou, S., Yao, C. et al. (2014). Reduction-resistant and reduction-catalytic double-crown nickel nanoclusters. *Nanoscale* 6: 14195–14199.

83 Joya, K.S., Sinatra, L., AbdulHalim, L.G. et al. (2016). Atomically monodisperse nickel nanoclusters as highly active electrocatalysts for water oxidation. *Nanoscale* 8: 9695–9703.

84 Touchton, A.J., Wu, G., and Hayton, T.W. (2020). Generation of a Ni_3 phosphinidene cluster from the Ni(0) synthon, $Ni(\eta^3\text{-}CPh_3)_2$. *Organometallics* 39: 1360–1365.

85 Moreau, L.M., Ha, D.-H., Bealing, C.R. et al. (2012). Unintended phosphorus doping of nickel nanoparticles during synthesis with TOP: a discovery through structural analysis. *Nano Lett.* 12: 4530–4539.

86 Zhang, H., Ha, D.-H., Hovden, R. et al. (2011). Controlled synthesis of uniform cobalt phosphide hyperbranched nanocrystals using tri-*n*-octylphosphine oxide as a phosphorus source. *Nano Lett.* 11: 188–197.

87 Touchton, A.J., Wu, G., and Hayton, T.W. (2021). $[Ni_8(CN^tBu)_{12}][Cl]$: a nickel isocyanide nanocluster with a folded nanosheet structure. *J. Chem. Phys.* 154: 211102.

88 Chini, P. (1967). Hexacobalt hexadecacarbonyl and its derivatives. *Chem. Commun.* 440–441.

89 Albano, V., Chini, P., and Scatturin, V. (1968). Crystal and molecular structures of new cobalt carbonyl clusters. *Chem. Commun.* 163–164.

90 Brunner, H., Lucas, D., Monzon, T. et al. (2000). Metal telluride clusters composed of niobocene carbonyl, telluride, and cobalt carbonyl units: syntheses, structures, and reactivity. *Chem. Eur. J.* 6: 493–503.

91 Ebner, A., Wachter, J., and Zabel, M. (2004). Synthesis and characterization of Cobalttelluride clusters with the cubic body-centered $Co_9Te_6(CO)_8$ Core. *J. Clust. Sci.* 15: 163–174.

92 Wachter, J. (2004). Metal telluride clusters – from small molecules to polyhedral structures. *Eur. J. Inorg. Chem.* 2004: 1367–1378.

93 Bencharif, M., Cador, O., Cattey, H. et al. (2008). Electron-sponge behavior, reactivity and electronic structures of cobalt-centered cubic $Co_9Te_6(CO)_8$ clusters. *Eur. J. Inorg. Chem.* 2008: 1959–1968.

94 Ebner, A., Meier, W., Mugnier, Y., and Wachter, J. (2007). Investigation into the reaction of $(t\text{-}BuC_5H_4)_2Nb(\eta^2\text{-}Te_2)H$ with CH_3Li: hydride abstraction versus telluride methylation. *Inorg. Chim. Acta* 360: 3330–3334.

95 Hull, A.W. (1921). X-ray crystal analysis of thirteen common metals. *Phys. Rev.* 17: 571–588.

96 Cook, A.W., Wu, G., and Hayton, T.W. (2018). A re-examination of the synthesis of monolayer-protected $Cox(SCH_2CH_2Ph)m$ nanoclusters: unexpected formation of a thiolate-protected Co(II) T3 Supertetrahedron. *Inorg. Chem.* 57: 8189–8194.

97 Cador, O., Cattey, H., Halet, J.-F. et al. (2007). Electron-sponge behavior and electronic structures in cobalt-centered pentagonal prismatic $Co_{11}Te_7(CO)_{10}$ and $Co_{11}Te_5(CO)_{15}$ cluster anions. *Inorg. Chem.* 46: 501–509.

98 Brunner, H., Cattey, H., Meier, W. et al. (2003). Preparative and electrochemical investigations on the electron sponge behavior of cobalt telluride clusters: CO substitution in $[Co_{11}Te_7(CO)_{10}]^{n-}$ ions ($n = 1, 2$) by PMe_2Ph and crystal structure of $[Co_{11}Te_7(CO)_5(PMe_2Ph)_5]$. *Chem. Eur. J.* 9: 3796–3802.

99 Seidel, R., Kliss, R., Weissgräber, S., and Henkel, G. (1994). $[Co_{11}Te_7(CO)_{10}]^{2-}$, a novel cobalt telluride cluster anion containing a body-centred pentagonal-prismatic cobalt framework, and $[Co_4Te_2(CO)_{11}]$, a possible precursor with a metal-deficient pentagonal-planar cobalt frame. *J. Chem. Soc. Chem. Commun.* 2791–2792.

100 Whitmire, K.H. and Eveland, J.R. (1994). Structural characterization of two large bismuth–cobalt carbonyl clusters: $(PPN)_2[Bi_4Co_9(CO)_8(\mu\text{-}CO)_8]\cdot 2THF$ and $(PPN)_2[Bi_8Co_{14}(CO)_{12}(\mu\text{-}CO)_8]\cdot 1.08THF$. *J. Chem. Soc. Chem. Commun.* 1335–1336.

101 Tachikawa, M. and Muetterties, E.L. (1981). *Progress in Inorganic Chemistry*, vol. 28 (ed. S.J. Lippard), 203–238. New York: John Wiley & Sons.

102 Lee, S.C., Lo, W., and Holm, R.H. (2014). Developments in the biomimetic chemistry of Cubane-type and higher Nuclearity Iron–sulfur clusters. *Chem. Rev.* 114: 3579–3600.

103 Braye, E.H., Dahl, L.F., Hubel, W., and Wampler, D.L. (1962). The preparation, properties and structure of the Iron carbonyl carbide $Fe_5(CO)_{15}C$. *J. Am. Chem. Soc.* 84: 4633–4639.

104 Churchill, M.R., Wormald, J., Knight, J., and Mays, M.J. (1971). Synthesis and crystallographic characterization of bis(tetramethylammonium) carbidohexadecacarbonylhexaferrate, a hexanuclear carbidocarbonyl derivative of iron. *J. Am. Chem. Soc.* 93: 3073–3074.

105 Churchill, M.R. and Wormald, J. (1974). Crystal and molecular structure of tetramethylammonium carbidohexadecacarbonylhexaferrate(2−), $[Me_4N]_2[Fe_6(CO)_{16}C]$, a hexanuclear iron cluster complex with an encapsulated six-co-ordinate carbon atom. *J. Chem. Soc. Dalton Trans.* 2410–2415.

106 Bailey, P.J., Johnson, B.F.G., and Lewis, J. (1994). The formation of carbides in hexaruthenium carbonyl clusters. *Inorg. Chim. Acta* 227: 197–200.

107 van Wonterghem, J., Mørup, S., Charles, S.W. et al. (1985). Formation of a metallic glass by thermal decomposition of $Fe(CO)_5$. *Phys. Rev. Lett.* 55: 410–413.

108 Farrell, D., Majetich, S.A., and Wilcoxon, J.P. (2003). Preparation and characterization of monodisperse Fe nanoparticles. *J. Phys. Chem. B* 107: 11022–11030.

109 Long, G.J., Hautot, D., Pankhurst, Q.A. et al. (1998). Mössbauer-effect and X-ray-absorption spectral study of sonochemically prepared amorphous iron. *Phys. Rev. B* 57: 10716–10722.

110 Hansen, M.F., Jönsson, P.E., Nordblad, P., and Svedlindh, P. (2002). Critical dynamics of an interacting magnetic nanoparticle system. *J. Phys. Condens. Matter* 14: 4901–4914.

111 Holm, R.H. and Lo, W. (2016). Structural conversions of synthetic and protein-bound Iron–sulfur clusters. *Chem. Rev.* 116: 13685–13713.

112 Eichhöfer, A., Olkowska-Oetzel, J., Fenske, D. et al. (2009). Synthesis and structure of an "iron-doped" copper selenide cluster molecule: $[Cu_{30}Fe_2Se_6(SePh)_{24}(dppm)_4]$. *Inorg. Chem.* 48: 8977–8984.

113 Deng, L. and Holm, R.H. (2008). Stabilization of fully reduced Iron—sulfur clusters by carbene ligation: the $[Fe_nS_n]^0$ oxidation levels ($n = 4, 8$). *J. Am. Chem. Soc.* 130: 9878–9886.

114 Noda, I., Snyder, B.S., and Holm, R.H. (1986). Trichlorotetrakis(triethylphosphine) hexathioheptairon ($Fe_7S_6(PEt_3)_4Cl_3$): a topological link between the prismatic and cube-octahedral structures of iron-sulfur clusters. *Inorg. Chem.* 25: 3851–3853.

115 Snyder, B.S., Reynolds, M.S., Noda, I., and Holm, R.H. (1988). Core conversion reactions of iron-sulfur clusters: topological isomer of the Fe_6S_6-prismane core. *Inorg. Chem.* 27: 595–597.

116 Synder, B.S. and Holm, R.H. (1988). Hexanuclear iron-sulfur basket clusters: topological isomers of prismanes. Synthesis, structure, and reactions. *Inorg. Chem.* 27: 2339–2347.

117 Rutchick, S., Kim, S., and Walters, M.A. (1988). Facile one-step synthesis of tetrachalcogenotetrairon(2+) (chalcogen = sulfur, selenium) cubane-like centers. *Inorg. Chem.* 27: 1513–1515.

118 Wong, G.B., Bobrik, M.A., and Holm, R.H. (1978). Inorganic derivatives of iron sulfide thiolate dimers and tetramers: synthesis and properties of the halide series $[Fe_2S_2X_4]^{2-}$ and $[Fe_4S_4X_4]^{2-}$ (X = chlorine, bromine, iodine). *Inorg. Chem.* 17: 578–584.

119 Cecconi, F., Ghilardi, C.A., Midollini, S., and Orlandini, A. (1997). Synthesis and crystal structure of the novel paramagnetic selenium/iron stellated octahedral cluster $[Fe_6(\mu_3\text{-}Se)_8(PEt_3)_6][PF_6]$. *Inorg. Chim. Acta* 254: 387–389.

120 Saak, W., Henkel, G., and Pohl, S. (1984). A new route to iron-sulfur clusters: synthesis and structure of $[(C_2H_5)_4N]_2Fe_6S_6I_6$. *Angew. Chem. Int. Ed.* 23: 150–151.

121 Wang, R., Xu, W., Zhang, J., and Li, L. (2010). A new hexanuclear iron—selenium nitrosyl cluster: primary exploration of the preparation methods, structure, and spectroscopic and electrochemical properties. *Inorg. Chem.* 49: 4814–4819.

122 Zhang, H., Ouyang, Z., Liu, Y. et al. (2014). (Aminocarbene)(Divinyltetramethyldisiloxane)Iron(0) compounds: a class of low-coordinate iron(0) reagents. *Angew. Chem. Int. Ed.* 53: 8432–8436.

123 Scott, T.A., Berlinguette, C.P., Holm, R.H., and Zhou, H.-C. (2005). Initial synthesis and structure of an all-ferrous analogue of the fully reduced $[Fe_4S_4]^0$ cluster of the nitrogenase iron protein. *Proc. Natl Acad. Sci. USA* 102: 9741–9744.

124 Jolly, W.L. (1970). *The Synthesis and Characterization of Inorganic Compounds*. Englewood Cliffs, NJ: Prentice-Hall.

125 Burger, B.J. and Bercaw, J.E. (1987). *Experimental Organometallic Chemistry*, ACS Symposium Series, vol. 357, 79–115. American Chemical Society.

126 Wu, L., Mendoza-Garcia, A., Li, Q., and Sun, S. (2016). Organic phase syntheses of magnetic nanoparticles and their applications. *Chem. Rev.* 116: 10473–10512.

127 Sun, S. (2006). Recent advances in chemical synthesis, self-assembly, and applications of FePt nanoparticles. *Adv. Mater.* 18: 393–403.

128 Rutledge, R.D., Morris, W.H., Wellons, M.S. et al. (2006). Formation of FePt nanoparticles having high coercivity. *J. Am. Chem. Soc.* 128: 14210–14211.

129 Seo, W.S., Lee, J.H., Sun, X. et al. (2006). FeCo/graphitic-shell nanocrystals as advanced magnetic-resonance-imaging and near-infrared agents. *Nat. Mater.* 5: 971–976.

10

Atomically Precise Heterometallic Rhodium Nanoclusters Stabilized by Carbonyl Ligands

Guido Bussoli[1], Cristiana Cesari[1], Cristina Femoni[1], Maria C. Iapalucci[1], Silvia Ruggieri[2], and Stefano Zacchini[1]

[1] *Department of Industrial Chemistry "Toso Montanari", University of Bologna, Viale del Risorgimento 4, 40136, Bologna, Italy*
[2] *Laboratory of Luminescent Materials, Department of Biotechnology, University of Verona, Strada Le Grazie 15, 37134, Verona, Italy*

10.1 Introduction

The intention of this chapter is to give the readers an overview of recent works on rhodium nanoclusters stabilized by carbonyl ligands that can be identified with atomic precision.

The first introductory part will be devoted to the state of the art of transition metal clusters in general, to set the premises, followed by a more specific section on rhodium-based cluster compounds. The second part will illustrate the synthesis employed to obtain rhodium nanoclusters and the strategy to grow their nuclearity, with the focus on those species containing interstitial heteroatoms. The description of their multivalence properties will be defined in the third part, where selected examples will be discussed in detail. The fourth section will conclude the chapter and present future perspectives.

With this chapter, authors would also like to honor those scientists that most greatly contributed to the growth of the carbonyl cluster chemistry after its very beginning, setting the premises for modern nanochemistry.

10.1.1 Metal Carbonyl Clusters: A Brief Historical Overview

Transition-metal carbonyl clusters carry a long tradition of chemical research that dates back to the beginning of the previous century. It all started in Germany with the pioneering work on iron carbonyls of Walter Otto Hieber, defined by Lawrence F. Dahl as the "Father of Metal Carbonyl Chemistry." Heiber identified the first polynuclear species [1, 2]. Dahl himself, in the United States, devoted his professional life to the synthesis and characterization of metal clusters and unraveled the first crystal structures of metal carbonyls [3], thus confirming the existence of direct metal bonds that are responsible for their stability.

On this basis, F. A. Cotton in 1966 proposed what is, at present, the most accredited definition of metal clusters. He described them as compounds "containing a finite group of metal atoms which are held together entirely, mainly, or at least to a significant extent, by bonds directly between the metal atoms even though some non-metal atoms may be associated intimately with the cluster" [4].

Atomically Precise Nanochemistry, First Edition. Edited by Rongchao Jin and De-en Jiang.

By also benefiting from the development of single-crystal X-ray techniques, which permitted the characterization of larger and larger species, otherwise too challenging for the first-generation diffractometers, Prof. Dahl's outstanding contribution to the chemistry of metal clusters lasted many decades [5], until his recent demise in 2021. The field greatly expanded in the 1970s and 1980s also thanks to the exceptional work of Brian F. G. Johnson and Lord Jack Lewis in Cambridge, and Paolo Chini in Milan. The former groups in Cambridge mainly concentrated their activities on osmium and ruthenium neutral carbonyl species [6–8], while the Milan school was mostly devoted to the synthesis of nickel, platinum, cobalt, and rhodium anionic clusters [9]. Notably, the platinum clusters based on $\{Pt_3(CO)_6\}$ units are referred as *Chini clusters*, in recognition of Chini's exceptional contribution. In the 1990s the scientific community's interest in cluster chemistry waned, but it regained some attention after the 2000s due to the growth of nanotechnology [10], which symbolically started a few years earlier with the appearance of the renowned paper published by Brust et al. in 1994, reporting a straightforward synthesis of thiolate-protected gold nanoparticles [11]. Large metal carbonyl clusters can, in fact, be described as metal nanoparticles stabilized by CO ligands. One of the most captivating aspects of transition metal carbonyls is that they have been among the first species of such kind to be characterized with atomic precision. This is due to their molecular nature and to their possibility of being obtained in single crystals suitable for X-ray analyses, allowing their structure and composition to be unambiguously unraveled.

Over the years, the nuclearity of metal carbonyl clusters (corresponding to the number of their metal atoms) has increased more and more, to the point that their size has reached a nanometric regime [12]. A symbolic example is represented by Mednikov and Dahl 's $(\mu_{12}\text{-}Pt)Pd_{164-x}Pt_x(CO)_{72}(PPh_3)_{20}$ ($x \approx 7$) heteroleptic cluster [13], whose diameter reaches ca. 3 nm. Large homoleptic nanoclusters are also known, both homo- and bimetallic, including $[Pt_{38}(CO)_{44}]^{2-}$ [14], $[H_{6-n}Ni_{38}Pt_6(CO)_{44}]^{n-}$ (n = 6–4) [15, 16], $[Ni_{32}Pt_{24}(CO)_{56}]^{6-}$ [17], and $[Au_{34}Fe_{14}(CO)_{50}]^{8-}$ [18].

10.1.2 State of the Art on Rhodium Carbonyl Clusters

The chemistry of rhodium carbonyl clusters was one of the subjects of Hieber's work. However, it was only in 1963 that $Rh_6(CO)_{16}$, the very first example of hexanuclear metal carbonyl species but mistakenly labeled as $Rh_4(CO)_{11}$ 20 years before, was finally structurally characterized [19].

Rhodium presents quite high Rh–Rh and Rh–CO binding energies [20, 21], similar to Pt, and this promotes the formation of large nuclearity compounds. Several rhodium nanoclusters have been synthetized over the past years, initially by the Milano's school, for instance the homometallic $[Rh_{15}(CO)_{27}]^{3-}$, $[Rh_{14}(CO)_{25}]^{4-}$ [22] and $[Rh_{22}(CO)_{37}]^{4-}$ [23], and the bimetallic $[Rh_{18}Pt_4(CO)_{35}]^{4-}$ [24], then also by other groups like the hydride $[H_{8-n}Rh_{22}(CO)_{35}]^{n-}$ (n = 4, 5) species [25, 26] and the series $[Ni_xRh_{14-x}(CO)_{25}]^{n-}$ (x = 1, 2, 5; n = 5, 4, 3, respectively) [27]. At the moment, to our knowledge, the highest nuclearity homoleptic carbonyl species is still represented by $[Rh_{33}(CO)_{47}]^{5-}$ [28]. Its structure, illustrated in Figure 10.1, is composed of an inner Rh_{19} core based on interpenetrated centered icosahedra that grow on both sides, giving rise to two arachno Rh_{11} moieties, and resulting in an overall $\{Rh_{27}\}$ rod. The remaining six Rh atoms symmetrically complete the metal skeleton.

Notably, structural icosahedral patterns can be found in several carbonyl species, irrespective of the nature of the metal atom, confirming the extra stability imparted by such a geometry. Among them, we can find $[Pt_{19}(CO)_{17}\{Cd_5(\mu\text{-}Br)_5Br_3(Me_2CO)_2\}\{Cd_5(\mu\text{-}Br)_5Br(Me_2CO)_4\}]^{2-}$ [29], $[Au_{21}\{Fe(CO)_4\}_{10}]^{5-}$ [30], $[Ni_{39}P_3(CO)_{44}]^{6-}$ [31], and the previously mentioned $(\mu_{12}\text{-}Pt)Pd_{164-x}Pt_x(CO)_{72}(PPh_3)_{20}$. Moreover, the icosahedral structure creates the base for the complex architectures of the large $Au_{133}(SR)_{52}$ [32], $Au_{144}(SR)_{60}$ [33] and $[Ag_{44}(p\text{-}MBA)_{30}]^{4-}$ [34] gold and silver nanoparticles. The relevance of the icosahedral arrangement within carbonyl clusters will be even more evident throughout the present chapter.

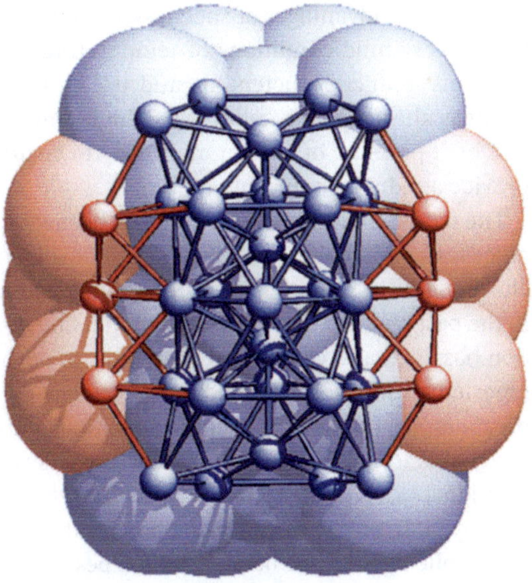

Figure 10.1 Metal skeleton of the $[Rh_{33}(CO)_{47}]^{5-}$ cluster anion.

Beside bimetallic species, Rh clusters can host second- and third-row *p* elements, such as C, N, P, and S, or can include in the metal skeleton heavier post-transition elements like As, Sn, and Sb. Whenever such heteroatoms are inserted in metal cavities, they are known to further strengthen the overall metal cluster core [35]. The chemistry of carbide and nitride Rh clusters has produced countless species, especially in the former case [36, 37], whereas only scattered examples of Rh clusters containing third-row *p* elements are known [38, 39]. Even less Rh compounds containing heavier post-transition metals have been reported [40, 41] until more recently. In fact, in the last 10 years or so, a more systematic synthesis of Rh clusters containing post-transition elements has been rationalized.

In this chapter, we review what has been thus far achieved on rhodium nanoclusters, highlighting the similarities and differences when combining post-transition elements with rhodium carbonyls. Moreover, we will present experimental evidence of some interesting electronic properties that such compounds may present, in particular how they possess an electron-sponge behavior by accepting or releasing electrons under electrochemical and/or chemical conditions.

10.2 Synthesis of Heterometallic Rhodium Carbonyl Nanoclusters

10.2.1 Synthesis of the $[Rh_{12}E(CO)_{27}]^{n-}$ Family of Nanoclusters

One of the best strategies for preparing large rhodium clusters containing post-transition elements is represented by the redox-condensation method [42, 43]. In this particular case, the method consists in reacting a rhodium cluster precursor in an anionic state, easily obtainable from consolidated preparation procedures, with a cationic/neutral complex of the desired element. By combining rhodium in a negative oxidation state and another metal in a positive one, a redox reaction occurs, during which several products, i.e. heterometallic clusters, are formed.

Within those new products, the rhodium atoms are in a less negative oxidation state than before, having been oxidized by the heteroatom which, in turn, is reduced to zero. Therefore, it is correct to state that heterometallic clusters thus obtained are the results of the oxidation of the cluster precursor.

One of the advantages of such a synthetic method lies in the possibility of carrying out the reaction in very mild conditions of temperature and pressure, because the reaction is driven by the reactants' reduction potentials and the carbonyl ligands are provided by the cluster precursor, whereas alternative procedures use high temperature to overcome the reaction activation energy and high pressure of carbon monoxide to ensure the formation of carbonyl species, as well as to provide a reducing atmosphere [44].

The redox condensation method has been systematically applied between the $[Rh_7(CO)_{16}]^{3-}$ anionic cluster and halides of Sn, Ge, Sb, Bi, to explore the possibility of preparing new heterometallic clusters. It is important to mention that heterometallic Rh—Sb clusters had already been explored in the past by Vidal and co-workers but using more drastic reaction conditions [41]. The investigation has been also carried out with indium derivatives; the results are very promising and will be reported elsewhere. The selection of the cluster precursor has been made based on its high-yield preparation [45], while the choice of the oxidant complex depends on the aim of the reaction. If the goal is to prepare heterometallic clusters where the heteroatom occupies a peripheral/ligand position [46, 47], then complex bearing strong ligands like alkyl- or aryl derivatives may be employed. If, on the contrary, the possibility of the insertion of the naked heteroatom is the target, then labile ligands like halides are the best choice.

It is quite remarkable that, when $[Rh_7(CO)_{16}]^{3-}$ and EX_n (E = Sn, Ge, Sb, Bi; X = Cl or Br; $n = 2$ or 3) react in similar stoichiometric ratios, in the same solvent and conditions of temperature (room) and atmosphere (CO or N_2 at around 1.5 bars), they give rise to the same compound, namely $[Rh_{12}E(CO)_{27}]^{n-}$ ($n = 4$ when E = Ge, Sn; $n = 3$ when E = Sb, Bi). These clusters are completely isostructural, being all based on an E-centered rhodium icosahedral metal skeleton stabilized by the same number of carbonyl ligands. The icosahedron defined by the 12 Rh atoms presents some structural deviations from the platonic polyhedron in terms of edge lengths, mainly due to the size of the interstitial heteroatom, and only marginally to the ligand's geometrical constraints. In fact, the least distorted geometry is that of $[Rh_{12}Ge(CO)_{27}]^{4-}$, being Ge the smallest heteroatom of the series (Rh—Rh bond length average of 2.935 Å), whereas the highest distortion is observed for $[Rh_{12}Bi(CO)_{27}]^{3-}$, as the largest Bi element requires further stretching of some Rh—Rh bond distances to be interstitially lodged (average Rh—Rh distance of 3.024 Å). The $[Rh_{12}Sn(CO)_{27}]^{4-}$ and $[Rh_{12}Sb(CO)_{27}]^{3-}$ congeners display intermediate distortions, with Rh—Rh average bond lengths of 2.978 and 2.982 Å, respectively, in consonance with their relative size. For the sake of comparison, the average Rh—Rh bond distance in $[Rh_7(CO)_{16}]^{3-}$, whose metal skeleton consists of a mono-capped octahedron, is 2.772 Å.

The overall molecular structure of $[Rh_{12}E(CO)_{27}]^{n-}$ is illustrated in Figure 10.2.

In addition to be isostructural, the $[Rh_{12}E(CO)_{27}]^{n-}$ species all possess 170 cluster valence electrons (CVE), resulting from the sum of the metal valence electrons (9 for the rhodium atom, 4 or 5 for E, depending on whether they belong to the 14- or the 15-group, respectively), the electrons given by the ligands (2 per each CO) and those provided by the negative charge (4 or 5, respectively depending on E). This "magic" number of 170 CVE is in total agreement with the electron-counting rules set by Wade-Mingos [48, 49], rationalized by exploiting the analogy with the borane chemistry, modeled with the support of semiempirical calculations [50], which predict $14N+2$ electrons (N = number of transition-metal atoms) for closo-deltahedral species. According to these rules for a given geometry and nuclearity, a defined number of valence electrons is expected, and for an icosahedron the predicted CVE are, indeed, 170 ($=14\times12+2$). Nonetheless, as we are going to show later in this chapter, compliance with these rules does not exclude the possibility of obtaining electron-rich or -poor carbonyl clusters, under suitable redox conditions.

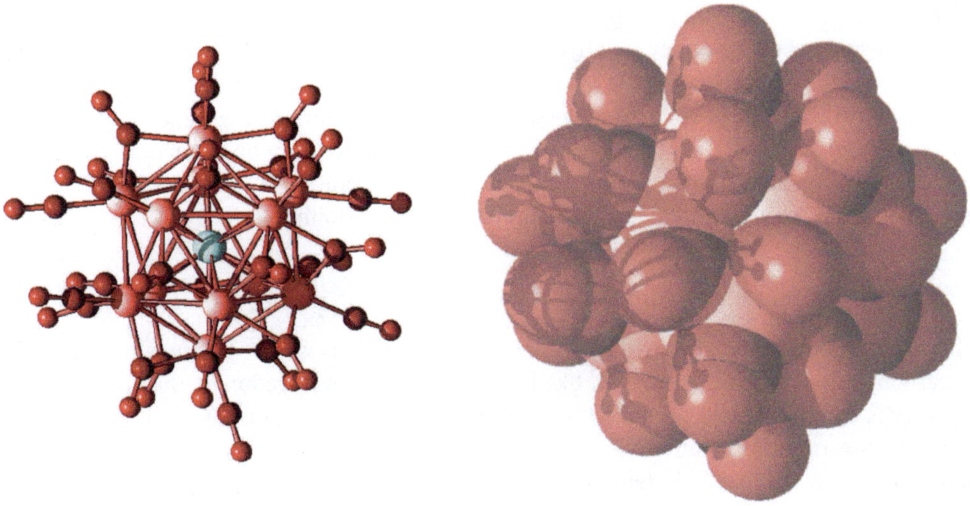

Figure 10.2 Molecular structure of the $[Rh_{12}E(CO)_{27}]^{n-}$ clusters (E = Ge, Sn, n = 4; E = Sb, Bi, n = 3).

The typical reaction between $[Rh_7(CO)_{16}]^{3-}$ and ECl_3 (E = Sb [51] or Bi [52]) to give $[Rh_{12}E(CO)_{27}]^{3-}$ is carried out in acetonitrile under CO atmosphere, and it proceeds according to the following stoichiometry (10.1):

$$4\left[Rh_7\left(CO\right)_{16}\right]^{3-} + 3ECl_3 \rightarrow 2\left[Rh_{12}E\left(CO\right)_{27}\right]^{3-} + \\ + 3\left[Rh\left(CO\right)_2Cl_2\right]^{-} + Rh + E + 4CO + 3Cl^{-} \tag{10.1}$$

To be exact, in the case of bismuth the same reaction would occur also under N_2 atmosphere.

In the case of the Rh—Sn derivative [53], the icosahedral cluster is obtained with a slightly different stoichiometry (10.2), but the reaction conditions remain those applied for the Rh—Sb congener:

$$2\left[Rh_7\left(CO\right)_{16}\right]^{3-} + 2SnCl_2 \rightarrow \left[Rh_{12}Sn\left(CO\right)_{27}\right]^{4-} + 2\left[Rh\left(CO\right)_2Cl_2\right]^{-} + Sn + CO \tag{10.2}$$

As for the $[Rh_{12}Ge(CO)_{27}]^{4-}$ cluster, its peculiarity is that it derives from exposure to CO atmosphere of $[Rh_{13}Ge(CO)_{25}]^{3-}$ in the presence of bromide. More specifically, a redox reaction favored by a carbon monoxide atmosphere causes the body-centered $[Rh_{13}Ge(CO)_{25}]^{3-}$ tri-anion to undergo a structural change into the Ge-centered icosahedral species, with the concurrent formation of $[Rh(CO)_2Br_2]^{-}$. Notably, this transformation is reversible under N_2 atmosphere [54].

All above reactions can be followed through infrared (IR) spectroscopy, thanks to the presence of the CO groups that allow detection of the progressive formation of the products and the concurrent disappearance of the starting material. There is an important remark to make about the IR spectra of carbonyl nanoclusters. Despite their high number of CO ligands, their infrared spectra are relatively invariant with respect to the number of CO groups and to the molecular geometry, and they appear to have an unexpected simplicity [55]. This phenomenon is useful for the identification of clusters in solution, as they do possess a spectral fingerprint that reveals their presence in the solution mixture.

Once the reaction is complete, the final solution is dried under vacuum and washed with water to eliminate halides of the alkylammonium cations which serve as the precursors' counterions. Owing to their different solubility, the obtained products can be separated by subsequent extractions with solvents of increasing polarity, namely tetrahydrofuran (THF), acetone, and acetonitrile. In details, $[Rh(CO)_2Cl_2]^-$ is soluble in THF, while the $[Rh_{12}E(CO)_{27}]^{n-}$ compounds can be extracted in acetone (E = Sb) or acetonitrile (E = Sn, Bi). Larger species are generally soluble only in acetonitrile. The $[Rh_{13}Ge(CO)_{25}]^{3-}$ too, from which $[Rh_{12}Ge(CO)_{27}]^{4-}$ can be prepared, can be extracted in acetone. Their crystallization is achieved by a gradual diffusion of their solution in a nonsolvent medium gently layered on top, thus allowing the slow precipitation of a crystalline solid.

10.2.2 Growth of Rhodium Heterometallic Nanoclusters

The synthesis of large carbonyl clusters can be performed in various ways – for instance, through reductive carbonylation and via thermal or, like in the presented cases, redox methods. While the former strategy is mostly used to produce low-nuclearity carbonyl compounds, with the notable exception of Pt clusters [56], the latter two are the best ones to obtain large species [57]. Moreover, a stable and fairly easily obtainable heterometallic cluster could serve, in turn, as precursor to grow larger species.

As mentioned before, the formation of the $[Rh_{12}E(CO)_{27}]^{n-}$ compounds is achieved with an approximately equimolar ratio between the reactants (between 0.75 and 1 equiv. of EX_n, depending on the nature of E), alongside some byproducts like Rh(I) carbonyl/halide complexes and metallic residue. How far the oxidation of the cluster precursor goes strongly depends on the amount of the added halide salts. More specifically, by adding more EX_n to $[Rh_7(CO)_{16}]^{3-}$ it is possible to further oxidize the original cluster precursor and, therefore, grow the icosahedral species to higher nuclearity nanoclusters. On the other side, an excess of the oxidant may have the opposite effect and break the cluster. It has been experimentally demonstrated that the addition of about 3 equiv. of EX_n (X = Cl or Br) represents an upper limit in all cases, as it causes the complete cluster degradation into the above-mentioned byproducts [58]. What happens between the close-to-equimolar reaction and further addition of the oxidant is where the different heterometallic systems' behavior diverges, giving rise to different higher-nuclearity nanoclusters. Cluster growth can go further, and it has also been performed by starting from the $[Rh_{12}E(CO)_{27}]^{n-}$ species when E = Sn, Sb, Bi, or $[Rh_{13}Ge(CO)_{25}]^{3-}$.

It is important to underline, however, that in this chemistry the prevision of what species are going to be formed in a specific reaction is not always feasible, as metal carbonyl clusters are able to exist in very different forms even when their nuclearity is similar. Like any compound in chemistry, their stability depends on the balance of energetic and steric factors, like the presence or absence of CO atmosphere, the size of the heterometallic elements, and the strength of the involved bonds. Yet, these factors appear to be rather close, so whether a reaction would take one direction or the other is not so easy to foresee, even with the input from theoretical calculations [59]. Moreover, only occasionally, and this seems the case for the $[Rh_{12}E(CO)_{27}]^{n-}$ clusters (E = Sn, Sb, Bi), a given species represents a relatively deep potential well in the reaction coordinate; yet, minor changes in the reagents or experimental conditions may result in completely different products [60].

10.2.2.1 Rh—Ge Nanoclusters

In the literature, there are some examples of Rh—Ge clusters where the germanium atoms are part of the stabilizing shell of the rhodium metal cluster [61]. When moving to the insertion of germanium atoms into the cluster, to our knowledge only the icosahedral $[Ni_{12}Ge(CO)_{22}]^{2-}$ and $[Ni_{10}Ge(CO)_{20}]^{2-}$ are known [62]. Notably, the former possesses, too, a valence electron counting of 170 CVE, in compliance with the Wade–Mingos rules.

The case of Rh—Ge system offers a different perspective in the herein presented scenario of heterometallic rhodium clusters, starting from the formation of the $[Rh_{13}Ge(CO)_{25}]^{3-}$ instead of the expected $[Rh_{12}Ge(CO)_{27}]^{4-}$ when $[Rh_7(CO)_{16}]^{3-}$ reacts with $GeBr_2$ [54]. The obtainment of $[Rh_{13}Ge(CO)_{25}]^{3-}$, which is based on a penta-capped Ge-centered cubic structure, is most likely due to the reduced dimensions of the Ge atom comparing with those of Sn, Sb, and Bi, more suitable to be lodged inside the smaller cavity of $[Rh_{13}Ge(CO)_{25}]^{3-}$ rather than the icosahedral one of $[Rh_{12}Ge(CO)_{27}]^{4-}$ [63].

When more $GeBr_2$ is added, the reaction gives rise to larger species. More specifically, the addition of around 2 equiv. of $GeBr_2$ to $[Rh_7(CO)_{16}]^{3-}$ under N_2 atmosphere, or 1.2 equiv. to $[Rh_{13}Ge(CO)_{25}]^{3-}$, leads to $[Rh_{14}Ge_2(CO)_{30}]^{2-}$, a species whose structure is based on two square antiprisms fused via one face and mono-capped on opposite sides, each centered by a germanium atom. Anything between the 1 : 1 and 1 : 2 stoichiometric ratio between $[Rh_7(CO)_{16}]^{3-}$ and $GeBr_2$, respectively, yields the same species, just in different amounts. Conversely, when the addition of 1.2 equiv. of $GeBr_2$ to $[Rh_7(CO)_{16}]^{3-}$ is done under CO atmosphere, but still in the same solvent and temperature conditions, the reaction produces the larger $[Rh_{23}Ge_3(CO)_{41}]^{5-}$ instead, albeit in very low yields.

The molecular structure of $[Rh_{23}Ge_3(CO)_{41}]^{5-}$ consists of three equivalent sphenocorona-like Ge-centered $Rh_{10}Ge$ polyhedra, sharing the inner Rh_2 edge and another Rh atom in pairs. The cluster overall contains three fully interstitial Ge atoms and two semi-interstitial Rh atoms, the latter capped by two hexagonal faces. The metal structure is stabilized by 41 carbonyl ligands. It is worth noting that, when reactions occur under CO atmosphere, the resulting nanoclusters adopt complex metal skeletons with large cavities, while under N_2 the experimental evidence has, so far, shown metal arrangements based on centered prismatic or antiprismatic architectures.

Figure 10.3 displays a synthetic scheme to obtain the so-far characterized Rh—Ge nanoclusters, as well as their metal structures as derived by single-crystal X-ray diffraction analyses.

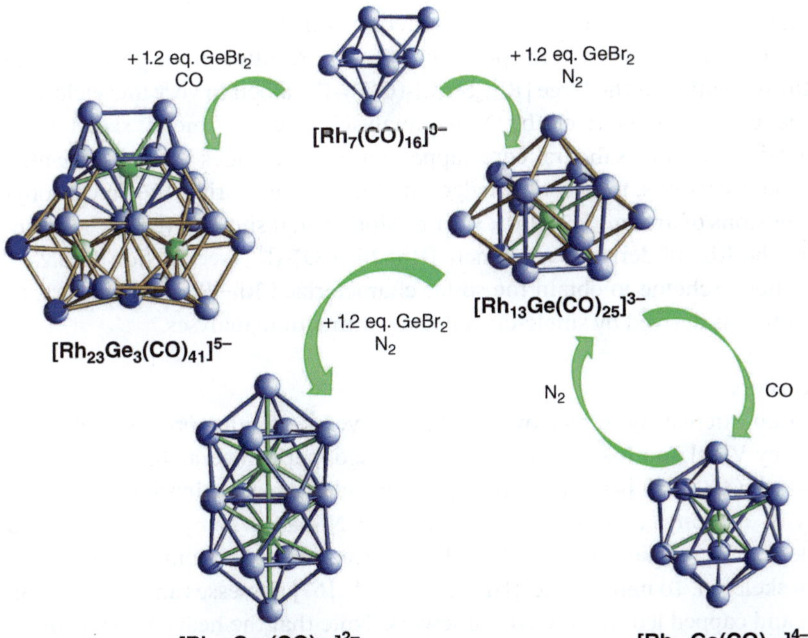

Figure 10.3 Reaction sketch and metal structures of the Rh—Ge nanoclusters. Rh atoms are in blue, Ge atoms in green.

10.2.2.2 Rh—Sn Nanoclusters

Many Rh-based compounds containing Sn ligands have been reported in the literature [64] whereas, to our knowledge, the insertion of tin in rhodium carbonyl clusters has only been reported in the species that are described herein. It is important, however, to mention that an analogous $[Ni_{12}Sn(CO)_{22}]^{2-}$ was reported over 30 years ago [62]. The Rh—Sn carbonyl cluster system so far characterized shows a similar variety of species with respect to the Rh—Ge one in terms of number, but the specific compounds are different with the sole exception of the icosahedral $[Rh_{12}Sn(CO)_{27}]^{4-}$ cluster. First of all, the reaction that leads to the latter can be carried out under either N_2 or CO atmosphere without significant difference in the observed products. Nonetheless, its carbon monoxide depletion must be performed at high temperature and under N_2, to yield the unsaturated species $[Rh_{12}Sn(CO)_{26}]^{4-}$ and $[Rh_{12}Sn(CO)_{25}]^{4-}$ [65].

The addition of 1 equiv. of $SnCl_2$ to the icosahedral species under N_2 in acetonitrile, corresponding to about 1.5 equiv. in total based on the initial amount of $[Rh_7(CO)_{16}]^{3-}$, results in the formation of the $[Rh_{12}SnCl_2(CO)_{23}]^{4-}$ adduct, where the chloride ligands not only partly replaced some carbonyls but also broke and deformed the initial polyhedron. If the same reaction is carried out under carbon monoxide, $[Rh_{12}Sn(RhCl)(CO)_{27}]^{4-}$ is produced instead. In this case, formally a {RhCl} neutral fragment is added onto the parent icosahedral species, which means that some cluster breaking must occur for this fragment to form. The reason for the different pathways that the two reactions take, depending on the employed atmosphere, has been attributed to the competing role of CO over the halide ligands. As a confirmation, when $[Rh_{12}SnCl_2(CO)_{23}]^{4-}$ is put under carbon monoxide atmosphere, it goes back to the parent $[Rh_{12}Sn(CO)_{27}]^{4-}$ compound.

Notably, very recent studies [66] showed that additions of $SnCl_2$ to the $[Rh_7(CO)_{16}]^{3-}$ precursor (ca. 1.3 : 1 ratio) under nitrogen but in acetone as opposed to acetonitrile, also give rise to the $[Rh_7Sn_4Cl_{10}(CO)_{14}]^{5-}$ species, which separates out from the final mixture by spontaneous precipitation. This compound is of a bright purple color, whereas all nanoclusters are black. Notably, the halides are retained by the tin atoms and the $\{SnCl_2\}$ and $\{SnCl_3\}$ fragments act as cluster ligands. Conversely, reactivity experiments performed on the icosahedral $[Rh_{12}Sn(CO)_{27}]^{4-}$, more specifically an addition of diluted sulfuric acid but in the presence of $SnCl_2$, resulted in the growth of the cluster precursor with the formation of the large $[Rh_{18}Sn_3Cl_2(CO)_{33}]^{4-}$, albeit in very low yields. Its molecular structure has been characterized by X-ray analysis, and the metal skeleton of $[Rh_{18}Sn_3Cl_2(CO)_{33}]^{4-}$ consists of an icosahedral core capped on opposite sides by two $\{Rh_3SnCl\}$ fragments. This compound represents, to our knowledge, the highest nuclearity Rh—Sn carbonyl cluster to date, with dimensions of around 1.5 × 1.0 × 1.2 nm. Moreover, it shows a very close structural similarity to one of the Rh—Bi derivatives, namely $[Rh_{17}Bi_3(CO)_{33}]^{4-}$ (see Section 10.2.2.4). Figure 10.4 shows a synthetic scheme to obtain the so-far characterized Rh—Sn nanoclusters, as well as their metal structures as derived by single-crystal X-ray diffraction analyses.

10.2.2.3 Rh—Sb Nanoclusters

The Rh—Sb system has been extensively studied over the last few years, after the preparation of the $[Rh_{12}Sb(CO)_{27}]^{3-}$ cluster by Vidal and his co-workers 40 years ago, performed at 140–160 °C and under elevated CO pressure (400 atm) [41]. If, however, we consider the combination of Sb with other transition metals in carbonyl clusters, there are several Ni-based compounds prepared through the same condensation reaction method described above, where Sb atoms contribute to the growing of the metal skeleton. To name some, $[Ni_{15}Sb(CO)_{24}]^{2-}$ [67] possesses an interstitial Sb atom in a very distorted and capped icosahedral Ni framework. Note that the heavier distortion of the icosahedral cage due to the interstitial Sb with respect to the one formed by Rh atoms depends on the smaller size of nickel atoms, which give rise to a smaller cavity. In fact, a much less distorted icosahedron is found for the Ni-centered, as opposed to Sb-centered, $[Ni_{11}Sb_2(CO)_{18}]^{3-}$ [63] and

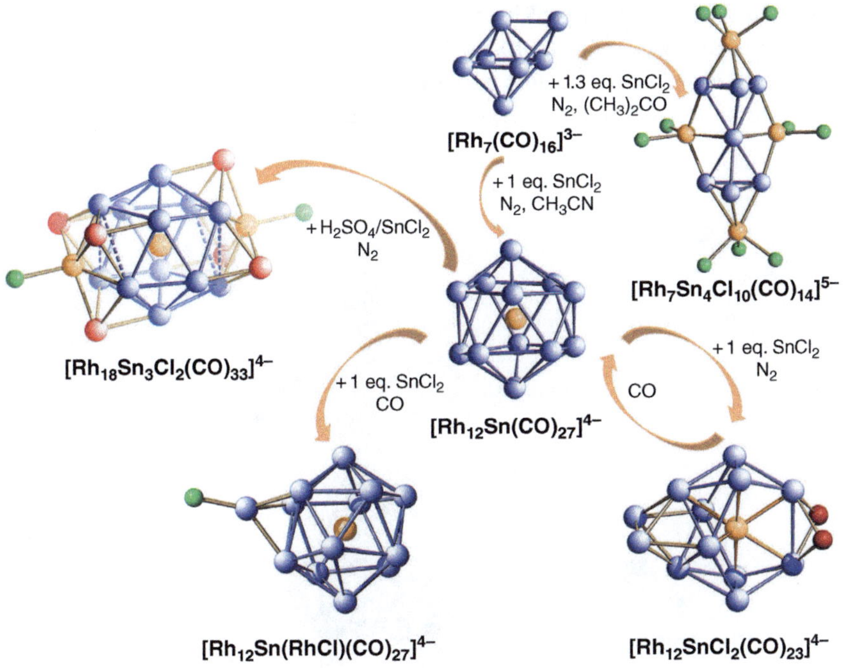

Figure 10.4 Reaction sketch and metal structures of Rh—Sn nanoclusters. Rh atoms are in blue (and light red in $[Rh_{18}Sn_3Cl_2(CO)_{33}]^{4-}$), Sn atoms in orange, Cl atoms in green. The icosahedral core, where present, has been highlighted.

$[Ni_{10}(SbR)_2(CO)_{18}]^{2-}$ (R = Me, Et, *i*Pr, *t*-Bu, *p*-FC_6H_4) [68]. The large $[Ni_{31}Sb_4(CO)_{40}]^{6-}$ presents only semi-interstitially lodged Sb atoms in a very geometrically complex architecture [69]. Other lower nuclearity, Sb-containing compounds are within the Os—Sb and Ru—Sb systems [70–72].

Once the possibility of obtaining $[Rh_{12}Sb(CO)_{27}]^{3-}$ in good yields through reactions carried out in mild conditions of temperature and pressure became feasible [51], the icosahedral cluster was also exploited as a starting material to grow larger heterometallic species, preferably under carbon monoxide atmosphere, as under N_2 the cluster is less stable. It is also worth mentioning that, during the synthesis of $[Rh_{12}Sb(CO)_{27}]^{3-}$ through the redox condensation method, the nido-icosahedral $[Rh_{11}Sb(CO)_{26}]^{2-}$ species was isolated, ideally representing the species just one step before the formation of the integer congener, as it is characterized by an incomplete Sb-centered icosahedral structure [73]. By the addition, under CO, of 0.5 equiv. of $SbCl_3$ to $[Rh_{12}Sb(CO)_{27}]^{3-}$, corresponding to an overall addition of 1.15 equiv. based on the original $[Rh_7(CO)_{16}]^{3-}$ precursor, the larger $[Rh_{20}Sb_3(CO)_{36}]^{3-}$ is formed. Conversely, when 0.7–0.8 equiv. of $SbCl_3$ is added under N_2 atmosphere, the overall result is a rather low-selectivity reaction where a mixture of nanoclusters is produced, namely $[Rh_{21}Sb_2(CO)_{38}]^{5-}$, in good yields, and traces of $[Rh_{28-x}Sb_x(CO)_{44}]^{6-}$ [74].

The metal structure of $[Rh_{20}Sb_3(CO)_{36}]^{3-}$ consists of a Rh-centered $Rh_{10}Sb_3$ icosahedron, with the two opposite Sb vertexes capped by pentagonal Rh_5 faces, and it is remarkably similar to that of $[Rh_{21}Sb_2(CO)_{38}]^{5-}$, the only difference being the substitution of the peripheral Sb atom with Rh. The carbonyl ligands and negative charge are also different, so to saturate their CVE. As for the $[Rh_{28-x}Sb_x(CO)_{44}]^{6-}$ species, the crystalline samples were not of good enough quality to allow a complete structural characterization. However, based on the analyses performed via electrospray ionization mass spectrometry (ESI-MS) and energy dispersive X-ray spectrometry (EDS) through scanning electron microscopy (SEM), it was tentatively formulated as $[Rh_{25}Sb_3(CO)_{44}]^{6-}$. Its metal structure can be described as made by an inner Rh-centered $Rh_{10}Sb_3$ icosahedron, capped on the three Sb atoms by pentagonal Rh faces. Figures 10.5 and 10.6 illustrate the synthetic scheme to obtain the so-far characterized Rh—Sb nanoclusters under CO and N_2 atmosphere, respectively, as well as their metal structures as derived by single-crystal X-ray diffraction analyses.

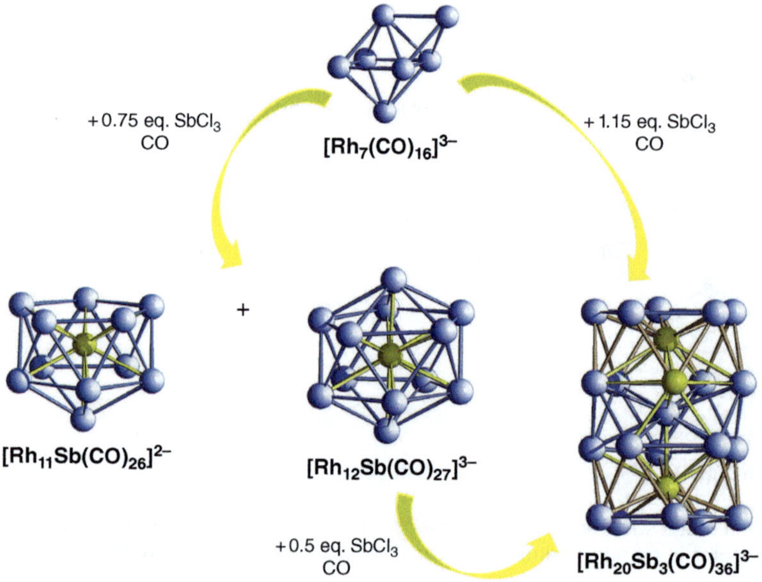

Figure 10.5 Reaction sketch under CO atmosphere and metal structures of Rh—Sb nanoclusters. Rh atoms are in blue, Sb atoms in yellow. The icosahedral core, where present, has been highlighted.

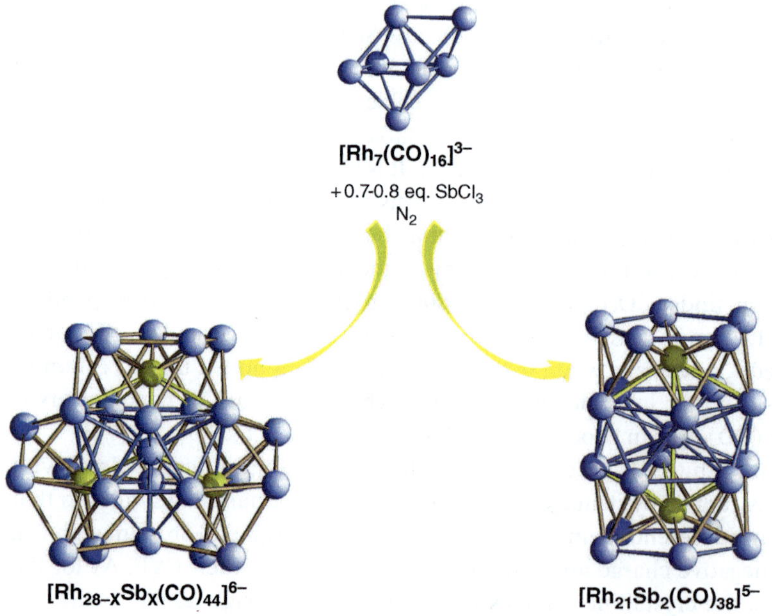

Figure 10.6 Reaction sketch under N_2 atmosphere and metal structures of Rh—Sb nanoclusters. Rh atoms are in blue, Sb atoms in yellow. The icosahedral core, where present, has been highlighted. For sake of scientific accuracy, $[Rh_{25}Sb_3(CO)_{44}]^{6-}$ has been indicated with the more general $[Rh_{28-x}Sb_x(CO)_{44}]^{6-}$ formula.

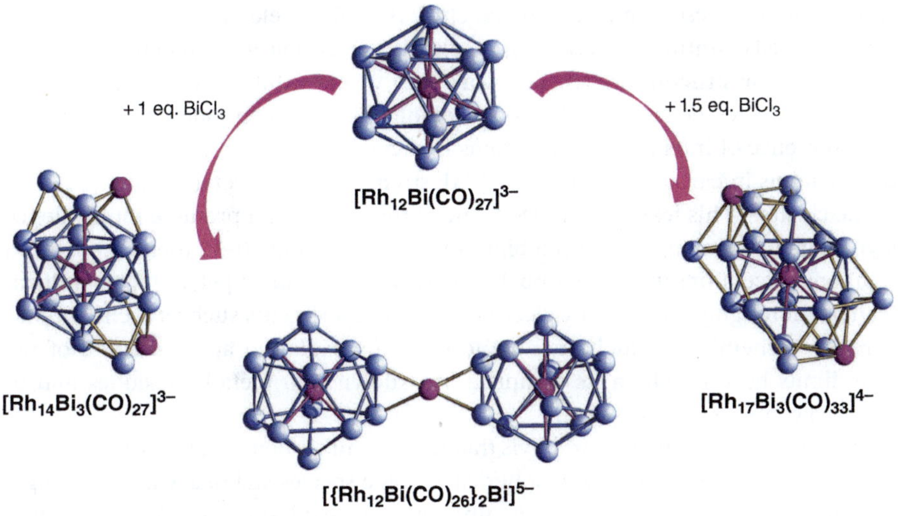

Figure 10.7 Reaction sketch under metal structures of Rh—Bi nanoclusters. Rh atoms are in blue, Bi atoms in magenta. The icosahedral core, where present, has been highlighted.

10.2.2.4 Rh—Bi Nanoclusters

In the field of heterometallic carbonyl clusters, there are a number of scattered examples of transition-metal compounds containing bismuth as reported in the literature – for instance, in combination with Fe [75], Co [76], Ni [77], and Ir [78, 79] – but none with rhodium until quite recently [52].

In the case of the Rh—Bi system, the icosahedral $[Rh_{12}Bi(CO)_{27}]^{3-}$ may again serve as the new precursor to grow larger species. However, the unique peculiarity with respect to the other systems is that the same products are obtained irrespective of the employed atmosphere: whether N_2 is used instead of CO, or vice-versa, it does not change the reaction path. After the addition of 1 equiv. of $BiCl_3$ to the icosahedral precursor $[Rh_{12}Bi(CO)_{27}]^{3-}$ (1.25 equiv. if based on the initial $[Rh_7(CO)_{16}]^{3-}$), both the dimeric $[\{Rh_{12}Bi(CO)_{26}\}_2Bi]^{5-}$ and the more compact $[Rh_{14}Bi_3(CO)_{27}]^{3-}$ nanoclusters are formed. If the addition is pushed to 0.5 more equivalent of $BiCl_3$ (i.e. 1.5 equiv. in total based on the initial amount of $[Rh_7(CO)_{16}]^{3-}$), then $[Rh_{17}Bi_3(CO)_{33}]^{4-}$ is produced. The latter, to our knowledge, represents the largest Rh—Bi carbonyl nanocluster to date.

The metal framework of $[\{Rh_{12}Bi(CO)_{26}\}_2Bi]^{5-}$ consists of two icosahedral moieties connected via a Bi atom, tilted by about 90° with respect to each other. Even $[Rh_{14}Bi_3(CO)_{27}]^{3-}$ has retained the parent structure, and it presents two additional Bi and Rh atoms capping the opposite triangular faces, in pairs. Conversely, $[Rh_{17}Bi_3(CO)_{33}]^{4-}$ shows a more compact expansion of the parent Bi-centered icosahedron of Rh atoms, as it is decorated by additional $\{Rh_3Bi\}$ and $\{Rh_2Bi\}$ fragments that, however, cause a further distortion from the ideal polyhedron. Figure 10.7 presents a synthetic scheme to obtain the so-far characterized Rh—Bi nanoclusters, as well as their metal structures as derived by single-crystal X-ray diffraction analyses.

10.3 Electron-Reservoir Behavior of Heterometallic Rhodium Nanoclusters

Transition metal carbonyl clusters usually adopt close-shell electronic configurations, and it has been a common belief that they should exhibit a perfect correspondence between the number of CVE and their metal structure [80, 81]. In most cases, this model remains true. Nonetheless, when certain ad-hoc conditions are present, metal carbonyl clusters could be electron-rich or -deficient with respect to the expected counting, by accepting or releasing electrons at given potentials while maintaining their molecular structure intact. This way, they would exhibit a so-called electron-sponge behavior and would act, de facto, as electron reservoirs [82]. As mentioned in the state-of-the-art section, the presence of interstitial heteroatoms in metal carbonyl compounds, but also of carbon and nitrogen atoms in carbides and nitrides [83], strengthens the metal core and imparts more stability to the clusters. This feature is, indeed, one of the factors that promote multivalence in metal carbonyl compounds, alongside with a high nuclearity (around 10–12 metal atoms, and beyond), although some exceptions with lower nuclearity species are known [84, 85]. The multivalence property can be thoroughly analyzed by electrochemical experiments such as cyclic voltammetry and spectroelectrochemistry, which are able to push the oxidation and reduction of the cluster toward the limits beyond which its complete breakdown into metallic residues and/or mononuclear carbonyl complexes occurs [86].

Among the less recently reported metal carbonyls that feature multivalence, it is worth mentioning the $[Ni_{32}C_6(CO)_{32}]^{6-}$ hexa-anion, a pseudo-spherical carbide species stabilized solely by edge-bridging carbonyl ligands, which can reversibly undergo one oxidation and four reduction monoelectronic steps [87]. Other notable examples of redox activity are represented by the $[Os_{18}Pd_3C_2(CO)_{42}]^{2-}$ carbide cluster [88] and the Co-centered $Co_{11}Te_7(CO)_{10}$ and $Co_{11}Te_5(CO)_{15}$ species [89]. More recently, similar properties have been reported for homoleptic Ni—Pd [90] and heteroleptic Au—Pd [91] and Au—Pt [92] nanoclusters. Notably, all the cited examples possess interstitial metallic atoms, including post-transition elements. It is important to underline, nonetheless, that the high nuclearity and the presence of interstitial atoms may not be per se sufficient to ensure multivalence, especially if the cohesive energy of the involved metals is not strong enough to support redox changes without metal-framework breakdown [93].

With those premises set, we would like to report the latest experimental evidence on the multivalence of the $[Rh_{12}E(CO)_{27}]^{n-}$ family of clusters. This feature was investigated via cyclic voltammetry, in situ infrared spectroelectrochemistry (IR SEC) conducted in an optically transparent thin-layer electrochemical (OTTLE) cell [94], and density functional theory (DFT) calculations [73].

Cyclic voltammetric (CV) profiles were registered at both Pt and glassy carbon (GC). As they showed low currents associated with the redox processes, it was not possible to obtain well-resolved peaks. However, the presence of multiple stable redox states of a carbonyl cluster could be inferred by in situ IR SEC studies, by the comparison between the sequence of IR spectra and the profile of the *i/E* curve [95–97]. Electrochemical experiments were performed in $CH_3CN/[N^nBu_4]PF_6$ solution, under either Ar or CO atmosphere, to ensure the cluster stability. It is important to underline that chemical reduction and oxidation reactions had also been performed on the $[Rh_{12}E(CO)_{27}]^{n-}$ icosahedral clusters, but they were mostly inconclusive because of the impossibility to crystallize the obtained products, which would otherwise have allowed their full characterization.

DFT calculations were performed with the Gaussian 16 package [98] and using LANL2DZ basis set with pseudopotential for transition metals [99], whereas 6-31G(d,p) basis set was used for the remaining atoms [100]. Geometry optimizations in vacuum were performed in combination with vibrational analysis to confirm the character of the stationary points and to simulate IR spectra, with vibrational frequencies computed analytically and rescaled using a 0.961 scaling factor [101].

The electrochemical investigation carried out on the four icosahedral clusters pointed out that $[Rh_{12}Ge(CO)_{27}]^{4-}$, $[Rh_{12}Sb(CO)_{27}]^{3-}$, and $[Rh_{12}Bi(CO)_{27}]^{3-}$ exhibit a rich redox activity and are able to reversibly accept or release electrons without significant alteration of their molecular structure, acting as electron reservoirs, while the behavior of $[Rh_{12}Sn(CO)_{27}]^{4-}$ appeared to be more complex.

In detail, the most redox active species is $[Rh_{12}Sb(CO)_{27}]^{3-}$, as it can sustain the addition of up to three further electrons and the depletion of one, reversibly, thus being stable with five different negative charges, from −2 to −6. The similarity of the IR spectra of the differently charged species, as well as the reversibility of the redox processes, suggests that no significant structural modifications occurred in the experimental conditions. This conclusion was also confirmed by DFT calculations, although it emerged that, upon reduction, all Rh—Rh and Rh—Sb bond distances systematically increase, even if the species remains intact. A similar expansion in the metal–metal bond lengths with the increasing of the cluster negative charge had been previously observed for multivalent carbonyl compounds, and it is likely due to the additional electrons populating metal–metal antibonding levels [89]. By sweeping the potential between +0.3 and −1.6 V (vs. Ag pseudo-reference electrode), at a scan rate of $1\,mV\,s^{-1}$, the IR spectrum species showed a ν_{CO} downshift related to the carbonyls of about $16–18\,cm^{-1}$ (for the terminally coordinated ones) for every increasing negative charge. Likewise, an upshift of the same magnitude was observed upon oxidation. When the potential was increased from +0.3 to +0.84 V, the oxidation resulted in relatively fast decomposition of the cluster. Figure 10.8 shows the selected infrared spectra of $[Rh_{12}Sb(CO)_{27}]^{n-}$ (n = 2–6) registered under CO in CH_3CN for the corresponding different cluster charges and its CV profile.

The in situ IR SEC experiments for $[Rh_{12}Bi(CO)_{27}]^{3-}$ were conducted under Ar atmosphere, where the cluster is perfectly stable, within a potential range between +0.4 and −2.1 V (vs. Ag pseudo-reference electrode), at the scan rate of $2\,mV\,s^{-1}$. The $[Rh_{12}Bi(CO)_{27}]^{3-}$ nanocluster seems to be more stable with an even number of CVE, at least in the more reduced form. In fact, while the $[Rh_{12}Bi(CO)_{27}]^{2-}$ could be fully isolated, the reduction process favored the formation of the penta- and hepta-anionic derivatives, and the IR spectrum of the intermediate tetra-anion could only be cleanly obtained by spectral deconvolution, even if its existence was predicted by DFT computational studies. The redox process was, in any case, perfectly reversible, and the cluster shows, nonetheless, a multivalence behavior as it is possible to isolate it with four different negative charges, namely −2, −3, −5, −7. DFT computations showed the same trend of bond elongations upon reduction than the one observed for the Sb congener, starting from the tetra-anion, but still with retention of the original icosahedral structure (Figure 10.9).

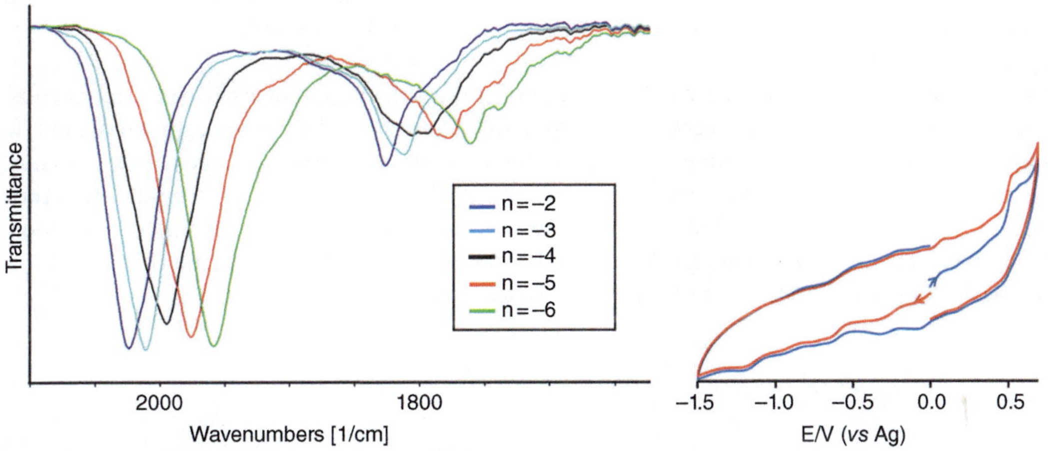

Figure 10.8 Selected infrared spectra of $[Rh_{12}Sb(CO)_{27}]^{n}$ as a function of the cluster charge (n) in CH_3CN (left); cyclic voltammetry profile of $[Rh_{12}Sb(CO)_{27}]^{3-}$ recorded at a platinum electrode in 0.1 M $[N^nBu_4]PF_6/CH_3CN$ solution (the arrows indicate the scan direction). Scan rate: $0.1\,V\,s^{-1}$ (right). *Source:* Reproduced with permission from Ref. [73]. © 2021 American Institute of Physics.

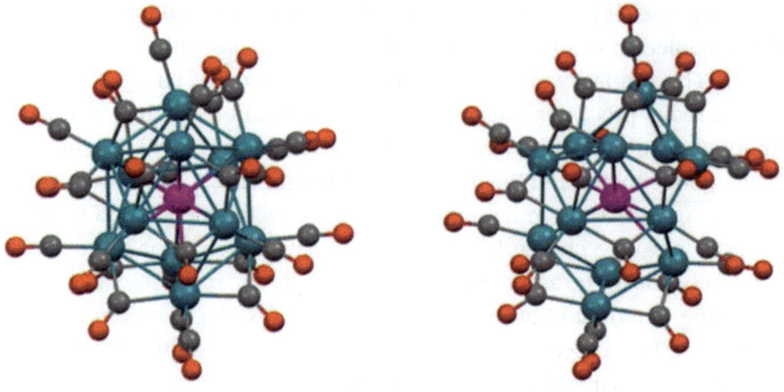

$[Rh_{12}Bi(CO)_{27}]^{2-}$ $[Rh_{12}Bi(CO)_{27}]^{4-}$

Figure 10.9 DFT optimized geometries of clusters $[Rh_{12}Bi(CO)_{27}]^{n-}$, with n = 2 and 4. Rh atoms are in turquoise, Bi atoms in magenta, C atoms in gray, and O atoms in red. *Source:* Reproduced with permission from Ref. [73]. © 2021 American Institute of Physics.

The in situ IR SEC measures for $[Rh_{12}Ge(CO)_{27}]^{4-}$ were performed under CO atmosphere and the IR spectra were recorded every 60 seconds during a slow scan (1 mV/s), sweeping the potential between +0.8 and −1.8 V. Similar to what was observed for the Rh—Bi congener, the oxidation process led to the stable tri- and di-anions, while under reduction it was possible to isolate only even-electron congeners, up to the hexa-anion, resulting in a total of four different cluster charges (−2, −3, −4, −6). Nonetheless, the existence of the fifth odd-electron $[Rh_{12}Ge(CO)_{27}]^{5-}$ was confirmed by theoretical calculations, and its IR spectrum could be obtained by spectral deconvolution analysis. Again, the oxidation and reduction processes were perfectly reversible.

Finally, the in situ IR SEC measurements for cluster $[Rh_{12}Sn(CO)_{27}]^{4-}$ were performed under Ar atmosphere, where the cluster is stable. The IR spectral changes were recorded every 60 seconds during the potential sweep between +0.2 and −1.9 V (vs. Ag pseudo-reference electrode) at a scan rate of 1 mV/s. In this case, only one stable oxidation process was observable, leading to the corresponding $[Rh_{12}Sn(CO)_{27}]^{3-}$, while the reduced congeners could not be properly isolated one from the other, even if the theoretical calculations predict the cluster stability in the gas phase up to the hexa-anion.

One notable outcome from the above comparative study is the importance of theoretical calculations in corroborating the hypothesis inferred from the experiments. For instance, the simulated IR spectra of all clusters bearing different negative charges maintain the same shape within the same system, even in the presence of significant structural modifications of the DFT optimized structures. Therefore, it can be concluded that whenever a reversible spectral modification in solution is observed, it is likely due to interactions with the solvent. Figure 10.10 shows the DFT simulated IR spectra for $[Rh_{12}E(CO)_{27}]^{n-}$ at various redox states (charges from −2 to −6).

10.4 Conclusions and Perspectives

In this chapter, authors aimed to give the readers an insight on the preparation and growth of heterometallic rhodium-based carbonyl nanoclusters, with a particular focus on the E-centered icosahedral carbonyl $[Rh_{12}E(CO)_{27}]^{n-}$ species, which have proven to be the common motif in the Rh—Sn, Rh—Ge, Rh—Sb, and Rh—Bi heterometallic systems. Moreover, when E = Sn, Bi, and Sb, these

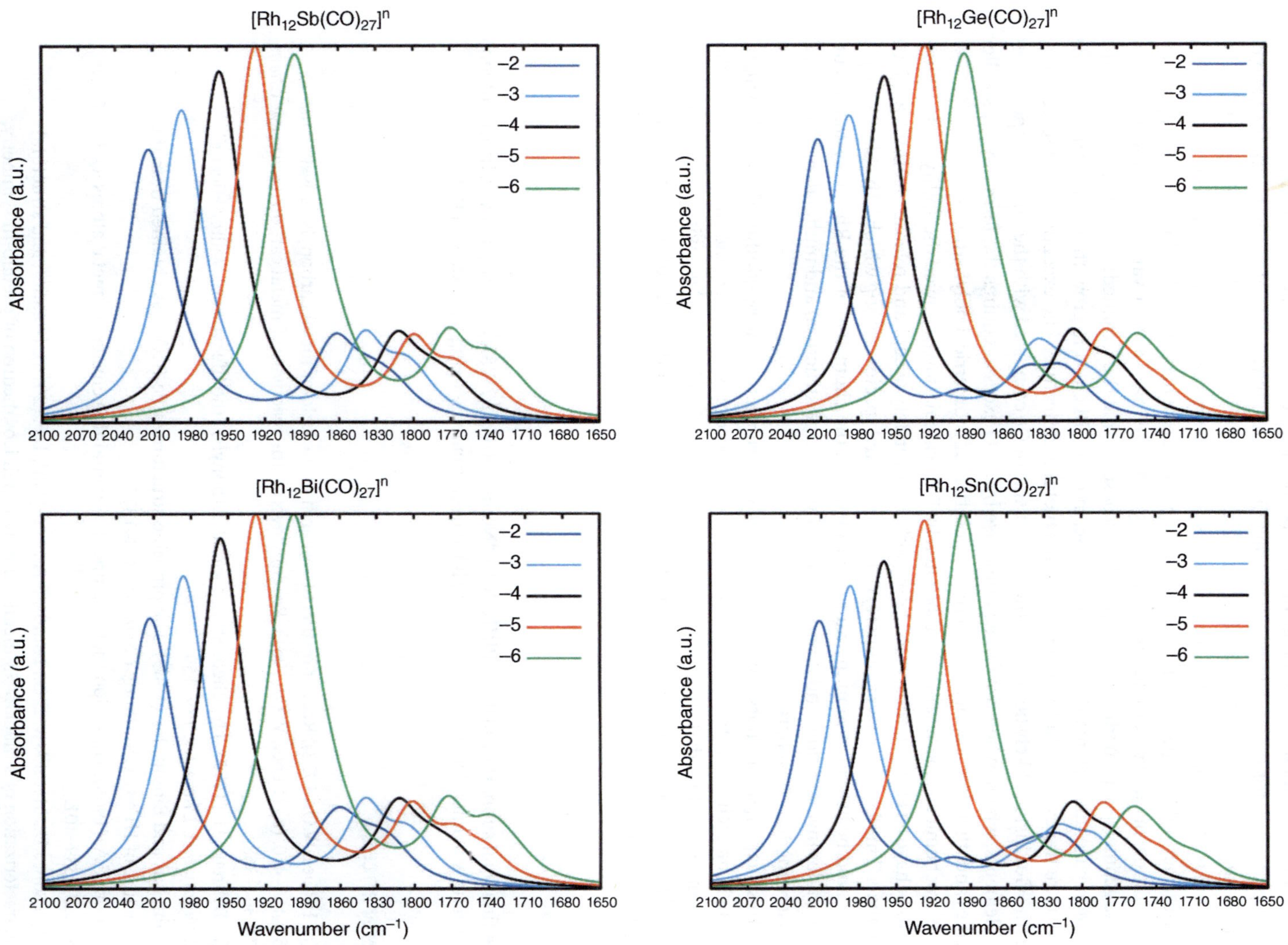

Figure 10.10 DFT simulated IR spectra for $[Rh_{12}E(CC)_{27}]^n$ at various redox states (n = −2 to −6). *Source:* Reproduced with permission from Ref. [73]. © 2021 American Institute of Physics.

specific nanoclusters appear to be a potential well in the reaction coordinate when $[Rh_7(CO)_{16}]^{3-}$ reacts with EX_n in a quasi-equimolar ratio. This, in turn, means that they can be exploited as new cluster precursors for growing larger heterometallic species. In the case of germanium, the yield of the icosahedral congener does not allow such a strategy and it is preferable to use $[Rh_{13}Ge(CO)_{25}]^{3-}$ instead. It is remarkable that the Rh-E systems herein described, despite sharing the isoelectronic and isostructural $[Rh_{12}E(CO)_{27}]^{n-}$ icosahedral species, do behave in such different ways upon further addition of the correspondent oxidant, e.g. EX_n, giving rise to a variety of nanoclusters of unique geometries. Nonetheless, it is possible to observe that the icosahedral geometry remains the common base upon which the Rh—Sn, Rh—Sb, and Rh—Bi clusters grow their nuclearity, proving once more the extra stability imparted by this metal architecture. The exception is represented by the Rh—Ge clusters, likely because of the smaller size of Ge compared with the ones of its neighboring elements, which, when not forced otherwise by a CO excess, requires the formation of smaller cavities and, therefore, results in the growth of different geometrical polyhedra.

Electrochemical experiments, corroborated by DFT calculations, showed the ability of such clusters to behave as electron reservoirs, being able to reversibly accept and release electrons at given potentials. This feature is also a property of other carbonyl species, among which $[Rh_{21}Sb_2(CO)_{38}]^{5-}$, and it could be extended to many more. For instance, the chemistry of the Rh—In system is currently under investigation, and it appears to share the same structural and multivalence properties as its icosahedral congeners.

In this chapter, the authors' intent was to show the readers that the possibility of preparing metal carbonyl nanoclusters through quasi-standard procedures has become a feasible reality. Furthermore, the fact that such nanoclusters can actually be characterized with atomic precision, can widen the study of the chemical, structural and electronic properties of metal nanoparticles of larger dimensions.

Acknowledgments

The financial support of MIUR (PRIN 2017 "Nemo" 20173L7W8K) and the University of Bologna is gratefully acknowledged. All drawings have been made by using Schakal 99 [102].

References

1 (a) Hieber, W.O. (1952). Recent results of research in the field of metal carbonyls. *Angew. Chem.* 64: 465–470. (b) Hieber, W.O. (1961). Recent views on the reactivity of metal carbonyls. *Angew. Chem.* 73: 364–368.

2 (a) Behrens, H. (1975). The chemistry of metal carbonyls: "the life work of Walter Hieber". *J. Organomet. Chem.* 94: 139–159.

3 Dahl, L.F. and Blount, J.F. (1965). The probable structure of $Fe_3(CO)_{12}$ as obtained from the structure of $HFe_3(CO)_{11}$. *Inorg. Chem.* 4: 1373–1375.

4 Cotton, F.A. (1966). Transition-metal compounds containing clusters of metal atoms. *Q. Rev. Chem. Soc.* 20: 389–401.

5 Erickson, J.D., Mednikov, E.G., Ivanov, S.A., and Dahl, L.F. (2016). Isolation and structural characterization of a Mackay 55-metal-atom two-shell icosahedron of pseudo-Ih symmetry, $Pd_{55}L_{12}(\mu_3\text{-}CO)_{20}$ (L = PR_3, R = isopropyl): comparative analysis with interior two-shell icosahedral geometries in capped three-shell Pd_{145}, Pt-centered four-shell Pd–Pt M_{165}, and four-shell Au_{133} nanoclusters. *J. Am. Chem. Soc.* 138 (5): 1502–1505.

6 McPartlin, M., Eady, C.R., Johnson, B.F.G., and Lewis, J. (1976). X-ray structures of the hexanuclear cluster complexes $[Os_6(CO)_{18}]^{2-}$, $[HOs_6(CO)_{18}]^-$, and $[H_2Os_6(CO)_{18}]$. *J. Chem. Soc., Chem. Commun.* 21: 883–885.

7 Braga, D., Johnson, B.F.G., Lewis, J. et al. (1982). The reaction of NO^+ with some anions of osmium and ruthenium: synthesis and X-ray characterization of $[H_3Os_4(CO)_{12}(\mu_2\text{-}NO)]$ and $[HRu_4N(CO)_{11}P(OMe)_3]$. *J. Chem. Soc., Chem. Commun.* 1081–1083.

8 Braga, D., Grepioni, F., Righi, S. et al. (1992). Synthesis of hexaruthenium bis(arene) clusters. Molecular and crystal structure of $[Ru_6C(CO)_{11}(\eta^6\text{-}1,3,5\text{-}C_6H_3Me_3)(\eta^6\text{-}C_6H_6)]$ and $[Ru_6C(CO)_{11}(\eta^6\text{-}1,3,5\text{-}C_6H_3Me_3)_2]$. *Organometallics* 11: 4042–40481.

9 Chini, P., Longoni, G., and Albano, V.G. (1976). High nuclearity metal carbonyl clusters. *Adv. Organomet. Chem.* 14: 285–344.

10 Hogarth, G., Kabir, S.E., and Nordlander, E. (2010). Cluster chemistry in the Noughties: new developments and their relationship to nanoparticles. *Dalton Trans.* 39: 6153–6174.

11 Brust, M., Walker, M., Bethell, D. et al. (1994). Synthesis of thiol-derivatised gold nanoparticles in a two-phase liquid–liquid system. *J. Chem. Soc., Chem. Commun.* 801–802.

12 Femoni, C., Cesari, C., Iapalucci, M.C. et al. (2022). Group 9 and 10 Carbonyl Clusters. In: *Comprehensive Organometallic Chemistry IV* (ed. G. Parkin, K. Meyer and D. O'Hare), 205–270. Elsevier https://doi.org/10.1016/B978-0-12-820206-7.00151-7.

13 Mednikov, E.G., Jewell, M.C., and Dahl, L.F. (2007). Nanosized $(\mu_{12}\text{-}Pt)Pd_{164-x}Pt_x(CO)_{72}(PPh_3)_{20}$ $(x \approx 7)$ containing Pt-centered four-shell 165-atom Pd–Pt core with unprecedented intershell bridging carbonyl ligands: comparative analysis of icosahedral shell-growth patterns with geometrically related $Pd_{145}(CO)_x(PEt_3)_{30}$ $(x \approx 60)$ containing capped three-shell Pd_{145} core. *J. Am. Chem. Soc.* 129: 11619–11630.

14 Ceriotti, A., Masciocchi, N., Macchi, P., and Longoni, G. (1999). $[Pt_{19}(CO)_{21}(NO)]^{3-}$ and $[Pt_{38}(CO)_{44}]^{2-}$: nitrosyl bending through intramolecular electron transfer as an intermediate step in the nucleation process from polydecker to CCP platinum carbonyl clusters. *Angew. Chem. Int. Ed.* 38: 3724–3727.

15 Ceriotti, A., Demartin, F., Longoni, G. et al. (1985). Synthesis and structure of the $[Ni_{38}Pt_6(CO)_{48}H_{6-n}]^{n-}$ (n = 5, 4) ions: Ni–Pt clusters as models for "cherry" crystallites. *Angew. Chem., Int. Ed. Engl.* 24: 697–698.

16 De Silva, N. and Dahl, L.F. (2006). Aprotic synthesis and structural determination of the nanosized nonprotonated ν_3-octahedral $[Pt_6Ni_{38}(CO)_{48}]^{6-}$ hexaanion stabilized as a cubic solvated $[NMe_4]^+$ salt. *Inorg. Chem.* 45: 8814–8816.

17 Femoni, C., Iapalucci, M.C., Longoni, G., and Svensson, P.H. (2004). New high-nuclearity Ni–Pt carbonyl clusters: synthesis and X-ray structure of the ordered $[HNi_{24}Pt_{17}(CO)_{46}]^{5-}$ and the substitutionally Ni/Pt disordered $[Ni_{32}Pt_{24}(CO)_{56}]^{6-}$ cluster anions. *Chem. Commun.* (20): 2274–2275.

18 Ciabatti, I., Femoni, C., Iapalucci, M.C. et al. (2018). The role of gold in transition metal carbonyl clusters. *Coord. Chem. Rev.* 355: 27–38.

19 Corey, E.R., Dahl, L.F., and Beck, W. (1963). $Rh_6(CO)_{16}$ and its identity with previously reported $Rh_4(CO)_{11}$. *J. Am. Chem. Soc.* 85: 1202–1203.

20 Wade, K. (1978). Metal–metal and metal–carbon bond energy terms for the rhodium carbonyl clusters $Rh_4(CO)_{12}$ and $Rh_6(CO)_{16}$. *Inorg. Nucl. Chem. Lett.* 14: 71–74.

21 Hughes, A.K. and Wade, K. (2000). Metal–metal and metal–ligand bond strengths in metal carbonyl clusters. *Coord. Chem. Rev.* 197: 191–229.

22 Martinengo, S., Ciani, G., Sironi, A., and Chini, P. (1978). Analogues of metallic lattices in rhodium carbonyl cluster chemistry. Synthesis and X-ray structure of the $[Rh_{15}(\mu\text{-}CO)_{14}(CO)_{13}]^{3-}$ and $[Rh_{14}(\mu\text{-}CO)_{16}(CO)_9]^{4-}$ anions showing a stepwise hexagonal close-packed/body-centered cubic interconversion. *J. Am. Chem. Soc.* 7096–7098.

23 Martinengo, S., Ciani, G., and Sironi, A. (1980). Synthesis and X-ray structural characterization of the $[Rh_{22}(\mu_3\text{-}CO)_7(\mu\text{-}CO)_{18}(CO)_{12}]^{4-}$ anion containing a large close-packed cluster with an ABAC sequence of compact layers. *J. Am. Chem. Soc.* 102: 7564–7565.

24 Fumagalli, A., Martinengo, S., Ciani, G. et al. (2012). Mixed platinum-rhodium carbonyl clusters. The isolation of $[Pt_4Rh_{18}(CO)_{35}]^{4-}$, an example of a cherry-like cluster with a semiexposed tetraplatinum core. *Inorg. Chem.* 31: 336–340.

25 Vidal, J.L., Schoening, R.C., and Troup, J.M. (1981). $[Cs_9(18\text{-crown-6})_{14}]^{9+}[Rh_{22}(CO)_{35}H_x]^{5-}$ $[Rh_{22}(CO)_{35}H_{x+1}]$. Synthesis, structure, and reactivity of a rhodium carbonyl cluster with body-centered and cubic arrangement of metal atoms. *Inorg. Chem.* 20: 227–238.

26 Collini, D., Fabrizi De Biani, F., Dolznikov, D.S. et al. (2011). Synthesis, structure, and spectroscopic characterization of $[H_{8-n}Rh_{22}(CO)_{35}]^{n-}$ (n = 4, 5) and $[H_2Rh_{13}(CO)_{24}\{Cu(MeCN)\}_2]^-$ clusters: assessment of CV and DPV as techniques to circumstantiate the presence of elusive hydride atoms. *Inorg. Chem.* 50: 2790–2798.

27 Collini, D., Femoni, C., Iapalucci, M.C. et al. (2002). Tuning electronic behavior of carbonyl metal clusters by substitution of interstitial and capping atoms. *Angew. Chem. Int. Ed.* 41: 3685–3688.

28 Dolzhnikov, D.S., Iapalucci, M.C., Longoni, G. et al. (2012). New high-nuclearity carbonyl and carbonyl-substituted rhodium clusters and their relationships with polyicosahedral carbonyl-substituted palladium and gold-thiolates. *Inorg. Chem.* 51: 11214–11216.

29 Femoni, C., Iapalucci, M.C., Longoni, G. et al. (2011). Icosahedral Pt-centered Pt_{13} and Pt_{19} carbonyl clusters decorated by $[Cd_5(\mu\text{-}Br)_5Br_{5-x}(\text{solvent})_x]^{x+}$ rings reminiscent of the decoration of Au–Fe–CO and Au-thiolate nanoclusters: a unifying approach to their electron counts. *J. Am. Chem. Soc.* 133: 2406–2409.

30 Femoni, C., Iapalucci, M.C., Longoni, G. et al. (2008). An organometallic approach to gold nanoparticles: synthesis and X-ray structure of CO-protected $Au_{21}Fe_{10}$, $Au_{22}Fe_{12}$, $Au_{28}Fe_{14}$ and $Au_{34}Fe_{14}$ clusters. *Angew. Chem. Int. Ed.* 47: 6666–6669.

31 Capacci, C., Cesari, C., Femoni, C. et al. (2020). Structural diversity in molecular nickel phosphide carbonyl nanoclusters. *Inorg. Chem.* 59: 16016–16026.

32 Zeng, C., Chen, Y., Kirschbaum, K. et al. (2015). Structural patterns at all scales in a nonmetallic chiral $Au_{133}(SR)_{52}$ nanoparticle. *Sci. Adv.* 1: e1500045.

33 Yan, N., Xia, N., Liao, L. et al. (2018). Unraveling the long-pursued Au144 structure by X-ray crystallography. *Sci. Adv.* 4: eaat7259.

34 Desireddy, A., Conn, B.E., Guo, J. et al. (2013). Ultrastable silver nanoparticles. *Nature* 501: 399–402.

35 Longoni, G., Femoni, C., Iapalucci, M.C., and Zanello, P. (1999). Electron-sink features of homoleptic transition-metal carbonyl clusters. In: *Metal Clusters in Chemistry*, vol. 2 (ed. P. Braunstein, L.A. Oro and P.R. Raithby), 1137–1158. Wiley-VCH.

36 (a) Fumagalli, A., Martinengo, S., Bernasconi, G. et al. (2000). Skeletal growth by condensation of small metal fragments on a carbido carbonyl cluster. Synthesis of the anions $[Rh_{15}C_2(CO)_{24}X_2]^{3-}$ (X = Cl, Br, I) and molecular structure of the bromo and iodo derivatives. *Organometallics* 19: 5149–5154. (b) Albano, V.G., Braga, D., Chini, P. et al. (1983). New carbide clusters in the cobalt sub-group. Part 10. Preparation and crystallographic characterization of dicarbido-octa-μ-carbonyl-hexadecacarbonyl-*polyhedro*-dodecarhodate(2-) as its bis(triphenyl-phosphine)iminium salt, $[N(PPh_3)_2]_2[Rh_{12}C_2(CO)_{24}]$. *J. Chem. Soc. Dalton Trans.* (2): 249–252.

37 Fumagalli, A., Martinengo, S., Bernasconi, G. et al. (1997). $[Rh_{28}N_4(CO)_{41}Hx]^{4-}$, a massive carbonyl cluster with four interstitial nitrogen atoms. *J. Am. Chem. Soc.* 119: 1450–1451.

38 Vidal, J.L., Walker, W.E., and Schoening, R.C. (1981). $[Rh_{10}P(CO)_{22}]^{3-}$. A transition-metal carbonyl cluster with a metal polyhedron based on the bicapped square antiprism as illustrated by the structural study of the benzyltriethylammonium salt. *Inorg. Chem.* 20: 238–242.

39 (a) Vidal, J.L., Fiato, R.A., Cosby, L.A., and Pruett, R.L. (1978). $[Rh_{17}(S)_2(CO)_{32}]^{3-}$. 1. An example of encapsulation of chalcogen atoms by transition-metal-carbonyl clusters. *Inorg. Chem.* 17: 2574–2582. (b) Garlaschelli, L., Fumagalli, A., Martinengo, S. et al. (1982). The synthesis and nuclear magnetic resonance spectroscopic study of $[Rh_{10}S(CO)_{22}]^{2-}$. *J. Chem. Soc. Dalton Trans.* (11): 2265–2267.

40 Vidal, J.L. (1981). $[Rh_{10}As(CO)_{22}]^{3-}$. Example of encapsulation of arsenic by transition-metal carbonyl clusters as illustrated by the structural study of the benzyltriethylammonium salt. *Inorg. Chem.* 20: 243–249.

41 Vidal, J.L. and Troup, J.M. (1981). $[Rh_{12}Sb(CO)_{27}]^{3-}$. An example of encapsulation of antimony by a transition metal carbonyl cluster. *J. Organomet. Chem.* 213: 351–363.

42 Hieber, W.O. and Schubert, E.H. (1965). Absorptionsmessungen an Carbonylferrat-Lösungen im Sichtbaren und UV-Gebiet. *Z. Anorg. Allg. Chem.* 338: 32–36.

43 Chini, P. (1980). Large metal carbonyl clusters (LMCC). *J. Organomet. Chem.* 200: 37–61.

44 Vidal, J.L. and Schoening, R.C. (1982). Rhodium carbonyl clusters chemistry under high pressures of carbon monoxide and hydrogen. 5. Synthesis and reactivity of $[Rh_{15}(CO)_{27}]^{3-}$. *Inorg. Chem.* 21: 438–441.

45 Martinengo, S. and Chini, P. (1972). Synthesis and characterization of the $[Rh_6(CO)_{15}]^{2-}$ and $[Rh_7(CO)_{16}]^{3-}$ anions. *Gazz. Chim. It.* 102: 344–354.

46 Adams, R.D., Chen, M., Trufan, E., and Zhang, Q. (2011). A new method for introducing tin ligands into tetrairidium dodecacarbonyl. *Organometallics* 30: 661–664.

47 Usher, A.J., Lucas, N.T., Dalton, G.T. et al. (2006). Mixed-metal cluster chemistry. 29. Core expansion and ligand-driven metal exchange at group 6-iridium clusters. *Inorg. Chem.* 45: 10859–10872.

48 Wade, K. (1971). The structural significance of the number of skeletal bonding electron-pairs in carboranes, the higher boranes and borane anions, and various transition-metal carbonyl cluster compounds. *J. Chem. Soc. D* (15): 792–793.

49 Mingos, D.M.P. (1984). Polyhedral skeletal electron pair approach. *Acc. Chem. Res.* 17: 311–319.

50 Teo, B.K. (1984). New topological electron-counting theory. *Inorg. Chem.* 23: 1251–1257.

51 Femoni, C., Ciabatti, I., Iapalucci, M.C. et al. (2016). Alternative synthetic route for the heterometallic CO-releasing $[Sb@Rh_{12}(CO)_{27}]^{3-}$ icosahedral carbonyl cluster and synthesis of its new unsaturated $[Sb@Rh_{12}(CO)_{24}]^{4-}$ and dimeric $[\{Sb@Rh_{12}Sb(CO)_{25}\}_2Rh(CO)_2PPh_3]^{7-}$ derivatives. *Prog. Nat. Sci.: Mater. Int.* 26: 461–466.

52 Femoni, C., Bussoli, G., Ciabatti, I. et al. (2017). Interstitial bismuth atoms in icosahedral rhodium cages: syntheses, characterizations, and molecular structures of the $[Bi@Rh_{12}(CO)_{27}]^{3-}$, $[(Bi@Rh_{12}(CO)_{26})_2Bi]^{5-}$, $[Bi@Rh_{14}(CO)_{27}Bi_2]^{3-}$, and $[Bi@Rh_{17}(CO)_{33}Bi_2]^{4-}$ carbonyl clusters. *Inorg. Chem.* 56: 6343–6351.

53 Femoni, C., Iapalucci, M.C., Longoni, G. et al. (2007). Sn-centred icosahedral Rh carbonyl clusters: synthesis and structural characterization and $^{13}C–\{^{103}Rh\}$ HMQC NMR studies. *Dalton Trans.* 35: 3914–3923.

54 Boccalini, A., Dyson, P.J., Femoni, C. et al. (2018). Insertion of germanium atoms in high-nuclearity rhodium carbonyl compounds: synthesis, characterization and preliminary biological activity of the heterometallic $[Rh_{13}Ge(CO)_{25}]^{3-}$, $[Rh_{14}Ge_2(CO)_{30}]^{2-}$ and $[Rh_{12}Ge(CO)_{27}]^{4-}$ clusters. *Dalton Trans.* 47: 15737–15744.

55 Kettle, S.F.A., Diana, E., Rossetti, R., and Stanghellini, P.L. (1997). Why are the terminal ν(CO) infrared spectra of metal cluster carbonyls so often so simple? *J. Am. Chem. Soc.* 119: 8228–8231.

56 Femoni, C., Kaswalder, F., Iapalucci, M.C. et al. (2005). High-yield one-step synthesis in water of $[Pt_{3n}(CO)_{6n}]^{2-}$ (n > 6) and $[Pt_{38}(CO)_{44}]^{2-}$. *Chem. Commun.* 46: 5769–5771.

57 Zacchini, S. (2011). Using metal carbonyl clusters to develop a molecular approach towards metal nanoparticles. *Eur. J. Inorg. Chem.* (27): 4125–4145.

58 Femoni, C., Iapalucci, M.C., Ruggieri, S., and Zacchini, S. (2018). From mononuclear complexes to molecular nanoparticles: the buildup of atomically precise heterometallic rhodium carbonyl nanoclusters. *Acc. Chem. Res.* 51: 2748–2755.

59 Qin, Z., Wang, J., Sharma, S. et al. (2021). Photo-induced cluster-to-cluster transformation of $[Au_{37-x}Ag_x(PPh_3)_{13}Cl_{10}]^{3+}$ into $[Au_{25-y}Ag_y(PPh_3)_{10}Cl_8]^+$: fragmentation of a trimer of 8-electron superatoms by light. *J. Phys. Chem. Lett.* 12: 10920–10926.

60 (a) Longoni, G. and Iapalucci, M.C. (1994). Low-valent organometallic clusters. In: *Clusters and Colloids: From Theory to Applications* (ed. G. Schmid), 91–177. New York: Wiley-VCH. (b) Fumagalli, A. and Della Pergola, R. (1999). Synthesis and properties of metal carbonyl clusters containing nitrido ligands. In: *Metal Clusters in Chemistry* (ed. P. Braunstein, L.A. Oro and P.R. Raithby), 323–347. New York: Wiley-VCH.(c) Lewis, J. and Raithby, P.R. (1999). High nuclearity osmium – gold clusters. In: *Metal Clusters in Chemistry* (ed. P. Braunstein, L.A. Oro and P.R. Raithby), 348–380. New York: Wiley-VCH.

61 Adams, R.D. and Smith, J.L. Jr. (2005). Rhodium cluster complexes containing bridging phenylgermanium ligands. *Inorg. Chem.* 44: 4276–4281.

62 Ceriotti, A., Demartin, F., Heaton, B.T. et al. (1989). Nickel carbonyl clusters containing interstitial carbon-congener atoms: synthesis and structural characterisation of the $[Ni_{12}(\mu_{12}\text{-E})(CO)_{22}]^{2-}$(E = Ge, Sn) and $[Ni_{10}(\mu_{10}\text{-Ge})(CO)_{20}]^{2-}$ dianions. *J. Chem. Soc. Chem. Commun.* (12): 786–787.

63 Pauling, L. (1947). Atomic radii and interatomic distances in metals. *J. Am. Chem. Soc.* 69: 542–553.

64 Adams, R.D., Captain, B., Smith, J.L. et al. (2004). Superloading of tin ligands into rhodium and iridium carbonyl cluster complexes. *Inorg. Chem.* 43: 7576–7578.

65 Femoni, C., Iapalucci, M.C., Longoni, G. et al. (2009). The loss of CO from $[Rh_{12}(\mu_{12}\text{-Sn})(CO)_{27}]^{4-}$: synthesis, spectroscopic and structural characterization of the electron-deficient, icosahedral $[Rh_{12}(\mu_{12}\text{-Sn})(CO)_{25}]^{4-}$ and $[Rh_{12}(\mu_{12}\text{-Sn})(CO)_{26}]^{4-}$ tetra-anions. *Dalton Trans.* (65): 2217–2223.

66 Bussoli, G., Cesari, C., Femoni, C. et al. (2022). Atomically precise rhodium nanoclusters: synthesis and characterization of the heterometallic $[Rh_{18}Sn_3Cl_2(CO)_{33}]^{4-}$ and $[Rh_7Sn_4Cl_{10}(CO)_{14}]^{5-}$ carbonyl compounds. *Results Chem.* 4: 100435.

67 Albano, V.G., Demartin, F., Femoni, C. et al. (2000). Synthesis and characterization of new paramagnetic nickel carbonyl clusters containing antimony atoms: X-ray structure of $[NEt_3CH_2Ph]_2[Ni_{15}(\mu_{12}\text{-Sb})(CO)_{24}]$ and $[NEt_4]_3[Ni_{10}Sb_2(\mu_{12}\text{-Ni})(CO)_{18}]$. *J. Organomet. Chem.* 593: 325–334.

68 Mlynek, P.D. and Dahl, L.F. (1997). New nickel-antimony carbonyl clusters: stereochemical analyses of the $[Ni_{10}(SbR)_2(CO)_{18}]^{2-}$ dianions (R = Me, Et, *i*Pr, *t*Bu, *p*-FC_6H_4) containing empty 1,12-$Ni_{10}Sb_2$ icosahedral cages and of the unprecedented stibinido-bridged 34-electron $Ni_2(CO)_4(\mu_2\text{-Sb}t\text{Bu}_2)_2$ dimer[1]. *Organometallics* 16: 1641–1654.

69 Femoni, C., Iapalucci, M.C., Longoni, G., and Svensson, P.H. (2000). A high-nuclearity Ni–Sb carbonyl cluster displaying unprecedented metal stereochemistries: synthesis and X-ray structure of $[NEt_4]_6[Ni_{31}Sb_4(CO)_{40}]\cdot 2\ Me_2CO$. *Chem. Commun.* 8: 655–656.

70 Li, Y.-Z., Ganguly, R., and Leong, W.K. (2016). Ligand substitution in the osmium-antimony rings $Os_3(\mu\text{-SbPh}_2)_2(CO)_{10}$ and $Os_3(\mu\text{-SbPh}_2)_3(Cl)(CO)_9$. *J. Organomet. Chem.* 820: 46–54.

71 Li, Y.-Z. and Leong, W.K. (2016). Raft-like osmium- and ruthenium-antimony carbonyl clusters. *J. Organomet. Chem.* 812: 217–225.

72 Chen, G. and Leong, W.K. (2006). Two group 8 carbonyl clusters containing a naked μ_5-Sb atom. *J. Clust. Sci.* 17: 111–118.

73 Cesari, C., Femoni, C., Funaioli, T. et al. (2021). Heterometallic rhodium clusters as electron reservoirs: chemical, electrochemical, and theoretical studies of the centered-icosahedral $[Rh_{12}E(CO)_{27}]^{n-}$ atomically precise carbonyl compounds. *J. Chem. Phys.* 155: 104301–104311.

74 Femoni, C., Funaioli, T., Iapalucci, M.C. et al. (2020). Rh–Sb nanoclusters: synthesis, structure, and electrochemical studies of the atomically precise $[Rh_{20}Sb_3(CO)_{36}]^{3-}$ and $[Rh_{21}Sb_2(CO)_{38}]^{5-}$ carbonyl compounds. *Inorg. Chem.* 59: 4300–4310.

75 Whitmire, K.H., Churchill, M.R., and Fettinger, J.C. (1985). Synthesis and crystal structure of $[Et_4N]_2[Bi_4Fe_4(CO)_{13}]$. Discovery of a hybrid Zintl-metal carbonyl cluster. *J. Am. Chem. Soc.* 107: 1056–1057.

76 Whitmire, K.H. and Eveland, J.R. (1994). Structural characterization of two large bismuth–cobalt carbonyl clusters: $(PPN)_2[Bi_4Co_9(CO)_8(\mu\text{-}CO)_8]\cdot 2THF$ and $(PPN)_2[Bi_8Co_{14}(CO)_{12}(\mu\text{-}CO)_8]\cdot 1.08THF$. *J. Chem. Soc., Chem. Commun.* (11): 1335–1336.

77 (a) Goicoechea, J.M., Hull, M.W., and Sevov, S.C. (2007). Heteroatomic deltahedral clusters: synthesis and structures of closo-$[Bi_3Ni_4(CO)_6]^{3-}$, closo-$[Bi_4Ni_4(CO)_6]^{2-}$, the open cluster $[Bi_3Ni_6(CO)_9]^{3-}$, and the intermetalloid closo -$[nix@\{Bi_6Ni_6(CO)_8\}]^{4-}$. *J. Am. Chem. Soc.* 129: 7885–7893. (b) Perla, L.G. and Sevov, S.C. (2015). $[Bi_{12}Ni_7(CO)_4]^{4-}$: aggregation of intermetalloid clusters by their thermal deligation and oxidation. *Inorg. Chem.* 54: 8401–8405.

78 Kruppa, W., Bläser, D., Boese, R., and Schmid, G. (1982). Heteronucleare clustersysteme, XX [1] μ_3-bismutio-cyclo-tris(tricarbonyliridium) (3Ir-Ir), $BiIr_3(CO)_9$ Darstellung und Strukturuntersuehung eines neuartigen Iridiumclusters. *Z. Naturforsch., B: J. Chem. Sci.* 37: 209–213.

79 Adams, R.D. and Elpitiya, G. (2016). Iridium-bismuth carbonyl cluster complexes. *J. Organomet. Chem.* 812: 115–122.

80 Schmid, G. (ed.) (1994). *Clusters and Colloids: From Theory to Applications*. Weinheim: VCH.

81 Braunstein, P., Oro, L.A., and Raithby, P.R. (ed.) (1999). *Metal Clusters in Chemistry*. Weinheim: Wiley-VCH.

82 Femoni, C., Kaswalder, F., Iapalucci, M.C. et al. (2006). The possible role of metal carbonyl clusters in nanoscience and nanotechnologies. *Coord. Chem. Rev.* 250: 1580–1604.

83 Fumagalli, A., Ulivieri, P., Costa, M. et al. (2004). Electron transfer and CO addition to polynitrido cobalt carbonyl clusters: parallel pathways for conversion of the $[Co_{10}N_2(CO)_{19}]^{4-}$ anion to the novel $[Co_{11}N_2(CO)_{21}]^{3-}$ anion. *Inorg. Chem.* 43: 2125–2131.

84 Longoni, G., Manassero, M., and Sansoni, M. (1980). Synthesis and structural characterization of bimetallic Fe–Pt carbonyl clusters: their relationship with bimetallic Fe–Pd carbonyl clusters. *J. Am. Chem. Soc.* 102: 7973–7974.

85 Adams, R.D., Arafa, I., Chen, G. et al. (1990). New platinum-iron carbonyl cluster complexes and their reactions with alkynes. *Organometallics* 9: 2350–2357.

86 Roth, J.D., Lewis, G.J., Safford, L.K. et al. (1992). Exploration of the ionizable metal cluster-electrode surface analogy: infrared spectroelectrochemistry of $[Pt_{24}(CO)_{30}]^n$, $[Pt_{26}(CO)_{32}]^n$, and $[Pt_{38}(CO)_{44}]^n$ ($n = 0$ to -10) and comparisons with potential-dependent spectra of carbon monoxide adlayers on platinum surfaces. *J. Am. Chem. Soc.* 114: 6159–6169.

87 Calderoni, F., Demartin, F., Fabrizi de Biani, F. et al. (1999). Electron-sink behaviour of the carbonylnickel clusters $[Ni_{32}C_6(CO)_{36}]^{6-}$ and $[Ni_{38}C_6(CO)_{42}]^{6-}$: synthesis and characterisation of the anions $[Ni_{32}C_6(CO)_{36}]^{n-}$ ($n = 5$–10) and $[Ni_{38}C_6(CO)_{42}]^{n-}$ ($n = 5$–9) and crystal structure of $[PPh_3Me]_6[Ni_{32}C_6(CO)_{36}]\cdot 4MeCN$. *Eur. J. Inorg. Chem.* (4): 663–671.

88 Yung, K.-F. and Wong, W.-T. (2003). $[\{N(PPh_3)_2\}_2\{Os_{18}Pd_3(\mu_6\text{-}C)_2(CO)_{42}\}]$: an osmium–palladium mixed-metal high-nuclearity carbonyl cluster. *Angew. Chem. Int. Ed.* 42: 553–555.

89 Olivier, C., Cattey, H., Halet, J.-F. et al. (2007). Electron-sponge behavior and electronic structures in cobalt-centered pentagonal prismatic $Co_{11}Te_7(CO)_{10}$ and $Co_{11}Te_5(CO)_{15}$ cluster anions. *Inorg. Chem.* 46: 501–509.

90 Berti, B., Cesari, C., Femoni, C. et al. (2020). Redox active Ni–Pd carbonyl alloy nanoclusters: syntheses, molecular structures and electrochemistry of $[Ni_{22-x}Pd_{20+x}(CO)_{48}]^{6-}$ ($x = 0.62$), $[Ni_{29-x}Pd_{6+x}(CO)_{42}]^{6-}$ ($x = 0.09$) and $[Ni_{29+x}Pd_{6-x}(CO)_{42}]^{6-}$ ($x = 0.27$). *Dalton Trans.* 49: 5513–5522.

91 Tran, N.T., Powell, D.R., and Dahl, L.F. (2004). Generation of $AuPd_{22}/Au_2Pd_{21}$ analogues of the high-nuclearity $Pd_{23}(CO)_{20}(PEt_3)_{10}$ cluster containing 19-atom centered hexacapped-cuboctahedral (ν_2-octahedral) metal fragment: structural-to-synthesis approach concerning formation of $Au_2Pd_{21}(CO)_{20}(PEt_3)_{10}$. *Dalton Trans.* 4: 209–216.

92 Ceriotti, A., Macchi, P., Sironi, A. et al. (2013). Cooperative effects of electron donors and acceptors for the stabilization of elusive metal cluster frameworks: synthesis and solid-state structures of $[Pt_{19}(CO)_{24}(\mu_4\text{-}AuPPh_3)_3]^-$ and $[Pt_{19}(CO)_{24}\{\mu_4\text{-}Au_2(PPh_3)_2\}]$. *Inorg. Chem.* 52: 1960–1964.

93 Kawano, M., Bacon, J.W., Campana, C.F. et al. (2001). High-nuclearity close-packed palladium-nickel carbonyl phosphine clusters: heteropalladium $[Pd_{16}Ni_4(CO)_{22}(PPh_3)_4]^{2-}$ and $[Pd_{33}Ni_9(CO)_{41}(PPh_3)_6]^{4-}$ containing pseudo-*Td* ccp $Pd_{16}Ni_4$ and pseudo-D_3h hcp $Pd_{33}Ni_9$ cores. *Inorg. Chem.* 40: 2554–2569.

94 Krejčik, M., Daněk, M., and Hartl, F. (1991). Simple construction of an infrared optically transparent thin-layer electrochemical cell: applications to the redox reactions of ferrocene, $Mn_2(CO)_{10}$ and $Mn(CO)_3$(3,5-di-t-butyl-catecholate)$^-$. *J. Electroanal. Chem. Interfacial Electrochem.* 317: 179–187.

95 Della Pergola, R., Bruschi, M., Sironi, A. et al. (2011). Structural variations, electrochemical properties and computational studies on monomeric and dimeric Fe–Cu carbide clusters, forming copper-based staple arrays. *Dalton Trans.* 40: 5464–5475.

96 Albinati, A., Balzano, F., Fabrizi de Biani, F. et al. (2010). Synthesis, structure, and electrochemistry of the dicluster molecular pincer $[Pt_3(\mu\text{-}PBut_2)_3(CO)_2]_2$(μ-1′,1‴-diethynylbiferrocene). *Inorg. Chem.* 49: 3714–3720.

97 Koentjoro, O.F., Low, P.J., Rousseau, R. et al. (2005). A combined spectroelectrochemical and computational study of the chemically reversible 2-electron reduction of $[Ru_4(\mu\text{-}RC_2R)_2(CO)_{11}]$ clusters. *Organometallics* 24: 1284–1292.

98 Frisch, M.J., Trucks, G.W., and Schlegel, H.B. (2016). *Gaussian Revision C.01*. Wallingford CT: Gaussian, Inc.

99 Hay, P.J. and Wadt, W.R. (1985). *Ab initio* effective core potentials for molecular calculations. Potentials for K to au including the outermost core orbitals. *J. Chem. Phys.* 82: 299–310.

100 Francl, M.M., Pietro, W.J., Hehre, W.J. et al. (1982). Self-consistent molecular orbital methods. XXIII. A polarization-type basis set for second-row elements. *J. Chem. Phys.* 77: 3654–3665.

101 NIST Computational Chemistry Comparison and Benchmark Database, NIST Standard Reference Database Number 101, Release 21 August 2020, Editor: Russell D. Johnson III, http://cccbdb.nist.gov

102 Keller, E. (1999). *SCHAKAL99*. Freiburg, Germany: University of Freiburg.

11

Endohedral Fullerenes: Atomically Precise Doping Inside Nano Carbon Cages

Yang-Rong Yao[1,2], Jiawei Qiu[1], Lihao Zheng[1], Hongjie Jiang[1], Yunpeng Xia[1], and Ning Chen[1]

[1] *College of Chemistry, Chemical Engineering and Materials Science, and State Key Laboratory of Radiation Medicine and Protection, Soochow University, Suzhou, Jiangsu 215123 China*
[2] *Department of Materials Science and Engineering, University of Science and Technology of China, Hefei 230026 China*

11.1 Introduction

The study of fullerenes has grown considerably in the past three decades [1]. One of the most attractive properties of fullerenes is their cage-like structures, capable of acting as robust nano-containers for other species such as metal ions or metallic clusters. Fullerenes with such host–guest structures are named as endohedral metallofullerenes (EMFs) [2]. The possibility of trapping a metal atom into a fullerene cage was proposed by Heath et al. in 1985 when they detected a mass signal of $C_{60}La$ [3]. This hypothesis was experimentally confirmed by the isolation of La@C_{82} in 1991 [4]. $C_{60}La$ was later determined to be La@C_{2v}(9)-C_{82} with a characteristic electronic structure of (La^{3+})@$(C_{2v}(9)\text{-}C_{82})^{3-}$ [5, 6], in which the description of "C_{2v}(9)" means the structural symmetry of this C_{82} cage with its isomeric identification code (spiral code) [7] in the bracket, and the signal "@" means that a lanthanum ion is endohedrally doped inside this C_{82} cage with a three-electron transfer from this lanthanum ion to the C_{82} cage. To date, various metal ions and metallic cluster have been successfully doped inside fullerene cages. Systematic studies of these novel nanocarbon compound revealed that, due to the metal-cage charge transfer and bonding interaction, the endohedral metal ion or metallic clusters have a major impact on the electronic structures and physicochemical properties of EMFs. Thus, the spectroscopic and chemical properties of EMFs can be atomically regulated by different dopants, giving this unique nanocarbon compound great potential applications in the fields such as magnetic materials [8], photovoltaics [9], and single-molecule electret devices [10]. In this chapter, we first summarize the synthetic methods of various EMFs as well as the impact of endohedral doping on their molecular structures. Then we discuss the physicochemical properties of EMFs, including spectroscopic, electrochemical, magnetic properties, and chemical reactivity, focusing on how precise endohedral doping inside fullerene cages can fine-tune these properties.

Atomically Precise Nanochemistry, First Edition. Edited by Rongchao Jin and De-en Jiang.

11.2 Synthesis of Endohedral Metallofullerenes

The first EMF was synthesized by Heath et al. in 1985 [3]. Briefly, a cluster beam apparatus was employed to generate a neodymium–yttrium–aluminum garnet (Nd : YAG) laser at 532 nm, which focused on a rotating graphite disk doped with $LaCl_3$ in a helium environment. The graphite disk was heated to a temperature higher than 10,000 K, and hence, carbon atoms, small carbon clusters, and La atoms were generated during the high-temperature evaporation. Empty fullerenes and La-doped fullerenes were formed under the carbon-rich vapor. A time-of-flight (TOF) mass spectrometer connected to a cluster beam apparatus was employed to probe the cluster-laden beam, resulting in mass spectra of two different series, empty fullerene C_{2n} and La-based fullerene La@ C_{2n}. This synthetic method is called laser vaporization or laser ablation. However, because of the expensive apparatus, high cost, and low productivity, this method has not been widely used for the preparation of EMFs.

Various strategies have been developed to prepare EMFs in the past three decades, including laser ablation [1], arc-discharge [11], radiofrequency furnace [12], high-pressure treatment [13, 14], ion bombardment [15–18], hot-atom incorporation [19–21], and chemical approaches [22–25]. Among these methods, the direct current (DC) arc-discharge method, invented by Krätschmer and Huffman et al., is the most popular one used to date.

Figure 11.1 shows a schematic diagram of the arc-discharge reactor. The arcing reaction is performed between two graphite electrodes with a specific current density and helium pressure. The cathode, generally made by a solid graphite-made rod or disc, is fixed in the arcing reactor. The anode is made by a graphite rod packed with a mixture of graphite powder and metal oxides with an optimized metal/C atomic ratio. Before the arcing reaction, the anode graphite rods are subjected to a high-temperature pretreatment, such as preliminary annealing in a tubular furnace or "in situ activation" [26] in the arcing reactor. The anode graphite rods are gradually consumed in the arcing reaction process, leading to the formation of carbon soot containing various empty fullerenes and EMFs. The yield of EMFs is sensitive to numerous factors, which can be roughly divided into two categories: the synthetic techniques and the intrinsic stability of EMFs. The synthetic parameters include the capacity of the arcing reactor, the arc-discharge

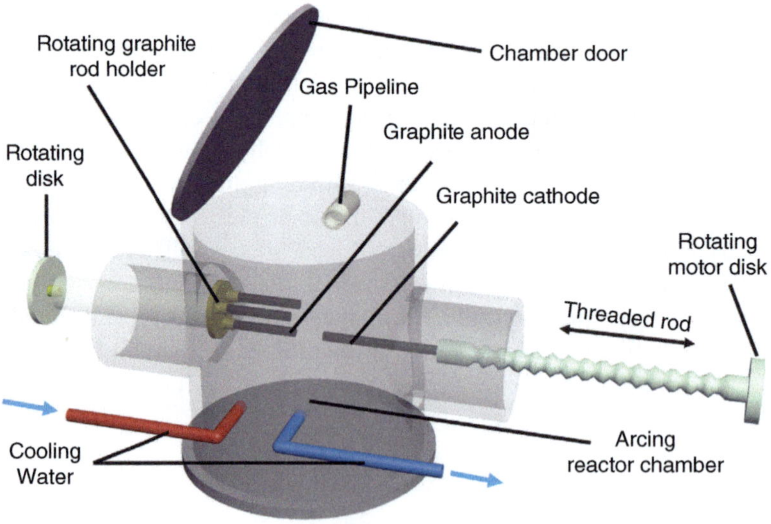

Figure 11.1 Schematic diagram of the arc-discharge reactor.

current density, the high-temperature pretreatment, the size of the graphite rod, the solid material composite and metal/C ratio, the pressure of the inert He gas, and the type and pressure of the reactive gases. The intrinsic properties of EMFs, including thermodynamic and kinetic stability, which are related to their electronic structures, are also dominant factors determining the yield of EMFs.

The arc-discharge method has unique advantages in synthesizing EMFs, as it is relatively simple, cost-effective, flexible, and productive compared to the other methods. To date, almost all rare earth metals have been encapsulated into fullerene cages, and some transition and actinide metals, such as Ti, V, Th, and U, are also capable of being enclosed inside fullerene cages using this method, as summarized in Figure 11.2 [27]. For example, recent advances in actinide EMFs revealed that U and Th can be encapsulated inside fullerene cages to form a large family of conventional mono-metallofullerenes U@C_{2n} [28–31] and Th@C_{2n} [31–36] as well as some di-metallofullerenes, such as U_2@C_{80} [37] and Th_2@C_{80} [38]. In addition to conventional mono-, di-, and tri-metallofullerenes, the flexibility of the arc-discharge method also allows the atomic endohedral doping of nonmetal elements (C, N, O, S, and H) into fullerene cages, which leads to the formation of various cluster EMFs, including carbide, nitride, cyanide, carbonitride, oxide, sulfide, and hydrocarbon clusterfullerenes (CCF, NCF, CYCF, CNCF, OCF, SCF, and HCCF).

Two strategies have been widely used to introduce nonmetal elements into the arcing reaction process. One is to use gaseous sources. In early studies of EMFs, only inert gases were used during the arcing process. Other gases were considered as harmful in the preparation of EMFs, until the discovery of the first clusterfullerene $Sc_3N@I_h(7)\text{-}C_{80}$, which was generated by an accidental leak of ambient air. This discovery opened a new research field of clusterfullerenes, which can be prepared by introducing nonmetallic elements during the arcing reaction [39]. For example, a mixed atmosphere of He and N_2 (i.e. 194 Torr He and 6 Torr N_2) is the most commonly used to synthesize NCFs. In addition to NCFs, CYCFs, and CNCFs are also available by introducing a small amount of N_2. In 2004, Dunsch et al. reported that highly selective synthesis of NCFs can be achieved by introducing NH_3 into the arc-discharging process. This strategy was called the *reactive gas*

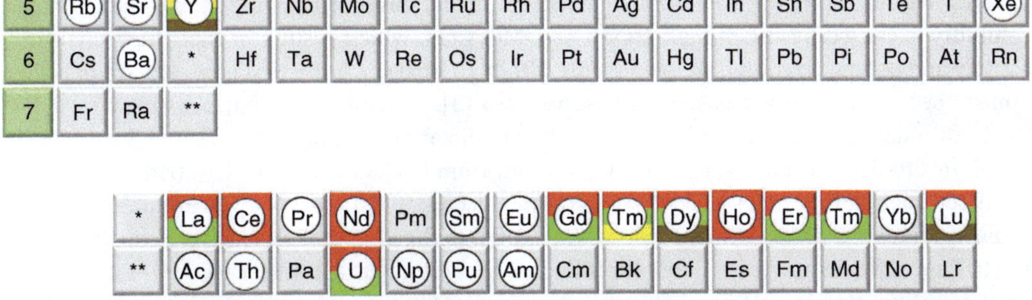

Figure 11.2 A periodic table illustrating the elements that have been encapsulated into fullerene cages. The metallic elements in circles are capable of forming conventional metallofullerenes. The element colors represent the types of clusterfullerenes that this element can form.

atmosphere method. With the introduction of *reactive gas*, the yield of NCFs is significantly increased, while the generation of empty fullerenes and other EMFs is largely suppressed [40]. Following this method, reactive gases, such as CH_4 [41], CO_2 [42], and SO_2 [43], are employed to synthesize clusterfullerenes such as HCCFs, OCFs, and SCFs, respectively. The only exception is CCFs. With carbide cluster inside fullerene cages, they can be produced without addition of any reactive gas into the He atmosphere [44].

In addition to gaseous sources, using solid reactive compound mixed with graphite powder and metal (or metal oxide) is also an effective way to introduce nonmetallic elements into fullerene cages. For example, Dunsch et al. employed an inorganic salt, CaNCN, as the nitrogen source for the synthesis of $Sc_3N@I_h(7)\text{-}C_{80}$, of which the relative yield was significantly increased up to 42% [40]. Later, Stevenson et al. developed a method named Chemically Adjusting Plasma Temperature, Energy, and Reactivity (CAPTEAR) for the selective synthesis of $Sc_3N@I_h(7)\text{-}C_{80}$ [45]. They used $Cu(NO_3)_2{\cdot}2.5H_2O$ as the solid nitrogen source, which can be decomposed into NO_x and Cu at high temperatures, adjusting the temperature and energy of the reactive plasma environments and thus selectively boosting the relative yield of $Sc_3N@I_h(7)\text{-}C_{80}$ to as high as 96%. Recently, Xin et al. employed Prussian blue ($Fe_4[Fe(CN)_6]_3$) as the solid cyanide/nitrogen dual source, resulting in selective simultaneous syntheses of Dy-based CYCFs and NCFs [46]. In addition, solid organic molecules, such as 3,5-diamino-1,2,4-triazole, thiosemicarbazide, and ammonium thiocyanate, have been used for the selective synthesis of clusterfullerenes [47].

11.3 Fullerene Structures Tuned by Endohedral Doping

11.3.1 Geometry of Empty and Endohedral Fullerene Cage Structures

In terms of geometrical structures, there are no drastic differences between the carbon cage structures of EMFs and those of empty fullerenes as both of them are generally composed of pentagons, hexagons. In particular, for nonmetallic endohedral fullerenes, such as $H_2@I_h\text{-}C_{60}$ [48] and $H_2O@I_h\text{-}C_{60}$ [24], their fullerene cages are identical to the corresponding empty fullerenes since there is no charge transfer between the interior small molecules and the exterior fullerene cages. These endohedral fullerenes are synthesized by a chemical approach figuratively called *molecular surgery* in which the empty fullerenes, i.e. $I_h\text{-}C_{60}$ (Figure 11.3a) and $D_{5h}\text{-}C_{70}$ are opened by organic synthetic methods and, after the insertion of H_2 or H_2O molecule (Figure 11.3b), they are restored to the original cage geometry.

However, when a trivalent lanthanide ion is capsulated inside $I_h\text{-}C_{60}$, the resulting $Ln@I_h\text{-}C_{60}$ shows a quite different solubility from empty $I_h\text{-}C_{60}$. It has been challenging to extract and purify $Ln@I_h\text{-}C_{60}$ because of their insolubility in common organic solvents, such as toluene and CS_2. This insolubility is caused by their small HOMO–LUMO gaps, which results in their high chemical reactivity and polymerization [49]. As a result, although $Ln@C_{60}$ is the first type of EMF detected by mass spectrometry in the gaseous phase in 1985 [3], its explicit structure was not revealed until 2018 when Nakagawa et al. reported the stabilization of $Ln@I_h\text{-}C_{60}$ (Ln = La and Gd) through in situ trifluoromethylation [50]. Different from $Gd@C_{60}$ and $La@C_{60}$, trifluoromethyl-derivatized metallofullerenes $Gd@C_{60}(CF_3)_n$ and $La@C_{60}(CF_3)_n$ (n = 3, 5) (Figure 11.3c) are soluble in toluene and CS_2, which could be attributed to the closed-shell electronic structures of these trifluoromethyl-derivatized metallofullerenes, leading to much wider HOMO–LUMO gaps than those of pristine $Ln@C_{60}$.

Essential differences between the electronic structures of EMFs and empty fullerenes play a key role on governing of their isomeric structures. Although a few isomeric cage structures are stable

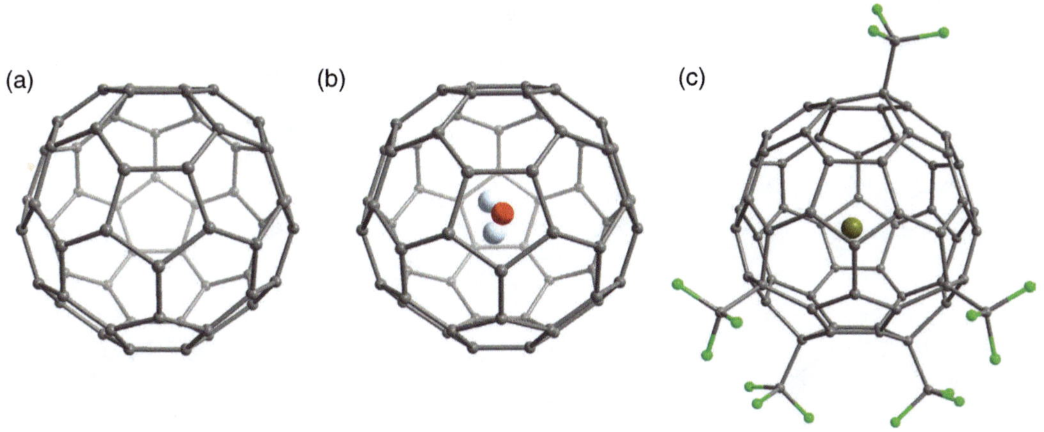

Figure 11.3 Molecular structures of (a) I_h-C_{60}; (b) $H_2O@I_h$-C_{60}; and (c) $La@I_h$-$C_{60}(CF_3)_5$.

in the form of both empty fullerene and EMF, such as $D_{2d}(23)$-C_{84} and $Sc_2C_2@D_{2d}(23)$-C_{84} [51], most of the EMFs bear isomeric cages different from the stable empty fullerenes. It is because that the isomeric cages of EMFs, which are thermodynamically not stable in its pristine form, can only be stabilized by specific charge transfer from the endohedral metal. The most symbolic case is $Sc_3N@I_h(7)$-C_{80}, in which neither the Sc_3N cluster nor the $I_h(7)$-C_{80} cage is thermodynamically stable [39]. However, the combination of these two unstable species gives rise to a stable EMF compound, in which six electrons are transferred from the encapsulated Sc_3N cluster to stabilize the $I_h(7)$-C_{80} cage. Due to internal charge transfer, the resulting $(Sc_3N)^{6+}@(I_h(7)$-$C_{80})^{6-}$ has a close-shell electronic structure with very high stability and yield, becoming the third most abundant fullerene behind I_h-C_{60} and D_{5h}-C_{70}.

Another representative case is the so-called IPR-violating or non-IPR fullerenes. Herein, IPR is abbreviated from "the isolated pentagon rule" proposed by Kroto, which states that stable fullerene cage structure only contain pentagons separated by five hexagons [52]. All the pristine fullerenes reported thus far strictly obey this rule. The destabilization of non-IPR fullerenes could be explained by the presence of fused-pentagon motifs, which result in increased local steric strain and reduced aromaticity [53]. Theoretical calculations suggested that the energy of fullerene increases by 19–24 kcal/mol with each increasing pair of fused pentagons [54]. However, benefiting from endohedral doping, some non-IPR fullerene cages are stabilized by the encapsulated species. $Sc_2@C_{2v}(4059)$-C_{66} [55] (Figure 11.4a) and $Sc_3N@D_3(6140)$-C_{68} [55] (Figure 11.5a) are the first two cases of EMFs violating the IPR principle, simultaneously reported by Wang et al. and Stevenson et al., respectively. The $C_{2v}(4059)$-C_{66} cage contains two pairs of triple sequentially fused pentagons (TSFP) distributed at the ends of the cage, and the two scandium ions are symmetrically located over the central pentagon of TSFP, each transferring three electrons to the IPR-violating motifs to make them obey Hückel's $(4n + 2)$ rule and thus stabilizing the whole EMF molecule [56]. In $Sc_3N@D_3(6140)$-C_{68}, three double-fused pentagons (DFPs) are triangularly distributed on the cage with three scandium ions located over the [5, 5] bond of the three DFPs. The stabilization of $D_3(6140)$-C_{68} is derived from the formal six-electron transfer from the HOMOs of the encapsulated Sc_3N cluster to the LUMOs of the cage. It should be noted that, in addition to the charger transfer between the encapsulated species and the fullerene cages, other factors, such as the electronic structures of the empty cage, the geometrical matching, and the interactions between the metal and fused pentagons, also play an essential role in the mechanism of endohedral stabilization of non-IPR fullerenes.

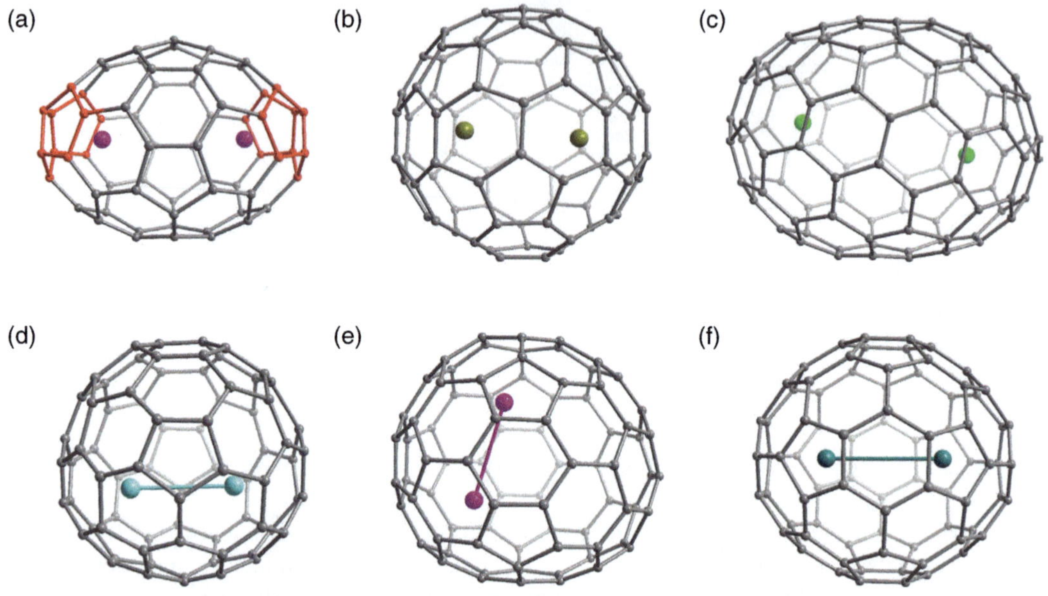

Figure 11.4 Molecular structures of (a) Sc_2@C_{2v}(4059)-C_{66}; (b) La_2@I_h(7)-C_{80}; (c) Sm_2@D_{3d}(822)-C_{104}; (d) Lu_2@T_d(2)-C_{76}; (e) Sc_2@C_{3v}(8)-C_{82}; and (f) Th_2@I_h(7)-C_{80}. The pentalene patterns in the fullerene cages are highlighted in red.

11.3.2 Conventional Endohedral Metallofullerenes

11.3.2.1 Mono-Metallofullerens

Starting from the observation of La@C_{60} and then the isolation of La@C_{82}, mono-metallofullerenes have been widely studied as prototypes of EMFs. Subsequent studies showed that most lanthanide metals could be encapsulated into the same cage, forming monometallofullerene M@C_{82} (M = La, Ce, Pr, Gd, Tb, Dy, Ho, Er, Lu). The encapsulated metal ions transfer three electrons to stabilize the C_{82} fullerene cage, forming a formal electronic structure of M^{3+}@${C_{82}}^{3-}$, and the whole molecule remains neutral. In addition to the trivalent oxidation state, metal ions with bivalent or tetravalent oxidation states, such as alkaline-earth metals [57], bivalent lanthanide [58, 59], and some actinide [28–36], have also been encapsulated into fullerene cages. An exceptional case is [(Li@I_h-C_{60})$^+$]·[($SbCl_6$)$^-$] salt, in which the whole (Li@I_h-C_{60})$^+$ acts as a cationic ion, different from the neutral form of the mono-metallofullerenes already mentioned [60].

It should be noted that for mono-metallofullerenes, different charge transfer from the endohedral metal ions to cages generally results in different isomeric fullerene structures, as summarized in Table 11.1. For example, the experimental and theoretical results of the first actinide endohedral fullerene Th@C_{3v}(8)-C_{82} suggested that the encapsulated thorium ion transferred four electrons to the C_{3v}(8)-C_{82} cage, an isomeric C_{82} structure never been reported for bivalent or trivalent lanthanide mono-metallofullerenes [36]. Interestingly, for metals with variable oxidation states, such as U, the metal oxidation states in mono-metallofullerenes show dependence on the carbon cage isomerism. For instance, Cai et al. reported experimental and theoretical studies on two U@C_{82} isomers, in which the encapsulated U ion transfers four electrons to the cage of C_2(5)-C_{82}, forming a tetravalent electronic configuration of (U^{4+})@(C_2(5)-C_{82})$^{4-}$, while only three electrons are transferred when the U ion is encapsulated inside C_{2v}(9)-C_{82}, suggesting the relevance between the metal oxidation states and isomeric cage structures [28].

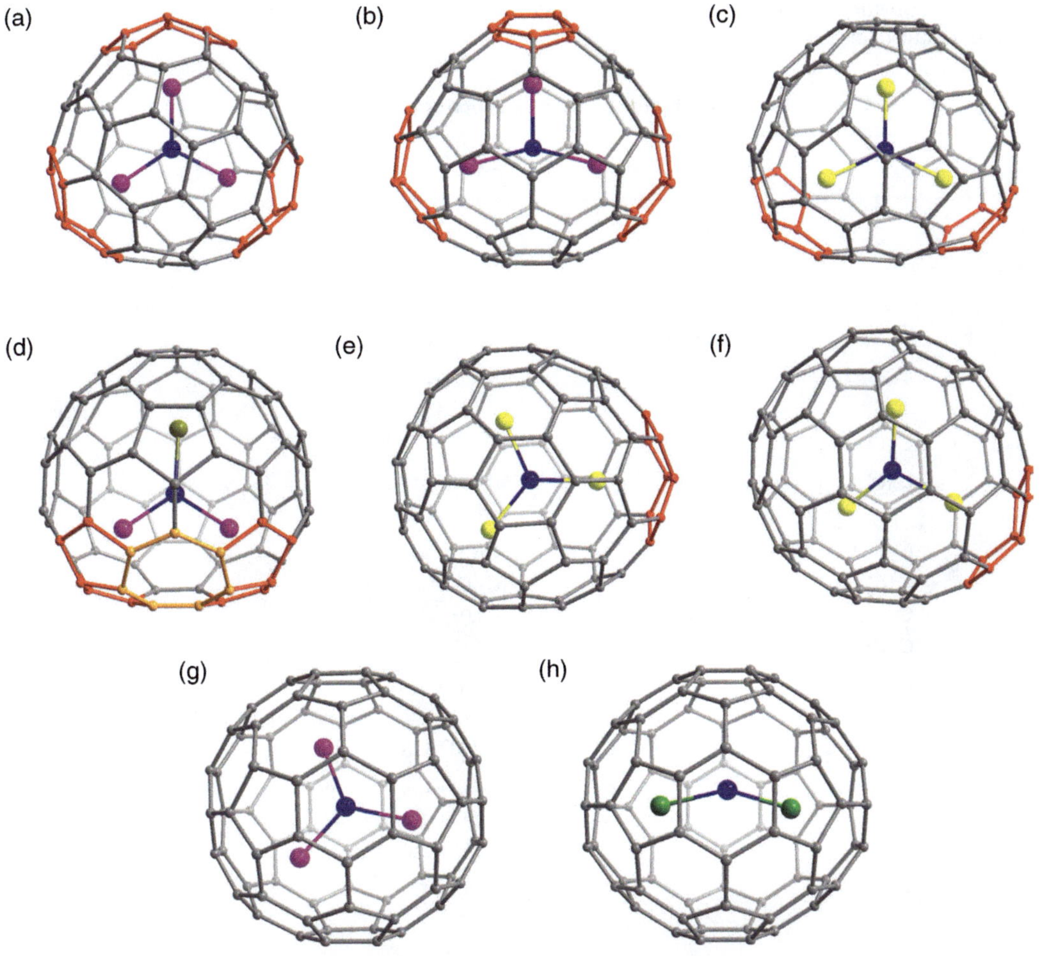

Figure 11.5 Molecular structures of (a) $Sc_3N@D_3(6140)$-C_{68}; (b) $Sc_3N@C_{2v}(7854)$-C_{70}; (c) $Gd_3N@C_2(22010)$-C_{78}; (d) $LaSc_2N@C_s$(hept)-C_{80}; (e) $Gd_3N@C_s(39663)$-C_{82}; (f) $Gd_3N@C_s(51365)$-C_{84}; (g) $Sc_3N@I_h(7)$-C_{80}; and (h) $U_2C@I_h(7)$-C_{80}. The pentalene and heptagon patterns in the fullerene cages are highlighted in red and orange, respectively.

11.3.2.2 Di-Metallofullerenes

Compared to mono-metallofullerenes, reports about well-characterized di-metallofullerenes are relatively rare. These EMFs only show tetravalent or hexavalent electronic configurations. $La_2@C_{80}$ was the first isolated di-metallofullerene and shared the same carbon cage with $Sc_3N@I_h(7)$-C_{80} [61]. The two encapsulated La ions, located over two parallel hexagons with a nonbonding La···La separation, totally transfer six electrons to stabilize $I_h(7)$-C_{80}. Later, the displacements of La to other trivalent lanthanides, such as Ce [62] and Dy [63], were reported, resulting in the analog $Ln_2@C_{80}$, which was found to be isostructural to $La_2@I_h(7)$-C_{80}. Notably, if the lanthanide ion tends to present a divalent oxidation state, the desirable fullerene cages are generally different. For example, Yang et al. reported four Sm-based di-metallofullerenes, including $Sm_2@D_2(35)$-C_{88} [64], $Sm_2@C_1(21)$-C_{90} [64], $Sm_2@D_3(85)$-C_{92} [64], and $Sm_2@D_{3d}(822)$-C_{104} [65]. Unlike La_2, the Sm_2 endo-unit favors large fullerene cages and only transfers four electrons to the cage.

Table 11.1 Structurally characterized monometallofullerenes classified by their metallic oxidation states.

	Cage isomers with different metallic oxidation states		
C_{2n}	+2	+3	+4
C_{60}		I_h-C_{60}	
C_{70}		I_h-C_{70}	
C_{72}		$C_2(10612)$-C_{72}	
C_{74}	D_{3h}-C_{74}	D_{3h}-C_{74}	D_{3h}-C_{74}
C_{76}	$C_{2v}(19138)$-C_{76}		$C_1(17418)$-C_{76} $T_d(2)$-C_{76}
C_{78}	$C_{2v}(3)$-C_{78}		
C_{80}	$C_{2v}(3)$-C_{80}	$C_{2v}(3)$-C_{80}	$C_1(28324)$-C_{80} $D_{5h}(6)$-C_{80}
C_{82}	$C_2(5)$-C_{82} $C_s(6)$-C_{82} $C_{3v}(7)$-C_{82} $C_{2v}(9)$-C_{82}	$C_s(6)$-C_{82} $C_{3v}(7)$-C_{82} $C_{2v}(9)$-C_{82}	$C_2(5)$-C_{82} $C_{3v}(8)$-C_{82} $C_{2v}(9)$-C_{82}
C_{84}	$C_2(11)$-C_{84} $C_2(13)$-C_{84} $D_{3d}(19)$-C_{84}		$C_2(8)$-C_{84} $C_s(15)$-C_{84} $D_2(21)$-C_{84}
C_{86}	$C_1(7)$-C_{86}		$C_1(11)$-C_{86} $C_1(12)$-C_{86} $C_s(15)$-C_{86}
C_{88}	$C_1(27)$-C_{88}		
C_{90}	$C_2(40)$-C_{90} $C_2(42)$-C_{90} $C_2(45)$-C_{90} $C_{2v}(46)$-C_{90}		
C_{92}	$C_s(24)$-C_{92} $C_1(42)$-C_{92}		
C_{94}	$C_{3v}(134)$-C_{94}		

In some cases, the two doped metals tend to form a rare lanthanide or actinide metal–metal bond inside the fullerene cage. For example, Lu et al. reported a series of lanthanide di-metallofullerenes $M_2@C_{2n}$ (M = Lu, Y, and Er, $2n = 76$–86) [66–69]. The common ground for these di-metallofullerenes is that the encapsulated ions adopt a low valent oxidation state of +2 and form a metal—metal bond inside the fullerene cages. The results reveal that a tetravalent electronic configuration for di-metallofullerenes is formed when the two La ions are replaced by two smaller metal ions, which tend to adopt a low valence state and thus form a metal–metal bond. Recently, Yao et al. reported the synthesis and characterization of a Sc_2 dimer trapped by a $C_{3v}(8)$-C_{82} cage, in which the Sc_2 dimer adopts a formal divalent state for each Sc ion, forming a Sc—Sc metal—metal bond and transferring four electrons to the fullerene cage [70]. The mechanism of the formation of the Sc—Sc bond has not been well elucidated since the Sc_2 dimer was also reported to be encapsulated inside $I_h(7)$-C_{80} and $C_{2v}(4059)$-C_{66} with a nonbonding Sc⋯Sc separation. Chen et al., on the other hand, reported actinide metal–metal bond inside two actinide dimetallofullerenes, $U_2@I_h(7)$-C_{80} [37] and $Th_2@I_h(7)$-C_{80} [38]. Similar to $La_2@I_h(7)$-C_{80}, the Ac_2 dimer transfers six electrons to the $I_h(7)$-C_{80} cage. A weak U—U bond and a strong Th—Th bond are formed inside the $I_h(7)$-C_{80}, which is caused by the significant overlap of hybrid actinide orbitals. The representative dimetallofullerenes are illustrated in Figure 11.4.

11.3.3 Clusterfullerenes

11.3.3.1 Nitride Clusterfullerenes

After the discovery of the first isolated clusterfullerene $Sc_3N@I_h(7)$-C_{80}, various clusterfullerenes have been systematically studied. Among them, the family of nitride clusterfullerenes, including lanthanide-based M_3N and mixed-metal $M_xM_{3-x}N$ (x = 1–2) NCFs, have become one of the most extensive groups [39]. For mixed-metal NCFs, the atomically precise doping of metallic ions can be achieved by replacing the three metal ions in the M_3N cluster one by one. This endohedral substitution alters the configuration of the nitride cluster and consequently the isomeric structures of the fullerene cages, because of the different ionic radii of the doped metals. For example, due to the larger Gd ionic radius and the longer Gd—N bond than those of Sc, the doping of one Gd ion into the Sc_3N cluster triggers the distortion of the cluster from a planar Sc_3N configuration in a pyramidal $GdSc_2N$ configuration, leading to the deviation of the N atom from the central position of the $I_h(7)$-C_{80} cage [71]. In 2014, Zhang et al. proved that endohedral doping could also change the isomeric structures of the fullerene cage [72]. They reported the first NCF with a mixed-metal $LaSc_2N$ cluster and a novel fullerene cage, C_s(hept)-C_{80}, containing 1 heptagon and 13 pentagons. Compared to Sc_3N encapsulated inside C_s(hept)-C_{80}, the stability of the C_s(hept)-C_{80} cage is dramatically enhanced through the La-doped $LaSc_2N$ cluster, which is rationalized by the strong interaction between the larger mixed-metal nitride cluster and the heptagon-containing fullerene cage.

Despite the diversity of the nitride cluster, all the subgroup-III metals of NCFs, such as Sc, Y, and lanthanides, unexceptionally form a trimetallic nitride cluster, following the well-known trimetallic nitride template (TNT) method. This phenomenon could be explained by the match between the inherent electronic features of $(M_3N)^{6+}$ and the hexavalent fullerene cage $(C_{2n})^{6-}$. However, the well-established TNT rule was unexpectedly broken when the metals in the nitride cluster were doped by actinide U. Recently, Li et al. reported a novel nitride clusterfullerene $U_2N@I_h(7)$-C_{80}, in which the central N atom is bridged by only two uranium ions [73]. This novel nitride cluster is unsymmetrical since the two U ions adopt different oxidation states of U(V) and U(IV), forming a hexavalent nitride cluster $(U_2N)^{6+}$ and thus transferring six electrons to stabilize the $I_h(7)$-C_{80} cage. It is evident that the flexible oxidation states of the U ions play an essential role in the formation of this novel nitride cluster. The representative nitride clusterfullerenes are illustrated in Figure 11.5.

11.3.3.2 Carbide Clusterfullerenes

Carbide clusterfullerenes, i.e. $M_xC_y@C_{2n}$ (x = 2–4, y = 1–3), have the most variable endohedral carbide species among the clusterfullerenes, as shown in Figure 11.6. The most common carbide cluster is M_2C_2, which applies to almost all subgroup-III metals, including Sc, Y, and lanthanides [44]. In such a carbide cluster, two encapsulated metal ions are bridged by a C_2 unit with a triple C—C bond, transferring four electrons to stabilize the fullerene cage. Interestingly, different from the encapsulated M_3N cluster with a rigid triangular configuration, the configuration of the M_2C_2 cluster is flexible, depending on the size and geometry of the carbon cage, varying from a butterfly like to orthogonal, planar, or linear shape [74]. On the other hand, the size of the host carbon cage can be somewhat dictated by the ionic radius of the encapsulated metals. For example, C_s(hept)-C_{88} is the largest reported cage for the small Sc_2C_2 cluster [75], while Y_2C_2 can even be

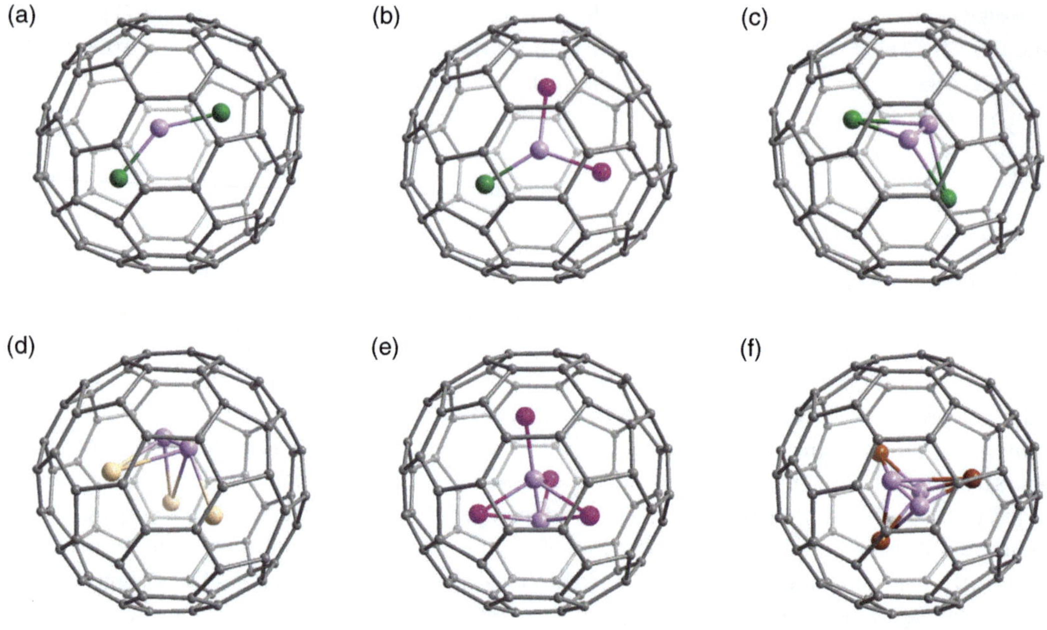

Figure 11.6 Molecular structures of endohedral metal carbide M_xC_y (x = 2–4, y = 1–3) inside the I_h(7)-C_{80} cage, including (a) $U_2C@I_h$(7)-C_{80}; (b) $USc_2C@I_h$(7)-C_{80}; (c) $U_2C_2@I_h$(7)-C_{80}; (d) $Dy_3C_2@I_h$(7)-C_{80}; (e) $Sc_4C_2@I_h$(7)-C_{80}; and (f) $Ti_3C_3@I_h$(7)-C_{80}. The endohedral C atoms are highlighted as lavender.

encapsulated into a giant C_{108} cage, forming the largest EMF $Y_2C_2@C_1$(1660)-C_{108} [76]. This phenomenon can also be explained by the geometrical match between the fullerene cage and the cluster, as Y_2C_2 is a much larger cluster than Sc_2C_2.

The ionic radius of the encapsulated metal plays a crucial role in determining the formation and structures of the carbide cluster. Recently, Jin et al. reported the synthesis and characterization of a novel carbide clusterfullerene $Dy_3C_2@I_h$(7)-C_{80}, in which the three Dy ions are distributed triangularly with the C_2 elevated above the Dy_3 plane, forming a bat ray configuration [77]. This carbide clusterfullerene has an electronic configuration of $(Dy_3)^{8+}(C_2)^{2-}@(I_h(7)\text{-}C_{80})^{6-}$ with a three-center single-electron Dy—Dy—Dy bond. Interestingly, when the three dysprosium ions are replaced by three smaller scandium ions, the resulting Sc_3C_2 cluster has an entirely different geometrical and electronic configuration, although they share the same host cage [78]. Although the small Sc_3C_2 cluster also adopts a bat ray configuration with three scandium ions triangularly distributed, the elevation of the C_2 unit above the Sc_3 plane is much smaller than that of Dy_3C_2. In addition, Sc_3C_2 has a different electronic configuration since the C_2 unit changes its electronic structures from $(C_2)^{2-}$ for Dy_3C_2 to $(C_2)^{3-}$ for Sc_3C_2, resulting in a hexavalent $((Sc_3)^{9+}(C_2)^{3-})^{6+}$ cluster and thus transferring six electrons to stabilize the I_h(7)-C_{80} cage. Furthermore, the small Sc_3C_2 cluster allows the doping of the fourth scandium ion to form a new carbide cluster, Sc_4C_2, in the same I_h(7)-C_{80} cage [79]. Due to the introduction of the additional scandium ion, the resulting Sc_4C_2 cluster adopts a tetrahedron configuration with the C_2 enveloped inside, with an altered electronic configuration of $((Sc_4)^{12+}(C_2)^{6-})^{6+}@(I_h(7)\text{-}C_{80})^{6-}$.

In addition, the endohedral doping of tetravalent metal ions, such as Ti^{4+} and U^{4+}, results in a hexavalent M_2C_2 cluster, different from the conventional lanthanide-based M_2C_2 cluster with a tetravalent electronic configuration [80, 81]. For example, Zhuang et al. reported the synthesis and characterization of two actinide carbide clusterfullerenes, $U_2C_2@D_{3h}$(7)-C_{78} and $U_2C_2@I_h$(7)-C_{80}. These two carbon cages have never been reported for Ln_2C_2-based carbide clusterfullerenes [81].

The stabilization of these two carbon cages can be ascribed to the six-electron charge transfer from the encapsulated U_2C_2 cluster with a $(U_2C_2)^{6+}$ electronic configuration, substantially different from the four-electron transfer between the conventional $(Ln_2C_2)^{4+}$ cluster and the fullerene cage.

The endohedral doping of Ti or U also gave rise to the formation of various types of novel carbide clusters, including $(Ti_3C_3)^{6+}$ [82], $(Ti(Lu/Dy/Sc)_2C)^{6+}$ [83–85], $(USc_2C)^6$ [86], and $(U_2C)^{6+}$ [87]. All these carbide clusters were hosted by the same cage of the $I_h(7)$-C_{80} cage. Inside the cage of Ti_3C_3@$I_h(7)$-C_{80}, the C_3 unit adopts a cyclopropane configuration with one weak C=C bond and two weak C—C bonds, while three Ti ions are triangularly distributed and bonded to the C_3 unit by two Ti—C bonds for each Ti ion, forming a unique Ti_3C_3 cluster and thus transferring six electrons to stabilize the $I_h(7)$-C_{80} cage [82]. For the clusters of USc_2C and $Ti(Lu/Dy/Sc)_2C$, the configuration of the USc_2C cluster is isostructural and isoelectronic to those of $Ti(Lu/Dy/Sc)_2C$ since they all have a triangle cluster configuration, and both U and Ti have an oxidation state of +4 and form an M=C bond with the central carbon [83–86]. For U_2C@$I_h(7)$-C_{80}, the first reported actinide cluster-fullerene, the encapsulated U_2C cluster adopts a symmetrically bent configuration with two equivalent U=C double bonds. The bend configuration can be ascribed to the highly electronegative central carbon, which results in a sp^1/sp^2-hybridized geometry. Such a carbide cluster has never been observed in lanthanide carbide clusterfullerenes, suggesting substantial differences in the electronic structures between lanthanide and actinide metal ions [87].

11.3.3.3 Oxide and Sulfide Clusterfullerenes

Oxygen atoms can also be doped into a di-metallofullerene, forming a new family of clusterfullerenes. Similar to those of the CCFs, the OCFs present variable configurations of endohedral oxide metallic clusters. Recently, Yang and Velkos et al. reported a family of oxide clusterfullerenes, Dy_2O@C_{2n} ($2n = 72$–82), with electronic configurations of $(Dy_2O)^{4+}$@$(C_{2n})^{4-}$ [88, 89]. The encapsulated oxygen atom serves as a bridge to connect two Dy ions, forming a bend or a linear Dy—O—Dy cluster, depending on the size and geometry of the fullerene cages. Interestingly, compared to large lanthanide ions, such as Dy^{3+} and Ho^{3+} [88–91], the small Sc^{3+} ion can not only form the isostructural Sc_2O@C_{2n} clusterfullerenes, including Sc_2O@$C_2(7854)$-C_{70} [42], Sc_2O@$T_d(2)$-C_{76} [92], Sc_2O@$C_{2v}(3)$-C_{78} [93], Sc_2O@$D_{3h}(5)$-C_{78} [93], Sc_2O@$C_{2v}(5)$-C_{80} [94], Sc_2O@$C_s(6)$-C_{82} [95], and Sc_2O@$C_{3v}(8)$-C_{82} [96], but also allow the doping of additional Sc and O to form larger oxide clusters, Sc_4O_2 and Sc_4O_3, both captured by the cage of $I_h(7)$-C_{80} [97, 98]. In the cage of Sc_4O_2@$I_h(7)$-C_{80}, the four scandium ions adopt a distorted tetrahedron configuration with two oxygen atoms asymmetrically located on two triangular faces of the Sc_4 tetrahedron. For Sc_4O_3@$I_h(7)$-C_{80}, the additional O atom is positioned on another triangular face based on the original configuration of the Sc_4O_2 cluster. Theoretical results revealed that two scandium ions of the Sc_4O_2 cluster adopt a low valent state of Sc^{2+} to form the clusterfullerene of Sc_4O_2@$I_h(7)$-C_{80} with a unique electronic structure of $(Sc_2)^{6+}(Sc_2)^{4+}(O_2)^{4-}$@$(I_h(7)$-$C_{80})^{6-}$. Interestingly, when an additional oxygen atom is doped into Sc_4O_2, the oxidation of the two low-valence scandium ions is altered, forming an oxide cluster-fullerene with an $(Sc_4)^{12+}(O_3)^{6-}$@$(I_h(7)$-$C_{80})^{6-}$ electronic structure.

Sulfide clusterfullerene, i.e. M_2S@C_{2n}, has an isoelectronic structure to the M_2O-based oxide clusterfullerene [99]. Similar to the Dy_2O and Sc_2O clusters, in the M_2S cluster, the Dy^{3+} and Sc^{3+} ions are also bridged by a sulfur atom [100–103]. Interestingly, because of the longer M—S bond than the M—O bond, the M_2S cluster shows a much smaller bond angle than those in the M_2O cluster in the same cage, as shown in Figure 11.7.

11.3.3.4 Carbonitride and Cyanide Clusterfullerenes

In 2010, Wang et al. reported the first carbonitride clusterfullerene, Sc_3CN@$I_h(7)$-C_{80} [104]. The Sc_3CN cluster adopts a planar configuration with three triangularly distributed scandium ions

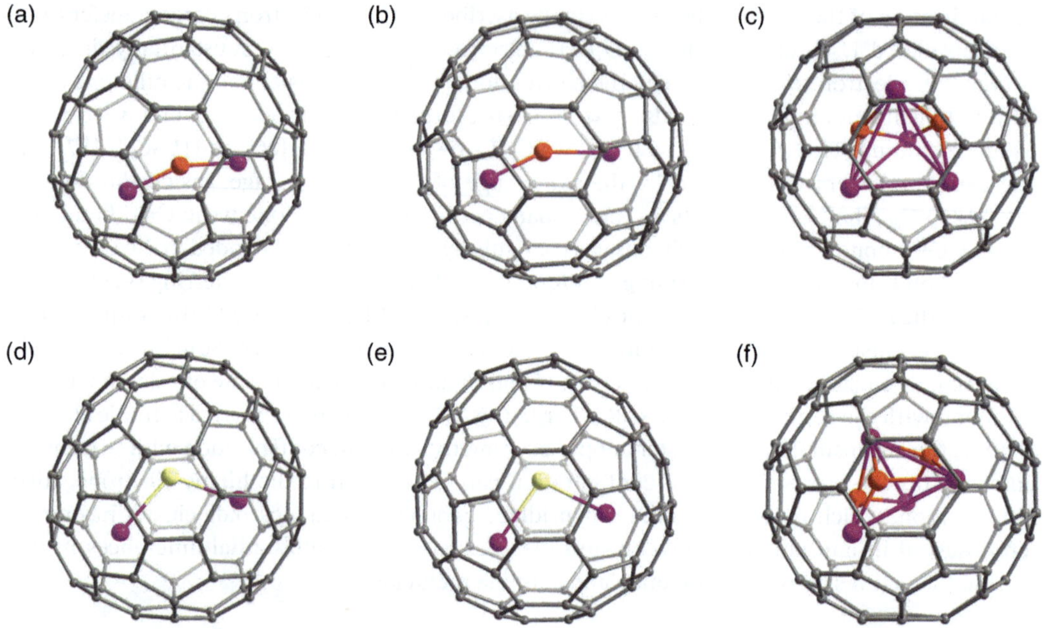

Figure 11.7 Molecular structures of oxide and sulfide clusterfullerenes, including (a) $Sc_2O@C_s(6)\text{-}C_{82}$; (b) $Sc_2O@C_{3v}(8)\text{-}C_{82}$; (c) $Sc_4O_2@I_h(7)\text{-}C_{80}$; (d) $Sc_2S@C_s(6)\text{-}C_{82}$; (e) $Sc_2O@C_{3v}(8)\text{-}C_{82}$; and (f) $Sc_4O_3@I_h(7)\text{-}C_{80}$.

connected by the central N atom and a side-located C atom. The calculation results suggest that the CN unit shows a unique valent state of $(CN)^{3-}$, combined with the three scandium ions to form a closed-shell electronic configuration, $(Sc_3)^{9+}(CN)^{3-}@(I_h(7)\text{-}C_{80})^{6-}$. Notably, when a C_2 unit is doped into the five-atom Sc_3CN cluster, a seven-carbonitride cluster Sc_3C_2CN is formed in the same cage. The C_2 unit and the CN unit are perpendicular to each other, distributed on the two sides of the triangle Sc_3 plane, forming a unique seven-number clusterfullerene with an electronic configuration of $(Sc_3)^{9+}(C_2)^{2-}(CN)^{-}@(I_h(7)\text{-}C_{80})^{6-}$ [105].

Unlike the small Sc^{3+} ion, Yang et al. reported that large lanthanide ions, such as Tb^{3+}, Dy^{3+}, and Y^{3+}, prefer to form a monometallic cluster with the CN unit [106–110]. The resulting cyanide clusterfullerenes have a triangle cluster configuration and an electronic structure of $(LnCN)^{2+}@(C_{2n})^{2-}$. The encapsulated CN unit contains a CN triple bond negatively charged by the encapsulated lanthanide ion. Recently, Meng et al. reported that the lanthanide metal in the cyanide cluster could be doped by a U ion to form a UCN cluster inside a $C_s(6)\text{-}C_{82}$ cage [111]. Although $UCN@C_s(6)\text{-}C_{82}$ has a geometrical structure similar to that of its analog $LnCN@C_s(6)\text{-}C_{82}$, the calculation results revealed that the encapsulated U ion shows ambiguous oxidation states between U(I) and U(III). This phenomenon can be explained by the substantial donation bonding from the cyanide unit and the fullerene to the uranium ion, resulting in the theoretical difficulty to distinguish from the neutral $(U^{+}CN^{-})$ cluster inside a neutral C_{82} cage. The representative carbonitride and cyanide clusterfullerenes are illustrated in Figure 11.8.

11.4 Properties Tuned by Endohedral Doping

11.4.1 Spectroscopic Properties

The spectroscopic properties of EMFs are sensitive to changes in their endohedral molecular structures. Although EMFs with similar electronic structures generally present similar spectra,

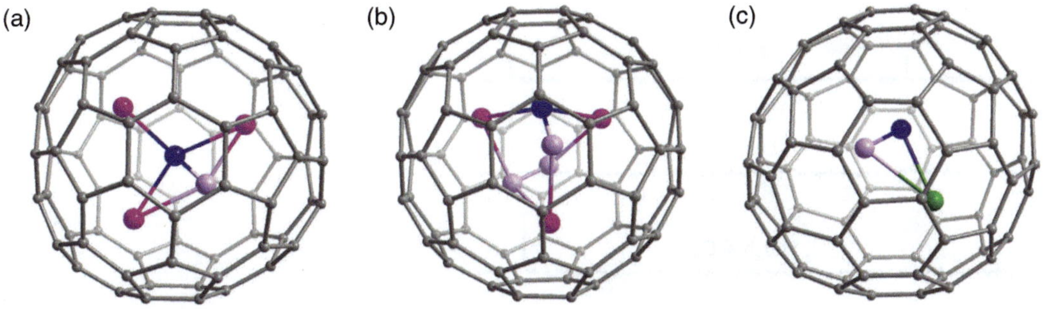

Figure 11.8 Molecular structures of (a) $Sc_3CN@I_h(7)\text{-}C_{80}$; (b) $Sc_3C_2CN@I_h(7)\text{-}C_{80}$; and (c) $UCN@C_s(6)\text{-}C_{82}$. The endohedral C atoms are highlighted as lavender.

endohedral doping can cause notable changes in their spectroscopic properties. In this section, we focus on the impact on the spectroscopic properties by endohedral doping of EMFs, including those of the nuclear magnetic resonance (NMR) spectroscopic, absorption spectroscopic, and vibrational spectroscopic properties.

11.4.1.1 NMR Spectroscopy

^{13}C NMR spectroscopy has been widely applied to obtain structural information of EMFs, such as the symmetry of the carbon cage and the structural differences of the endo-units. For example, mixed-metal NCFs $M_xM_{3-x}N@C_{80}$ (M = Sc, Lu, Y, x = 0–3) were systematically studied by ^{13}C NMR spectroscopy. All the spectra present only two peaks, typical for the $I_h(7)\text{-}C_{80}$ cage with high symmetry, as shown in Figure 11.9 [112]. However, these characteristic ^{13}C NMR peaks show a systematic upshift with endohedral doping with larger metallic ions, which can be explained by the increased π-orbital axis vector (POAV) pyramidalization angles of carbon atoms. With the increased size of the nitride cluster, the carbon atoms adjacent to the encapsulated metal ions are slightly extruded due to the limited $I_h(7)\text{-}C_{80}$ cage cavity. As a result, the corresponding carbon atoms are distorted, and the POAV pyramidalization angles are increased, leading to a systematic upshift of the ^{13}C NMR spectra.

Another case is the ^{13}C NMR spectra of $U_2C@I_h(7)\text{-}C_{80}$ and $U_2C_2@I_h(7)\text{-}C_{80}$ [81, 87]. These two EMFs show two signals with a ca. 3 : 1 intensity ratio, consistent with the two sets of equivalent carbon atoms of the $I_h(7)\text{-}C_{80}$ fullerene cage. In addition, both of the ^{13}C NMR spectra show significant temperature dependence, which is caused by the paramagnetic nature of the two U-based clusters and the weak covalent bonding between the uranium's 5f electron shell and the carbon cage. However, despite these similarities, the two ^{13}C NMR signals are located at 138.53 and 125.14 ppm for $U_2C@I_h(7)\text{-}C_{80}$ and at 132.23 and 106.42 ppm for $U_2C_2@I_h(7)\text{-}C_{80}$. Such a significant difference is probably related to the different sizes of encaged clusters and different U oxidation states in the two EMFs (U(V) in $U_2C@I_h(7)\text{-}C_{80}$ and U(IV) in $U_2C_2@I_h(7)\text{-}C_{80}$).

Metal NMR spectroscopy, such as ^{45}Sc [99, 101, 104], ^{89}Y [113], and ^{139}La [114, 115], is also informative for understanding the chemical environment, endo-unit structure, and oxidation states of the encapsulated metal ions. For example, the ^{45}Sc NMR spectrum of $Sc_3N@I_h(7)\text{-}C_{80}$ presents a single line at 200 ppm (Table 11.2), suggesting a homogeneous electronic environment for the three Sc ions of the Sc_3N cluster [39]. In contrast, after being doped with an additional C atom, $Sc_3NC@I_h(7)\text{-}C_{80}$ shows a different ^{45}Sc NMR spectrum in which two signals with a 1 : 2 intensity ratio were observed [104]. Although all three Sc ions in the Sc_3NC cluster have formal trivalent electronic configurations, they are divided by the NC unit into two types: one connected linearly with the NC unit and the other two distributed on the side of the NC unit. Accordingly, the Sc_3NC cluster has a unique C_{2v} symmetry and exhibits an inhomogeneous ^{45}Sc NMR spectrum.

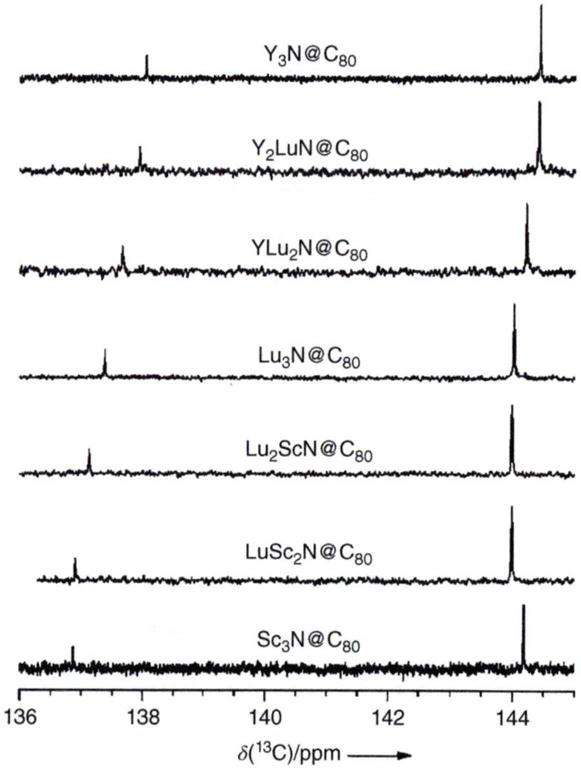

Figure 11.9 The 125 MHz ^{13}C NMR spectra of mixed-metal NCFs $M_xM_{3-x}N@I_h(7)$-C_{80} (M = Sc, Lu, Y, x = 0–3) at room temperature. *Source:* Copyright 2008, WILEY-VCH Verlag GmbH & Co. KGaA, Weinheim.

Table 11.2 ^{45}Sc NMR chemical shifts in three Sc-based EMFs.

EMFs	δ_1	δ_2
$Sc_3N@I_h(7)$-C_{80}		200
$Sc_3NC@I_h(7)$-C_{80}	360	280
$Sc_4O_2@I_h(7)$-C_{80}	285	135

For $Sc_4O_2@I_h(7)$-C_{80}, with four Sc doped inside the $I_h(7)$-C_{80} cage, the ^{45}Sc NMR spectrum also exhibits two peaks. In this case, the different chemical shifts can be ascribed to the mixed-valence state of the four Sc ions, two of which are assigned to typical Sc(III), and the other two are assigned to lower valent Sc(II) [116].

11.4.1.2 Absorption Spectroscopy

Ultraviolet–visible–near-infrared (UV–Vis–NIR) absorption spectroscopy is a facile method for the characterization of the electronic structures of EMFs. The advantage of UV–Vis–NIR absorption spectroscopy includes its broad availability and high structural sensitivity originating from the $\pi \rightarrow \pi^*$ excitation of the fullerene cage π-system. It has been well recognized that EMFs with the same carbon isomer and charge state share similar UV–Vis–NIR absorption spectra, regardless of the encapsulated metal. Therefore, UV–Vis–NIR absorption spectroscopy has been widely employed to identify the isomeric structures and metallic oxidation states of isolated EMFs.

For example, Yao et al. reported the comparison of endo-units of Sc_2C_2 and Sc_2, both captured by a C_{82} fullerene cage [70]. These two EMFs have identical UV–Vis–NIR absorption spectra, suggesting that $Sc_2C_2@C_{82}$ and $Sc_2@C_{82}$ share the same isomeric cage, which was confirmed as the $C_{3v}(8)$-C_{82} cage by single-crystal X-ray crystallographic studies.

On the other hand, even with the same isomeric cage structure, the UV–Vis–NIR absorption spectra could be substantially different if the charge states of endo-units defer. For instance, when trivalent metal ions are encapsulated inside the cage of $C_{2v}(9)$-C_{82}, their absorption spectra are almost identical [117–120]. However, the same $C_{2v}(9)$-C_{82}, when doped with divalent metal ions, such as Eu [121], Yb [122], and Sm [123], leads to the $M^{II}@C_{2v}(9)$-C_{82} electronic configuration and demonstrates dramatically different absorption spectra from that of $M^{III}@C_{2v}(9)$-C_{82}. Recently, Meng et al. reported structural and spectroscopic studies of $Th@C_{2v}(9)$-C_{82}, which exhibits a unique absorption spectrum with little resemblance to those of $M^{III}@C_{2v}(9)$-C_{82} or $M^{II}@C_{2v}(9)$-C_{82}, as shown in Figure 11.10 [35]. This result could be partially explained by the four-electron transfer between the encapsulated Th ion and the carbon cage, which results in a different $Th^{IV}@C_{2v}(9)$-C_{82} electronic structure. This is also related to the fact that Th ions have a stronger metal-cage interaction than lanthanide-based EMFs, which greatly affects the energy distribution of molecular orbitals and the corresponding absorption spectrum.

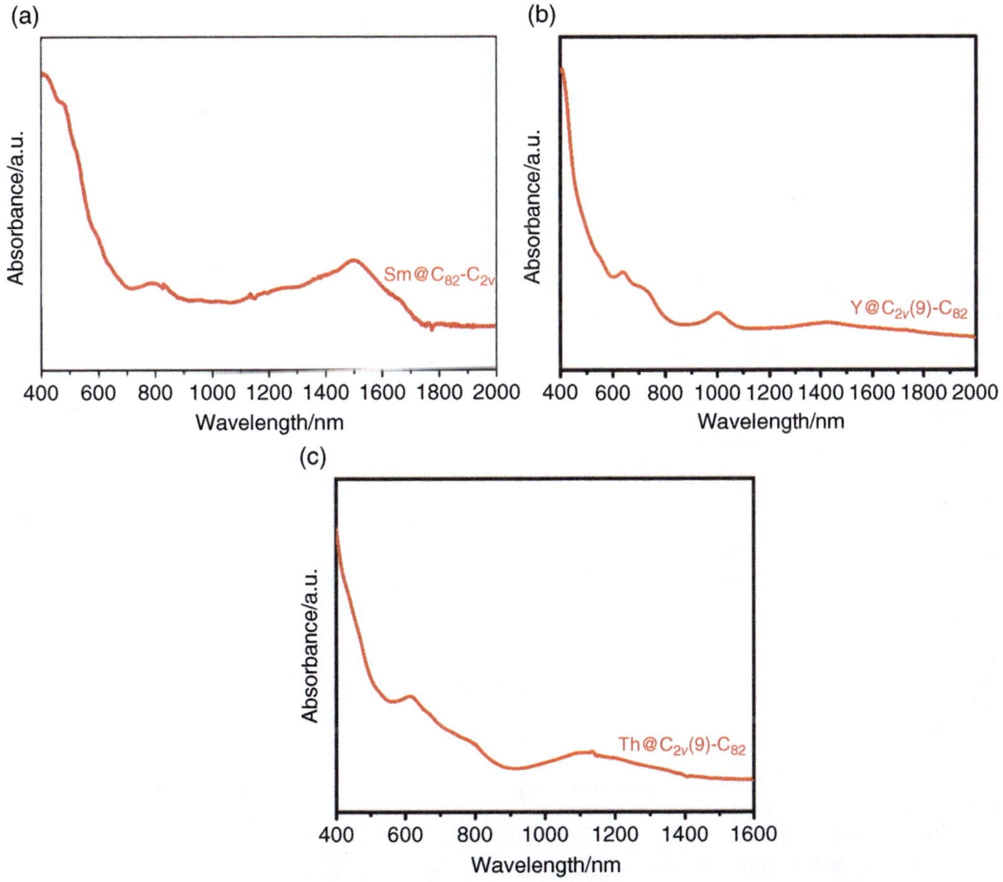

Figure 11.10 Absorption spectra of (a) $Sm^{II}@C_{2v}(9)$-C_{82}; (b) $Y^{III}@C_{2v}(9)$-C_{82}, and (c) $Th^{IV}@C_{2v}(9)$-C_{82}. *Source:* Copyright 2015, American Chemical Society; Copyright 1994, American Chemical Society; Copyright 2021, American Chemical Society.

In fact, the similarity of the absorption spectra of EMFs with the same cage structure can be observed only when the endo-unit does not play an essential role in the frontier orbitals of the whole EMF molecule. The absorption spectra of EMFs will show significant differences if the endo-units have vital contributions to the molecular orbitals, even though the cage isomerism and charge transfer remain the same. For instance, $Sc_2O@T_d(2)\text{-}C_{76}$ exhibits a virtually identical absorption spectrum to $Lu_2@T_d(2)\text{-}C_{76}$ due to the same four-electron charge transfer [68, 92].

However, as shown in Figure 11.11, the features of the absorption spectrum of $Sc_2O@T_d(2)\text{-}C_{76}$ are substantially different from those of $Th@T_d(2)\text{-}C_{76}$, especially in the range of 600–1000 nm, despite the same four-electron transfer from the endo-units of the Th ion and the Sc_2O cluster to the same $T_d(2)\text{-}C_{76}$ cage [33]. Similar differences are also observed between $Th@C_{3v}(8)\text{-}C_{82}$ and $Sc_2O@C_{3v}(8)\text{-}C_{82}$ [36, 96]. Such phenomena strongly indicate that the endohedral doping of actinide-based endo-units indeed substantially changes the electronic structures of the EMF molecule with the deep involvement of the actinide frontier orbitals in the resulting EMF molecular orbitals.

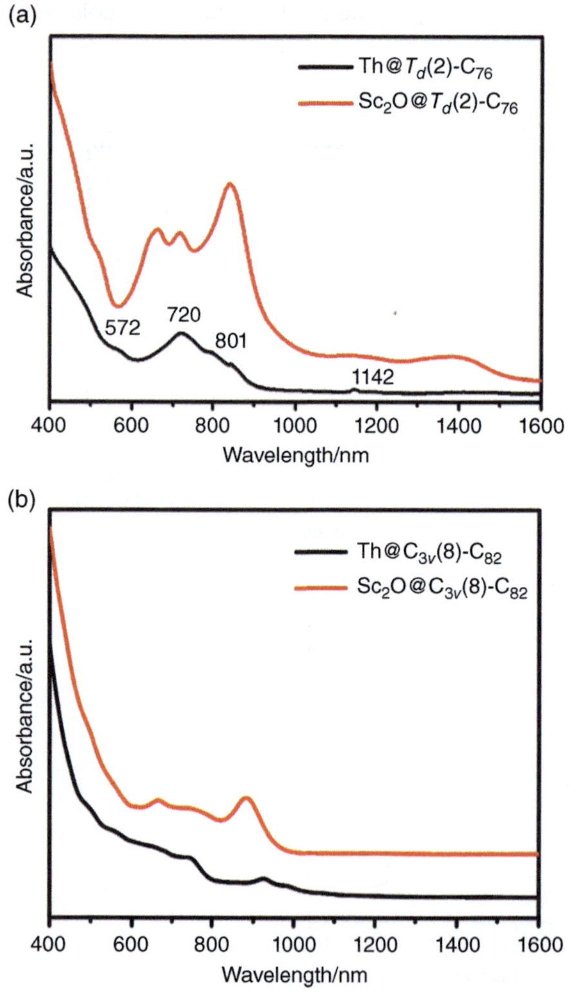

Figure 11.11 Absorption spectra of (a) $Th@T_d(2)\text{-}C_{76}/Sc_2O@T_d(2)\text{-}C_{76}$; and (b) $Th@C_{3v}(8)\text{-}C_{82}/Sc_2O@C_{3v}(8)\text{-}C_{82}$. *Source:* Copyright 2019, American Chemical Society; Copyright 2017, American Chemical Society.

11.4.1.3 Vibrational Spectroscopy

In addition to absorption spectroscopy, information on the molecular and electronic structures of EMFs can also be obtained from IR and Raman vibrational spectra. In general, because of the significant differences in the atomic masses between the encapsulated metal atoms and the fullerene carbon atoms, the metal-based vibrations are effectively separated from those of the carbon-based vibrations. For example, in the recently reported $U_2C_2@D_{3h}(5)\text{-}C_{78}$ and $U_2C_2@I_h(7)\text{-}C_{80}$, the IR and Raman vibrational spectra have been well characterized and assigned [81]. The IR vibrational modes of the carbon cages are mainly dominated in high frequencies ranging from 1000 to $1600\,cm^{-1}$, while the encapsulated U_2C_2 clusters exhibit unique U-cage vibration modes mainly located in low-frequency regions at approximately $600\text{–}900\,cm^{-1}$. Similarly, in the Raman spectra, the peaks at high frequencies of 200–250 and $400\text{–}600\,cm^{-1}$ are contributed by cage vibrations, and the peaks at low frequencies of 148 and $123\,cm^{-1}$ can be assigned to U-cage vibrations in the cages of $D_{3h}(5)\text{-}C_{78}$ and $I_h(7)\text{-}C_{80}$, respectively.

Notably, although the vibrational modes can be roughly distinguished into two types, i.e. metal-based and cage-based vibration modes, in some cases, the combination of these two vibration modes should also be considered. For instance, a systematic comparison of the IR spectra of $Sc_3N@D_3(6140)\text{-}C_{68}$ between the experimental and theoretical results was performed by Yang et al. This work showed that without the consideration of the encapsulated Sc_3N cluster, the vibration mode of neither neutral $D_3(6140)\text{-}C_{68}$ nor hexavalent $(D_3(6140)\text{-}C_{68})^{6-}$ agrees with the experimental spectrum (Figure 11.12) [124]. The simulated spectrum can be consistent with the experimental spectrum only when the influence of the Sc_3N cluster is fully considered. These results suggest that the vibration spectra of EMFs are not simply contributed by the spectral superposition of the individual cage anion and the endo-unit cation. The contribution from the metal–cage interaction should also be considered.

The vibrational spectra of EMFs are sensitive to the structures of both the carbon cages and the encapsulated species, as well as the extent of the charge transfer. Accordingly, vibrational spectroscopy of EMFs has been widely employed to identify isomeric cage structures and metal–cage

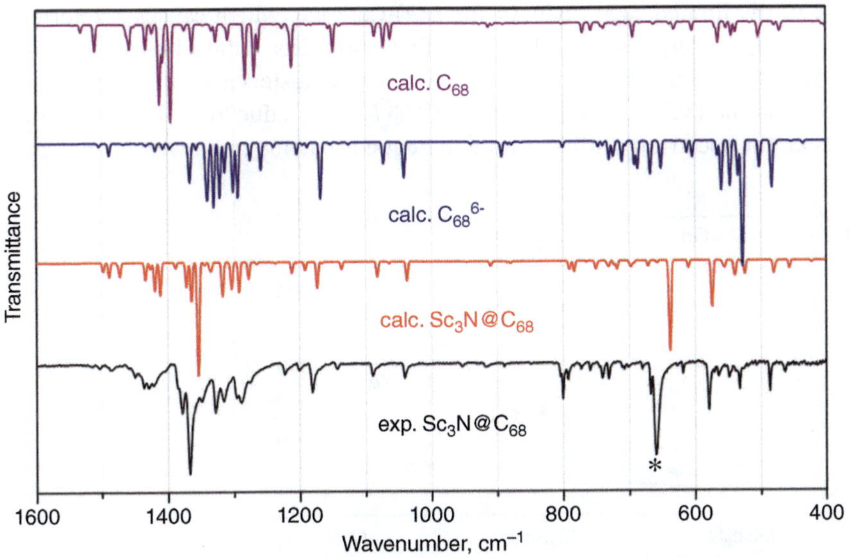

Figure 11.12 Comparison of the DFT-optimized IR spectra of $D_3(6140)\text{-}C_{68}$, $(D_3(6140)\text{-}C_{68})^{6-}$, $Sc_3N@D_3(6140)\text{-}C_{68}$, and the experimental IR spectrum of $Sc_3N@D_3(6140)\text{-}C_{68}$. An asterisk marks the Sc–N antisymmetric stretching mode. *Source:* Copyright 2006, WILEY-VCH Verlag GmbH & Co. KGaA, Weinheim.

interactions. In an early study in 1998, Lebedkin et al. employed Raman spectroscopy for the structural analysis of M@C_{2v}(9)-C_{82} (M = Y, La, Ce, and Gd) [125]. All the mono-metallofullerenes have virtually identical vibrational modes in the range above 200 cm^{-1}, indicating the same cage isomer, which was confirmed by subsequent X-ray crystallography studies. These mono-metallofullerenes also exhibit similar longitudinal metal-cage vibration modes when the encapsulated metal ions have an oxidation state of +3. However, the frequencies of the metal-cage vibration modes significantly shift toward lower wavenumbers due to the endohedral doping of divalent metal ions. This phenomenon could be explained by the frequency equation $\nu = (f/\mu)^{0.5}$ in the harmonic approximation, where ν, f, and μ represent the metal-based vibrational frequency, metal-cage force constant, and reduced mass of the metal-cage system, respectively [125]. The parameter f is highly related to the metal-cage charge transfer, suggesting that the metal-cage mode is expected at a higher frequency for metal ions with higher oxidation states.

According to this equation, the metal radius and mass effect also play an essential role in affecting the vibrational spectra, which was verified by systematic studies on the IR spectra of $M_xSc_{3-x}N$ @I_h(7)-C_{80} (x = 0–3, M = Nd, Dy, Gd, Er, and Lu) [126]. As shown in Figure 11.13, the mid-IR region in the range of 600–800 cm^{-1} is dominated by the antisymmetric stretching vibrational modes of ν_{as}(M—N), which is caused by the in-plane vibration of the nitrogen atom. The frequency of the ν_{as}(M—N) vibrational mode shows a strong metal dependence, such as the metallic ion radius, atomic mass, and thus the cluster geometry. For example, Sc_3N@I_h(7)-C_{80} exhibits a characteristic frequency at 599 cm^{-1}, which is related to the force constant of the Sc—N bonds. The doping of other lanthanide metals, leading to a mixed-metal cluster with reduced symmetry, splits the ν_{as}(M—N) vibrational mode into two components due to the different metal-based force constants. The mixed ν_{as}(M—N) vibrational mode shows a remarkable correlation between the ν_{as}(M—N) frequency and the number of doped lanthanide metals. Hence, the ν_{as}(M—N) frequency upshifts significantly with the increasing number of lanthanide metals. In addition, the mixed ν_{as}(M—N) vibrational mode is also affected by the ionic radius of doped metal ions. From Lu to Nd, with increasing ionic radius, the ν_{as}(M—N) frequency increases with the gradually distorted cluster geometry. With increasing ionic radius, the cluster structure changes from a planar to a pyramidal configuration because of the limited cage cavity of the I_h(7)-C_{80} cage, which causes a deformed local cage surface over the doped large lanthanide ion, leading to upshifts of the vibrational modes.

It should be noted that the nonmetallic atoms in the endohedral clusters also affect the cluster vibrational modes. In the Raman vibrational mode of Sc_3CN@I_h(7)-C_{80}, due to the compact structure of the Sc_3CN cluster, the spectrum exhibits a unique frequency at 468 cm^{-1}, which belongs to

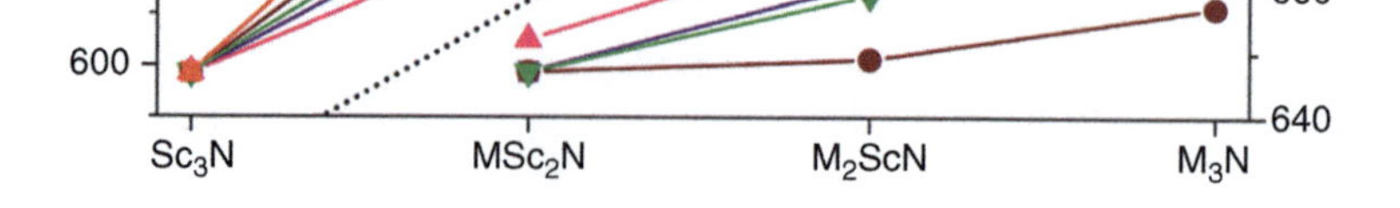

Figure 11.13 Dependence of ν_{as}(Sc−N) and ν_{as}(M−N) frequencies in $Sc_xM_{3-x}N$@I_h(7)-C_{80} (M = Nd, Gd, Dy, Er, and Lu, x = 0−3) on the cluster compositions. The sales for ν_{as}(Sc−N) and ν_{as}(M−N) are divided by a dotted line. *Source:* Copyright 2009, American Chemical Society.

the characteristic Sc-NC vibration mode and is not present in $Sc_3N@I_h(7)$-C_{80} [104]. Similarly, the metal-based vibration mode for the Sc_4O_2 cluster has a redshift compared to that of the Sc_3N cluster inside the same $I_h(7)$-C_{80} cage [116]. Furthermore, due to the doping of the oxygen atoms, characteristic vibrations related to the Sc-O vibrational mode were identified in the range of 400–600 cm^{-1}.

11.4.2 Electrochemical Properties

Electrochemical studies of EMFs on their redox behaviors, mainly determined by the frontier molecular orbitals (MOs) of EMFs, are facile tools to investigate the electronic properties and charge transfer processes of EMFs. As mentioned above, both the carbon cages and the endo-units can make significant contributions to the frontier MOs of EMFs, which could be probed by studying their redox behaviors. The precise atomic doping of the endo-units often has a major impact on the LUMO orbitals of the corresponding EMFs. In this section, we summarize the impact of endohedral doping of EMFs on the electronic structures of EMFs, investigated by electrochemical measurements, such as cyclic voltammetry (CV), differential pulse voltammetry (DPV), and square-wave voltammetry (SWV).

11.4.2.1 Conventional Endohedral Metallofullerenes

For mono-metallofullerenes, their frontier MOs are centered on the carbon cages. Hence, the electrochemical properties of these EMFs are mainly dominated by the electronic structures of the carbon cages. Similar to absorption spectroscopy, the electrochemical properties of monometallofullerenes are also widely employed to characterize cage isomerism because monometallofullerenes present virtually identical electrochemical patterns when they share the same cage isomer with the same electronic configuration. The conventional monometallofullerenes $M@C_{2v}(9)$-C_{82} (M = La, Y, Ce, and Gd) with trivalent metal ions are classic examples, as shown in Figure 11.14 [127]. These lanthanide metal atoms transfer three electrons to the $C_{2v}(9)$-C_{82} cage, leading to an open-shell electronic structure. With one unpaired electron in the HOMO, $M^{III}@C_{2v}(9)$-C_{82} presents a

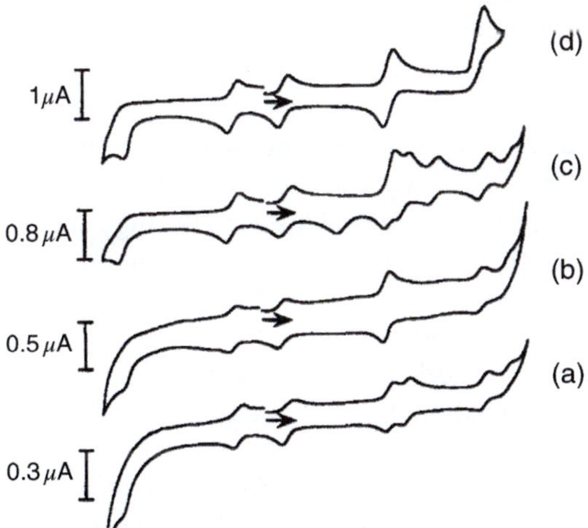

Figure 11.14 Cyclic voltammograms of (a) $La@C_{2v}(9)$-C_{82}; (b) $Y@C_{2v}(9)$-C_{82}, (c) $Ce@C_{2v}(9)$-C_{82}; and (d) $Gd@C_{2v}(9)$-C_{82}. *Source:* Copyright 1996, Published by Elsevier Ltd.

considerably positively shifted first reduction potential (${}^{red}E_1$) and a negatively shifted first oxidation potential (${}^{ox}E_1$) compared to those of the empty fullerenes. Consequently, $M^{III}@C_{2v}(9)$-C_{82} exhibits a very small electrochemical bandgap ($\Delta E = {}^{ox}E_1 - {}^{red}E_1$) of ca. 0.5 V. Such small bandgaps are also observed in another trivalent open-shell mono-metallofullerene, $M^{III}@C_s(6)$-C_{82}, suggesting the negligible influence of cage isomerism on the electrochemical activity of open-shell mono-metallofullerenes [120, 128].

Nevertheless, endohedral doping of different trivalent metal ions can also slightly alter the electrochemical activity of open-shell mono-metallofullerenes. For example, Sc@$C_{2v}(9)$-C_{82} shows the first oxidation potential at 0.15 V, which is 0.05–0.08 V larger than those of other lanthanide-based M@$C_{2v}(9)$-C_{82}, while the reduction potential is slightly smaller (0.02–0.06 V) [129]. Such discrepancies are related to the ionic radii of the metal ions and the contributions of the molecular orbitals of the metals to the frontier MOs of the EMF molecule. However, the endohedral doping of divalent metal ions, which results in closed-shell electronic structures of the EMFs, brings about significant changes in their electrochemical properties. For example, $M^{II}@C_{2v}(9)$-C_{82} has oxidation steps at much higher potentials (0.5–0.6 V) than those of trivalent $M^{III}@C_{2v}(9)$-C_{82}, while the reduction potentials are similar or even more cathodically shifted [122, 123]. As a result, the electrochemical bandgap of $M^{II}@C_{2v}(9)$-C_{82} is much larger than that of $M^{III}@C_{2v}(9)$-C_{82}.

Another typical example is the Th@C_{2n} family reported by Chen et al. These mono-metallofullerenes demonstrate considerable differences in their redox processes due to the doping of the actinide Th ion [31–36]. For instance, Th@$C_{2v}(9)$-C_{82} employs a unique tetravalent electronic configuration because the encapsulated Th ion generally prefers to present an oxidation state of +4 rather than +3 [35]. Hence, the $C_{2v}(9)$-C_{82} cage possesses a closed-shell electronic structure due to four-electron transfer. Compared to the lanthanide-based M@$C_{2v}(9)$-C_{82}, the first reduction step (−1.05 V) of Th@$C_{2v}(9)$-C_{82} is negatively shifted, and the first oxidation potential (0.20 V) is slightly positively shifted due to the contribution of the Th ion to the HOMO of Th@$C_{2v}(9)$-C_{82}. As a result, the electrochemical bandgap of Th@$C_{2v}(9)$-C_{82} (1.25 V) is significantly larger than that of lanthanide-based M@$C_{2v}(9)$-C_{82}.

It should be noted that although except for U@$C_{2v}(9)$-C_{82}, most U@C_{2n} also adapts an oxidation state of +4, similar to that of Th@C_{2n}, the U@C_{2n} family displays quite different electrochemical properties from those of Th@C_{2n} [28–31]. For example, U@$C_1(28324)$-C_{80} and Th@$C_1(28324)$-C_{80} share similar oxidation processes, but the first reduction for Th@$C_1(28324)$-C_{80} (−1.22 V) is much more negative than that for U@$C_1(28324)$-C_{80} (−0.57 V), leading to a significant difference in the electrochemical bandgap between these two actinide monometallofullerenes with the same carbon cage [31]. The phenomenon can be rationalized by the difference in the frontier MOs of these two EMFs. The LUMO of Th@$C_1(28324)$-C_{80} is located on the carbon cage due to the high energy of the Th 5f empty orbitals, while the U 5f orbitals are comparable to that of the carbon cage. Consequently, while the reduction processes occur on the carbon cage for Th@$C_1(28324)$-C_{80}, as typically observed for mono-metallofullerenes, the reduction process of U@$C_1(28324)$-C_{80} occurs on encapsulated U ions due to the low energies of the U 5f orbitals, a rare exception for mono-metallofullerenes. The redox potentials of the mono-metallofullerenes discussed are listed in Table 11.3.

For di-metallofullerenes, the localization of frontier MOs is different from that of mono-metallofullerenes. For example, in $La_2@I_h(7)$-C_{80}, the two La ions are separated from each other because of Coulomb repulsion and individually transfer three electrons to the carbon cage, and hence, the molecule's electronic structure can be described as $(La_2)^{6+}@I_h(7)$-$C_{80}{}^{6-}$ [130]. The first oxidation and reduction potentials of $La_2@I_h(7)$-C_{80} are 0.56 and −0.31 V (Table 11.4), respectively, presenting its ability as an electron acceptor rather than an electron donor. Theoretical calculations reveal that the LUMO of $La_2@I_h(7)$-C_{80} is localized between two La ions and that its relatively

Table 11.3 Redox potentials of the representative monometallofullerenes.

$M@C_{2v}(9)$-C_{82}	$^{ox}E_2$	$^{ox}E_1$	$^{red}E_1$	$^{red}E_2$	$^{red}E_3$	ΔE_{gap}
$Sc@C_{2v}(9)$-C_{82}	/	0.15	−0.35	−1.29	/	0.50
$Y@C_{2v}(9)$-C_{82}	1.07[a]	0.10	−0.37	−1.34	/	0.47
$La@C_{2v}(9)$-C_{82}	1.07[a]	0.07	−0.42	−1.37	−1.53	0.49
$Ce@C_{2v}(9)$-C_{82}	1.08[a]	0.08	−0.41	−1.41	−1.53	0.49
$Gd@C_{2v}(9)$-C_{82}	1.08[a]	0.09	−0.39	−1.38	/	0.48
$Yb@C_{2v}(9)$-C_{82}	/	0.61	−0.46	−0.78	−1.60	1.07
$U@C_{2v}(9)$-C_{82}	0.92	0.10	−0.43	−1.42	−1.76[a]	0.53
$Sm@C_{2v}(9)$-C_{82}	/	0.52	−0.42	−0.77	−1.60	0.94
$Yb@C_{2v}(9)$-C_{82}	/	0.61	−0.46	−0.78	−1.60	1.07
$Th@C_{2v}(9)$-C_{82}	0.64[a]	0.20	−1.05	−1.36	−1.72	1.25
$Th@C_1(28324)$-C_{80}	−1.11[a]	0.24	−1.22	−1.50	−2.05	1.46
$U@C_1(28324)$-C_{80}	/	0.28	−0.57	−1.50	−1.91	0.85

[a] Irreversible process.

Table 11.4 Redox potentials of $M_2@I_h(7)$-C_{80} (M = La, Ce, and U).

$M_2@I_h(7)$-C_{80}	$^{ox}E_2$	$^{ox}E_1$	$^{red}E_1$	$^{red}E_2$	$^{red}E_3$	ΔE_{gap}
$La_2@I_h(7)$-C_{80}	0.95	0.56	−0.31	−1.71	−2.13[a]	0.87
$Ce_2@I_h(7)$-C_{80}	0.95	0.57	−0.39	−1.71	/	0.96
$U_2@I_h(7)$-C_{80}	1.16[a]	0.40	−0.44	−1.58[a]	−1.78[a]	0.84

[a] Irreversible process.

low energy level gives rise to the low potential of the first reduction step. An identical redox process is also observed for another lanthanide-based $M_2@I_h(7)$-C_{80}, $Ce_2@I_h(7)$-C_{80}, which has a similar LUMO distribution to $La_2@I_h(7)$-C_{80} [131]. In contrast, despite the identical charge transfer, di-metallofullerene, $U_2@I_h(7)$-C_{80}, shows redox properties somewhat different from those of $La_2@I_h(7)$-C_{80} [37]. The first oxidation and reduction steps proceed at 0.40 and -0.44 V, cathodically shifting approximately 0.15 V compared to those for $La_2@I_h(7)$-C_{80}. The other oxidation and reduction steps also show some differences. The results suggest that the endohedral doping of two encapsulated actinide ions influences the formation of the HOMO and LUMO of the EMF molecule, which is likely related to the strong interaction between U and the carbon cage as well as the weak bonding interaction between the two U ions.

11.4.2.2 Clusterfullerenes

The electrochemical properties of clusterfullerenes have been well studied, especially for $M_3N@I_h(7)$-C_{80} (M = Sc, Y, and lanthanides). The representative molecule is $Sc_3N@I_h(7)$-C_{80}, the first reported clusterfullerene with the typical hexavalent electronic structure of $(Sc^{3+})_3N^{3-}@I_h(7)$-$C_{80}{}^{6-}$ [132]. Compared to the isoelectronic $La_2@I_h(7)$-C_{80}, $Sc_3N@I_h(7)$-C_{80} has a similar first oxidation potential (0.62 V), whereas the first reduction potential (−1.24 V) is negatively shifted, resulting in a bandgap of 1.86 V, which is significantly larger than that of most conventional EMFs.

The redox behaviors of NCFs can be roughly classified into two groups, i.e. nitride clusters with and without Sc [133–135].

Generally, the first reduction potential of non-Sc $M_3N@I_h(7)$-C_{80} is cathodic shifted compared to $Sc_3N@I_h(7)$-C_{80} due to the different MO distributions. The theoretical studies indicate that the reduction of $Sc_3N@I_h(7)$-C_{80} proceeds at the nitride cluster due to the low energy level of the 3d orbitals of the encapsulated Sc ion. In contrast, the reduction of non-Sc $M_3N@I_h(7)$-C_{80} mainly takes place on the LUMO of the carbon cage.

The electrochemical properties of NCFs are significantly affected by atomic doping of the nitride cluster. A representative example is $CeLu_2N@I_h(7)$-C_{80}, which exhibits peculiar oxidation processes with the first oxidation potential of only 0.01 V, drastically cathodically shifted compared to those of other $M_3N@I_h(7)$-C_{80} with potentials generally larger than 0.6 V [136]. Such a significant discrepancy is mainly caused by the engagement of the Ce ion with the HOMO of the EMF molecule, in which the 4f electron of Ce is involved in the oxidation process, suggesting that Ce^{3+} is much easier to oxidize than the cage carbon atoms. Another example is the doping of transition metals, such as Ti and V, which also substantially alter the electrochemical properties of NCFs [137–139]. The $TiSc_2N$ cluster shows a unique electrochemical behavior in both reduction and oxidation processes. The first oxidation and reduction steps of $TiSc_2N@I_h(7)$-C_{80} proceed at 0.16 and -0.94 V, respectively, which suggests that it is much easier to oxidize or reduce than $Sc_3N@I_h(7)$-C_{80} [137]. Similar results can also be observed in the electrochemical study of $VSc_2N@I_h(7)$-C_{80}, leading to a significant positive shift of the first reduction potential to -0.42 V, thus suggesting the considerable contributions of doped transition metals to the electrochemical properties of NCFs [139].

The nonmetal unit of the encapsulated cluster also exhibits nonnegligible contributions to the electrochemical properties of clusterfullerenes. The endohedral doping of C into the trimetallic nitride cluster results in a trimetallic μ_3-carbide clusterfullerene, such as $TiLu_2C@I_h(7)$-C_{80}, which shows unique electrochemical properties [83]. Similar to most Ti-based clusterfullerenes, the reduction step of $TiLu_2C@I_h(7)$-C_{80} occurs on the localized orbital of doped Ti atom. However, $TiLu_2C@I_h(7)$-C_{80} exhibits excellent reversibility in the reduction steps, different from those of $TiLu_2N@I_h(7)$-C_{80} with irreversible reduction processes.

$Sc_3C_2@I_h(7)$-C_{80} is also a representative case to present the crucial influence of the endohedral nonmetal unit on the electrochemical properties [140]. $Sc_3C_2@I_h(7)$-C_{80} can be considered the product of the endohedral doping of the C_2 unit into $Sc_3N@I_h(7)$-C_{80} but presents notable differences in the electrochemical properties from that of $Sc_3N@I_h(7)$-C_{80}. The first oxidation and reduction of $Sc_3C_2@I_h(7)$-C_{80} proceed at −0.03 and −0.50 V, respectively, making it electrochemically active in both reduction and oxidation processes with a very narrow bandgap, dramatically smaller than that of $Sc_3N@I_h(7)$-C_{80} (1.86 V). This result is caused by the doping of the C_2 unit, which leaves one unpaired electron on the frontier MOs and thus results in singly occupied MOs with a small bandgap. Interestingly, the doping of an additional Sc into the Sc_3C_2 cluster leads to the formation of one of the largest endohedral carbide clusters, Sc_4C_2. The resulting $Sc_4C_2@I_h(7)$-C_{80} presents three irreversible reduction processes at −1.49, −1.96, and −2.19 V along with two oxidation steps at 0.06 and 0.74 V, resembling neither $Sc_3C_2@I_h(7)$-C_{80} nor $Sc_3N@I_h(7)$-C_{80} [116]. The discrepancy results from the unique electronic structure of the $(C_2)^{6-}@(Sc_4)^{12+}$ cluster, where the HOMO of $Sc_4C_2@I_h(7)$-C_{80} is localized. During the oxidation processes, the $(C_2)^{6-}$ unit is the unit that loses electrons due to the difficulty of oxidizing the Sc^{3+} ions. Furthermore, compared to the single oxidation step of $Sc_3N@I_h(7)$-C_{80}, $Sc_4C_2@I_h(7)$-C_{80} possesses two oxidation steps since $(C_2)^{6-}$ can be easily transformed into $(C_2)^{5-}$ and then to $(C_2)^{4-}$, suggesting the unique role of the encapsulated $(C_2)^{6-}$ on the very low first oxidation potential of the molecule (0.06 V). The redox potentials of the clusterfullerenes just mentioned are listed in Table 11.5.

Table 11.5 Redox potentials of the representative clusterfullerenes.

Clusterfullerenes	${}^{ox}E_2$	${}^{ox}E_1$	${}^{red}E_1$	${}^{red}E_2$	${}^{red}E_3$	ΔE_{gap}
$Sc_3N@I_h(7)\text{-}C_{80}$	/	0.62	−1.24	−1.62	/	1.86
$Y_3N@I_h(7)\text{-}C_{80}$	/	0.63	-1.39^a	/	/	2.02
$Tb_3N@I_h(7)\text{-}C_{80}$	/	0.59	-1.38^a	/	/	1.97
$Ho_3N@I_h(7)\text{-}C_{80}$	/	0.60	-1.45^a	/	/	2.05
$Lu_3N@I_h(7)\text{-}C_{80}$	/	0.60	-1.48^a	/	/	2.08
$TiSc_2N@I_h(7)\text{-}C_{80}$	1.08^a	0.16	−0.94	−1.58	−2.21	1.10
$VSc_2N@I_h(7)\text{-}C_{80}$	/	0.44^a	−0.42	−0.66	−1.33	0.86
$CeLu_2N@I_h(7)\text{-}C_{80}$	/	0.01	−1.39	−1.88	/	1.40
$TiLu_2C@I_h(7)\text{-}C_{80}$	/	0.64	−0.91	/	/	1.55
$Sc_3C_2@I_h(7)\text{-}C_{80}$	/	−0.03	−0.50	−1.64	−1.84	0.47
$Sc_4C_2@I_h(7)\text{-}C_{80}$	0.74	0.06	-1.49^a	-1.96^a	-2.19^a	1.55

[a] Irreversible process.

11.4.3 Magnetic Properties

When the encapsulated metal ions have unpaired electrons or transfer an odd number of electrons to the carbon cage, the resulting EMF molecules present interesting magnetic properties due to the paramagnetic state of the molecule. Early studies on the magnetic properties of EMFs focused on paramagnetic behaviors measured by magnetic technologies, such as X-ray magnetic circular dichroism (XMCD) and superconducting quantum interference devices (SQUIDs). Recently, potential applications as single-molecule magnets (SMMs) have been explored, especially for EMFs containing lanthanide ions with localized partially filled 4f shells. Accordingly, the SMM behavior of EMFs can be precisely regulated by the endohedral doping of different lanthanide ions. In this section, we focus on the significant impact of the endohedral doping of different lanthanide ions on the magnetic properties of EMFs as SMMs.

11.4.3.1 Dimetallofullerenes

Dimetallofullerenes exhibit pronounced SMM performances when the two encapsulated metal ions feature a single-electron bond [63, 141]. However, such dimetallofullerenes with a single-electron metal–metal bond are unstable and tend to aggregate to form polymers, making it difficult to extract from soot [142–144]. Recently, two strategies have been explored to synthesize stable di-metallofullerenes containing single-electron metal–metal bond. The most common strategy is to stabilize the di-metallofullerene radical by a free-radical addition reaction, which results in a stable EMF monoadduct [145, 146]. The other method is the in situ doping of N into the carbon cage of the dimetallofullerene [147–149]. The resulting N-substituted azafullerenes, such as $M_2@C_{79}N$ (M = Y, Tb, Gd, and Dy), are stable compounds. Very recently, inspired by the SMM behaviors inside fullerenes, mixed-valence dilanthanide complexes, $(Cp^{iPr5})_2Ln_2I_3$ (Ln = Gd, Tb, or Dy, Cp^{iPr5} = pentaisopropylcyclopentadienyl), containing a singly occupied Ln–Ln σ-bonding, have been synthesized and characterized [150]. These dilanthanide complexes exhibit excellent magnetic SMM properties, and $(Cp^{iPr5})_2Dy_2I_3$ shows a blocking temperature of 72 K, the highest value among the SMMs reported to date.

The first N-substituted endohedral aza[80]fullerenes $M_2@C_{79}N$ (M = Y and Tb), which contain the fullerene cage of $I_h(7)\text{-}C_{80}$ doped by an N atom, were reported in 2008 by Dorn and coworkers, although their SMM properties have not been explored at that time [148]. In 2019, Velkos et al. reported a study of the magnetic properties of $Tb_2@C_{79}N$, which exhibits a large Tb-electron exchange coupling constant of 40–45 cm^{-1} with a blocking temperature (T_B) of 28 K (Figure 11.15) [151]. The authors further proved that the excellent SMM properties of $Tb_2@C_{79}N$ are induced by the unpaired [Tb^{3+}-*e*-Tb^{3+}] spin system, suggesting the potential applications of dimetallofullerenes as SMMs. Recently, Wang et al. reported the SMM study of another dimetalloazafullerene, $Dy_2@C_{79}N$ [149]. Similar to $Tb_2@C_{79}N$, the SMM properties of $Dy_2@C_{79}N$ are caused by ferromagnetic coupling between the two Dy ions via the one-electron radical bridge. Notably, the effective energy (669 $\pm$ 7 K) and blocking temperature (24 K) are slightly lower than those of $Tb_2@C_{79}N$ (757 $\pm$ 4 K and 28 K, respectively), suggesting the essential role of the encapsulated ions in the SMM performance.

In 2017, Liu et al. reported the successful stabilization of $M_2@I_h(7)$-C_{80} with a single-electron metal–metal bond through the reaction of $M_2@I_h(7)$-C_{80} with benzyl bromide in DMF [63]. The resulting monoadduct $Dy_2@I_h(7)$-$C_{80}(CH_2Ph)$ exhibits an observable hysteresis loop from 1.8 to 21 K (Figure 11.15). The blocking temperature of $Dy_2@I_h$-$C_{80}(CH_2Ph)$ was measured up to 21.9 K, which was the highest blocking temperature reported for lanthanide-based SMMs at the time. Such excellent SMM performance comes from the strong ferromagnetic coupling between two Dy ions through the single-electron bridge. Later, Liu et al. further synthesized and characterized an array of $Ln_2@I_h(7)$-$C_{80}(CH_2Ph)$ (Ln_2 = Y_2, Gd_2, Tb_2, Dy_2, Ho_2, Er_2, TbY, TbGd), among which $Tb_2@I_h(7)$-$C_{80}(CH_2Ph)$ has the highest blocking temperature, up to 28.9 K (Figure 11.15), which is one of the highest blocking temperatures in fullerene SMMs [152]. Notably, the blocking temperatures of $TbGd@I_h(7)$-$C_{80}(CH_2Ph)$ and $TbY@I_h(7)$-$C_{80}(CH_2Ph)$ were measured to be 14.4 K and 5.0 K, respectively. This phenomenon could be rationalized by the differences in the lanthanide ions, for which the former is attributed to the isotropy of the Gd^{3+} ion, while the latter can be explained by the nonmagnetism of the Y^{3+} ion. These results also reveal that the presence of two uniaxially anisotropic magnetic lanthanides seems necessary for excellent dimetallofullerene SMMs.

11.4.3.2 Clusterfullerenes

In 2012, Westerström et al. reported the magnetic properties of $DySc_2N@I_h(7)$-C_{80}, which was the first reported clusterfullerene exhibiting SMM behaviors [153]. The element-specific magnetization curve measured by XMCD is similar to the curve of the magnetic moment versus the field measured by SQUID, indicating the magnetic moments of $DySc_2N@I_h(7)$-C_{80} as a mononuclear SMM. The butterfly-shaped hysteresis loop originating from the quantum tunneling of magnetization (QTM) suggests slow magnetic relaxation with a blocking temperature T_B of 6.9 K and the SMM nature of $DySc_2N@I_h(7)$-C_{80}, which is induced by the encapsulated Dy^{3+} ion. Furthermore, the hysteresis remained below 6 K when the sample was diluted with 10–20 equiv. of C_{60}, suggesting that the SMM nature is substantially brought about by the encapsulated Dy^{3+} ion instead of the intermolecular magnetic interactions.

In 2014, Westerström et al. further investigated the SMM properties of $Dy_nSc_{3-n}N@I_h(7)$-C_{80} (n = 1–3) and showed how precise atomical doping of additional Dy^{3+} ions into the $DySc_2N$ cluster can alter the magnetic properties of the clusterfullerenes [154]. As shown in Figure 11.16, compared to the butterfly-shaped hysteresis loop of $DySc_2N@I_h(7)$-C_{80}, $Dy_2ScN@I_h(7)$-C_{80} exhibited a pronounced stepped hysteresis loop with the largest remanence among the three compounds, which can be explained by the solid ferromagnetic coupling between the two Dy^{3+} ions overcoming a high barrier of the magnetic moments and thus stabilizing the zero-field magnetization. Popov et al. further pointed out that $Dy_2ScN@I_h(7)$-C_{80} exhibits slow magnetic relaxation below 76 K with a blocking temperature of 8 K. Furthermore, the effective barrier for the thermally driven

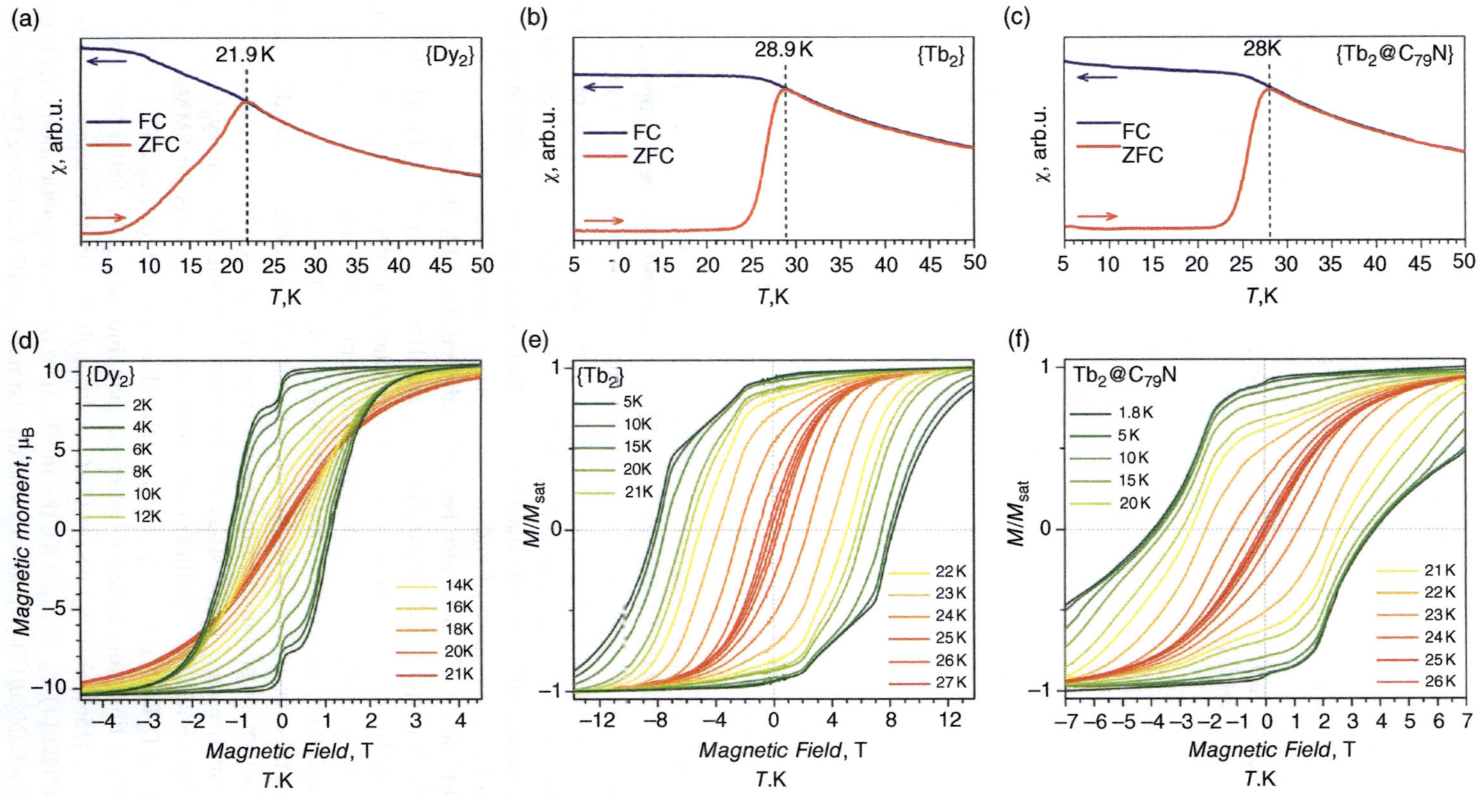

Figure 11.15 (a–c) Magnetization blocking temperatures of (a) $Dy_2@I_h(7)\text{-}C_{80}(CH_2Ph)$; (b) $Tb_2@I_h(7)\text{-}C_{80}(CH_2Ph)$; and (c) $Tb_2@C_{79}N$; (d–f) Magnetic hysteresis of (d) $Dy_2@I_h(7)\text{-}C_{80}(CH_2Ph)$; (e) $Tb_2@I_h(7)\text{-}C_{80}(CH_2Ph)$; and (f) $Tb_2@C_{79}N$. *Source:* Copyright 2019 American Chemical Society.

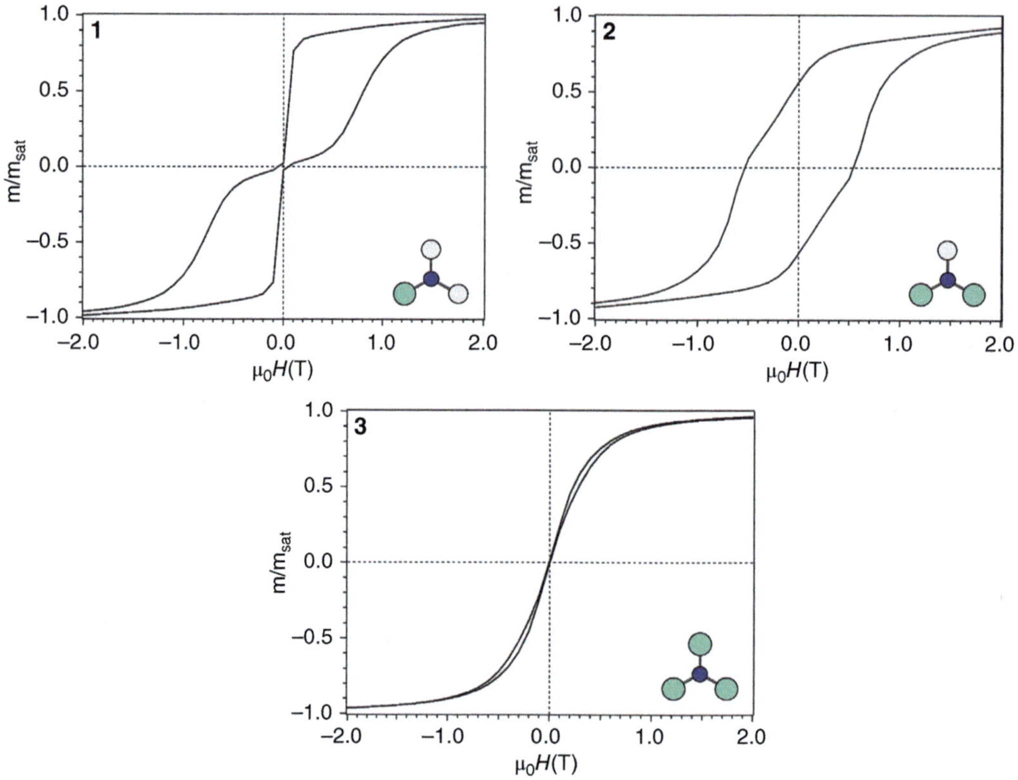

Figure 11.16 Magnetic hysteresis of $DySc_2N@I_h(7)\text{-}C_{80}$, $Dy_2ScN@I_h(7)\text{-}C_{80}$, and $Dy_3N@I_h(7)\text{-}C_{80}$ (left to right). *Source:* Copyright 2014 by the American Physical Society.

relaxation of $Dy_2ScN@I_h(7)\text{-}C_{80}$ was estimated to be 1735 ± 24 K, which is much higher than that of $DySc_2N@I_h(7)\text{-}C_{80}$ (24 ± 0.5 K). For $Dy_3N@I_h(7)\text{-}C_{80}$, a narrow hysteresis loop with vanishing zero-field magnetization was observed. This phenomenon can be explained by the magnetic frustration of the ground state of $Dy_3N@I_h(7)\text{-}C_{80}$, in which the magnetic moments of three Dy^{3+} ions are counteracted due to the symmetrical D_{3h} configuration of the Dy_3N cluster, resulting in silent Dy–Dy ferromagnetic couplings and thus hindering the remanence of the EMF molecule.

In addition to Dy^{3+}, the effects of endohedral doping of nitride clusters with other lanthanide ions, such as Ho^{3+} and Lu^{3+}, have also been investigated. In 2014, Dreiser et al. symmetrically studied magnetic behaviors of $DySc_2N@I_h(7)\text{-}C_{80}$ and $HoSc_2N@I_h(7)\text{-}C_{80}$ [155]. Different from $DySc_2N@I_h(7)\text{-}C_{80}$, $HoSc_2N@I_h(7)\text{-}C_{80}$ exhibits a slow relaxation of magnetization with timescales on the order of milliseconds, revealing the strong magnetic anisotropy of Ho^{3+} and the nature of a field-induced single-ion magnet (SIM) for $HoSc_2N@I_h(7)\text{-}C_{80}$. The large disparity in the magnetization relaxation times is attributed to the low-symmetry ligand field generated by the C_{80} cage, which has a much more profound influence on $HoSc_2N@I_h(7)\text{-}C_{80}$ than on $DySc_2N@I_h(7)\text{-}C_{80}$. Another representative case is the study on the magnetic behaviors of $Dy_nLu_{3-n}N@I_h(7)\text{-}C_{80}$ (n = 1, 2), which shows the significant effects of diamagnetic ions on SMMs. Compared to $DySc_2N@I_h(7)\text{-}C_{80}$, $DyLu_2N@I_h(7)\text{-}C_{80}$ has a higher blocking temperature (T_B = 9.5 K), longer magnetic relaxation time, and broader hysteresis loop, indicating the better SMM properties of $DyLu_2N@I_h(7)\text{-}C_{80}$ [156]. For $Dy_2LuN@I_h(7)\text{-}C_{80}$, although the blocking temperature is identical to that of $Dy_2ScN@I_h(7)\text{-}C_{80}$ (T_B = 8 K), the mechanism of the magnetic moments is essentially different since $Dy_2LuN@I_h(7)\text{-}C_{80}$ shows a solo dipolar nature with vanishing Dy—Dy interactions, much smaller than that of $Dy_2ScN@I_h(7)$-C_{80}. Such differences can be explained by the variation in the coupling constant caused by the endohedral substitution of Sc^{3+} by Lu^{3+}, resulting in a smaller exchange interaction and thus a smaller energy difference between the ferromagnetically and antiferromagnetically coupled states.

It should be noted that in addition to the encapsulated metal ions, other factors, such as cage isomerism, cage size, and nonmetallic units, also affect the SMM properties of EMFs. In 2017, Chen et al. studied the magnetic properties of $Dy_2S@C_s(6)$-C_{82}, $Dy_2S@C_{3v}(8)$-C_{82}, $Dy_2C_2@C_s(6)$-C_{82}, and $Dy_2S@C_s(10528)$-C_{72} [103]. Hysteresis loops were observed from the magnetization curves for all four EMFs, confirming their SMM properties. Nevertheless, the blocking temperature of $Dy_2S@C_{3v}(8)$-C_{82} was higher than that of $Dy_2S@C_s(6)$-C_{82}, indicating that carbon cage isomerism had a considerable influence on the SMM properties of EMFs. On the other hand, the relaxation time of $Dy_2S@C_{82}$ is longer than that of $Dy_2C_2@C_{82}$, suggesting that the replacement of C_2 unit by S atom in the endohedral cluster brings about better SMM properties. In 2019, Chen, Popov, and their coworkers reported the magnetic properties of three isomers of $Dy_2O@C_{82}$ [88]. Compared to $Dy_2S@C_{82}$, the $Dy_2O@C_{82}$ isomers show broader hysteresis, longer relaxation times, and higher blocking temperatures (Figure 11.17), suggesting that the doping of S by O results in superior SMM properties. In fact, although the Dy–Dy exchange interaction in $Dy_2O@C_{82}$ is similar to that in $Dy_2S@C_{82}$, their magnetic behaviors proceed at different states, i.e. ferromagnetically coupled excited state for $Dy_2O@C_{82}$ and ground state for $Dy_2S@C_{82}$, which determines their different SMM performances.

In addition to the endohedral doping, the cage structure and the cluster configuration also exert notable impact the SMM properties of clusterfullerenes though explicit mechanism hasn't been fully explored thus far. In the studies of the $Dy_2O@C_{82}$ isomers mentioned above, the three isomers of $Dy_2O@C_{82}$ show different blocking temperatures (4.4 K for $Dy_2O@C_s(6)$-C_{82}, 5.8 K for $Dy_2O@C_{2v}(9)$-C_{82}, and 7.4 K for $Dy_2O@C_{3v}(8)$-C_{82}), demonstrating the impact of cage isomerism on the SMM properties of EMFs [88]. Recently, Chen, Popov, and their coworkers reported the magnetic properties of two small non-IPR Dy_2O-based EMFs, $Dy_2O@C_s(10528)$-C_{72} and $Dy_2O@C_2(13333)$-C_{74} [89]. The results reveal that $Dy_2O@C_s(10528)$-C_{72} shows an open hysteresis curve with a higher blocking temperature (7 K) than that of the isostructural $Dy_2S@C_s(10528)$-C_{72} (3 K), in agreement with the theoretical prediction of the higher crystal-field splitting energy for

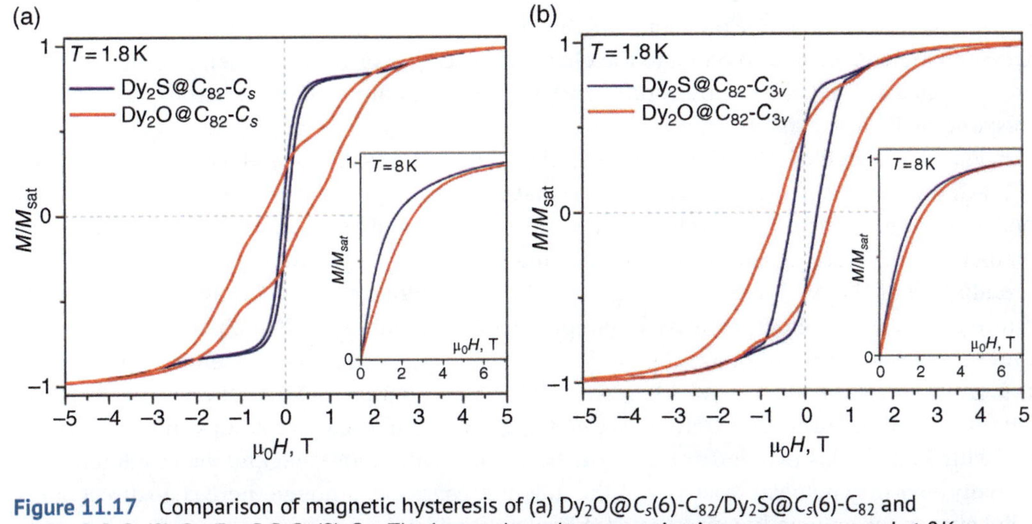

Figure 11.17 Comparison of magnetic hysteresis of (a) $Dy_2O@C_s(6)$-C_{82}/$Dy_2S@C_s(6)$-C_{82} and (b) $Dy_2O@C_{3v}(8)$-C_{82}/$Dy_2S@C_{3v}(8)$-C_{82}. The insets show the magnetization curves measured at 8 K. *Source:* Copyright 2019 WILEY-VCH Verlag GmbH & Co. KGaA, Weinheim.

Dy_2O-based EMFs. Notably, the hysteresis curve for $Dy_2O@C_2(13333)$-C_{74} is up to 14 K, the highest blocking temperature among all reported Dy-clusterfullerenes. The excellent SMM properties of $Dy_2O@C_2(13333)$-C_{74} originate from the counterbalanced contributions from the Dy—Dy ferromagnetic dipolar coupling and the antiferromagnetic exchange due to the linear Dy—O—Dy configuration, suggesting that the configuration of endohedral cluster also plays an essential role in the SMM performance of EMFs.

11.5 Chemical Reactivity Tune by Endohedral Doping

Chemical functionalization is essential for EMF studies. On one hand, it helps introduce external functional groups to exquisitely modulate the physical and chemical properties of EMFs to engender diversely applicable functional materials. On the other hand, it assists in increasing the solubility of EMFs by obtaining water-soluble derivatives as well as in crystallization, contributing to structural characterization of the parent EMFs by using X-ray crystallography. Studies show that after the doping of metal or metal clusters, the chemical characteristics of the resulting EMFs differ considerably from those of their parent empty fullerenes. Moreover, various factors can impact the reactivity of EMFs from the perspective of metal clusters, such as the metal cluster type and size, the metal-to-cage charge transfer, and the strain energy of the carbon framework caused by encapsulation of metal clusters. Consequently, atomically precise doping, which results in subtle changes in the encapsulated metal ions or metallic clusters, can also cause notable differences in their chemical reactivity as well as regioselectivity. In this section, we summarize recent developments in the chemical reactivity of EMFs, mainly focusing on how the atomically precise regulation of the endo-units can affect the chemical reactivity of EMFs.

11.5.1 Impact of Endohedral Doping on the Reactivity of Fullerene Cages

The above mentioned researches have shown that by encapsulating metal ions or metallic clusters, the molecular and electronic structure of fullerene can be fundamentally changed. In particular, electrochemical studies show that the redox properties of EMFs differ significantly from those of empty fullerenes due to charge transfer from the metal to the cage. Thus, endohedral doping could also be an efficient method to regulate the chemical reactivity of EMFs. Systematic studies have been conducted to study the reactivity difference between the empty fullerene carbon cage and the corresponding EMFs.

Yamada et al. studied the differences in the chemical reactivity and selectivity between $D_{2d}(23)$-C_{84} and $Sc_2C_2@D_{2d}(23)$-C_{84}, a rare case that fullerene cage can be stable in both empty and endohedral form. Comparative studies were carried on two kind of reaction: photoreactions with 2-adamantane-2,3′-[3*H*]-diazirine (**1**) [51, 157] and 5′,5′-dimethoxyspiro[adamantane]-2,2′-[Δ^3-1, 3,4-oxadiazoline] (**3**). As illustrated in Figure 11.18, the photoreaction of C_{84} and **1** at room temperature generated two mono-adducts, major product **2a** and minor product **2b**. In contrast to that of the empty fullerene, the photoreaction of $Sc_2C_2@D_{2d}(23)$-C_{84} with **1** at room temperature yielded four mono-adducts, **4a**, **4b**, **4c**, and **4d,** in an 8 : 7 : 3 : 1 ratio after one minute of reaction. As to the reactions with 3, the results show that $Sc_2C_2@D_{2d}(23)$-C_{84} is not reactive at all while $D_{2d}(23)$-C_{84} reacts with 3 and yields two derivative products. These results show that endohedral doping can essentially alter the chemical reactivity of the fullerene cages even though there is no direct interaction between the endohedral moiety and the external addend.

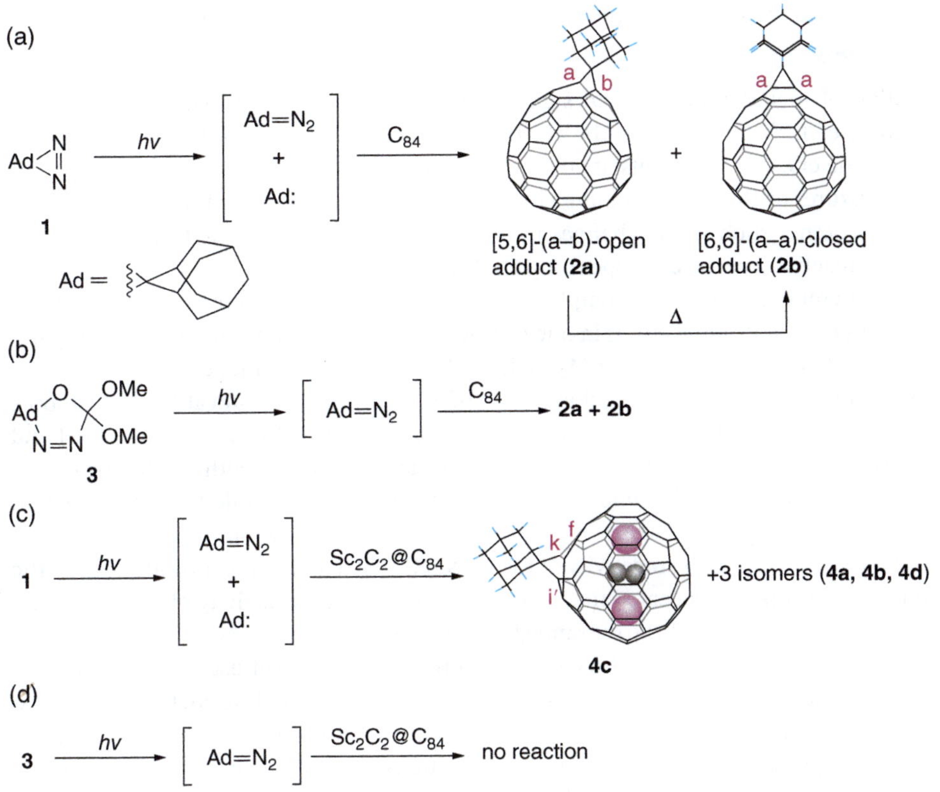

Figure 11.18 Photochemical reactions of (a) $D_{2d}(23)$-C_{84} and **1**; (b) $D_{2d}(23)$-C_{84} and **3**; (c) $Sc_2C_2@D_{2d}(23)$-C_{84} and **1**; and (d) $Sc_2C_2@D_{2d}(23)$-C_{84} and **3**. *Source:* Copyright 2020 MDPI.

Further mechanism studies suggest that the photochemical reaction of $Sc_2C_2@D_{2d}(23)$-C_{84} and **1** proceeded via a carbene addition mechanism, unlike the corresponding empty fullerene through [3+2] cycloaddition at the (a–a) bond. An inspection of the molecular orbital (MO) diagrams shows that the LUMO of $Sc_2C_2@D_{2d}(23)$-C_{84} is higher in energy than that of $D_{2d}(23)$-C_{84}. As a result, the energy mismatch between their frontier orbitals can be used to rationalize $Sc_2C_2@D_{2d}(23)$-C_{84}'s inertness toward diazoadamantane. However, the higher level of HOMO in $Sc_2C_2@D_{2d}(23)$-C_{84} compared to $D_{2d}(23)$-C_{84} may promote reactivity toward the electrophilic carbene. Throughout, the endohedral doping elevates the carbon cage's HOMO and LUMO levels, which give rise to the different chemical reactivities of $D_{2d}(23)$-C_{84} and $Sc_2C_2@D_{2d}(23)$-C_{84}.

In addition to experimental studies, several computational studies were also conducted to compare the reactivity of D_{3h}-C_{78} and $M_3N@D_{3h}$-C_{78} [158–160]. In 2015, Bickelhaupt et al. investigated the DA reactions between 1,3-butadiene and free C_{78} and the metallofullerene $Sc_3N@C_{78}$ [161]. The results indicate that, after the encapsulation of Sc_3N, $Sc_3N@C_{78}$ presents lower reactivity than the empty C_{78}. Computational analysis suggests that this lower reactivity was caused mainly by the less stabilizing interaction between the deformed reactants induced by the Sc_3N moiety. These studies suggest that metal cluster doping changes the HOMO and LUMO levels of the carbon cage and causes deformation, thus altering the reactivity in different kinds of reactions compared to the empty carbon cage.

11.5.2 Chemical Reactivity of Endohedral Fullerenes Altered by Atomically Endohedral Doping

The encapsulation of metallic clusters can significantly alter the electronic structures and the chemical reactivity of EMFs. Studies on the electronic structures of clusterfullerenes show that subtle changes in the encaged cluster itself could cause differences in the front orbitals as well as the electronic structures of the corresponding clusterfullerenes. Thus, research has also been carried out to study whether such subtle changes in the atoms of the encapsulated cluster can also alter the chemical reactivity of the corresponding EMFs.

Nitride clusterfullerenes provide ideal templates for such studies due to their capacity to capture three homogeneous or mixed-metal atoms in one nitride cluster. For example, based on a comparative study of the 1,3-dipole cycloaddition of $M_3N@C_{80}$ (M = Sc, Y) and the Bingel–Hirsch reaction, Cardona et al. found that the 1,3-dipole cycloaddition of $Y_3N@C_{80}$ only occurs at the [6,6] double bond, which is in markable contrast to the previously reported results for the [5,6] double bond addition of $Sc_3N@C_{80}$ [162]. In addition, the cyclopropanation reaction with diethyl bromomalonate occurred at the [6,6] double bond of $Y_3N@C_{80}$ and $Er_3N@C_{80}$, while the same reaction performed on $Sc_3N@C_{80}$ failed to yield any recognizable product, as shown in Figure 11.19 [163]. These studies demonstrated that the reactivity of the $M_3N@C_{80}$ NCF is significantly affected by the encapsulated metal clusters and revealed the dependence of the regioselectivity of the cycloaddition reactions on the composition of the entrapped metal nitride cluster.

In addition to the replacement of the whole metallic nitride cluster, the effect of the substitution of the individual lanthanide metal ions on the reactivity of the corresponding NCFs was also systematically investigated. For example, Chen et al. studied the regioselectivity of a series of mixed NCFs $Sc_xGd_{3-x}N@C_{80}$ (x = 0–3) in 1,3-dipolar cycloadditions and found that as the size of the $Sc_xGd_{3-x}N$ cluster increases, the regioselectivity of fullerenes gradually changes from [5,6]-cycloaddition to [6,6]-cycloaddition [164]. Later, they investigated the regioselectivity of $Sc_xY_{3-x}N@C_{80}$ (x = 0–3) in the cycloaddition reaction of the N-enthylazomethine ylide [165]. The results showed that the regioselectivity follows the same trend of electronic properties as the size of the endohedral moiety increases. For $Sc_3N@C_{80}$ and $Sc_2YN@C_{80}$, only [5,6]-pyrrolidine regioisomers are observed, while notably in $ScY_2N@C_{80}$, the [6,6]-pyrrolidine regioisomer appears as a minor regioisomer and finally becomes the major regioisomer in the reaction of $Y_3N@C_{80}$. These studies show that even though the geometric structures are similar, the size of the endohedral moiety significantly affects the electronic and chemical properties of the NCFs and the chemical reactivity of the whole EMF molecule can be manipulated by atomically precise altering the endohedral clusters.

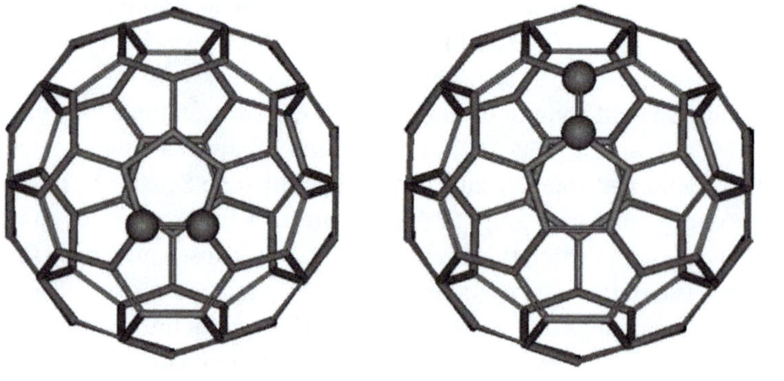

Figure 11.19 Two possible sites of addition to the $I_h(7)$-C_{80} cage: a [5,6] ring junction (on the left) and a [6,6] ring junction (on the right). *Source:* Copyright 2006 American Chemical Society.

In addition to the doping of different lanthanide ions on the endohedral M_3N cluster, other metal ions, such as Ti, can also be doped into an M_3N cluster. Yang and Popov et al. found that when doped by Ti, the resulting TiM_2N cluster demonstrated a drastically altered electronic structure, which changed the reactivity of the corresponding $TiM_2N@C_{80}$ endohedral fullerenes [137, 166]. In the study of the Bingel–Hirsch reactivity of $TiSc_2N@I_h(7)$-C_{80} in 2013, the results showed that while $Sc_3N@C_{80}$ shows no reactivity to this reaction, after the endohedral doping of Ti, the resulting $TiSc_2N@$ $I_h(7)$-C_{80} demonstrated a significantly improved reactivity, and the corresponding reaction yielded two unconventional single-bonded adducts [167]. The authors suggested that one Sc atom substitution by Ti dramatically changed the rotation dynamics of the endohedral nitride clusters and hindered the rotation of $TiSc_2N$, which should account for the improved reactivity of $TiSc_2N@I_h(7)$-C_{80}. A subsequent computational study was conducted discussing the role of endohedral titanium nitride in the reaction mechanisms on $TiSc_2N@I_h(7)$-C_{80} [168]. They suggested that the higher reactivity can be attributed to the stronger positive charge of the 6–6–6 carbon atom, which is induced by the substitution of Sc by Ti and is located above the Ti–Sc edge of the inner cluster.

In addition, a similar study on the Bingel–Hirsch reactivity of $TiY_2N@I_h(7)$-C_{80} and $Y_3N@I_h(7)$-C_{80} was performed later in 2016 [169]. They found that with one metal ion of Ti doped on the Y_3N cluster, the reactivity of $TiY_2N@I_h(7)$-C_{80} is dramatically changed. Both $TiY_2N@I_h(7)$-C_{80} and $Y_3N@I_h(7)$-C_{80} were reactive toward the Bingel-Hirsch reaction. However, the addition patterns were different. Upon substituting one endohedral yttrium (Y) atom of $Y_3N@I_h$-C_{80} with titanium (Ti), the Bingel-Hirsch derivative changes from the cyclopropane to the singly bonded monoadduct. This result reveals that the precise doping can alter not only the reactivity but also the addition pattern on the endohedral metallic clusters.

On the other hand, while the Bingel-Hirsch reaction of $TiY_2N@I_h(7)$-C_{80} demonstrates high selectivity with only one monoadduct, the same reaction on $TiSc_2N@I_h(7)$-C_{80} results in two monoadducts of $TiSc_2N@I_h(7)$-C_{80}. The commonly accepted formation mechanism of the conventional cyclopropane Bingel–Hirsch derivative, including $TiSc_2N@I_h(7)$-C_{80}, involves the stabilization of the singly bonded intermediate anion by oxidization, thus forming a second bond between the malonate and the carbon cage [168]. Moreover, the intrinsic reason for the high selectivity of $TiY_2N@I_h(7)$-C_{80} is that the formation mechanism differs: the nucleophilic reaction occurs at a carbon atom to form an intermediate $[TiY_2N@I_h(7)$-C_{80}-$CBr(COOC_2H_5)_2]^-$. The intermediate is then oxidized to give the final singly bonded monoadduct. This indicates that not only the doping of Ti but also the change of the rare earth metal from Y to Sc has a significant impact on the reactivity of the mixed nitride clusterfullerenes upon the Bingel-Hirsch reaction. The cluster-dominated regioselectivity was also found for the multiaddition reaction of $M_3N@C_{80}$ fullerenes. In 2015, a study on the regioselectivity of bis-1,3-dipolar cycloadditions of $M_3N@I_h(7)$-C_{80} (M = Sc, Lu) was carried out by Cerón and coworkers [170]. They concluded that the regiochemistry of the multi-addition of EMFs is strongly determined by the encapsulated clusters rather than the exohedral tether.

In 2020, a study on the regioselectivity of tris- and tetra-Prato adducts of $M_3N@I_h(7)$-C_{80} (M = Y, Gd) was reported by Semivrazhskaya and coworkers [171]. Both $Y_3N@I_h(7)$-C_{80} and $Gd_3N@I_h(7)$-C_{80} demonstrate high selectivity in this multiaddition reaction. Out of a large number of possible isomers (1190 for tris isomers), only a very limited number of multiaddition products were found for both nitride cluster fullerenes, suggesting that the doping of endohedral nitride clusters considerably reduced possible addition sites on the fullerene cages. Nevertheless, $Gd_3N@I_h$-C_{80} demonstrates much higher regioselectivity in the Prato reaction than $Y_3N@I_h$-C_{80}. The electron spin resonance studies of pristine, bis-, and tris-adducts of $Gd_3N@C_{80}$ suggested that the rotation of the endohedral metal cluster slowed on the increase of the additional numbers to the C_{80} cage, which is favored for accommodating the Gd atoms of the relatively large Gd_3N cluster inner space at the sp^3 addition sites.

In conclusion, due to the unique host-guest molecular structures and the complex metal-cage interaction, atomically precise doping of the endohedral clusters can significantly alter the chemical reactivity and the regioselectivity of the EMFs. Several aspects are found to be relevant to the reactivity of EMFs: (i) the type of endohedral metal cluster; (ii) the size of the metal cluster; (iii) the charge state of the EMF; (iv) the reaction types and reagents; and (v) the pyramidalization angles of fullerene cage.

11.6 Conclusions and Perspectives

These chapter presents a well-established class of zero-dimension carbon nanostructures, endohedral fullerenes (EMFs), focusing on the impact of atomically precise doping on their electronic structures and physicochemical properties. Different from nanotubes and graphene, EMFs are molecular compound that can be isolated in chemically pure form. Thus, atomic doping can be achieved by synthetic methods, and the fine tuning of their molecular and electronic structures can be precisely monitored by single crystal X-ray crystallography as we as spectroscopic methods. These studies show that, on one hand, endohedral doping can stabilize fullerene structures that would not be stable independently and form a new host-guest compound due to the strong interaction between the metal ion and the fullerene cage. On the other hand, due to the substantial involvement of endohedral metal orbitals into the EMF molecular orbitals, the doping of endohedral metal ion can substantially alter the electronic structures, spectroscopic properties, and the chemical reactivity of the EMF compounds, while the geometry of the fullerene cage remains unchanged. In many cases, substitution of one metal inside cages can cause notable changes on their physicochemical properties. Moreover, the chemical reactivity of EMFs can be gradually altered by the gradual substitution of the metal ions in the metal clusters. Thus, advantages of precise doping inside a nanocarbon cage compound, which can be purified and fine characterized, give rise to many potential applications of the EMFs. For example, with the different combination of metal ions inside di-metallofullerenes, the nature of the metal–metal bond that forms inside cages could be fundamentally different, which facilitates the fine tuning of their superior SMM properties and their application as magnetic carbon materials. $Gd@C_{2n}$ water-soluble derivatives show great potential in the treatment of cancer, and encapsulating of the radioactive isotopes will contribute to their applications in nuclear medicine. In addition, tubular fullerene can be considered as carbon nanotube with precise molecular structure, the doping of metal ions inside this kind of fullerenes will likely alter their single-molecular conductivities and provide wide applications in the fields of integrated circuit chips as well as nano-electronics. Though showing great potentials in the abovementioned research field, currently, more efficient synthesis techniques are urgently needed to improve the low product yield and costly and time-consuming purification process, which somewhat hinder further fundamental and applied exploration of EMFs. Nevertheless, the flexibility of precise doping into various nanocarbon cages, which offers many possibilities in the construction of molecular structures and tunable physicochemical properties, will continue to provide unlimited research scope for the future creative work of EMFs.

References

1 Kroto, H.W., Heath, J.R., O'Brien, S.C. et al. (1985). C_{60}: buckminsterfullerene. *Nature* 318: 162–163.
2 Popov, A.A., Yang, S., and Dunsch, L. (2013). Endohedral fullerenes. *Chem. Rev.* 113: 5989–6113.
3 Heath, J.R., O'Brien, S.C., Zhang, Q. et al. (1985). Lanthanum complexes of spheroidal carbon shells. *J. Am. Chem. Soc.* 107: 7779–7780.

4 Chai, Y., Guo, T., Jin, C. et al. (1991). Fullerenes with metals inside. *J. Phys. Chem.* 95: 7564–7568.

5 Kobayashi, K. and Nagase, S. (1998). Structures and electronic states of $M@C_{82}$ [M = Sc, Y, La and lanthanides]. *Chem. Phys. Lett.* 282: 325–329.

6 Maeda, Y., Matsunaga, Y., Wakahara, T. et al. (2004). Isolation and characterization of a carbene derivative of $La@C_{82}$. *J. Am. Chem. Soc.* 126: 6858–6859.

7 Fowler, P.W. and Manolopoulos, D.E. (1995). *An Atlas of Fullerenes*. Oxford; New York: Clarendon Press; Oxford University Press.

8 Shu, C.-Y., Wang, C.-R., Zhang, J.-F. et al. (2008). Organophosphonate functionalized $Gd@C_{82}$ as a magnetic resonance imaging contrast agent. *Chem. Mater.* 20: 2106–2109.

9 Ross, R.B., Cardona, C.M., Guldi, D.M. et al. (2009). Endohedral fullerenes for organic photovoltaic devices. *Nat. Mater.* 8: 208–212.

10 Zhang, K., Wang, C., Zhang, M. et al. (2020). A $Gd@C_{82}$ single-molecule electret. *Nat. Nanotechnol.* 15: 1019–1024.

11 Krätschmer, W., Lamb, L.D., Fostiropoulos, K., and Huffman, D.R. (1990). Solid C_{60}: a new form of carbon. *Nature* 347: 354–358.

12 Peters, G. and Jansen, M. (1992). A new fullerene synthesis. *Angew. Chem. Int. Ed.* 31: 223–224.

13 Saunders, M., Jiménez-Vázquez, H.A., Cross, R.J., and Poreda, R.J. (1993). Stable compounds of helium and neon: $He@C_{60}$ and $Ne@C_{60}$. *Science* 259: 1428.

14 Saunders, M., Jimenez-Vazquez, H.A., Cross, R.J. et al. (1994). Incorporation of helium, neon, argon, krypton, and xenon into fullerenes using high pressure. *J. Am. Chem. Soc.* 116: 2193–2194.

15 Tellgmann, R., Krawez, N., Lin, S.H. et al. (1996). Endohedral fullerene production. *Nature* 382: 407–408.

16 Weiske, T., Böhme, D.K., Hrušák, J. et al. (1991). Endohedral cluster compounds: inclusion of helium within ${C_{60}}^{+}$ and ${C_{70}}^{+}$ through collision experiments. *Angew. Chem. Int. Ed.* 30: 884–886.

17 Almeida Murphy, T., Pawlik, T., Weidinger, A. et al. (1996). Observation of atomlike nitrogen in nitrogen-implanted solid C_{60}. *Phys. Rev. Lett.* 77: 1075–1078.

18 Suetsuna, T., Dragoe, N., Harneit, W. et al. (2002). Separation of $N_2@C_{60}$ and $N@C_{60}$. *Chem. Eur. J.* 8: 5079–5083.

19 Kikuchi, K., Kobayashi, K., Sueki, K. et al. (1994). Encapsulation of radioactive ^{159}Gd and ^{161}Tb atoms in fullerene cages. *J. Am. Chem. Soc.* 116: 9775–9776.

20 Cagle, D.W., Thrash, T.P., Alford, M. et al. (1996). Synthesis, characterization, and neutron activation of holmium metallofullerenes. *J. Am. Chem. Soc.* 118: 8043–8047.

21 Diener, M.D., Alford, J.M., Kennel, S.J., and Mirzadeh, S. (2007). $^{212}Pb@C_{60}$ and its water-soluble derivatives: synthesis, stability, and suitability for radioimmunotherapy. *J. Am. Chem. Soc.* 129: 5131–5138.

22 Vougioukalakis, G.C., Roubelakis, M.M., and Orfanopoulos, M. (2010). Open-cage fullerenes: towards the construction of nanosized molecular containers. *Chem. Soc. Rev.* 39: 817–844.

23 Morinaka, Y., Sato, S., Wakamiya, A. et al. (2013). X-ray observation of a helium atom and placing a nitrogen atom inside $He@C_{60}$ and $He@C_{70}$. *Nat. Commun.* 4: 1554.

24 Kurotobi, K. and Murata, Y. (2011). A single molecule of water encapsulated in fullerene C_{60}. *Science* 333: 613.

25 Krachmalnicoff, A., Bounds, R., Mamone, S. et al. (2016). The dipolar endofullerene $HF@C_{60}$. *Nat. Chem.* 8: 953–957.

26 Liu, B.-b., Xu, W.-g., Liu, Z.-y. et al. (1996). High-yield synthesis and extraction of $La@C_{2n}$. *Solid State Commun.* 97: 407–410.

27 Yang, S., Wei, T., and Jin, F. (2017). When metal clusters meet carbon cages: endohedral clusterfullerenes. *Chem. Soc. Rev.* 46: 5005–5058.

28 Cai, W., Morales-Martínez, R., Zhang, X. et al. (2017). Single crystal structures and theoretical calculations of uranium endohedral metallofullerenes [$U@C_{2n}$, $2n = 74, 82$] show cage isomer dependent oxidation states for U. *Chem. Sci.* 8: 5282–5290.

29 Cai, W., Alvarado, J., Metta-Magaña, A. et al. (2020). Interconversions between uranium mono-metallofullerenes: mechanistic implications and role of asymmetric cages. *J. Am. Chem. Soc.* 142: 13112–13119.

30 Yao, Y.-R., Roselló, Y., Ma, L. et al. (2021). Crystallographic characterization of U@C_{2n} [2_n= 82–86]: insights about metal–cage interactions for mono-metallofullerenes. *J. Am. Chem. Soc.* 143: 15309–15318.

31 Cai, W., Abella, L., Zhuang, J. et al. (2018). Synthesis and characterization of non-isolated-pentagon-rule actinide endohedral metallofullerenes U@C_1[17418]-C_{76}, U@C_1[28324]-C_{80}, and Th@C_1[28324]-C_{80}: low-symmetry cage selection directed by a tetravalent ion. *J. Am. Chem. Soc.* 140: 18039–18050.

32 Wang, Y., Morales-Martínez, R., Cai, W. et al. (2019). Th@C_1[11]-C_{86}: an actinide encapsulated in an unexpected C_{86} fullerene cage. *Chem. Commun.* 55: 9271–9274.

33 Jin, M., Zhuang, J., Wang, Y. et al. (2019). Th@T_d[19151]-C_{76}: a highly symmetric fullerene cage stabilized by a tetravalent actinide metal ion. *Inorg. Chem.* 58: 16722–16726.

34 Yan, Y., Morales-Martínez, R., Zhuang, J. et al. (2021). Th@D_{5h}[6]-C_{80}: a highly symmetric fullerene cage stabilized by a single metal ion. *Chem. Commun.* 57: 6624–6627.

35 Meng, Q., Morales-Martínez, R., Zhuang, J. et al. (2021). Synthesis and characterization of two isomers of Th@C_{82}: Th@C_{2v}[9]-C_{82} and Th@C_2[5]-C_{82}. *Inorg. Chem.* 60: 11496–11502.

36 Wang, Y., Morales-Martínez, R., Zhang, X. et al. (2017). Unique four-electron metal-to-cage charge transfer of Th to a C_{82} fullerene cage: complete structural characterization of Th@C_{3v}[8]-C_{82}. *J. Am. Chem. Soc.* 139: 5110–5116.

37 Zhang, X., Wang, Y., Morales-Martínez, R. et al. (2018). U_2@I_h[7]-C_{80}: crystallographic characterization of a long-sought dimetallic actinide endohedral fullerene. *J. Am. Chem. Soc.* 140: 3907–3915.

38 Zhuang, J., Morales-Martínez, R., Zhang, J. et al. (2021). Characterization of a strong covalent Th^{3+}–Th^{3+} bond inside an I_h[7]-C_{80} fullerene cage. *Nat. Commun.* 12: 2372.

39 Stevenson, S., Rice, G., Glass, T. et al. (1999). Small-bandgap endohedral metallofullerenes in high yield and purity. *Nature* 401: 55–57.

40 Dunsch, L., Krause, M., Noack, J., and Georgi, P. (2004). Endohedral nitride cluster fullerenes: formation and spectroscopic analysis of $L_{3-x}M_xN$@C_{2n} [$0 \leq x \leq 3$; $N = 39, 40$]. *J. Phys. Chem. Solids* 65: 309–315.

41 Krause, M., Ziegs, F., Popov, A.A., and Dunsch, L. (2007). Entrapped bonded hydrogen in a fullerene: the five-atom cluster Sc_3CH in C_{80}. *ChemPhysChem* 8: 537–540.

42 Zhang, M., Hao, Y., Li, X. et al. (2014). Facile synthesis of an extensive family of Sc_2O@C_{2n} [n = 35–47] and chemical insight into the smallest member of Sc_2O@C_2[7892]–C_{70}. *J. Phys. Chem. C* 118: 28883–28889.

43 Chen, N., Chaur, M.N., Moore, C. et al. (2010). Synthesis of a new endohedral fullerene family, Sc_2S@C_{2n} [n = 40–50] by the introduction of SO_2. *Chem. Commun.* 46: 4818–4820.

44 Wang, C.-R., Kai, T., Tomiyama, T. et al. (2001). A scandium carbide endohedral metallofullerene: [Sc_2C_2]@C_{84}. *Angew. Chem. Int. Ed.* 40: 397–399.

45 Stevenson, S., Thompson, M.C., Coumbe, H.L. et al. (2007). Chemically adjusting plasma temperature, energy, and reactivity [CAPTEAR] method using NO_x and combustion for selective synthesis of Sc_3N@C_{80} metallic nitride fullerenes. *J. Am. Chem. Soc.* 129: 16257–16262.

46 Xin, J., Jin, F., Guan, R. et al. (2021). Ancient pigment to treasure: Prussian blue as a cheap solid cyanide/nitrogen dual-source affording the high-yield syntheses of pricey endohedral clusterfullerenes. *Inorg. Chem. Front.* 8: 1719–1726.

47 Jiao, M., Zhang, W., Xu, Y. et al. (2012). Urea as a new and cheap nitrogen source for the synthesis of metal nitride clusterfullerenes: the role of decomposed products on the selectivity of fullerenes. *Chem. Eur. J.* 18: 2666–2673.

48 Komatsu, K., Murata, M., and Murata, Y. (2005). Encapsulation of molecular hydrogen in fullerene C_{60} by organic synthesis. *Science* 307: 238.

49 Diener, M.D. and Alford, J.M. (1998). Isolation and properties of small-bandgap fullerenes. *Nature* 393: 668–671.

50 Nakagawa, A., Nishino, M., Niwa, H. et al. (2018). Crystalline functionalized endohedral C_{60} metallofullerides. *Nat. Commun.* 9: 3073.

51 Yamada, M., Tanabe, Y., Dang, J.-S. et al. (2016). D_{2d}[23]-C_{84} versus Sc_2C_2@D_{2d}[23]-C_{84}: impact of endohedral Sc_2C_2 doping on chemical reactivity in the photolysis of diazirine. *J. Am. Chem. Soc.* 138: 16523–16532.

52 Kroto, H.W. (1987). The stability of the fullerenes C_n, with $n = 24, 28, 32, 36, 50, 60$ and 70. *Nature* 329: 529–531.

53 Tan, Y.-Z., Xie, S.-Y., Huang, R.-B., and Zheng, L.-S. (2009). The stabilization of fused-pentagon fullerene molecules. *Nat. Chem.* 1: 450–460.

54 Campbell, E.E.B., Fowler, P.W., Mitchell, D., and Zerbetto, F. (1996). Increasing cost of pentagon adjacency for larger fullerenes. *Chem. Phys. Lett.* 250: 544–548.

55 Wang, C.-R., Kai, T., Tomiyama, T. et al. (2000). C_{66} fullerene encaging a scandium dimer. *Nature* 408: 426–427.

56 Schmalz, T.G., Seitz, W.A., Klein, D.J., and Hite, G.E. (1986). C_{60} carbon cages. *Chem. Phys. Lett.* 130: 203–207.

57 Reich, A., Panthöfer, M., Modrow, H. et al. (2004). The structure of Ba@C_{74}. *J. Am. Chem. Soc.* 126: 14428–14434.

58 Bao, L., Li, Y., Yu, P. et al. (2020). Preferential formation of mono-metallofullerenes governed by the encapsulation energy of the metal elements: a case study on Eu@C_{2n} [$2n$ = 74–84] revealing a general rule. *Angew. Chem. Int. Ed.* 59: 5259–5262.

59 Bao, L., Yu, P., Pan, C. et al. (2019). Crystallographic identification of Eu@C_{2n} [$2n$ = 88, 86 and 84]: completing a transformation map for existing metallofullerenes. *Chem. Sci.* 10: 2153–2158.

60 Aoyagi, S., Nishibori, E., Sawa, H. et al. (2010). A layered ionic crystal of polar Li@C_{60} superatoms. *Nat. Chem.* 2: 678–683.

61 Alvarez, M.M., Gillan, E.G., Holczer, K. et al. (1991). Lanthanum carbide [La_2C_{80}]: a soluble dimetallofullerene. *J. Phys. Chem.* 95: 10561–10563.

62 Feng, L., Suzuki, M., Mizorogi, N. et al. (2013). Mapping the metal positions inside spherical C_{80} cages: crystallographic and theoretical studies of Ce_2@D_{5h}-C_{80} and Ce_2@I_h-C_{80}. *Chem. Eur. J.* 19: 988–993.

63 Liu, F., Krylov, D.S., Spree, L. et al. (2017). Single molecule magnet with an unpaired electron trapped between two lanthanide ions inside a fullerene. *Nat. Commun.* 8: 16098.

64 Yang, H., Jin, H., Hong, B. et al. (2011). Large endohedral fullerenes containing two metal ions, Sm_2@D_2[35]-C_{88}, Sm_2@C_1[21]-C_{90}, and Sm_2@D_3[85]-C_{92}, and their relationship to endohedral fullerenes containing two gadolinium ions. *J. Am. Chem. Soc.* 133: 16911–16919.

65 Mercado, B.Q., Jiang, A., Yang, H. et al. (2009). Isolation and structural characterization of the molecular nanocapsule Sm_2@D_{3d}[822]-C_{104}. *Angew. Chem. Int. Ed.* 48: 9114–9116.

66 Pan, C., Shen, W., Yang, L. et al. (2019). Crystallographic characterization of Y_2C_{2n} [$2n$ = 82, 88–94]: direct Y–Y bonding and cage-dependent cluster evolution. *Chem. Sci.* 10: 4707–4713.

67 Shen, W., Bao, L., Wu, Y. et al. (2017). Lu_2@C_{2n} [$2n$ = 82, 84, 86]: crystallographic evidence of direct Lu–Lu bonding between two divalent lutetium ions inside fullerene cages. *J. Am. Chem. Soc.* 139: 9979–9984.

68 Shen, W., Bao, L., Hu, S. et al. (2019). Crystallographic characterization of Lu_2C_{2n} [$2n$ = 76–90]: cluster selection by cage size. *Chem. Sci.* 10: 829–836.

69 Hu, S., Shen, W., Yang, L. et al. (2019). Crystallographic and theoretical investigations of Er_2@C_{2n} [$2n$ = 82, 84, 86]: indication of distance-dependent metal–metal bonding nature. *Chem. Eur. J.* 25: 11538–11544.

70 Yao, Y.-R., Shi, X.-M., Zheng, S.-Y. et al. (2021). Atomically precise insights into metal–metal bonds using comparable endo-units of Sc_2 and Sc_2C_2. *CCS Chem.* 3: 294–302.

71 Stevenson, S., Chancellor, C.J., Lee, H.M. et al. (2008). Internal and external factors in the structural organization in cocrystals of the mixed-metal endohedrals [$GdSc_2N@I_h$-C_{80}, $Gd_2ScN@I_h$-C_{80}, and $TbSc_2N@I_h$-C_{80}] and Nickel[II] octaethylporphyrin. *Inorg. Chem.* 47: 1420–1427.

72 Zhang, Y., Ghiassi, K.B., Deng, Q. et al. (2015). Synthesis and structure of $LaSc_2N@C_s$[hept]-C_{80} with one heptagon and thirteen pentagons. *Angew. Chem. Int. Ed.* 54: 495–499.

73 Li, X., Roselló, Y., Yao, Y.-R. et al. (2021). $U_2N@I_h$[7]-C_{80}: fullerene cage encapsulating an unsymmetrical U[IV]=N=U[V] cluster. *Chem. Sci.* 12: 282–292.

74 Chen, C.-H., Ghiassi, K.B., Cerón, M.R. et al. (2015). Beyond the Butterfly: $Sc_2C_2@C_{2v}$[9]-C_{86}, an endohedral fullerene containing a planar, twisted Sc_2C_2 unit with remarkable crystalline order in an unprecedented carbon cage. *J. Am. Chem. Soc.* 137: 10116–10119.

75 Chen, C.-H., Abella, L., Cerón, M.R. et al. (2016). Zigzag Sc_2C_2 carbide cluster inside a [88]fullerene cage with one heptagon, $Sc_2C_2@C_s$[hept]-C_{88}: a kinetically trapped fullerene formed by C_2 insertion? *J. Am. Chem. Soc.* 138: 13030–13037.

76 Pan, C., Bao, L., Yu, X. et al. (2018). Facile access to Y_2C_{2n} [$2n = 92$–130] and crystallographic characterization of $Y_2C_2@C_1$[1660]-C_{108}: a giant nanocapsule with a linear carbide cluster. *ACS Nano* 12: 2065–2069.

77 Jin, F., Xin, J., Guan, R. et al. (2021). Stabilizing a three-center single-electron metal–metal bond in a fullerene cage. *Chem. Sci.* 12: 6890–6895.

78 Iiduka, Y., Wakahara, T., Nakahodo, T. et al. (2005). Structural determination of metallofullerene Sc_3C_{82} revisited: a surprising finding. *J. Am. Chem. Soc.* 127: 12500–12501.

79 Wang, T.-S., Chen, N., Xiang, J.-F. et al. (2009). Russian-doll-type metal carbide endofullerene: synthesis, isolation, and characterization of $Sc_4C_2@C_{80}$. *J. Am. Chem. Soc.* 131: 16646–16647.

80 Yu, P., Bao, L., Yang, L. et al. (2020). Crystallographic characterization of $Ti_2C_2@D_{3h}$[5]-C_{78}, $Ti_2C_2@C_{3v}$[8]-C_{82}, and $Ti_2C_2@C_s$[6]-C_{82}: identification of unsupported Ti_2C_2 cluster with cage-dependent configurations. *Inorg. Chem.* 59: 9416–9423.

81 Zhuang, J., Abella, L., Sergentu, D.-C. et al. (2019). Diuranium[IV] carbide cluster U_2C_2 stabilized inside fullerene cages. *J. Am. Chem. Soc.* 141: 20249–20260.

82 Yu, P., Shen, W., Bao, L. et al. (2019). Trapping an unprecedented Ti_3C_3 unit inside the icosahedral C_{80} fullerene: a crystallographic survey. *Chem. Sci.* 10: 10925–10930.

83 Svitova, A.L., Ghiassi, K.B., Schlesier, C. et al. (2014). Endohedral fullerene with μ_3-carbido ligand and titanium–carbon double bond stabilized inside a carbon cage. *Nat. Commun.* 5: 3568.

84 Junghans, K., Schlesier, C., Kostanyan, A. et al. (2015). Methane as a selectivity booster in the arc-discharge synthesis of endohedral fullerenes: selective synthesis of the single-molecule magnet $Dy_2TiC@C_{80}$ and its congener $Dy_2TiC_2@C_{80}$. *Angew. Chem. Int. Ed.* 54: 13411–13415.

85 Junghans, K., Ghiassi, K.B., Samoylova, N.A. et al. (2016). Synthesis and isolation of the titanium–scandium endohedral fullerenes-$Sc_2TiC@I_h$-C_{80}, $Sc_2TiC@D_{5h}$-C_{80} and $Sc_2TiC_2@I_h$-C_{80}: metal size tuning of the Ti[IV]/Ti[III] redox potentials. *Chem. Eur. J.* 22: 13098–13107.

86 Li, X., Yao, Y.-R., Yang, W. et al. (2020). Crystallographic and spectroscopic characterization of a mixed actinide–lanthanide carbide cluster stabilized inside an I_h[7]-C_{80} fullerene cage. *Chem. Commun.* 56: 3867–3870.

87 Zhang, X., Li, W., Feng, L. et al. (2018). A diuranium carbide cluster stabilized inside a C_{80} fullerene cage. *Nat. Commun.* 9: 2753.

88 Yang, W., Velkos, G., Liu, F. et al. (2019). Single molecule magnetism with strong magnetic anisotropy and enhanced Dy⋅⋅Dy coupling in three isomers of Dy-oxide clusterfullerene $Dy_2O@C_{82}$. *Adv. Sci.* 6: 1901352.

89 Velkos, G., Yang, W., Yao, Y.-R. et al. (2020). Shape-adaptive single-molecule magnetism and hysteresis up to 14 K in oxide clusterfullerenes $Dy_2O@C_{72}$ and $Dy_2O@C_{74}$ with fused pentagon pairs and flexible Dy–[μ_2-O]–Dy angle. *Chem. Sci.* 11: 4766–4772.

90 Liu, A., Nie, M., Hao, Y. et al. (2019). $Ho_2O@C_{74}$: Ho_2O cluster expands within a small non-IPR fullerene cage of C_2[13333]-C_{74}. *Inorg. Chem.* 58: 4774–4781.

91 Yu, Y., Slanina, Z., Wang, F. et al. (2020). $Ho_2O@D_3$[85]-C_{92}: highly stretched cluster dictated by a giant cage and unexplored isomerization. *Inorg. Chem.* 59: 11020–11027.

92 Yang, T., Hao, Y., Abella, L. et al. (2015). $Sc_2O@T_d$[19151]-C_{76}: hindered cluster motion inside a tetrahedral carbon cage probed by crystallographic and computational studies. *Chem. Eur. J.* 21: 11110–11117.

93 Hao, Y., Tang, Q., Li, X. et al. (2016). Isomeric $Sc_2O@C_{78}$ related by a single-step Stone–Wales transformation: key links in an unprecedented fullerene formation pathway. *Inorg. Chem.* 55: 11354–11361.

94 Tang, Q., Abella, L., Hao, Y. et al. (2015). $Sc_2O@C_{2v}$[5]-C_{80}: dimetallic oxide cluster inside a C_{80} fullerene cage. *Inorg. Chem.* 54: 9845–9852.

95 Mercado, B.Q., Stuart, M.A., Mackey, M.A. et al. (2010). Sc_2[μ_2-O] trapped in a fullerene cage: the isolation and structural characterization of Sc_2[μ_2-O]@C_s[6]-C_{82} and the relevance of the thermal and entropic effects in fullerene isomer selection. *J. Am. Chem. Soc.* 132: 12098–12105.

96 Tang, Q., Abella, L., Hao, Y. et al. (2016). $Sc_2O@C_{3v}$[8]-C_{82}: a missing isomer of $Sc_2O@C_{82}$. *Inorg. Chem.* 55: 1926–1933.

97 Mercado, B.Q., Olmstead, M.M., Beavers, C.M. et al. (2010). A seven atom cluster in a carbon cage, the crystallographically determined structure of Sc_4[μ_3-O]$_3$@I_h-C_{80}. *Chem. Commun.* 46: 279–281.

98 Stevenson, S., Mackey, M.A., Stuart, M.A. et al. (2008). A distorted tetrahedral metal oxide cluster inside an icosahedral carbon cage. Synthesis, isolation, and structural characterization of Sc_4[μ_3-O]$_2$@I_h-C_{80}. *J. Am. Chem. Soc.* 130: 11844–11845.

99 Dunsch, L., Yang, S., Zhang, L. et al. (2010). Metal sulfide in a C_{82} fullerene cage: a new form of endohedral clusterfullerenes. *J. Am. Chem. Soc.* 132: 5413–5421.

100 Chen, N., Mulet-Gas, M., Li, Y.-Y. et al. (2013). $Sc_2S@C_2$[7892]–C_{70}: a metallic sulfide cluster inside a non-IPR C_{70} cage. *Chem. Sci.* 4: 180–186.

101 Chen, N., Beavers, C.M., Mulet-Gas, M. et al. (2012). $Sc_2S@C_s$[10528]-C_{72}: a dimetallic sulfide endohedral fullerene with a non isolated pentagon rule cage. *J. Am. Chem. Soc.* 134: 7851–7860.

102 Mercado, B.Q., Chen, N., Rodríguez-Fortea, A. et al. (2011). The shape of the Sc_2[μ_2-S] unit trapped in C_{82}: crystallographic, computational, and electrochemical studies of the isomers, Sc_2[μ_2-S]@C_s[6]-C_{82} and Sc_2[μ_2-S]@C_{3v}[8]-C_{82}. *J. Am. Chem. Soc.* 133: 6752–6760.

103 Chen, C.-H., Krylov, D.S., Avdoshenko, S.M. et al. (2017). Selective arc-discharge synthesis of Dy_2S-clusterfullerenes and their isomer-dependent single molecule magnetism. *Chem. Sci.* 8: 6451–6465.

104 Wang, T.-S., Feng, L., Wu, J.-Y. et al. (2010). Planar quinary cluster inside a fullerene cage: synthesis and structural characterizations of $Sc_3NC@C_{80}$-I_h. *J. Am. Chem. Soc.* 132: 16362–16364.

105 Wang, T., Wu, J., and Feng, Y. (2014). Scandium carbide/cyanide alloyed cluster inside fullerene cage: synthesis and structural studies of Sc_3[μ_3-C_2][μ_3-CN]@I_h-C_{80}. *Dalton Trans.* 43: 16270–16274.

106 Yang, S., Chen, C., Liu, F. et al. (2013). An improbable monometallic cluster entrapped in a popular fullerene cage: YCN@C_s[6]-C_{82}. *Sci. Rep.* 3: 1487.

107 Liu, F., Wang, S., Gao, C.-L. et al. (2017). Mononuclear clusterfullerene single-molecule magnet containing strained fused-pentagons stabilized by a nearly linear metal cyanide cluster. *Angew. Chem. Int. Ed.* 56: 1830–1834.

108 Guan, R., Chen, M., Xin, J. et al. (2021). Capturing the missing carbon cage isomer of C_{84} via mutual stabilization of a triangular monometallic cyanide cluster. *J. Am. Chem. Soc.* 143: 8078–8085.

109 Liu, F., Gao, C.-L., Deng, Q. et al. (2016). Triangular monometallic cyanide cluster entrapped in carbon cage with geometry-dependent molecular magnetism. *J. Am. Chem. Soc.* 138: 14764–14771.

110 Liu, F., Wang, S., Guan, J. et al. (2014). Putting a terbium-monometallic cyanide cluster into the C_{82} fullerene cage: TbCN@C_2[5]-C_{82}. *Inorg. Chem.* 53: 5201–5205.

111 Meng, Q., Abella, L., Yang, W. et al. (2021). UCN@C_s[6]-C_{82}: an encapsulated triangular UCN cluster with ambiguous U oxidation state [U[III] versus U[I]]. *J. Am. Chem. Soc.* 143: 16226–16234.

112 Yang, S., Popov, A.A., and Dunsch, L. (2008). Carbon pyramidalization in fullerene cages induced by the endohedral cluster: non-scandium mixed metal nitride clusterfullerenes. *Angew. Chem. Int. Ed.* 47: 8196–8200.

113 Fu, W., Xu, L., Azurmendi, H. et al. (2009). ^{89}Y and ^{13}C NMR cluster and carbon cage studies of an yttrium metallofullerene family, $Y_3N@C_{2n}$ [$n = 40$–43]. *J. Am. Chem. Soc.* 131: 11762–11769.

114 Nishibori, E., Takata, M., Sakata, M. et al. (2001). Pentagonal-dodecahedral La_2 charge density in [80-I_h]fullerene: $La_2@C_{80}$. *Angew. Chem. Int. Ed.* 40: 2998–2999.

115 Wakahara, T., Yamada, M., Takahashi, S. et al. (2007). Two-dimensional hopping motion of encapsulated La atoms in silylated $La_2@C_{80}$. *Chem. Commun.* 2680–2682.

116 Popov, A.A., Chen, N., Pinzón, J.R. et al. (2012). Redox-active scandium oxide cluster inside a fullerene cage: spectroscopic, voltammetric, electron spin resonance spectroelectrochemical, and extended density functional theory study of $Sc_4O_2@C_{80}$ and its ion radicals. *J. Am. Chem. Soc.* 134: 19607–19618.

117 Kikuchi, K., Nakao, Y., Suzuki, S. et al. (1994). Characterization of the isolated $Y@C_{82}$. *J. Am. Chem. Soc.* 116: 9367–9368.

118 Suzuki, M., Lu, X., Sato, S. et al. (2012). Where does the metal cation stay in Gd@C_{2v}[9]-C_{82}? A single-crystal X-ray diffraction study. *Inorg. Chem.* 51: 5270–5273.

119 Suzuki, M., Yamada, M., Maeda, Y. et al. (2016). The unanticipated dimerization of Ce@C_{2v}[9]-C_{82} upon co-crystallization with Ni[octaethylporphyrin] and comparison with monomeric M@C_{2v}[9]-C_{82} [M = La, Sc, and Y]. *Chem. Eur. J.* 22: 18115–18122.

120 Hu, S., Liu, T., Shen, W. et al. (2019). Isolation and structural characterization of Er@C_{2v}[9]-C_{82} and Er@C_s[6]-C_{82}: regioselective dimerization of a pristine endohedral metallofullerene induced by cage symmetry. *Inorg. Chem.* 58: 2177–2182.

121 Sun, B.-Y., Inoue, T., Shimada, T. et al. (2004). Synthesis and characterization of Eu-metallofullerenes from $Eu@C_{74}$ to $Eu@C_{90}$ and their nanopeapods. *J. Phys. Chem. B* 108: 9011–9015.

122 Lu, X., Slanina, Z., Akasaka, T. et al. (2010). $Yb@C_{2n}$ [$n = 40, 41, 42$]: new fullerene allotropes with unexplored electrochemical properties. *J. Am. Chem. Soc.* 132: 5896–5905.

123 Hu, Z., Hao, Y., Slanina, Z. et al. (2015). Popular C_{82} fullerene cage encapsulating a divalent metal ion Sm^{2+}: structure and electrochemistry. *Inorg. Chem.* 54: 2103–2108.

124 Yang, S., Kalbac, M., Popov, A., and Dunsch, L. (2006). A facile route to the non-IPR fullerene $Sc_3N@C_{68}$: synthesis, spectroscopic characterization, and density functional theory computations [IPR = isolated pentagon rule]. *Chem. Eur. J.* 12: 7856–7863.

125 Lebedkin, S., Renker, B., Heid, R. et al. (1998). A spectroscopic study of $M@C_{82}$ metallofullerenes: Raman, far-infrared, and neutron scattering results. *Appl. Phys. A* 66: 273–280.

126 Yang, S., Popov, A.A., Chen, C., and Dunsch, L. (2009). Mixed metal nitride clusterfullerenes in cage isomers: $Lu_xSc_{3-x}N@C_{80}$ [$x = 1, 2$] as compared with $M_xSc_{3-x}N@C_{80}$ [M = Er, Dy, Gd, Nd]. *J. Phys. Chem. C* 113: 7616–7623.

127 Suzuki, T., Kikuchi, K., Oguri, F. et al. (1996). Electrochemical properties of fullerenolanthanides. *Tetrahedron* 52: 4973–4982.

128 Akasaka, T., Okubo, S., Kondo, M. et al. (2000). Isolation and characterization of two $Pr@C_{82}$ isomers. *Chem. Phys. Lett.* 319: 153–156.

129 Hachiya, M., Nikawa, H., Mizorogi, N. et al. (2012). Exceptional chemical properties of $Sc@C_{2v}[9]\text{-}C_{82}$ probed with adamantylidene carbene. *J. Am. Chem. Soc.* 134: 15550–15555.

130 Suzuki, T., Maruyama, Y., Kato, T. et al. (1995). Electrochemistry and ab initio study of the dimetallofullerene $La_2@C_{80}$. *Angew. Chem. Int. Ed.* 34: 1094–1096.

131 Yamada, M., Mizorogi, N., Tsuchiya, T. et al. (2009). Synthesis and characterization of the D_{5h} isomer of the endohedral dimetallofullerene $Ce_2@C_{80}$: two-dimensional circulation of encapsulated metal atoms inside a fullerene cage. *Chem. Eur. J.* 15: 9486–9493.

132 Krause, M. and Dunsch, L. (2004). Isolation and characterisation of two $Sc_3N@C_{80}$ isomers. *ChemPhysChem* 5: 1445–1449.

133 Elliott, B., Yu, L., and Echegoyen, L. (2005). A simple isomeric separation of D_{5h} and Ih $Sc_3N@C_{80}$ by selective chemical oxidation. *J. Am. Chem. Soc.* 127: 10885–10888.

134 Valencia, R., Rodríguez-Fortea, A., Clotet, A. et al. (2009). Electronic structure and redox properties of metal nitride endohedral fullerenes $M_3N@C_{2n}$ [M = Sc, Y, La, and Gd; $2n$ = 80, 84, 88, 92, 96]. *Chem. Eur. J.* 15: 10997–11009.

135 Chaur, M.N., Aparicio-Anglès, X., Mercado, B.Q. et al. (2010). Structural and electrochemical property correlations of metallic nitride endohedral metallofullerenes. *J. Phys. Chem. C* 114: 13003–13009.

136 Zhang, L., Popov, A.A., Yang, S. et al. (2010). An endohedral redox system in a fullerene cage: the Ce based mixed-metal cluster fullerene $Lu_2CeN@C_{80}$. *Phys. Chem. Chem. Phys.* 12: 7840–7847.

137 Popov, A.A., Chen, C., Yang, S. et al. (2010). Spin-flow vibrational spectroscopy of molecules with flexible spin density: electrochemistry, ESR, cluster and spin dynamics, and bonding in $TiSc_2N@C_{80}$. *ACS Nano* 4: 4857–4871.

138 Chen, C., Liu, F., Li, S. et al. (2012). Titanium/yttrium mixed metal nitride clusterfullerene $TiY_2N@C_{80}$: synthesis, isolation, and effect of the group-III metal. *Inorg. Chem.* 51: 3039–3045.

139 Wei, T., Wang, S., Lu, X. et al. (2016). Entrapping a group-VB transition metal, vanadium, within an endohedral metallofullerene: $V_xSc_{3-x}N@I_h$ C_{80} [$x = 1, 2$]. *J. Am. Chem. Soc.* 138: 207–214.

140 Wakahara, T., Sakuraba, A., Iiduka, Y. et al. (2004). Chemical reactivity and redox property of $Sc_3@C_{82}$. *Chem. Phys. Lett.* 398: 553–556.

141 Liu, F., Velkos, G., Krylov, D.S. et al. (2019). Air-stable redox-active nanomagnets with lanthanide spins radical-bridged by a metal–metal bond. *Nat. Commun.* 10: 571.

142 Wang, Z., Kitaura, R., and Shinohara, H. (2014). Metal-dependent stability of pristine and functionalized unconventional dimetallofullerene $M_2@I_h\text{-}C_{80}$. *J. Phys. Chem. C* 118: 13953–13958.

143 Tagmatarchis, N. and Shinohara, H. (2000). Production, separation, isolation, and spectroscopic study of dysprosium endohedral metallofullerenes. *Chem. Mater.* 12: 3222–3226.

144 Sun, B. and Gu, Z. (2002). Solvent-dependent anion studies on enrichment of metallofullerene. *Chem. Lett.* 31: 1164–1165.

145 Bao, L., Chen, M., Pan, C. et al. (2016). Crystallographic evidence for direct metal–metal bonding in a stable open-shell $La_2@I_h\text{-}C_{80}$ derivative. *Angew. Chem. Int. Ed.* 55: 4242–4246.

146 Yamada, M., Kurihara, H., Suzuki, M. et al. (2015). Hiding and recovering electrons in a dimetallic endohedral fullerene: air-stable products from radical additions. *J. Am. Chem. Soc.* 137: 232–238.

147 Fu, W., Zhang, J., Fuhrer, T. et al. (2011). $Gd_2@C_{79}N$: isolation, characterization, and monoadduct formation of a very stable heterofullerene with a magnetic spin state of S = 15/2. *J. Am. Chem. Soc.* 133: 9741–9750.

148 Zuo, T., Xu, L., Beavers, C.M. et al. (2008). $M_2@C_{79}N$ [M = Y, Tb]: isolation and characterization of stable endohedral metallofullerenes exhibiting M–M bonding interactions inside Aza[80]fullerene cages. *J. Am. Chem. Soc.* 130: 12992–12997.

149 Wang, Y., Xiong, J., Su, J. et al. (2020). $Dy_2@C_{79}N$: a new member of dimetalloazafullerenes with strong single molecular magnetism. *Nanoscale* 12: 11130–11135.

150 Gould Colin, A., McClain, K.R., Reta, D. et al. (2022). Ultrahard magnetism from mixed-valence dilanthanide complexes with metal–metal bonding. *Science* 375: 198–202.

151 Velkos, G., Krylov, D.S., Kirkpatrick, K. et al. (2019). High blocking temperature of magnetization and giant coercivity in the azafullerene $Tb_2@C_{79}N$ with a single-electron terbium–terbium bond. *Angew. Chem. Int. Ed.* 58: 5891–5896.

152 Liu, F., Spree, L., Krylov, D.S. et al. (2019). Single-electron lanthanide–lanthanide bonds inside fullerenes toward robust redox-active molecular magnets. *Acc. Chem. Res.* 52: 2981–2993.

153 Westerström, R., Dreiser, J., Piamonteze, C. et al. (2012). An endohedral single-molecule magnet with long relaxation times: $DySc_2N@C_{80}$. *J. Am. Chem. Soc.* 134: 9840–9843.

154 Westerström, R., Dreiser, J., Piamonteze, C. et al. (2014). Tunneling, remanence, and frustration in dysprosium-based endohedral single-molecule magnets. *Phys. Rev. B* 89: 060406.

155 Dreiser, J., Westerström, R., Zhang, Y. et al. (2014). The metallofullerene field-induced single-ion magnet $HoSc_2N@C_{80}$. *Chem. Eur. J.* 20: 13536–13540.

156 Spree, L., Schlesier, C., Kostanyan, A. et al. (2020). Single-molecule magnets $DyM_2N@C_{80}$ and $Dy_2MN@C_{80}$ [M = Sc, Lu]: the impact of diamagnetic metals on Dy^{3+} magnetic anisotropy, Dy. . .Dy coupling, and mixing of molecular and lattice vibrations. *Chem. Eur. J.* 26: 2436–2449.

157 Yamada, M., Liu, M.T.H., Nagase, S., and Akasaka, T. (2020). New horizons in chemical functionalization of endohedral metallofullerenes. *Molecules* 25 (16): 3626.

158 Osuna, S., Swart, M., Campanera, J.M. et al. (2008). Chemical reactivity of D_{3h} C_{78} [metallo]fullerene: regioselectivity changes induced by Sc_3N encapsulation. *J. Am. Chem. Soc.* 130: 6206–6214.

159 Garcia-Borràs, M., Osuna, S., Luis, J.M. et al. (2012). The exohedral Diels–Alder reactivity of the titanium carbide endohedral metallofullerene $Ti_2C_2@D_{3h}$-C_{78}: comparison with D_{3h}-C_{78} and $M_3N@D_{3h}$-C_{78} [M = Sc and Y] reactivity. *Chem. Eur. J.* 18: 7141–7154.

160 Garcia-Borràs, M., Osuna, S., Luis, J.M. et al. (2013). A complete guide on the influence of metal clusters in the Diels–Alder regioselectivity of I_h-C_{80} endohedral metallofullerenes. *Chem. Eur. J.* 19: 14931–14940.

161 Bickelhaupt, F.M., Solà, M., and Fernández, I. (2015). Understanding the reactivity of endohedral metallofullerenes: C_{78} versus $Sc_3N@C_7$. *Chem. Eur. J.* 21: 5760–5768.

162 Cardona, C.M., Kitaygorodskiy, A., and Echegoyen, L. (2005). Trimetallic nitride endohedral metallofullerenes: reactivity dictated by the encapsulated metal cluster. *J. Am. Chem. Soc.* 127: 10448–10453.

163 Cardona, C.M., Elliott, B., and Echegoyen, L. (2006). Unexpected chemical and electrochemical properties of $M_3N@C_{80}$ [M = Sc, Y, Er]. *J. Am. Chem. Soc.* 128: 6480–6485.

164 Chen, N., Zhang, E.-Y., Tan, K. et al. (2007). Size effect of encaged clusters on the exohedral chemistry of endohedral fullerenes: a case study on the pyrrolidino reaction of $Sc_xGd_{3-x}N@C_{80}$ [x = 0–3]. *Org. Lett.* 9: 2011–2013.

165 Chen, N., Fan, L.-Z., Tan, K. et al. (2007). Comparative spectroscopic and reactivity studies of $Sc_{3-x}Y_xN@C_{80}$ [x = 0–3]. *J. Phys. Chem. C* 111: 11823–11828.

166 Yang, S., Chen, C., Popov, A.A. et al. (2009). An endohedral titanium[III] in a clusterfullerene: putting a non-group-III metal nitride into the C_{80}-I_h fullerene cage. *Chem. Commun.* 6391–6393.

167 Yang, S., Chen, C., Li, X. et al. (2013). Bingel–Hirsch monoadducts of $TiSc_2N@I_h$-C_{80}versus $Sc_3N@I_h$-C_{80}: reactivity improvement via internal metal atom substitution. *Chem. Commun.* 49: 10844–10846.

168 Zhao, P., Dang, J.-S., and Zhao, X. (2016). Bingel–Hirsch reaction mechanisms on $TiSc_2N@I_h$-C_{80}: the role of endohedral titanium nitride. *Phys. Chem. Chem. Phys.* 18: 9709–9714.

169 Wang, S., Huang, J., Gao, C. et al. (2016). Singly bonded monoadduct rather than methanofullerene: manipulating the addition pattern of trimetallic nitride clusterfullerene through one endohedral metal atom substitution. *Chem. Eur. J.* 22: 8309–8315.

170 Cerón, M.R., Izquierdo, M., Garcia-Borràs, M. et al. (2015). Bis-1,3-dipolar cycloadditions on endohedral fullerenes $M_3N@I_h$-C_{80} [M = Sc, Lu]: remarkable endohedral-cluster regiochemical control. *J. Am. Chem. Soc.* 137: 11775–11782.

171 Semivrazhskaya, O., Aroua, S., Yulikov, M. et al. (2020). Regioselective synthesis and characterization of tris- and tetra-prato adducts of $M_3N@C_{80}$ [M = Y, Gd]. *J. Am. Chem. Soc.* 142: 12954–12965.

12

On-Surface Synthesis of Polyacenes and Narrow Band-Gap Graphene Nanoribbons

Hironobu Hayashi and Hiroko Yamada

Division of Materials Science, Nara Institute of Science and Technology, 8916-5 Takayama-cho, Ikoma, 630-0192, Japan

12.1 Introduction

The objective of this chapter is to briefly introduce the recent synthetic achievements of polyacenes and narrow-bandgap graphene nanoribbons (GNRs) using surface-assisted and on-surface reactions under ultra-high vacuum (UHV) conditions. The molecular structure was observed at an atomic level with cutting-edge technology to investigate the physical properties of a single molecule to encourage researchers to synthesize unexplored molecules and nanocarbon materials. Moore's law strongly indicates that innovative materials should be developed for next-generation devices. For example, devices fabricated using nanocarbon materials, such as graphene and carbon nanotubes (CNTs), are expected to exhibit excellent properties as compared to silicon-based devices; however, the properties strongly depend on their structural purity. Therefore, an atomically precise structure is required for the as-prepared nanocarbon materials.

The top-down approach enables the large-scale production of nanocarbon materials, whereas the bottom-up approach could potentially control the structure of nanocarbon materials at an atomic level. This chapter provides an overview of recent achievements in this field, focusing specifically on polyacenes and GNRs. Following the introduction, a summary of graphene-based materials is provided, along with the relationship between small acene molecules and graphene-based materials. The second section outlines the bottom-up synthesis of GNRs using conventional in-solution synthesis and advanced surface-assisted reactions. The third part of this chapter illustrates the on-surface synthesis of armchair-type GNRs (AGNRs) with narrow bandgaps, which could be useful in fabricating GNR-based field effect transistors (GNR-FETs). The fourth section focuses on the on-surface synthesis of polyacenes that are important precursors of GNRs and models for investigating the intrinsic physical properties of zigzag-type GNRs. The final section presents a conclusion and perspective, including recognition of the tailor-made synthesis of GNRs as key materials for next-generation devices. The aim of this chapter is to emphasize the importance of molecular design for obtaining desired carbon nanostructures using on-surface synthesis.

Atomically Precise Nanochemistry, First Edition. Edited by Rongchao Jin and De-en Jiang.

12.1.1 Nanocarbon Materials

The development of integrated circuit fabrication technology consisting of silicon semiconductors, which play a central role in electronic products, has facilitated the miniaturization, high performance, high functionality, and low power consumption of products. In the near future, an incredible processing speed will be required for next-generation devices, because electronic devices will be integrated with other fields, such as communication functions. The miniaturization of large integrated circuits, which was achieved by following Moore's law, strongly indicates that innovative materials that can overcome the performance limitations of silicon-based devices should be developed.

Nanocarbon materials, such as fullerene, CNTs, and graphene, have garnered much interest as organic semiconductors (Figure 12.1). Fullerenes [1] are composed entirely of carbon in the form of hollow spheres, and their derivatives play an important role in organic photovoltaics as electron acceptors [2–5]. The discovery of fullerenes in 1985 [1] was the beginning of the study on carbon nanostructures. The small reorganization energy of fullerenes arising from the delocalized π system on the sphere, along with the rigid structure, contributed to their unique electron transport properties. Subsequently, CNTs, which have a tubular structure formed by rolling up a graphene sheet, were discovered in 1991 [6]. The single-walled CNTs (SWCNTs) are expected to be used in various applications because of their long-range, one-dimensional shape, toughness, and high electrical and thermal conductivities [7–9]. In addition, their most significant feature is the electronic structure, which can be semiconducting or metallic, depending on their chirality [10]. Finally, graphene, a flat monolayer of carbon atoms tightly packed into a two-dimensional honeycomb lattice, was isolated in 2004 by the micromechanical cleavage of bulk graphite [11]. As observed in CNTs, graphene also exhibits an ultra-high carrier mobility. Therefore, nanocarbon materials are expected to play an important role in next-generation electronic devices.

12.1.2 Graphene Nanoribbons

Although the charge-carrier mobility of graphene is a promising feature for its application in transistor channels, the lack of a bandgap concludes that graphene-based transistors do not turn off appropriately. This presents a significant difference between graphene and CNTs. Accordingly, the feasibility of inducing a bandgap in the electronic structure of graphene without impairing the high charge-carrier mobility is of considerable interest. One possible way is to cut graphene into narrow ribbons, known as graphene nanoribbons (GNRs), where the lateral confinement of electrons on the nanometer scale creates a bandgap.

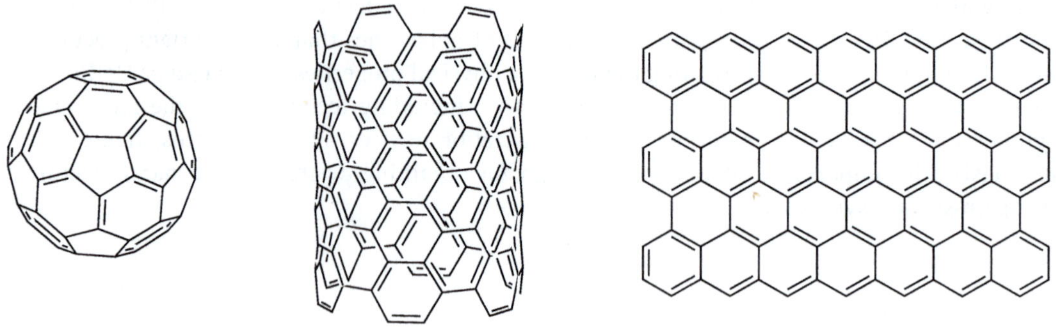

Figure 12.1 Fullerene, CNT, and graphene.

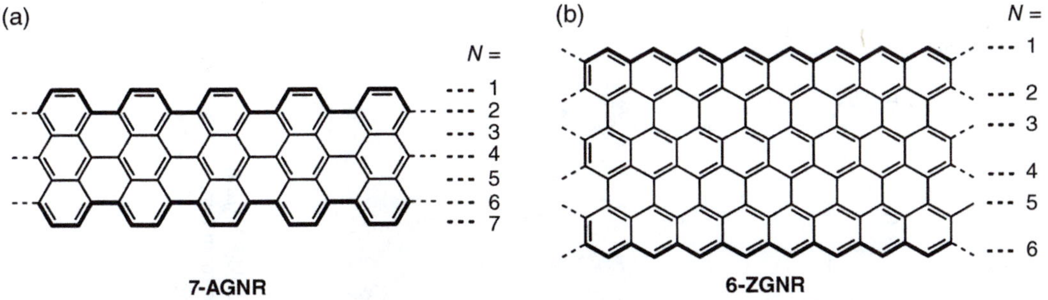

Figure 12.2 Structure of (a) armchair-type GNR (7-AGNR) and (b) ZGNR (6-ZGNR).

Recently, quasi-one-dimensional strips of graphene, GNRs, have drawn considerable attention as promising materials for fabricating nanoelectronic devices. Their electronic properties are determined by width and edge topology [12–14]. Depending on the edge structure, armchair *N*-GNRs (*N*-AGNRs) (*N* is the width measured by the number of rows of carbon atoms across the GNRs) and zigzag GNRs (*N*-ZGNRs) can be synthesized (Figure 12.2a,b). Theoretical and experimental investigations indicate that AGNRs are categorized into three subfamilies with $N = 3p$, $3p + 1$, and $3p + 2$ (p is a natural number), in which the electronic structure varies, depending on the structural boundary conditions [15, 16]. According to the first-principle calculations using the many-body perturbation theory, the gap size Δ decreases with increasing N (width) within each subfamily, and among subfamilies with the same p, the gap size is in the order of $\Delta^{3p+1} > \Delta^{3p} > \Delta^{3p+2}$. Because a $\Delta < 1$ eV is desirable for the application of GNRs in electronic devices instead of silicon-based semiconductors, as observed in the case of CNTs [8, 17], a synthetic method for GNRs with a low bandgap should be developed. In addition, ZGNRs are predicted to exhibit a smaller bandgap than AGNRs, and theoretical calculations have indicated that ZGNRs exhibit metallic properties [15, 16, 18]. The electronic states are localized along the zigzag edge. This "edge state" can be spin-polarized, making ZGNRs promising materials for spintronic applications.

12.2 Bottom-Up Synthesis of Graphene Nanoribbons

Thus far, several top-down approaches have been applied to prepare GNRs from graphite and/or graphene [18–23]; however, they have encountered two major problems. First, theoretical calculations indicate that GNRs must have a width of less than 10 nm to exhibit sufficiently large bandgaps. The production of such a narrow GNR using the standard lithography method remains a challenge. Second, the representative top-down methods use ion beams and/or lasers, which generally cause serious damage to the edge of GNRs. This might hamper the achievement of excellent physical properties of GNRs as their electronic structures is significantly affected by the edge topology.

In contrast, bottom-up approaches have the potential to produce atomically precise GNRs. Structurally well-defined GNRs can be obtained by combining appropriate building blocks and methods [12–14]. Accordingly, the development of GNR synthesis via solution-based organic chemistry has been considered [24–32]. The first report of GNR synthesis using solution-based organic chemistry was achieved in 2008 by Yang et al. (Figure 12.3) [24]. The Suzuki-Miyaura polymerization followed by the intramolecular Scholl reaction with $FeCl_3$ provided GNRs with

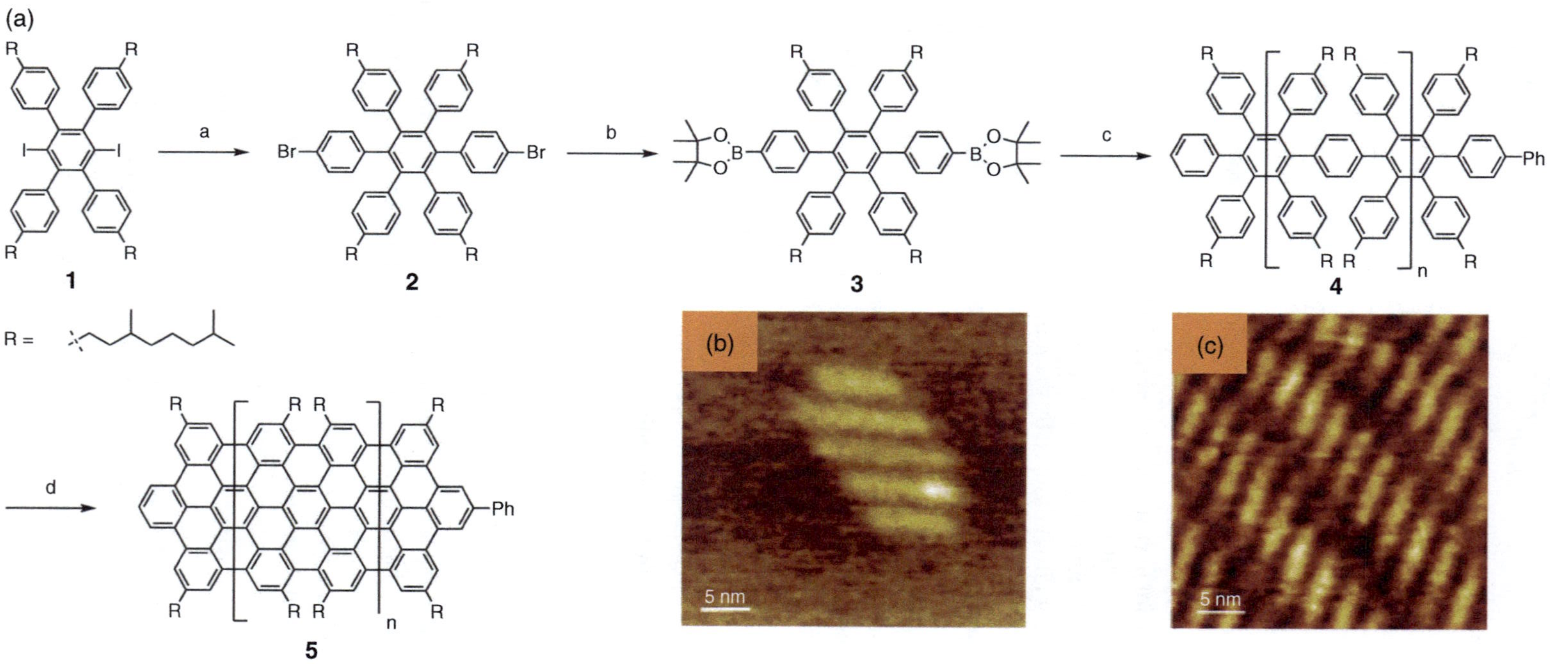

Figure 12.3 (a) Synthetic scheme of GNR **5** via solution-based organic synthesis. Reaction conditions: (a) 4-bromophenylboronic acid, $Pd(PPh_3)_4$, aliquat 336, K_2CO_3, toluene, 80 °C, 24 hours, 93%. (b) n-BuLi, tetrahydrofuran, −78 °C, 1 hour; 2-isopropoxy-4,4,5,5-tetramethyl [1, 3, 2] dioxaborolane, room temperature, 2 hours, 82%. (c) Compound **1**, $Pd(PPh_3)_4$, aliquat 336, K_2CO_3, toluene/H_2O, reflux, 72 hours, 75%. (b, c) Scanning tunneling microscopy (STM) images of GNR **5** at the solid-liquid interface on highly oriented pyrolytic graphite. *Source:* Adapted with permission [24]. Copyright 2008, American Chemical Society.

a length of up to 12 nm. Subsequently, several types of GNRs have been synthesized using solution-based organic chemistry. Specifically, the introduction of long alkyl chains and bulky substituents to the edges of GNRs significantly improves their solubility, enabling the characterization and processability of the solution. In many cases, graphitization, that is, the oxidative aryl–aryl bond formation of the polymer, is the key to their synthesis.

The bottom-up synthesis of GNRs with atomic precision has also been accomplished oxidatively on a metal surface, such as Au(111). Bottom-up synthesis based on surface-assisted reactions is an emerging approach to GNR fabrication [12–14, 32]. Figure 12.4 illustrates the first report of an on-surface synthesis of GNRs [33] in which the molecular building block is 10,10′-dibromo-9,9′-bianthryl (DBBA, in Figure 12.4), and the on-surface reaction yields 7-AGNR. The reaction mechanism is as follows (Figure 12.4) [33, 34]. First, DBBA molecules, which are precursors of 7-AGNR, are sublimated on the Au(111) surface under UHV conditions. Thermally induced debromination occurs, creating a biradical species, which undergoes polymerization to form a linear polymer on the Au(111) surface. The subsequent annealing of the polymers at higher temperatures afforded 7-AGNR via surface-assisted intramolecular cyclodehydrogenation. The precursor molecular design defines the ultimate GNR structure with atomic-level precision in spite of the immense production of GNRs being impossible using on-surface synthesis. The surface-assisted reaction does not encounter solubility issues for synthesis and characterization processes, which can be observed in the case of solution-based organic chemistry. Furthermore, the strong catalytic property of the Au(111) surface smoothly induces the graphitization of linear polymers compared with the solution-based organic synthesis [35]. In addition, the development of cutting-edge technology to explore the molecular structure at an atomic level and to investigate the physical property of a single molecule have significantly contributed to the research field of GNR synthesis [36–39]. High-resolution scanning tunneling microscopy (STM) together with noncontact atomic force microscopy (nc-AFM) under UHV conditions clearly visualize the molecular structure with atomic precision [33, 40]. Because the structure and reactivity of molecular building blocks determine the ultimate GNR structure, the molecular design of acene and its derivative evidently plays an important role in the synthesis of GNRs. The synthesis of AGNRs exhibiting a narrow band gap essentially relies on the construction of benzene rings in the desired structure, and hence, the structural design of the molecular building block, the GNR precursor, is crucial. The subsequent section introduces the recent achievements in the bandgap modulation of AGNRs using different molecular building blocks.

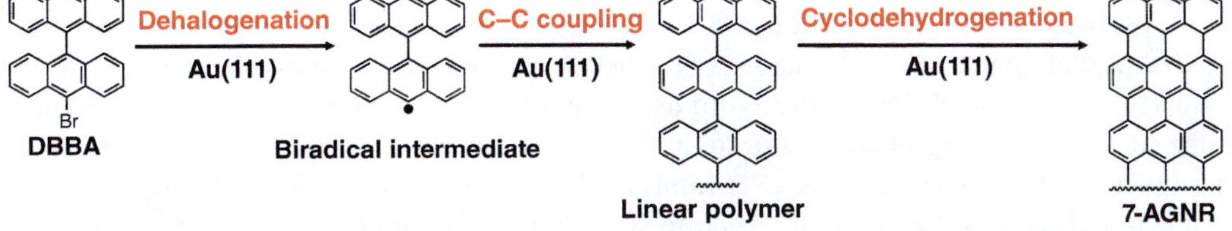

Figure 12.4 Basic steps for surface-supported 7-AGNR synthesis by using 10,10′-dibromo-9,9′-bianthryl monomers (DBBA) as the precursor [33].

12.3 On-Surface Synthesis of Narrow Bandgap Armchair-Type Graphene Nanoribbons

As described in Section 12.1, the precursor molecular design defines the ultimate GNR structure with atomic-level precision. Therefore, the accurate molecular design of acene-based precursors plays an important role in the preparation of AGNRs with narrow bandgaps. Specifically, the diffusion of molecules on the Au(111) surface during polymerization is crucial for obtaining long-range GNRs. For instance, if the precursors possess a flat structure, they are strongly adsorbed onto the substrate after deposition. Eventually, efficient polymerization does not occur because of low diffusion on the substrate. This issue can be overcome using precursors with a cross-shaped structure, which is essential for efficient radical step-growth polymerization on the Au(111) surface. DBBA has a simple and well-considered structure for the polymerization step of diradical intermediates (Figure 12.4) [12]. After the deposition of DBBA on the Au(111) surface, the anthryl units in DBBA are both tilted out of plane owing to steric hindrance between their inner hydrogens. Therefore, this conformation allows for an unhindered approach of the two bianthryl radicals at the polymerization step because of the reduced interaction between the molecule and the Au(111) surface. Once polyanthrylene is formed via an Ullmann-type reaction, [41–44] cyclodehydrogenation takes place efficiently, owing to the strong catalytic property of the Au(111) surface. The resulting anthracene nanoribbons (7-AGNRs) exhibited a band gap of $\Delta = 2.3$ eV on the Au(111) surface.

The chemical modification of DBBA is expected to tune the GNR structure [45–53]. In fact, 2,2′-di((1,1′-biphenyl)-2-yl)-10,10′-dibromo-9,9′-bianthracene (**6**) was prepared through the Suzuki cross-coupling reaction of 2,2′-dibromo-9,9′-bianthracene with (1,1′-biphenyl)-2-ylboronic acid, followed by the selective bromination of the bianthracene core with Br_2 (Figure 12.5) [48]. Similar to DBBA, the precursor was evaporated onto the Au(111) surface, and the subsequent annealing of the substrate at 200 °C under UHV conditions cleaved the C—Br bond, resulting in the formation of linear chains via step-growth polymerization (Figure 12.5b,c). Further annealing at 400 °C induced intramolecular cyclodehydrogenation, yielding hexacene nanoribbons (13-AGNNRs) (Figure 12.5d,e). Scanning tunneling spectroscopy (STS) measurement reveals the band gap Δ of 13-AGNRs to be 1.4 ± 0.1 eV, which is 1.2 eV smaller than that of 7-AGNRs, while 13-AGNRs and 7-AGNRs belong to the same subfamily ($N = 3p+1$). Therefore, the difference in the obtained bandgap reflects the effect on the GNR width.

GNRs belonging to the subfamily $N = 3p$ are expected to exhibit a lower band gap than the subfamily $N = 3p+1$. Considering this theory, a dibromo-*o*-terphenyl precursor was designed and synthesized through multiple organic syntheses to prepare 9-AGNR (Figure 12.6a) [54]. The key to synthesize this precursor is to obtain 1,4-dibromo-2,3-diiodobenzene (DBTP), which enables Suzuki–Miyaura cross-coupling using the different reactivities of iodine and bromine. The precursor was evaporated onto the Au(111) surface, and extended one-dimensional polymers were formed upon annealing at 250 °C (Figure 12.6b). However, covalent bonds between the precursors can only be formed if they are rotated by 180° with respect to each other. STM measurements indicated that the poly(*p*-phenylene) backbone of the polymer possessed an almost planar structure on the Au(111) surface, while the sterically hindered phenyl side groups rotated out of the plane. Finally, further annealing of the substrate at 350 °C transformed the polymers into tetracene nanoribbons (9-AGNRs) (Figure 12.6c). STM and nc-AFM measurements using a CO-functionalized tip explicitly visualized the 9-AGNR structure with atomic precision (Figure 12.6d,e). The key results of the electronic structure of 9-AGNRs are the low effective mass $m^* \approx 0.1\ m_e$ for both electrons and holes and the low band gap $\Delta = 1.4$ eV on the Au(111) surface. Reflecting over the features of AGNRs belonging to the subfamily $N = 3p$, the band gap of 9-AGNR was slightly improved compared to 13-AGNRs, [48] even though 9-AGNRs possessed narrower widths than 13-AGNRs.

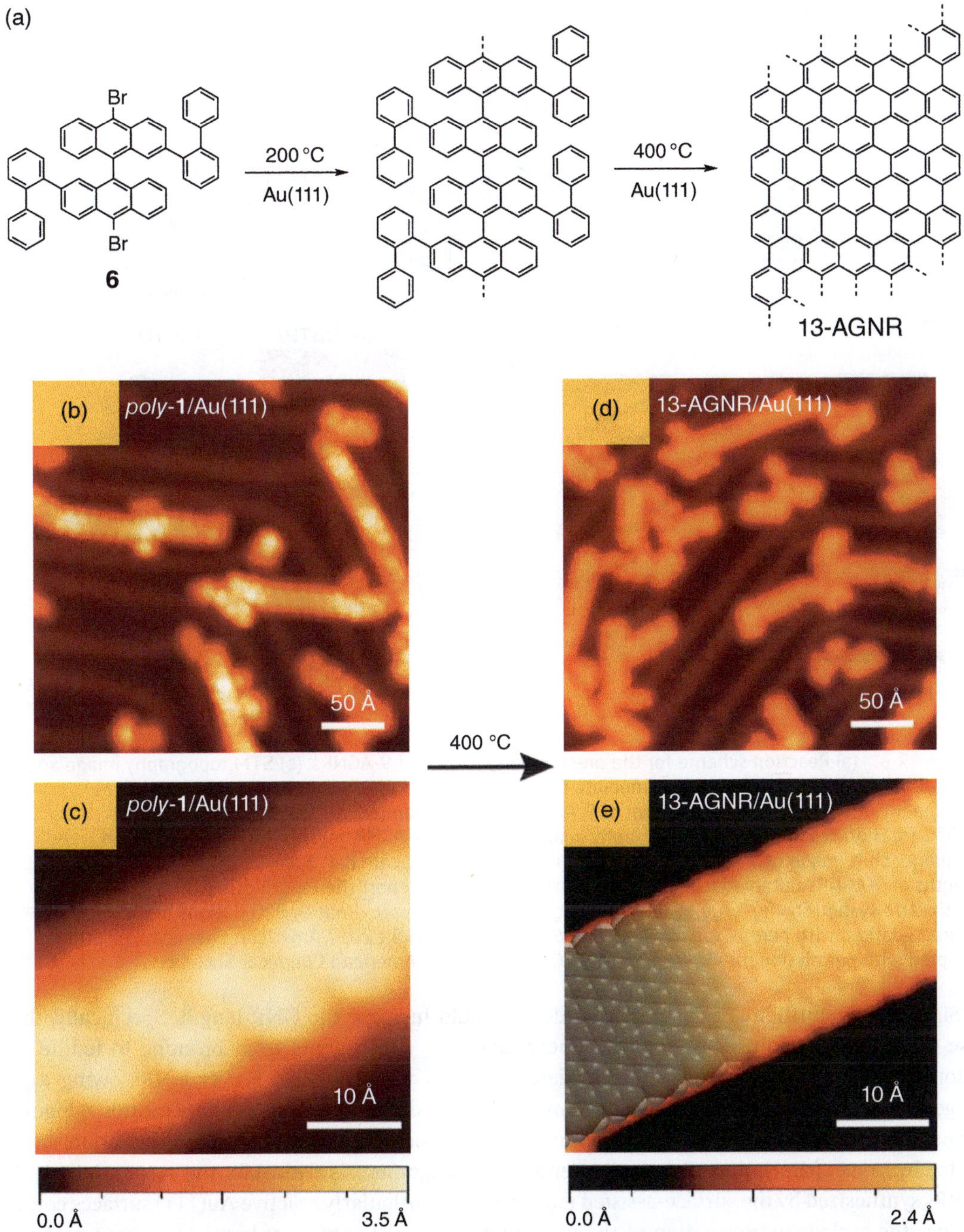

Figure 12.5 (a) Schematic representation of the synthesis of 13-AGNRs from the molecular building block (**6**). (b) STM image of the polymer formed after the deposition of precursor onto the Au(111) surface maintained at 200 °C. (c) High-resolution STM image of the polymer. (d) STM image of 13-AGNRs formed after annealing polymer at 400 °C. (e) Close-up STM image of 13-AGNR. Structural model of 13-AGNR has been overlaid onto the STM image. *Source:* Adapted with permission [48]. Copyright 2012, American Chemical Society.

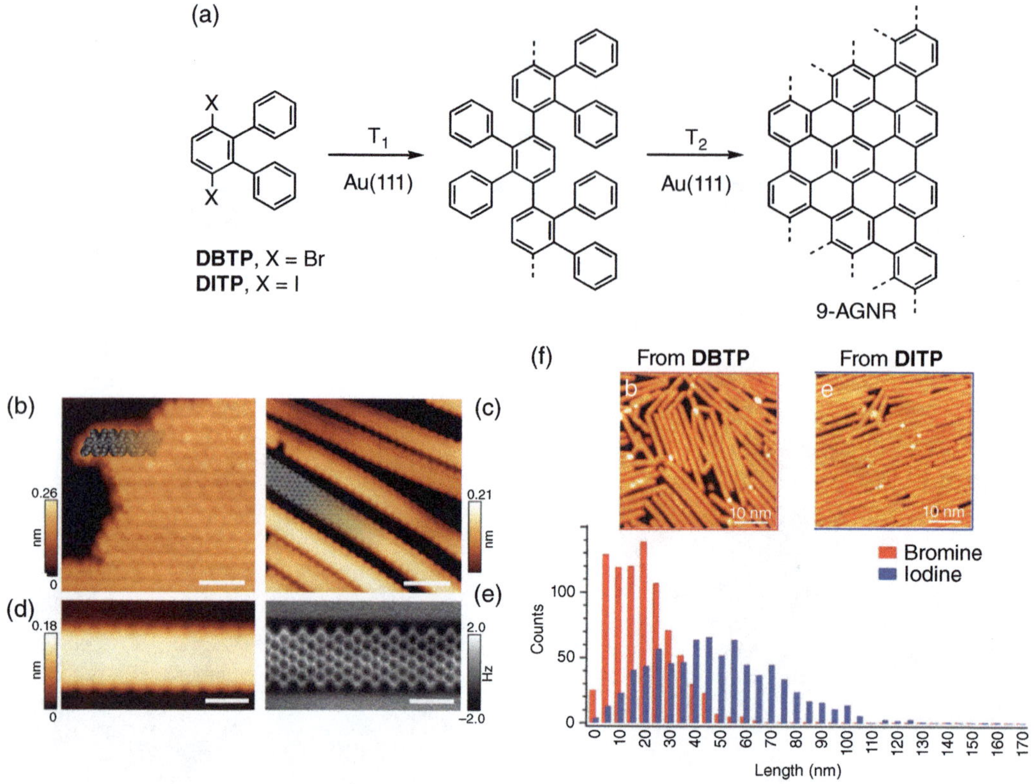

Figure 12.6 (a) Reaction scheme for the on-surface synthesis of 9-AGNRs. (b) STM topography image and structural model of the polymer intermediate from DBTP formed after annealing to 250 °C. (c) STM topography image and structural model of 9-AGNRs formed by annealing the polymer sample to 350 °C. (d) High-resolution STM topography image of a single 9-AGNR. (e) High-resolution nc-AFM frequency shift image of 9-AGNR using a CO-functionalized tip. Scale bar: 1 nm. (f) STM showing different lengths of 9-AGNRs obtained from DBTP and DITP. Histograms report the length distribution determined for the two systems using large-scale STM images, measuring approximately 800 ribbons in each case. *Source:* Adapted with permission for Figure 12.6b–e [54]. Copyright 2017, American Chemical Society. Adapted with permission for Figure 12.6f [55]. Copyright 2018, American Chemical Society.

Significantly, further careful molecular design could increase the GNR length. Specifically, the use of iodine-containing monomers, namely, changing substituents from bromine to iodine to afford surface-assisted diradical intermediates, induced the growth of longer GNRs owing to a larger separation of the polymerization and cyclodehydrogenation temperatures (Figure 12.6a,f) [55]. Here, the undesired termination of the sequential growth during polymerization, such as the hydrogen passivation of the radical ends of the oligomers, is critical for the average length of GNRs synthesized by the surface-assisted reaction [56]. Particularly reactive Au(111) surfaces could induce the cyclodehydrogenation of the monomers and/or oligomers at lower temperatures than expected, generating hydrogen atoms, which could be the origin of the passivation and would be effective for decoupling these two steps, thereby producing fewer undesired hydrogen atoms. Here, the bond enthalpy with phenyl is 67 kcal mol^{-1} for I and 84 kcal mol^{-1} for Br [57]. Therefore, 3′,6′-diiodo-1,1′:2′,1′′-terphenyl (DITP) was synthesized. Fast X-ray photoemission spectroscopy measurements (XPS) measurements revealed a lower dehalogenation temperature for DITP compared with that of DBTP, reducing crosstalk between the polymerization and cyclodehydrogenation steps and therefore limiting the passivation of growing chains by atomic hydrogen. These features clearly affected the length of 9-AGNRs: the average 9-AGNR length significantly increased from 15 to 45 nm when DITP was used as the precursor instead of DBTP (Figure 12.6f).

AGNRs with a subfamily of $N = 3p + 2$ are theoretically predicted to be metallic with zero bandgap at the tight-binding level of theory and semiconducting with very small band gaps using density functional theory (DFT) calculations [16]. The first successful example of AGNRs belonging to the subfamily of $N = 3p + 2$ was achieved by using 1,4,5,8-tetrabromonaphthalene (TBN) as the precursor [58]. The structure of TBN does not possess a cross-shape, and therefore, it first forms Au-naphthalene organometallic complexes. Because no polymerization step is required, further annealing yielded naphthalene nanoribbons (5-AGNRs). STS measurements indicated that the band gap of 5-AGNR on the Au(111) surface was 2.8 ± 0.1 eV, which was significantly larger than the value (1.32 eV) predicted by GW approximation. The interaction between 5-AGNR and the Au(111) substrate or a local electric field introduced by the STM tip is the origin of this finding [59, 60]. Concurrently, 5-AGNRs were also synthesized using 3,9 (10)-dibromoperylene [50]. In this case, a bandgap of ~100 meV was realized in short ribbons of only 5 nm in length [61].

The simple chemical modification of DBTP enabled us to prepare a precursor for 17-AGNRs, which belong to the subfamily $N = 3p + 2$, together with 13-AGNRs. Guided by this idea, 1,2-bis-(2-anthracenyl)-3,6-dibromobenzene (BADBB) for the 17-AGNRs and 1,2-bis-(2-naphthale nyl)-3,6-dibromobenzene (BNDBB) for 13-AGNRs were successfully synthesized via multi-step organic synthesis by Yamaguchi et al. (Figure 12.7a) [62]. Annealing Au(111) surfaces with a submonolayer of BADBB or BNDBB up to 250 °C induced dehalogenation and formed extended

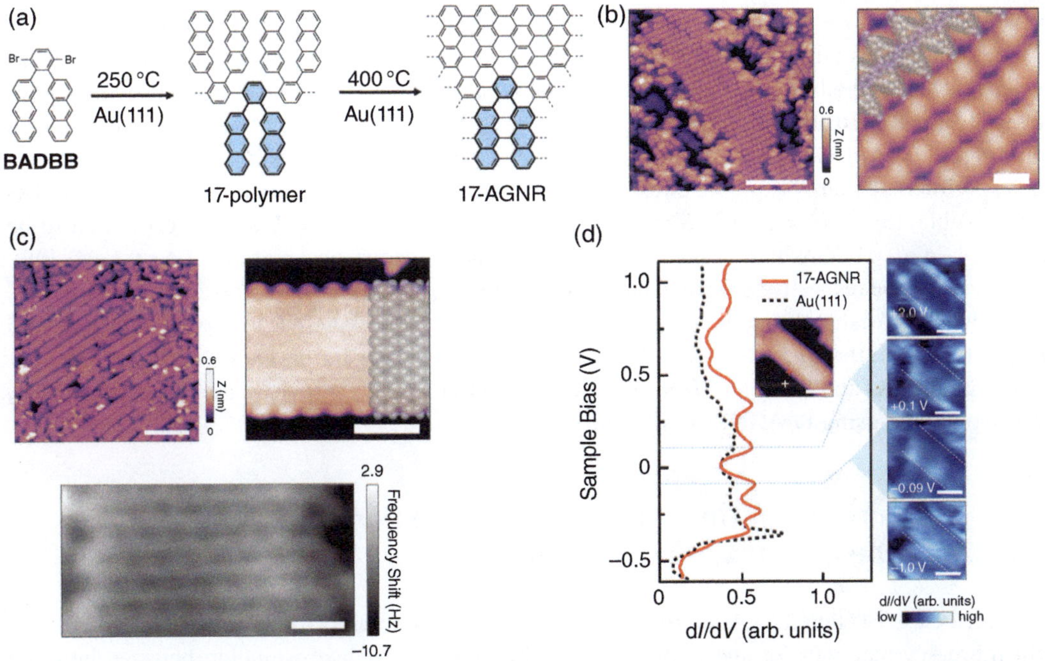

Figure 12.7 (a) Schematic drawing of an on-surface reaction for 17-AGNR on the Au(111) surface. (b) Large-scale STM image of 17-polymer on the Au(111) surface (left) and small-scale image (right). Structural model of single 17-polymer is superimposed in right image. (c) Overview STM image of 17-AGNRs on the Au(111) surface (left); high-resolution STM image of single 17-AGNR together with structural model (right), and constant-height frequency shift image of single 17-AGNR measured by nc-AFM with a CO-functionalized tip (bottom). (d) Differential conductance (dI/dV) point spectra recorded on the edge of single 17-AGNR (red line) and the Au(111) surface (black dotted line) (left). Crosses in the inset of the STM image indicate tip positions for STS. Constant-height dI/dV maps of the 17-AGNR obtained at energies indicated in each map (right). Dashed lines indicate outer edges of the AGNR. *Source:* Reproduced under the terms of the CC-BY Creative Commons Attribution 4.0 International license (http://creativecommons.org/licenses/by/4.0) [62]. Copyright 2020, The Authors, published by Springer Nature.

one-dimensional 17-polymer and 13-polymer by aryl–aryl coupling (Figure 12.7a, b for 17-polymer). As observed in the synthesis of 9-AGNRs from DBTP, the monomers can be covalently bridged only when they are rotated 180° with respect to each other. A large-scale STM topographic image of 17-polymers on the Au(111) surface showed that the 17-polymers are assembled into extended islands (Figure 12.7b). A small-scale STM image of 17-polymers showed the alignment of protrusions, which were derived from the sterically hindered anthracene units. Further annealing of the substrate at 400 °C resulted in the complete planarization of the polymers and led to fully conjugated AGNRs with an apparent height of 0.18 nm in 17-AGNRs (Figure 12.7c). The apparent height of 13-AGNRs was similar to that observed in the case synthesized with a different precursor (vide supra) [48]. Significantly, ex-situ nc-AFM imaging using CO-functionalized tips clearly visualized the bond-resolved structure of AGNRs (Figure 12.7c), although the ex-situ nc-AFM imaging is often a challenging task because of contaminations on the surface adsorbed during air exposure. Both 17-AGNRs and 13-AGNRs possess some defects along their edges compared with 9-AGNRs. This is because of the larger steric hindrance between the two acene units (anthracene for 17-AGNR or naphthalene for 13-AGNR vs. benzene for 9-GNR), resulting in the breaking of the attached acene units during the cyclodehydrogenation process. A differential conductance (dI/dV) obtained by STS for 17-AGNR indicated that the experimental energy gap is estimated to be $\Delta_{STS} = 0.19 \pm 0.03$ eV for the 17-AGNR on the Au(111) surface (Figure 12.7d). In the GW calculations, the quasiparticle gap of the freestanding 17-AGNR is predicted to be $\Delta_{GW} = 0.63$ eV. In general, the STS values determined by STS in GNRs on metal surfaces are significantly underestimated compared with those of Δ_{GW} for freestanding GNRs. The reduction in Δ_{STS} compared to Δ_{GW} is caused by a substrate-induced weakening of the electrostatic potential owing to long-range screening effects [63]. An advanced image-charge model that includes the substrate screening as well as internal screening of the GNRs could correct the Δ_{GW} values [64]. Therefore, the renormalized quasiparticle gap of 17-AGNRs is predicted to be $\Delta_{GW}' = 0.20$ eV, which is in good agreement with the experimental energy gap. The energy gap of 17-AGNRs clearly reflects the features of AGNRs with a subfamily of $N = 3p+2$. Similarly, the band gap of the 13-AGNRs was estimated to be $\Delta_{STS} = 1.34 \pm 0.03$ eV, which was consistent with the previous observation (1.4 ± 0.1 eV) [48]. Furthermore, the respective effective masses (m^*_{CB} and m^*_{VB} for the conduction and valence bands, respectively) are estimated to be $m^*_{CB} = m^*_{VB} = 0.06m_e$ in the 17-AGNR and $m^*_{CB} = 0.14m_e$ and $m^*_{VB} = 0.13m_e$ in the 13-AGNR. The electron/hole effective mass in the 17-AGNR is even smaller than that in GaAs and InP [65]. These results indicate that excellent devices, including transistors, can be obtained using 17-AGNRs in near future.

12.4 On-Surface Synthesis of Polyacenes as Partial Structure of Zigzag-Type Graphene Nanoribbons

As described in Section 12.1.2, ZGNRs are expected to host spin-polarized electronic edge states, which are key materials for spintronic devices. However, cyclodehydrogenation between benzene rings in precursors or polymers for GNRs cannot form a zigzag structure. Therefore, a completely different molecular design for the ZGNR precursors is required. In 2016, Ruffieux et al. designed a U-shaped precursor and successfully synthesized 6-ZGNR by surface-assisted reaction of this precursor (Figure 12.8) [66]. The following are the key to their success: (i) this precursor possesses a short zigzag structure, which would be a partial zigzag edge of ZGNR, and (ii) benefiting from the strong catalytic property of the Au(111) surface, surface-assisted reaction between the methyl and benzene groups also leads to a zigzag edge, linking the short zigzag edge of precursor. A similar synthetic strategy successfully prepared nitrogen-substituted 6-ZGNRs (N-6-ZGNRs), where every sixth C—H group

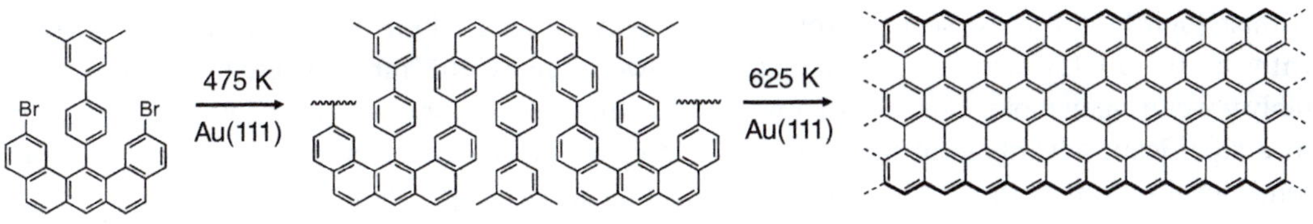

Figure 12.8 Synthetic scheme for 6-ZGNR on the Au(111) surface by using U-shaped precursor [66].

along the zigzag edge of a 6-ZGNR was replaced by a nitrogen atom [67]. Despite these efforts, the synthetic difficulty for ZGNRs hinders the investigation of the intrinsic properties of various ZGNRs.

Polyacenes, composed of multiple linearly fused benzene rings, are an important class of polycyclic aromatic hydrocarbons [68]. One of the most important features of polyacenes is their ultra-narrow highest occupied molecular orbital (HOMO)–lowest unoccupied molecular orbital (LUMO) gaps. As the length of the acene increases, the energy gap between the HOMO and LUMO decreases rapidly. This unique electronic property stems from its zigzag-edged polyacene structure, and hence, they are not only regarded as synthetic precursors of GNRs but also as partial structures of ZGNRs. The evaluation of their electronic properties significantly contributes to our understanding of the edge states of ZGNRs. However, the stability of polyacenes decreases as the acene length increases. Furthermore, polyacenes are poorly soluble in common organic solvents owing to their uncomplicated structures. These features have prevented the synthesis of polyacenes and investigation of their intrinsic properties.

One promising way to synthesize polyacenes is by their functionalization with stabilizing and protecting groups [69–72]. The stable and soluble precursors could be converted into the corresponding polyacenes at the final stage under appropriate conditions. The most successful example is the Strating–Zwanenburg reaction [73], where α-diketone groups undergo visible-light-induced photodecarbonylation to generate polyacenes. Quantitative conversion from precursors to polyacenes can be achieved by simple photoirradiation with only gaseous byproducts if the purity of the precursor is sufficient. In addition, thermally induced decarbonylation and retro-Diels-Alder reaction are representative examples of precursor methods [69]. Despite these synthetic approaches, the characterization of polyacenes, namely, the confirmation of polyacene generation, is mostly achieved by measuring absorption spectra under inert conditions at ultra-low temperatures or by evaluating using mass spectrometry alone [72–76]. Indeed, hexacene, which has six linearly fused benzene rings, is the longest acene that is isolated and characterized by X-ray single crystal analysis [77]. Hexacene single crystals were obtained via the sublimation of the monoketone precursor. In situ single-crystal formation during sublimation was achieved via thermally induced decarbonylation of the precursor (Figure 12.9). The hexacene single-crystal field-effect transistor exhibited a maximum hole mobility of 4.28 cm^2/Vs, and these devices could function for more than 19 days, while the mobility showed a 32% reduction even in a nitrogen atmosphere, suggesting the oxidation of the surface of the single crystal.

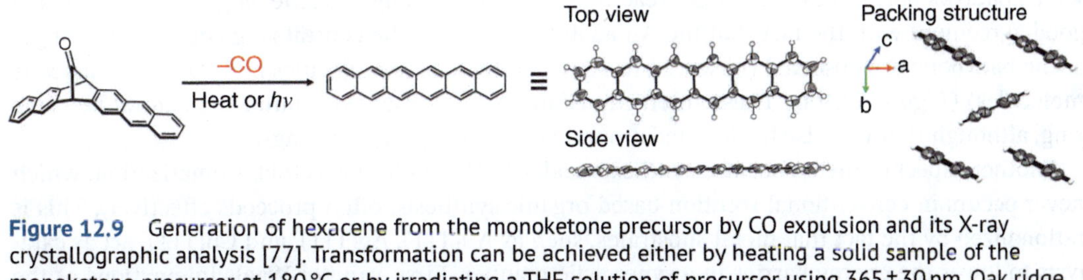

Figure 12.9 Generation of hexacene from the monoketone precursor by CO expulsion and its X-ray crystallographic analysis [77]. Transformation can be achieved either by heating a solid sample of the monoketone precursor at 180 °C or by irradiating a THF solution of precursor using 365 ± 30 nm. Oak ridge thermal ellipsoid plot (ORTEP) drawing of hexacene and its packing structure are also shown.

Consequently, the most severe problem with polyacenes is their oxidation. The on-surface synthesis under UHV conditions with pressures lower than ca. 10^{-9} mbar, where there are considerably fewer oxygen molecules, together with in situ high-resolution STM and nc-AFM techniques, has a significant advantage in the research field of polyacenes. The molecular structure at the atomic level – along with the physical properties such as bandgap, – could be evaluated.

Polyacenes have a partial structure at the edge of ZGNRs; therefore, they are regarded as the narrowest members in the family of ZGNRs. Although it is generally accepted that short acenes (up to pentacene or hexacene) have a closed-shell electronic structure, the ground state of longer acenes has been gaining interest. Recent theoretical studies have indicated that longer acenes, such as nonacene, possessed an open-shell singlet ground state [78]. These studies also show that this trend is accomplished by decreasing singlet-triplet and electronic gaps, significantly increasing chemical reactivity and increasing the difference between ladder and rung carbon–carbon bonds, structurally rendering longer acenes as weakly coupled oligoacetylene chains. Therefore, polyacenes are good counterparts to study the edge states of ZGNRs.

Urgel et al. experimentally investigated the chemical and electronic structures of nonacene and heptacene through on-surface synthesis and characterization under UHV conditions, [79] and in this case, α-diketone-type nonacene and heptacene precursors were provided (Figure 12.10a). The precursors with α-diketone groups attracted attention because of their thermal stability. Although the precursors must be evaporated on a metal surface without decomposition, the α-diketone groups are thermally stable up to ca. 350–400 °C, resulting in the successful deposition of precursors on the Au(111) surface. Constant-height STM images of submonolayers of nonacene precursors indicate mostly intact *syn*–bisdiketone nonacene precursors, where the two bright protrusions are assigned to the bridged α-diketone groups pointing upward (Figure 12.10b). After light exposure ($\lambda = 470$ nm, $\phi_{\text{photon}} = 2.5\times10^{18}$ cm^{-2} s^{-1}), most of the bright protrusions disappeared, revealing smooth, rod-like structures with a homogeneous apparent height of 1.6 Å (Figure 12.10b). Ultimate structural information was obtained by constant-current STM and nc-AFM measurements with a CO-functionalized tip (Figure 12.10c), indicating that the nonacene precursor was converted to nonacene via the Strating–Zwanenburg reaction. The α-bisdiketone-type heptacene precursor was also converted to heptacene in the same manner (Figure 12.10d), although heptacene was also successfully obtained from α-monodiketone-type heptacene precursor on the Ag(111) surface [80]. STS measurements revealed their band gaps on the Au(111) surface: 1.25 eV for nonacene and 1.50 eV for heptacene. To elucidate the relationship between the HOMO-LUMO gaps of polyacenes and their radical character, quasiparticle GW calculations for acenes, including screening effects from the underlying Au(111) substrate, were performed (Figure 12.10e). As a result, the singlet open-shell (SOS) GW gaps are in good agreement with the experimental spectroscopic values, both in trend and absolute values, for longer acenes, whereas the SOS ground state and closed-shell (CS) from DFT calculations are identical for pentacene and hexacene. Hence, polyacenes larger than hexacene clearly possess an open-shell character on the Au(111) surface, which endows polyacenes with larger reactivity as the increasing of acene length. This result is in good agreement with the fact that the Au adatoms bound to the central ring edges of the longer acene backbone are observed (observed in ~ 60% of heptacene molecules and ~95% of nonacene molecules) (Figure 12.10f). These interactions are mostly observed around the central benzene ring, although they can also be detected around neighboring benzene rings.

Another aspect of the on-surface synthesis under UHV conditions is that aromatization, which never occurs in conventional solution-based organic synthesis, often proceeds effectively. This is rationalized by the fact that metal substrates, such as Au(111), Ag(111), and Cu(111), act as catalysts to lower the energy barrier in a synergistic manner. First, van der Waals interactions of the molecule with the metal surface tend to planarize the adsorbed molecule to increase the binding

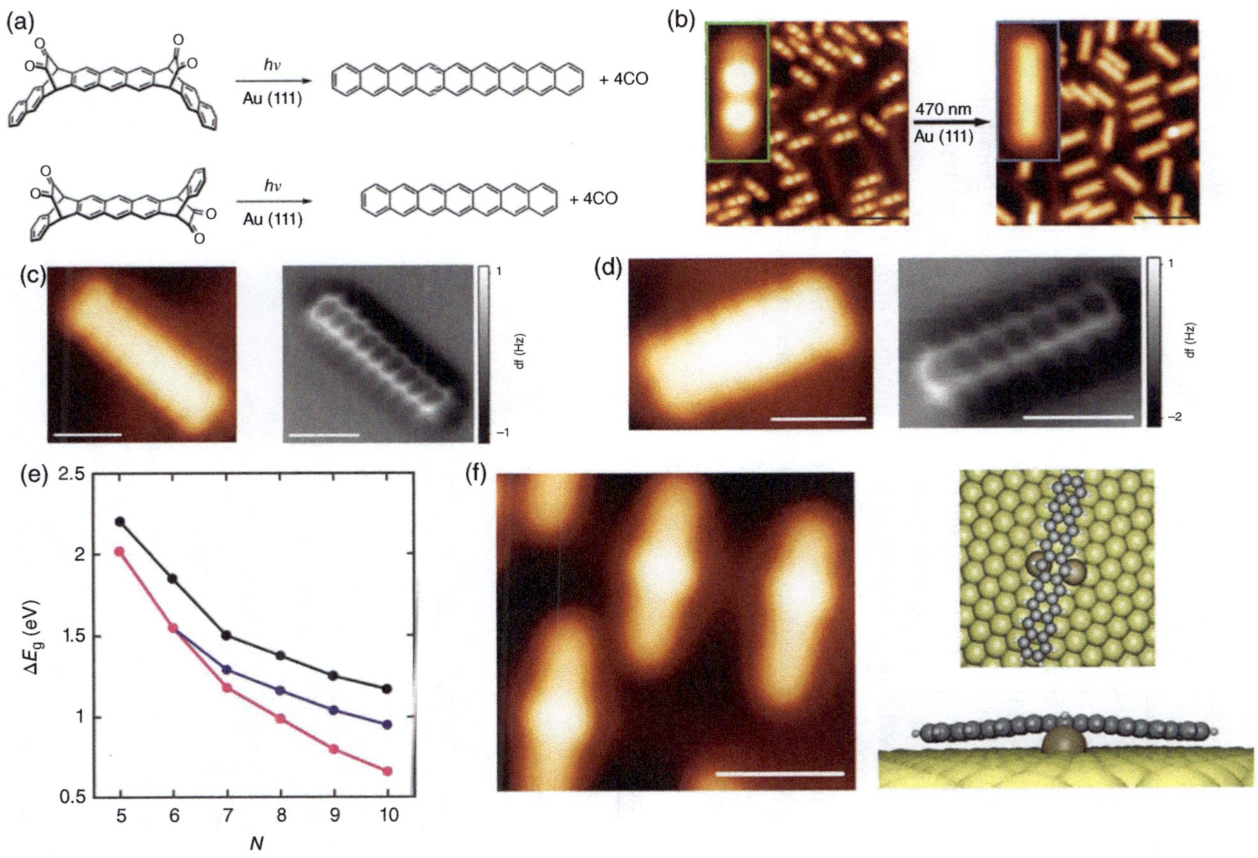

Figure 12.10 (a) Light-induced on-surface formation of nonacene and heptacene. (b) Nonacene formation via photodissociation. Left: STM image after the room temperature desorption of the α-diketone-type nonacene precursor on the Au(111) surface. Scale bar: 5 nm. Inset shows a high-resolution STM image of an individual α-diketone-type nonacene precursor. Right: STM image of nonacene species after light exposure (λ = 470 nm, 3 hours). Scale bar: 5 nm. Inset shows a high-resolution STM image of an individual nonacene. (c, d) Nonacene (c) and heptacene (d) on the Au(111) surface. High-resolution STM image (left) and constant-height frequency-shift nc-AFM image acquired with a CO-functionalized tip (right). Scale bars: 1 nm. (e) Experimental electronic gaps (black curve) together with computed quasiparticle energy gaps for the singlet open-shell (blue curve) and closed-shell (pink curve) configurations, including screening effects from the Au(111) substrate. (f) High-resolution STM image showing the Au-nonacene interaction as two protrusions at certain Vbias. Scale bar: 2 nm. Top and side views of the corresponding DFT equilibrium geometry are also shown. *Source:* Reproduced under the terms of the CC-BY Creative Commons Attribution 4.0 International license (http://creativecommons.org/licenses/by/4.0) [79]. Copyright 2019, The Authors, published by Springer Nature.

energy to the substrate. Furthermore, the metal surface provides the electronic stabilization of the intermediates (e.g. radicals). These features often provide a much lower energy barrier for on-surface reactions than for the corresponding reactions in the solution or gas phase [35].

Indeed, tetracene [81], hexacene [82], decacene [83], and dodecacene [84] precursors protected by epoxy groups were successfully deposited on the Cu(111) or Au(111) surfaces, followed by efficient deoxygenation to afford the corresponding polyacenes (Figure 12.11a). For instance, STS measurements for decacene provided the energy gap of 1.17 eV on the Au(111) surface, though the energy gap of dodecacene unexpectedly increased to 1.4 eV on Au(111) surface. In addition, hydropolyacene precursors were also converted to the corresponding polyacenes by surface-assisted reaction (Figure 12.11b) [85, 86]. Notably, undecacene [85] exhibited a HOMO–LUMO band gap of 1.09 eV on the Au(111) surface. The significance of these findings is that these precursors are thermally inert molecules and therefore possess sufficient stability for sublimation.

Wider AGNRs, which exhibit a narrow bandgap, are expected to be synthesized from halogenated polyacenes using on-surface synthesis. Guided by this idea, the α-diketone-type heptacene precursor was brominated at the 7,16-positions to ensure a controlled reaction on the Au(111) surface (Figure 12.12a) [87]. The annealing of a submonolayer of 7,16-dibrominated heptacene precursors on the Au(111) surface (Figure 12.12b) at 435 K induced debromination and the subsequent formation of an organometallic intermediate, where heptacene precursors were linked together via Au adatoms at the 7,16-positions (Figure 12.12c). Subsequent

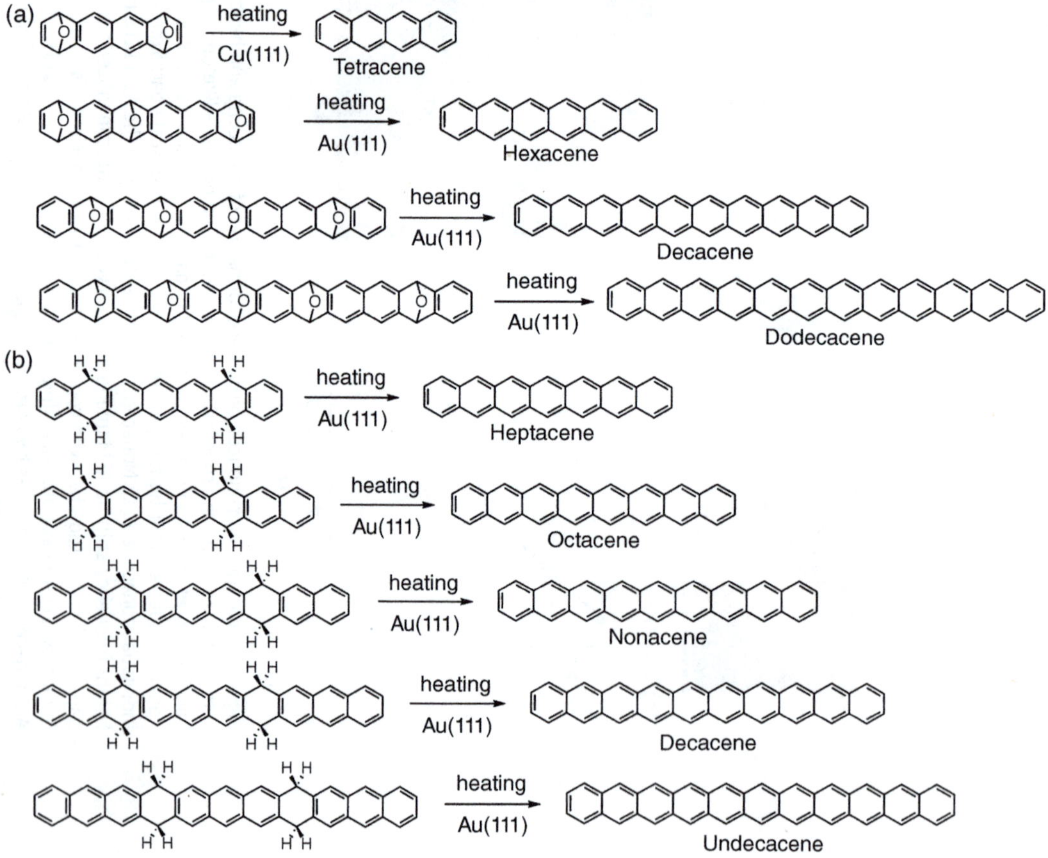

Figure 12.11 (a) Surface-assisted deoxygenation of epoxy precursors of polyacene [81–83]. (b) On-surface dehydrogenation of hydropolyacenes [84, 85].

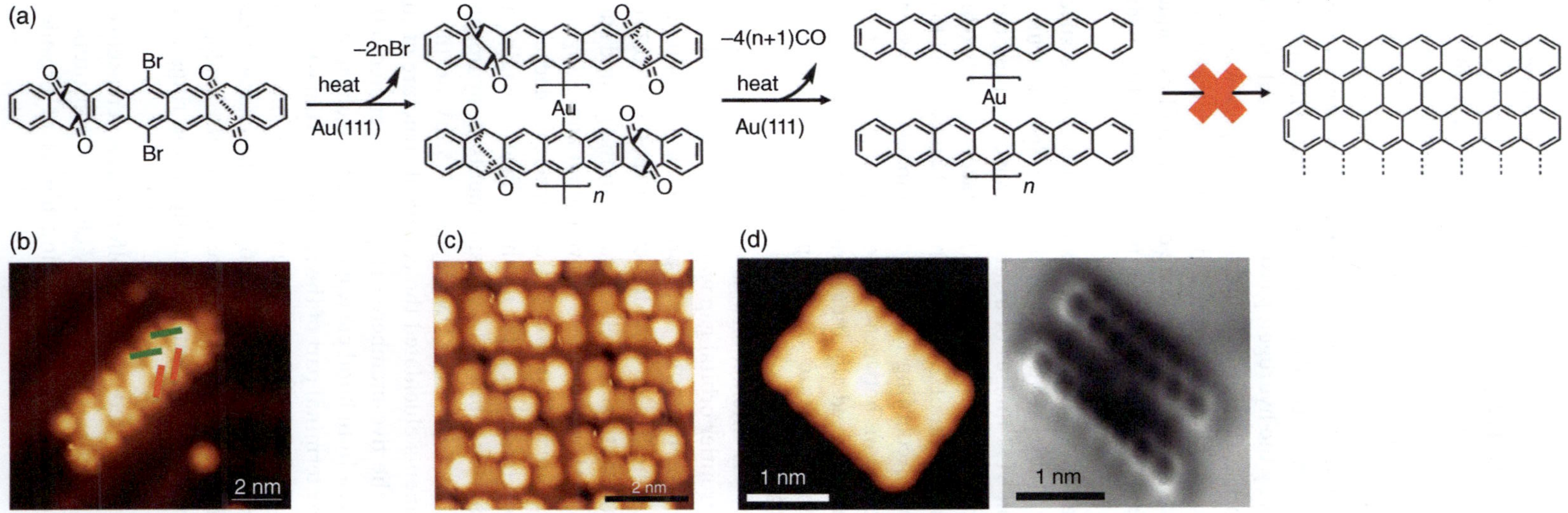

Figure 12.12 (a) On-surface synthetic strategy toward 15-AGNRs from the 7,16-diboromoheptacene precursor. (b) STM topography image of precursor after deposition on the Au(111) surface. Green and red lines are depicted as a guide to the eye for molecule identification. (c) High-resolution STM image of a Au-directed tetramer of heptacene precursors adsorbed on the Au(111) surface. (d) High-resolution STM image (left) and constant-height frequency-shift nc-AFM image acquired with a CO-functionalized tip (right) of heptacene-Au-heptacene organometallic complex. *Source:* Adapted with permission [86]. Copyright 2017, American Chemical Society.

annealing at 535 K cleaved the α-diketone moieties from the intermediate to provide heptacene organometallic complexes (Figure 12.12d). STM images indicated that two seven-lobed species were linked by a single bright protrusion, and the C—Au bond length was 2.5 ± 0.2 Å in a heptacene–Au–heptacene organometallic complex. nc-AFM measurements were performed using a CO-functionalized tip to confirm the chemical structure. Significantly, the length of the complex can vary from two to nine heptacene units, and 76% of the complexes present are two to four units in length. This is tentatively associated with the presence of residual hydrogen gas in the vacuum chamber, even under UHV conditions (vide supra), inhibiting the formation of longer organometallic chains by the hydrogen passivation of radical sites. Even though the length of the complex was short, further annealing of the complex was expected to induce aryl–aryl coupling between the heptacene species, providing heptacene nanoribbons (15-AGNRs). However, the graphitization of the heptacene organometallic complexes did not proceed at all, probably because of the steric hindrance between the hydrogen atoms of heptacene [88].

Furthermore, 6,12,19,25-tetraazaundecacene and its derivatives were successfully synthesized via on-surface synthesis, [89] in which an ethano-bridged precursor was prepared (Figure 12.13a). High-resolution STM images of a submonolayer of the precursors indicated the intramolecular features of precursor on the Au(111) surface, where two bright protrusions per molecule of similar apparent height of 2.3 Å, which could be assigned to two out-of-plane ethano bridges pointing upward. According to structural models, self-assembled molecules are stabilized by the N•••H interactions between adjacent molecules. The tetraazaundecacene precursor on the Au(111) surface showed significant reactivity, depending on the conditions (Figure 12.13a). First, tip-induced release of the protecting group afforded the target tetraazaundecacene (Figure 12.13b). The obtained tetraazaundecacenes remained self-assembled and stabilized by N•••H interactions, as observed in the case of the precursor. Significantly, the experimental frontier orbital gap was 1.35 eV. A GW calculation with image charge corrections for the open-shell ground state provides a gap value of 0.97 eV, which matches better with the experimental value than the closed-shell gap (0.74 eV), thereby providing further evidence for significant open-shell character in the ground state of tetraazaundecacene on the Au(111) surface. Meanwhile, the conventional annealing of the precursor at 280 °C on the Au(111) surface provided hydrogenated tetraazaundecacene and its analog (Figure 12.13a). The deprotection reaction occurs on the surface, providing a significant density of mobile hydrogen adatoms (vide supra). STS measurements for hydrogenated tetraazaundecacene provided an experimental HOMO–LUMO gap of 1.55 eV on the Au(111) surface. Additionally, nc-AFM measurements elucidated the chemical structure of the lateral protrusions, which were attributed to the five-membered rings (hydrogenated tetraazaundecacene analog). This hypothesis is derived from bond cleavage at the bridgehead position of the bicyclo[2.2.2]octadieno unit, with the terminal part of the resulting partial structure re-bonding to the neighboring nitrogen atom. These steps lead to the removal of the two protons of the ethano group via dehydrogenative aromatization. Additionally, the characterization of the aromaticity shows that hydrogenated tetraazaundecacene and its analog with edge-fused five-membered rings exhibit unique sharp antiaromaticity in the rings containing nitrogen atoms (Figure 12.13c).

These results indicate that the presented procedure offers a new perspective for investigating the complex electronic correlation effects in acenes by chemical substitution and opens up a new path toward obtaining precisely tailored building blocks of organic electronics and spintronics.

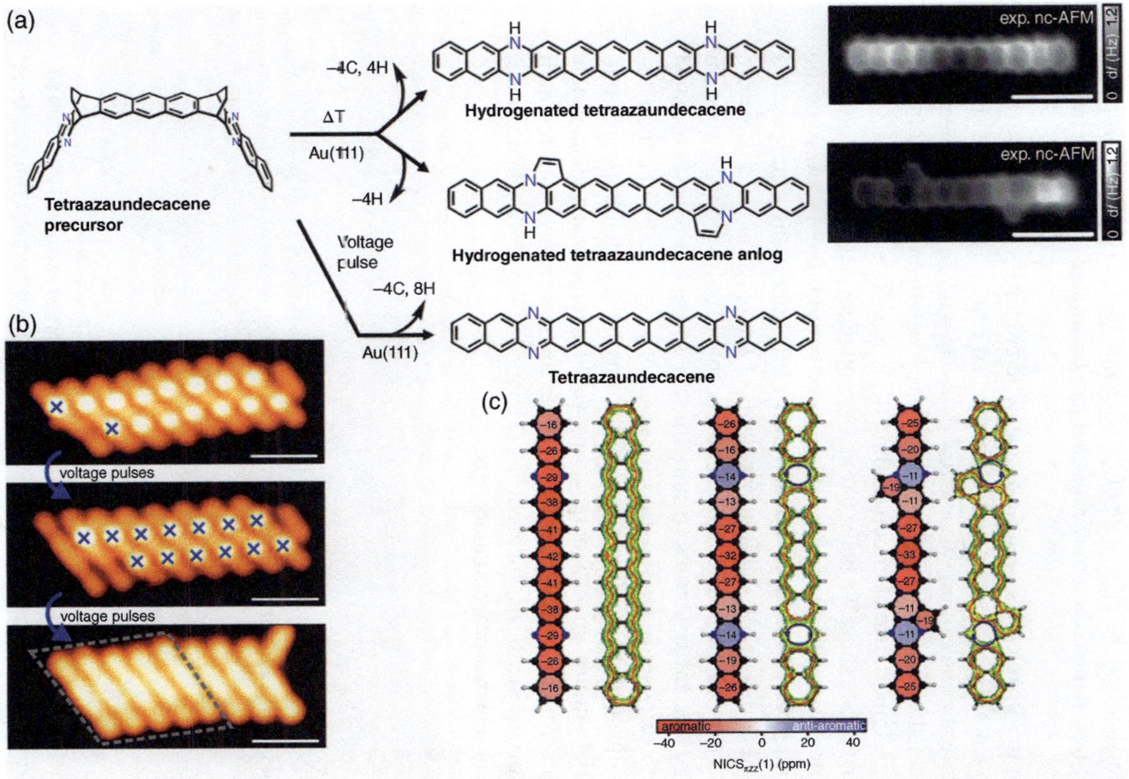

Figure 12.13 (a) On-surface synthetic strategy toward the synthesis of tetraazaundecacene and its derivatives from the tetraazaundecacene precursor. Constant-height frequency-shift nc-AFM images acquired with a CO-functionalized tip are also shown. Scale bar: 1 nm. (b) On-surface synthesis of tetraazaundecacene via the STM tip-induced cleavage of ethano protecting groups of precursor. Successive STM topography images illustrating the cleavage process. Scale bar: 5 nm. (c) Computational characterization of the aromaticity of tetraazaundecacene (left), hydrogenated tetraazaundecacene (middle), and its analog with two edge-fused five-membered rings (right). NICSπ_{ZZ} (1) patterns (each cycle is represented by the NICSπ_{ZZ} (1) evaluated at the center) and π-ACID plot (magnetic field is applied perpendicular to the plane of the system, pointing toward the reader; red and blue arrows indicate aromatic diatropic and anti-aromatic paratropic ring currents, respectively) for the closed-shell solution of tetraazaundecacene (left), hydrogenated tetraazaundecacene (middle), and its analog with two edge-fused five-membered rings.

12.5 Conclusion and Perspectives

Combining the on-surface synthesis of GNRs with cutting-edge observation techniques by STM and nc-AFM enables the synthesis of various types of GNRs (AGNRs and ZGNRs) and small molecules (e.g. polyacenes) that could not otherwise be obtained. The investigation of the relationship between the physical properties and structures (e.g. length and substitution) of polyacenes would significantly contribute to understanding the edge state of ZGNRs and discovering an appropriate molecular design to synthesize new nanocarbon materials that can be applied to semiconducting and/or spintronic devices.

Meanwhile, as introduced in this chapter, the significant catalytic properties of metal surfaces, such as Au(111), Ag(111), and Cu(111), sometimes cause unexpected reactions. Even the simple anthracene-based precursor molecule can exhibit variable reactivity. On-surface synthesis features hierarchical reactions, such as dehalogenation, polymerization, and cyclodehydrogenation [45–47]. These results strongly suggest the importance of understanding the reaction mechanism and transient structural role in the intermediate state. We believe that a thorough investigation of the reaction mechanism using experimental and theoretical methods together with the discovery of the critical intermediate state for unexpected reactions substantially contribute to the synthesis of a wide range of essential nanocarbon materials and improvements of the reaction yield. Recently, the rational molecular design of edge-fluorinated GNR precursors, that could avoid the formation of undesired and destabilizing transient structures, resulted in the success of edge-fluorinated GNR synthesis [90], while in an early study, carbon–fluorine bond cleavage, despite its high dissociation energy, was observed during the cyclodehydrogenation of partially edge-fluorinated polyanthrylenes to form GNRs [53]. In future, on-surface reactions would emerge as an indispensable pathway to create attractive nanostructures. Collaboration with researchers from various research fields (organic synthesis, on-surface synthesis, theoretical calculations, and device fabrication) would accelerate the synthesis of new nanocarbon materials.

Acknowledgments

H.H. and H.Y. acknowledge JST CREST (No. JPMJCR15F1 [HY]), JST PRESTO (No. JPMJPR21AC [HH]), and Grants-in-Aid for Scientific Research (Nos. JP20H02816 (HH), JP20H00379 (HY), and JP20H05833 (HY)).

References

1 Kroto, H.W., Heath, J.R., O'Brien, S.C. et al. (1985). C_{60}: Buckminsterfullerene. *Nature* 318: 162–163.

2 Sariciftci, N.S., Smilowitz, L., Heeger, A.J., and Wudl, F. (1992). Photoinduced electron transfer from a conducting polymer to buckminsterfullerene. *Science* 258: 1474–1476.

3 Yu, G., Gao, J., Hummelen, J.C. et al. (1995). Polymer photovoltaic cells: enhanced efficiencies via a network of internal donor-acceptor heterojunctions. *Science* 270: 1789–1791.

4 Hummelen, J.C., Knight, B.W., LePeq, F. et al. (1995). Preparation and characterization of fulleroid and methanofullerene derivatives. *J. Org. Chem.* 60: 532–538.

5 Wienk, M.M., Kroon, J.M., Verhees, W.J.H. et al. (2003). Efficient methano[70]fullerene/MDMO-PPV bulk heterojunction photovoltaics cells. *Angew. Chem. Int. Ed.* 42: 3371–3375.

6 Iijima, S. (1992). Helical microtubules of graphitic carbon. *Nature* 354: 56–58.

7 Yu, M.-F., Lourie, O., Dyer, M.J. et al. (2000). Strength and breaking mechanism of multiwalled carbon nanotubes under tensile load. *Science* 287: 637–640.

8 Javey, A., Guo, J., Wang, Q. et al. (2003). Ballistic carbon nanotube field-effect transistors. *Nature* 424: 654–657.

9 Pop, E., Mann, D., Wang, Q. et al. (2006). Thermal conductance of an individual single-wall carbon nanotube above room temperature. *Nano Lett.* 6: 96–100.

10 Krupke, R., Hennrich, F., Löhneysen, H.V., and Kappes, M.M. (2003). Separation of metallic from semiconducting single-walled carbon nanotubes. *Science* 301: 344–347.

11 Novoselov, K.S., Geim, A.K., Morozov, S.V. et al. (2004). Electric field effect in atomically thin carbon films. *Science* 306: 666–669.

12 Talirz, L., Ruffieux, P., and Fasel, R. (2016). On-surface synthesis of atomically precise graphene nanoribbons. *Adv. Mater.* 28: 6222–6231.

13 Narita, A., Wang, X.-Y., Feng, X., and Müllen, K. (2015). New advances in nanographene chemistry. *Chem. Soc. Rev.* 44: 6616–6643.

14 Zhou, X. and Yu, G. (2020). Modified engineering of graphene nanoribbons prepared via on-surface synthesis. *Adv. Mater.* 32: 1905957.

15 Son, Y.-W., Cohen, M.L., and Louie, S.G. (2006). Energy gaps in graphene nanoribbons. *Phys. Rev. Lett.* 97: 216803.

16 Yang, L., Park, C.-H., Son, Y.-W. et al. (2007). Quasiparticle energies and band gaps in graphene nanoribbons. *Phys. Rev. Lett.* 99: 186801.

17 Dürkop, T., Getty, S.A., Cobas, E., and Fuhrer, M.S. (2004). Extraordinary mobility in semiconducting carbon nanotubes. *Nano Lett.* 4: 35–39.

18 Fujita, M., Wakabayashi, K., Nakada, K., and Kusakabe, K. (1996). Peculiar localized state at zigzag graphite edge. *J. Phys. Soc. Jpn.* 65: 1920–1923.

19 Li, X., Wang, X., Zhang, L. et al. (2008). Chemically derived, ultrasmooth graphene nanoribbon semiconductors. *Science* 319: 1229–1232.

20 Campos-Delgado, J., Romo-Herrera, J.M., Jia, X. et al. (2008). Bulk production of a new form of sp^2 carbon: crystalline graphene nanoribbons. *Nano Lett.* 8: 2773–2778.

21 Jiao, L., Zhang, L., Wang, X. et al. (2009). Narrow graphene nanoribbons from nanotubes. *Nature* 458: 877–880.

22 Kosynkin, D.V., Higginbotham, A.L., Sinitskii, A. et al. (2009). Longitudinal unzipping of carbon nanotubes to form graphene nanoribbons. *Nature* 458: 872–877.

23 Jiao, L., Wang, X., Diankov, G. et al. (2010). Facile synthesis of high-quality graphene nanoribbons. *Nat. Nanotechnol.* 5: 321–325.

24 Yang, X., Dou, X., Rouhanipor, A. et al. (2008). Two-dimensional graphene nanoribbons. *J. Am. Chem. Soc.* 130: 4216–4217.

25 Narita, A., Feng, X., Hernandez, Y. et al. (2014). Synthesis of structurally well-defined and liquid-phase-processable graphene nanoribbons. *Nat. Chem.* 6: 126–132.

26 Jordan, R.S., Wang, Y., McCurdy, R.D. et al. (2016). Synthesis of graphene nanoribbons via the topochemical polymerization and subsequent aromatization of a diacetylene precursor. *Chem* 1: 78–90.

27 Jordan, R.S., Li, Y.L., Lin, C.-W. et al. (2017). Synthesis of $N = 8$ armchair graphene nanoribbons from four distinct polydiacetylenes. *J. Am. Chem. Soc.* 139: 15878–15890.

28 Yang, W., Lucotti, A., Tommasini, M., and Chalifoux, W.A. (2016). Bottom-up synthesis of soluble and narrow graphene nanoribbons using alkyne benzannulations. *J. Am. Chem. Soc.* 138: 9137–9144.

29 Li, G., Yoon, K.-Y., Zhong, X. et al. (2018). A modular synthetic approach for band-gap engineering of armchair graphene nanoribbons. *Nat. Commun.* 9: 1687.

30 Daigle, M., Miao, D., Lucotti, A. et al. (2017). Helically coiled graphene nanoribbons. *Angew. Chem. Int. Ed.* 56: 6213–6217.

31 Jolly, A., Miao, D., Daigle, M., and Morin, J.-F. (2020). Emerging bottom-up strategies for the synthesis of graphene nanoribbons and related structures. *Angew. Chem. Int. Ed.* 132: 4652–4661.

32 Narita, A., Chen, Z., Chen, Q., and Müllen, K. (2019). Solution and on-surface synthesis of structurally defined graphene nanoribbons as a new family of semiconductors. *Chem. Sci.* 10: 964–975.

33 Cai, J., Ruffieux, P., Jaafar, R. et al. (2010). Atomically precise bottom-up fabrication of graphene nanoribbons. *Nature* 466: 470–473.

34 Blankenburg, S., Cai, J., Ruffieux, P. et al. (2012). Intraribbon heterojunction formation in ultranarrow graphene nanoribbons. *ACS Nano* 6: 2020–2025.

35 Treier, M., Pignedoli, C.A., Laino, T. et al. (2011). Surface-assisted cyclodehydrogenation provides a synthetic route towards easily processable and chemically tailored nanographenes. *Nat. Chem.* 3: 61–67.

36 Gross, L., Mohn, F., Moll, N. et al. (2009). The chemical structure of a molecule resolved by atomic force microscopy. *Science* 325: 1110–1114.

37 Gross, L., Mohn, F., Moll, N. et al. (2010). Organic structure determination using atomic resolution scanning prove microscopy. *Nat. Chem.* 2: 821–825.

38 Gross, L. (2011). Recent advances in submolecular resolution with scanning prove microscopy. *Nat. Chem.* 3: 273–278.

39 Sugimoto, Y., Pou, P., Abe, M. et al. (2007). Chemical identification of individual surface atoms by atomic force microscopy. *Nature* 446: 64–67.

40 van der Lit, J., Boneschanscher, M.P., Vanmaekelbergh, D. et al. (2023). Suppression of electron-vibron coupling in graphene nanoribbons contacted via a single atom. *Nat. Commun.* 2013: 4.

41 Ullmann, F. and Bielecki, J. (1901). Ueber Syntheses in der Biphenylreihe. *Ber. Dtsch. Chem. Ges.* 34: 2174.

42 Dong, L., Liu, P.N., and Lin, N. (2015). Surface-activated coupling reactions confined on a surface. *Acc. Chem. Res.* 48: 2765–2774.

43 Eichhorn, J., Nieckarz, D., Ochs, O. et al. (2014). On-surface Ullmann coupling: the influence of kinetic reaction parameters on the morphology and quality of covalent networks. *ACS Nano* 8: 7880–7889.

44 Clair, S. and de Oteyza, D.G. (2019). Controlling a chemical coupling reaction on a surface: tools and strategies for on-surface synthesis. *Chem. Rev.* 119: 4717–4776.

45 Sánchez-Sánchez, C., Dienel, T., Deniz, O. et al. (2016). Purely armchair or partially chiral: noncontact atomic force microscopy characterization of dibromo-bianthryl-based graphene nanoribbons grown on Cu(111). *ACS Nano* 10: 8006–8011.

46 de Oteyza, D.G., García-Lekue, A., Vilas-Varela, M. et al. (2016). Substrate-independent growth of atomically precise chiral graphene nanoribbons. *ACS Nano* 10: 9000–9008.

47 Jacobse, P.H., van den Hoogenband, A., Moret, M. et al. (2016). Aryl radical geometry determines nanographene formation on Au(111). *Angew. Chem. Int. Ed.* 55: 13052–13055.

48 Chen, Y.-C., de Oteyza, D.G., Pedramrazi, Z. et al. (2013). Tuning the band gap of graphene nanoribbons synthesized from molecular precursors. *ACS Nano* 7: 6123–6128.

49 Kawai, S., Saito, S., Osumi, S. et al. (2015). Atomically controlled substitutional boron-doping of graphene nanoribbons. *Nat. Commun.* 6: 8098.

50 Cloke, R.R., Marangoni, T., Nguyen, G.D. et al. (2015). Site-specific substitutional boron doping of semiconducting armchair graphene nanoribbons. *J. Am. Chem. Soc.* 137: 8872–8875.

51 Carbonell-Sanromà, E., Hieulle, J., Vilas-Varela, M. et al. (2017). Doping of graphene nanoribbons *via* functional group edge modification. *ACS Nano* 11: 7355–7361.

52 Nguyen, G.D., Toma, F.M., Cao, T. et al. (2016). Bottom-up synthesis of $N = 13$ sulfur-doped graphene nanoribbons. *J. Phys. Chem. C* 120: 2684–2687.

53 Hayashi, H., Yamaguchi, J., Jippo, H. et al. (2017). Surface-assisted graphene nanoribbon synthesis featuring carbon–fluorine bond cleavage. *ACS Nano* 11: 6204–6210.

54 Talirz, L., Söde, H., Dumslaff, T. et al. (2017). On-surface synthesis and characterization of 9-atom wide armchair graphene nanoribbons. *ACS Nano* 11: 1380–1388.

55 Giovannantonio, M.D., Deniz, O., Urgel, J.I. et al. (2017). On-surface growth dynamics of graphene nanoribbons: the role of halogen functionalization. *ACS Nano* 11: 1380–1388.

56 Talirz, L., Söda, H., Cai, J. et al. (2013). Termini of bottom-up fabrication graphene nanoribbons. *J. Am. Chem. Soc.* 135: 2060–2063.

57 Blanksby, S.J. and Ellison, G.B. (2003). Bond dissociation energies of organic molecules. *Acc. Chem. Res.* 36: 255–263.

58 Zhang, H., Lin, H., Sun, K. et al. (2015). On-surface synthesis of rylene-type graphene nanoribbons. *J. Am. Chem. Soc.* 137: 4022–4025.

59 Binnig, G., Frank, K.H., Fuchs, H. et al. (1985). Tunneling spectroscopy and inverse photoemission: image and field states. *Phys. Rev. Lett.* 55: 991.

60 Bombis, C., Ample, F., Lafferentz, L. et al. (2009). Single molecular wires connecting metallic and insulating surface areas. *Angew. Chem. Int. Ed.* 48: 9966–9970.

61 Kimouche, A., Ervasti, M.M., Drost, R. et al. (2015). Ultra-narrow metallic armchair graphene nanoribbons. *Nat. Commun.* 6: 10177.

62 Yamaguchi, J., Hayashi, H., Jippo, H. et al. (2020). Small bandgap in atomically precise 17-atom-wide armchair-edged graphene nanoribbons. *Commun. Mater.* 1: 36.

63 Deniz, O., Sánchez-Sánchez, C., Dumslaff, T. et al. (2017). Revealing the electronic structure of silicon intercalated armchair graphene nanoribbons by scanning tunneling spectroscopy. *Nano Lett.* 17: 2197–2203.

64 Kharche, N. and Meunier, V. (2016). Width and crystal orientation dependent band gap renormalization in substrate-supported graphene nanoribbons. *J. Phys. Chem. Lett.* 7: 1526–1533.

65 Sze, S.M. and Ng, K.K. (2007). *Physics of Semiconductor Devices*, 3e. Hoboken, NJ: Wiley.

66 Ruffieux, P., Wang, S., Yang, B. et al. (2016). Roman Fasel, on-surface synthesis of graphene nanoribbons with zigzag edge topology. *Nature* 531: 489–492.

67 Blackwell, R.E., Zhao, F., Brooks, E. et al. (2021). Spin splitting of dopant edge state in magnetic zigzag graphene nanoribbons. *Nature* 600: 647–652.

68 Anthony, J.E. (2007). The larger acenes: versatile organic semiconductors. *Angew. Chem. Int. Ed.* 47: 452–483.

69 Yamada, H., Okujima, T., and Ono, N. (2008). Organic semiconductors based on small molecules with thermally or photochemically removable groups. *Chem. Commun.* 2957–2974.

70 Suzuki, M., Aotake, T., Yamaguchi, Y. et al. (2014). Synthesis and photoreactivity of α-diketone-type precursors of acenes and their use in organic-device fabrication. *J. Photochem. Photobiol. C-Photochem. Rev.* 18: 50–70.

71 Yamada, H., Kuzuhara, D., Suzuki, M. et al. (2020). Synthesis and morphological control of organic semiconducting materials using the precursor approach. *Bull. Chem. Soc. Jpn.* 93: 1234–1267.

72 Yamada, H., Yamashita, Y., Kikuchi, M. et al. (2005). Photochemical synthesis of pentacene and its derivatives. *Chem. Eur. J.* 11: 6212–6220.

73 Strating, J., Zwanenburg, B., Wagenaar, A., and Udding, A.C. (1969). Evidence for the expulsion of bis-CO from bridged α-diketones. *Tetrahedron Lett.* 10: 125–128.

74 Mondal, R., Shah, B.K., and Neckers, D.C. (2006). Photogeneration of heptacene in a polymer matrix. *J. Am. Chem. Soc.* 128: 9612–9613.

75 Tönshoff, C. and Bettinger, H.F. (2010). Photogeneration of octacene and nonacene. *Angew. Chem. Int. Ed.* 49: 4125–4128.

76 Shen, B., Tatchen, J., Sanchez-Garcia, E., and Bettinger, H.F. (2018). Evolution of the optical gap in the acene series: Undecacene. *Angew. Chem. Int. Ed.* 57: 10506–10509.

77 Watanabe, M., Chang, Y.J., Liu, S.-W. et al. (2012). The synthesis, crystal structure and charge transport properties of hexacene. *Nat. Chem.* 4: 574–578.

78 Jiang, D. and Dai, S. (2008). Electronic ground state of higher acenes. *J. Phys. Chem. A* 112: 332–335.

79 Urgel, J.I., Mishra, S., Hayashi, H. et al. (2019). On-surface light-induced generation of higher acenes and elucidation of their open-shell character. *Nat. Commun.* 10: 861.

80 Zugermeier, M., Gruber, M., Schmid, M. et al. (2017). On-surface synthesis of heptacene and its interaction with a metal surface. *Nanoscale* 9: 12461–12469.

81 Krüger, J., Pavliček, N., Alonso, J.M. et al. (2016). Tetracene formation by on-surface reduction. *ACS Nano* 10: 4538–4542.

82 Krüger, J., Eisenhut, F., Alonso, J.M. et al. (2017). Imaging the electronic structure of on-surface generated hexacene. *Chem. Commun.* 53: 1583–1586.

83 Krüger, J., García, F., Eisenhut, F. et al. (2017). Decacene: on-surface generation. *Angew. Chem. Int. Ed.* 56: 11945–11948.

84 Eisenhut, F., Kühne, T., García, F. et al. (2020). Dodecacene generated on surface: Reopening of the energy gap. *ACS Nano* 14 (1): 1011–1017.

85 Zuzak, R., Dorel, R., Krawiec, M. et al. (2017). Nonacene generated by on-surface dehydrogenation. *ACS Nano* 11: 9321–9329.

86 Zuzak, R., Dorel, R., Kolmer, M. et al. (2018). Higher acenes by on-surface dehydrogenation: from heptacene to undecacene. *Angew. Chem. Int. Ed.* 57: 10500–10505.

87 Urgel, J.I., Hayashi, H., Giovannantonio, M.D. et al. (2017). On-surface synthesis of heptacene organometallic complexes. *J. Am. Chem. Soc.* 139: 11658–11661.

88 Urgel, J.I., Giovannantonio, M.D., Gandus, G. et al. (2019). Overcoming steric hindrance in aryl-aryl homocoupling via on-surface copolymerization. *ChemPhysChem* 20: 2360–2366.

89 Eimre, K., Urgel, J.I., Hayashi, H. et al. (2021). On-surface synthesis and characterization of nitrogen-substituted undecacenes. *Nat. Commun.* 13: 511.

90 Panighel, M., Quiroga, S., Brandimarte, P. et al. (2020). Stabilizing edge fluorination in graphene nanoribbons. *ACS Nano* 14: 11120–11129.

13

A Branch of Zintl Chemistry: Metal Clusters of Group 15 Elements

Yu-He Xu[1], Nikolay V. Tkachenko[2], Alvaro Muñoz-Castro[3], Alexander I. Boldyrev[2] and Zhong-Ming Sun[1]

[1] State Key Laboratory of Elemento-Organic Chemistry, School of Materials Science and Engineering, Nankai University, Tianjin, 300350 China

[2] Department of Chemistry and Biochemistry, Utah State University, 0300 Old Main Hill, Logan, UT, 84322-0300, USA

[3] Grupo de Química Inorgánica y Materiales Moleculares, Facultad de Ingeniería, Universidad Autonoma de Chile, El Llano Subercaseaux, Santiago 2801 Chile

13.1 Introduction

Zintl clusters of group 15, characterized by fruitful structures and fascinating bonding patterns, play a special role in Zintl chemistry. Unlike group 14 elements, which easily form cage compounds with borane-like structures, the coordination modes of group 15 elements are more challenging and their aggregations tend to be constructed from small building blocks into larger clusters. It has been nearly one century since Eduard Zintl first recognized the presence of As_7^{3-} and Sb_7^{3-} in liquid ammonia by potentiometric titrations [1, 2]. The two anions are isostructural and are the most basic units in group 15 clusters acting as versatile precursors. With the introduction of cation-sequestering agents and polar solvents, a number of homoatomic polyanions have been discovered and provided access to construct a wider range of Zintl clusters. In this chapter, some representative Zintl-type clusters containing group 15 elements will be presented briefly, including their syntheses, structural characterizations, and bonding analyses. Particularly, the concept of aromaticity was applied to metalloid and metal cluster systems, which allowed a more comprehensive understanding of the stability of some Zintl-type clusters and aroused extensive interest among chemists. Some clusters involving aromaticity and anti-aromaticity will be highlighted here as well.

13.1.1 Homoatomic Group 15 Clusters

Most mixed clusters consisting of group 15 elements and transition metals were synthesized from homoatomic Zintl anions. As noted, Pn_7^{3-} cages (Pn = P-Bi) with nortricyclane-like configuration are one of the common homoatomic group 15 clusters [3–7]. In addition, Pn_2^{2-} (Pn = Sb-Bi) with a naked double bond [8, 9], Pn_4^{2-} (Pn = P-Bi) with square-planar structure [3, 10, 11], cyclic Pn_6^{4-} (Pn = P-As) [12–14], and polycyclic Pn_{11}^{3-} (Pn = P-Bi) anions have been observed and employed for further reactions as well [15–18]. Many of these anions could be readily isolated from the solutions of corresponding binary Zintl phase composed of pnictogens and alkali metals, whereas a few anions only exist in liquid ammonia such as Pn_4^{2-} and Pn_6^{4-} (Pn = P-As), which limits their

Atomically Precise Nanochemistry, First Edition. Edited by Rongchao Jin and De-en Jiang.

further applications. Additionally, two unusual antimony clusters – crown-like Sb_8^{8-} and Sb_{10}^{2-} fused by two Sb_7 units through the four Sb atoms – were discovered later and have not been used widely as precursors until now [8, 19].

Among the elements in group 15, bismuth as a heavier congener, is known to have a tendency to form chain oligomers Bi_2^{2-} and Bi_4^{2-} instead of polyanions. It was not until recent years that Bi_7^{3-} was synthesized by Sevov and coworkers using K_5Bi_4 with (C_6H_6) $Cr(CO)_3$ in pyridine [6]. Bi_{11}^{3-} was obtained by Dehnen's group in 2013 by dissolving crystals of $[K(2.2.2\text{-crypt})]_2(GaBi_3){\cdot}en$ in pyridine solution [18]. Since then, the missing family members of group 15 polyanions have been discovered by employing uncommon synthetic approaches. Some selected homoatomic group 15 clusters are shown in Figure 13.1, and many of the intermetalloid and heterometallic clusters described in this chapter are formed with these anions as precursors. Additionally, the P-containing Zintl clusters have been less reported in recent years and the reactivity of P_4 with transition metals has been reviewed, so phosphorus clusters will not be discussed in this chapter [20, 21].

13.1.2 Bonding Concepts

Early explorations on the relationship between structure and chemical bonding of Zintl anions and related species were originated by Eduard Zintl in 1939 and then developed by Wilhelm Klemm in 1958, together combined into the Zintl-Klemm concept [22]. The Zintl-Klemm concept addresses electron-precise structures composed of two-center two-electron (2c–2e) bonds, in which each atom satisfies a full electron octet. Consequently, the concept was extended to the Mooser-Pearson (8-N) rule, where N refers to the valence number of more electronegative atoms in the cluster [23]. The majority of homoatomic group 15 anions just mentioned are electron-precise structures, suitable for the Zintl-Klemm concept. Taking Pn_7^{3-} as an example, the geometry is made of four 3-connected and three 2-connected Pn atoms. The three negative charges carried by the whole cluster are distributed among the three 2-connected atoms so that the outermost shell of each atom is filled with eight electrons. However, the limitations of these simple models emerged for the lack of accounting for polyhedral clusters dominated by multicenter bonds. At that point, Wade-Mingos rules – which were generated initially to rationalize the valence electron counting and chemical

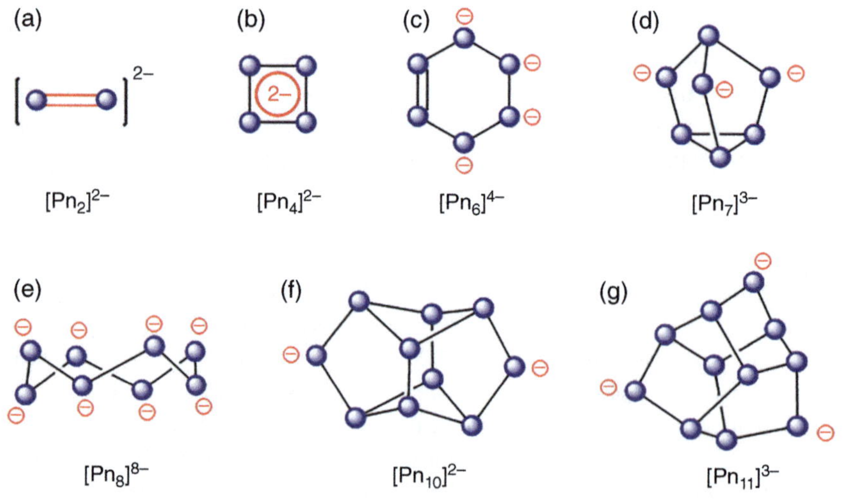

Figure 13.1 Selected examples of known homoatomic group 15 clusters; (a) Pn = Sb-Bi [8, 9]; (b) Pn = P-Bi [3, 10, 11]; (c) Pn = P-As [12–14]; (d) Pn = P-Bi [3–7]; (e) Pn = Sb [19]; (f) Pn = Sb [8]; (g) Pn = P-Bi [15–18].

structure of polyhedral boranes – were introduced into Zintl chemistry [24–27]. According to Wade-Mingos rules, a *closo*-type cluster with n vertices has 2n + 2 skeletal electrons (SE), and a *nido* cluster is derived from the *closo* parent by removal of one vertex with SE counts of 2n + 4. By analogy, *arachno* and *hypho* cages possess 2n + 6 and 2n + 8 SEs, respectively, with two and three missing vertices. Actually, many cage compounds of group 14 fit into Wade-Mingos rules very well, such as *closo*-E_{10}^{2-} (E = Ge, Pb) and *nido*-E_9^{4-} (E = Si-Pb) [28–33], whereas the situation is a little bit more complicated in the group 15 clusters with complex coordination. A minority of heterometallic group 15 clusters with formal polyhedral structure can be explained by Wade-Mingos rules, such as pentagonal bipyramidal *closo*-$[Bi_3Ni_4(CO)_6]^{3-}$ and bisdisphenoidal *closo*-$[Bi_4Ni_4(CO)_6]^{2-}$ [34]. Some clusters are related to vertex, edge, and face fusion, while others with multicenter bonds can be rationalized by the aromaticity concept prominently.

13.1.3 Aromaticity in Zintl Chemistry

The term *aromaticity*, was originated in 1865 to explain the special stability in cyclic hydrocarbon molecules. To date, the concept has been developed in the field of inorganic chemistry and numerous all-metal aromatic rings, including Ga_3^{2-} and Al_4^{2-}, have been reported, which has aroused extensive interest [35–38]. It was found that the geometries and electron structures of some Zintl-type clusters also can be clarified in terms of aromaticity. For instance, the square anions Pn_4^{2-} (Pn = P-Bi) and the planar pentagonal rings As_5^-, Sn_5^{6-}, and Pb_5^{6-} can be regarded as inorganic analogues of aromatic hydrocarbons $[C_4H_4]^{2-}$ and $[C_5H_5]^-$ [39–41]. Furthermore, with the spatial delocalization of electrons, the concept of aromaticity can be extended from two-dimensional planar hydrocarbons to three-dimensional polyhedral clusters. It has been reported that the primary origin of the stability of polyhedral boranes lies in their 3D aromaticity. In particular, $[B_{12}H_{12}]^{2-}$ and $[B_6H_6]^{2-}$ possess a significant degree of aromaticity [42, 43]. Among the known homoatomic clusters, some group 14 anions are isoelectronic with *closo* polyhedral boranes, such as E_{10}^{2-} (E = Ge, Pb) and E_{12}^{2-} (E = Sn, Pb) and have been classified into the same 3D aromatic system [28, 29, 44–46]. It follows that the aromaticity concept may be further applied in heterometallic and intermetalloid clusters and some representative examples of group 15 anions with aromaticity or anti-aromaticity will be introduced in this chapter.

A very useful tool in the study of aromatic properties of Zintl clusters is the adaptive natural density partitioning (AdNDP) algorithm [47, 48]. The AdNDP is a localization algorithm that can represent a chemical bonding pattern in terms of both Lewis bonding elements (lone pairs (1c—2e) and two-center two-electron (2c—2e) bonds) as well as delocalized bonding elements (such as nc—2e [n > 2] bonds), allowing to describe highly delocalized electrons associated with the concept of aromaticity. This technique has been widely used in describing chemical bonding patterns of various metal-containing clusters [49–51]. Using the AdNDP method, we can build chemically intuitive and vivid chemical bonding patterns employing the concept of electron pair as the main element of chemical bonding models. For instance, the chemical bonding in the previously mentioned Sn_5^{6-} species can be represented in terms of classical Lewis 2c—2e bonds and three delocalized 5c—2e bonds responsible for the aromatic properties of the compound (Figure 13.2a). Moreover, we can notice a striking similarity between chemical bonding elements of Sn_5^{6-} and $C_5H_5^-$ species (Figure 13.2b), agreeing with the aromatic description of the flat Zintl ion.

For the theoretical analysis of Zintl clusters, the ionic limit model is often used [49–51]. The model assumes that electrons from the donating alkali-metals or alkali-metal cryptand complexes are completely transferred on the anionic species. This model allows researchers not to consider bulk counterions in the quantum chemical calculation which significantly reduces the computational complexity of the calculation. This model produces reliable and meaningful results, and the

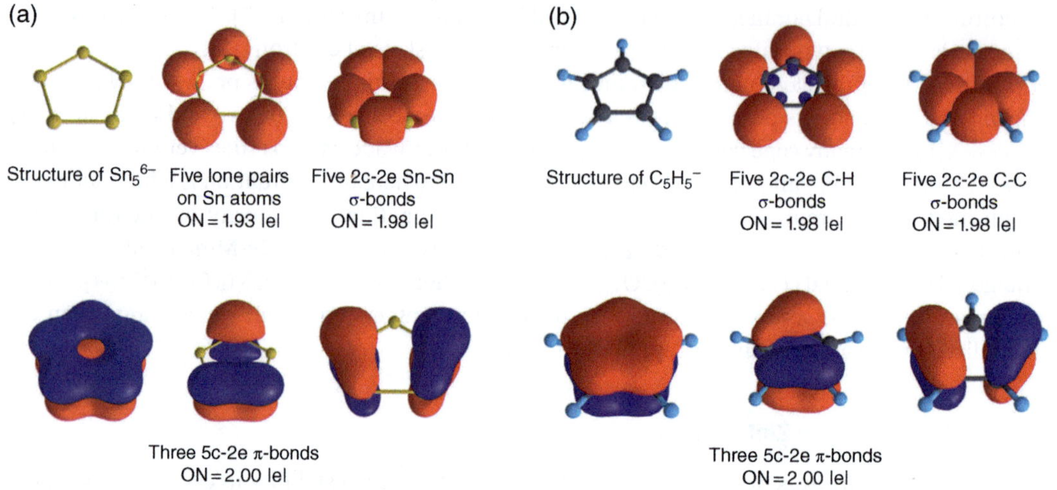

Figure 13.2 Chemical bonding pattern of Sn_5^{6-} (a) and $C_5H_5^-$ (b) obtained from the AdNDP analysis. ON stands for occupation number and is equal to 2.0|e| in an ideal case.

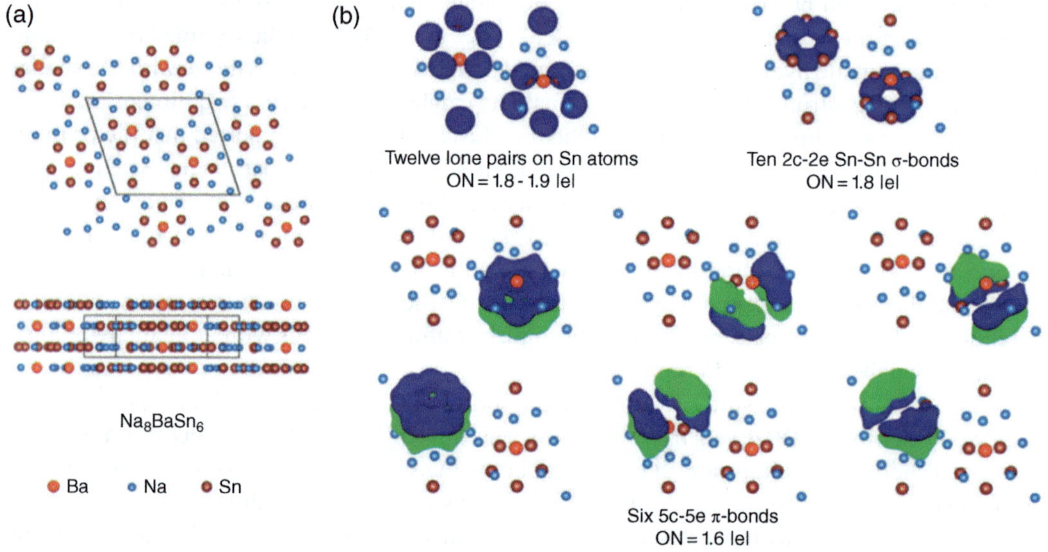

Figure 13.3 Structure (a) and SSAdNDP (b) chemical bonding pattern for the Na_8BaSn_6 Zintl phase. The unit cell is shown in black. *Source:* Reproduced from Ref. [52].

chemical bonding analysis for isolated anion coincides with the chemical bonding of a periodic structure surrounded by the counterions. To illustrate that, we can consider the example of Na_8BaSn_6 Zintl phase that was synthesized by Todorov and Sevov in 2004 [41]. The title compound contains flat pentagonal rings of Sn_5 surrounded by the alkali and alkaline-earth metals. The chemical bonding pattern of the Na_8BaSn_6 was investigated with the solid state adaptive natural density partitioning (SSAdNDP) algorithm [52], which is the extension of the AdNDP method to periodic systems. The chemical bonding pattern of Sn_5 units was found to be similar to the chemical bonding pattern of the isolated Sn_5^{6-} (Figure 13.3) showing the reliability of the ionic limit model for chemical bonding analysis in Zintl clusters.

13.2 Complex Coordination Modes in Arsenic Clusters

Arsenic resides in the midway position of the pnictogen family, combining the properties of both the nonmetallic and metallic members of group 15 and conferred by this position, the range of its bond energies lies in an intriguing situation. The bond energies of arsenic are significantly higher than those of phosphorus, which always produce energy barriers to thermal conversions that prevent the separation of intermediates. Simultaneously, they are much lower than the bond energies of heavier analogies in which almost only thermodynamically favorable products will be isolated [53]. Therefore, not only is the structural chemistry varied in metal-coordinated arsenic, but also oligomer fragments are easily produced in the reaction process related to polyarsenic systems. In Zintl chemistry, polyarsenic anions combined with transition metals create many unprecedented structures and have enriched arsenic chemistry greatly.

As mentioned in the introduction, nortricyclane-like Pn_7^{3-} anions act as the most widely used Zintl precursors of group 15. Generally, two identical Pn_7^{3-} tend to be linked by one transition metal in which the central ion coordinated by the two Pn_7^{3-} cages in η^4 mode, such as $[(Pn_7)M(Pn_7)]^{4-}$ (Pn = P/As, M = Zn/Cd; Pn = Sb, M = Zn; Pn = Bi, M = Cd) [54–57]. However, the situation is more complicated in arsenic-containing clusters. In addition to the above configuration, two As_7^{3-} cages have been also observed to be respectively connected by two mutually perpendicular units, M-As (M = Zn, Hg) and Hg—Hg [55, 56, 58]. Taking $[ZnAs_{15}]^{3-}$ shown in Figure 13.4a as an example, the Zn—As unit combined with four As atoms, like an open book, acts as a junction of two As_7 cages and the whole structure is the same as P_{16}^{2-}. As for $[Hg_2As_{14}]^{4-}$ [58], the Hg—Hg bond (2.68 Å) is like a rope holding two As_7^{3-} units together with slightly shorter As—As bonds (2.355–2.488 Å) than those of $[ZnAs_{15}]^{3-}$ (2.367–2.495 Å).

In 1998, Eichhorn and co-workers reported the first isolated free binary transition metal pnictide ion $[MoAs_8]^{2-}$, which can be viewed as a molybdenum atom in the center of a regular S_8-like As_8^{8-} ring as shown in Figure 13.4b [59]. Previously, an isostructural unit $[NbAs_8]^{3-}$ had been found in a one-dimensional chain complex ${}^{1}_{\infty}[Rb\cdot NbAs_8]^{3-}$ but not a free anion [64]. Later, free

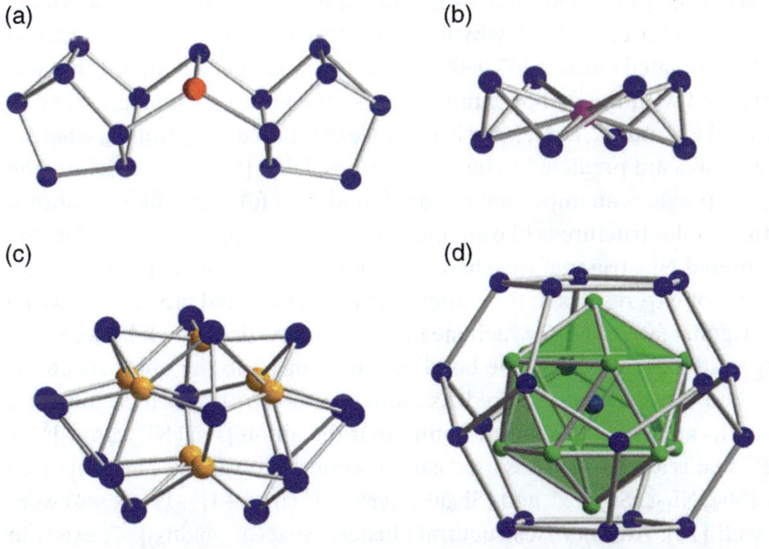

Figure 13.4 (a) $[MAs_{15}]^{3-}$ (M = Zn, Hg) [55, 56]; (b) $[M@As_8]^{q-}$ (M/q = Mo/2, Cr/3, Nb/3) [59–61]; (c) $[Pd_7As_{16}]^{4-}$ [62]; (d) $[As@Ni_{12}@As_{20}]^{3-}$ [63]. The global colors are as follow: As = blue, Zn/Hg = red, Mo/Cr/Nb = purple, Pd = orange, Ni = green.

$[NbAs_8]^{3-}$ and $[CrAs_8]^{3-}$ were also synthesized, both of which contain an eight-coordinate metal atom in the center [60, 61]. The As—As bonds in $[MoAs_8]^{2-}$ and $[CrAs_8]^{3-}$ (2.423–2.437 Å) are slightly shorter than those in $[NbAs_8]^{3-}$ (2.445–2.453 Å). It is worth noting that $[MoAs_8]^{2-}$ is a 16 e^- diamagnetic complex, whereas $[CrAs_8]^{3-}$ adopts a 17 e^- paramagnetic electron structure. The $As_8{}^{8-}$ ring is a novel exception to the hydrocarbon coordination analogy and its heavier analogy $Sb_8{}^{8-}$ was observed independently in liquid ammonia by the direct reduction of antimony. Similarly, isostructural $[MSb_8]^{q-}$ (M/q = Mo/3, Nb/3) clusters were obtained as well, with the average Sb—Sb bond length of 2.783 Å.

Besides $[NbAs_8]^{3-}$ and $[MoAs_8]^{2-}$ (Figure 13.4b), $[Pd_7As_{16}]^{4-}$ (Figure 13.4c) and $[Pd_2As_{14}]^{4-}$ were the second examples for transition metal binary anions of group 15 elements [62, 65]. The two As_7 units here have made some subtle changes, i.e. they were reduced to $As_7{}^{5-}$ with norbornadiene-like configuration and bound to a $Pd_2{}^{6+}$ dimetal center in a μ, η^2, η^2 mode. The formal Pd^{3+} ions reside in distorted square planar coordination situations and are linked by a Pd—Pd bond of 2.714 Å. $[Pd_7As_{16}]^{4-}$ is another product of the same material as $[Pd_2As_{14}]^{4-}$ ($As_7{}^{3-}$ and $Pd[PCy_3]_2$) consisting of one distorted capped trigonal prismatic Pd_7 core, two As_5 rings, two As_2 units, and two single As atoms. The Pd—As bonds fall in the range of 2.435–2.651 Å and As—As bond lengths vary from 2.410 to 2.495 Å, both kinds of bonds being slightly longer than those of $[Pd_2As_{14}]^{4-}$. The electron attributions are as follows: 2 $As_5{}^{1-}$, 2 $As_2{}^{2-}$, and 2 As^{3-} with 6 trigonal prismatic Pd^+ ions and one capped Pd^{2+}. Four years later, a similar structure of heavier analogy $[Ni_5Sb_{17}]^{4-}$ was reported [66], which was intermixed crystalized with an unclear compound $[K(2.2.2\text{-crypt})]_4[(Sb_7)_2Ni_x(Ni_2Sb_2)]$ ·2en. In $[Ni_5Sb_{17}]^{4-}$, a Ni(*cyclo*-Ni_4Sb_4) ring sits inside a Sb_{13} bowl and the Ni_5 unit can also be regarded as the removal of one trigonal prismatic edge from the Pd_7 unit [66].

$[As@Ni_{12}@As_{20}]^{3-}$ is a landmark cluster in Zintl chemistry [63]. Highly connected and symmetric compounds generate great interest among chemists due to their wonderful structures, unusual stability, and reactivity. "Fullerene" carbon clusters, which originated from C_{60} and later covered hollow cages of carbon with globe, cylinder, or tube, are one of the representatives of such species. In the past period of time, the synthesis of inorganic fullerenes received extensive interest, but only a few experimental results have been published, such as In_{74} in the $Na_{96}In_{97}Z_2$ phases (Z = Ni, Pd, Pt), and Si_{20} in the $Cs_8Na_{16}Si_{136}$ phase, both of which are components in three-dimensional extended solids but not solution-isolated clusters [67, 68]. Among the pnictogen elements, the situation is much different contrasted with the carbon family because the group 15 analogies tend to form 2c—2e bonds. The observed P_{20} and N_{20} are relatively unstable compared to P_4 and N_2, whereas the heavier analogies As_{20} and Sb_{20} are predicted to be stable and isolable [69–71]. Therefore, the synthesis of $[As@Ni_{12}@As_{20}]^{3-}$ provides an important research model for fullerene-like structures. As shown in Figure 13.4d, the whole structure is like an onion-skin: an As_{20} pentagonal dodecahedron shell packing an As-centered Ni_{12} trigonal icosahedron, together exhibiting I_h point symmetry. Interestingly, every Ni atom of Ni_{12} resides in the center of every pentagonal face of As_{20} while each As atom of As_{20} caps a trigonal face of Ni_{12}, which means there is a vertice switch between the icosahedron and the pentagonal dodecahedron. The bond distances of Ni to the outer As atoms (2.388–2.404 Å) are shorter than those to the centered As atom (2.546–2.570 Å), which indicates strong interactions between Ni_{12} and As_{20}. It is worth noting that the anion $[As@Ni_{12}@As_{20}]^{3-}$ is crystallized as a salt of Bu_4P^+ that is a relatively less used cation sequestering agent. Later reported similar onion-skin clusters $[Sb@Ni_{12}@Sb_{20}]^{-/+}$ and $[Sb@Pd_{12}@Sb_{20}]^n$ (n = +1, −1, −3, −4) were crystallized with Bu_4P^+ as well [72]. Another isostructural cluster, $[Sn@Cu_{12}@Sn_{20}]^{12-}$, exists in the ternary phase $A_{12}Cu_{12}Sn_{21}$ (A = Na, K) [73].

The publication of these works aroused extensive interest among theoretical chemists. Both Zhao and Baruah et al. reanalyzed the structural as well as electronic properties of $[As@Ni_{12}@As_{20}]^{3-}$

and obtained a similar conclusion on the results of geometry optimization, total binding energy, and the LUMO-HOMO gap, in which they confirmed the stabilization on the formation of $[As@Ni_{12}@As_{20}]^{3-}$ [74, 75]. Furthermore, due to the extended Hückel theory with low-level accuracy used by original authors, Zhao et al. recalculated the symmetry of HOMO and LUMO employing plane-wave based DFT. It turned out that the HOMO and LUMO of $[As@Ni_{12}@As_{20}]^{3-}$ are not threefold-degenerate with t_{1u} symmetry as predicted by Moses et al., but fivefold-degenerate with h_u and h_g symmetries, respectively. From another perspective, Alvaro et al. applied the concept of superatom to understand the bonding and antibonding interaction between the concentric shells for such a three-layered cluster. The charge distribution of $[As@Ni_{12}@As_{20}]^{3-}$ reveals that the central atom and the outer dodecahedron display a closed-shell configuration that is given by the $[\{As@Ni_{12}\}^{3-}@\{As_{20}\}^{0}]$ form, with the inner-core displaying a $1S^21P^62S^2$ superatomic electron configuration [76].

Analysis of the magnetic response of $[As@Ni_{12}@As_{20}]^{3-}$ allows to evaluate its aromatic characteristics, similar to $[Sn@Cu_{12}@Sn_{20}]^{12-}$ [77]. From the magnetic criteria of aromaticity, aromatic species enable a shielding region at the center of the structure under an external magnetic field. The use of a single nuclear-independent-chemical-shift probe (NICS) for evaluating aromaticity cannot be employed on endohedral clusters with a central atom, where the analysis of the magnetic behavior outside the cage is a useful approach [78, 79]. In this sense, aromatic clusters sustain a shielding surface following the cage backbone for the orientational averaged term (Iso), which is a three-dimensional array of NICS points. In addition, under a specific orientation of the external field, the shielding cone property is revealed as a direct connection between planar and three-dimensional aromatic compounds [79–81]. For $[As@Ni_{12}@As_{20}]^{3-}$, a homogenous shielding region following the icosahedral cage is obtained, suggesting an aromatic character for the overall cluster. Interestingly, under a specific orientation of the external field, a long-ranged shielding cone property is enabled along the z-, x- and y-axis as representative orientations (Figure 13.5a). At 5.3 Å from the center of the structure, the shielding effect amounts to −20 ppm, i.e. about 1.0 Å over the outer As_{20} dodecahedron radius, which is of −10 ppm at 7.2 Å from the center −5 ppm at 9.2 Å (Figure 13.5b), and retaining the strength of −1 ppm at 15 Å. Hence, it is a spherical aromatic multilayered cluster of ~4 Å radius leading to a large aromatic spot.

13.3 Antimony Clusters with Aromaticity and Anti-Aromaticity

The polyantimonides catch the attention of chemists due to their tendency to undergo disproportionation processes with transition metals and facile structural rearrangements, compared with their lighter congeners [7]. In the field of Zintl chemistry, many reported Sb-containing clusters are involved in the terms of aromaticity and anti-aromaticity, so the application of the aromaticity concept in Zintl related clusters will be elucidated in this section. A representative all-metal aromatic sandwich $[Sb_3Au_3Sb_3]^{3-}$ was reported by Sun and coworkers in 2015 [82]. One Au_3 sheet is jammed between two Sb_3 rings, building up a ligand-free triangular prismatic structure with pseudo-D_{3h} symmetry (Figure 13.6a). The Au—Au bond distances fall in a narrow range of 2.918–2.954 Å, comparable well to typical Au—Au bonds but shorter than the observed Au_3 rings in $[(LAu)_6(N_2)]^{2+}$ (3.244 Å *a.v.*; L = $PPh_2{}^iPr$), which indicates a more "solid" golden triangle [87]. The Sb—Sb interactions span a narrow range between 2.849 and 2.882 Å and are almost equilateral triangles. Six Au—Sb bonds acting as pillars of the triangular prism connect three sheets and the bond lengths are close to each other (2.592–2.609 Å), significantly shorter than common Au—Sb bonds. The publication of $[Sb_3Au_3Sb_3]^{3-}$ received extensive attention among theoretical scientists, and its electronic structure, as well as chemical bonding, still remains greatly controversial. In the original

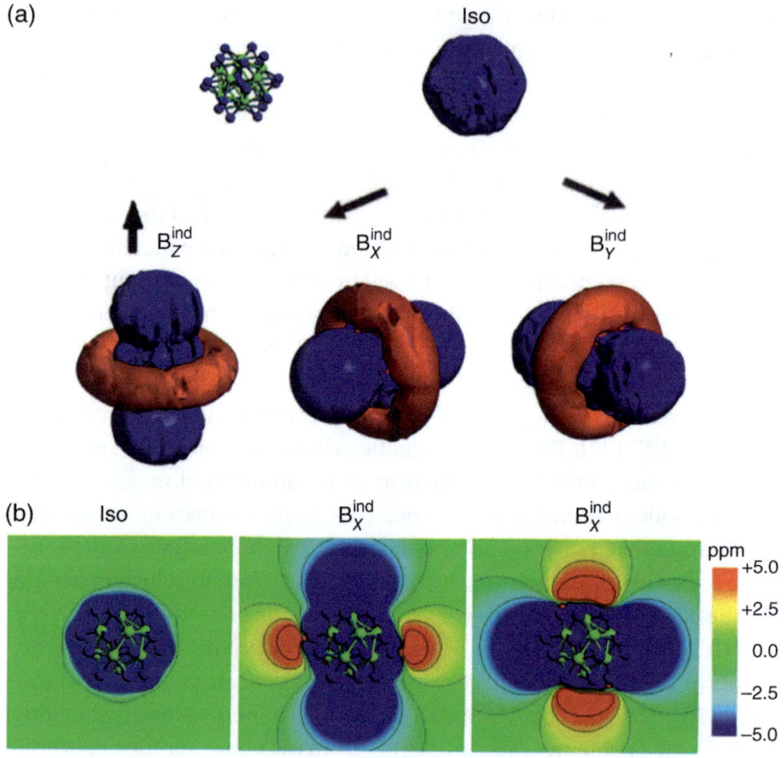

Figure 13.5 Magnetic response properties obtained for $[As@Ni_{12}@As_{20}]^{3-}$, given by the isotropic term (Iso), and from specific orientations of the external field (B_z^{ind}, B_x^{ind}, and B_y^{ind}), as (a) isosurfaces at ±5 ppm and (b) contour plot. Blue surface: Shielding; Red surface: Deshielding.

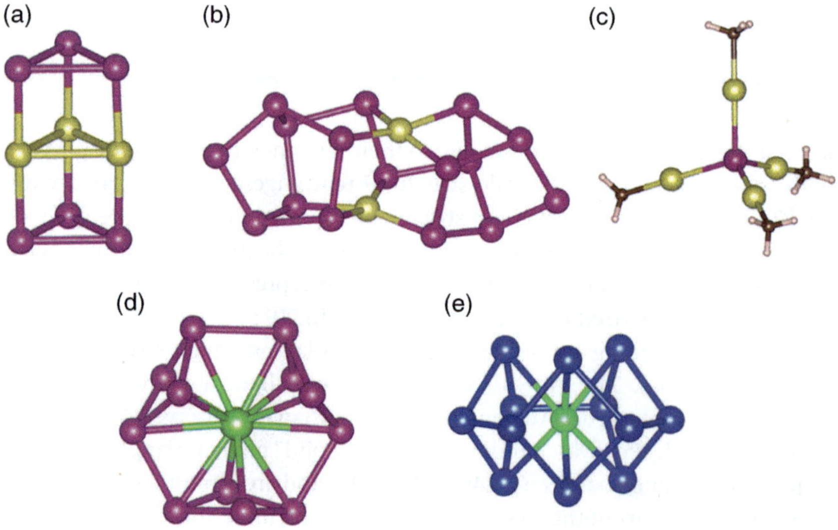

Figure 13.6 (a) $[Sb_3Au_3Sb_3]^{3-}$ [82]; (b) $[Au_2Sb_{16}]^{4-}$ [83]; (c) $[Sb(AuMe)_4]^{3-}$ [84]; (d) $[Ln(\eta^4\text{-}Sb_4)_3]^{3-}$ (Ln = La, Y, Ho, Er, Lu) [85]; (e) $[Th@Bi_{12}]^{4-}$ [86]. Global colors are as follows: Sb = dark pink, Au = gold, Ln = green, Bi = blue, Th = light green, C = brown, H = light pink.

article, DFT calculations revealed the π-aromaticity of $[Sb_3Au_3Sb_3]^{3-}$. According to the results, HOMO/HOMO' and HOMO-3 exhibit delocalized π-characters, which are mainly attributed to the *cyclo*-Sb_3 units. These π MOs are delocalized over the two Sb_3 rings carrying the three extra charges of the whole cluster, which exactly offset the Sb to Au σ-donation. Consequently, each Sb_3 ring achieves a total of 15 valence electrons and has a set of delocalized three-center three-electron (3 c—3 e) π-bonds. To some extent, the 3c—3e π-bonds are equivalent to a typical 3c—4e ππ* triplet system, and the latter is known to be aromatic on the basis of the reversed 4n Hückel rule for aromaticity in a triplet state, thus $[Sb_3Au_3Sb_3]^{3-}$ can also be considered as π aromatic [88]. The year after the publication of this article, Li et al. performed theoretical calculations on $[Sb_3Au_3Sb_3]^{3-}$ using the AdNDP method [89]. They considered that the primary stability of $[Sb_3Au_3Sb_3]^{3-}$ comes from the interlayer σ interactions between the Sb 5p π orbitals of the Sb_3 rings and the vertical Au 5d6s hybrid orbitals of Au_3. As a new perspective to be put forward, they held the opinion that there are three 3c—2e σ-delocalized bonds on each Sb_3 ring, resulting in a dual σ-aromaticity on $[Sb_3Au_3Sb_3]^{3-}$.

The favorable bonding pattern from the multidecker $[Sb_3Au_3Sb_3]^{3-}$ cluster denotes three aromatic rings fragments, which exhibits an interesting magnetic behavior given by the cumulative of three shielding cones [80, 81], as denoted by the B_z^{ind} term (Figure 13.7a). This behavior results from the aromatic characteristics of Sb_3^{3-} rings and their bonding contribution toward the 5d6s orbitals of Au_3. From the contour plot (Figure 13.7b) it is possible to identify three complementary in-plane deshielding short-ranged regions centered at each stacked ring, supporting the view of three aromatic rings within the $[Sb_3Au_3Sb_3]^{3-}$ cluster.

One year after $[Sb_3Au_3Sb_3]^{3-}$ was reported, another all-metal aromatic cluster $[Au_2Sb_{16}]^{4-}$ was observed [83]. As shown in Figure 13.6b, two Au atoms inlay on the outer Sb_{16} rod-like cage in η^4-coordination fashion. Two Au atoms combined with respective surrounding four Sb atoms form

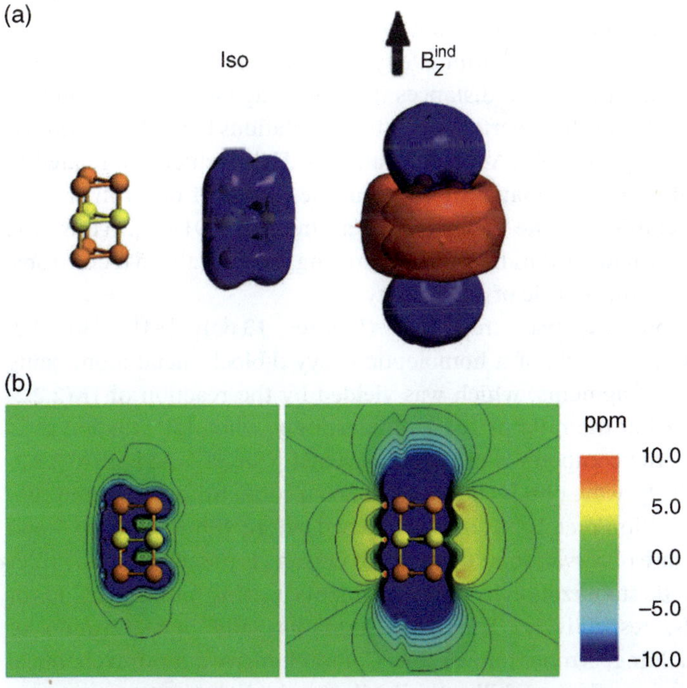

Figure 13.7 Magnetic response properties obtained for $[Sb_3Au_3Sb_3]^{3-}$, given by the isotropic term (Iso), and from a perpendicular to the rings external field (B_z^{ind}), as (a) isosurfaces at ±5 ppm and (b) contour plot.

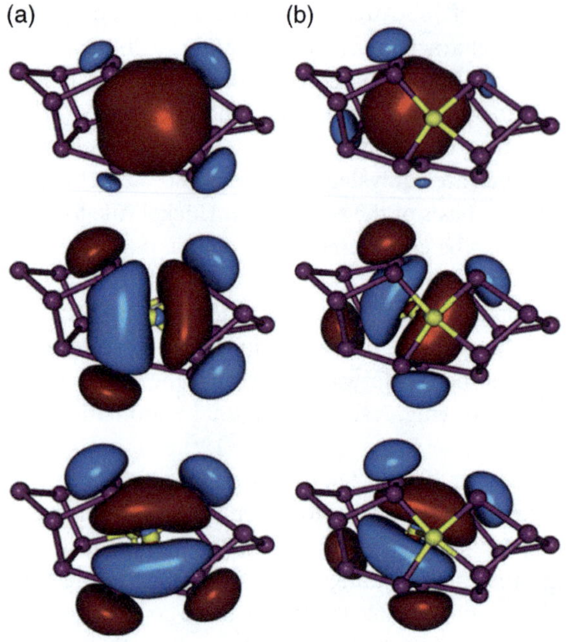

Figure 13.8 Delocalized chemical bonding elements of $[Au_2Sb_{16}]^{4-}$. Six 5c-2e delocalized σ-bonds with ON = 1.86–1.99 |e| found at the upper (a) and lower (b) $AuSb_4$ fragments. *Source:* Reproduced from Ref. [83].

two $AuSb_4$ planes, which are almost parallel to each other. Eight Au—Sb bonds fall in a markedly narrow range of (2.69 ± 0.02 Å), implying the possibility of some special interactions. According to the results of chemical bonding analysis performed by AdNDP method, 19 classical 2c—2e σ-bonds are found on Sb—Sb interactions and three 5c—2e σ-bonds are found within each $AuSb_4$ plane (Figure 13.8). The three delocalized bonds on each $AuSb_4$ unit are distributed as one complete bonding and two occupying one nodal plane. Such a bonding pattern is similar to typical σ-aromatic bonds in gas-phase clusters [90]. Therefore, local aromaticity exists in the $[Au_2Sb_{16}]^{4-}$ cluster. However, the question comes from that the Sb—Sb distances in $AuSb_4$ fragment are not equal as expected for the aromatic systems, so the authors performed extra calculations to verify the aromaticity within the $AuSb_4$ unit. Two simple models $[AuSb_4H_8]^-$ and $[Sb_4H_8]^{2-}$ which are related to $AuSb_4$, were established and found to be σ-aromatic. In addition, the results of electron-sharing indices (I_{ring} and MCIs) further confirmed the existence of aromaticity [91–93]. Thus, the $[Au_2Sb_{16}]^{4-}$ anion possesses two σ-aromatic $AuSb_4$ fragments featuring six delocalized σ electrons, each of which separately satisfies the Hückel rule of aromaticity.

In 2021, another Au—Sb complex was reported (Figure 13.6c) [84]. $[K(2.2.2\text{-crypt})]_3[Sb(AuMe)_4]$·py represents an example of a homoleptic heavy *p*-block metal atom being surrounded by four transition metal fragments, which was yielded by the reaction of $[K(2.2.2\text{-crypt})]_2(Sn_2Sb_2)$·en with $[AuMePPh_3]$ at the ratio of 1 : 2.2. Interestingly, when the reaction ratio changes from 1 : 2.2 to 2 : 1 and 1 : 1, the compounds of $[K(2.2.2\text{-crypt})]_3[(Sn_2Sb_2)_2Au]$ or $[K(2.2.2\text{-crypt})]_4[(Sn_5Sb_3Au)_2]$ ·2py, respectively, were obtained [94, 95]. Such a configuration that the central Sb atom is attached by four transition metals has been described before, $[Sb\{Fe(CO)_4\}_4]^{3-}$ and $[Sb\{Co(CO)_3PPh_3\}_4]^+$ [96, 97]. The rare observation of this kind of clusters is possible due to rather labile bonding, which required kinetic stabilization by bulky fragments bearing four CO ligands or three CO and one PPh_3 ligands, respectively. Thus, the formation and crystallization of $[Sb(AuMe)_4]^{3-}$, with little steric shielding, is more remarkable. Comprehensive quantum chemical studies showed that electron correlation effects stabilize this labile tetrahedral structure.

$[Ln(\eta^4\text{-}Sb_4)_3]^{3-}$ (Ln = La, Y, Ho, Er, Lu) is another important series of antimony clusters, reported in the same year with $[Au_2Sb_{16}]^{4-}$ [85]. It was obtained by the reaction of $Ln(benzyl)_3(THF)_3$ with

K_5Sb_4 in pyridine solution. As shown in Figure 13.6d, the lanthanide ion is surrounded by three *cyclo*-Sb_4 units. There are two types of Ln-Sb bonds, six in equatorial positions ranging from 3.4338(5) to 3.4735(5) Å and another six with shorter distances of 3.2386(5) to 3.2634(5) Å. Sb—Sb distances can be divided into two species as well: the bonds in Sb_4 rings are relatively short with the range of 2.8088–2.8339 Å and the other Sb—Sb bonds linking *cyclo*-Sb_4 fall in the range between 3.0179 and 3.0517 Å. A simple electron count indicated that the Ln atom carries a positive charge of 3+ and, spontaneously, each Sb_4 unit carries a negative charge of 2^-.

To enhance the understanding of the bonding in $[Ln(\eta^4\text{-}Sb_4)_3]^{3-}$, AdNDP analysis was performed. It revealed that there are 12 localized 2c—2e σ bonds (Figure 13.9b) on the three Sb_4 rings, three 3c—2e σ bonds (Figure 13.9c) responsible for the bonding between the three separate *cyclo*-Sb_4 units as well as for the interaction of La with the equatorial Sb atoms and six 5c—2e π bonds on three $LaSb_4$ fragments (Figure 13.9d). The two 5c—2e π bonds on each Sb_4 unit are akin to the two 4c—2e π bonds in the neutral Sb_4 cluster with π-antiaromaticity according to Breslow's 4n rule [98]. Therefore, it can be concluded that there are three π-antiaromatic Sb_4 units in $[Ln(\eta^4\text{-}Sb_4)_3]^{3-}$ cluster. As for the two charges carried by each *cyclo*-Sb_4 already mentioned, each Sb_4 unit approximately provides two electrons to form 3c—2e σ bonds, thus leaving only four electrons for the π framework on the Sb_4 unit, which is not in conflict with the 4n rule. Therefore, the $[Ln(\eta^4\text{-}Sb_4)_3]^{3-}$ clusters are the first all-metal antiaromatic compounds observed in the condensed phase.

In 2021, Dehnen et al. reported a topologically related heavy-metal cluster $[Th@Bi_{12}]^{4-}$ shown in Figure 13.6e [86], which was obtained by $[ThCp^{\#}{}_3Cl]$ ($Cp^{\#} = C_5Me_4H$) reacting with ternary solid-phase $K_5Ga_2Bi_4$ in ethylenediamine (en) solution. Interestingly, according to their results of DFT calculations, $[Th@Bi_{12}]^{4-}$ exhibits substantial π-aromaticity with a ring current much stronger

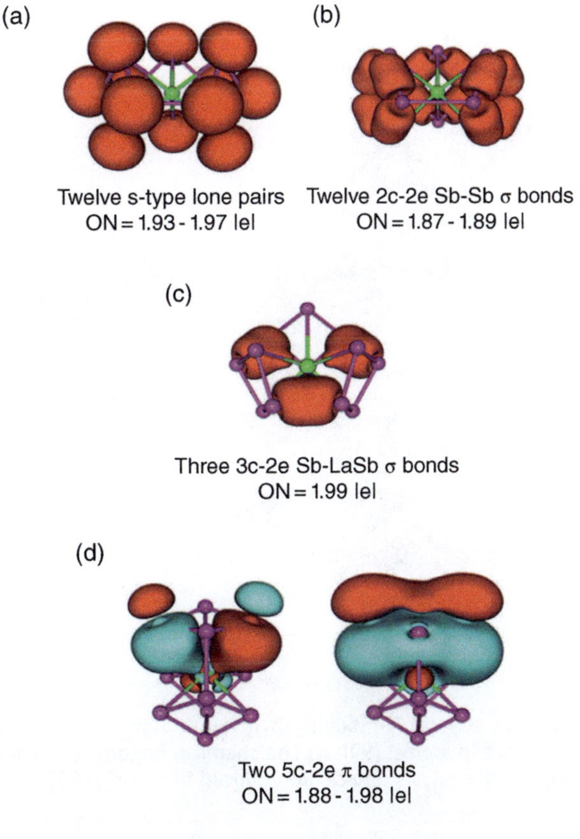

Figure 13.9 Chemical bonding pattern of $[Ln(\eta^4\text{-}Sb_4)_3]^{3-}$ obtained from the AdNDP analysis, (a) twelve s-type lone pairs (1c−2e bonds) on twelve Sb atoms; (b) twelve 2c−2e Sb−Sb σ bonds; (c) three 3c−2e σ bonds; (d) two 5c−2e π bonds in one *cyclo*-Sb_4 unit. *Source:* Reproduced from ret. [85].

than that of benzene (6π) and equivalent to that of porphine (26π), which is completely different from the results of theoretical calculations on $[Ln(\eta^4\text{-}Sb_4)_3]^{3-}$. There are two types of Bi—Bi bonds observed as well, slightly shorter Bi—Bi distances between Bi_4 rings, and longer Bi—Bi bonds within Bi_4 rings, which are opposite to the structural properties in $[Ln(\eta^4\text{-}Sb_4)_3]^{3-}$. Furthermore, all Bi—Bi bond lengths span a narrower range of 3.0402–3.132 Å than that of $[Ln(\eta^4\text{-}Sb_4)_3]^{3-}$, which can be connected to the equalization of bond lengths for aromatic molecules. To probe the aromaticity in $[Th@Bi_{12}]^{4-}$, the authors carried out a localization procedure for the 34 valence molecular orbitals, and calculated the magnetically induced current density and NICS values, all of which collectively confirm the existence of π-aromaticity.

Among the Zintl related clusters, organic isomers are an important class. As mentioned in Section 1.1, many homoatomic group 15 clusters are isostructural with cycloalkanes, such as nortricyclane-like $Sb_7{}^{3-}$ and trishomocubane-like $Sb_{11}{}^{3-}$. A heterometallic compound $[K(2.2.2\text{-crypt})]_4[In_8Sb_{13}]$ was reported in 2019, which contained a 1 : 1 mixture of $[Sb@In_8Sb_{12}]^{3-}$ and $[Sb@In_8Sb_{12}]^{5-}$ [99]. The former anion displays a perfect T_h symmetry (Figure 13.10a) and is isomorphic with dodecatetrahedrene $C_{20}H_{12}$, while the penta-anion is C_s-symmetric (Figure 13.10b). Different from most intermetalloid clusters embedded with transition metals, $[K(2.2.2\text{-crypt})]_4[In_8Sb_{13}]$ is composed of the shell with two main group elements and one Sb atom in the center of the cluster. The synthetic method is different from most mixed main-group clusters. Many heterometallic clusters made up of main group atoms are obtained by directly extracting Zintl phase solids. For instance, $[Sn_5Sb_3]^{3-}$ is extracted from K_2SnSb and $[Ge_4Bi_{14}]^{4-}$ is crystallized in the solution of K_2GeBi [101, 102]. However, $[K(2.2.2\text{-crypt})]_4[In_8Sb_{13}]$ is prepared by the reaction of K_5Sb_4 and $In(benzyl)_3$, which provides a new idea for the synthesis of Zintl clusters

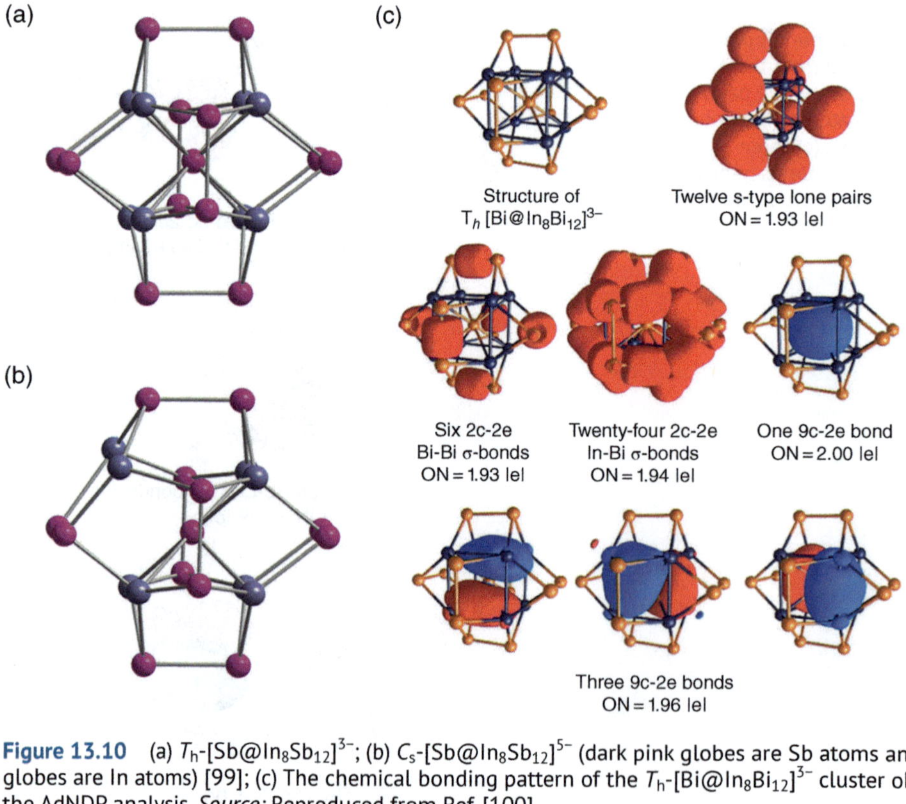

Figure 13.10 (a) T_h-$[Sb@In_8Sb_{12}]^{3-}$; (b) C_s-$[Sb@In_8Sb_{12}]^{5-}$ (dark pink globes are Sb atoms and gray purple globes are In atoms) [99]; (c) The chemical bonding pattern of the T_h-$[Bi@In_8Bi_{12}]^{3-}$ cluster obtained from the AdNDP analysis. *Source:* Reproduced from Ref. [100].

comprising main group elements. The $[Sb@In_8Sb_{12}]^{3-}$ anion possesses a single Sb atom in the core of the In_8 cube, whereas the $[Sb@In_8Sb_{12}]^{5-}$ cluster exhibits a distortion in the In_8 cube, leading to a near-planar In_2Sb_3 five-membered ring. Noticeably, the In—In interactions as well as the distances of In atoms to the central Sb atom in T_h-$[Sb@In_8Sb_{12}]^{3-}$ both span a very narrow range of 3.433–3.476 Å and 2.977–2.995 Å, respectively, which may imply the existence of aromaticity.

Subsequently, a heavier analogy $[Bi@In_8Bi_{12}]^{3-/5-}$ was reported by Sun and co-workers, and the spherical aromaticity was revealed in the cubic $Bi@In_8$ unit [100]. According to the results of AdNDP analysis (Figure 13.10c), there are 12 1c—2e s-type lone pairs on Bi atoms and 30 2c—2e Bi—Bi and In—Bi σ-bonds. The remaining 8 electrons form four delocalized 9c—2e bonds, responsible for the binding interaction in the cubic $Bi@In_8$ unit, which indicates spherical aromaticity on the basis of $2^* (N+1)^2$ electron counting rule with N = 1. Furthermore, the NICS value and induced magnetic field were studied, both of which agree with the overall concept of spherically aromatic description of the investigated clusters. This year, Dehnen et al. presented a similar compound, $[K(2.2.2\text{-crypt})]_{3.67}[Bi@Ga_8Bi_{12}]$, by reacting $K_5Ga_2Bi_4$ with $[La(C_5Me_4H)_3]$ [103].

To support the spherical aromaticity depicted by the bonding pattern in $[Bi@In_8Bi_{12}]^{3-/5-}$ clusters, their magnetic behavior was obtained (Figure 13.11), which sustain a shielding surface for the orientational averaged term (Iso), ascribed to the cage backbone. Moreover, under specific orientations of the external field, a shielding cone property is enabled [79–81], supporting the spherical aromatic behavior of $[Bi@In_8Bi_{12}]^{3-/5-}$. In contrast to planar aromaticity, which exhibits a shielding cone only under a perpendicular applied field, spherical aromatic species are able to sustain the

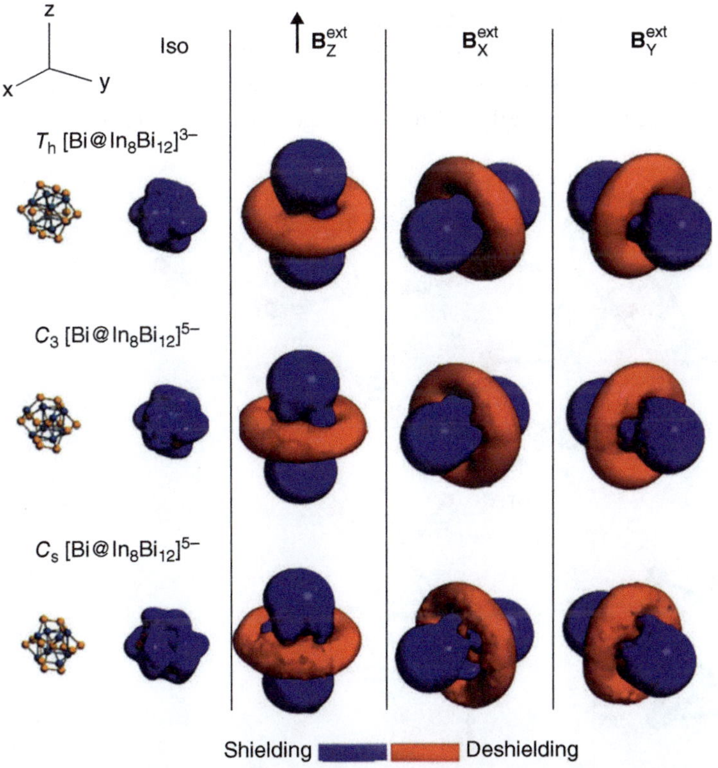

Figure 13.11 Magnetic response properties obtained for $[Bi@In_8Bi_{12}]^{3-/5-}$ clusters, given by the isotropic term (Iso), and from specific orientations of the external field (B_z^{ind}, B_x^{ind}, and B_y^{ind}), as isosurfaces at ±2 ppm. *Source:* Reproduced from Ref. [100].

shielding cone property in any orientation according to the external magnetic field [104]. In addition, the obtained magnetic behavior for $[Bi@In_8Bi_{12}]^{3-/5-}$ is similar to $[As@Ni_{12}@As_{20}]^{3-}$, owing to their spherical aromatic characteristics, despite their different symmetries and architectures.

13.4 Recent Advances in Bismuth-Containing Compounds

Compared with the lighter analogies of pnictogen elements, much less activity has emerged in the synthesis of metal-rich clusters of bismuth. Most assemblies of Bi atoms with transition metals were found in polycations in previous years, such as $[CuBi_8]^{3+}$, $[(Bi_8)Au(Bi_8)]^{5+}$, and $[Rh@Bi_9]^{4+}$ [105–107]. However, access to Bi-rich polyanions is relatively blocked, as the common precursors $Bi_7{}^{3-}$ and $Bi_{11}{}^{3-}$ were not observed until 2014, more than 20 years after their lighter analogies were reported. Until recently, a number of Bi-rich polyanions were synthesized and characterized, greatly improving the research progress of Bi clusters. Xu et al. carried out a series of works reacting K_5Bi_4 with carbonyl compounds of group 9 elements (Co, Rh, Ir) and obtained four heterometallic clusters, $[Bi_7M_3(CO)_3]^{2-}$ (M = Co, Rh), $[(\eta^3\text{-}Bi_3)_2(IrCO)_6(\mu^4\text{-}Bi)_3]^{3-}$, $[Rh@Bi_{10}(RhCO)_6]^{3-}$, and $[Rh@Bi_9(RhCO)_5]^{3-}$ from 2019 to 2020. As exhibited in Figure 13.12a, $[Bi_7Co_3(CO)_3]^{2-}$ and $[Bi_7Rh_3(CO)_3]^{2-}$ can be described as the three five-membered ring of $Bi_7{}^{3-}$ capped by three MCO carbonyl fragments, leading to the formation of 10-vertex deltahedral hybrids [108]. Obviously, due to the insertion of MCO fragments in η^5-coordination, the Bi_7 unit is significantly expanded with the apical-linking Bi—Bi distances (av. 3.395(1) Å) and the triangular basal separations (av. 3.4008(9) Å) in $[Bi_7Rh_3(CO)_3]^{2-}$ markedly longer than those in $Bi_7{}^{3-}$. With an eye to the whole cluster carrying −2 charge, the three MCO fragments collectively account for +1 positive charge considering the presence of $Bi_7{}^{3-}$. So it is concluded that in the reaction progress of K_5Bi_4 and monovalent MCO^+ species ($CpCo(CO)_2$ and $Rh(CO)_2(acac)$), $Bi_4{}^{5-}$ is oxidized to $Bi_7{}^{3-}$ and simultaneously $M(CO)^+$ is reduced to $[M(CO)]_3{}^+$ with an average oxidation state of +0.333.

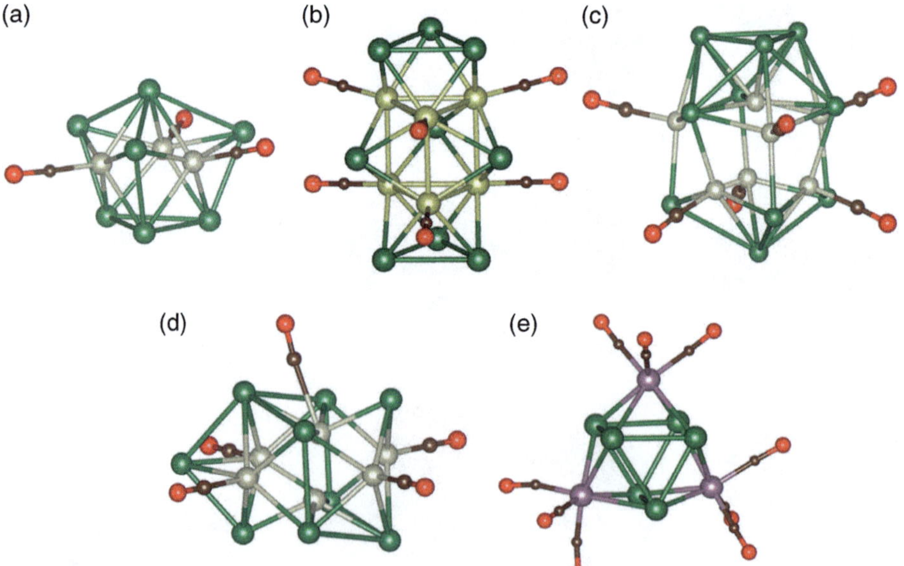

Figure 13.12 (a) $[Bi_7M_3(CO)_3]^{2-}$ (M = Co, Rh) [108]; (b) $[(\eta^3\text{-}Bi_3)_2(IrCO)_6(\mu^4\text{-}Bi)_3]^{3-}$ [109]; (c) $[Rh@Bi_{10}(RhCO)_6]^{3-}$; (d) $[Rh@Bi_9(RhCO)_5]^{3-}$ [110]; (e) $[Bi_6Mo_3(CO)_9]^{4-}$ [111]. The global colors are as follows: Bi = blackish green, Co/Rh = light gray, Ir = light blonde, Mo = light purple, C = brown, O = red.

DFT calculations were performed on $[Bi_7Rh_3(CO)_3]^{2-}$, which revealed that there are two types of nine 3c—2e bonds. Three of them consist of two basal Bi atoms and one Rh atom, and the other six bonds are formed by the apical Bi atom, one linking Bi atom and Rh atoms. The normalized 3c—2e bond orders are comparable to that of benzene, indicating correspondingly strong interactions and their stabilization toward the deltahedral hybrids.

The clusters comprising Ir atoms are quite rare in Zintl chemistry; only a *closo* type cluster $[E_9Ir(cod)]^{3-}$ (E = Sn, Pb) and an intermetalloid $[Ir@Sn_{12}]^{3-}$ have been reported in 2010 [112, 113]. However, in $[(\eta^3\text{-}Bi_3)_2(IrCO)_6(\mu^4\text{-}Bi)_3]^{3-}$ (Figure 13.12b), a trigonal prism composed of Ir-CO fragments is stabilized by coordinated Bi triangles and separate Bi atoms, using K_5Bi_4 and $Ir(CO)_2(acac)$ [109]. The compounds combining pnictide elements (Pn_x) with transition metal carbonyl complexes $M_y(CO)_z$ have been reported more than 400 times, such as $[(\mu^4\text{-}Bi)_4Co_9(CO)_{16}]^{2-}$ and $[(\mu^3\text{-}Bi)\,Ir_3(CO)_8(PPh_3)]$ [114, 115]. In contrast to neutral naked pnictide atoms, those composed of Zintl anions of pnictide and polyhedral $M_y(CO)_z$ cluster units are extremely rare, which may be due to the lack of cations to stabilize the highly negatively charged Zintl ions. Taking these into consideration, the employment of $Ir(CO)_2(acac)$ with a monovalent $[Ir(CO)_2]^+$ moiety is necessary so the positive charges can balance the surrounding negatively charged Bi_x^{x-}; the electron-deficient property makes it tend to form polyhedral $[M(CO)]_y^{y+}$ to share electrons further. As expected, the trigonal prismatic $[Ir(CO)]_6^{6+}$ subunit is aggregated from replacing $acac^-$ and carbonyl ligands of $Ir(CO)_2(acac)$ with three capping $\mu^4\text{-}Bi^-$ and two *cyclo*-Bi_3^{3-} and the entire cluster carries a total of −3 charges. According to Wade–Mingos rules and Polyhedral Skeletal Electron Pair Theory (PSEPT), for a mixed main group (E)–transition metal (M) *closo*-deltahedron $E_{n1}M_{n2}$ (n1: the number of E vertices, n2: the number of M vertices), each time a transition metal atom is substituted with the main group atom, the total cluster valence electron (CVE) count falls by 10, which leads to a CVE formula of $14(n1+n2) - 10n1+2 = 4n1+14n2+2$. Furthermore, a condensed triangular face is considered to give a decrease of 48 CVE since each shared metal atom contributes 16 electrons. $[(\eta^3\text{-}Bi_3)_2(IrCO)_6(\mu^4\text{-}Bi)_3]^{3-}$ can be regarded as a condensation of two $[Bi_3(IrCO)_3]$ octahedra and one middle $[(IrCO)_6(\mu^4\text{-}Bi)_3]$ tricapped trigonal prism by sharing two Ir_3 triangular faces. Thus, theoretically, the number of its CVE is $(14\times3+4\times3+2)\times2+(14\times6+4\times3+2) - 2\times48 = 114$, which is exactly in accord with the number of available CVE: 9×5 (Bi)$+6\times9$ (Ir)$+6\times2$ (CO)$+3 = 114$.

Inspired by the work of $[(\eta^3\text{-}Bi_3)_2(IrCO)_6(\mu^4\text{-}Bi)_3]^{3-}$, Xu et al. used a similar organometallic compound $Rh(CO)_2(acac)$ at 60 °C to yield a 15-vertex $[Rh@Bi_9(RhCO)_5]^{3-}$ and $Rh_2(CO)_4Cl_2$ at room temperature to yield a 16-vertex $[Rh@Bi_{10}(RhCO)_6]^{3-}$ [110]. When these two $^+$Rh-CO precursors were reacted with K_5Bi_4, the K^+-Cl^- ion pairing effect prefers the Cl-substitution with Bi_n^{n-} compared with the K^+-$acac^-$, so the $Rh(CO)_2(acac)$ precursor needs a higher reaction temperature to break the chelating Rh—O bonds. In $[Rh@Bi_{10}(RhCO)_6]^{3-}$ (Figure 13.12c), one crown-like Bi_6 embedding a Rh atom and one pyramidal Bi_4 unit are combined by five Rh-CO fragments. As for $[Rh@Bi_9(RhCO)_5]^{3-}$ (Figure 13.12d), it can be described as a Bi_9 boat composed of a trimer Bi_3, two Bi_2 dinuclear fragments, and two weakly bonded Bi atoms in which a Rh_6 cluster is inlaid. According to the results of Löwdin population analysis [116], in both compounds, each Rh-CO unit gains approximately one electron and the naked Rh atoms gain slightly more than one electron. Hence, it is rational to attribute the charge of −1 to the naked Rh atoms, which precisely achieve the d^{10} electron configuration for the Rh- ions. In addition, the AdNDP as well as the Mayer bond order (MBO) and Fuzzy bond order (FBO) analyses were carried out. There are obviously short Bi—Bi and Rh—Bi bonds in both clusters, and in support of these observations, the calculated bond orders are substantially positive, implying the presence of substantial bonding. In the cluster $[Rh@Bi_{10}(RhCO)_6]^{3-}$, the short Bi—Rh distances (2.64 Å) as well as the Bi_6 crown are associated with 3c—2e bonds and relatively longer ones (2.83 Å) with 4c—2e bonds. While in the

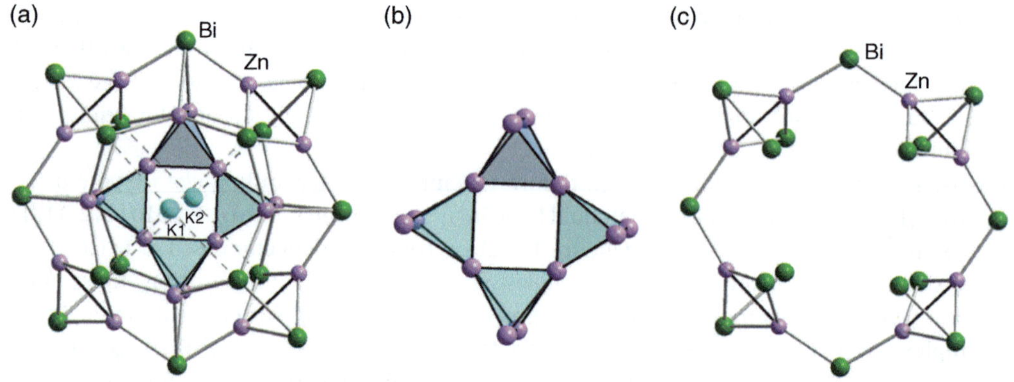

Figure 13.13 (a) $[K_2Zn_{20}Bi_{16}]^{6-}$; (b) $[Zn_{12}]$ unit core; (c) $[Zn_8Bi_{16}]^{q-}$ shell [117]. The global colors are as follows: K = blue, Bi = green, Zn = purple.

cluster $[Rh@Bi_9(RhCO)_5]^{3-}$, the short Rh—Bi bond with 2.54 Å has a 2c—2e bond and the one with 2.50 Å is surrounded by one 4c—2e bond and one 5c–2e bond.

In 2020, a ligand-free Bi-rich nanocluster $[K_2Zn_{20}Bi_{16}]^{6-}$ (Figure 13.13a) was discovered, in which one 24-atom polymetallide ring embeds a $[Zn_{12}]$ unit (Figure 13.13b) [117]. The formation of low-valent zinc compounds has long been of great interest to chemists, for the purpose of breaking the high stability of +II oxidation state zinc complexes to enhance their chemical reactivity. Up to now, a few low-valent and mixed-valent zinc compounds have been reported, including Cp*Zn–ZnCp* with Zn (I) atoms, Cp*Zn–Zn–ZnCp* with mixed Zn (0) and Zn (I) atoms and an inverse sandwich-type cluster dimer $\{[K_2ZnSn_8(ZnMes)]_2\}^{4-}$ with mixed-valence Zn (I)/Zn (II) [49, 118, 119]. Cluster in this work simultaneously exhibits a zero-valent situation at the central four Zn atoms and aggregation of 12 Zn atoms. The entire cluster can be divided into one 24-atom Zn—Bi shell and one assembled zinc tetrahedron core. The four zinc tetrahedra are corner-sharing that form a nearly undistorted inner Zn_4 square with a short inner edge of 2.544 Å. And it is revealed by the natural population analysis (NPA) that there are charges of +0.05 for the inner four Zn atoms and + 0.71 for the other eight Zn atoms. The latter is a typical value for an oxidation state of +I, so this unit as a whole is clearly low valent. As for the $[Zn_8Bi_{16}]^{q-}$ unit (Figure 13.13c), its topology is reminiscent of the organic macrocycle porphine, both of which possess a 24-atom ring and have a similar number of *s* and *p* valence electrons. To probe if there is aromaticity in $[Zn_8Bi_{16}]^{q-}$ units being the same as porphine, the authors calculated the magnetically induced current density of the $[Zn_8Bi_{16}]^{8-}$ cluster based on the magnetic criterion (the negative charges of 8- referred to a heavy metal inorganic mimics of porphine $[Hg_8Te_{16}]^{8-}$ and the structure optimization consequences of $[Zn_{12}]$ unit [120]). The cluster keeps a net diatropic ring current of 0.43–7.0 nA/T, which is about a fifth of the ring current calculated for porphine (25.4 nA/T at the same computational level) and about a third of the value calculated for benzene (11–15 nA/T). Furthermore, the NICS analysis was employed on $[Zn_8Bi_{16}]^{8-}$ and $[K_2Zn_8Bi_{16}]^{6-}$, revealing negative NICS values of −4.4 and − 4.2 ppm, respectively. Thus, the weak aromaticity of $[Zn_8Bi_{16}]^{q-}$ was confirmed.

In contrast to the diverse polybismuth structures existing in the intermetalloid cluster cations, most reported Bi-rich heterometallic cluster anions are composed of common Bi_7^{3-} cage or Bi_3, Bi_2 fragments. Recently, a novel $[Bi_6Mo_3(CO)_9]^{4-}$ cluster containing a distorted Bi_6 triangular prism was reported by Sun and co-workers [111]. By adjusting the amount of organometallic compound $(MeCN)_3Mo(CO)_3$, two different clusters were obtained, $[Bi_6Mo_3(CO)_9]^{4-}$ shown in Figure 13.12e with 2 eq. and $[Bi_3Mo_2(CO)_6]^{3-}$ which has been reported in 2001 with 1 eq. [121]. The Bi_6^{4-} in $[Bi_6Mo_3(CO)_9]^{4-}$ exhibits a markedly distorted trigonal prismatic geometry with two elongated prism heights (3.3087(6) and 3.5554(7) Å), indicating four weakly bonded Bi atoms. Three $Mo(CO)_3$ fragments capping on the near-square planes of Bi_6^{4-} form a Ge_9^{2-}- like cage. However, compared to 20-electron Ge_9^{2-}, the $[Bi_6Mo_3(CO)_9]^{4-}$ cluster has 22 skeleton electrons ($3\times6+4 = 22$), the same as Ge_9^{4-} cluster. As the DFT calculations performed on the three clusters, the HOMO of $[Bi_6Mo_3(CO)_9]^{4-}$ (or Ge_9^{4-}) and the LUMO of Ge_9^{2-} are π-bonding within the triangular bases but σ-antibonding between them. The two more electrons in $[Bi_6Mo_3(CO)_9]^{4-}$ (or Ge_9^{4-}) occupy the LUMO of Ge_9^{2-} leading to the two prolonged prism heights. Hence, the reported cluster can be considered as a nonclassical "*closo*-type" cluster with 22 skeleton electrons. In addition, analyses of the principal interacting orbitals (PIOs) [122] manifested that each Bi atom participates in the formation of a 3c–2e σ bond, while each Mo atom forms three 5c–2e σ bonds with four Bi atoms of a Bi_4Mo unit. Consequently, every Bi_3 triangle carries two electrons, and each Bi_4Mo unit is assigned with six electrons, both indicating the presence of σ-aromaticity on the basis of Hückel rule with $n = 0$ and $n = 1$, respectively. The combination of electron density of delocalized bonds (EDDB) analysis, induced magnetic field (Bind), and $NICS_{iso}$ indices calculation further confirm the multiple local σ-aromatic characteristic of $[Bi_6Mo_3(CO)_9]^{4-}$.

13.5 Ternary Clusters Containing Group 15 Elements

Ternary clusters are usually obtained from the reaction of binary precursors with the transition metal, lanthanide, or actinide organometallic compounds. Up to now, the most widely used binary Zintl anion precursors consist of group 15 elements. Zintl phase complexes, such as K_8SnSb_4 or KPbBi, are dissolved in ethylenediamine solution to crystalize the salt of $[Sn_2Sb_2]^{2-}$ or $[Pb_2Bi_2]^{2-}$ as materials for further reactions [123–125]. Binary 9-atom Zintl anions, like $[Sn_7Bi_2]^{2-}$ and $[Pb_7Bi_2]^{2-}$, are generally considered as byproducts in the oxidative process from 4-atom Zintl precursors to larger ternary clusters [126, 127]. Because they contain more kinds of atom species and diverse coordination modes, the configurations of ternary clusters possess more abundant variations than binary clusters. But there is a common occupational disorder problem in the crystal structure since the existence of two *p*-block elements in ternary clusters often contributes to several energetically similar isostructural isomers. Thus, electrospray ionization mass spectrometry (ESI-MS) and energy dispersive X-ray (EDX) are used to determine the formula of the cluster, while quantum chemical calculations must therefore be employed to identify both the correct atomic positions and the lowest energy isomers that contribute to the averaged structure. Since the first attempt to expand the intermetalloid cluster concept to the ternary system, the pentacle-like cluster $[Zn_6Sn_3Bi_8]^{4-}$ was obtained, and later a number of ternary clusters were observed successively which greatly enriched structural models in Zintl chemistry [126]. For instance, an icosahedron with two tetrahedra $[Pd@Pd_2Pb_{10}Bi_6]^{4-}$, $[Pd_3Sn_8Bi_6]^{4-}$ embedding a Pd_3 triangle, an inhomogeneous superatom dimer $\{[CuSn_5Sb_3]^{2-}\}_2$, a trimeric cluster anion $[Cd_3(Ge_3P)_3]^{3-}$ and so on [128–131]. The structural properties and chemical bonding of these clusters have been summarized before, so our focus here is on the recently reported ternary clusters.

In 2018, Dehnen and co-workers reported an asymmetric occupation of a single-metal centered 12-vertex cluster $[Co@Sn_6Sb_6]^{3-}$ (Figure 13.14a), as well as double-metal centered $[Co_2@Sn_5Sb_7]^{3-}$ (Figure 13.14b) and $[Ni_2@Sn_7Sb_5]^{3-}$ [132]. The $[Co@Sn_6Sb_6]^{3-}$ and $[Co_2@Sn_5Sb_7]^{3-}$ co-crystallized in the salt of $[K(2.2.2\text{-crypt})]_3[Co@Sn_6Sb_6]_{0.83}[Co_2@Sn_5Sb_7]_{0.17}\cdot2dmf\cdot2tol$ by reacting $[K(2.2.2\text{-crypt})]_2(Sn_2Sb_2)\cdot en$ with $[K(thf)_x][Co(cod)_2]$ at 5 °C. However, at a higher reaction temperature, only the crystals of $[Co_2@Sn_5Sb_7]^{3-}$ were isolated. All three cluster shells have a total of 56 valence electrons, which is supported by the results of calculated natural atomic charges (for the Co atoms in $[Co@Sn_6Sb_6]^{3-}$ and $[Co_2@Sn_5Sb_7]^{3-}$ are −1.32 to −1.35, respectively, whereas the value for the

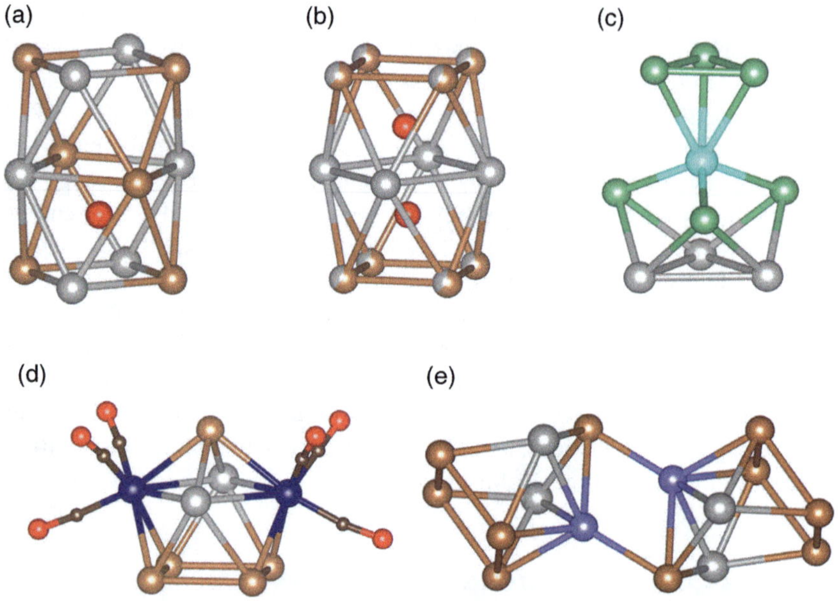

Figure 13.14 (a) $[Co@Sn_6Sb_6]^{3-}$ (the disorder has been omitted and atoms are colored to represent the most probable isomer); (b) $[Ni_2@Sn_7Sb_5]^{3-}$ (the atoms with mixed colors indicate statistic atomic disorder of Sn and Sb) [132]; (c) $[As_3M(As_3Pb_3)]^{3-}$ (M = Nb, Ta) [133]; (d) $[M_2(CO)_6Sn_2Sb_5]^{3-}$ (M = Cr, Mo); (e) $[(MSn_2Sb_5)_2]^{4-}$ (M = Cu, Ag) [134]. Global colors are as follows: Sn/Pb = gray, Sb = light brown, As = light green, Co/Ni = red, Nb/Ta = sky blue, Cr/Mo = dark blue, Cu/Ag = light purple, C = brown, O = red.

Ni atoms in $[Ni_2@Sn_7Sb_5]^{3-}$ is −0.33, indicating the existence of formally Co^- and Ni^0). $[Co@Sn_6Sb_6]^{3-}$ exhibits an unusual configuration with the inner atom being offset from the cluster center to hold the center of one prismatic "pocket" and leave the other empty. The formation of such unexpected topology can be understood by the outcomes of energy calculations. The first is to satisfy the demands to fit with a total − 3 charge. Second, reasonably stable cages for $[Co@Sn_xSb_y]^{3-}$ result with 50 (I_h-symmetry), 54 (distorted C_{2v}-symmetry) and 56 valence electrons (C_{4v}-symmetry) realized with Sn:Sb compositions of 12 : 0, 8 : 4, or 6 : 6, respectively. Obviously, the maximization of heteroatomic bonds leads to a strong preference for the formation of the 6 : 6 variant.

Among ternary intermetalloid clusters, arsenic-bearing clusters are a little bit special in that they can be extracted from quaternary Zintl phases when arsenic element reacts with group five metals at temperatures above 600 °C, resulting in their incorporation into the solid [64]. Sun et al. reported a series of works such as $[As_3Nb(As_3Sn_3)]^{3-}$ and $[As_3M(As_3Pb_3)]^{3-}$ (M = Nb, Ta), which are extracted from the ethylenediamine solution of $K_8NbSnAs_5$, $K_8NbPbAs_5$ and $K_8TaPbAs_5$, respectively [133, 135]. The former was reported in 2016, and the latter two clusters were published this year, so $[As_3M(As_3Pb_3)]^{3-}$ (M = Nb, Ta) will be primarily introduced here. It is shown in Figure 13.14c that the entire cluster exhibits a flower-vase shape with a Nb/Ta neck. The bottom half of $[Pb_3As_3]^{5-}$ is isostructural with the known *hypho*-type cluster formed in liquid ammonia $[Sn_3Bi_3]^{5-}$ [136] and the upper As_3 ring carries the negative charges of −3, hence the medium Nb/Ta displays a formally +5 positive charges. The bonding features of the three clusters are similar, taking $[As_3Nb(As_3Pb_3)]^{3-}$ as an example. The total 50 valence electrons can be assigned to 9 s-type lone pairs on the 6 As atoms and 3 Pb atoms as well as 6 two-center two-electron (2c—2e) Nb—As σ bonds and 3 2c—2e As—As σ bonds. The remaining 14 electrons (7 pairs) can be partitioned into 6 3c—2e Nb-As-Pb σ bonds representing linear combinations of the 4p and 6p orbitals of the As and Pb atoms with directed 4d orbitals on the Nb center, and delocalized 3c—2e σ bond on the *cyclo*-Pb_3 ring. Thus, the two delocalized electrons on the Pb_3 ring are considered to lead to σ-aromaticity following the Hückel rule ($4n+2$, $n = 0$).

As mentioned in Section 1.2, Wade-Mingos rules are widely used in polyhedral clusters dominated by multicenter bonds now. In the four types of clusters defined by Wade-Mingos rules, *closo* and *nido*-type clusters are the most stable and common ones, while *hypho*-type clusters are difficult to exist under conventional experimental conditions due to their high degree of structural incompleteness, which limits the further research on their structure and reactivity. Recently, a series of ternary clusters based on *hypho*-$[Sn_2Sb_5]^{3-}$ were reported [134]. The *hypho*-$[Sn_2Sb_5]^{3-}$ can be regarded as removing three vertices from a 10-vertex *closo*-type cluster with a bicapped square antiprismatic topology. As shown in Figure 13.14d, two $CrCO_3$ ($MoCO_3$) carbonyl fragments occupied two missing five-connected vertices of *hypho*-$[Sn_2Sb_5]^{3-}$ to form *nido*-type anions $[M_2(CO)_6Sn_2Sb_5]^{3-}$ (M = Cr, Mo). Additionally, in Figure 13.14e, one Cu (Ag) atom capped one 5-connected vertex constructing an *arachno*-type unit then dimerized to clusters $[(MSn_2Sb_5)_2]^{4-}$ (M = Cu, Ag). So, it can be inferred that coordinating to transition metal fragments may capture unstable structure units that is reminiscent of previous works $[Pb_5(MoCO_3)_2]^{4-}$ and $[Bi_3Mo_2(CO)_6]^{3-}$ in which Pb_5 ring and ozone-like Bi_3 were both stabilized by $MoCO_3$ fragments [121, 137]. All four clusters fit into Wade-Mingos and *mno* rules very well [138], in *hypho*-$[Sn_2Sb_5]^{3-}$, every naked Sn atom bearing a lone pair provides two electrons for cluster bonding (2×2), and each Sb atom owns three skeleton electrons (3×5), plus three negative charges, leading to a total of 22 skeleton electrons ($2n+8$). The cluster contains one polyhedron ($m = 1$), seven atoms ($n = 7$), and three missing vertices ($o = 3$), totally generating 11 orbitals available for cluster bonding. In light of this, Ag^+ or Cu^+ provides one electron for cluster bonding, leading to 22 skeleton electrons ($2n+6$) and followed by $m+n+o$ rule as well. The cases of Cr^0 and Mo^0 occupations are similar to those described above.

To further understand the chemical bonding of the four clusters, the authors carried out the AdNDP analysis. As the consequence, there are four 3c-2e σ bonds over Mo–Sb–Sn units in $[Mo_2(CO)_6Sn_2Sb_5]^{3-}$ and extra six electrons form three 5c–2e bonds, which may lead to locally σ-aromatic behavior of the Mo_2Sn_2Sb cap fragment (Figure 13.15a). Things are totally different in $[(CuSn_2Sb_5)_2]^{4-}$, on the diamond plane of Cu_2Sb_2, four electrons form two 4c–2e σ Sb—Cu—Cu—Sb bonds whose combination can give two 3c-2e σ Cu—Cu—Sb bonds indicating the anti-aromatic character of this plane (Figure 13.15b). The calculated positive $NICS_{zz}$ values at the center of Cu_2Sb_2 confirm the anti-aromatic feature of this unit. What's more, it is noted that the anti-aromatic fragment with four atoms contributes to the formation of two locally σ-aromatic 3c–2e islands which is denoted by their negative $NICS_{zz}$ values at the center of each Cu_2Sb triangle. As a result, the σ-aromaticity on Mo_2Sn_2Sb cap and divisional aromaticity over Cu_2Sb_2 plane provide extra stabilization for *hypho*-$[Sn_2Sb_5]^{3-}$ cluster.

Moreover, to further support the aromatic/antiaromatic characteristics of $[Mo_2(CO)_6Sn_2Sb_5]^{3-}$, $[(AgSn_2Sb_5)_2]^{4-}$, and $[(CuSn_2Sb_5)_2]^{4-}$, their magnetic response properties were evaluated accounted by the orientational averaged term (Iso) from different representative orientations of the external field (Figure 13.16). For $[Mo_2(CO)_6Sn_2Sb_5]^{3-}$, as a single cluster unit, the isotropic term shows a spherical-like shielding region ascribed to the cluster cage, where under a field along the *z*-, *x*-, and *y*-axis, delocalized electrons lead to a shielding cone property centered at the cage, similar to $[As@Ni_{12}@As_{20}]^{3-}$ and $[Bi@In_8Bi_{12}]^{3-/5-}$ clusters. However, for the dimerized clusters $[(MSn_2Sb_5)_2]^{4-}$ (M = Cu, Ag), two spherical-like shielding surfaces are obtained following the cage backbone from each cluster unit (MSn_2Sb_5), which in turn enables two adjacent shielding cones under specific orientations of the external field (*z*- and *x*-axis). Interestingly, for a field along with the intercluster bonding (*y*-axis), the two shielding cones are overlapped through the same axis, enhancing the shielding response at the center of the inter-cluster boning array. Hence, this shows that multiple spherical aromatic clusters can be aggregated, retaining spherical aromatic units within the same molecular structure.

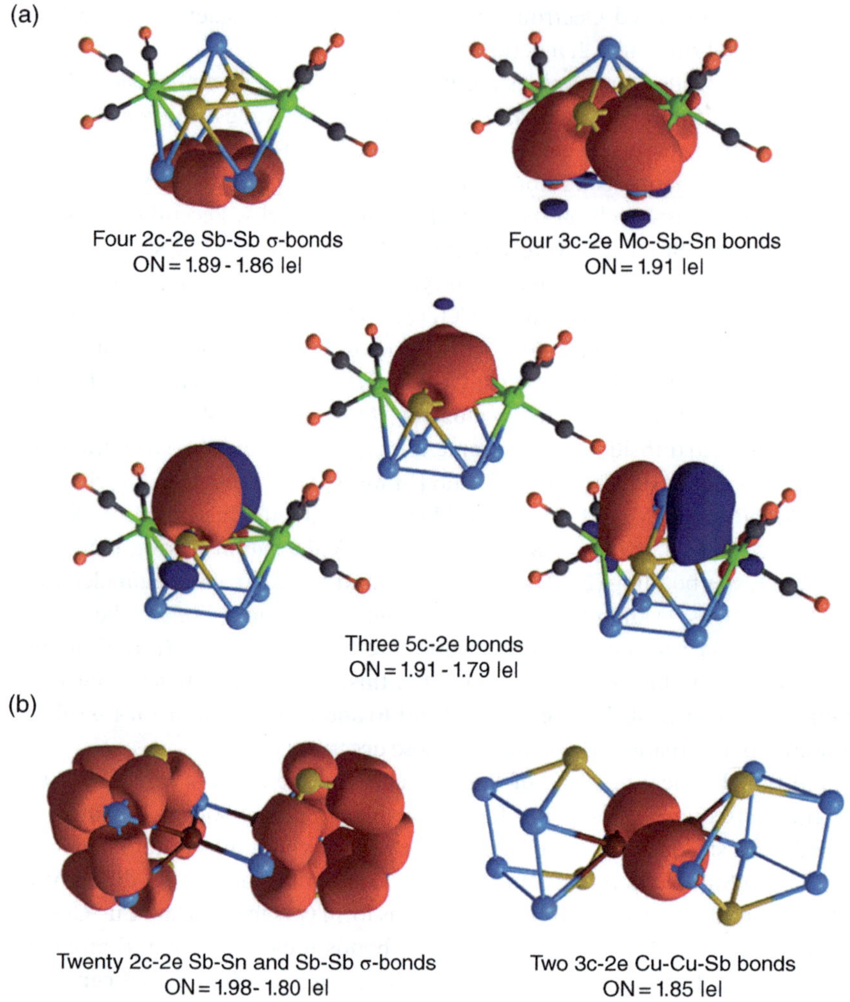

Figure 13.15 Chemical bonding pattern of the cage fragments of $[Mo_2(CO)_6Sn_2Sb_5]^{3-}$ (a) and $[(CuSn_2Sb_5)_2]^{4-}$ (b) obtained from the AdNDP analysis. *Source:* Reproduced from Ref. [134].

13.6 Conclusion and Perspectives

In this chapter, we summarized some classical group 15 clusters based on heavier arsenic, antimony, and bismuth reported in the past six years. Ternary clusters dominated by group 15 elements were involved likewise, and the configurations of ternary clusters have more variations compared with binary heterometallic clusters due to the introductions of group 13/14 elements. No matter to which main family of Zintl clusters, building blocks play an important role, in that they are the most fundamental units of structures and the basis to assemble larger clusters. From the view of the latest reported group 15 anions, the building blocks not only display the common cage configuration but also tend to appear as discrete fragments, which lead to the coordination modes of the whole cluster being more complex and may guide a breakthrough from the existing classical geometries. Besides, most current building blocks of group 15 clusters are restricted to those consisting of only main group elements, so what happens if one contains both main group and transition metal elements? Taking clusters of group 14 elements, for example, lately, a family of lead clusters

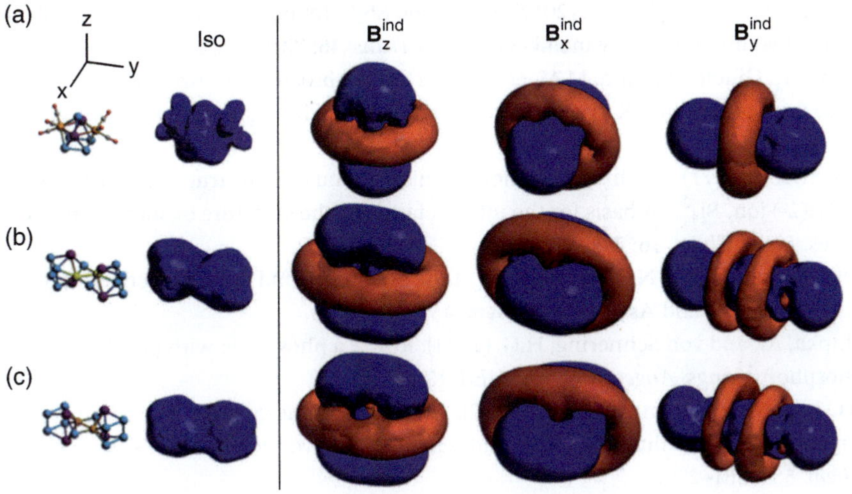

Figure 13.16 Magnetic response properties obtained for (a) $[Mo_2(CO)_6Sn_2Sb_5]^{3-}$; (b) $[(AgSn_2Sb_5)_2]^{4-}$; (c) $[(CuSn_2Sb_5)_2]^{4-}$ clusters, given by the isotropic term (Iso), and from specific orientations of the external field (B_z^{ind}, B_x^{ind}, and B_y^{ind}), as isosurfaces at ±2 ppm. *Source:* Reproduced from Ref. [134].

with Au cores was observed in which the two sizable $[Au_8Pb_{33}]^{6-}$ as well as $[Au_{12}Pb_{44}]^{8-}$ can be considered as being aggregated by $[Au@Pb_{11}]^{3-}$ unit, and isostructural $[Ag@Pb_{11}]^{3-}$ has been isolated as a confirmation [139]. So, it is expected that, as for group 15 Zintl-type clusters, heterometallic clusters can be employed as building blocks for further assembly. Presumably, $[Nb@As_8]^{3-}$ or other new heterometallic cages can serve in this manner. With a deeper exploration of Zintl-ion chemistry, it is anticipated that many unexpected new structures will be discovered in the future, and this will open up new windows for synthetic chemistry.

References

1 Zintl, E. and Dullenkopf, W. (1932). Polyantimonide, Polywismutide und ihr Übergang in Legierungen. *Z. Phys. Chem. Abt. B* 16: 183–194.

2 Zintl, E., Goubean, J., and Dullenkopf, W. (1931). Salzartige Verbindungen und intermetallische Phasen des Natriums in flüssigem Ammoniak. *Z. Phys. Chem. Abt. A* 154: 1–46.

3 Critchlow, S.C. and Corbett, J.D. (1984). Homopolyatomic anions of the post transition elements. Synthesis and structure of potassium-crypt salts of the tetraantimonide (2-) and heptaantimonide (3-) anions, Sb_4^{2-} and Sb_7^{3-}. *Inorg. Chem.* 23: 770–774.

4 Kraus, F. and Korber, N. (2005). The chemical bond in polyphosphides: crystal structures, the electron localization function, and a new view of aromaticity in P_4^{2-} and P_5. *Chem. Eur. J.* 11: 5945–5959.

5 Adolphson, D., Corbett, J.D., and Merryman, D. (1976). Stable homopolyatomic anions of the post-transition metals. "Zintl ions". The synthesis and structure of a salt containing the heptantimonide (3-) anion. *J. Am. Chem. Soc.* 98: 7234–7239.

6 Perla, G.L., Oliver, G.A., and Sevov, S.C. (2015). Bi_7^{3-}: the missing family member, finally isolated and characterized. *Inorg. Chem.* 54: 872–875.

7 Turbervill, R.S.P. and Goicoechea, J.M. (2014). From clusters to unorthodox pnictogen sources: solution-phase reactivity of $[E_7]^{3-}$ (E = P–Sb) anions. *Chem. Rev.* 114: 10807–10828.

8 Ruan, H.P., Wang, L.L., Li, Z.Y., and Xu, L. (2017). Sb_{10}^{2-} and Sb_2^{2-} found in [K(18-crown-6)]$_6$[Sb_{10}][Sb_4{Mo(CO)$_3$}$_2$]·2en: two missing family members. *Dalton Trans.* 46: 7219–7222.

9 Sun, H.T., Yonezawa, T., Gillett-Kunnath, M.M. et al. (2012). Ultra-broad near-infrared photoluminescence from crystalline (K-crypt)$_2$Bi$_2$ containing $[Bi_2]^{2-}$ dimers. *J. Mater.Chem.* 22: 20175–20178.

10 Cisar, A. and Corbett, J.D. (1977). Polybismuth anions. Synthesis and crystal structure of a salt of the Tetrabismuthide(2-) ion, Bi_4^{2-}. A basis for the interpretation of the structure of some complex intermetallic phases. *Inorg. Chem.* 16: 2482–2487.

11 Kraus, F., Hanauer, T., and Korber, N. (2006). Nature of the chemical bond in polypnictides: the lone pair aromatic anions P_4^{2-} and As_4^{2-}. *Inorg. Chem.* 45: 1117–1123.

12 Schmettow, W., Lipka, A., and von Schnering, H.G. (1974). Rb_4P_6, a phosphide with planar six-membered phosphorus rings. *Angew. Chem. Int. Ed.* 13: 345–345.

13 von Schnering, H.G., Meyer, T., HRnle, W. et al. (1987). Tetrarubidiumhexaphosphid und Tetracäsiumhexaphosphid: Darstellung, Struktur und Eigenschaften von Rb_4P_6 und Cs_4P_6. *Z. Anorg. Allg. Chem.* 553: 261–279.

14 Kraus, F., Hanauer, T., and Korber, N. (2005). Chemical bond in the cyclic anions P_6^{4-} and As_6^{4-}: synthesis, crystal structure, and electron localization function of {Rb[18]crown-6)}$_2$Rb$_2$As$_6$·6NH$_3$. *Angew. Chem. Int. Ed.* 44: 7200–7204.

15 Korber, N. and von Schnering, H.G. (1996). Alkali metal polyphosphides in low dimensions: synthesis and structure of $[(CH_3)_4N]_2RbP_7(NH_3)$ and $[(C_2H_5)_4N]\ Cs_2P_{11}$. *Chem. Ber.* 129: 155–159.

16 Hanauer, T. and Korber, N. (2006). $[Sb_{11}]^{3-}$ and $[As_{11}]^{3-}$: synthesis and crystal structure of two new ammoniates containing Trishomocubane-like polyanions. *Z. Anorg. Allg. Chem.* 632: 1135–1140.

17 Belin, C.H.E. (1980). Polyarsenide anions. Synthesis and structure of a salt containing the Undecaarsenide(3-) ion. *J. Am. Chem. Soc.* 102: 6036–6040.

18 Weinert, B., Eulenstein, A.R., Ababei, R., and Dehnen, S. (2014). Formation of $[Bi_{11}]^{3-}$, a homoatomic, polycyclic bismuth polyanion, by pyridine-assisted decomposition of $[GaBi_3]^{2-}$. *Angew. Chem. Int. Ed.* 53: 4704–4708.

19 Reil, M. and Korber, N. (2007). A S_8-like $[Sb_8]^{8-}$ Zintl anion in the ammoniate $[K_{17}(Sb_8)_2(NH_2)]\cdot 17.5NH_3$. *Z. Anorg. Allg. Chem.* 633: 1599–1602.

20 Caporali, M., Gonsalvi, L., Rossin, A., and Peruzzini, M. (2010). P_4 activation by late-transition metal complexes. *Chem. Rev.* 110: 4178–4235.

21 Cossairt, B.M., Prio, N.A., and Cummins, C.C. (2010). Early-transition-metal-mediated activation and transformation of white phosphorus. *Chem. Rev.* 110: 4164–4177.

22 Nesper, R. (2014). The Zintl-Klemm concept – a historical survey. *Z. Anorg. Allg. Chem.* 640: 2639–2648.

23 Mooser, E. and Pearson, W.B. (1956). Chemical bond in semiconductors. *Phys. Rev.* 101: 1608.

24 Wade, K. (1971). Structural significance of number of skeletal bonding electron-pairs in carboranes, the higher boranes and borane anions, and various transition-metal carbonyl cluster compounds. *J. Chem. Soc. D* 792–793.

25 Wade, K. (1976). Structural and bonding patterns in cluster chemistry. *Adv. Inorg. Chem. Radiochem.* 18: 1–66.

26 Mingos, D.M.P. (1984). Polyhedral skeletal electron pair approach. *Acc. Chem. Res.* 17: 311–319.

27 Mingos, D.M.P., Slee, T., and Lin, Z.Y. (1990). Bonding models for ligated and bare clusters. *Chem. Rev.* 90: 383–402.

28 Bentlohner, M.M., Fischer, C., and Fässler, T.F. (2016). Synthesis and characterization of pristine *closo*-$[Ge_{10}]^{2-}$. *Chem. Commun.* 52: 9841–9843.

29 Spiekermann, A., Hoffmann, S.D., and Fässler, T.F. (2006). The Zintl ion $[Pb_{10}]^{2-}$: a rare example of a homoatomic *closo* cluster. *Angew. Chem. Int. Ed.* 45: 3459–3462.

30 Quéneau, V., Todorov, E., and Sevov, S.C. (1998). Synthesis and structure of isolated silicon clusters of nine atoms. *J. Am. Chem. Soc.* 120: 3263–3264.

31 von Schnering, H.G., Baitinger, M., Bolle, U. et al. (1997). Binary alkali metal compounds with the Zintl anions $[Ge_9]^{4-}$ and $[Sn_9]^{4-}$. *Z. Anorg. Allg. Chem.* 623: 1037–1039.

32 Corbett, J.D. and Edwards, P.A. (1977). The Nonastannide (4-) anion Sn_9^{4-}, a novel capped antiprismatic configuration (C_{4v}). *J. Am. Chem. Soc.* 99: 3313–3317.

33 Yong, L., Hoffmann, S.D., and Fässler, T.F. (2006). $^2_\infty[K4Pb9]$ – A low-dimensional arrangement of $[Pb_9]^{4-}$ clusters in $[K(18\text{-crown-}6)]_2K_2Pb_9{\cdot}(en)_{1.5}$. *Inorg. Chim. Acta* 359: 4774–4778.

34 Goicoechea, J.M., Hull, M.W., and Sevov, S.C. (2007). Heteroatomic deltahedral clusters: synthesis and structures of *closo*-$[Bi_3Ni_4(CO)_6]^{3-}$, *closo*-$[Bi_4Ni_4(CO)_6]^{2-}$, the open cluster $[Bi_3Ni_6(CO)_9]^{3-}$, and the Intermetalloid *closo*-$[Ni_x@\{Bi_6Ni_6(CO)_8\}]^{4-}$. *J. Am. Chem. Soc.* 129: 7885–7893.

35 Li, X.W., Pennington, W.T., and Robinson, G.H. (1995). Metallic system with aromatic character. Synthesis and molecular structure of $Na_2[[(2,4,6\text{-}Me_3C_6H_2)_2C_6H_3]Ga]_3$ the first Cyclogallane. *J. Am. Chem. Soc.* 117: 7578–7579.

36 Xie, Y., Schreiner, P.R., Schaefer, H.F. et al. (1996). Are cyclogallenes $[M_2(GaH)_3]$ (M = Li, Na, K) aromatic? *J. Am. Chem. Soc.* 118: 10635–10639.

37 Li, X., Kuznetsov, A.E., Zhang, H.F. et al. (2000). Observation of all-metal aromatic molecules. *Science.* 291: 859–861.

38 Zhan, C.G., Zheng, F., and Dixon, D.A. (2002). Electron affinities of Al_n clusters and multiple-fold aromaticity of the square Al_4^{2-} structure. *J. Am. Chem. Soc.* 124: 14795–14803.

39 Zhai, H.J., Wang, L.S., Kuznetsov, A.E., and Boldyrev, A.I. (2002). Probing the electronic structure and aromaticity of pentapnictogen cluster anions Pn_5^- (Pn = P, As, Sb, and Bi) using photoelectron spectroscopy and ab initio calculations. *J. Phys. Chem. A* 106: 5600–5606.

40 Todorov, I. and Sevov, S.C. (2005). Heavy-metal aromatic and conjugated species: rings, oligomers, and chains of tin in $Li_{9-x}EuSn_{6+x}$, $Li_{9-x}CaSn_{6+x}$, $Li_5Ca_7Sn_{11}$, $Li_6Eu_5Sn_9$, $LiMgEu_2Sn_3$, and $LiMgSr_2Sn_3$. *Inorg. Chem.* 44: 5361–5369.

41 Todorov, I. and Sevov, S.C. (2004). Heavy-metal aromatic rings: cyclopentadienyl anion analogues Sn_5^{6-} and Pb_5^{6-} in the Zintl phases Na_8BaPb_6, Na_8BaSn_6, and Na_8EuSn_6. *Inorg. Chem.* 43: 6490–6494.

42 King, R.B. and Rouvray, D.H. (1977). Chemical applications of group theory and topology. 7. A graph-theoretical interpretation of the bonding topology in polyhedral boranes, carboranes, and metal clusters. *J. Am. Chem. Soc.* 99: 7834–7840.

43 Aihara, J. (1978). Three-dimensional aromaticity of polyhedral boranes. *J. Am. Chem. Soc.* 100: 3339–3342.

44 King, R.B. (2001). Three-dimensional aromaticity in polyhedral boranes and related molecules. *Chem. Rev.* 101: 1119–1152.

45 Cui, L.F., Huang, X., Wang, L.M. et al. (2006). Sn_{12}^{2-}: Stannaspherene. *J. Am. Chem. Soc.* 128: 8390–8391.

46 Cui, L.F., Huang, X., Wang, L.M. et al. (2006). Pb_{12}^{2-}: Plumbaspherene. *J. Phys. Chem. A* 110: 10169–10172.

47 Zubarev, D.Y. and Boldyrev, A.I. (2008). Developing paradigms of chemical bonding: adaptive natural density partitioning. *Phys. Chem. Chem. Phys.* 10: 5207–5217.

48 Tkachenko, N.V. and Boldyrev, A.I. (2019). Chemical bonding analysis of excited states using the adaptive natural density partitioning method. *Phys. Chem. Chem. Phys.* 21: 9590–9596.

49 Xu, H.L., Tkachenko, N.V., Muñoz-Castro, A. et al. (2021). $[Sn_8]^{6-}$-bridged mixed-valence Zn^I/Zn^{II} in $\{[K_2ZnSn_8(ZnMes)]_2\}^{4-}$ inverse Sandwich-type cluster supported by a $Zn^I{-}Zn^I$ bond. *Angew. Chem. Int. Ed.* 60: 9990–9995.

50 Xu, H.L., Popov, I.A., Tkachenko, N.V. et al. (2020). σ-Aromaticity-induced stabilization of heterometallic supertetrahedral clusters $[Zn_6Ge_{16}]^{4-}$ and $[Cd_6Ge_{16}]^{4-}$. *Angew. Chem. Int. Ed.* 59: 17286–17290.

51 Tkachenko, N.V. and Boldyrev, A.I. (2019). Multiple local σ-aromaticity of nonagermanide clusters. *Chem. Sci.* 10: 5761–5765.

52 Galeev, T.R., Dunnington, B.D., Schmidt, J.R., and Boldyrev, A.I. (2013). Solid state adaptive natural density partitioning: a tool for deciphering multi-center bonding in periodic systems. *Phys. Chem. Chem. Phys.* 15: 5022–5029.

53 Dimaio, A.J. and Rheingold, A.L. (1990). Structural chemistry of transition-metal complexes containing arsenic-arsenic bonds. *Chem. Rev.* 90: 169–190.

54 Knapp, C., Zhou, B.B., Denning, M.S. et al. (2010). Reactivity studies of group 15 Zintl ions towards homoleptic post-transition metal organometallics: a 'bottom-up' approach to bimetallic molecular clusters. *Dalton Trans.* 39: 426–436.

55 Kaas, M. and Korber, N. (2017). Synthesis and characterization of $[Rb([18]crown\text{-}6)]_3As_{15}Zn\cdot 4NH_3$ and $[Rb([2.2.2]crypt)]_4Sb_{14}Zn\cdot 8NH_3$. *Z. Anorg. Allg. Chem.* 643: 1331–1334.

56 Mandal, S., Reber, A.C., Qian, M.C. et al. (2012). Synthesis, structure and band gap energy of covalently linked cluster-assembled materials. *Dalton Trans.* 41: 12365–12377.

57 Lichtenberger, N., Massa, W., and Dehnen, S. (2019). Polybismuthide anions as ligands: the homoleptic complex $[(Bi_7)Cd(Bi_7)]^{4-}$ and the ternary cluster $[(Bi_6)Zn_3(TlBi_5)]^{4-}$. *Angew.Chem.Int. Ed.* 58: 3222–3226.

58 Mandal, S., Reber, A.C., Qian, M.C. et al. (2012). On the stability of an unsupported mercury–mercury bond linking group 15 Zintl clusters. *Dalton Trans.* 41: 5454–5457.

59 Eichhorn, B.W., Mattamana, S.P., Gardner, D.R., and Fettinger, J.C. (1998). Synthesis, structure and coordination chemistry of $MoAs_8{}^{2-}$. *J. Am. Chem. Soc.* 120: 9708–9709.

60 Kesanli, B., Fettinger, J., Scott, B., and Eichhorn, B.W. (2004). Gas phase, solution, and solid state alkali ion binding by the $[NbE_8]^{3-}$ ($E = As, Sb$) complexes: synthesis, structure, and spectroscopy. *Inorg. Chem.* 43: 3840–3846.

61 Kesanli, B., Fettinger, J., and Eichhorn, B.W. (2003). Controlled aggregation of $ME_8{}^{n-}$ binary anions ($M = Cr, Mo$; $E = As, Sb$) into one-dimensional arrays: structures, magnetism and spectroscopy. *J. Am. Chem. Soc.* 125: 7367–7376.

62 Moses, M.J., Fettinger, J.C., and Eichhorn, B.W. (2002). Charged molecular alloys: synthesis and characterization of the binary anions $Pd_7As_{16}{}^{4-}$ and $Pd_2As_{14}{}^{4-}$. *J. Am. Chem. Soc.* 124: 5944–5945.

63 Moses, M.J., Fettinger, J.C., and Eichhorn, B.W. (2003). Interpenetrating As_{20} fullerene and Ni_{12} icosahedra in the onion-skin $[As@Ni_{12}@As_{20}]^{3-}$ ion. *Science* 300: 778–780.

64 von Schnering, H.G., Wolf, J., Weber, D. et al. (1986). $[NbAs_8]^{3-}$, a novel type of complex and an unexpected one-dimensional chain structure: ${}^{1}_{\infty}\left[Rb\cdot NbAs_8\right]^{3-}$. *Angew. Chem. Int. Ed. Engl.* 25: 353–354.

65 Reber, A.C., Mandal, S., Qian, M.C. et al. (2012). Palladium in the gap: cluster assemblies with band edges localized on linkers. *J. Phys. Chem. C* 116: 10207–10214.

66 Moses, M.J., Fettinger, J.C., and Eichhorn, B.W. (2007). $[Ni_5Sb_{17}]^{4-}$ transition-metal Zintl ion complex: crossing the Zintl border in molecular intermetalloid clusters. *Inorg. Chem.* 46: 1036–1038.

67 Sevov, S.C. and Corbett, J.D. (1993). Carbon-free fullerenes: condensed and stuffed anionic examples in indium systems. *Science* 262: 880–883.

68 Bobev, S. and Sevov, S.C. (1999). Synthesis and characterization of stable stoichiometric clathrates of silicon and germanium: $Cs_8Na_{16}Si_{136}$ and $Cs_8Na_{16}Ge_{136}$. *J. Am. Chem. Soc.* 121: 3795–3796.

69 Scherer, O.J. (2000). Small Neutral P_n Molecules. *Angew. Chem. Int. Ed.* 39: 1029–1030.

70 Ha, T.K., Suleimenov, O., and Nguyen, M.T. (1999). A quantum chemical study of three isomers of N_{20}. *Chem. Phys. Lett.* 315: 327–334.

71 Shen, M. and Schaefer, H.F. (1994). Dodecahedral and smaller arsenic clusters: As_n, n=2, 4, 12, 20. *J. Chem. Phys.* 101: 2261–2266.

72 Wang, Y., Moses-DeBusk, M., Stevens, L. et al. (2017). $Sb@Ni_{12}@Sb_{20}^{-/+}$ and $Sb@Pd_{12}@Sb_{20}^{n}$ cluster anions, where n = +1, −1, −3, −4: multi-oxidation-state clusters of interpenetrating platonic solids. *J. Am. Chem. Soc.* 139: 619–622.

73 Stegmaier, S. and Fässler, T.F. (2011). A bronze matryoshka: the discrete intermetalloid cluster $[Sn@Cu_{12}@Sn_{20}]^{12-}$ in the ternary phases $A_{12}Cu_{12}Sn_{21}$ (A = Na, K). *J. Am. Chem. Soc.* 133: 19758–19768.

74 Zhao, J.J. and Xie, R.H. (2004). Density functional study of onion-skin-like $[as@Ni_{12}As_{20}]^{3-}$ and $[Sb@Pd_{12}Sb_{20}]^{3-}$ cluster ions. *Chem. Phys. Lett.* 396: 161–166.

75 Baruah, T., Zope, R.R., Richardson, S.L., and Pederson, M.R. (2003). Electronic structure and rebonding in the onion like $as@Ni_{12}@As_{20}$ cluster. *Phys. Rev. B* 68: 241404(R).

76 Carey, D.M., Morales-Verdejo, C., and Muñoz-Castro, A. (2015). $[as@Ni_{12}@As_{20}]^{3-}$ and $[Sn@Cu_{12}@Sn_{20}]^{12-}$ clusters. Related structures with different construction philosophy. *Chem. Phys. Lett.* 638: 99–102.

77 Kulichenko, M., Fedik, N., Boldyrev, A.I., and Muñoz-Castro, A. (2020). Expansion of magnetic aromaticity criteria to multilayer structures: magnetic response and spherical aromaticity of matryoshka-like cluster $[Sn@Cu_{12}@Sn_{20}]^{12-}$. *Chem. Eur. J.* 26: 2263–2268.

78 Muñoz-Castro, A. and King, R.B. (2017). Aromatic and antiaromatic spherical structures: use of long-range magnetic behavior as an aromatic indicator for bare icosahedral $[Al@Al_{12}]^{-}$ and $[Si_{12}]^{2-}$ clusters. *Phys. Chem. Chem. Phys.* 19: 15667–15670.

79 Tkachenko, N.V., Popov, I.A., Kulichenko, M. et al. (2021). Bridging aromatic/antiaromatic units: recent advances in aromaticity and antiaromaticity in main-group and transition-metal clusters from bonding and magnetic analyses. *Eur. J. Inorg. Chem.* 41: 4239–4250.

80 von Schleyer, P.R. and Jiao, H.J. (1996). What is aromaticity? *Pure Appl. Chem.* 68: 209–218.

81 Islas, R., Heine, T., and Merino, G. (2012). The induced magnetic field. *Acc. Chem. Res.* 45: 215–228.

82 Pan, F.X., Li, L.J., Wang, Y.J. et al. (2015). An all-metal aromatic Sandwich complex $[Sb_3Au_3Sb_3]^{3-}$. *J. Am. Chem. Soc.* 137: 10954–10957.

83 Popov, I.A., Pan, F.X., You, X.R. et al. (2016). Peculiar all-metal σ-aromaticity of the $[Au_2Sb_{16}]^{4-}$ anion in the solid state. *Angew. Chem. Int. Ed.* 55: 15344–15346.

84 Pan, F.X., Lukanowski, M., Weigend, F., and Dehnen, S. (2021). Tetrahedral $[Sb(AuMe)_4]^{3}$ occurring in multimetallic cluster syntheses: about the structure-directing role of methyl groups. *Angew. Chem. Int. Ed.* 60: 2–8.

85 Min, X., Popov, I.A., Pan, F.X. et al. (2016). All-metal antiaromaticity in Sb_4-type Lanthanocene anions. *Angew. Chem. Int. Ed.* 55: 5531–5535.

86 Eulenstein, A.R., Franzke, Y.J., Lichtenberger, N. et al. (2021). Substantial π-aromaticity in the anionic heavy-metal cluster $[Th@Bi_{12}]^{4-}$. *Nat. Chem.* 13: 149–155.

87 Shan, H., Yang, Y., James, A.J., and Sharp, P.R. (1997). Dinitrogen bridged gold clusters. *Science* 275: 1460–1462.

88 Baird, N.C. (1972). Quantum organic photochemistry. II. Resonance and aromaticity in the lowest $^3\pi\pi^*$ state of cyclic hydrocarbons. *J. Am. Chem. Soc.* 94: 4941–4948.

89 Li, W.L., Xu, C.Q., Hua, S.X., and Li, J. (2016). Theoretical studies on the bonding and electron structures of a $[Au_3Sb_6]^{3-}$ complex and its oligomers. *Dalton Trans.* 45: 11657–11667.

90 Alexandrova, A.N., Koyle, E., and Boldyrev, A.I. (2006). Theoretical study of hydrogenation of the doubly aromatic B_7^- cluster. *J. Mol. Model.* 12: 569–576.

91 Giambiagi, M., de Giambiagi, M.S., dos Santos Silva, C.D., and de Figuereido, A.P. (2000). Multicenter bond indices as a measure of aromaticity. *Phys. Chem. Chem. Phys.* 2: 3381–3392.

92 Bultinck, P., Ponec, R., and Van Damme, S. (2005). Multicenter bond indices as a new measure of aromaticity in polycyclic aromatic hydrocarbons. *J. Phys. Org. Chem.* 18: 706–718.

93 Cioslowski, J., Matito, E., and Solà, M. (2007). Properties of aromaticity indices based on the one-electron density matrix. *J. Phys. Chem. A* 111: 6521–6525.

94 Pan, F.X., Li, L.J., and Sun, Z.M. (2016). [au(η^2-Sn_2Sb_2)$_2$]$^{3-}$: a mixed group 14/15 tetrahedral dimer bridged by Au. *Chin. J. Struct. Chem.* 35: 1099–1106.

95 Pan, F.X., Guggolz, L., Weigend, F., and Dehnen, S. (2020). Atom exchange versus reconstruction: $(Ge_xAs_{4-x})^{x-}$ (x = 2, 3) as building blocks for the Supertetrahedral Zintl cluster $[Au_6(Ge_3As)(Ge_2As_2)_3]^{3-}$. *Angew. Chem. Int. Ed.* 59: 16638–16643.

96 Luo, S. and Whitmire, K.H. (1989). Synthesis and characterization of a series of antimony-containing iron carbonyl complexes: $[Et_4N]_3[SbFe_4(CO)_{16}]$, $[Et_4N]_2[HSbFe_4(CO)_{13}]$, $[Et_4N][H_2SbFe_4(CO)_{13}]$, and $[Et_4N]_2[ClSbFe_3(CO)_{12}]$. *Inorg. Chem.* 28: 1424–1431.

97 Cobbledick, R.E. and Einstein, F.W.B. (1979). The crystal structure of μ^4-Antimonio-Tetrakis[tricarbonyl(triphenylphosphine)cobalt] tetraphenylborate-dichloromethane, $[\{CO(CO)_3PPh_3\}_4Sb][BPh_4].CH_2Cl_2$. *Acta Crystallogr. Sect. B* 35: 2041–2044.

98 Ramos-Cordoba, E., Postils, V., and Salvador, P. (2015). Oxidation states from wave function analysis. *J. Chem. Theory Comput.* 11: 1501–1508.

99 Liu, C., Tkachenko, N.V., Popov, I.A. et al. (2019). Structure and bonding in $[Sb@In_8Sb_{12}]^{3-}$ and $[Sb@In_8Sb_{12}]^{5-}$. *Angew. Chem. Int. Ed.* 58: 8367–8371.

100 Tkachenko, N.V., Zhang, X.W., Qiao, L. et al. (2020). Spherical aromaticity of all-metal $[Bi@In_8Bi_{12}]^{3-/5-}$ clusters. *Chem. Eur. J.* 26: 2073–2079.

101 Wilson, R.J., Weigend, F., and Dehnen, S. (2020). A new facet of the *pseudo*-element concept: the *Arachno*-Zintl ion $(Sn_5Sb_3)^{3-}$ and the effects of element composition on the structures of isoelectronic clusters. *Angew. Chem. Int. Ed.* 59: 14251–14255.

102 Wilson, R.J. and Dehnen, S. (2017). $(Ge_4Bi_{14})^{4-}$: a case of "element segregation" on the molecular level. *Angew. Chem. Int. Ed.* 56: 1–6.

103 Pan, F.X., Wei, S.X., Guggolz, L. et al. (2021). Insights into formation and relationship of multimetallic clusters: on the way toward Bi-rich nanostructures. *J. Am. Chem. Soc.* 143: 7176–7188.

104 Muñoz-Castro, A. (2017). The shielding cone in spherical aromatic structures: insights from models for spherical $2(N + 1)^2$ aromatic fullerenes. *Phys. Chem. Chem. Phys.* 19: 12633–12636.

105 Knies, M., Kaiser, M., Isaeva, A. et al. (2018). The intermetalloid cluster cation $(CuBi_8)^{3+}$. *Chem. Eur. J.* 24: 127–132.

106 Müller, U., Isaeva, A., Richter, J. et al. (2016). Polyhedral bismuth polycations coordinating gold(I) with varied hapticity in a homoleptic heavy-metal cluster. *Eur. J. Inorg. Chem.* 2016: 3580–3584.

107 Groh, M.F., Müller, U., Isaeva, A., and Ruck, M. (2017). Ionothermal syntheses, crystal structures, and chemical bonding of the rhodium-centered clusters $[RhBi_9]^{4+}$ and $[(RhBi_7)I_8]$. *Z. Anorg. Allg. Chem.* 643: 1482–1490.

108 Li, Z., Ouyang, D., and Xu, L. (2019). $[Bi_7M_3(CO)_3]^{2-}$ (M= Co, Rh): a new prototype of 10-vertex deltahedral hybrids from unprecedented polycyclic η^5–coordination addition of Bi_7^{3-} and trimetallic fragments. *Chem. Commun.* 55: 6783–6786.

109 Li, Z.Y., Liu, C.P., Wu, J. et al. (2019). $[(\eta^3\text{-}Bi_3)_2(IrCO)_6(\mu^4\text{-}Bi)_3]^{3-}$: a new archetype of a 15-vertex deltahedral hybrid from Bi_x^{x-}-coordination aggregation of cationic $[IrCO]^+$ units. *Dalton Trans.* 48: 12013–12017.

110 Chen, S., Li, Z.Y., Yuan, B.B. et al. (2020). Aggregation of Polybismuthide anions in a single compound using $^+$Rh-CO units: heterometallic cluster ions $[Rh@Bi_{10}(RhCO)_6]^{3-}$ and $[Rh@Bi_9(RhCO)_5]^{3-}$. *Inorg. Chem.* 59: 10628–10633.

111 Qiao, L., Chen, D.D., Zhu, J. et al. (2021). $[Bi_6Mo_3(CO)_9]^{4-}$: a multiple local σ-aromatic cluster containing a distorted Bi_6 triangular prism. *Chem. Commun.* 57: 3656–3659.

112 Downing, D.O., Zavalij, P., and Eichhorn, B.W. (2010). The *closo*-$[Sn_9Ir(cod)]^{3-}$ and $[Pb_9Ir(cod)]^{3-}$ Zintl ions: isostructural Ir^{I} derivatives of the *nido*-E_9^{4-} anions (E = Sn, Pb). *Eur. J. Inorg. Chem.* 2010: 890–894.

113 Wang, J.Q., Stegmaier, S., Wahl, B., and Fässler, T.F. (2010). Step-by-step synthesis of the endohedral stannaspherene $[Ir@Sn_{12}]^{3-}$ via the capped cluster anion $[Sn_9Ir(cod)]^{3-}$. *Chem. Eur. J.* 16: 1793–1798.

114 Zouchoune, B., Ogliaro, F., Halet, J.-F. et al. (1998). Bonding analysis in inorganic transition-metal cubic clusters. 3. Metal-centered tetracapped $M_9(\mu^5\text{-}E)_4Ln$ species with a tetragonal distortion. *Inorg. Chem.* 37: 865–875.

115 Adams, R.D. and Elpitiya, G. (2016). Iridium–bismuth carbonyl cluster complexes. *J. Organomet. Chem.* 812: 115–122.

116 Bruhn, G., Davidson, E.R., Mayer, I., and Clark, A.E. (2006). Löwdin population analysis with and without rotational invariance. *Int. J. Quantum Chem.* 106: 2065–2072.

117 Eulenstein, A.R., Franzke, Y.J., Bügel, P. et al. (2020). Stabilizing a metalloid $\{Zn_{12}\}$ unit within a polymetallide environment in $[K_2Zn_{20}Bi_{16}]^{6-}$. *Nat Commun* 11: 5122–5129.

118 Resa, I., Carmona, E., Gutierrez-Puebla, E., and Monge, A. (2004). Decamethyldizincocene, a stable compound of Zn(I) with a Zn-Zn bond. *Science* 305: 1136–1138.

119 Hicks, J., Underhill, E.J., Kefalidis, C.E. et al. (2015). A mixed-valence tri-zinc complex, [LZnZnZnL] (L = bulky amide), bearing a linear chain of two-coordinate zinc atoms. *Angew. Chem. Int. Ed.* 54: 10000–10004.

120 Donsbach, C., Reiter, K., Sundholm, D. et al. (2018). $[Hg_4Te_8(Te_2)_4]^{8-}$: a heavy metal porphyrinoid embedded in a lamellar structure. *Angew. Chem. Int. Ed.* 57: 8770–8774.

121 Xu, L., Ugrinov, A., and Sevov, S.C. (2001). Stabilization of ozone-like $[Bi_3]^{3-}$ in the heteroatomic *closo*-clusters $[Bi_3Cr_2(CO)_6]^{3-}$ and $[Bi_3Mo_2(CO)_6]^{3-}$. *J. Am. Chem. Soc.* 123: 4091–4092.

122 Zhang, J.X., Sheong, F.K., and Lin, Z.Y. (2018). Unravelling chemical interactions with principal interacting orbital analysis. *Chem. Eur. J.* 24: 9639–9650.

123 Lips, F., Schellenberg, I., Pöttgen, R., and Dehnen, S. (2009). The subtle influence of binary versus homoatomic Zintl ions: the phenyl-ligated trimetallic cage $[Sn_2Sb_5(ZnPh)_2]^{3-}$. *Chem. Eur. J.* 15: 12968–12973.

124 Ababei, R., Heine, J., Holyńska, M. et al. (2012). Making practical use of the *pseudo*-element concept: an efficient way to ternary intermetalloid clusters by an isoelectronic Pb^--Bi combination. *Chem. Commun.* 48: 11295–11297.

125 Pan, F.X., Guggolz, L., and Dehnen, S. (2021). Cluster chemistry with (pseudo-)tetrahedra involving group 13–15 (semi-)metal atoms. *CCS Chem.* 3: 2969–2984.

126 Lips, F. and Dehnen, S. (2009). $[Zn_6Sn_3Bi_8]^{4-}$: expanding the intermetalloid Zintl anion concept to ternary systems. *Angew. Chem. Int. Ed.* 48: 6435–6438.

127 Wilson, R.J., Lichtenberger, N., Weinert, B., and Dehnen, S. (2019). Intermetalloid and heterometallic clusters combining p-block (semi)metals with d- or f-block metals. *Chem. Rev.* 119: 8506–8554.

128 Ababei, R., Massa, W., Harms, K. et al. (2013). Unusual 14-electron fragments $[Pd(\eta^3\text{-}Bi_{3-x}Pb_x)]^{(x+1)-}$ as *pseudo* Lead atoms in *closo*-$[Pd@Pd_2Pb_{10}Bi_6]^{4-}$. *Angew. Chem. Int. Ed.* 52: 13544–13548.

129 Lips, F., Clérac, R., and Dehnen, S. (2011). $[Pd_3Sn_8Bi_6]^{4-}$: a 14-vertex Sn/Bi cluster embedding a Pd_3 triangle. *J. Am. Chem. Soc.* 133: 14168–14171.

130 Wilson, R.J., Broeckaert, L., Spitzer, F. et al. (2016). $\{[CuSn_5Sb_3]^{2-}\}_2$: a dimer of inhomogeneous Superatoms. *Angew. Chem. Int. Ed.* 55: 1–7.

131 Mitzinger, S., Bandemehr, J., Reiter, K. et al. (2018). $(Ge_2P_2)^{2-}$: a binary analogue of P_4 as a precursor to the ternary cluster anion $[Cd_3(Ge_3P)_3]^{3-}$. *Chem. Commun.* 54: 1421–1424.

132 Wilson, R.J., Hastreiter, F., Reiter, K. et al. (2018). $[Co@Sn_6Sb_6]^{3-}$: an off-Center endohedral 12-vertex cluster. *Angew. Chem. Int. Ed.* 57: 15359–15363.

133 Shu, C.C., Qiao, L., Muñoz-Castro, A., and Sun, Z.M. (2021). $[As_3M(As_3Pb_3)]^{3-}$ (M = Nb, Ta): ternary heterometallic clusters with early transition metal atoms and aromatic $[Pb_3]^{2-}$. *Chin. J. Chem.* 39: 1953–1957.

134 Xu, Y.H., Tkachenko, N.V., Popov, I.A. et al. (2021). Ternary aromatic and anti-aromatic clusters derived from the *hypho* species $[Sn_2Sb_5]^{3-}$. *Nat Commun* 12: 4465–4472.

135 Pan, F.X., Xu, C.Q., Li, L.J. et al. (2016). A niobium-necked cluster $[As_3Nb(As_3Sn_3)]^{3-}$ with aromatic $Sn_3{}^{2-}$. *Dalton Trans.* 45: 3874–3879.

136 Friedrich, U. and Korber, N. (2016). A step in between: $[Sn_3Bi_3]^{5-}$ and its structural relationship to $[Sn_3Bi_5]^{3-}$ and $[Sn_4Bi_4]^{4-}$. *Chemistry Open* 5: 306–310.

137 Yong, L., Hoffmann, S.D., Fässler, T.F. et al. (2005). $[Pb_5\{Mo(CO)_3\}_2]^{4-}$: a complex containing a planar Pb_5 unit. *Angew. Chem. Int. Ed.* 44: 2092–2096.

138 Jemmis, E.D., Balakrishnarajan, M.M., and Pancharatna, P.D. (2001). A unifying electron-counting rule for macropolyhedral boranes, metallaboranes, and metallocenes. *J. Am. Chem. Soc.* 123: 4313–4323.

139 Shu, C.C., Morgan, H.W.T., Qiao, L. et al. (2020). A family of Lead clusters with precious metal cores. *Nat. Commun.* 11: 3477–3485.

14

Exploration of Controllable Synthesis and Structural Diversity of Titanium–Oxo Clusters

Mei-Yan Gao[1,2], *Lei Zhang*[1,3], *and Jian Zhang*[1]

[1] *State Key Laboratory of Structural Chemistry, Fujian Institute of Research on the Structure of Matter, Chinese Academy of Sciences, Fuzhou, Fujian, P. R. China 350002*
[2] *Department of Chemical Sciences, Bernal Institute, University of Limerick, Limerick, Limerick County, Republic of Ireland V94 T9PX*
[3] *Institute of Modern Optics, College of Electronic Information and Optical Engineering, Nankai University, Tianjin, P. R. China 300350*

14.1 Introduction

Titanium-oxo clusters (abbreviated as TOCs; also known as polyoxo-titanium clusters, PTCs), form a significant branch of metal-oxo cluster chemistry, often featuring atomically well-defined structures [1–3]. Compared with the intrinsically anionic polyoxometalate (POM) compounds, the majority of TOCs are neutral, consisting of cationic inorganic Ti–O cores wrapped by organic functional ligands (e.g. carboxylate, alkoxide) [4]. A small number of TOCs are anionic in nature and capped with inorganic ligands (e.g. sulfates) or other components (as encapsulated in POMs) [5, 6]. Due to significant similarities with the pigment titanium dioxide (TiO_2) in terms of both composition and structure, atomically precise TOCs have been considered as high-quality structural and reactivity models for TiO_2 [7]. Structural information can be accurately visualized by solving data obtained from single crystal X-ray diffraction (SCXRD), which gives a chance for theoretical research and mechanistic explanations for their properties at the molecular level. It is worth mentioning that the O/Ti ratio ("degree of condensation," dc) in TOCs ranges from $0 < dc < 2$, whereby 2 is the maximum value found in pure TiO_2 [8]. Moreover, the coordinating organic or inorganic ligands and inorganic Ti–O cores are highly tunable, allowing for significant development in their applications, typically as spectroscopic or electronic properties. Additionally, the functionalized ligands used to coordinate with the Ti–O cores allow TOCs to be dissolved in organic or aqueous solvents, resulting in easy handling compared to TiO_2 nanoparticles. Such solubility endows them with significant benefits for solution-based techniques, such as purification and preparation of composite thin films with MOFs by the liquid-phase epitaxy (LPE) layer-by-layer approach [9, 10].

Since single-crystal X-ray analysis was used to determine the crystal structure of the initial crystalline hydrolysis product of titanium tetraethoxide [$Ti_7O_4(OEt)_{20}$] in 1967 [11], atomically precise TOCs have drawn significant attention due to their low toxicity, appealing structures, intriguing properties, and extremely promising applications. So far, more than 300 homometallic examples of TOCs have been reported and characterized by SCXRD, spanning from 3 to 52 titanium atoms [8]. However, compared with well-explored silver nanoclusters (with the largest reported cluster being

Atomically Precise Nanochemistry, First Edition. Edited by Rongchao Jin and De-en Jiang.

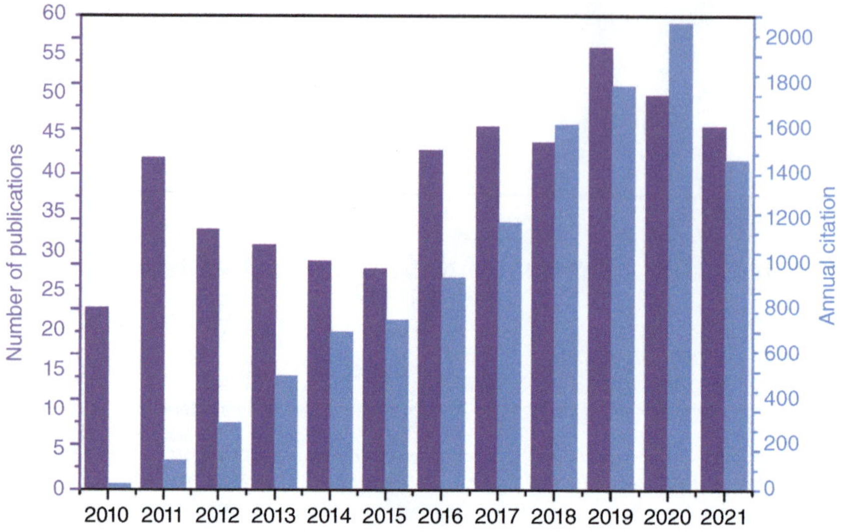

Figure 14.1 The number of TOCs-related research articles published each year and the related annual citations from 2010 to 2021. Web of Science data by searching keywords "polyoxo-titanium cluster" or "polyoxotitanium cluster" or "titanium-oxo cluster" or "titanium-oxide cluster" on August 11, 2021.

Ag_{490} [12]), as well as POMs (with the largest reported cluster being Mo_{368} [13]), the nuclearity and number of publications, together with annual citations, are still relatively scarce (Figure 14.1), mainly owing to the difficulty in the kinetic control of tetravalent Ti^{4+} cations in the reaction system. Titanium (IV) ions are prone to rapidly hydrolyze and spontaneously generate a stubborn precipitation of TiO_2 due to the high polarizing power of Ti^{4+} and the low electronegativity of Ti. Most recently, coordination delayed hydrolysis (CDH) strategies have paved the way to fabricating novel TOCs by controlling the degree of Ti^{4+} hydrolysis to facilitate their crystallization. Both cluster nuclearity and structural types have made significant advances, from fullerene-like Ti_{42} [10] to gigantic two-halves of Ti_{44} [14] and Ti_{52} [15]. For further comprehensive information on the structures of TOCs, please refer to earlier reviews on titanium-oxo clusters.

TiO_2 is a significantly important photocatalyst with particular uses in water splitting for H_2 production. However, its large bandgap (ca. 3.2 eV) unfortunately restricts the photocapture to the UV light range (making up a small portion of solar light, <5%) [16, 17]. In addition, the risk assessment committee of the European Chemicals Agency (ECHA) has recently proposed that the white pigment TiO_2 in worldwide use be classified as a potential carcinogen [18, 19]. Based on the tunable structures and excellent optical responsitivities, TOCs are arguably regarded as promising materials to replace TiO_2 in many practical applications and are expected to show superior performance. To this end, tremendous research has been devoted to clearly understand the relationship between bandgap, photo-related properties, and crystal structures. By metal/nonmetal doping and ligand modification, the bandgap of TOCs has been adjusted to visible-light adsorption, and a series of black TOCs were isolated through further changing synthetic conditions [20–22]. Furthermore, researchers also avidly investigate their applications in photocatalytic H_2 production, CO_2 reduction, photochromic effects, and nonlinear optical (NLO) effects, for example [7, 23, 24].

In this chapter, we provide a brief overview of the state-of-the-art advances in this area. We will start with a discussion of CDH strategies to prepare titanium-oxo clusters, followed by an analysis on recent progress concerning cluster core diversity, highlighting some of the most compelling cases of TOCs to illustrate different cluster core types. Section 14.4 outlines the ligand diversity and its advantages in synthetic control. Section 14.5 illustrates the diversity of metal-doping strategies, including transition metal doping and rare earth metal doping. Subsequently, we describe the bandgap engineering of TOCs and potential applications in photo-related fields. Finally, we conclude with a summary of the major challenges and opportunities associated with titanium-oxo nanoclusters.

In this chapter, we restrict our analysis to homometallic TOCs, except for the section on metal-doping strategies. Metal-organic frameworks (MOFs) with TOCs as building blocks are not covered. Low nuclearity TOCs are occasionally mentioned, though a complete summary of these materials may be found elsewhere.

14.2 Coordination Delayed Hydrolysis Strategy

The titanium(IV) ion, a strong Lewis acid, is known to be highly reactive, tending to uncontrollably and rapidly hydrolyze into a TiO_2 precipitate [25–27]. Thus, intense research efforts have been devoted to the controllable synthesis of a marvelous diversity of crystallinity in the form of TOCs. Currently, the highly unpredictable reactivity is "tamed" by the CDH strategy to some extent, which slows down the hydrolysis dynamics, i.e. the nucleation and crystallization process (Figure 14.2). Specifically, the Ti^{4+} ion is pre-protected by complexing ligands or solvent under specific reaction conditions to decrease the activity of Ti^{4+} [4]. In this regard, it is extremely crucial to choose synthetic conditions judiciously, including titanium precursors, protecting ligands, solvents, pH, reaction time, and temperature. For titanium precursors, the coordination number is vital with both $TiCl_4$ and $Ti(OR)_4$ being 4-coordinate, lower than that of TiO_2 (6-coordinate). As such, $TiCl_4$ and $Ti(OR)_4$ display significant reactivity toward H_2O, creating a driving force for reaction. It is often observed that the structures of the obtained TOCs are greatly dependent on the applied synthetic methods. In this section, we discuss four synthetic methods (solvothermal synthesis, aqueous sol-gel synthesis, ionothermal synthesis and solid-state-like synthesis), which are illustrated by related examples.

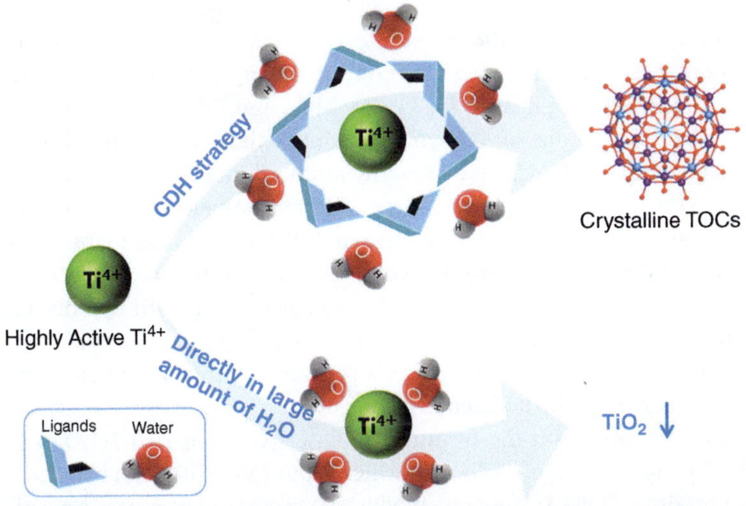

Figure 14.2 Illustration of the coordination delayed hydrolysis strategy for preparing crystalline TOCs.

14.2.1 Solvothermal Synthesis

Solvothermal synthesis has been found to be the most reported and effective method [28, 29]. A large part of crystalline TOCs were prepared via this approach, due to its simple operation, wide solute selection, high yield, and a rich variety of solvent species, including conventional solvents of alcohols, acetonitrile, amides, toluene, amino alcohols, or combinations thereof [20, 30, 31]. These solvents can serve as reaction media and dissolve the reactants. Additionally, they may be used as structure-directing reagents, greatly influencing the nuclearity and type of Ti–O cores. Moreover, they typically impact the stability of the final products, working as coordinating terminal groups or bridging ligands. Targeted TOCs with strong coordinating solvents (new type of solvents, e.g. acids) generally show better stability against moisture than those linking with the weakly coordinating solvents (traditional solvents) mentioned above. One impressive example is the Ti_{44}-oxo cluster totally surrounded with propionate groups and coordinating H_2O [14].

The principal aim in the preparation of TOCs is limiting the uncontrollable growth of TiO_2 precipitation and the controllable synthesis of highly crystalline TOCs by the CDH strategy. Consequently, it is crucial to control the H_2O/Ti^{4+} molar ratio, to ensure the reaction remains controllable. One of the significant approaches is through the use of trace water generated by in-situ esterification of alcohol and carboxylic acid and keeping the reaction in balance (i.e. the amount of water in reaction). For example, when using the solvent isopropanol, the use of auxiliary solvents from formic acid to propionic acid under the same temperature conditions may be used to generate TOCs from a Ti_{42}-oxo cluster to a Ti_{52}-oxo cluster, due to the slower water-generation rate (more delayed time) for the aggregation of larger cluster cores [10, 15]. Also, this deduction is confirmed by the presented ester groups in the final products of this type of reaction [32, 33].

14.2.2 Aqueous Sol-Gel Synthesis

A wide range of crystalline TOCs have been isolated from reactions in nonaqueous solutions. Typically, in this kind of reactions, $Ti(OR)_4$ is more preferable to $TiCl_4$, attributed to its reduced readiness to hydrolysis due to steric hindrance and the hydrophobic character of the alkoxy groups, instead of the comparatively risky manipulation of $TiCl_4$, which may hydrolyze and release hydrochloric acid in a fierce reaction. Fortunately, this problem may be treated with aqueous sol-gel synthesis in highly acidic aqueous systems with the assistance of organic ammonium cations (OACs), where TOCs may be spontaneously obtained via the controllable hydrolysis and condensation of Ti^{4+} without any additional/chelating ligands [5, 34, 35]. $TiCl_4$ and $TiOSO_4$ are the most used precursors to prepare TOCs by this aqueous sol-gel synthesis, due to their readier dissolution in water, giving highly acidic clear solutions. OACs, like tetraethylammonium (TEA) and tetrabutylammonium (TBA) cations, are essential to change the chemical environment of the solution and particularly to promote the construction of target structures of TOCs, despite not being coordinating ligands and not enhancing the condensation degree of Ti^{4+}. In the highly acidic Ti^{4+}/SO_4^{2-} reaction, SO_4^{2-} increases the condensation degree of Ti^{4+} by coordinating with hydrolyzed species (also possibly delays the hydrolysis and avoids the production of TiO_2 particles). Unlike the uncharged clusters obtained by solvothermal approaches, TOCs prepared by this method are primarily polycationic compounds and are soluble in aqueous media.

The first aqueous TOC, a cubic octamer with the formula $[Ti_8O_{12}(H_2O)_{24}]Cl_8{\cdot}HCl{\cdot}7H_2O$, was structurally determined in 1987, by the reaction of $TiCl_4$ with water vapor [36]. Although considerable efforts have been made to produce TOCs in aqueous media, the reported structures are still particularly rare. Wang and coworkers judiciously improved on this method, developing a series of TOCs such as $[Ti_4O_4(OH_2)_6Cl_2(SO_4)_4]^{2-}$ $\{Ti_4\}$ [37], $[Ti_6(O\mu)_8(O_tH_2)_{20}]^{8+}$ $\{Ti_6\}$ [35], $[Ti_8O_{12}(OH_2)_{24}]^{8+}$

$\{Ti_8\}$ [35], $[Cs@Ti_{12}O_{18}(SO_4)_6]^+$ $\{Ti_{12}O_{18}\}$ [38], and the triple-decked pentagonal prism $[Ti_{18}O_{27}(OH_2)_{30}(SO_4)_6]^{6+}$ $\{Ti_{18}O_{27}\}$ [5], etc. This method provides a new perspective to prepare TOCs in an eco-friendly, safe, and economical way.

14.2.3 Ionothermal Synthesis

Ionothermal synthesis is based on ionic liquids (ILs) and is attracting extensive research attention with more than 6000 reported articles [39]. The first carbon nanotube (CNT)-like titanium oxide (Ti—O) nanotube was assembled via this approach [40]. Deep eutectic solvents (DESs) are considered a new category of ILs, owing to their similar physical characteristics. Compared with conventional ILs, the study of DESs is still in its infancy, with the first example reported in 2001 [41]. The melting point of the eutectic mixture largely relies on a reciprocal interaction, lower than that of each component. DESs can not only work as solvents and reaction media but also promote the controllable hydrolysis of Ti^{4+} to facilitate the formation of crystalline TOCs. A series of crystalline titanium-oxo clusters with nuclearity up to 28 were prepared using ionothermal synthesis with a eutectic mixture of choline chloride and phenol (molar ratio = 1 : 2) [42].

14.2.4 Solid-State-Like Synthesis

Zhang and coworkers developed a new method for preparing TOCs, based on solids of organic compounds with low melting points [43, 44]. When the reaction was put in an oven with the temperatures above the organic compound's melting point, the melting solid served as a solvent. Thus far, the solids used to successfully prepare crystalline TOCs include phenol (melting point: 43 °C) or its derivatives (also denoted as phenol-thermal synthesis), pyrazole (melting point: 66–70 °C), or its derivatives (also termed as the pyrazole-thermal approach). Some other ligands (solids) may also be promising choices, such as imidazole (88–91 °C), urea (132.7 °C), benzoic anhydride (43 °C), and phenylacetic acid (76.5 °C) or their derivatives. Compared with conventional solvothermal synthesis, the melting ligands act not only as solvents and reaction media but also as coordinating ligands to delay the hydrolysis of Ti^{4+} ions. Additionally, phenol can functionalize TOCs by improving their stability and reducing their bandgap, while pyrazole-modified TOCs show fantastic stability and potential separation promotion of photogenerated carriers due to its chelating ability and abundant nitrogen doping, respectively [43, 44]. Attributed to this novel approach, more structural types of stable TOCs are expected in the future.

14.3 Ti—O Core Diversity

Thus far, numerous TOCs have been reported, and we expand on the rich structural diversity of viable species in this section. High nuclearity TOCs have been updated with Ti_{42}, Ti_{44}, and Ti_{52} clusters reported by the Zhang group, with the cluster size increasing to 36 Å [10, 14, 15]. TOCs with the same nuclearity may also have distinct core structures. A large portion of chemists concentrate on the preparation, properties, and (potential) applications of TOCs, while a comprehensive and systematic understanding of the structural diversity of Ti—O cores is still necessary.

Many factors endow TOCs with both core structure and shape diversity. First, the diverse synthetic strategies regularly affect the core structures. As explained, these reaction procedures are not limited to organic solvents or solutions, with crystalline TOCs also being isolated in aqueous solutions or ionothermal methods or in solid-statelike reactions through the CDH strategy. In addition, the inorganic Ti—O cores are influenced by various ligands, including O-donor ligands and N-donor ligands in diverse terminal and/or bridging coordination modes. In aqueous sol-gel synthesis, the choice of Ti sources can also be expanded beyond $Ti(OR)_4$ to $TiCl_4$ and $TiOSO_4$. Additionally, lower or higher coordination numbers of Ti^{4+} cations have been observed (ranging from 4 to 7), with the six-coordinated octahedral TiO_6 being the most frequent one. TiO_4, TiO_5, TiO_6, TiO_7 polyhedra, or combinations thereof, form TOCs by edge-sharing, face-fused, or corner-linking modes [8]. Moreover, basic structural units have proven to be particularly diverse. Among the reported TOCs, triangular $\{Ti_3\}$ is found to be the most basic subunit, existing in almost all the structures. Pentagonal $\{(Ti)Ti_5\}$ fragments are also very common, existing in examples such as the fullerene-like Ti_{42}, Ti_{20} isomers, water-soluble Ti_{18}, and even larger Ti_{52} units, which consist of a seven-coordinate Ti atom and 5 five-coordinate Ti atoms [5, 10, 15, 45]. Furthermore, the connecting modes of O-bridges are also varied, including μ_2, μ_3, or μ_4 linking modes. Last but not least, TOCs can not only be neutral but also polycationic compounds. The majority of TOCs are known to be uncharged and commonly generated from the partial hydrolysis and condensation of titanium alkoxo sources. In summary, by virtue of the diversities of synthetic strategies, various ligands, Ti sources, coordination modes of Ti^{4+} cations, structural units, connecting modes of O-bridges, and so on, Ti—O cores are responsible for considerable diversity allowing substantial progress in TOC development.

In this section, the known TOCs are mainly categorized according to the shape of the cluster cores into dense structures, wheel-shaped structures, sphere-shaped structures, and multicluster structures, which have scarcely been discussed in previous reviews. Meanwhile, representative TOCs (with nuclearity more than six) are listed in Figures 14.3–14.6 according to nuclearity to give a systematic and comprehensive understanding of the Ti—O cores diversity.

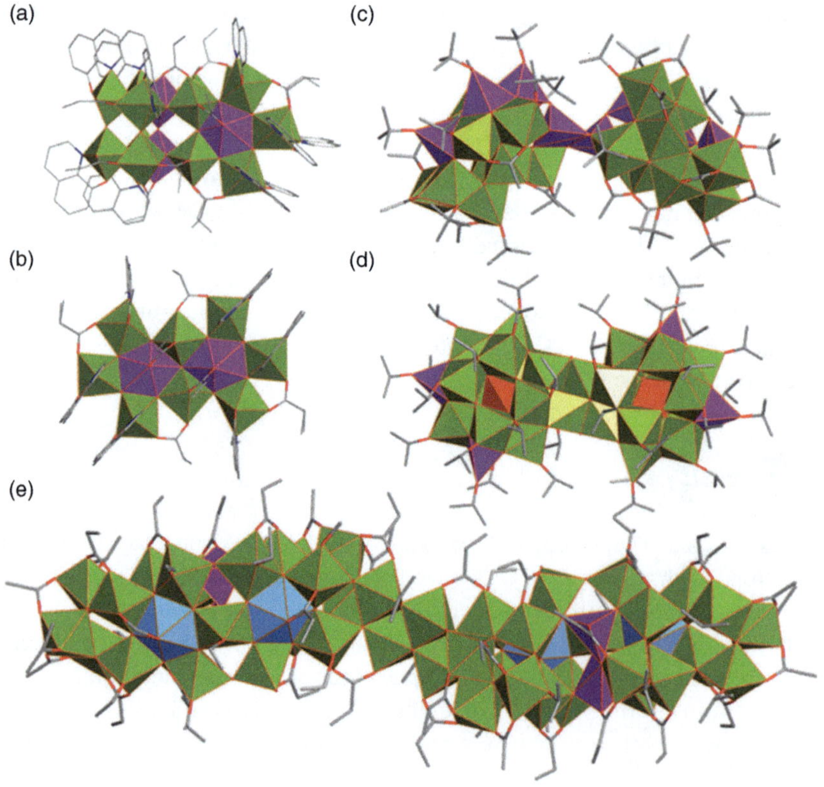

Figure 14.3 Representative examples of TOCs with dense structures. (a, b) Ti_{20} isomers [45]; (c) Ti_{28} [46]; (d) Ti_{34} [47]; (e) Ti_{52} [15]. H atoms are omitted for clarity. Color code: TiO_4, red; TiO_5, purple, TiO_6, green, TiO_7, blue.

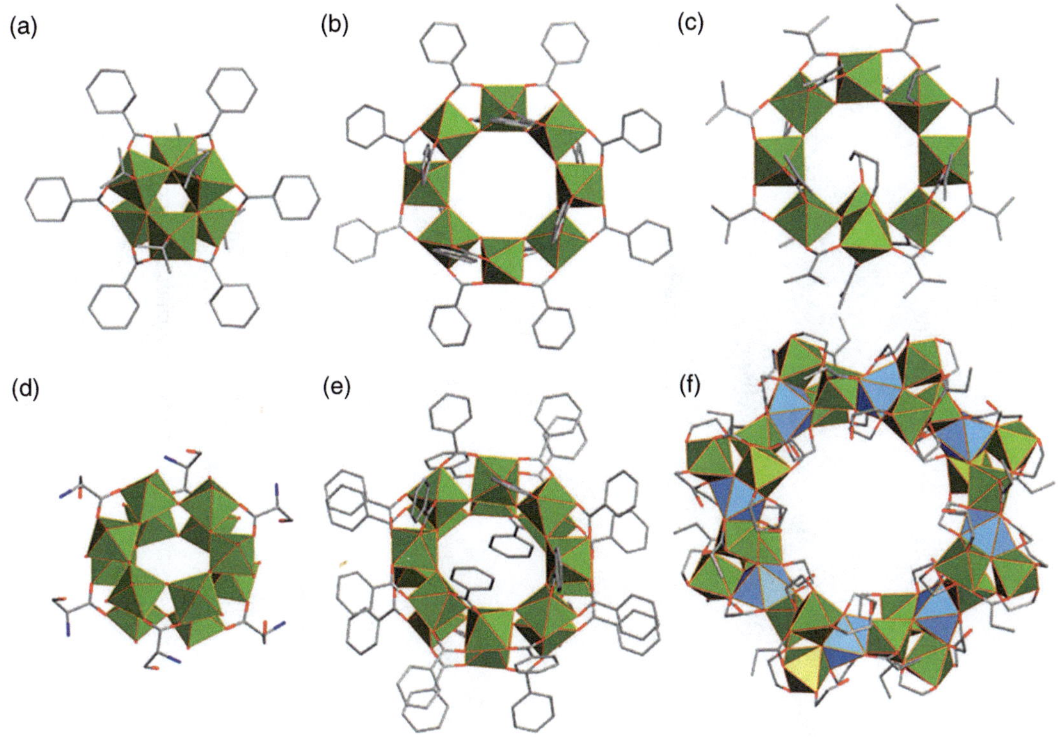

Figure 14.4 Representative examples of TOCs with ring-like structures. (a), Ti_6 [30]; (b) Ti_8 [48]; (c) Ti_9 [49]; (d) Ti_{12} [38]; (e) Ti_{16} [50]; (f) Ti_{32} [51]. C and H atoms are omitted for clarity.

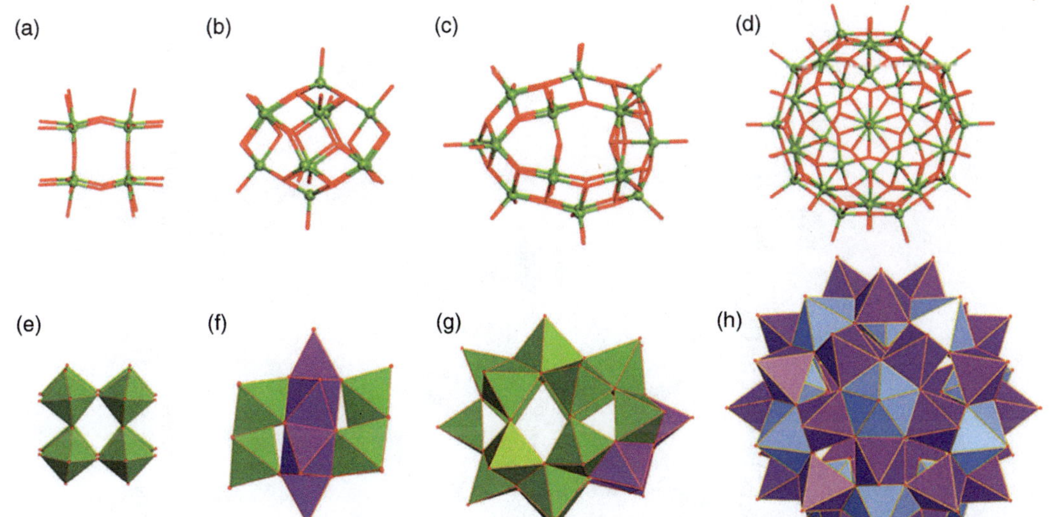

Figure 14.5 Representative examples of Ti–O sphere-shaped structures, (a, e) Ti_8 [36]; (b, f) Ti_{12} [52]; (c, g) Ti_{22} [53]; (d, h) Ti_{42} [10]. C and H atoms are omitted for clarity.

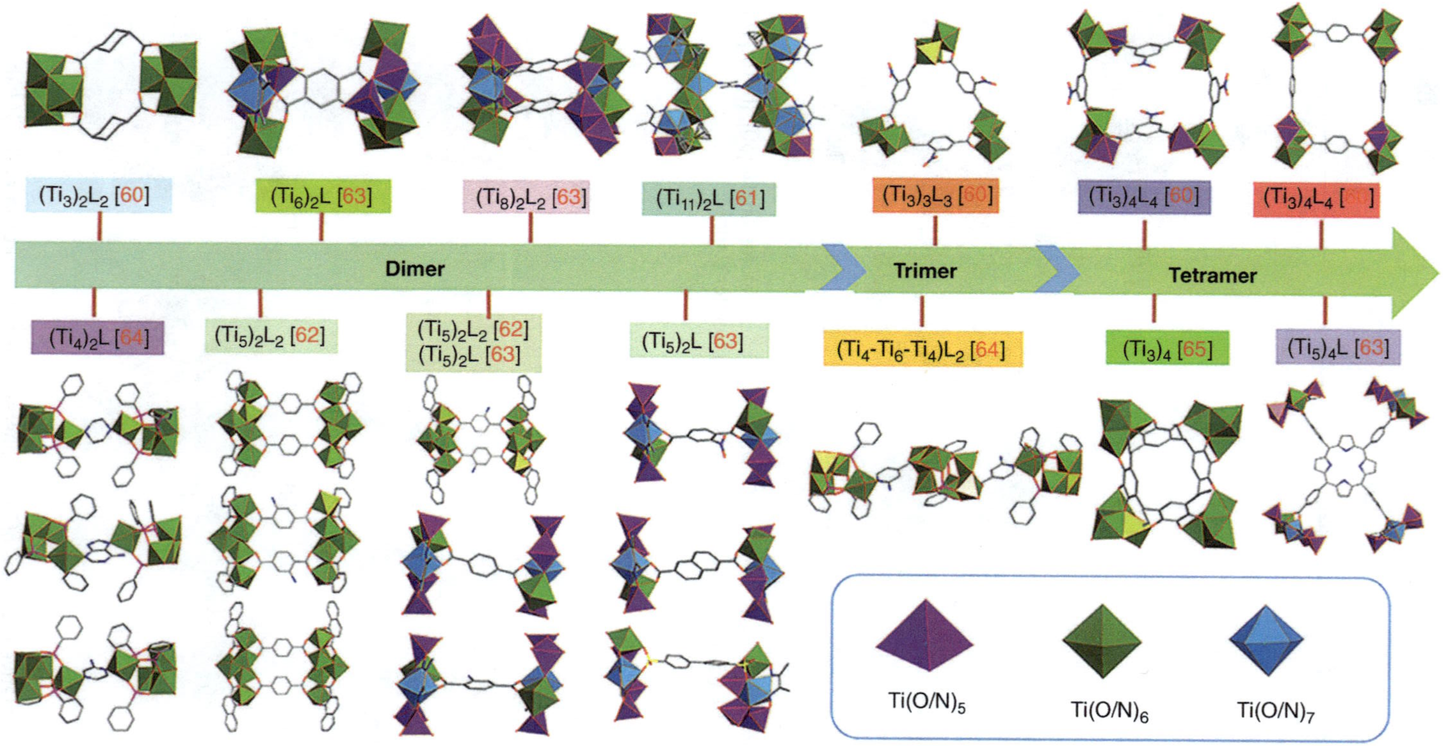

Figure 14.6 Representative examples of multicluster Ti—O structures. H atoms and alkoxides are omitted for clarity; L = ligands between/among cluster units.

14.3.1 Dense Structures

Recently, tremendous progress has been made in emerging categories of Ti–O cores. Dense structures are found to be most reported, especially for large TOCs, like Ti_{20} [45], Ti_{28} [46], Ti_{29} [3], Ti_{34} [47], Ti_{44} [14], Ti_{52} [15], and so forth. Some representative compounds are summarized in Figure 14.3. These dense structures also contain diverse shapes, such as saddles, butterfly, pentagonal-prismatic, two halves, baskets, and dumbbells. As for why dense structures are commonly presented, we deduce that dense structural features show better stability due to the protection by ligands to a larger degree, which prohibits moisture damage. In addition, we propose that high-nuclearity TOCs are prone to form dense structures according to the gradual accumulation of slowly hydrolyzed Ti^{4+} ions. Further investigations into the deeper formation mechanism are required, which could benefit the controllable construction of large TOCs.

14.3.2 Wheel-Shaped Structures

At present, numerous wheel-shaped coordination complexes are published based on transition metals, or lanthanides, however, wheel-shaped (or ring-shaped, or cyclic) TOCs with "donut" shapes and open cavities are still scarcely reported, much less than dense TOCs. Typically, cyclic TOCs are modified by carboxylate ligands on the inside or outside of the rings to stabilize the whole backbone, sometimes with alkoxides as auxiliary ligands. Thus far, the largest wheel-type TOC is $\{Ti_{32}O_{16}\}$, initially reported by Zheng and coworkers, with the cyclic skeleton modified by mono-deprotonated and doubly deprotonated ethylene glycolates, as well as carboxylate ligands [51]. The carboxylates can be propionic acid, methacrylic acid, or pivalic acid. Additionally, the mono-deprotonated ethylene glycol ligands were found to be exchangeable with other alcohols. Wang et al. also prepared a Ti_{32} nanoring decorated by 4-aminobenzoic acid and ethylene glycol, which underwent single-crystal-to-single-crystal transformation owing to the desolvation process, allowing the ring size to shrink [54]. Other known cyclic TOCs include Ti_6 rings (double layers) [30], Ti_8 rings [48], Ti_9 rings [49], Ti_{12} rings (double layers) [38], and Ti_{16} rings (double layers) [50] (Figure 14.4), among which Ti_{16} rings can often be regarded as double-layer Ti_8 rings. Among the series, Ti_6 rings and Ti_8 rings are the most studied wheel-shaped structures, and their peripheral carboxylates ligands can be replaced by a large range of carboxybenzene derivatives and fatty acid ligands, and even sulfate ligands for Ti_8 [34, 55]. There is also another kind of novel single-layer Ti_6 ring, with di-tert-butylphosphate as the functionalized ligand [56]. These structures not only display an architectural beauty but also show interesting supramolecular applications. These rings can stack into supramolecular nanotubes, which can be used to selectively adsorb gas molecules, encapsulate toluene molecules, and separate isomers, as well as further assemble interlocked structures, etc. [54, 55, 57, 58].

14.3.3 Sphere-Shaped Structures

The adjacent Ti atoms of sphere-shaped TOC structures (or cage structures) may be linked by oxides or bridges to form a sphere. Different from wheel-shaped TOCs, the cavity is closed with a comparatively dense "shell." The inside is usually found to be hollow or contain guest molecules or solvents. Attributed to the copious efforts of researchers, cage TOCs have been created, including the most common sphere-shaped Ti_8 [36], Ti_{12} [52], Ti_{22} [53], and Ti_{42} [10] (Figure 14.5). Among them, Ti_{42} has proven to be the largest sphere-like structure and the only TOC to compare favorably with fullerene [10]. It is worth noting that it contains 12 pentagonal $\{(Ti)Ti_5\}$ fragments, with the central $\{TiO_7\}$ surrounded by five $\{TiO_5\}$ units in an edge-sharing mode, similar to pentagonal {(Mo)Mo_5} fragments shown in giant molybdenum-oxide clusters [59]. The {Ti_{12}} icosahedron formed by 12 seven-coordinating Ti atoms, is encapsulated by the {Ti_{30}} icosidodecahedron constructed by 30 five-coordinated Ti atoms. Their organic ligands in the outer sphere are usually alkoxides, except in the case of Ti_8 [3, 52].

14.3.4 Multicluster Structures

Multicluster compounds $\{Ti_m\}_n$ ($m \geq 3$; $n \geq 2$) are composed of two or more TOCs (Ti_m) connected by bidentate or multidentate ligands (Figure 14.6). The nuclearity of subcluster (m) greatly depends on the type of linking ligands and auxiliary ligands. In general, Ti_3 subclusters are predominant when no additional ligand is used, Ti_5 is common with salicylic acid or derivatives as secondary ligands, Ti_5, Ti_6, or Ti_8 appear with dimethylglyoxime, and Ti_{11} form with auxiliary ligands of phosphoric acid and 1,1′-ferrocenedicarboxylic acid [60–63]. Thus far, dimers, trimers, and tetramers have been found, but no higher oligomer is yet known. Among these, dimers are the most prevalent type. In known cases, the linking ligands have been varied, including carboxylates, nitrogen heterocyclic ligands, and dimethylglyoxime ($DmgH_2$) [60, 61, 64, 65]. It is noteworthy that the geometric characteristics of the prepared multicluster TOCs strongly rely on the configuration of the used ligands, typically showing linear, triangular, or rectangular architectures.

In 2017, Zhang's group published three supramolecular structures with double Ti_4-Ti_4 and triple Ti_4-Ti_6-Ti_4 configurations [64]. Similarly, in 2018, a series of multicluster TOCs were isolated by this group, bridged by four dicarboxylates with {Ti_3} building units [60]. A linear dimeric $(Ti_3)_2$ complex was fabricated by a flexible cyclohexane dicarboxylic acid, while a triangular $[Ti_3]_3$ was constructed by changing the flexible dicarboxylate ligand to the rigid 2-nitro-1,4-benzenedicarboxylate. Two interesting $(Ti_3)_4$ rectangles were also formed when the ligands were further tuned to the more symmetric terephthalate or 5-nitro-1,3-benzenedicarboxylate [60]. After that, another series of multicluster structures were constructed, including $(Ti_5)_2$, $(Ti_6)_2$, $(Ti_8)_2$, and $(Ti_5)_4$ [62, 63]. These oligomers are compiled in Figure 14.6. More empirical principles for the synthesis of these materials may be acquired as more and more TOC-based supramolecular structures are reported.

14.4 Ligand Diversity

In addition to the development of new Ti—O cores, considerable research endeavors have also been devoted to the design and construction of TOCs modified with diverse functionalized ligands. According to the soft and hard acid-base theory (HSAB) in coordination chemistry, Ti^{4+} cation, as a hard acid, has a strong coordination ability with hard bases or borderline bases, including alkoxides, carboxylates, sulfates, phosphonates, polyphenolic groups, nitrogen heterocyclic ligands, or any combination of these in a chelating or a bridging binding mode. In the early stages of this research, apart from the ubiquitous alkoxide groups, the structures of TOCs were mostly modified with carboxylate and phosphonate ligands, due to the highly oxyphilic nature of Ti^{4+} cations [3, 8, 29, 31]. With the synthetic strategies of TOCs improving, more and more O-donor ligands have been used to prepare TOCs with high crystallinity, especially phenolic ligands and inorganic sulfate ligands [5, 66]. In addition to O-donor ligands, N-donor ligands were successfully introduced in the system of TOCs, along with their capability in the assembly of multicluster TOCs [42].

There have been several reviews about TOCs anchored with different ligands. In a critical 2011 review by Rozes and Sanchez, TOCs were classified into three categories based on the capping ligands [27]. Zhang and coworkers systematically reviewed and categorized TOCs into four types according to the supporting ligands [29].

In this section, the representative peripheral ligands are introduced in five parts for clarity: (i) carboxylate ligands; (ii) phosphonate ligands; (iii) polyphenolic ligands; (iv) sulfate ligands; and (v) nitrogen heterocyclic ligands. The majority of TOCs contain alkoxide groups due to prevalent use of alcohols as solvent or from residual OR groups of $Ti(OR)_4$ precursor. Thus, we will not discuss alkoxide ligands individually.

14.4.1 Carboxylate Ligands

Carboxylate-containing TOCs account for almost two thirds of all TOCs, constituting the greatest subset of TOCs (with the exception of alkoxide-based TOCs). The primary reason is that the Ti^{4+} cation has a high affinity for O-donor ligands, especially carboxylate linkers. Second, there are various options for carboxylate ligands, such as fatty acids and aromatic acids, including mono-, di-, tri-, and poly-carboxylic acids. Other than their use as linkers, the carboxylate ligands are also used to adjust the acidity of the reaction or even serve as reaction media. Moreover, the coordinating mode of carboxylate ligands is diverse, including dangling and bridging modes. Finally, the in situ generated trace amount of water from esterification reactions between carboxylic acid and liberated alcohol in the reaction mixture can control the growth of highly crystalline TOCs, although small amounts of water can occasionally be introduced into the reaction system by other methods, like the residual moisture of the starting materials [26, 27]. This turns out to be a tried-and-true strategy proven by the TOCs with ester groups [32, 33]. This reaction can happen even in the absence of alcohol solvent (but with $Ti(OR)_4$ precursors) and be kept in equilibrium. At present, carboxylate ligands are frequently used to prepare high nuclearity TOCs by this kind of self-limiting manner, although they do not always appear in the obtained structures [10].

Ti-oxo-carboxylate clusters have witnessed the rapid evolution of high nuclearity structures in recent years, including the Ti_{42}-oxo cluster formed by isopropanol and formic acid (Figure 14.5d) [10, 14, 15]. Despite the formic acid not appearing in the resulting crystal structure, it is indispensable for the synthetic process. Zhang and coworkers also reported a series of TOCs using propionic acid with or without the tert-butanol (^tBuOH) solvent (Figure 14.7) [14]. With the amount of propionic acid in the reaction increasing, the nuclearity grows gradually, up to a gigantic Ti_{44} cluster. Compared to other TOCs decorated by alkoxide or mixed alkoxide and carboxylate groups, the Ti_{44}-oxo cluster is totally surrounded by carboxylate groups in dangling and bridging modes, except for a pair of coordinating water molecules. Two other Ti_{44}-oxo clusters with the same type were obtained when the formic/propionic ligands were pairwise replaced by two ligands of glutaric acid or 3-methylglutaric acid. These were found to be much less sensitive to moisture than TOCs decorated by alkoxide or mixed alkoxide and carboxylate ligands, contributing to the total protection by carboxylate groups. Also, a 3.6 nm Ti_{52}-oxo cluster with the highest nuclearity to date was constructed using isopropanol and propionic acid (Figure 14.3e) [15]. The growth of this structure is based on a stepwise interlayer assembly approach from smaller Ti_6-oxo and Ti_{17}-oxo clusters.

14.4.2 Phosphonate Ligands

Phosphonate ligands based on phosphoric acid and its derivates RPO_3H_2 (R = Ph, Me, 4-CNPh, ^tBu, etc.), as O-donor ligands with up to three available oxygen atoms, are frequently employed to fabricate TOCs. At present, phosphonate-stabilized TOCs make up around one third of all the TOCs (excluding alkoxide-based TOCs), with sizes ranging from Ti_3 to Ti_{26}, due to the great affinity of the phosphonate linkers to titanium oxide surfaces [3, 67]. Among them, phenylphosphonate

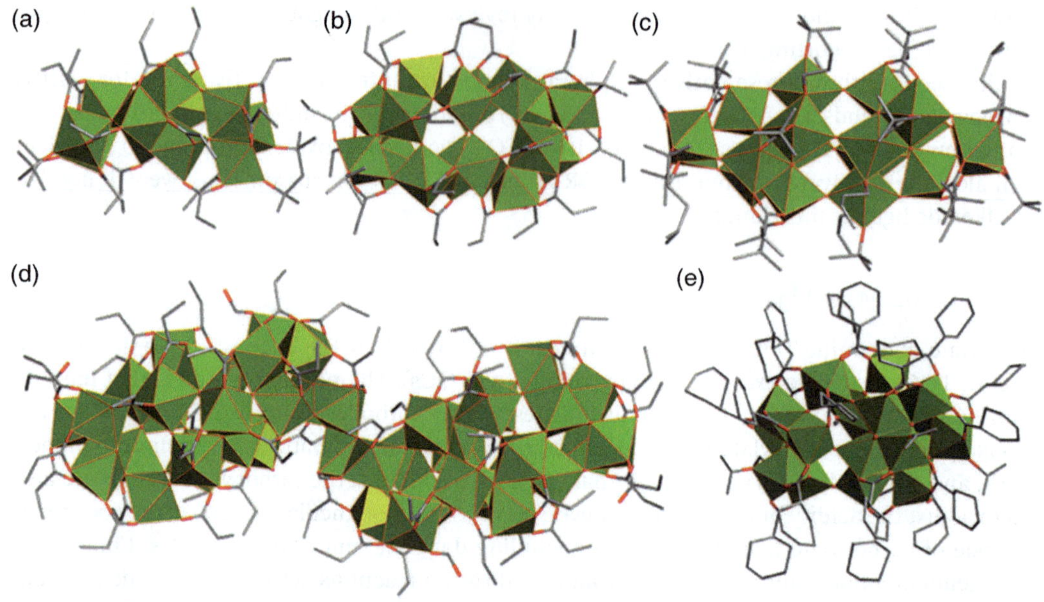

Figure 14.7 Polyhedral views of representative carboxylate-substituted TOCs. (a) Ti_{11} [14]; (b) Ti_{16} [14]; (c) Ti_{18} [14]; (d) Ti_{44} [14]; (e) Ti_{18} [30]. H atoms are omitted for clarity.

and its derivative-modified structures are some of the most researched. Helpfully, this kind of ligand reveals a more complex coordination chemistry, with not just edge- or corner-sharing modes. Diprotic phosphonate ligands $RP(O)(OH)_2$ can produce two anions, $RPO_2(OH)^-$ and RPO_3^{2-}, which can both work as ligands coordinating with Ti^{4+} cations. $RPO_2(OH)^-$ only possesses one possible coordination mode, i.e. the bridging 2.11. There are four possible coordination modes for RPO_3^{2-}, two kinds of bridging 2.11, one bridging 3.111, and one doubly bridging chelation 3.112. Among these, tripodal 3.111, like a claw, is the most frequently applied coordination mode in the structures of TOCs. In the w.xyz designation, *w* signals the number of Ti atoms binding to phosphonate ligand and each of *x*, *y*, *z* denotes the number of metal atoms coordinating to each oxygen of phosphonate ligand. These coordination modes of phosphonate ligands are illustrated in the paper reported by Coppens and coworkers [3]. The various coordination modes are also a significant factor for the expansion of the phosphonate-substituted TOC family.

According to the review of Schubert, the humidity in the feed reactants is probably the major source of the oxide ions in the final phosphonate-supported TOCs, which is comparable to the case of carboxylate ligands [8]. Moreover, large phosphonate-substituted clusters are scarce, with the highest nuclearity being 26. And the majority of phosphonate-substituted clusters are homometallic, possibly due to the coordination characteristics of phosphonate ligands. The most typical structures are the series of phosphonate-stabilized Ti_6—O clusters with labile sites, which can coordinate with a large range of carboxylates (Figure 14.8a) [20]. Zhang and coworkers also isolated a phenylphosphonate-stabilized Ti_7-oxo cluster, whose phenylphosphonate could be replaced by its derivatives, including (4-bromophenyl)methylphosphonate and methyl phenylphosphonate (Figure 14.8b) [31, 68, 69]. In 2013, Coppens and coworkers reported a list of TOCs decorated by phenyl phosphonate ligands, including the largest phosphonate-supported TOC, Ti_{26}, by far (Figure 14.8d) [67]. The average length of Ti—O bonds linking the phosphonate groups is calculated to be 2.010 Å, considerably shorter than the average value (2.087 Å) of Ti—O bonds in carboxylate-functionalized TOCs, which reveals its stability to some degree.

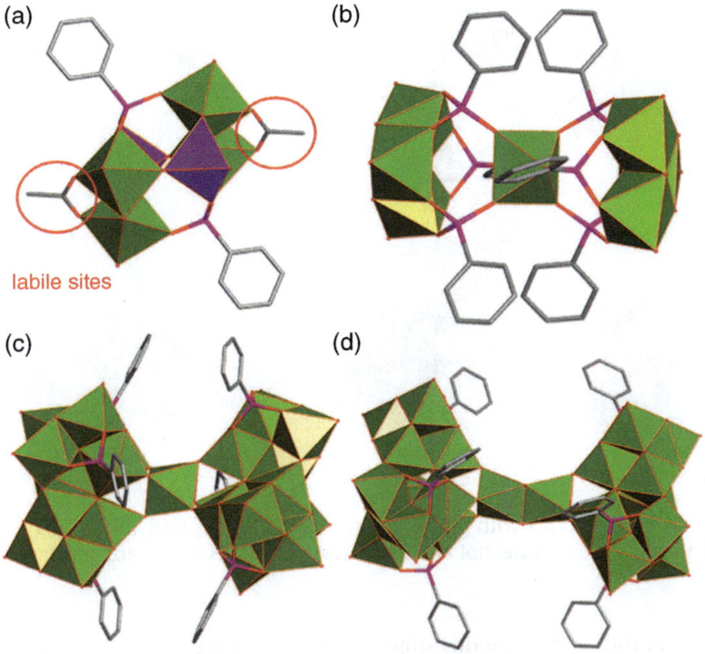

Figure 14.8 Polyhedral views of representative phosphonate-substituted TOCs. (a) Ti_6 [20]; (b) Ti_7 [31]; (c) Ti_{25} [67]; (d) Ti_{26} [67]. H atoms and alkoxides are omitted for clarity.

14.4.3 Polyphenolic Ligands

Polyphenolic ligands (catechol, salicylic acid, 2,2′-biphenol, calix[n]arenes, and their derives), as a group of O-donor ligands, also show high affinity with the Ti^{4+} cations and reveal various coordination modes [70]. Coppens and coworkers reviewed five coordinating modes of the ligands with two linking oxygen atoms, like catechol (Cat) in 2012 [47]. These contained (i) monodentate μ_1-(O), (ii) chelating bidentate μ_1-(O,O′), (iii) bridging bidentate μ_2-(O,O′), (iv) bridging chelating μ_2-(O,O′,O′), and (v) doubly bridging chelating μ_3-(O,O′,O,O′) modes. In spite of these diverse binding modes, the molecular multi-chelating phenolic modified TOCs are extremely scarce and also of low nuclearity, only ranging from Ti_3 to Ti_{17}, due to the low solubility of these ligands and lack of water molecules in the reactions [32, 71, 72]. In this section, we take catechol (abbrev. Cat) as an example. It is widely reported that TOCs with Ti^{4+} cations coordinated to O-donor ligands featuring $C(sp^2)$-O^- or $N(sp^2)$-O^- commonly display narrowed band gaps [73]. Therefore, Cat-functionalized TOCs generally show a red or dark red color, because of the charge transfer (CT) from the organic ligand to the Ti—O inorganic core [2]. The absorption edges of Cat-substituted TOCs are effectively extended to the visible light region and they are regarded as model compounds for dyed-TiO_2. Up to now, Cat-functionalized TOCs have attracted extensive attention in the pursuit of photo-related applications [66, 74]. The number of Cat ligands coordinating with TOCs is generally less than or equal to that of Ti^{4+} cations [21, 22, 75, 76].

In 2001, Henry's group obtained Cat-supported homometallic TOCs for the first time, $[Ti_{10}O_6(OBu^n)_{12}(HOBu^n)_2(Cat)_8]$ (Figure 14.9a) [70]. There are two Cat ligands using the less common doubly bridging μ_3-(O,O,O′,O′) coordination mode in this structure. In 2010, Coppens and coworkers reported the largest Cat-functionalized TOC, the Ti_{17}-oxo cluster (Figure 14.9b) [71]. Four Cat groups in this cluster apply the same type of binding mode, chelating bidentate μ_1-(O,O′). The Ti—O bond lengths in TiO_2 phases are shorter than those in the Ti_{17}-oxo cluster, exhibiting the

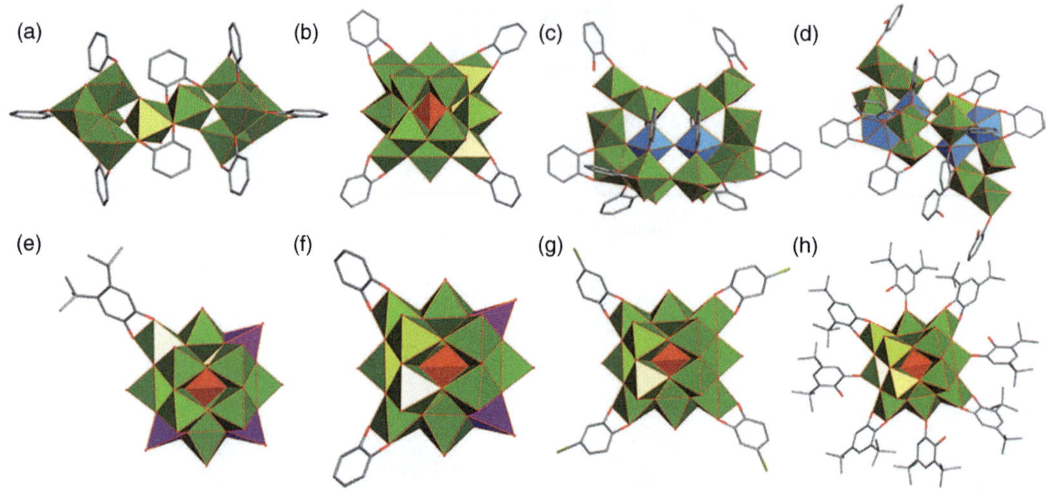

Figure 14.9 Polyhedral views of representative TOCs with polyphenolic ligands. (a) Ti_{10} [70]; (b) Ti_{17} [71]; (c, d) Ti_{14} [21]; (e, f). (g, h) $Ti_{17}Ln$ [72]. Ligands except catechol and its derivatives, and H atoms are omitted for clarity.

importance of surface effects. Whether the binding modes affect the bandgaps of Cat-functionalized TOCs needs to be further researched. In 2019, Wang and coworkers performed ligand exchange reactions using Cat derivatives without damaging the core structure, and a series of $Ti_{17}Ln$ ($n = 1$, 2, 4, and 8; $L =$ Cat or its derivatives) clusters were formed (Figures 14.9e–h) [72]. The findings implied that the degree of functionality largely depends on the ratio of Ti_{17} and Cat. Very recently, Zhang and coworkers prepared a series of Cat-functionalized Ti_{14}-oxo clusters in acid solvent, which present the lowest bandgap in this class of materials (Figures 14.9c,d) [21].

14.4.4 Sulfate Ligands

SO_4^{2-}, as a flexible ligand with hydrophilic properties, has various coordination modes and can link with 1-5, or even more metal ions [77]. Also, transition metal sulfates tend to produce low dimensional materials, owing to the low charge density of SO_4^{2-}. However, few examples of sulfate-substituted TOCs are known, with the majority being prepared in nonaqueous solvents, using less readily hydrolyzed sources of $Ti(OR)_4$ [10, 14, 78]. As already discussed, the nuclearity is largely dependent on the $[H_2O]/[Ti(OR)_4]$ ratio in the reactants. A too-large ratio generally leads to uncontrollable and spontaneous formation of TiO_2 precipitate. Typically, the trace of water coming from the esterification reaction or the humidity in the feed reactants provides controllable preparations. Two-low nuclearity complexes, $Ti_6(\mu_2\text{-}OH)_2(\mu_3\text{-}O)_2(SO_4)_4(O^iPr)_{10}$ (Figure 14.10a) [14] and $Ti_6O_4(O^iPr)_{10}(O_3PPhen)_2(HSO_4)_2$ (Figure 14.10b) [79], were obtained with organic ligands and inorganic sulfate ligands through the solvothermal method in isopropanol. It remains a great challenge to acquire higher nuclearity TOCs in aqueous solution.

Nevertheless, the highly acidic aqueous solution system can restrict the hydrolysis of Ti^{4+} cations. This opens the door to isolate sulfate supported TOCs without organic ligands. Fu et al. initially obtained two new wheel-like Ti-oxo sulfates $(C_4N_3H_{16})_5(H_3O)[TiO(SO_4)_2]_8 \cdot 15H_2O$ and $(C_4N_2H_{12})_8[TiO(SO_4)_2]_8 \cdot 23H_2O$ using a highly acidic aqueous sulfate solution as solvent in the presence of organic amines (Figure 14.10c) [34]. Wang and coworkers made significant contributions to the field of sulfate-functionalized TOCs [5, 37, 80]. Polycationic clusters of $[Ti_{18}O_{27}(OH_2)_{30}(SO_4)_6]^{6+}$ $\{Ti_{18}O_{27}\}$ can be considered as notable examples (Figure 14.10d) [5]. The triple-decked

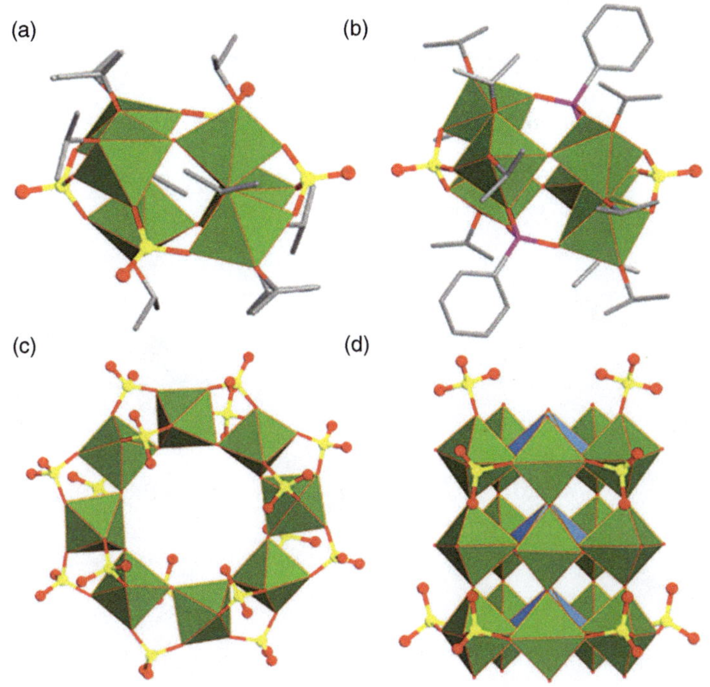

Figure 14.10 Polyhedral views of representative sulfate-substituted TOCs. (a) Ti_6 [14]; (b) Ti_6 [79]; (c) Ti_8 [34]; (d) Ti_{18} [5]. H atoms are omitted for clarity.

pentagonal prism structures are surrounded and stabilized by external sulfate anions and coordinating H_2O molecules, exhibiting good solubilities and stabilities in a number of polar solvents like acetonitrile and water.

14.4.5 Nitrogen Heterocyclic Ligands

The great majority of published TOCs are based on O-donor ligands, and those with Ti—N bonds are extremely rare. N-doping of TiO_2 has previously been prepared and displays great impact on its properties, with enhanced separation and extraction efficiency for the photogenerated carriers. Greatly motivated by N-doping of TiO_2, considerable research efforts have been taken to synthesize N-doped TOCs. Isolating crystalline N-containing TOCs using nitrogen heterocyclic ligands, which are typically nonacidic ligands, is deeply challenging. Thanks to the improvements in the CDH strategy, several N-doped TOCs have been successfully isolated. A few prominent nitrogen heterocyclic ligands, including 1,10-phenanthroline and 8-hydroxyquinoline, were employed to prepare TOCs [42, 44, 45, 81]. These nitrogen heterocyclic ligands and the Cambridge Structural Database (CSD) Refcode of corresponding TOCs are summarized in Figure 14.11 to give a systematic and comprehensive understanding.

In 2017, Zhang and coworkers reported a series of multicluster TOCs modified by nitrogen heterocyclic ligands, combined with phenyl-phosphonic acid (mentioned above in the Section 14.3.4) [64]. Three dumbbell-like Ti_4-Ti_4 dimeric structures were obtained using the N-ligands of piperazine (PIP), adenine (ADN), and aminopyrazine (APZ), respectively (Figure 14.11). When the ligand was replaced by the O/N-donor ligand 2-aminoisonicotinic acid, a triple Ti_4-Ti_6-Ti_4 complex was fabricated with the diameter of up to 2.8 nm. In the same year, Zhang and coworkers

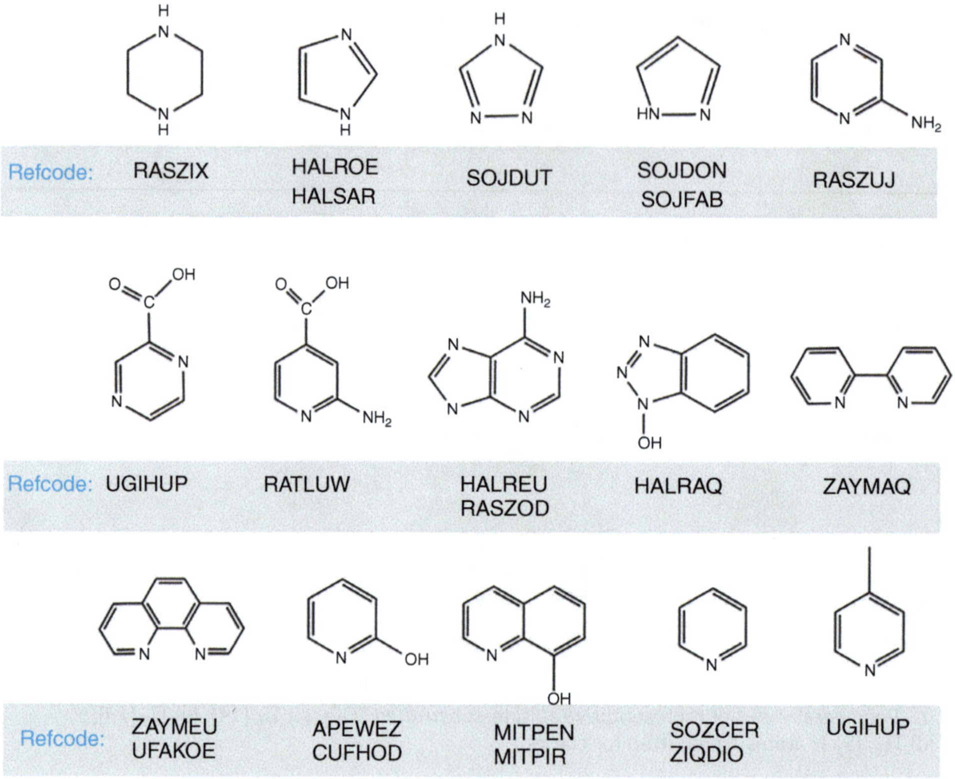

Figure 14.11 The representative nitrogen heterocyclic ligands and the CSD Refcode of corresponding TOCs.

also published a list of N-doped TOCs via ionothermal synthesis using DESs of choline chloride/phenol (molar ratio of 1 : 2) for the first time [81]. These were found to allow complete protection of the cluster core by nitrogen heterocyclic ligands (1,10-phenanthroline or 2,2′-bipyridine) and electron-donating phenols with a nuclearity of up to 28 and cluster-core diameter as high as 1.6 nm. The π-conjugated shell not only protected the Ti—O cores, preventing decomposition, but also affected their dispersible behavior in a water/triethanolamine system. The amount of doped nitrogen was subsequently increased in this system. Three N-rich TOCs were prepared via solid-state-like synthesis using nitrogen heterocyclic ligands (pyrazole and/or 1, 2, 4-triazole) as both ligands and reaction medium [44]. The ratio of N to Ti in the N-rich TOCs was as high as 24 : 10.

14.5 Metal-Doping Diversity

Up to now, considerable efforts have been devoted to heterometallic TOCs (M-TOCs), with a large quantity of them being prepared successfully with various doping metal ions, dopant numbers, and dopant sites [50, 82–90]. The majority of metal ions in the periodic table of elements have been incorporated into TOCs, including main group ions [82, 88], transition ions [23, 61, 87, 89, 91], lanthanide ions [92, 93], and so on. Metal doping brings considerable benefits in terms of their synthesis and performance. First, M-TOCs have been used as structural and reactivity models for bulk nanoscale metal-doped TiO_2 [TiO_2(M)] to investigate the correlation between structures and properties, as well as to understand the mechanisms at the molecular level, overcoming the imprecise and inhomogeneous features of [TiO_2(M)] nanomaterials. Second, metal doping can effectively reduce the band gaps, with the absorption edge shifting to the visible light region, which makes it possible to enhance solar light harvesting and photocatalytic applications. Additionally, the stability of doped polymetallic components may be improved if they are wrapped by TOCs. Moreover, M-TOCs may be more reactive than bare TOCs, with a subsequent efficiency improvement in photoelectrochemical (PEC) performance, luminescent properties, and so on, owing to their polarity, asymmetry, as well as the different electronic and steric demands of the different metal centers [94]. Finally, doped metal ions can act as cationic guests of host-guest systems and counter ions to stabilize the titanium-oxo anionic fragments.

There are two main doping modes for M-TOCs, one of which is by forming a cluster directly by coordination with Ti^{4+} cations through bridging-O atoms, [$Ti_xO_y(OR)_zM_nX_m$] (Type I, R = organic species; M = dopant metal ion; X = inorganic anion) [86, 91, 92, 95, 96] and the other is through producing a cocrystal of two separated components, [$Ti_xO_y(OR)_z$][M_nX_m] (Type II) [3, 97, 98]. In this section, we mainly describe transition metal and rare earth metal doped TOCs, in a Type I doping mode.

14.5.1 Transition Metal Doping

The great majority of transition metal ions have been integrated into TOCs including noble metal ions, forming transition metal-doped TOCs (TM-TOCs) [87, 91, 95, 99]. These transition metals typically exhibit variable coordination with Ti^{4+} ions. Silver-doped TOCs (Ag-TOCs) are relevant examples. Based on the hard and soft acid-base theory (HSAB), Ag^+ cations, as a soft acid, are apt to combine with soft bases or borderline bases, like thiolates, alkynyls, phosphines, pyridines, or acetonitrile, etc., which is distinct from the coordination "habit" of Ti^{4+} ions (hard acid with highly oxyphilic characteristics). Consequently, it is extremely challenging to fabricate Ag-TOCs considering such different coordination "habits." This challenge was solved by the discovery of negatively charged Ti—O modules with rich coordination sites, which not only induce the formation of positive Ag clusters, but also provide protection for Ag clusters [99]. By carefully analyzing the synthetic procedures, with the exception of O-donor ligands (benzoic acid, salicylic acid, etc.), soft bases or borderline bases (thiolates, alkynyls, phosphines, pyridines, or acetonitrile, etc.) are necessary to make both Ti^{4+} and Ag^+ cations "comfortable," which is consistent with the above analysis. The prepared Ag-TOCs typically demonstrate enhanced stability and exceptional applications, such as optical limiting effects, catalysis, and electricity. Along these lines, Zhang and coworkers obtained a family of Ag-TOCs, which constitute notable progress in this field (Figure 14.12) [91, 95, 100, 101, 104].

In 2018, two unprecedented $Ag_6@Ti_{16}$-oxo nanoclusters were isolated via a solvothermal reaction of $Ag(CH_3COO)_2$, $Ti(O^iPr)_4$, and benzoic acid in acetonitrile [100]. These two atomically precise Ag-TOCs revealed core-shell arrangements with a positively charged multimetallic core wrapped by a negatively charged semiconductive Ti_{16}-oxo shell. The wrapped Ag cluster was also capped by acetonitrile groups, and the $[Ag_6]^{4+}$ cation was linked with $[Ti_{16}]^{4-}$ anions through Ag-O-Ti coordination interactions. Among these, the octahedral Ag_6 core showed a mixture of valences, 0 and + 1. Interestingly, both compounds displayed intriguing NLO properties in dichloromethane solutions toward 532 nm laser light, making them promising candidates for the construction of optical limiting devices.

Subsequently, a range of increasingly larger silver nanoclusters were fabricated applying a flexible trifurcate TiL_3 (L = salicylic acid or 5-fluorosalicylic acid) module through the solvothermal approach: $Ag_4@Ag_4@Ti_4$, $Ag_{12}@Ti_4$, $Ag_4@Ag_6@Ag_{12}@Ti_4$, $Ag_6@Ag_{24}@Ag_{12}@Ti_4$, and $Ag_{12}@Ag_{24}@Ti_4$ (Figure 14.12d) [91]. These flexible Ti—O modules were found to "grasp" the $\{Ag_3\}$ vertex units with different sizes and further induce a series of encapsulated silver nanoclusters to form a varied combination of Archimedean and/or Platonic solid arrangements with average edge lengths

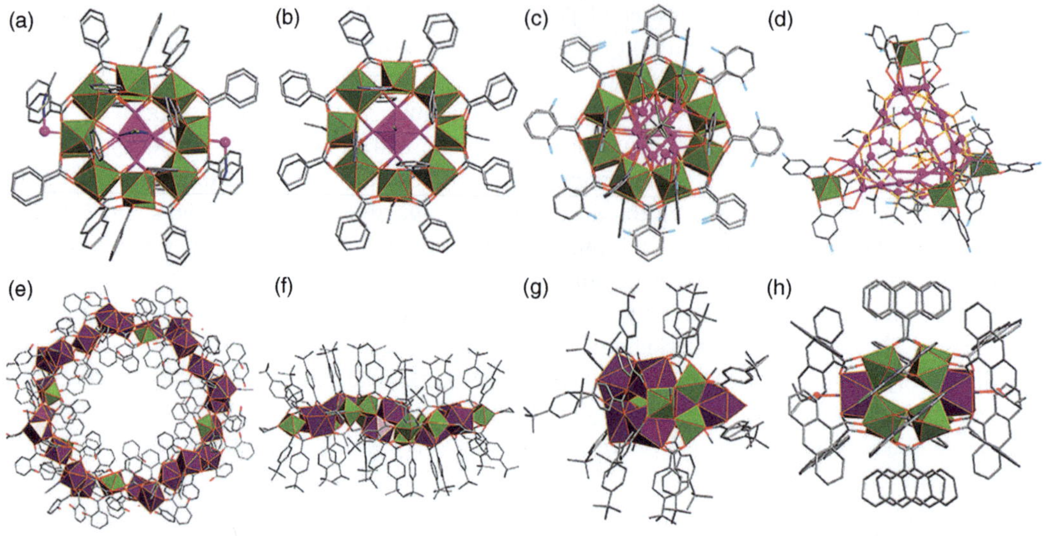

Figure 14.12 Examples of transition metal doping (upper row) and rare earth metal (RE) doping (lower row) TOCs in the polyhedral mode. (a, b) $Ti_{16}Ag_6$ [100]; (c) $Ti_{16}Ag_{14}$ [101]; (d) Ti_4Ag_{36} [91]; (e) Ti_8RE_{24} [92]; (f) $Ti_{10}RE_8$ [93]; (g) Ti_8RE_6 [102]; (h) Ti_8RE_2 [103]. H atoms are omitted for clarity. Atom color code: green Ti; pink Ag; blue N; red O; light orange S; sky blue F; gray C. Polyhedral color code: green TiO_6; violet REO_x.

varying from ~8.12 Å to ~17.37 Å. The encapsulated silver nanoclusters with tetrahedral geometry showed an overall valence state of +8 from each +1 Ag ion, due to the stabilization of the four −2 Ti—O modules. Remarkably, these clusters were proven to be ultra-stable in air for several months, and in water for more than three days, attributed to the protecting effects of the TiL_3 modules on the periphery. Additionally, the prepared clusters demonstrated NLO effects in optical limiting measurements and temperature-dependent photoluminescent performances.

Most recently, two Ag-TOCs of $Ag_5@Ti_8$ and $Ag_{14}@Ti_{16}$ were synthesized by a solvothermal reaction of $Ag(CH_3COO)_2$, $Ti(O^iPr)_4$ and 2-fluorobenzoic acid in the presence of 3,3-dimethyl-1-butyne in acetonitrile or toluene [101]. These two Ag-TOCs revealed similar arrangements with $Ag_6@Ti_{16}$-oxo nanoclusters, and the positive multimetallic core was enveloped by negatively charged semiconductive Ti_8 or Ti_{16}-oxo shells. The $[Ag_5]^{5+}$ or $[Ag_{14}]^{14+}$ cations were bound with $[Ti_8]^{5-}$ or $[Ti_{16}]^{14-}$ anions, respectively, through Ag-O-Ti coordination interactions. In $Ag_{14}@Ti_{16}$, apart from the Ti—O module, the Ag cluster was also anchored by 3,3-dimethyl-1-butyne, significantly increasing its stability. Interestingly, these clusters presented solid-state optical limiting properties toward 532 nm laser light.

14.5.2 Rare Earth Metal Doping

Lanthanide-titanium oxo clusters (Ln-TOCs) were first published in 1994, with the formula $[Sm_4Ti(\mu_5\text{-}O)(\mu_3\text{-}OR)_2(\mu\text{-}OR)_6(OR)_6]$ ($R = {}^iPr$) [105]. Since then, extensive research interest has arisen from rare earth metal-doped TOCs, with the primary reasons being the pursuit of their unique applications in the fields of magnetic susceptibility, luminescence, photocatalysis, and photoelectrocatalysis. It is thought that the formation of Eu-Ti-O species may lower the energy demands for rare earth RE^{3+} photoluminescence sensitization, compared with pure lanthanide-based metal clusters. Correspondingly, Ln-TOCs are expected to show distinct and appealing luminescence properties and have been proposed as more suitable candidates for luminescent materials [102]. In addition, the well-defined Ln-TOCs are acknowledged as theoretical and experimental models for how rare earth metals are integrated into the host lattice of TiO_2.

The existing Ln-TOCs, including mono-substituted and polymetallic-doped compounds, demonstrate a general formula [$Ti_xO_y(L)_zRE_n$] (L = ligand, RE = rare earth metal dopant) with certain characteristics. The main doping modes include surface doping (forming heterometallic cages) [106] and inner doping [92], where the doped rare earth metals connect with Ti^{4+} cations by Ti-O-RE bonds. In contrast to Ag-TOCs, there is no example of polymetallic rare earth metal clusters existing in Ln-TOCs [100]. In comparison with the multiple valence states of transition metals in TM-TOCs, the RE dopant usually possesses only one valence state (+3) in Ln-TOCs, except for the two valence states for Ce (+3 and +4). Thus far, cage structures form ranging from Ti_2 to Ti_{29} and have incorporated up to 24 RE dopants ions [3, 92]. In fitting the condensation number (y/[x+n]) and total metal nuclearity (x+n), a surprising linear relationship has been observed, which is also suitable for other metal-doped TOCs [26].

In 2012, Wright and coworkers made the highest nuclearity heterometallic cages of Ln-TOCs reported at that time, mono-substituted $Ti_{28}RE$ (RE = La and Ce) [106]. The mixtures were obtained in the same reaction, and they revealed "pear-shaped" cage arrangements. There is another known RE-doped mono-substituted cluster of $Ti_{29}Pr$, which is the largest heterometallic cages of Ln-TOCs so far [26]. The RE dopant atoms are located on the periphery close to the pseudo threefold axis of the Ti—O framework, in a configuration reminiscent of crown ethers. The high nuclearity of these species was deduced to be driven by the demand for high coordination number of the trivalent RE cations. In heterometallic cages of Ln-TOCs, doped RE ions are similarly located near the surface of the clusters. Except for $Ti_{28}RE$ and $Ti_{29}Pr$, a range of RE-containing cages are known, such as $Ti_{10}RE_2$ (RE = Nd and Eu) [107] and $Ti_{11}RE$ (RE = Sm, Eu, and Gd) [108] published by Dai's group, as well as Ti_8RE_2 (RE = Eu and Tb) [103] and Ti_8Eu_6 [102] reported by Long and coworkers.

However, to date, high-nuclearity Ln-TOCs are still rare due to the synthetic difficulty. The Long's group made a particular contribution in the preparation of large Ln-TOCs [92, 93]. [$RE_8Ti_{10}(\mu_3\text{-}O)_{14}(tbba)_{34}(Ac)_2(H_2O)_4(THF)_2$] (RE = Eu, Sm, and Gd; Htbba = 4-tert-butylbenzoic acid; Ac = acetate) is a representative example [93]. The structure, with C_2 symmetry, consists of two Ti_5RE_4 subunits fused together by two $\mu_3\text{-}O^{2-}$ ions and four tbba groups. These chemically stable Ln-TOCs demonstrate highly efficient photoelectrochemical water oxidation, which is more efficient than that of commercial TiO_2 (P25). The reasons are deduced to be lower overpotential with the incorporation of RE and oxidative cleavage of the first O—H bond. Recently, they reported the highest nuclearity Ln-TOCs, [$RE_{24}Ti_8(L)_31(HL)_{42}(CH_3CN)_{11}(H_2O)_8$] (RE = Eu and Gd, H_2L = salicylic acid) [92]. This giant structure showed a diameter of ca. 4.1 nm and was constructed based on a building block strategy. The wheel-like octamer of $RE_{24}Ti_8$ was described as eight tiny Eu_3Ti fragments linked together by rotating 180° in the head to tail mode by bridging HL^- ligands. The assembly process of the building blocks was investigated through time- and temperature-dependent high-resolution electro-spray ionization mass spectrometry (HRESI-MS).

14.6 Structural Influence on Properties and Applications

The above structural diversity has also greatly affected the physiochemical behaviors of TOCs. Inspired by band-gap regulation of TiO_2, atomically precise TOCs have applied similar methods to narrow the band gaps, i.e. metal and/or nonmetal doping and functional ligand modification, due to modifiability of the structures. Modification of functional ligands, especially photoactive ligands, is an often-used strategy to shift the absorption to a longer wavelength. In this part, we mainly describe bandgap adjustment of TOCs with functional ligands, particularly photoactive ligands.

This branch of TOC research was pioneered by Coppens and coworkers in 2010 [71]. From then on, the design and synthesis of TOCs functionalized with photoactive groups has been rapidly expanding. Dai's and Coppens's group made a significant contribution to dye-sensitized TOCs [2, 47, 109, 110]. According to the review by Zhu et al. [2] and our understanding, the photoactive ligands used to modify TOCs are divided into three categories: (i) polyphenolic ligands and coumarin derivatives, including catechol and calix[n]arene; (ii) arylamine dyes; (iii) other photoactive ligands (π-conjugated ligands), like 1,10-phenanthroline, 9-carboxyl anthracene, carboxyl ferrocene, mono- or bi-carboxyl tetrathiafulvalene, tetrakis(4-carboxyphenyl)porphyrin (TCPP), and so on [2]. Among them, the polyphenolic ligands and coumarin derivatives were mostly studied, followed by arylamine dyes, which commonly show low nuclearity. More recently, π-conjugated ligands have been introduced. One of the most prominent TOCs in this area is the kegging-like mother compound of $Ti_{17}O_{24}(O^iPr)_{20}$ (Ti_{17}) reported by Sanchez and coworkers [111]. It was found that various photoactive ligands can be used to sensitize this Ti_{17} cluster, such as cat and its derivatives, real dye of coumarin 343, p-nitrophenyl acetylacetone (NPA—H), and so on [72, 109, 112]. TOCs modified by photoactive ligands are representatively summarized in Table 14.1, where TOCs of low nuclearity (<Ti10) are not included, except TOCs with arylamine dyes.

The diversity in the structures of TOCs has also prompted an exploration of their applications, including in photocatalytic H_2 production, photocatalytic CO_2 reduction, photochromic effects, NLO effects, water oxidation, and so forth. The application of TOCs has been considerably influenced by the synergistic effect between the Ti—O cores and the functional groups [2, 3, 117]. In this section, we mainly describe the photo-related applications.

Unfortunately, the stability of some TOCs are doubtful, particularly in aqueous environments, substantially confining their wider applications. Thus, the research on increasing stability is extraordinarily essential prior to the study of their applications. One simple strategy to improve the stability of TOCs is using hydrophobic peripheral groups. Another method of enhancing stability is applying strong complexing ligands like carboxylates or pyrazoles, instead of alkoxides [14, 44]. In addition to stability, solubility is the second influential factor to be considered when using TOCs as thin-film devices. Compared with bare nano TiO_2, TOCs stabilized by protecting organic ligands can be possibly dissolved in organic solvents, while TOCs with the inorganic sulfate ligands synthesized by the aqueous sol-gel method have an indication of water-solubility and TOCs modified by POMs are found to be soluble in aqueous solution [6, 28, 118].

In the early stage, photocatalytic H_2 production was mostly studied because of the composition similarity with TiO_2 [20, 30]. The representative examples are two isomeric Ti_{20}-oxo clusters composed of pentagonal {Ti(Ti5)} fragments in corner-sharing and edge-sharing modes, separately, showing distinct H_2 evolution activities [45]. The photochromic effect is also often investigated as the Ti(IV) cations in TOCs can be reduced to Ti(III) cations under irradiation, with the color turning blue, which is then recovered by air oxidation. There are many TOCs presenting such an effect, like the first examples of dicarboxylate-substituted TOCs by Dai's group [33], linear, trimeric and rectangular TOCs with dicarboxylate ligands reported by the Zhang's group [60], carboxylate-stabilized gigantic TOCs with the nuclearity of 44 [14], and N-rich TOCs prepared via pyrazole-thermal synthesis [44].

The application diversity is also dependent on the metal doping and functionalized photoactive ligands of TOCs. Metal-doping usually generates some additional properties, like NLO effect, water oxidation, and CO_2 photoreduction. For example, Ag-TOCs showed good solution and solid NLO performance [91, 99–101]. RE-doped TOCs of $[RE_8Ti_{10}(\mu_3\text{-}O)_{14}(tbba)_{34}(Ac)_2(H_2O)_4(THF)_2]$ 2Htbba (RE = Eu, Sm, and Gd; Htbba = 4-tert-butylbenzoic acid; Ac— = acetate) exhibited outstanding water oxidation properties [93]. Another case is the largest noble metal-doped TOC, $Ag_{10}Ti_{28}$ [95], as a new type of single Ag site doped Ti—O material displayed unique CO_2 reduction potential as confirmed by DFT calculations (Figures 14.13a,b). TOCs anchored with photoelectric-active ligands could exhibit interesting photocatalytic CO_2 reduction applications, owing to the

Table 14.1 The representative TOCs functionalized by photoactive ligands.[a]

Formula	Color	Bandgap (eV)	Solvent	CCDC Number	CSD Refcode	References
Polyphenolic Ligands and Coumarin Derivatives						
$Ti_{10}O_6(^nBuO)_{12}(^nBuOH)_2(Cat)_8$	Red-orange	–	^nBuOH	167909	ICUDIT	[70]
$Ti_{10}O_{12}(Cat)_8(py)_8$	Dark red	1.97	Py	953666	ZIQDIO	[75]
$Ti_{10}O_{12}(Cat)_8(pico)_8$	Dark red	1.88	$CHCl_3$	1413512	UMEPAF	[76]
$Ti_{10}O_{12}(Cat)_8(4\text{-}Phpy)_8$	Dark red	1.86	$CHCl_3$	1413510	UMEPEJ	[76]
$Ti_{10}O_{12}(Cat)_8(pyrald)_8$	Dark red	1.85	$CHCl_3$	1413511	UMEPIN	[76]
$Ti_{17}O_{24}(Cat)_4(^iPrO)_{16}$	Red	–	Toluene	779644	LUVHIU	[71]
$Ti_{17}O_{24}(^iPrO)_{19}(DTBC)$	Red	–	^iPrOH	1843138	QOSTUQ	[72]
$Ti_{17}O_{24}(^iPrO)_{19}(^tBuCat)\cdot 2(^iPrOH)$	Red	–	^iPrOH	1843140	QOSVAY	[72]
$Ti_{17}O_{24}(^iPrO)_{18}(Cat)_2\cdot 2(^iPrOH)$	Red	–	^iPrOH	1843141	QOSVEC	[72]
$Ti_{17}O_{24}(^iPrO)_{16}(BrCat)_4\cdot 7(^iPrOH)$	Red	–	^iPrOH	1843149	QOSVOM	[72]
$Ti_{17}O_{24}(^iPrO)_{12}(DTBC)_8\cdot 13(^iPrOH)$	Red	–	^iPrOH	1843150	QOSVUS	[72]
$[I@Ti_{22}O_{30}(\mu_2\text{-}OH)_2(^iPrO)_{20}(DTBC)_2]\ (I_3)$	Dark red	3.50	^iPrOH	1500118	OSULUL	[53]
$Ti_{10}O_3(TBC[8])(BAA)_{12}(^iPrO)_6]\cdot 2CH_3CN$	Deep red	1.757	CH_3CN	2019356	SAHMIB	[113]
$Ti_{10}O_3(TBC[8])(PFBA)_{12}(^iPrO)_6$	Deep red	1.737	CH_3CN	2019357	SAHNOI	[113]
$Ti_{12}O_4(PgC_3)(O^iPr)_{28}$	Red	–	^iPrOH	1554693	LECHEJ	[65]
$Ti_{17}O_{24}(O^iPr)_{18}(C343)_2$	Orange	–	^iPrOH	898869	QAYCUQ	[47]
$Ti_{14}O_{12}(PA)_{10}(Cat)_{10}(HCat)_2$	Black	1.54	PA	2057536	TAYQUJ	[21]
$Ti_{14}O_{11}(PA)_{10}(Cat)_{10}(HCat)_4$	Black	1.51	PA	2057537	TAYRAQ	[21]
Arylamine Dyes						
$Ti_{17}O_{24}(^iPrO)_{16}(NPA)_4$	Pale yellow	–	Benzene	–	–	[112]
$Ti_4O_2(^iPrO)_4(DPAPA)_2(DDMO)_2$	Orange	2.62	Acetone	1946008	XOWDAR	[114]
$Ti_4O_2(^iPrO)_4(DEAPA)_2(DDMO)_2$	Orange	2.66	Acetone	1946009	XOWDEV	[114]
$Ti_3O(TPA)_2(^iPrO)_8$	Orange	2.2	^iPrOH	1884294	LOJZAO	[115]

(Continued)

Table 14.1 (Continued)

Formula	Color	Bandgap (eV)	Solvent	CCDC Number	CSD Refcode	References
$Ti_3O(DEPA)_2(^iPrO)_8$	Orange	2.46	^iPrOH	1884295	LOKFEZ	[115]
$Ti_6O_4(TPA)_2(O_3PPh)_2(^iPrO)_{10}$	Yellow	2.4	^iPrOH	1884297	LOKFID	[115]
$Ti_6O_4(DEPA)_2(O_3PPh)_2(^iPrO)_{10}$	Yellow	2.55	^iPrOH	1884298	LOJZIW	[115]
$Ti_6O_4ABA_2(O_3PPh)_2(^iPrO)_{10}$	Yellow	3.22	^iPrOH	1909379	JONZIY	[116]
$Ti_6O_4DABA_2(O_3PPh)_2(^iPrO)_{10}$	Yellow	3.23	^iPrOH	1909380	JONZOE	[116]
$Ti_6O_4(TPA\text{-}ABA)_2(O_3PPh)_2(^iPrO)_{10}$	Yellow	2.77	^iPrOH	1909381	JONZUK	[116]
$Ti_6O_4(DEPA\text{-}ABA)_2(O_3PPh)_2(^iPrO)_{10}$	Yellow	2.79	^iPrOH	1909382	JOPBAU	[116]
$Ti_6O_4(TPA\text{-}DABA)_2(O_3PPh)_2(^iPrO)_{10}$	Yellow	2.81	^iPrOH	1909383	JOPBEY	[116]
$Ti_6O_4(DEPA\text{-}DABA)_2(O_3PPh)_2(^iPrO)_{10}$	Yellow	2.84	^iPrOH	1909384	JOPCUP	[116]
Other Photoactive Ligands						
$[Ti_{28}O_{48}(PhenO)_{14}(1,10\text{-}phn)_{14}]Cl_2 \cdot 6PhenOH$	Red	1.9	DES	1542117	ZAYMEU	[42]
$[Ti_{14}O_{19}(1,10\text{-}phn)_8(Ac)_{12}(O^iPr)_2] \cdot 2^iPrOH \cdot 4Ac^-$	Colorless	3.15	^iPrOH THF	1849117	UFAKOE	[7]
$Ti_{11}O_9(^iPrO)_{12}(Fdc)_2(Dmg)_5$	Yellow	~1.9	^iPrOH	2053564	AMANUA	[61]
$Ti_{22}O_{18}(^iPrO)_{30}(Fdc)_4(Dmg)_7$	Yellow	~1.8	^iPrOH	2053565	AMAPAI	[61]
$Ti_{20}O_{12}(Dmg)_4(TCPP)(O^iPr)_{44}$	Crimson	1.56	^iPrOH	1914961	GOGKUL	[63]

[a] **A**bbreviations: Cat = catechol or pyrocatecholate; pico = 4-picoline; 4-Phpy = 4-phenylpyridine; pyrald = 2-methoxy-5-(pyridin-4-yl)benzaldehyde; H_2DTBC = 3,5-di-tert-butylcatechol; ^tBuCat = 4-tert-butylcatecholate; BrCat = 4-bromocatecholate; $PhPO_3H_2$ = phenylphosphonate; H_3DDMO = 2,6-dihydroxy-2,6-dimethylheptan-4-one; H_2DTBC = 3,5-di-tert-butylcatechol; PgC_3 = C-propylpyrogallol[4] arene; TBC[8] = methylene bridged t-butylcalix[8] arene; C_{343} = coumarin-343; NPA-H = p-nitrophenyl acetylacetone; DAPA = 3-(4-(diphenylamino)phenyl)acrylic acid; DEAPA = 3-(4-(diethylamino)phenyl)acrylic acid; TPA = (E)-2-cyano-3-(4-(diphenylamino)phenyl)acrylic acid; DEPA = (E)-2-cyano-3-(4-(diethylamino)phenyl)acrylic acid; ABA = 4-aminobenzoic acid; DABA = 3,5-diaminobenzoic acid; TPA-ABA = (E)-4-((4-(diphenylamino) benzylidene)amino)benzoic acid; DEPA-ABA = (E)-4-((4-(diethylamino)benzylidene)amino)benzoic acid; TPA-DABA = 3,5-bis(((E)-4-(diphenylamino)benzylidene)amino) benzoic acid; DEPA-DABA = 3,5-bis(((E)-4-(diethylamino)benzylidene)amino)benzoic acid; 1,10-phn = 1,10-phenanthroline; PhenOH = phenol; $DmgH_2$ = dimethylglyoxime; $FdcH_2$ = 1,1'-ferrocenedicarboxylic acid; TCPP = meso-Tetra(4-carboxyphenyl)porphine; BAA = bromoacetic acid; PFBA = pentafluorobenzoic acid; py = pyridine; CH_3CN = acetonitrile; ^iPrOH = isopropanol; PA = propionic acid; Ac^- = acetate; DES = choline chloride/phenol (in this review); DMF = dimethylformamide; THF = Tetrahydrofuran.

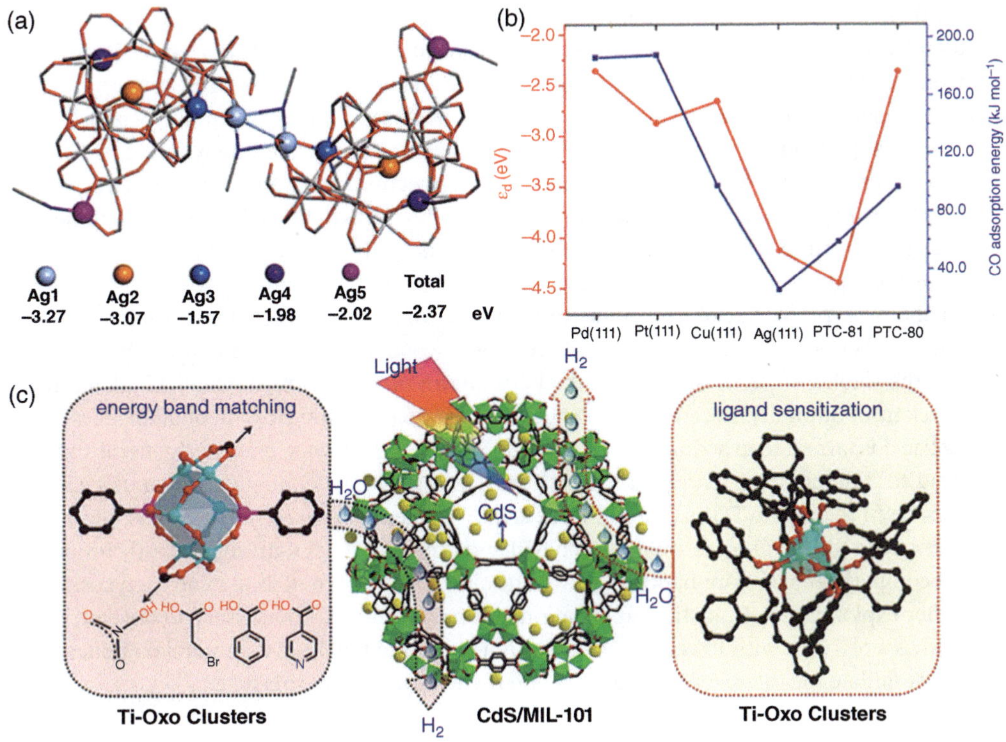

Figure 14.13 (a) DFT calculation results of the d-band center (ε_d) for inequivalent Ag atoms in $Ag_{10}Ti_{28}$. The Fermi energy was shifted to zero [95]. (b) The ε_d and CO adsorption energy for different metallic surfaces, atom-clusters [95]. (c) Strategy for constructing ternary TOC/CdS/MIL-101 photocatalysts with efficient H_2-evolution activities [119].

large number of photoelectric active groups with conjugated π-electron structures that may also act as efficient and stable electron donors. For instance, Lan and coworkers reported two Ti_6 and Ti_8 oxide clusters functionalized by 1,1-ferrocene dicarboxylic acid (Fcdc) in 2021 [120]. The introduction of Fcdc into TOCs improved the activity of photoreduction of CO_2 to $HCOO^-$, since the Fcdc ligand has the capability to rapidly transfer electrons.

Other than the above applications, TOCs can also be used to prepare composite materials with enhanced performance, attributed to the synergistic effect of each component [9, 10, 119]. For example, Jiang et al. developed a stepwise solvothermal strategy to load a range of TOCs into preformed CdS/MIL-101 composites to form new ternary PTC/CdS/MIL-101 materials (Figure 14.13c) [119]. The ternary catalysts displayed wonderful H_2 production activities in the visible-light range. Moreover, the catalytic performance could be further adjusted through modifying the functionalized ligands on TOCs. Regarding the structural diversity of TOCs, huge prospects for new TOC-based composite materials are expected.

14.7 Conclusion and Perspectives

In this chapter, we systematically outline the CDH strategy for the preparation of TOCs, the Ti–O core diversity, the ligands and metal-doping diversity, and the influence of structural diversity on applications. In the following, some conclusions and perspectives have been drawn according to our understanding of the state of this field, informed by the work of our group in recent years.

Although there are still no general methods currently available to design particular structures of TOCs, we endeavor to uncover the strategies to fabricate high-nuclearity TOCs. By "taming" the highly uncontrollable reactivity of the strong Lewis acid Ti^{4+} cations using the CDH strategy, significant advances have already been achieved both in the nuclearity and structural types of TOCs. However, we must point out that research on TOCs is still in its infancy, and many foreseeable challenges need to be emphasized and properly tackled. From a structural perspective, both nuclearity and number of the existing TOCs are still far less than the well-developed POMs field. It is worth noting that the nuclearity of most known TOCs is of even numbers, with only two odd-number examples known among the high nuclearity titanium-oxo clusters (>Ti_{20}), Ti_{25} [67], and Ti_{29} [3]. A better understanding of the bulk properties of TiO_2 may be achieved when larger and more varied TOCs are attained and characterized. Regarding the stability, most existing TOCs are modified by alkoxide ligands from the alcohol solvents or from residual OR groups of $Ti(OR)_4$ precursors, which are then prone to attack from humidity. Recent reports have demonstrated the substitution of alkoxide solvent by propionic acid is favorable to acquire TOCs with high stability. However, it is not a general strategy and may lead to different core structures. Therefore, it is highly challenging to judiciously choose ligands and solvent to prepare TOCs that fulfill the stability criteria. As to the applications, the synergistic effect between the Ti–O cores and functional ligands endows TOCs with significantly improved potential, in comparison with bare nano TiO_2. Nevertheless, the tough scalability and high costs hinder the further explorations of practical applications. Consequently, it is necessary to develop more scalable synthesis of TOCs with lower cost. It is our hope that the potential commercial applications of TOCs will stimulate the interest of researchers and invite new ideas to this field.

Acknowledgment

This work was supported by National Natural Science Foundation of China (21922111 and 91961108).

References

1 Liu, C. and Wang, Y. (2021). Supramolecular chemistry of titanium oxide clusters. *Chem. Eur. J.* 27: 4270–4282.

2 Zhu, Q.-Y. and Dai, J. (2021). Titanium oxo/alkoxyl clusters anchored with photoactive ligands. *Coord. Chem. Rev.* 430: 213664.

3 Coppens, P., Chen, Y., and Trzop, E. (2014). Crystallography and properties of polyoxotitanate nanoclusters. *Chem. Rev.* 114: 9645–9661.

4 Zhang, Y.J., de Azambuja, F., and Parac-Vogt, T.N. (2021). The forgotten chemistry of group(IV) metals: a survey on the synthesis, structure, and properties of discrete Zr(IV), Hf(IV), and Ti(IV) oxo clusters. *Coord. Chem. Rev.* 438: 213886.

5 Zhang, G., Liu, C., Long, D.L. et al. (2016). Water-soluble pentagonal-prismatic titanium-oxo clusters. *J. Am. Chem. Soc.* 138: 11097–11100.

6 Chakraborty, B. and Weinstock, I.A. (2019). Water-soluble titanium-oxides: complexes, clusters and nanocrystals. *Coord. Chem. Rev.* 382: 85–102.

7 Hong, Z.-F., Xu, S.-H., Yan, Z.-H. et al. (2018). A large titanium oxo cluster featuring a well-defined structural unit of rutile. *Cryst. Growth Des.* 18: 4864–4868.

8 Schubert, U. (2021). Titanium-oxo clusters with bi- and tridentate organic ligands: gradual evolution of the structures from small to big. *Chem. Eur. J.* 27: 11239–11256.

9 Gu, Z.G., Fu, H., Neumann, T. et al. (2016). Chiral porous metacrystals: employing liquid-phase epitaxy to assemble enantiopure metal-organic nanoclusters into molecular framework pores. *ACS Nano* 10: 977–983.

10 Gao, M.Y., Wang, F., Gu, Z.G. et al. (2016). Fullerene-like polyoxotitanium cage with high solution stability. *J. Am. Chem. Soc.* 138: 2556–2559.

11 Watenpaugh, K. and Caughlan, C.N. (1967). The crystal and molecular structure of the first hydrolysis product ($Ti_7O_{24}Et_{19}$) of titanium tetraethoxide. *Chem. Commun.* 76–77.

12 Anson, C.E., Eichhöfer, A., Issac, I. et al. (2008). Synthesis and crystal structures of the ligand-stabilized silver chalcogenide clusters $[Ag_{154}Se_{77}(dppxy)_{18}]$, $[Ag_{320}(StBu)_{60}S_{130}(dppp)_{12}]$, $[Ag_{352}S_{128}(StC_5H_{11})_{96}]$ and $[Ag_{490}S_{188}(StC_5H_{11})_{114}]$. *Angew.Chem. Int. Ed.* 120: 1346–1351.

13 Muller, A., Beckmann, E., Bogge, H. et al. (2002). Inorganic chemistry goes protein size: a Mo_{368} nano-hedgehog initiating nanochemistry by symmetry breaking. *Angew. Chem. Int. Ed.* 41: 1162–1167.

14 Gao, M.Y., Zhang, L., and Zhang, J. (2019). Acid-controlled synthesis of carboxylate-stabilized Ti_{44}-oxo clusters: scaling up preparation, exchangeable protecting ligands and photophysical properties. *Chem. Eur. J.* 25: 1–7.

15 Fang, W.H., Zhang, L., and Zhang, J. (2016). A 3.6 nm Ti_{52}-oxo nanocluster with precise atomic structure. *J. Am. Chem. Soc.* 138: 7480–7483.

16 Linsebigler, A.L., Lu, G., and Yates, J.T. (1995). Photocatalysis on TiO_2 surfaces: principles, mechanisms, and selected results. *Chem. Rev.* 95: 735–758.

17 Kment, S., Riboni, F., Pausova, S. et al. (2017). Photoanodes based on TiO_2 and alpha-Fe_2O_3 for solar water splitting – superior role of 1D nanoarchitectures and of combined heterostructures. *Chem. Soc. Rev.* 46: 3716–3769.

18 Nguyen, H.L. (2017). The chemistry of titanium-based metal–organic frameworks. *New J. Chem.* 41: 14030–14043.

19 Bomgardner, M.M. (2017). Europe may call TiO_2 a carcinogen. *C&EN News* 95: 13.

20 Liu, J.X., Gao, M.Y., Fang, W.H. et al. (2016). Bandgap engineering of titanium-oxo clusters: labile surface sites used for ligand substitution and metal incorporation. *Angew. Chem. Int. Ed.* 55: 5160–5165.

21 Gao, M.Y., Wang, Z., Li, Q.H. et al. (2022). Black titanium-oxo clusters with ultralow band gaps and enhanced nonlinear optical performance. *J. Am. Chem. Soc.* 144: 8153–8161.

22 Yu, Y.Z., Zhang, Y.R., Geng, C.H. et al. (2019). Precise and wide-ranged band-gap tuning of Ti_6-Core-based titanium oxo clusters by the type and number of chromophore ligands. *Inorg. Chem.* 58: 16785–16791.

23 Wang, C., Liu, C., Li, L.J., and Sun, Z.M. (2019). Synthesis, crystal structures, and photochemical properties of a family of heterometallic titanium oxo clusters. *Inorg. Chem.* 58: 6312–6319.

24 Li, N., Liu, J.-J., Sun, J.-W. et al. (2020). Calix[8]arene-constructed stable polyoxo-titanium clusters for efficient CO_2 photoreduction. *Green Chem.* 22: 5325–5332.

25 Li, L., Wang, X.S., Liu, T.F., and Ye, J.H. (2020). Titanium-based MOF materials: from crystal engineering to photocatalysis. *Small Methods* 4: 2000486.

26 Matthews, P.D., King, T.C., and Wright, D.S. (2014). Structure, photochemistry and applications of metal-doped polyoxotitanium alkoxide cages. *Chem. Commun.* 50: 12815–12823.

27 Rozes, L. and Sanchez, C. (2011). Titanium oxo-clusters: precursors for a Lego-like construction of nanostructured hybrid materials. *Chem. Soc. Rev.* 40: 1006–1030.

28 Gao, M., Zhang, L., and Zhang, J. (2018). Preparation and properties of polyoxo-titanium clusters. *Chin. Sci. Bull.* 63: 2731–2744.

29 Fang, W.H., Zhang, L., and Zhang, J. (2018). Synthetic strategies, diverse structures and tuneable properties of polyoxo-titanium clusters. *Chem. Soc. Rev.* 47: 404–421.

30 Gao, M.Y., Chen, S., Hu, L.X. et al. (2017). Synthesis and photocatalytic H_2 evolution properties of four titanium-oxo-clusters based on a cyclohex-3-ene-1-carboxylate ligand. *Dalton Trans.* 46: 10630–10634.

31 Gao, M.Y., Liu, X.X., Fang, W.H. et al. (2018). Synthesis and structural characterization of a dumbbell-like phenylphosphonate-stabilized Ti_7-oxide cluster. *Acta Crystallogr. C* 74: 1248–1251.

32 Gao, M.Y., Sun, Y., Wang, F. et al. (2021). Synthesis and structure of a series of Ti_6-oxo clusters functionalized by in situ esterified dicarboxylate ligands. *Chin. J. Chem.* 39: 1259–1264.

33 Wu, Y.Y., Luo, W., Wang, Y.H. et al. (2012). Titanium-oxo-clusters with dicarboxylates: single-crystal structure and photochromic effect. *Inorg. Chem.* 51: 8982–8988.

34 Fu, Y., Xu, Z., and Zhang, F. (2008). New inorganic discrete 8-membered wheel clusters: organically directed titanium sulfate supermolecules. *J. Mol. Struct.* 873: 168–172.

35 Zhang, G., Hou, J., Li, M. et al. (2016). Counteranion-stabilized titanium(IV) Isopolyoxocationic clusters isolated from water. *Inorg. Chem.* 55: 4704–4709.

36 Reichmann, M.G., Hollander, F.J., and Bell, A.T. (1987). Structure of $[Ti_8O_{12}(H_2O)_{24}]Cl_8$. HCl. $7H_2O$. *Acta Crystallogr. C* 43: 1681–1683.

37 Liu, W., Liu, C., Wu, W. et al. (2020). Effects of organic ammonium cations on the isolation of $\{Ti_4\}$ cyclic clusters from water: an ^{17}O NMR study. *Dalton Trans.* 49: 5957–5964.

38 Zhang, G.Y., Li, W.Y., Liu, C.Y. et al. (2018). Titanium-oxide host clusters with exchangeable guests. *J. Am. Chem. Soc.* 140: 66–69.

39 Smith, E.L., Abbott, A.P., and Ryder, K.S. (2014). Deep eutectic solvents (DESs) and their applications. *Chem. Rev.* 114: 11060–11082.

40 Chintakrinda, K., Narayanam, N., Li, Y.-Z. et al. (2020). Atomically precise titanium–oxo nanotube with selective water adsorption and semiconductive behaviors. *CCS Chemistry* 2: 209–215.

41 Abbott, A.P., Capper, G., Davies, D.L. et al. (2001). Preparation of novel, moisture-stable, Lewis-acidic ionic liquids containing quaternary ammonium salts with functional side chains. *Chem. Commun.* 2010–2011.

42 Narayanam, N., Fang, W.H., Chintakrinda, K. et al. (2017). Deep eutectic-solvothermal synthesis of titanium-oxo clusters protected by pi-conjugated chromophores. *Chem. Commun.* 53: 8078–8080.

43 Fan, X., Fu, H., Gao, M.Y. et al. (2019). One-pot and postsynthetic phenol-thermal synthesis toward highly stable titanium-oxo clusters. *Inorg. Chem.* 58: 13353–13359.

44 Fan, X., Fu, H., Zhang, L., and Zhang, J. (2019). Pyrazole-thermal synthesis: a new approach towards N-rich titanium-oxo clusters with photochromic behaviors. *Dalton Trans.* 48: 8049–8052.

45 Fan, X., Wang, J., Wu, K. et al. (2019). Isomerism in titanium-oxo clusters: molecular anatase model with atomic structure and improved photocatalytic activity. *Angew. Chem. Int. Ed.* 58: 1320–1323.

46 Benedict, J.B., Freindorf, R., Trzop, E. et al. (2010). Large polyoxotitanate clusters: well-defined models for pure-phase TiO_2 structures and surfaces. *J. Am. Chem. Soc.* 132: 13669–13671.

47 Sokolow, J.D., Trzop, E., Chen, Y. et al. (2012). Binding modes of carboxylate- and acetylacetonate-linked chromophores to homodisperse polyoxotitanate nanoclusters. *J. Am. Chem. Soc.* 134: 11695–11700.

48 Frot, T., Cochet, S., Laurent, G. et al. (2010). $Ti_8O_8(OOCR)_{16}$, a new family of titanium-oxo clusters: potential NBUs for reticular chemistry. *Eur. J. Inorg. Chem.* 2010: 5650–5659.

49 Czakler, M., Artner, C., and Schubert, U. (2015). Influence of the $Ti(O\mathit{i}Pr)_4$: methacrylic acid ratio on the formed oxo/alkoxo clusters. *Monatsh. Chem.* 146: 897–902.

50 Yuan, L.B., Li, Q.H., Zhang, J., and Zhang, L., Stepwise Bi and Bi/Cu doping of molecular titanium-oxo clusters: Solution behaviors, electronic structures, and assembly pathway unraveling by mass spectrometry. Submitted.

51 Zhao, C., Han, Y.Z., Dai, S. et al. (2017). Microporous cyclic titanium-oxo clusters with labile surface ligands. *Angew. Chem. Int. Ed.* 56: 16252–16256.

52 Day, V.W., Eberspacher, T.A., Klemperer, W.G., and Park, C.W. (1993). Dodecatitanates – a new family of stable polyoxotitanates. *J. Am. Chem. Soc.* 115: 8469–8470.

53 Hou, J., Hu, J., Sun, Q. et al. (2016). A post-functionalizable iso-polyoxotitanate cage cluster. *Inorg. Chem.* 55: 7075–7078.

54 Wang, X., Gui, D., Zhai, F. et al. (2018). Single-crystal-to-single-crystal desolvation in a Ti_{32} nanoring cluster. *CrystEngComm* 20: 7062–7065.

55 Hong, K. and Chun, H. (2013). Nonporous titanium-oxo molecular clusters that reversibly and selectively adsorb carbon dioxide. *Inorg. Chem.* 52: 9705–9707.

56 Verma, S., Bhat, G.A., and Murugavel, R. (2021). Cyclopentadienyl removal assisted nuclearity expansion in thermolabile titanium and zirconium organophosphates sourced from metallocene dichlorides. *J. Organomet. Chem.* 932: 121642.

57 Gao, M.Y., Song, B.Q., Sensharma, D., and Zaworotko, M.J. (2020). Crystal engineering of porous coordination networks for C3 hydrocarbon separation. *SmartMat* 2: 38–55.

58 Barrow, H., Brown, D.A., Alcock, N.W. et al. (1995). An octameric titanium oxo metallacycle with host-guest interactions. *J. Chem. Soc. Chem. Commun.* 1231–1232.

59 Müller, A. and Gouzerh, P. (2012). From linking of metal-oxide building blocks in a dynamic library to giant clusters with unique properties and towards adaptive chemistry. *Chem. Soc. Rev.* 41: 7431–7463.

60 Gao, M.Y., Fan, X., Zhang, L., and Zhang, J. (2018). Dicarboxylate ligands oriented assembly of $\{Ti_3(\mu_3\text{-}O)\}$ units: from dimer to coordination triangles and rectangles. *Inorg. Chem.* 57: 5642–5647.

61 Han, E.M., Yu, W.D., Li, L.J. et al. (2021). Accurate assembly of ferrocene-functionalized $\{Ti_{22}Fc_4\}$ clusters with photocatalytic amine oxidation activity. *Chem. Commun.* 57: 2792–2795.

62 Ding, Q.R., Liu, J.X., Narayanam, N. et al. (2017). Construction of molecular rectangles with titanium-oxo clusters and rigid aromatic carboxylate ligands. *Dalton Trans.* 46: 16000–16003.

63 Ding, Q.R., Xu, G.L., Zhang, L., and Zhang, J. (2019). Ligand-directed assembly engineering of trapezoidal $\{Ti_5\}$ building blocks stabilized by dimethylglyoxime. *Dalton Trans.* 48: 9916–9919.

64 Gao, M.-Y., Fang, W.-H., Wen, T. et al. (2017). Connecting titanium-oxo clusters by nitrogen heterocyclic ligands to produce multiple cluster series with photocatalytic H_2 evolution activities. *Cryst. Growth Des.* 17: 3592–3595.

65 Su, K., Wu, M., Tan, Y. et al. (2017). A monomeric bowl-like pyrogallol[4]arene Ti12 coordination complex. *Chem. Commun.* 53: 9598–9601.

66 Lv, H.T., Li, H.M., Zou, G.D. et al. (2018). Titanium-oxo clusters functionalized with catecholate-type ligands: modulating the optical properties through charge-transfer transitions. *Dalton Trans.* 47: 8158–8163.

67 Chen, Y., Trzop, E., Sokolow, J.D., and Coppens, P. (2013). Direct observation of the binding mode of the phosphonate anchor to nanosized polyoxotitanate clusters. *Chem. Eur. J.* 19: 16651–16655.

68 Hayami, R., Sagawa, T., Tsukada, S. et al. (2018). Synthesis, characterization and properties of titanium phosphonate clusters. *Polyhedron* 147: 1–8.

69 Czakler, M., Artner, C., and Schubert, U. (2013). Influence of the phosphonate ligand on the structure of phosphonate-substituted titanium oxo clusters. *Eur. J. Inorg. Chem.* 2013: 5790–5796.

70 Gigant, K., Rammal, A., and Henry, M. (2001). Synthesis and molecular structures of some new titanium(IV) aryloxides. *J. Am. Chem. Soc.* 123: 11632–11637.

71 Benedict, J.B. and Coppens, P. (2010). The crystalline nanocluster phase as a medium for structural and spectroscopic studies of light absorption of photosensitizer dyes on semiconductor surfaces. *J. Am. Chem. Soc.* 132: 2938–2944.

72 Liu, C., Hu, J., Zhu, F. et al. (2019). Functionalization of titanium oxide cluster $Ti_{17}O_{24}(O^iC_3H_7)_{20}$ with Catechols: structures and ligand-exchange reactivities. *Chem. Eur. J.* 25: 14843–14849.

73 Passadis, S.S., Hadjithoma, S., Kalampounias, A.G. et al. (2019). Synthesis, structural and physicochemical characterization of a new type Ti_6-oxo cluster protected by a cyclic imide dioxime ligand. *Dalton Trans.* 48: 5551–5559.

74 Wu, Y.Y., Lu, X.W., Qi, M. et al. (2014). Titanium-oxo cluster with 9-anthracenecarboxylate antennae: a fluorescent and photocurrent transfer material. *Inorg. Chem.* 53: 7233–7240.

75 Chaumont, C., Mobian, P., and Henry, M. (2014). An unprecedented high nuclearity catecholato-based Ti(IV)-architecture bearing labile pyridine ligands. *Dalton Trans.* 43: 3416–3419.

76 Chaumont, C., Chaumont, A., Kyritsakas, N. et al. (2016). Titanium oxo-clusters derivatized from the $Ti_{10}O_{12}(cat)_8(py)_8$ complex: structural investigation and spectroscopic studies of light absorption. *Dalton Trans.* 45: 8760–8769.

77 Hfidhi, N., Kammoun, O., Pelka, R. et al. (2018). Structure features, thermal analysis, model and experiment magnetic behavior of tetranuclear oxo and sulfato bridged copper (II) aggregate. *Inorg. Chim. Acta* 469: 431–439.

78 Zhu, B.C., Fang, W.H., Emayavaramban, P. et al. (2018). Structures and photophysical performances of (fluoro)salicylate stabilized polyoxo-titanium clusters. *CrystEngComm* 20: 5964–5968.

79 Zheng, A.P., Gao, M.Y., Fang, W.H., and Kang, Y. (2021). Two novel $\{Ti_6P_2\}$ clusters decorated with inorganic acids((1)). *Chin. J. Struc. Chem.* 40: 277–282.

80 Zhang, G., Hou, J., Tung, C.H., and Wang, Y. (2016). Small titanium oxo clusters: primary structures of titanium(IV) in water. *Inorg. Chem.* 55: 3212–3214.

81 Narayanam, N., Chintakrinda, K., Fang, W.H. et al. (2016). Azole functionalized Polyoxo-titanium clusters with sunlight-driven dye degradation applications: synthesis, structure, and photocatalytic studies. *Inorg. Chem.* 55: 10294–10301.

82 Molina, P.I., Kozma, K., Santala, M. et al. (2017). Aqueous bismuth titanium-oxo sulfate cluster speciation and crystallization. *Angew. Chem. Int. Ed.* 56: 16277–16281.

83 Yang, S., Su, H.-C., Hou, J.-L. et al. (2017). The effects of transition-metal doping and chromophore anchoring on the photocurrent response of titanium-oxo-clusters. *Dalton Trans.* 46: 9639–9645.

84 Liu, Y.-J., Fang, W.-H., Zhang, L., and Zhang, J. (2020). Recent advances in heterometallic polyoxotitanium clusters. *Coord. Chem. Rev.* 404: 213099.

85 Jin, W.T., Yang, F., Deng, L. et al. (2018). Wheel-like icosanuclear peroxotitanate – a stable water-soluble catalyst for oxygen transfer reactions. *Inorg. Chem.* 57: 14116–14122.

86 Liu, S., Hu, W., Nath, J.K. et al. (2017). $[Ti_{12}In_6O_{18}(OOCC_6H_5)_{30}]$: a multifunctional hetero-polyoxotitanate nanocluster with high stability and visible photoactivity. *Dalton Trans.* 46: 678–684.

87 Fu, H., Lin, Q., Wang, F., and Zhang, J. (2021). Construction of titanium-based metal-organic frameworks based on the Ti/Cu heteronuclear cluster. *Inorg. Chem.* 60: 24–27.

88 Zhou, S., Li, C., Fu, H. et al. (2020). Lead-doped titanium-oxo clusters as molecular models of perovskite-type $PbTiO_3$ and Electron-transport material in solar cells. *Chem. Eur. J.* 26: 6894–6898.

89 Fu, H., Zhou, S., Fan, X. et al. (2021). Assembly and packing models of $[Ti_6Co_{12}]$ ring based on the titanium-capped cobalt clathrochelates. *Chin. Chem. Lett.* 32: 923–925.

90 Artner, C., Kronister, S., Czakler, M., and Schubert, U. (2014). Ion-size-dependent formation of mixed titanium/lanthanide oxo clusters. *Eur. J. Inorg. Chem.* 2014: 5596–5602.

91 Gao, M.Y., Wang, K., Sun, Y. et al. (2020). Tetrahedral geometry induction of stable Ag-Ti nanoclusters by flexible trifurcate TiL_3 metalloligand. *J. Am. Chem. Soc.* 142: 12784–12790.

92 Zheng, H., Du, M.H., Lin, S.C. et al. (2018). Assembly of a wheel-like $Eu_{24}Ti_8$ cluster under the guidance of high-resolution electrospray ionization mass spectrometry. *Angew. Chem. Int. Ed.* 57: 10976–10979.

93 Lu, D.F., Kong, X.J., Lu, T.B. et al. (2017). Heterometallic lanthanide-titanium oxo clusters: a new family of water oxidation catalysts. *Inorg. Chem.* 56: 1057–1060.

94 Yang, L., Shu, X.P., Fu, M.Y. et al. (2020). Molybdenum-titanium oxo-cluster, an efficient electrochemical catalyst for the facile preparation of black titanium dioxide film. *Dalton Trans.* 49: 10516–10522.

95 Chen, S., Chen, Z.N., Fang, W.H. et al. (2019). $Ag_{10}Ti_{28}$ -oxo cluster containing single-atom silver sites: atomic structure and synergistic electronic properties. *Angew. Chem. Int. Ed.* 58: 10932–10935.

96 Son, J.H. and Casey, W.H. (2016). Titanium-substituted Polyoxotantalate clusters exhibiting wide pH stabilities: $[Ti_2Ta_8O_{28}]^{8-}$ and $[Ti_{12}Ta_6O_{44}]^{10-}$. *Chem. Eur. J.* 22: 14155–14157.

97 Fang, W.H., Zhang, L., and Zhang, J. (2017). Assembly of titanium-oxo cations with copper-halide anions to form supersalt-type cluster-based materials. *Chem. Commun.* 53: 3949–3951.

98 Gao, M.Y., Bai, H., Cui, X. et al. (2022). Precisely tailoring heterometallic polyoxotitanium clusters for the efficient and selective photocatalytic oxidation of hydrocarbons. *Angew. Chem. Int. Ed.* 61: e202215540.

99 Fan, X., Chen, S., Zhang, L., and Zhang, J. (2021). Protection of Ag clusters by metal-oxo modules. *Chem. Eur. J.* 27: 15563–15570.

100 Chen, S., Fang, W.H., Zhang, L., and Zhang, J. (2018). Atomically precise multimetallic semiconductive nanoclusters with optical limiting effects. *Angew. Chem. Int. Ed.* 57: 11252–11256.

101 Fan, X., Yuan, F., Li, D. et al. (2021). Threefold collaborative stabilization of Ag_{14}-nanorods by hydrophobic Ti_{16}-oxo clusters and alkynes: designable assembly and solid-state optical-limiting application. *Angew. Chem. Int. Ed.* 60: 12949–12954.

102 Zhao, Y.R., Zheng, H., Chen, L.Q. et al. (2019). The effect on the luminescent properties in lanthanide-titanium OXO clusters. *Inorg. Chem.* 58: 10078–10083.

103 Chen, R., Hong, Z.F., Zhao, Y.R. et al. (2019). Ligand-dependent luminescence properties of lanthanide-titanium oxo clusters. *Inorg. Chem.* 58: 15008–15012.

104 Hou, S., Li, Q., Ding, Q.R. et al. (2020). Heterometallic Ag_2Ti_{10} and Ag_4Ti_8-oxo clusters with different silver doping models: synthesis, structure, and theoretical studies. *Dalton Trans.* 49: 11005–11009.

105 Daniele, S., Hubert-Pfalzgraf, L.G., Daran, J.-C., and Halut, S. (1994). Synthesis and molecular structure of $[Sm_4Ti(\mu_5\text{-}O)(\mu_3\text{-}OR)_2(\mu\text{-}OR)_6(OR)_6]$ ($R = Pr^i$): a novel framework for heteronuclear alkoxides with a 1:4 stoichiometry. *Polyhedron* 13: 927–932.

106 Lv, Y., Willkomm, J., Leskes, M. et al. (2012). Formation of $Ti_{28}Ln$ cages, the highest nuclearity polyoxotitanates (Ln = La, Ce). *Chem. Eur. J.* 18: 11867–11870.

107 Wang, S., Su, H.-C., Yu, L. et al. (2015). Fluorescence and energy transfer properties of heterometallic lanthanide-titanium oxo clusters coordinated with anthracenecarboxylate ligands. *Dalton Trans.* 44: 1882–1888.

108 Zhang, G.-L., Wang, S., Hou, J.-L. et al. (2016). A lanthanide-titanium ($LnTi_{11}$) oxo-cluster, a potential molecule based fluorescent labelling agent and photocatalyst. *Dalton Trans.* 45: 17681–17686.

109 Negre, C.F., Young, K.J., Oviedo, M.B. et al. (2014). Photoelectrochemical hole injection revealed in polyoxotitanate nanocrystals functionalized with organic adsorbates. *J. Am. Chem. Soc.* 136: 16420–16429.

110 Hou, J.L., Huo, P., Tang, Z.Z. et al. (2018). A titanium oxo cluster model study of synergistic effect of co-coordinated dye ligands on photocurrent responses. *Inorg. Chem.* 57: 7420–7427.

111 Steunou, N., Kickelbick, G., Boubekeur, K., and Sanchez, C. (1999). A new polyoxo–alkoxo titanium cluster of the Keggin family: synthesis and characterization by X-ray diffraction and NMR spectroscopy. *J. Chem. Soc. Dalton Trans.* 3653–3655.

112 Snoeberger, R.C. 3rd, Young, K.J., Tang, J. et al. (2012). Interfacial electron transfer into functionalized crystalline polyoxotitanate nanoclusters. *J. Am. Chem. Soc.* 134: 8911–8917.

113 Yang, X.X., Yu, W.D., Yi, X.Y. et al. (2020). Monocarboxylate-driven structural growth in calix[n]arene-polyoxotitanate hybrid systems: utility in hydrogen production from water. *Chem. Commun.* 56: 14035–14038.

114 Liu, P.Y., Cui, L.N., Yang, L. et al. (2019). Bio-compatible fluorescent nano TiO materials prepared from titanium-oxo-cluster precursors. *Chem. Commun.* 55: 12360–12363.

115 Zou, D.H., Cui, L.N., Liu, P.Y. et al. (2019). Molecular model of dye sensitized titanium oxides based on aryl-amine dye anchored titanium oxo clusters. *Inorg. Chem.* 58: 9246–9252.

116 Cui, L.N., Liu, P.Y., Yang, L. et al. (2019). A series of Ti_6 oxo clusters anchored with arylamine dyes: effect of dye structures on photocurrent responses. *Chem. Asian J.* 14: 3198–3204.

117 Li, N., Matthews, P.D., Luo, H.K., and Wright, D.S. (2016). Novel properties and potential applications of functional ligand-modified polyoxotitanate cages. *Chem. Commun.* 52: 11180–11190.

118 Nomiya, K., Sakai, Y., and Matsunaga, S. (2011). Chemistry of group IV metal ion-containing polyoxometalates. *Eur. J. Inorg. Chem.* 2011: 179–196.

119 Jiang, Z., Liu, J., Gao, M. et al. (2017). Assembling polyoxo-titanium clusters and CdS nanoparticles to a porous matrix for efficient and tunable H_2-evolution activities with visible light. *Adv. Mater.* 29: 1603369.

120 Liu, J.-J., Li, N., Sun, J.-W. et al. (2021). Ferrocene-functionalized Polyoxo-titanium cluster for CO_2 photoreduction. *ACS Catal.* 11: 4510–4519.

15

Atom-Precise Cluster-Assembled Materials: Requirement and Progresses

Sourav Biswas[1], Panpan Sun[2], Xia Xin[2], Sukhendu Mandal[1], and Di Sun[2]

[1] *School of Chemistry, Indian Institute of Science Education and Research Thiruvananthapuram, Kerala 69551, India*
[2] *School of Chemistry and Chemical Engineering, State Key Laboratory of Crystal Materials, Shandong University, Ji'nan 250100, China*

15.1 Introduction

In this chapter, we briefly introduce cluster and cluster-assembled materials. Then we discuss the prospects and the classifications of the nanocluster assembling processes. In the subsequent subsections, we summarize the advancement in this area with variable dimensionality (one-, two-, and three-dimensional cluster assembly) of the assembling approach and consider their key beneficial application-oriented properties. Finally, we discuss all the plausible future possibilities for creating new cluster-assembled materials and outline the direction in which the upcoming research will continue to flourish.

Encouraged by the concept of atom-precision, the research in nanotechnology is set to explore the opportunities in every new possible direction to counteract the inherent disadvantages of conventional metal nanoparticles. The prosperous effect of electronic confinement and unmatched properties with great control over the atomic homogeneity in size, shape, and composition make atom-precise nanocluster (NC) materials more alluring [1–4]. Such metal NCs are composed of several metal atoms in the sub-nanometer regime and tens to hundreds in the nanometer regime, both with a definite protecting ligand shell as well as the interface between them, providing unique opportunities for investigating the structure–property correlation. So, these newly synthesized metal NCs have emerged to bridge the gap between discrete metal complexes and plasmonic nanoparticles [1, 5]. The structural determination of such a type of nanomaterials based on single-crystal X-ray diffraction (SCXRD) analysis has come up just a decade before, which gives the further leverage of upsurge.

In recent years, a massive advancement of this area is observed with a profound understanding of interfacial connectivity between the metal and ligands, spatial arrangement of the kernel, and staple motifs, along with the structure–property correlation, which is completely inaccessible for conventional imprecise nanoparticles [6–11]. Several mono- and bi-metallic NCs are already synthesized and studied for their exotic properties; among them, noble metal nanoclusters are highly commended [1, 2, 12–14].

To promote such atom-precise NCs in the practical application as the next generation of materials, recent research has come up with a maiden solution of fortifying the impact of assembling the component NC nodes with tailored properties [15–20]. Since then, this nanocluster field has flourished in a new direction of cluster-assembled material with improved stability and photophysical

Atomically Precise Nanochemistry, First Edition. Edited by Rongchao Jin and De-en Jiang.

properties. Multiple techniques have already been developed to form the NC assembly in various dimensionality, which is increasing day by day [21]. This chapter summarizes the advancement of these newly generated techniques of assembling metal NCs and to find their potential applications with tailored structure—property correlation.

This chapter also attempts to produce a logical guide to the different methodologies embraced in yesteryears in the design of atom-precise cluster-assembled materials, which features the crucial logical necessities and points of interest of utilizing such materials.

15.2 Prospect of Cluster-Assembling Process and Their Classification

Metal NCs have already established their niche as promising functional nanomaterials due to their unmatched properties. However, the insufficient stability (e.g. Ag, Cu), electronic property, and the small-scale fabrication methods limit the application of these novel nanomaterials in modern electronic devices [21]. Due to the strong quantum confinement effect, noble metal NCs are particularly attractive, owing to their photoluminescence property [9]. It has been observed that the photoluminescence property of NCs varies with the number of the noble metal atoms in the kernel, size, and geometry of the kernel, electronic effect of the ligands, overall geometry of the NCs, and even the rigidity of the ligands. Both silver and gold have almost indistinguishable atomic radii (2.89 vs. 2.88Å); however, silver NCs have gained more attention due to their high effectual photoluminescence property under ambient conditions. But, silver is much more reactive in its zerovalent state compared to gold and is easily susceptible to oxidization, which also obstructs its future applicability [11].

So, a strategic replacement is necessary to grow these NCs to a size that allows their combination with structures manufactured by top-down methods favorable for the cutting-edge applications and the long-term stability in the ambient environment. The smaller size of the individual NCs also appears as a boundary for the wide-band practical applicability. So, to counteract these problems, assembling of the individual NCs to form a sheet-like architecture has come into play. It has been observed that such assembly of the individual NCs improves the overall properties of the parent NCs in various ways.

Intercluster interactions (such as chemical bonding, electrostatic interaction, hydrogen bonding, van der Waals interaction, π–π interaction) are playing a pivotal role in the assembling of NC building blocks [22]. A common method of the assembly process of atomically precise metal NCs is crystallization, as the cluster nodes tend to organize in a specific pattern due to the presence of specific interactions between them, which might also account for the assembly of the individual NCs into long-range ordered arrays, coined single crystals [23–25]. However, the geometrical architecture and the size of the cluster nodes along with the other physical states influenced mostly such assembling processes [17]. To overcome the limitation of this narrow window of assembly through crystal lattice stacking, the connectivity between the cluster nodes through metal–metal bonding took the place immediately [26]. Introduction of the organic linkers is also helpful to assemble the parent NCs through covalent interactions between the cluster nodes, and N terminals of organic linkers were vastly utilized to form porous assemblies, which are reminiscent of metal–organic frameworks [27, 28]. Furthermore, the discrete NC nodes can also be assembled in the solution through an aggregation process, which restricts the molecular vibrations and strengthens the luminescence [29–31]. There have been a lot of plausible assembling approaches in the literature, which are summarized in Figure 15.1. We shall discuss these approaches and explain their superiority over the individual NCs in the following subsections. We recommend that readers who want to know about the details of synthetic procedures for each example refer to each original article.

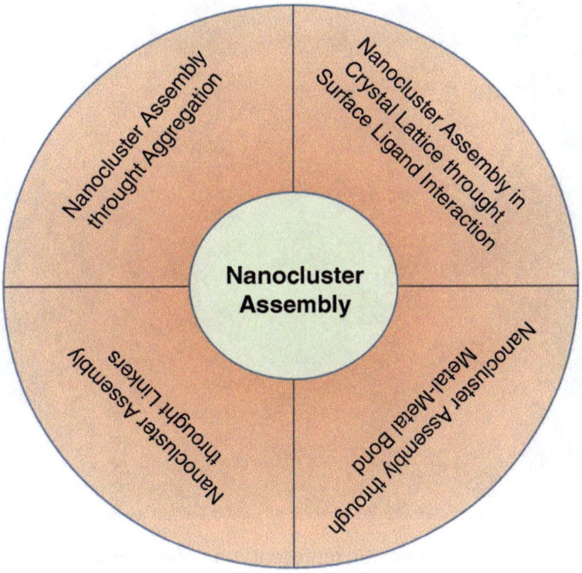

Figure 15.1 Schematic representation of the nanocluster assembling process.

15.2.1 Nanocluster Assembly in Crystal Lattice through Surface Ligand Interaction

The interactions between any individual cluster depend on their size, charge, and electronic or geometric configuration, and play a pivotal role in entropy-driven packing of the clusters in the crystal lattice that dictates its ultimate packing manner and properties. So, the determination of the assembly mode of any NCs and their dominating atomically precise interactions will be influential. Zeng et al. has shown the complex structure of $[Au_{246}(\text{p-methylbenzenethiolate})_{80}]$, which undergoes the self-assembly packing of the metal cores through surface ligands interaction [32]. The (C—H$\cdots\pi$) interaction between the surface protecting ligands directly influenced the packing of the NCs with orientational, rotational, and translational orders. They showed that the surface-protecting ligands of any particular NCs interact with other ligands from three adjacent clusters nodes through this specific (C—H$\cdots\pi$) interaction, and five pairs of interacted ligands contribute to each unit cell formation (Figure 15.2). They explained that the crystal lattice of this NC in {001} plane was organized in a square configuration, similar to the packing mode in the {100} plane of the face-centered cubic (fcc) lattice. However, interestingly, the measured intercluster distance was lower than the actual fcc lattice, which was due to the formation of interlocking conformation of the protecting ligands. As a result, these two-dimensional lattice planes were stacked up diagonally along the z-direction to maximize the inter-cluster interactions that lead to the formation of a three-dimensional lattice with an ···*ABAB*··· pattern, which led to a deviation from the standard fcc lattice structure. The calculated packing density also had a detrimental effect in comparison with a perfect fcc lattice. So, it was well observed that specific inter-particle interactions favor the entropy-controlled packing mode of the gold NCs. Furthermore, the rotational patterns of ligands during the stacking of the clusters induced chirality in the interfacial layer and transitional layer.

Further research activity showed that adequate inter-particle interactions and the nature of the surface protecting ligands led to crystallizing some more self-assembled NCs. In an example, Yan et al. described the structural stability and the availability of the crystalline state of the $[Au_{144}(\text{phe nylmethanethiolate})_{60}]$ and their packing mode in *ABAB* pattern through weak inter-cluster (C—H$\cdots\pi$) and (H$\cdots$H) interactions of surface ligands [33]. In another work, Higaki et al. showed

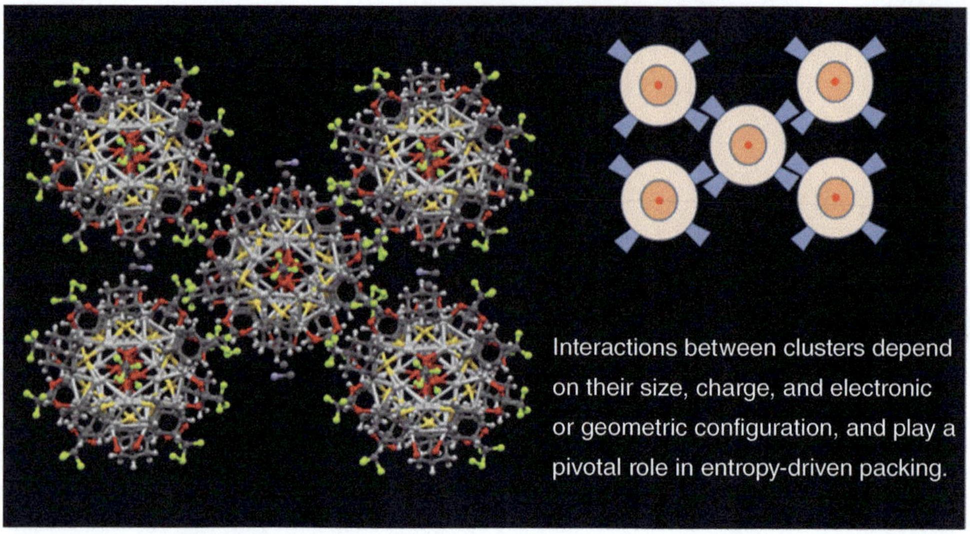

Figure 15.2 Schematic representation showing the directional packing of inter-cluster assembly and the list of influential parameters. White: Silver; red: oxygen; green: fluorine; blue: nitrogen; grey: carbon; light gray: hydrogen.

the unique herringbone-pattern-like, compact packing of [$Au_{103}S_2$(2-napthalenethiol)$_{41}$], due to the presence of both (C—H···π) and (H···H) interactions between the surface protecting ligands (Figure 15.3) [34]. The surface protecting ligands of neighboring NCs were cross-linked with T-shaped (C—H···π) interactions that accounted for the assembling process with atomic precision. Such an orderly T-shape arrangement gave rise to a "herringbone pattern," ubiquitous in crystal structures of polycyclic aromatic hydrocarbons, and the stability was increased by forming a different electronic structure.

15.2.2 Nanocluster Assembly through Metal–Metal Bonds

De Nardi et al. first discovered the metal–metal linked gold NC assembled one-dimensional chain. They synthesized [Au_{25}(n-butanethiolate)$_{18}$] NC where the Au—Au distance between the two NCs is 3.15 Å, which is in the range of aurophilic interaction (Figure 15.4) [26]. They mentioned that the choice of the surface protecting ligand was crucial for the formation of the Au—Au bond within this short distance. However, due to the specific interaction between the surface protecting ligands, the adjacent NCs were twisted and approached each other to form the one-dimensional chain. Then they observed that in the solution state, isolated [Au_{25}(n-butanethiolate)$_{18}$] NCs showed the paramagnetic behavior due to the presence of unpaired electrons whereas after forming the one-dimensional chain it showed anti-ferromagnetic behavior. This striking transformation was mainly attributed to the formation of the one-dimensional assembled network that led to the proximity of NCs, allowing the unpaired electrons of adjacent NCs to form electron pairs. Due to the formation of such electron pairs, a filled conduction band was observed with an empty valence band. So, the assembled one-dimensional framework was predicted to have semiconducting properties. Later on, the same group succeeded to synthesize metal–metal connected one-dimensional chains of [Au_{25}(pentanethiolate)$_{18}$] [35], [$Au_{24}Cd$(n-butanethiolate)$_{18}$] [36], and [$Au_{24}Hg$(n-butanethiolate)$_{18}$] [36] NCs. For all these NCs, they compared the physical properties and electronic structures with their zero-dimensional molecular clusters and found a distinct difference due to the formation of one-dimensional chains.

Later on, Chen et al. discovered two more one-dimensional alloy NCs [Au_4Pd_2(2-phenylethanethiolate)$_8$] [37], [$Au_4Pt_2(SR)_8$] (SR = 2-chlorobenzenemethanethiolate, and

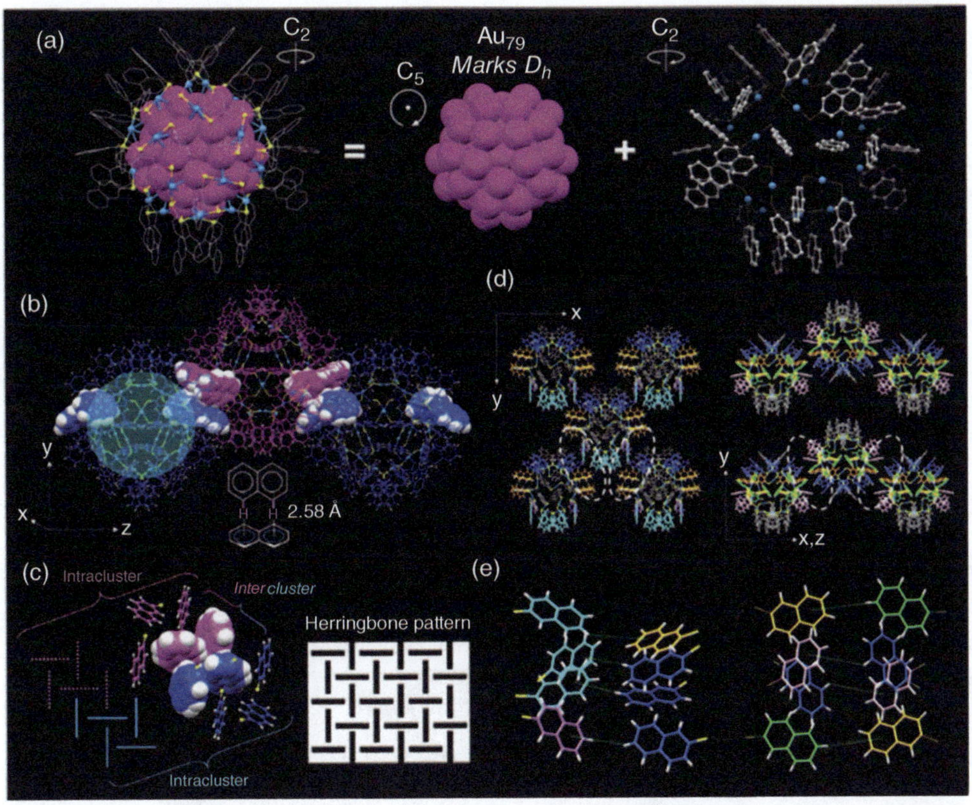

Figure 15.3 (a) Crystallographic structure of the [$Au_{103}S_2$(2-napthalenethiol)$_{41}$] NC, total structure of the NC (left), kernel (middle), and surface staple motifs (right). Magenta: Au in the kernel, blue: Au in the staple motifs. (b) Inter-cluster ligand–ligand interactions in the crystal lattice via (C—H···π) interaction. (c) Combination of inter-cluster and intra-cluster ligand–ligand interactions to form "herringbone pattern." Light blue: Au in the staple motifs, yellow: S, white: H. (d) Inter-cluster interactions between neighboring NC nodes. Sky blue: Au in the staple motifs; yellow: S; white: H; and all the other colors are used for C in different positions. (e) (C—H···π) interactions between ligands on adjacent nanoclusters. *Source:* Reproduced from Ref. [34] with permission from the ACS, copyright 2017.

2-phenylethanethiolate) [38], respectively, via Au—Au bond connectivity. It has been observed that Au and Pt form a stronger bond in comparison with Au and Pd, which facilitated the formation of a more stable one-dimensional chain of AuPt alloy NCs. So, in the case of AuPt alloy NCs, they explained that inter-cluster interactions depend exclusively on the ligand structures, and they varied the ligand structure to identify it. They observed that in the case of 2-phenylethanethiol as a ligand, the π-π interactions between the ligands oriented the clusters through rotation and form a zigzag, one-dimensional chain, whereas in the case of 2-chlorobenzenemethanethiol as a ligand, the rigidity of the ligand is crucial in the formation of the nearly straight-chain (Figure 15.5). However, when they used [$SCH(CH_3)_2$], [$SCH_2Ph(CH_3)_3$], and [SCH_2Ph^tBu], the chains were not formed properly due to the absence of any specific interactions between the ligands. Finally, they studied the electronic structure of these chains and compared them with the molecular cluster unit, and confirmed the formation of a narrow bandgap due to the assembled networks of the molecular cluster units.

Very recently, Yuan et al. the synthetic protocol of the formation of a one-dimensional chain of [$(AuAg)_{34}$(adamantanethiolate)$_{20}$] alloy NC with an (Ag—Au—Ag) bond connectivity (Figure 15.6) [39]. By the theoretical calculation and experiments, they showed the bandgap of this material is 1.3 eV, which is in the semiconducting range. Further, they utilized its semiconducting

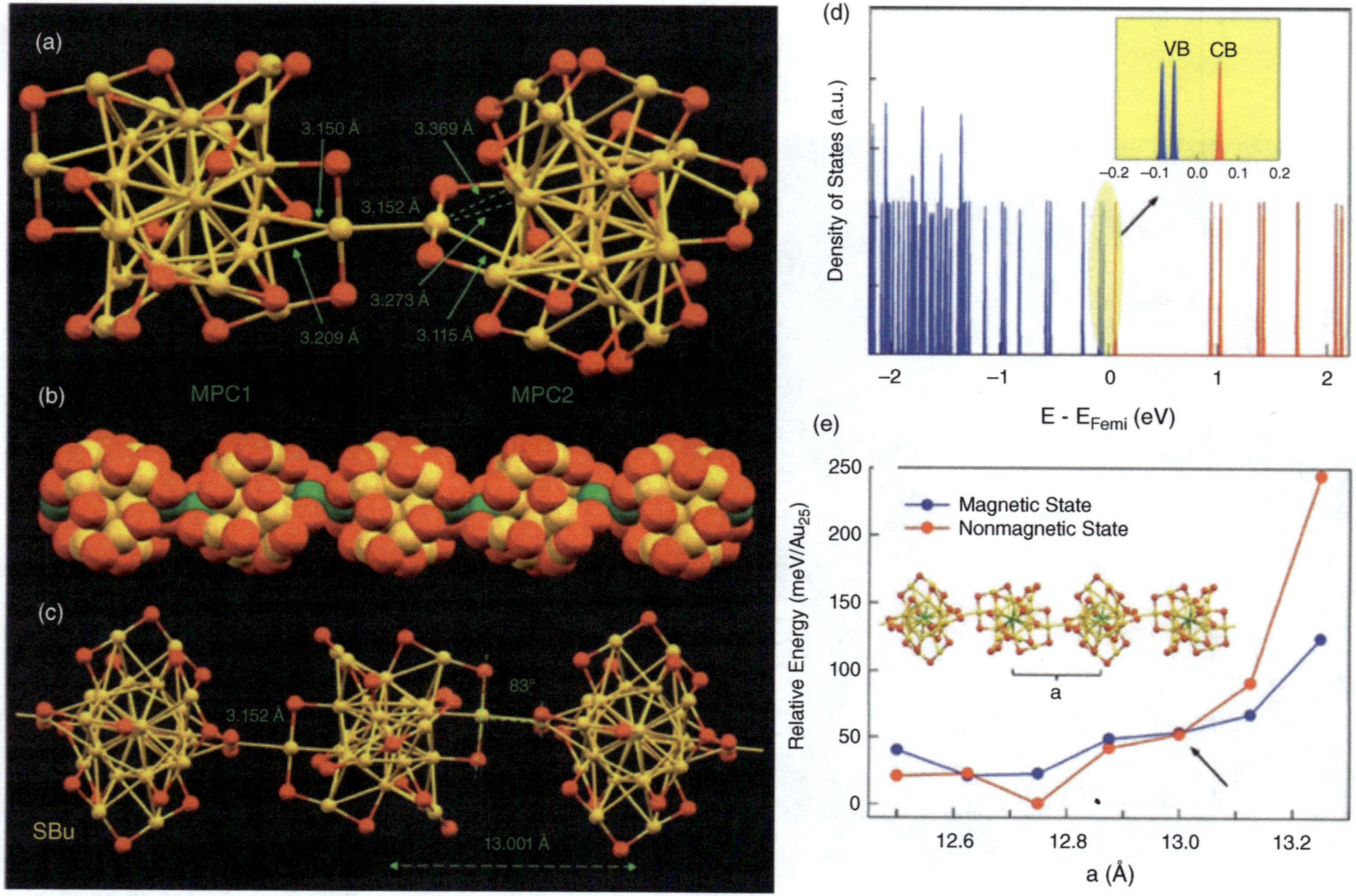

Figure 15.4 (a) Stick-and-ball view of the two-cluster node of [Au_{25}(n-butanethiolate)$_{18}$] that are connected through aurophilic interaction. (b) View of the 1D chain evidenced along the (111) direction; connecting Au—Au bonds are in green. (c) One-dimensionally assembled long chain of the [Au_{25}(n-butanethiolate)$_{18}$] NC with relevant distances and torsional angles. Yellow: Au, red: S (C and H atoms are removed for clarity). (d) Electronic density of states of the one-dimensional chain structure optimized further by density functional theory in the nonmagnetic state at 0 K. Inset shows a zoom-in of the region near the Fermi energy, with the valence band (VB) and conduction band (CB) labeled. (e) Relative energy of the nonmagnetic and magnetic states of the one-dimensional chain a function of the distance *a* between the central atoms of two neighboring clusters.
Source: Reproduced from Ref. [26] with permission from the ACS, copyright 2014.

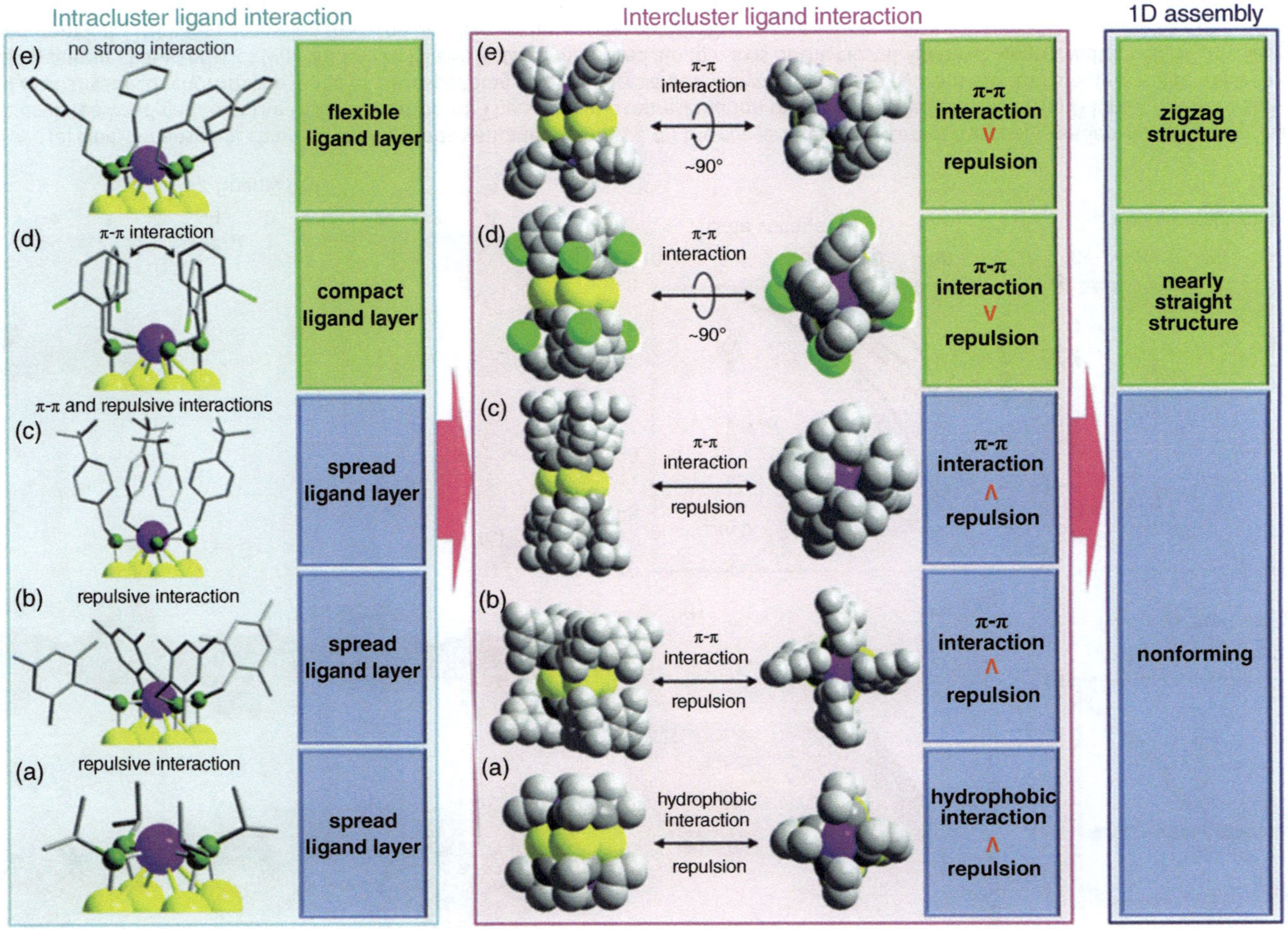

Figure 15.5 Relationship between the intracluster ligand interaction, which is related to the distribution of the ligands within each cluster, intercluster ligand interaction, and one-dimensional assembly for different ligands (a) $[SCH(CH_3)_2]$, (b) $[SCH_2Ph(CH_3)_3]$, (c) $[SCH_2Ph^tBu]$, (d) $[SCH_2PhCl]$, and (e) $[SC_2H_4Ph]$. *Source:* Reproduced from Ref. [38] with permission from the RSC, copyright 2019.

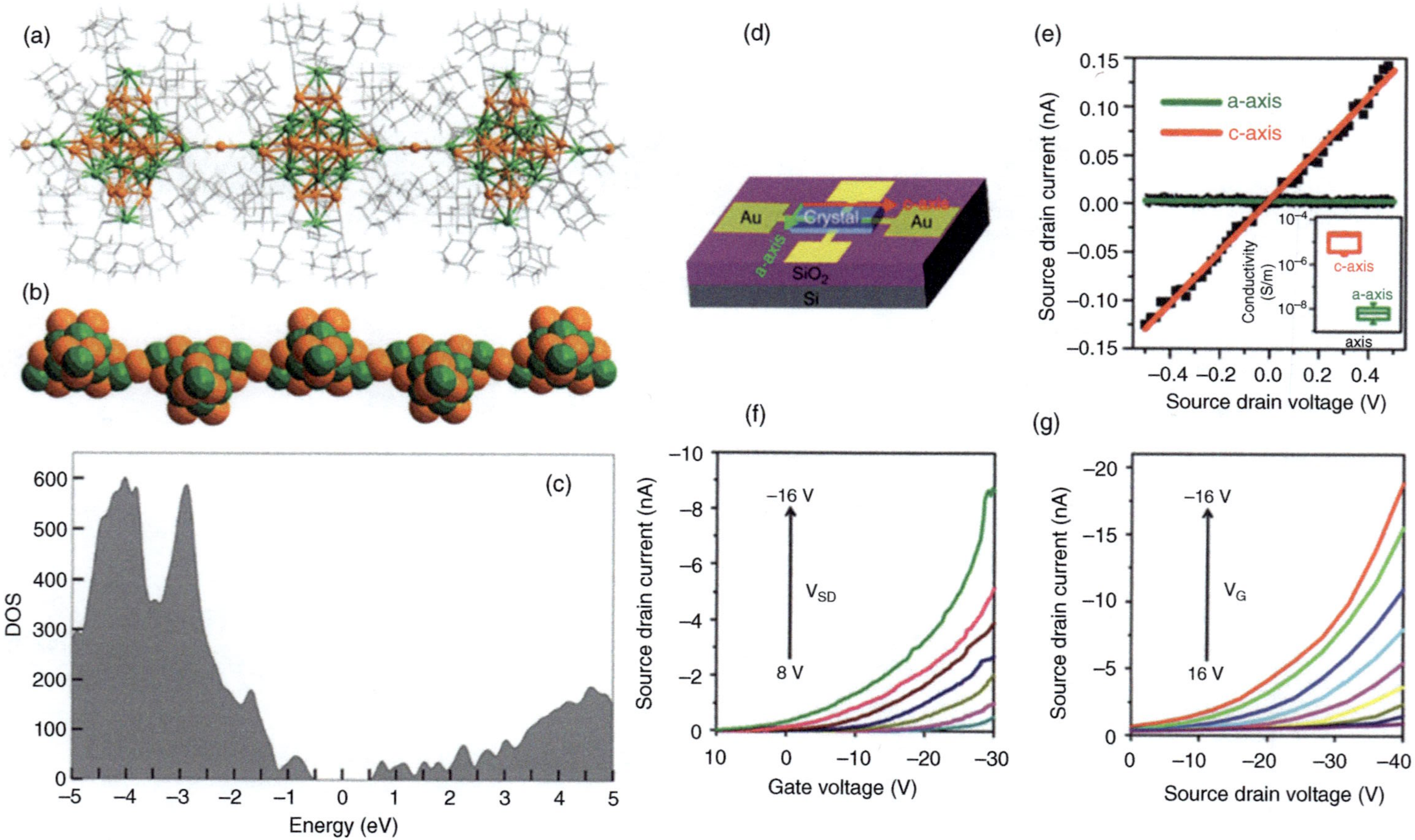

Figure 15.6 (a) One-dimensional chain of [(AuAg)$_{34}$(adamantanethiolate)$_{20}$], (b) the space-filled structure of the chain, golden: Au, green: Ag, gray: C. (c) the calculated band gap form the DOS calculation. (d) The schematic representation of the field-effect transistor. (e) *I*–*V* plot of the crystal along *a*-axis and *c*-axis, respectively, with the range of corresponding conductivity values shown in the inset. (f) Transistor transfer (V_{SD} = 8 to −16 *V* in −4 *V* steps) and (g) output characteristics (V_G = 16 to −16 *V* in −4 *V* steps) measured along *c*-axis of the crystal. Channel length, width = 80, 50 μm. *Source:* Reproduced from Ref. [39] with permission from the Springer-Nature, copyright 2020.

properties by fabricating a field-effect transistor device using the single crystal and they obtained a promising ON/OFF ratio of 4000 with hole mobility of $0.02\,cm^2/Vs$. They also discovered the anisotropic electrical conducting behavior in its single crystal. They noted that the conductivity ($1.49 \times 10^{-5}\,S\,m^{-1}$) of this crystal in the crystallographic c-axis was one-to-three orders of magnitude higher than the values reported for one-dimensional NC assembly consisting of Au_{21} clusters, where one-dimensional NC assembly was formed by modulating the weak interactions in the ligand layers. They stated that this striking enhancement in conductivity and charge carrier mobility was obtained in the one-dimensional chain due to the direct linking of each cluster node by (Ag—Au—Ag) connectivity in the crystal.

15.2.3 Nanocluster Assembly through Linkers

Although the assembly of the NCs through lattice packing via surface ligand interactions and metal–metal bond formations are favorable for the electronic properties of the assembled network, the availability of such assembled materials is very limited due to the complex synthetic protocols. On the other hand, the organic linkers are already highlighted due to their well-known unique architectural stabilizing ability in metal–organic framework designing, so, it quickly takes the place as a superior linker molecule here for forming the periodic framework of NCs. The improved chemical stability, high structural tunability, large channel with improved photophysical properties make this attractive since its exploration in comparison with their molecular clusters. To date, silver nanoclusters are extensively used as the building blocks for this type of assembly with improved structural stability and photophysical properties. The covalent interaction between Ag and N is the prime factor here that introduces the N-based organic linker molecules mostly [28].

In a pioneering work, Wang et al. reported the cluster node of Ag_{14} superatoms protected by face-capping [1,2-dithiolate-*o*-carborane] ligands [40]. They found that the corner N-donor ligands contribute predominately to the stability and luminescence property of the Ag_{14} superatom. As the first-formed zero-dimensional $[Ag_{14}(1,2\text{-dithiolate-}o\text{-carborane})_6(CH_3CN)_8]\cdot 4CH_3CN$ NC with labile vertex-coordinated CH_3CN ligands is highly unstable, monodentate pyridine linker was used to replace these CH_3CN species site-specifically, giving $[Ag_{14}(1,2\text{-dithiolate-}o\text{-carborane})_6(\text{pyridine}/p\text{-methylpyridine})_8]$ in gram scale with its core structure intact, which featured high thermal stability up to 150 °C in the air (Figure 15.7). However, when they replaced the monodentate linker with a bidentate linker to bridge the superatomic Ag_{14} cluster nodes, an unprecedented hierarchical three-dimensional superatomic silver cluster-assembled material was formed, which was stable up to 220 °C in air. Furthermore, this series of stable solid-state superatomic-NC materials exhibited tunable dual emission with wide-range thermochromism. So, this study constituted a major step toward the development of ligand-modulation of the structure, stability, assembly, and functionality of superatomic silver NCs.

Later on, Alhilaly et al. demonstrated an approach of synthesizing atomically precise anion-templated silver thiolate NC and its assembly in one- and two-dimensional frameworks through [4,4′-bipyridine] linker (Figure 15.8) [41]. They first synthesized a zero-dimensional $[Cl@Ag_{16}(\textit{tert}\text{-butylthiolate})_8(CF_3COO)_7(DMF)_4(H_2O)]\cdot 1.5(DMF)$ NC and used this as a node for further assembly process and finally made one-dimensional assembly of $[Cl@Ag_{15}(\textit{tert}\text{-butylthiolate})_8(CF_3COO)_{5.67}(NO_3)_{0.33}(4,4'\text{-bipyridine})_2(DMF)_2]\cdot 4.3(DMF)\cdot H_2O$ and two-dimensional assembly of $[Cl@Ag_{14}(\textit{tert}\text{-butylthiolate})_8(CF_3COO)_5(4,4'\text{-bipyridine})_2(DMF)]\cdot 2(DMF)$. They identified that these three cluster units shared notable similarities. In particular, the units possessed a similar geometrical core structure and the protecting ligand shell of each cluster core contains a combination of the thiolate and trifluoroacetate ligands as well as coordinated solvent molecules. As the ratio of the ligands to the Ag thiolate complex increased in the presence of a linker, clusters began to distort more to accommodate

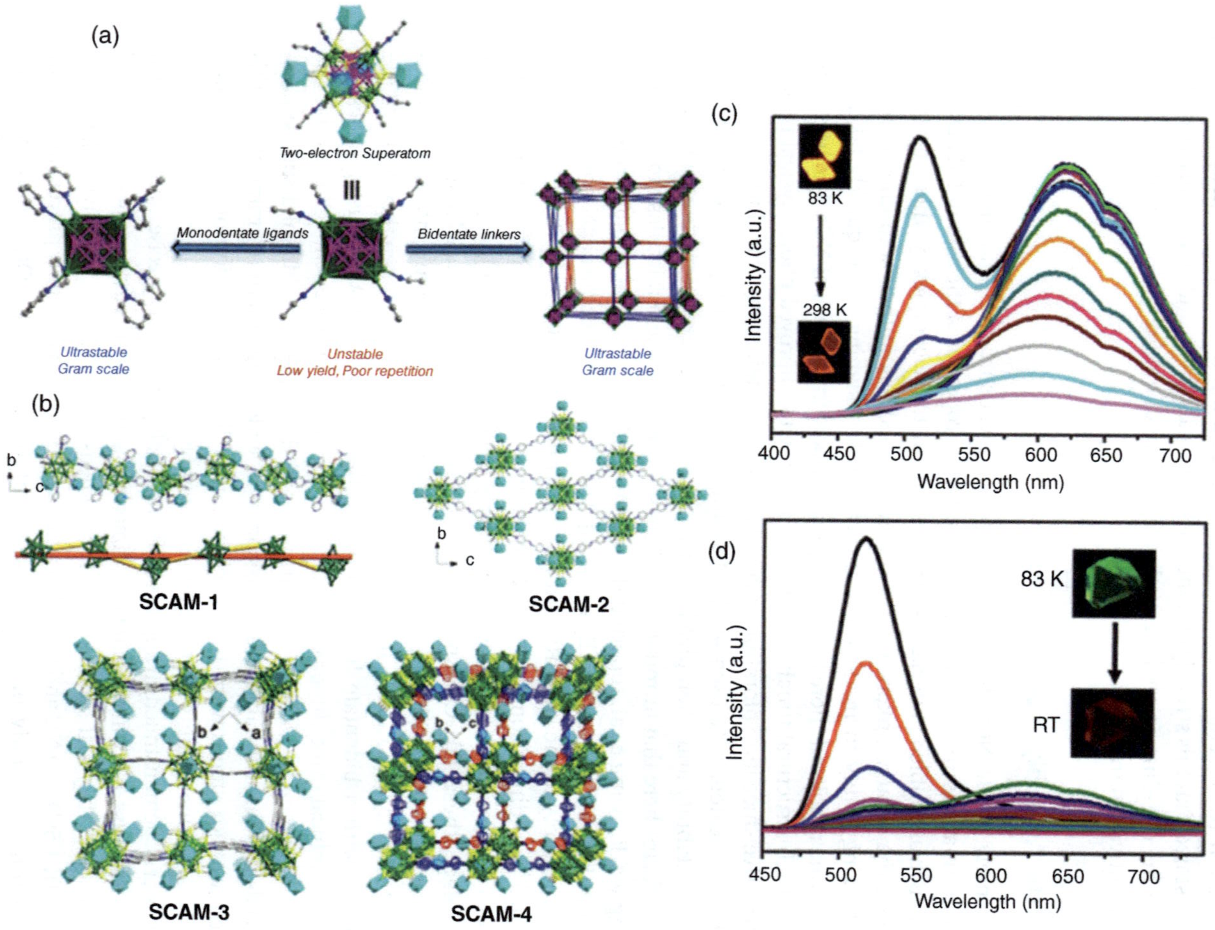

Figure 15.7 (a) Schematic representation of site-specific tailoring and directional assembly of Ag_{14} superatom NC, (b) structural representation of one-dimensional helix of SCAM-1, two-dimensional grid network of SCAM-2, three-dimensional porous framework of SCAM-3, and twofold interpenetrated porous three-dimensional frameworks of SCAM-4, green and pink: silver, yellow: sulfur; gray: carbon, blue: nitrogen, turquoise: carborane. Variable-temperature emission spectra of (c) one-dimensional chain and (d) twofold interpenetrated porous three-dimensional frameworks (excited at 380 nm) from 83 K to 298 K in air. Inset: the corresponding photographs UV irradiation. *Source:* Reproduced from Ref. [40] with permission from the ACS, copyright 2018.

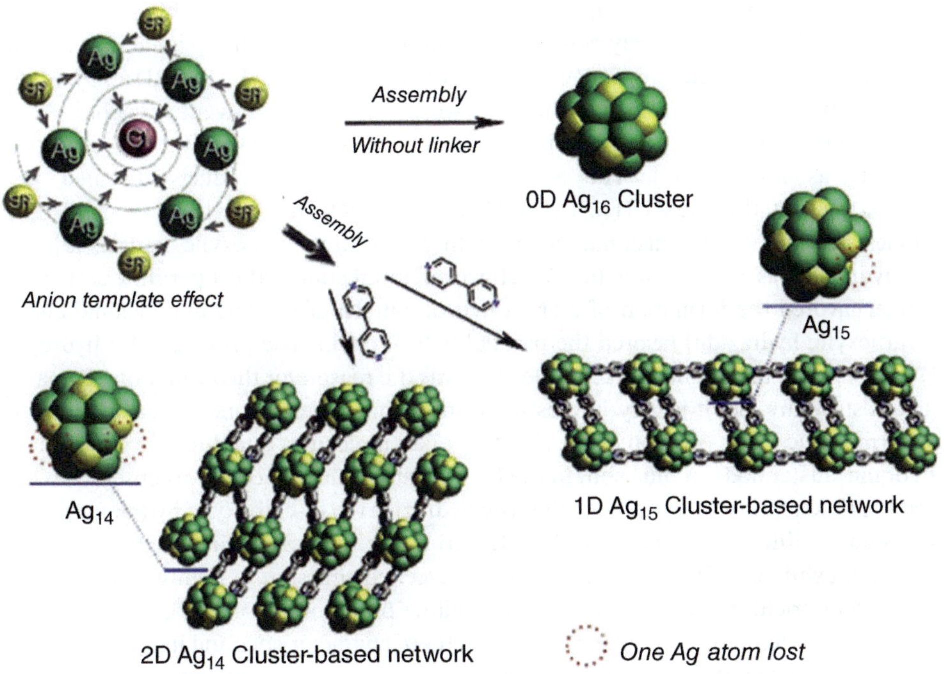

Figure 15.8 Synthesis of anion template nanocluster assembly of molecular Ag_{16} nanocluster in one and two-dimensional framework. *Source:* Reproduced from Ref. [41] with permission from the ACS, copyright 2019.

structures of higher connectivity and thus confined spacing. These distortions led to a sequential loss of a single surface Ag atom as the NC adopted one- and two-dimensional assembled networks of Ag_{15} and Ag_{14}, respectively. They also identified that these three structures exhibited qualitatively distinct characteristics in photoluminescence and thermal stability. Specifically, higher-dimensional structures showed enhanced photoluminescence and higher thermal stability. So, there is a vast library of these nanocluster nodes and their assembling process through linker molecules available in the literature, which dictates the superiority of the assembling process over the zero-dimensional cluster nodes. The following subsections address their dimensionality and explain their properties.

15.2.3.1 One-Dimensional Nanocluster Assembly

Ma et al. reported a one-dimensional silver cluster based framework of $[Ag_{18}(PhPO_3)$(*tert*-butylthiolate)$_{10}(CF_3COO)_2(PhPO_3H)_4]$ with [3-amino-4,4′-bipyridine] as a linker [42]. The obtained assembled network was stable up to 110 °C in a nitrogen atmosphere, which is far improved from the parent cluster. They identified that when mechanical stimulation was applied to the assembled network, the photoluminescence emission wavelength was changed. When the assembled network sample is subjected to mechanical stimulation and recrystallization happened, its photoluminescence emission wavelength returned to the previous value. Thus, the photoluminescence emission of the assembled network composed of $[Ag_{18}(PhPO_3)$(*tert*-butylthiolate)$_{10}]$ NCs exhibited reversible mechanochromism behavior. By showing the emission property at two wavelengths of this assembled network and its temperature-dependent photoluminescence emission intensity ratio, the authors suggested that it could be used in a thermometer.

Later on, Li et al. reported another one-dimensional chain of $[Ag_{10}$(*tert*-butylthiolate)$_6(CF_3CF_2COO)_4]$ with the linker of [4,4′-bipyridine] [43]. Due to the rigid and bi-dentate nature of the bipyridine, they utilized it to coordinate the silver cluster nodes at a specific distance that is favorable for the formation of the excimer or exciplex emission. They discovered that this silver cluster assembled network exhibits a selectively erasable photoluminescent switch triggered by a coordinated CH_3CN molecule. The switching mechanism was distinctly resolved by SCXRD analysis and the reverse process was achieved by treating the framework with CH_3CN solvent or vapor.

Lu et al. reported two different, one-dimensional silver cluster assembled networks by using the pyridine derivate as linkers [44]. The SCXRD analysis of the products revealed that [$Ag_{10}(CF_3COO)_4$(*tert*-butylthiolate)$_6(CH_3CN)_2$] and [$Ag_{10}(CF_3COO)_4$(*tert*-butylthiolate)$_6(CH_3CN)$] were the nodes of the one-dimensional chain with [4-pyridine carboxylic hydrazide] and [2-carboxylic hydrazide] as N containing linkers. However, interestingly, the 4-pyridine carboxylic hydrazide linker favored the formation of a cross helical pattern of the one-dimensional chain, whereas [2-carboxylic hydrazide] favored the parallel helical chain. The presence of a hydrogen bond (N—H···O) between the parallel linker molecules was the reason for the formation of a parallel helical chain structure. So, it is very interesting to note that different one-dimensional chain architecture formed due to the position of the N at the linker molecules that directly coordinate to the Ag atom of the cluster node. A blue shift in the UV—vis absorbance spectra was observed when they changed the linkers from [4-pyridine carboxylic hydrazide] to [2-carboxylic hydrazide]; however, the emission maximum was the same. The [2-carboxylic hydrazide] linker contained one-dimensional chain exhibited a weak green photoluminescence emission in highly polar solvents (dichloromethane or trichloromethane) and strong yellow photoluminescence emission in solvents with low polarity (tetrachloromethane). Based on these characteristics and impressive stability of the fluorescence property, the authors suggested that the one-dimensional parallel helical fashioned silver assembled network with [2-carboxylic hydrazide] as a linker could be used to measure the concentration of dichloromethane or trichloromethane, and tetrachloromethane.

Xu and co-workers reported the trans-1,2-bis(4-pyridyl)ethylene linker contained zigzag, one-dimensional chain of [Cd_6Ag_4(benzene thiolate)$_{16}$(DMF)][(H_2O)] alloy NC assembly [45]. Interestingly, the N atoms of the linkers are attached herewith the Cd, not with the Ag atom. They compared the electronic structures of the resulting assembled alloy NC and individual alloy [Cd_6Ag_4(benzene thiolate)$_{16}$(DMF)$_3$][(CH_3OH)] NC. The results revealed that the bandgap of the NC was narrowed by the formation of the one-dimensional assembled network. In the assembled network, the optical absorption onset was redshifted to the visible region. They utilized the one-dimensional assembled network as a visible light (>420 nm)-responsive photocatalyst for decomposition of the organic dye Rhodamine B in water. The one-dimensional assembled network exhibited higher photocatalytic activity toward rhodamine B degradation than the molecular cluster counterpart and high stability during the photocatalytic reaction.

Recently, Das et al. synthesized a new one-dimensional chain of [Ag_{11}(adamantanethiolate)$_6(CF_3COO)_4$(4,4′-azopyridine)$_2$(DMF)$_2$][NO_3] cluster assembled network [46]. Here, intramolecular parallel-displaced $\pi\cdots\pi$ interaction between the two linkers [4,4′-azopyridine] played a pivotal role in growing the one-dimensional unit. The interlayer (C—H···F) and (C—H···π) interactions of the surface-protecting ligands also favored the supramolecular assembly in the three-dimensional network (Figure 15.9). They have shown that the high thermal stability, up to 140 °C of this n-type semiconducting material with a bandgap of 1.98 eV, is achieved due to the linear conjugation of the silver cluster nodes. They have also explained the stability of this material in the solution state and the solvent polarity-dependent photoluminescence property, where the maximum photoluminescence intensity was observed in acetonitrile solution.

To date, only thiolate-protected hydrophobic silver nanoclusters and their fruitful assembly have been extensively synthesized and studied; however, Biswas et al. recently showed the synthetic pathway of water-soluble thiol free, smallest number of silver atom contained nanocluster assembly of [$Ag_2(PhPO_3H)_2$(4,4′-azopyridine)$_2$], through [4,4′-azopyridine] as a linker [47]. They have shown that the smallest size of the cluster building blocks along with the choice of organic linker molecules

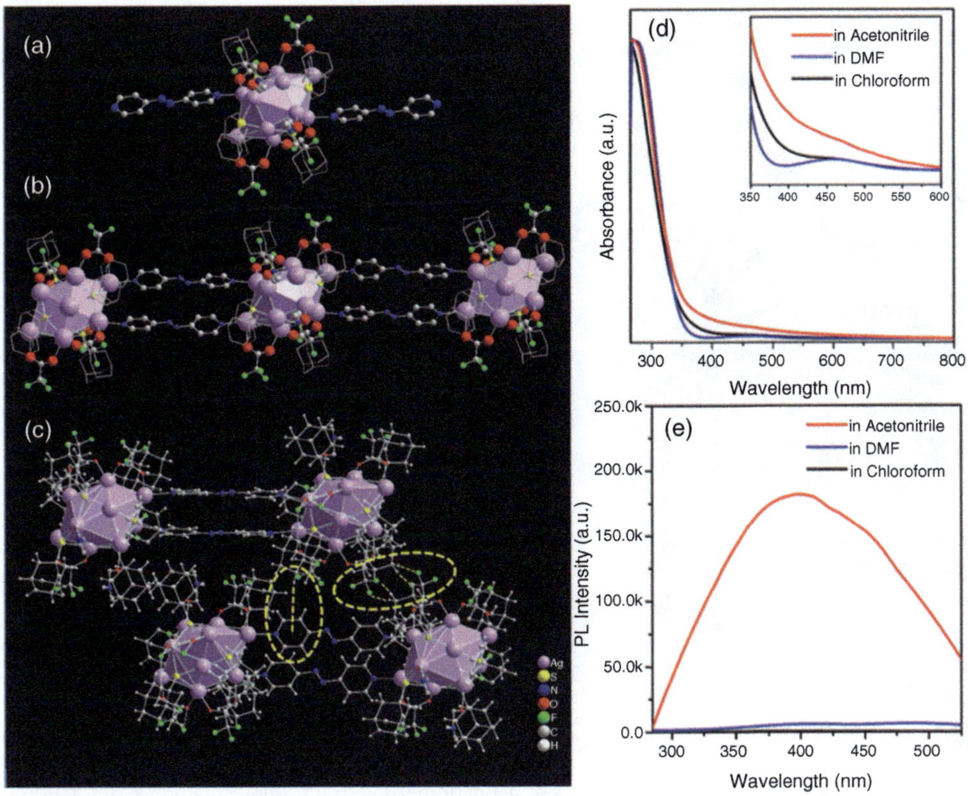

Figure 15.9 (a) The repeating unit of the Ag_{11} NC, (b) the one-dimensional chain assembled from Ag_{11} NC. For clarity all H atoms are omitted from first two figures. (c) Inter-chain (C—H···F) and (C—H···π) interactions where C—H, F, and π are originating from the adamantane group, CF_3COO ligand, and azopyridine linker of the Ag_{11} NC, respectively (d) UV–vis absorbance spectra of Ag_{11} NC in different solvents, and (e) solvent polarity dependent PL emission spectra of Ag_{11} NC. *Source:* Reproduced from Ref. [46] with permission from the ACS, copyright 2021.

efficiently stabilized the one-dimensional structure via intrachain π···π stacking and further chains were assembled by the interchain T-shaped π···π interactions (Figure 15.10). The advantageous semi-conducting band structure associated with the charge transfer phenomenon and high structural and thermal stability of the material was guided them to explore the photoresponsive character of this assembled network. Additionally, the unprecedented water solubility, which is very rare for this class of materials, was utilized for photoacoustic imaging with chicken breast tissue as a model. They measured the photoacoustic signal strength and confirmed the blood vessel mimicking capability of the obtained one-dimensional chain at a depth of ~3 mm inside the chicken breast tissue.

15.2.3.2 Two-Dimensional Nanocluster Assembly

Huang et al. reported the first two-dimensional assembly of [Ag_{12}(*tert*-butylthiolate)$_6$(CF_3COO)$_6$] cluster nodes via [4,4′-bipyridine] linker [48]. They showed that the symmetric two-dimensional sheets were formed when each cluster node was linked with six other cluster nodes and each layer was separated by weak non-covalent interaction with a distance of 7.23 Å. However, interestingly, they showed that the structural deformation happened due to the solvent interactions and they transformed this two-dimensional structure into a three-dimensional structure. They also compared the photoluminescence property of both structural architecture and identified the different emission properties. Later, the emission property of this two-dimensional structure was further improved by amino-bipyridine as a linker.

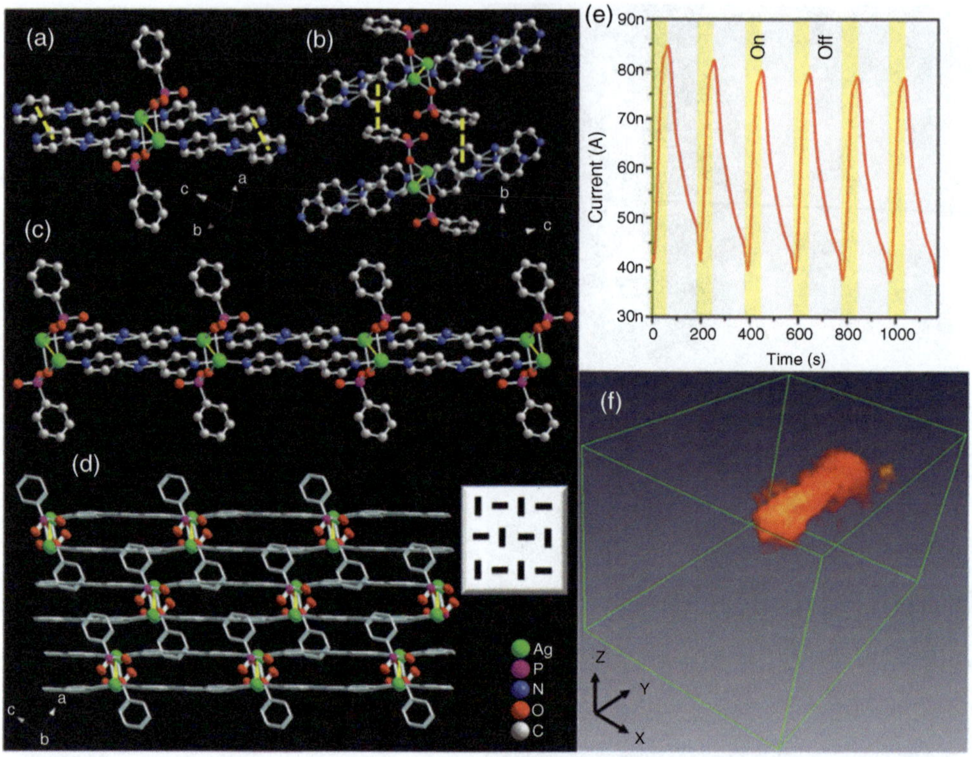

Figure 15.10 (a) Monomer unit of $[Ag_2(PhPO_3H)_2(4,4'\text{-azopyridine})_2]$ with intramolecular $\pi\ldots\pi$ interaction between the linkers, (b) T-shaped $\pi\ldots\pi$ interaction showing in between two adjacent monomer unit, (c) one-dimensional chain-like architecture of one chain unit, (d) stacking pattern of one-dimensional chain of Ag_2 CAM, inset showing the herringbone pattern of growth, (e) photocurrent response of the material, and (f) three-dimensional photoacoustic image of the inserted tube containing sample at a depth of 2 mm chicken tissue, confirming the image-producing capability of the material. *Source:* Reproduced from Ref. [47] with permission from the RSC, copyright 2021.

Afterward, Du et al. reported an interesting two-dimensional assembly of $[Ag_{12}$(*tert*-butylthiolate)$_6(CF_3COO)_6]$ by using [*tris*(4-pyridylphenyl)-amine] as a linker, where the distance between the layers can be modulated [49]. In this two-dimensional assembly, dimethylacetamide was used as a reactant solvent and trapped between layers immediately after the synthesis of the cluster when the layers were overlapping each other. Interestingly, when they tried to remove this trapped solvent from the structure, an architectural disorder by shortening the inter-planar distances along with varied emission properties was observed (Figure 15.11). However, after redissolving the structure in the mother liquid, they observed another different characteristic with a larger inter-planar distance. Further, they also reported another two-dimensional assembled network of $[Ag_{12}$(*tert*-butylthiolate)$_6(CF_3COO)_3]$ NCs by using [5,10,15,20-tetra(4-pyridyl)porphyrin] as a linker [50]. As the used linker has a photosensitizing effect, they showed the superiority of this assembly toward the photodegradation of toxic [2-chloroethyl ethyl sulphide] (mustard gas) (Figure 15.11). The synthesized two-dimensional structure showed better photocatalytic activity than the reported metal–organic framework due to the synergistic effect of both the silver cluster node and the linker molecules that produced the singlet oxygen species that was used for the degradation.

Das et al. reported a two-dimensional assembly of $[Ag_{14}$(*tert*-butylthiolate)$_{10}(CF_3COO)_4]$ cluster nodes through [4,4′-azopyridine] linker (Figure 15.12) [51]. They represented a new way of

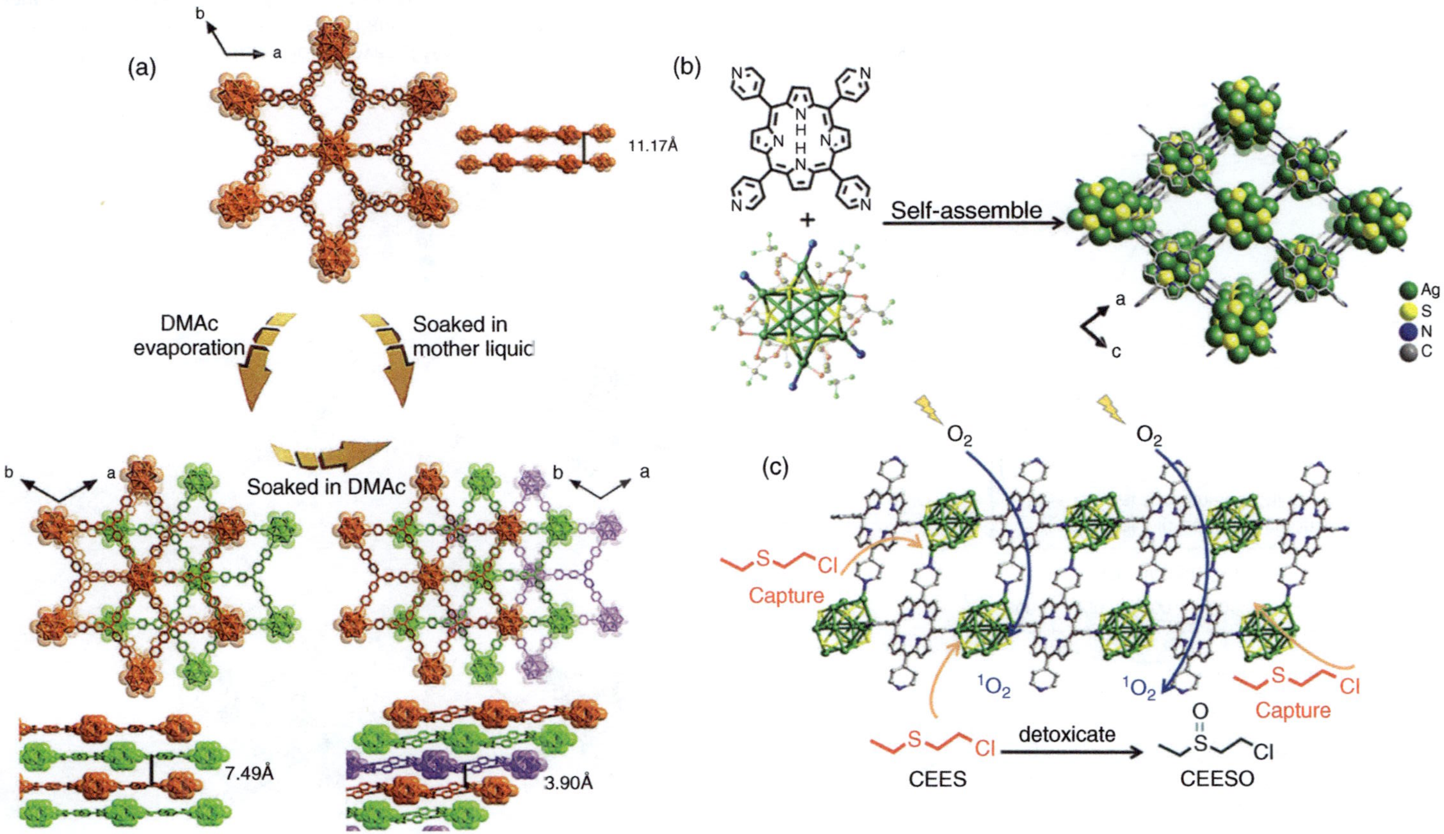

Figure 15.11 (a) Two-dimensional assembly of [Ag_{12}(*tert*-butylthiolate)$_6$(CF_3COO)$_6$] with [*tris*(4-pyridylphenyl)-amine] as a linker showing that trapped solvent evaporation and re-desorption change the inter-planar layer distances. *Source:* Reproduced from Ref. [49] with permission from the RSC, copyright 2018. (b) Synthetic process of the self-assembled two-dimensional [Ag_{12}(*tert*-butylthiolate)$_6$(CF_3COO)$_3$] NCs by using [5,10,15,20-tetra(4-pyridyl)porphyrin] as a linker molecule, (c) schematic illustration of the capture and photodetoxification of mustard gas. *Source:* Reproduced from Ref. [50] with permission from the ACS, copyright 2019.

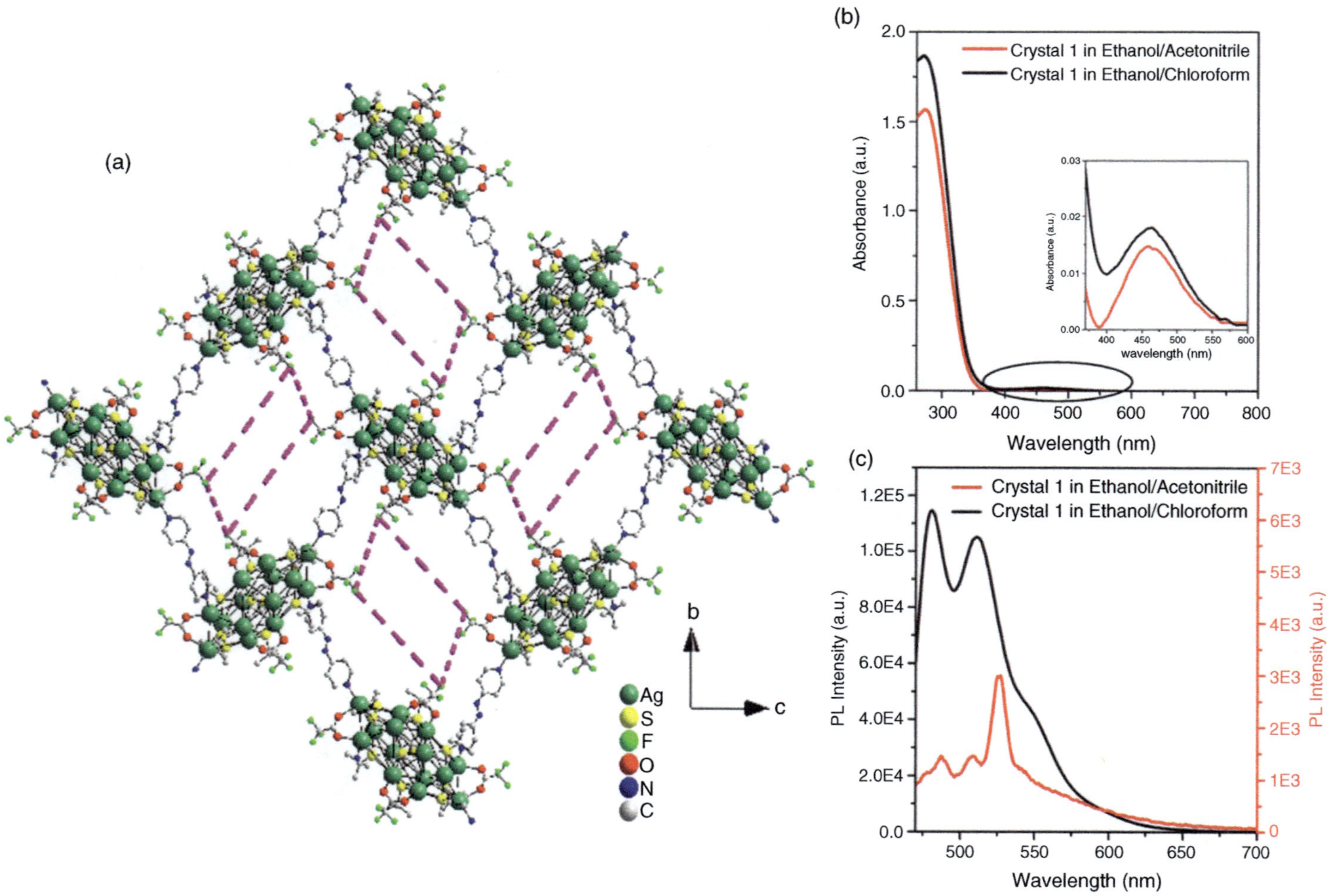

Figure 15.12 (a) Two-dimensional layer of $[Ag_{14}$(*tert*-butylthiolate)$_{10}(CF_3COO)_4]$ cluster nodes through [4,4′-azopyridine] linker denoted as crystal 1 for next two figures, (b) UV–vis absorbance and (c) photoluminescence spectra of this two-dimensional framework in two different solvent mediums. *Source:* Reproduced from Ref. [51] with permission from the RSC, copyright 2021.

controlling photophysical properties of assembled cluster nodes via interchangeable cluster–solvent interactions. From the structural understanding and theoretical modeling, they showed that additional interlayer noncovalent interaction that confines the linker molecule with an unexpected stimuli-responsive frontier molecular orbital arrangement and influenced the charge transfer phenomenon. This structural transformation was reflected in photoluminescence properties with ~650 times enhancement of PL quantum yield at room temperature. Further, to manifest this restricted molecular vibration in the nonradiative phonon emission process, photoacoustic signal strength was measured, which suggests a concomitant photon and phonon emission pathway. Further, they introduced an innovative technique, *pre-illumination,* to amplify the photoacoustic signal strength (>85%) at 650 nm frequency region for stepping toward the opening of biomedical imaging applicability. They further used this technique to confirm the blood-vessel mimicking capability of the portrayed material while the sample was embedded inside the chicken breast tissue at a depth of ~2 mm for finding a new route of wide applicability.

15.2.3.3 Three-Dimensional Nanocluster Assembly

Huang et al. has reported the first three-dimensional nanocluster assembly of $[Ag_{12}$(*tert*-butylthiolate)$_8(CF_3COO)_4$(4,4′-bipyridine)$_4]$, where the cluster node was bridged by the [4,4′-bipyridine] linkers [48]. The obtained assembled framework was a bilayer structural unit that markedly improved the stability of the molecular cluster. The formation of three-dimensional network helped to build the porous network inside the structure, which was further utilized for inspecting the gas-adsorption behavior. The photoluminescence property of this material was changed greatly by the formation of the three-dimensional network in comparison with the parent molecular cluster. In general, the molecular cluster showed a weak red photoluminescence emission whereas the assembled network showed strong green emission under vacuum with an enhancement of PLQY by 60% due to the suppression of the nonradiative emission pathway in the three-dimensional rigid network. However, in the presence of oxygen, the emission is quenched drastically, which is absent for the molecular counterpart. With the help of this observation, the authors have identified the superiority of this three-dimensional framework toward the sensing of oxygen gas. In addition, they have also discovered the ability of absorption of volatile organic compounds, which is detected by the different emission colors.

Later, Dong et al. reported two-dimensional layers of $[Ag_{10}$(*tert*-butylthiolate)$_6(CF_3COO)_2(PhPO_3H)_2]$ NCs linked via [4,4′-bipyridine], which were further stacked through hydrogen bond (O—H···O) and (C—H···O) interactions to form the three-dimensional framework (Figure 15.13) [52]. So, the two-dimensional layers were thus packed together by weak interactions, which facilitated the sliding of the layers, allowing the three-dimensional framework to undergo structural deformation in response to guest organic molecules. They have discovered that this framework exhibited green photoluminescence in the air. Upon inclusion of guest organic molecules, it exhibited photoluminescence with an emission color depending on the guest organic molecule, and that actually helped in finding the potential application as a volatile organic compound sensor designing. Later, they reported another three-dimensional silver cluster assembled framework of $[Ag_{12}$(*tert*-butylthiolate)$_6(CF_3COO)_6]$ with [2,5-bis(4-cyanophenyl)-1,4-bis(4-(pyridine-4-yl)-phenyl)-1,4-dihydropyrrolo[3,2-b]pyrrole] as a linker [53]. This three-dimensional framework exhibited a 66% higher PLQY than the linker molecules due to the suppression of aggregation-induced quenching of the linker molecules while bonded with the cluster nodes. Later, Wu et al. synthesized a three-dimensional metal–organic-framework consisting of two types of silver cluster nodes connected via [1,1,2,2-tetrakis(4-(pyridin-4-yl)phenyl)-ethene] linker [54]. So, in the three-dimensional framework, they observed that one $[Ag_{12}$(*tert*-butylthiolate)$_6(CF_3COO)_6]$ NC node and three $[Ag_8$(*tert*-butylthiolate)$_4(CF_3COO)_4]$ NC nodes were connected to the four N atoms of

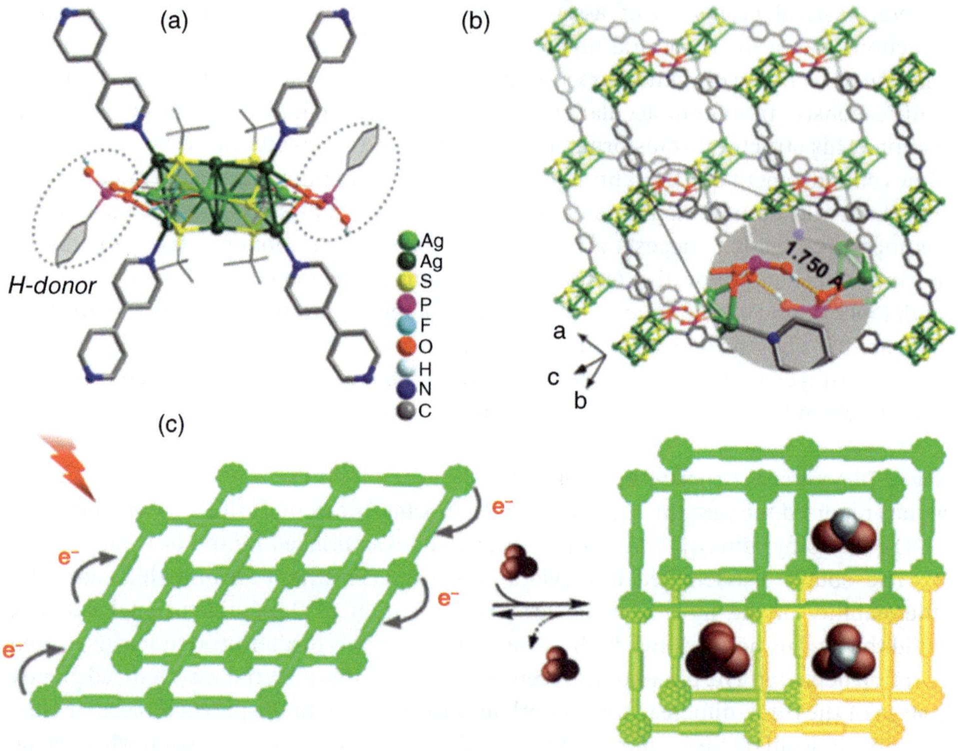

Figure 15.13 (a) Perspective view of the coordination environment of the [Ag_{10}(*tert*-butylthiolate)$_6$(CF_3COO)$_2$($PhPO_3H$)$_2$] core in the assembled structure, (b) two-layer stack of the host framework of Ag_{10} NC with complementary hydrogen bonding (O—H···O, H···O distance is 1.750 Å) between interlayer —PO_3H moieties, (c) illustration of CH_2Cl_2, $CHCl_3$, and CCl_4 represented in space-filling mode induced reversible pore-open/closed structural transformation and switchable solvatochromism. *Source:* Reproduced from Ref. [52] with permission from the ACS, copyright 2018.

the linker molecule. They obtained a highly porous framework with a highly intense photoluminescence property when the pores are empty. However, when the solvent is present inside the pores, the molecular rotation of the linker molecule is suppressed, and that changes the excited-state-dynamic of the whole framework.

15.2.4 Nanocluster Assembly through Aggregation

Luo et al. reported using aggregation-induced assembly in nanocluster assembly for the first time to obtain the enhanced photoluminescence property [55]. They showed that the photoluminescence intensity of [$Au_n(SR)_m$] NC was enhanced with the increasing polarity of the mixed solvent (ethanol/water) (Figure 15.14). They observed that the gold NCs loosely aggregated in aqueous solution is nonemissive but weakly red emissive when they used the ratio of mixed solvents up to 75%. However, further improvement in the emission intensity was observed when the ratio of the mixed solvents was increased up to 95%, and that induced the dense aggregation of these gold NCs. They also observed that this aggregation influenced the position of the photoluminescence peak of these NCs, which was blue-shifted from the red to yellow region. Later, they found that the heat treatment also promoted the aggregation of nonemissive gold complexes and formed a highly emissive metallic kernel containing silver and gold NCs assemblies.

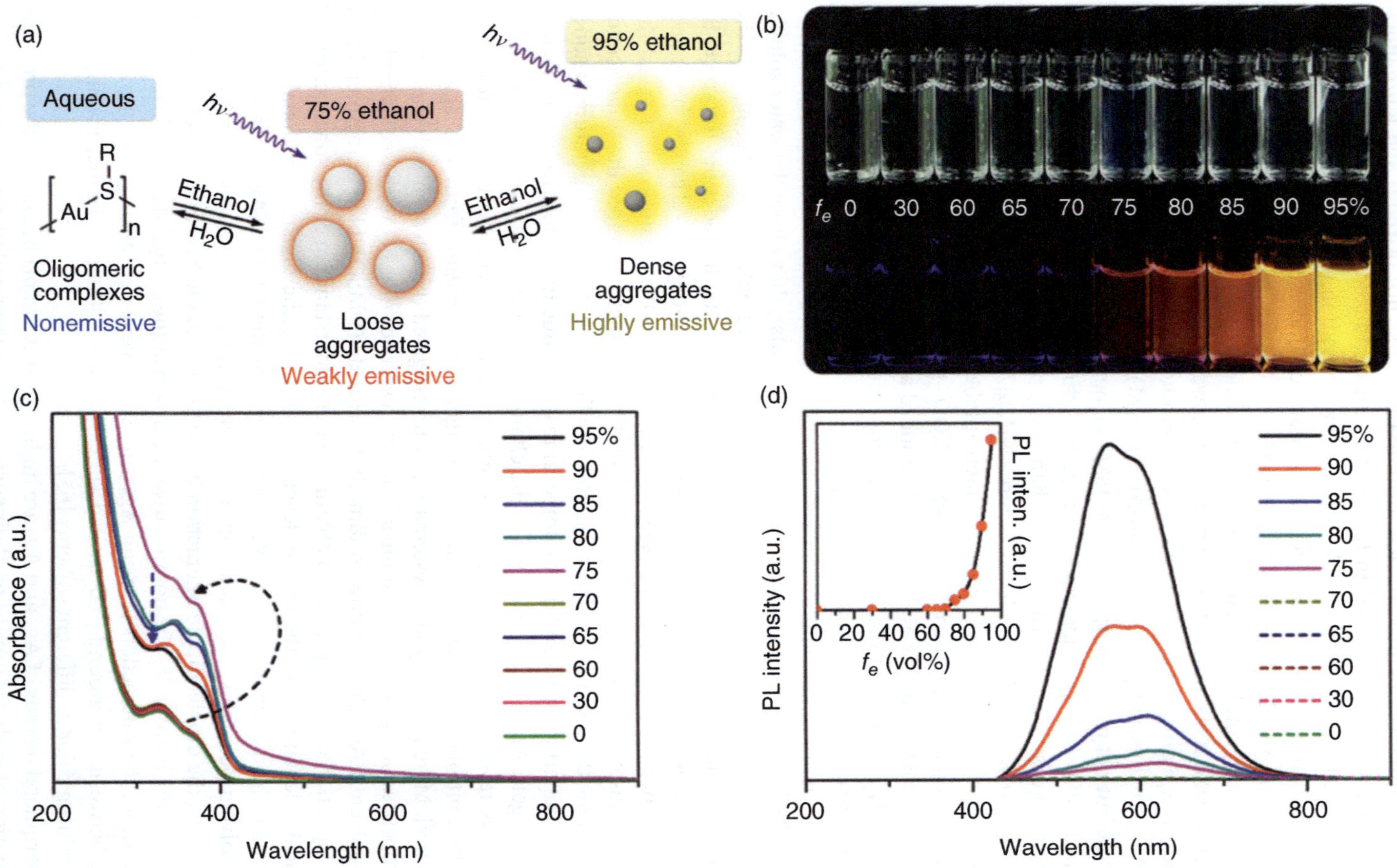

Figure 15.14 (a) Schematic illustration of solvent-induced aggregation-induced emission properties of $[Au_n(SR)_m]$ NC. (b) Digital photos of $[Au_n(SR)_m]$ NC in mixed solvents of ethanol and water with different f_e under visible (top row) and UV (bottom row) light, (c) UV–vis absorption, and (d) photoemission spectra of $[Au_n(SR)_m]$ NC in mixed solvents with different f_e. (Inset) Relationship between the luminescence intensity and f_e. The spectra were recorded 30 minutes after the sample preparation. *Source:* Reproduced from Ref. [55] with permission from the ACS, copyright 2012.

Jin et al. reported the highly intense luminescence nature of the aggregated [$Au_4Ag_5(DPPM)_2$ (1-adamantanethiolate)$_6$] NCs in methanol/water medium, while the individual NC is nonemissive [56]. The crystal structure of this NC demonstrated the existence of right- and left-handed enantiomers, which were closely packed in the crystal lattice induced by the staggered surface structures. Such a close packing is responsible for the enhancement in the emission property during aggregation. Sugiuchi et al. showed an aggregation-induced fluorescence-to-phosphorescence switching based on the [$Au_8(DPPP)_4L_2$] nanoclusters (L=Cl, $C{\equiv}CC_6H_5$, or $C{\equiv}C(CH_3)_2CH_3$) [57]. These materials are highly soluble in dichloromethane, in which these materials displayed a single photoluminescence peak at 600 nm with a short lifetime; however, in poor solvents these NCs tend to aggregate and show a different emission band at 700 nm with a very long lifetime. Chen et al. showed the enhanced emission property of [$Au_4Ag_{13}(DPPM)_3$(2,5-dimethylbenzenethiolate)$_9$] in its crystalline state but not in its amorphous state [58]. So, here the crystallization-induced emission property was visible due to the aggregation of the NCs through compact C—H⋯π interaction.

To address the high cost of gold clusters and enhance the possible biological applications, Sun et al. synthesized two water-soluble silver clusters $(NH_4)_6[Ag_6(mna)_6]$ and $(NH_4)_9[Ag_9(mba)_9]$ (H_2mna = 2-mercaptonicotinic acid, H_2mba = 2-mercaptobenzoic acid, hereafter referred to as Ag_6 and Ag_9, respectively). Due to the rich peripheral $-COO^-$ groups, Ag_6 and Ag_9 have the potential to bind secondary metal ions [59–61], or form H-bonded aggregates. Whereafter, Sun and Xin groups collaboratively investigated the fluorescence properties of the assembled materials built from these two water-soluble silver clusters [62–70].

First, the modulation of the optical properties of Ag_6 and Ag_9 was achieved through the solvent effect to construct long-range, ordered nanostructures. In particular, they chose the water-soluble Ag_9 cluster as the assembly unit, which has nine incompletely coordinated carboxyl groups on the surface of the silver cluster. Therefore, it can be well dissolved in water but does not fluoresce. By adding ethanol to its aqueous solution, a strongly luminescent metal-organic gel can be fabricated. The addition of ethanol enhanced the hydrogen bonding between the clusters and induced the aggregation of Ag_9 clusters to form highly ordered nanofibers, which enhanced the fluorescence intensity and overcame the disadvantage that Ag_9 does not fluoresce at room temperature in aqueous solution. More importantly, it can be monitored in real time due to the relatively slow formation process of Ag_9 gels. An in-depth analysis of the changes in spectroscopic properties during the gelation process revealed an interesting conversion from fluorescence to phosphorescence, which is mainly controlled by the structural rearrangements bursting from the aggregates (Figure 15.15a) [69]. Moreover, they also systematically investigated the effects of solvents on the fluorescence properties of Ag_6-based assemblies. The fluorescence color of Ag_6-based assemblies varies with the solvents: Ag_6 glows yellow in dimethyl sulfoxide (DMSO), blue in methanol and acetonitrile, and green in ethylene glycol. To obtain a better understanding of the solvatochromic effects, theoretical calculations on the energy levels of Ag_6 using the density functional theory (DFT) demonstrated the existence of intramolecular charge transfer (ICT) in Ag_6 clusters. As the polarity of the solvent decreases, the maximum emission wavelength of Ag_6 is blue-shifted. In highly polar solvents, the dipole–dipole interaction between the Ag_6 clusters and the solvent molecules is enhanced, which promotes the nonradiative transition of the excited state and leads to the emission. In addition, the interaction between Ag_6 and the solvent may also change the energy level difference between the ground and excited states of the emitting molecule, leading to a change in the emitting color in different solvents [65].

Second, the peripheral ligands of Ag_6 and Ag_9 are highly sensitive to pH because of the protonation or deprotonation of carboxyl groups on their surfaces. They can regulate the degree of protonation of the carboxyl groups on the surfaces of Ag_6 and Ag_9 by using inorganic [63] and organic acids [62, 68], respectively, and the modulation of the aggregate morphology can be achieved by

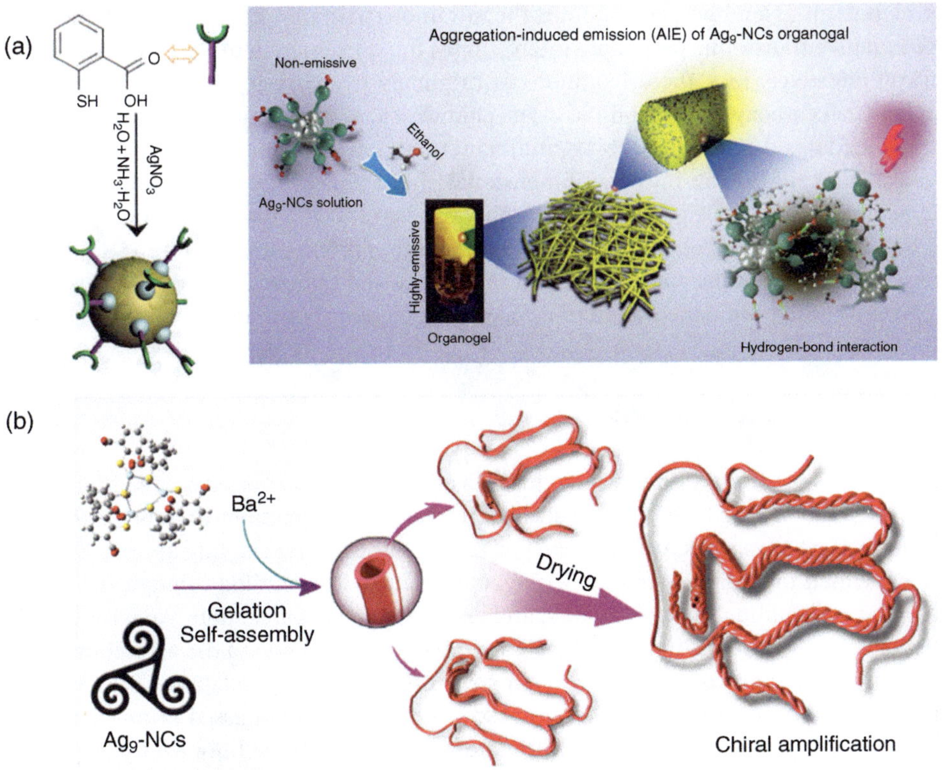

Figure 15.15 (a) Schematic illustration of solvent-induced gelation of Ag nanoclusters with atomic precision. *Source:* Reproduced from Ref. [69] with permission from the Wiley-VCH, copyright 2020. (b) Illustration of the chiral self-assembly of Ag_9 induced by coordination of Ba^{2+}. *Source:* Reproduced from Ref. [70] with permission from the ACS, copyright 2021.

regulating the protonation degree of peripheral $-COO^-$ group of the silver clusters, which results in abundant nanostructures (e.g. nanorods [63], nanoribbons [62, 63] and nanohollow tubes [68]), and these highly ordered nanostructures can effectively limit the rotations and vibrations of the protonated ligands, which is favorable for achieving ligand-to-metal charge transfer (^{3}LMCT) or ligand-to-metal–metal charge transfer (^{3}LMMCT), thus realizing the modulation of Ag_6 and Ag_9 aggregate morphology and fluorescence properties. Finally, they obtained novel amphiphilic molecules $Ag_6@C_{16}$mim-NCs based on metal nanoclusters (NCs) by selecting hydrophobic cation 1-hexadecyl-3-methylimidazole ($C_{16}mim^+$) to modify hydrophilic Ag_6. $Ag_6@C_{16}$mim-NCs possess surfactant properties with unique thermotropic liquid crystal and thermofluorescent properties. In addition, the ratio of water/DMSO binary solvent can be adjusted to modulate the aggregation morphology to obtain highly ordered nanorods and nanosheets due to the amphiphilic nature of $Ag_6@C_{16}$mim-NCs, and these ordered structures can effectively limit the intramolecular vibrations of the capping ligands, displaying aggregation-induced emission properties.

In this system, they can also select suitable templates to increase the synergy of their non-covalent interactions, thus promoting the supramolcular assembly of Ag_6 and Ag_9 for the modulation of optical properties (e.g. fluorescence and chirality). They used the polyethyleneimine (PEI) as a soft template to co-assemble with Ag_6 and formed a tightly ordered assembly through the electrostatic interaction between Ag_6 and PEI. Functional colloid aggregates of Ag_6 such as nanospheres and nanovesicles were formed along with the enhanced emission because of the formation of compact-ordered self-assemblies, which limits the intramolecular vibration of Ag_6 with increasing chance of radiative transitions [64]. More interestingly, the introduction of Ba^{2+} into Ag_9 aqueous solution can effectively lock the orientations of originally free carboxyl groups through the coordination between carboxyl group and Ba^{2+}. The enantioselective orientation of the peripheral carboxyl group facilitates the assembly of Ag_9 into nanotubes with a chiral cubic (I^*) lattice. The nanotubes can further intertwine into one-dimensional chiral nanobraids with a preferred left-handed arrangement. These multiple levels of chirality can be tuned by drying, during which the I^* phase is missing but the chiral entanglement of the nanotubes is enhanced (Figure 15.15b) [70]. The above results suggest that the conformation of metal NCs and their supramolecular assembly can be fine-tuned by designing noncovalent or covalent interactions between individuals.

15.3 Conclusions and Outlook

In this chapter, we have summarized the intercluster assembling processes where the individual NC building blocks were stacked due to some specific intercluster interactions to improve the stability and photophysical properties. We have categorized all these NC assembling approaches in a few subsections, which include assembling in the crystal lattice, assembling through the direct metal—metal bond, assembling through linkers, and assembling via aggregation. We have identified that most of the properties of individual NCs are greatly affected due to the assembling and that is well researched by the community through structure–property correlation. We have also observed that in most of the cases the photoluminescence property is discussed primarily; however, there is a lack of guidance to reach the ultimate applicability. In addition to the aesthetic photoluminescence property, these materials also have a semiconducting electronic characteristic, which is almost unrevealed for practical application. So, although there have been great advances, much work remains to be done in the future:

- Compared with the recent developments for controlling crystal lattices of small-sized metal complexes, research on dictating the cell units of nanoclusters is still lacking.
- Although the introduction of the organic linker molecules is a good move for the NCs assembling approach, only N-based organic linkers have emerged enormously. There is a void of M—O and M—S robust NCs frameworks, which should be pursued in future work.
- Assembly of water-soluble NCs are still not much achieved and most of them were not characterized by SCXRD. Future work along this line will be beneficial in biomedical applications.
- Studying the electronic properties of the assembled networks with their single layer architecture will be helpful for modern electronic device designing.
- Beyond gold and silver, NCs of other elements and alloying material as building blocks in the assembling process will also be worthy of future research, especially for various applications and structure–property correlation study.

As a whole, a more complete understanding of both cluster nodes and their assembling process will offer more tailor-made approaches to functional, cluster-based nanomaterials with desired properties. This chapter allowed us to obtain a common understanding of the NCs assembling process reported to date and their functions. We hope that the knowledge thus clarified will lead to clear design guidelines for developing new assembling approaches with desired functions in the future.

Notes

The authors declare no competing financial interest.

Acknowledgments

S. B. acknowledges the NPDF fund from SERB (PDF/2020/001085). D. S. thanks the fund from the National Natural Science Foundation of China (Grant Nos. 22171164, 91961105, 52261135637).

References

1 Jin, R., Zeng, C., Zhou, M., and Chen, Y. (2016). Atomically precise colloidal metal nanoclusters and nanoparticles: fundamentals and opportunities. *Chem. Rev.* 116 (18): 10346–10413.

2 Chakraborty, I. and Pradeep, T. (2017). Atomically precise clusters of noble metals: emerging link between atoms and nanoparticles. *Chem. Rev.* 117 (12): 8208–8271.

3 Qian, H., Zhu, M., Wu, Z., and Jin, R. (2012). Quantum sized gold nanoclusters with atomic precision. *Acc. Chem. Res.* 45 (9): 1470–1479.

4 Sun, C., Teo, B.K., Deng, C. et al. (2021). Hydrido-coinage-metal clusters: rational design, synthetic protocols and structural characteristics. *Coord. Chem. Rev.* 427: 213576.

5 Negishi, Y., Nobusada, K., and Tsukuda, T. (2005). Glutathione-protected gold clusters revisited: bridging the gap between gold (I)–thiolate complexes and thiolate-protected gold nanocrystals. *J. Am. Chem. Soc.* 127 (14): 5261–5270.

6 Du, Y., Sheng, H., Astruc, D., and Zhu, M. (2019). Atomically precise noble metal nanoclusters as efficient catalysts: a bridge between structure and properties. *Chem. Rev.* 120 (2): 526–622.

7 Yamazoe, S., Koyasu, K., and Tsukuda, T. (2014). Nonscalable oxidation catalysis of gold clusters. *Acc. Chem. Res.* 47 (3): 816–824.

8 Yao, Q., Wu, Z., Liu, Z. et al. (2021). Molecular reactivity of thiolate-protected noble metal nanoclusters: synthesis, self-assembly, and applications. *Chem. Sci.* 12 (1): 99–127.

9 Kang, X. and Zhu, M. (2019). Tailoring the photoluminescence of atomically precise nanoclusters. *Chem. Soc. Rev.* 48 (8): 2422–2457.

10 Antoine, R. (2020). Supramolecular gold chemistry: from atomically precise thiolate-protected gold nanoclusters to gold-thiolate nanostructures. *Nanomaterial* 10: 377.

11 Yang, J. and Jin, R. (2019). New advances in atomically precise silver nanoclusters. *ACS Mater. Lett.* 1 (4): 482–489.

12 Yuan, X., Dou, X., Zheng, K., and Xie, J. (2015). Recent advances in the synthesis and applications of ultrasmall bimetallic nanoclusters. *Part. Part. Syst. Charact.* 32 (6): 613–629.

13 Yang, S. and Zhu, M. (2020). Insight of the photoluminescence of atomically precise bimetallic nanoclusters with free electrons. *J. Chin. Chem. Soc.* 67 (12): 2171–2181.

14 Zhao, J. and Jin, R. (2018). Heterogeneous catalysis by gold and gold-based bimetal nanoclusters. *Nano Today* 18: 86–102.

15 Claridge, S.A., Castleman, A. Jr., Khanna, S.N. et al. (2009). Cluster-assembled materials. *ACS Nano* 3 (2): 244–255.

16 Chakraborty, P., Nag, A., Chakraborty, A., and Pradeep, T. (2018). Approaching materials with atomic precision using supramolecular cluster assemblies. *Acc. Chem. Res.* 52 (1): 2–11.

17 Kang, X. and Zhu, M. (2019). Intra-cluster growth meets inter-cluster assembly: the molecular and supramolecular chemistry of atomically precise nanoclusters. *Coord. Chem. Rev.* 394: 1–38.

18 Wang, Z., Li, X.Y., Liu, L.W. et al. (2016). Beyond clusters: supramolecular networks self-assembled from nanosized silver clusters and inorganic anions. *Chem. A Eur. J.* 22 (20): 6830–6836.

19 Daniel, M.-C. and Astruc, D. (2004). Gold nanoparticles: assembly, supramolecular chemistry, quantum-size-related properties, and applications toward biology, catalysis, and nanotechnology. *Chem. Rev.* 104 (1): 293–346.

20 Xie, Y.-P., Jin, J.-L., Duan, G.-X. et al. (2017). High-nuclearity silver (I) chalcogenide clusters: a novel class of supramolecular assembly. *Coord. Chem. Rev.* 331: 54–72.

21 Ebina, A., Hossain, S., Horihata, H. et al. (2020). One-, two-, and three-dimensional self-assembly of atomically precise metal nanoclusters. *Nanomaterial* 10 (6): 1105.

22 Grzybowski, B. and Whitesides, G. (2002). Self-assembly at all scales. *Science* 295 (5564): 2418–2421.

23 Wu, Z., Yao, Q., Zang, S., and Xie, J. (2019). Directed self-assembly of ultrasmall metal nanoclusters. *ACS Mater. Lett.* 1 (2): 237–248.

24 Rival, J.V., Mymoona, P., Lakshmi, K.M. et al. (2021). Self-assembly of precision noble metal nanoclusters: hierarchical structural complexity, colloidal superstructures, and applications. *Small* 2005718.

25 Häkkinen, H. (2012). The gold–sulfur interface at the nanoscale. *Nat. Chem.* 4 (6): 443–455.

26 De Nardi, M., Antonello, S., Jiang, D.-e. et al. (2014). Gold nanowired: a linear $(Au_{25})_n$ polymer from Au_{25} molecular clusters. *ACS Nano* 8 (8): 8505–8512.

27 Lei, Z., Pei, X.L., Jiang, Z.G., and Wang, Q.M. (2014). Cluster linker approach: preparation of a luminescent porous framework with NbO topology by linking silver ions with gold (I) clusters. *Angew. Chem.* 126 (47): 12985–12989.

28 Jin, Y., Zhang, C., Dong, X.-Y. et al. (2021). Shell engineering to achieve modification and assembly of atomically-precise silver clusters. *Chem. Soc. Rev.* 50 (4): 2297–2319.

29 Goswami, N., Yao, Q., Luo, Z. et al. (2016). Luminescent metal nanoclusters with aggregation-induced emission. *J. Phys. Chem. Lett.* 7 (6): 962–975.

30 Bera, D. and Goswami, N. (2021). Driving forces and routes for aggregation-induced emission-based highly luminescent metal nanocluster assembly. *J. Phys. Chem. Lett.* 12 (37): 9033–9046.

31 Wu, Z., Yao, Q., Zang, S.-q., and Xie, J. (2021). Aggregation-induced emission in luminescent metal nanoclusters. *Natl. Sci. Rev.* 8 (6): nwaa208.

32 Zeng, C., Chen, Y., Kirschbaum, K. et al. (2016). Emergence of hierarchical structural complexities in nanoparticles and their assembly. *Science* 354 (6319): 1580–1584.

33 Yan, N., Xia, N., Liao, L. et al. (2018). Unraveling the long-pursued Au_{144} structure by X-ray crystallography. *Sci. Adv.* 4 (10): eaat7259.

34 Higaki, T., Liu, C., Zhou, M. et al. (2017). Tailoring the structure of 58-electron gold nanoclusters: $Au_{103}S_2(S\text{-nap})_{41}$ and its implications. *J. Am. Chem. Soc.* 139 (29): 9994–10001.

35 Antonello, S., Dainese, T., Pan, F. et al. (2017). Electrocrystallization of monolayer-protected gold clusters: opening the door to quality, quantity, and new structures. *J. Am. Chem. Soc.* 139 (11): 4168–4174.

36 Fei, W., Antonello, S., Dainese, T. et al. (2019). Metal doping of Au25 (SR) 18–clusters: insights and hindsights. *J. Am. Chem. Soc.* 141 (40): 16033–16045.

37 Chen, J., Liu, L., Liu, X. et al. (2017). Gold-doping of double-crown Pd nanoclusters. *Chem. Eur. J.* 23 (72): 18187–18192.

38 Hossain, S., Imai, Y., Motohashi, Y. et al. (2020). Understanding and designing one-dimensional assemblies of ligand-protected metal nanoclusters. *Mater. Horiz.* 7 (3): 796–803.

39 Yuan, P., Zhang, R., Selenius, E. et al. (2020). Solvent-mediated assembly of atom-precise gold–silver nanoclusters to semiconducting one-dimensional materials. *Nat. Commun.* 11 (1): 1–8.

40 Wang, Z.-Y., Wang, M.-Q., Li, Y.-L. et al. (2018). Atomically precise site-specific tailoring and directional assembly of superatomic silver nanoclusters. *J. Am. Chem. Soc.* 140 (3): 1069–1076.

41 Alhilaly, M.J., Huang, R.-W., Naphade, R. et al. (2019). Assembly of atomically precise silver nanoclusters into nanocluster-based frameworks. *J. Am. Chem. Soc.* 141 (24): 9585–9592.

42 Ma, X.H., Wang, J.Y., Guo, J.J. et al. (2019). Reversible wide-range tuneable luminescence of a dual-stimuli-responsive silver cluster-assembled material. *Chin. J. Chem.* 37 (11): 1120–1124.

43 Li, Y.-H., Huang, R.-W., Luo, P. et al. (2019). 1D silver cluster-assembled materials act as a platform for selectively erasable photoluminescent switch of acetonitrile. *Sci. Chin. Chem.* 62 (3): 331–335.

44 Lu, S.-H., Li, Y., Yang, S.-X. et al. (2019). Three silver coordination polymers with diverse architectures constructed from pyridine carboxylic hydrazide ligands. *Inorg. Chem.* 58 (17): 11793–11800.

45 Xu, C., Hedin, N., Shi, H.-T., and Zhang, Q.-F. (2014). A semiconducting microporous framework of $Cd_6Ag_4(SPh)_{16}$ clusters interlinked using rigid and conjugated bipyridines. *Chem. Commun.* 50 (28): 3710–3712.

46 Das, A.K., Biswas, S., Manna, S.S. et al. (2021). Solvent-dependent photophysical properties of a semiconducting one-dimensional silver cluster-assembled material. *Inorg. Chem.* 60 (23): 18234–18241.

47 Biswas, S., Das, A.K., Nath, A. et al. (2021). Novel class of water-soluble phosphonate silver cluster assembled material for efficient photoelectric sensing and photoacoustic imaging. *Nanoscale* 13 (41): 17325–17330.

48 Huang, R.-W., Wei, Y.-S., Dong, X.-Y. et al. (2017). Hypersensitive dual-function luminescence switching of a silver-chalcogenolate cluster-based metal–organic framework. *Nat. Chem.* 9 (7): 689–697.

49 Du, X.-S., Yan, B.-J., Wang, J.-Y. et al. (2018). Layer-sliding-driven crystal size and photoluminescence change in a novel SCC-MOF. *Chem. Commun.* 54 (42): 5361–5364.

50 Cao, M., Pang, R., Wang, Q.-Y. et al. (2019). Porphyrinic silver cluster assembled material for simultaneous capture and photocatalysis of mustard-gas simulant. *J. Am. Chem. Soc.* 141 (37): 14505–14509.

51 Das, A.K., Biswas, S., Thomas, A. et al. (2021). Switchable photon and phonon emission properties of an atomically precise Ag_{14} core-based two-dimensional silver cluster-assembled material. *Mater. Chem. Front.* 5 (24): 8380–8386.

52 Dong, X.-Y., Huang, H.-L., Wang, J.-Y. et al. (2018). A flexible fluorescent SCC-MOF for switchable molecule identification and temperature display. *Chem. Mater.* 30 (6): 2160–2167.

53 Wei, Z., Wu, X.H., Luo, P. et al. (2019). Matrix coordination induced emission in a three-dimensional silver cluster-assembled material. *Chem. Eur. J.* 25 (11): 2750–2756.

54 Wu, X.H., Luo, P., Wei, Z. et al. (2019). Guest-triggered aggregation-induced emission in silver chalcogenolate cluster metal–organic frameworks. *Adv. Sci.* 6 (2): 1801304.

55 Luo, Z., Yuan, X., Yu, Y. et al. (2012). From aggregation-induced emission of Au(I)–thiolate complexes to ultrabright Au(0)@Au(I)–thiolate core–shell nanoclusters. *J. Am. Chem. Soc.* 134 (40): 16662–16670.

56 Jin, S., Liu, W., Hu, D. et al. (2018). Aggregation-induced emission (AIE) in Ag–Au bimetallic nanocluster. *Chem. A Eur. J.* 24 (15): 3712–3715.

57 Sugiuchi, M., Maeba, J., Okubo, N. et al. (2017). Aggregation-induced fluorescence-to-phosphorescence switching of molecular gold clusters. *J. Am. Chem. Soc.* 139 (49): 17731–17734.

58 Chen, T., Yang, S., Chai, J. et al. (2017). Crystallization-induced emission enhancement: a novel fluorescent Au–Ag bimetallic nanocluster with precise atomic structure. *Sci. Adv.* 3 (8): e1700956.

59 Sun, D., Zhang, L., Lu, H. et al. (2013). Bright-yellow to orange-red thermochromic luminescence of an Ag^{I}_{6}-Zn^{II}_{2} heterometallic aggregate. *Dalton Trans.* 42: 3528–3532.

60 Sun, D., Zhang, L., Yan, Z., and Sun, D. (2012). Stepwise construction of a Ag^{I}_{9}-Cu^{II}_{4} heterometallic cluster incorporating two unusual vertex-shared trigonal-bipyramidal silver polyhedra. *Chem. Asian J.* 7 (7): 1558–1561.

61 Wang, Z., Zhuang, G., Deng, Y. et al. (2016). Near-infrared emitters: stepwise assembly of two heteropolynuclear clusters with tunable Ag^{I}:Zn^{II} ratio. *Inorg. Chem.* 55 (10): 4757–4763.

62 Sun, P., Wang, Z., Bi, Y. et al. (2020). Self-assembly-driven aggregation-induced emission of silver nanoclusters for light conversion and temperature sensing. *ACS Appl. Nano Mater.* 3 (2): 2038–2046.

63 Sun, P., Wang, Z., Sun, D. et al. (2020). pH-guided self-assembly of silver nanoclusters with aggregation-induced emission for rewritable fluorescent platform and white light emitting diode application. *J. Colloid Interface Sci.* 567 (1): 235–242.

64 Shen, J., Wang, Z., Sun, D. et al. (2018). pH-responsive nanovesicles with enhanced emission co-assembled by Ag(I) nanoclusters and polyethyleneimine as a superior sensor for Al^{3+}. *ACS Appl. Mater. Interfaces* 10 (4): 3955–3963.

65 Shen, J., Wang, Z., Sun, D. et al. (2017). Self-assembly of water-soluble silver nanoclusters: superstructure formation and morphological evolution. *Nanoscale* 9: 19191–19200.

66 Shen, J., Wang, Z., Xia, C. et al. (2019). Amphiphilicity regulation of Ag(I) nanoclusters: self-assembly and its application as a luminescent probe. *Chem. A Eur. J.* 25 (18): 4713–4721.

67 Zhang, Y., Sun, D., Shen, J., and Xin, X. (2020). Dynamic self-assembly of silver nanoclusters into luminescent nanotubes with controlled surface roughness: scaffold of superhydrophobic materials. *Appl. Surf. Sci.* 514 (1): 145913.

68 Cheng, X., Sun, P., Zhang, N. et al. (2021). Self-assembly of silver nanoclusters and phthalic acid into hollow tubes as a superior sensor for Fe^{3+}. *J. Mol. Liq.* 323 (1): 115032.

69 Xie, Z., Sun, P., Wang, Z. et al. (2020). Metal–organic gels from silver nanoclusters with aggregation-induced emission and fluorescence-to-phosphorescence switching. *Angew. Chem. Int. Ed.* 59 (25): 9922–9927.

70 Bi, Y., Wang, Z., Liu, T. et al. (2021). Supramolecular chirality from hierarchical self-assembly of atomically precise silver nanoclusters induced by secondary metal coordination. *ACS Nano* 15 (10): 15910–15919.

16

Coinage Metal Cluster-Assembled Materials

Zhao-Yang Wang and Shuang-Quan Zang

Henan Key Laboratory of Crystalline Molecular Functional Materials, Henan International Joint Laboratory of Tumor Theranostical Cluster Materials, Green Catalysis Center, and College of Chemistry, Zhengzhou University, Zhengzhou 450001, P. R. China

16.1 Introduction

Atomically precise, monolayer-protected noble metal nanoclusters (NCs), which fill the gap between discrete metal complexes and plasmonic nanoparticles, emerged as a fascinating area of nanoscience research for the past two decades [1–3]. The high uniformity, precise atomic compositions, and well-defined atom-packing modes of NCs facilitate the study of structure-reactivity correlations. Besides the well-studied solution chemistry of metal nanoclusters focusing on their discrete molecule form, the aggregated state is also very attractive. For instance, intercluster aggregations induced by the addition of poor solvents or cations to the solution of metal nanoclusters may display enhanced PL intensities – a phenomenon known as aggregation-induced emission (AIE) [4, 5]. As a common aggregated form, the solid-state metal cluster-assembled materials (MCAMs) show promising applications in many fields (e.g. fluorescence, catalysis, chirality, magnetism, electrochemistry, etc.), [6] and their properties highly depend on the packing modes and the inter-cluster interactions. The self-assembly of NCs are usually driven by hydrogen bonding, electrostatic interactions, van der Waals interactions (including dipolar, C–H$\cdots\pi$, and $\pi\cdots\pi$ interactions) facilitated by the surface ligands between neighboring NCs, and in some cases, the metallophilic and amphiphilic interactions [6–10]. Recently, the externally directed assembly of NCs by adding external reagents such as metal ions and bridging ligands have been explored [11]. In this chapter, we focus on the coordination-assisted MCAMs with atom-precise structure determined by X-ray single-crystal diffraction.

The materials based on metal clusters bridged by linkers (e.g. simple inorganic anions or polyoxometalates (POM), bi-/multi-topic organic linkers) are known as MCAMs, and the ones with high dimensional porous structures are also known as metal cluster-based metal–organic frameworks (MC-MOFs). MCAMs show some advantages over discrete metal clusters:

- Metal clusters in MCAMs show significantly improved stability. Instability is a deadly shortcoming of metal clusters, particularly for silver or copper clusters. One major reason is that on the surface of the metal clusters lie volatile molecules via weak coordination bonds. Replacing them with selected linkers could effectively solve this problem.

Atomically Precise Nanochemistry, First Edition. Edited by Rongchao Jin and De-en Jiang.

- MCAMs integrating the metal cluster nodes and linkers show richer photophysical or other properties benefited from the multiple charge transfer paths and synergistic effect.
- Self-assembled metal clusters usually adopt a dense packing manner to maintain stability. In contrast, the exposed metal cluster nodes in the highly porous MC-MOFs could effectively interact with the small molecules or substrates captured, which endows the MC-MOFs with promising sensing and catalysis properties. Therefore, the investigation of MCAMs is of great significance in both fundamental research and applications.

In this chapter, we discuss the crystal structures of MCAMs sorted out by metal type, linker type and nuclearity of metal cluster nodes, and analyze the opportunities and challenges in the fabrication of such materials with different metal and linker components basically following the hard-soft-acid-base (HSAB) theory. The potential applications of MCAMs in many fields, including catalysis, optical sensing, and biomedicine, are also introduced with several attractive representatives.

16.2 Structures of Metal Cluster-Assembled Materials

16.2.1 Silver Cluster-Assembled Materials (SCAMs)

16.2.1.1 Simple Ion Linker

Simple inorganic anions from the metal salts employed in the synthesis of silver clusters usually act as co-protective agent in the ligand-shell, structure-directing templates inside the clusters, and counterions in the lattice, nevertheless, in some cases they also play roles as small bridges linking large clusters to form cluster-based polymeric materials. For instance, the Zang group reported an extended layer with a formula of $[Ag_{31}S_3(S^tBu)_{16}(NO_3)_9]_n$ based on nest-like Ag_{31} clusters bridged by NO_3^- anions in 2014 [12]. Later on, two anion-templated silver clusters, a drum-like $CO_3@Ag_{20}$ and a ball-shaped $SO_4@Ag_{22}$ with a twisted truncated tetrahedral geometry, were incorporated into a two-dimensional (2D) *sql* lattice and an unprecedented three-dimensional (3D) twofold interpenetrated *dia* network assisted by NO_3^- linkers, respectively, under mild solvothermal conditions by the Sun group [13]. In another work, the same group reported a novel 7-connected 3D framework with a rare *kwh* topology, comprising Ag_{19} clusters template by one μ_3-η^1:η^1:η^1 CrO_4^{2-} and one $^iBuS^-$ and CN^- as bridges, where the cyanide anion is in situ generated from the decomposition of acetonitrile, and a 3D MOF with a chiral *qtz* topology constructed from egg-like Ag_{20} clusters encapsulating one μ_8-η^1:η^1:η^3:η^3 CrO_4^{2-} as a template and μ_6-$Cr_2O_7^{2-}$ as bridges [14]. Besides these oxyacidates, halide ions can also serve as bridges in silver cluster assembly. Zhang et al. reported a novel water-stable 3D *pcu-h* metal-organic open framework, $[Cl@Ag_{14}(cPrC{\equiv}C)_{10}Cl_2\cdot(p\text{-}TOS)\cdot 1/3H_2O]_n$ (cPrC–CH = cyclopropylacetylene; p-TOS = p-toluenesulfonate), wherein the chloride acts not only as template to shape the tetradecanuclear silver cluster from the interior, but also bridges the cluster to an extended superstructure [15].

Very recently, the Zhu group reported a new type of framework material, termed a superatom complex inorganic framework (SCIF), in which a chiral nanocluster/superatom complex ($[Au_1Ag_{22}(SR)_{12}]^{3+}$, SC) serves as the building block and SbF_6^- serves as the linker [16]. Two frameworks with different topologies and different channel sizes were obtained by the assembly of the SC with adjusted amounts of SbF_6^- (Figure 16.1). SCIF-1 is racemic and showed no CPL response. SCIF-2 displayed a pair of mirror-image CPL signals for the left- and right-handed crystals.

Similarly, cations also act as counterions of negative-charged clusters, which can not only control the syntheses of nanoclusters but also assemble these nanoclusters into cluster-based supracrystals. Wei et al. reported the hierarchical assembly of $Ag_{29}(BDT)_{12}$ clusters into linear chains

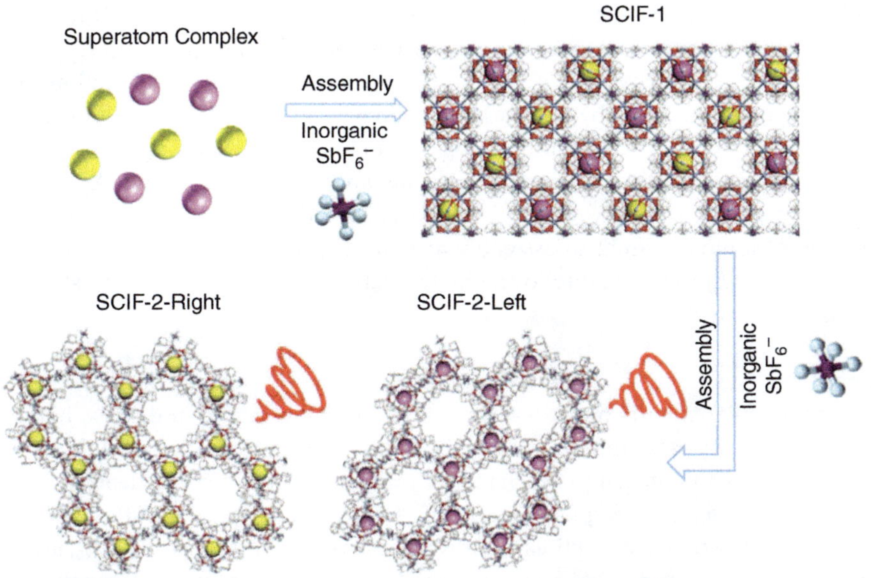

Figure 16.1 Schematic illustration of SCIFs assembly with SbF_6^- linkers. *Source:* Redrawn with permission from Ref. [16]. © 2020 Wiley-VCH GmbH.

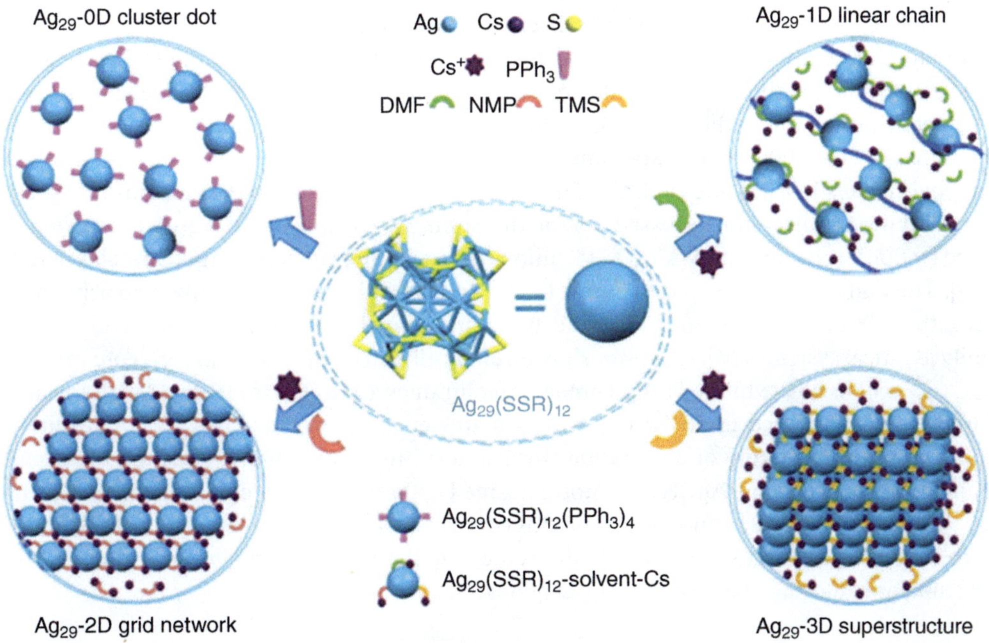

Figure 16.2 Scheme illustration of the 1D-3D assemblies of $[Ag_{29}(BDT)_{12}]^{3-}$. Ag_{29}-0D represents the discrete $[Ag_{29}(BDT)_{12}(TPP)_4]^{3-}$ clusters, linear chains of Ag_{29}-1D, grid networks of Ag_{29}-2D and 3D array of Ag_{29}-3D. *Source:* Reproduced with permission from Ref. [17]. © 2020 Oxford University Press.

$[Cs_3Ag_{29}(BDT)_{12}(DMF)_x]_{2n}$ (Ag_{29}-1D, DMF = dimethylformamide), 2D grid networks $[Cs_3Ag_{29}(BDT)_{12}(NMP)_x]_n$, ($Ag_{29}$-2D; NMP = N-methyl-2-pyrrolidone), and 3D superstructures $[Cs_2Ag_{29}(BDT)_{12}(TMS)_x]^-_n$ (Ag_{29}-3D; TMS = tetramethylene sulfone) in which the Cs^+ cations are surrounded by solvent molecules (Figure 16.2) [17]. The interactions among the cluster framework, the Cs^+ cations, and the O-containing solvent molecules (including DMF, NMP, and TMS) assemble the anionic clusters into an extended structure in the crystalline phase. In Ag_{29}-1D, two adjacent Ag_{29} clusters are bridged by two Cs^+ cations through Cs–C and Cs–π interactions. In addition, intercluster assembly is induced by outward interactions from four Cs^+ conjunction sites. For Ag_{29}-2D, each $[Cs@Ag_{29}(BDT)_{12}(NMP)_x]_2$ unit is adjacent to four identical units through four Cs^+ conjunction sites, producing the two-dimensional grid network. For Ag_{29}-3D, each Ag_{29} cluster is surrounded by six adjacent cluster molecules, and each Cs^+ cation links two clusters (Figure 16.2). This kind of ionic hierarchical assembly of silver clusters holds great promise for superstructures with collective properties due to the higher capacity of interactions and shorter interaction distances.

16.2.1.2 POMs Linker

POMs are a class of inorganic oxo-enriched clusters with controllable shapes and sizes, highly negative charges, and universally structural topologies. Benefiting from the strong oxophilicity of silver, POMs were employed as anionic templates to induce the formation of high-nuclearity silver nanoclusters. The Sun group and others prepared a series of POM-templated silver clusters, including $Ag_{44}@Mo_6O_{22}$ and $Ag_{50}@Mo_8O_{28}$, as well as three-shell structures of $Ag_{10}@(W_7O_{26})_2@Ag_{74}$ and $Ag_{10}@(Mo_7O_{26})_2@Ag_{70}$, to name a few, with different size, geometry, and compositions. Notably, in some cased the POMs could undergo a structure transformation in the synthesis to generate unprecedented POMs in situ, which conversely enriches the library of POMs [18, 19]. Moreover, silver cluster-POMs hybrids with inverted structures, that is, silver nanoclusters encapsulated in hollow POM framework, have also been reported [20].

Not unnaturally, multidentate POMs can also act as bulk linkers to bridge the silver clusters. Gruber et al. synthesized the first chain-like intercluster compounds containing large, novel silver alkynyl clusters as cationic building blocks, which are linked to the polyanions by direct silver-oxygen contacts [21]. Later on, the Mak and Lu groups contributed in the study of POM-based silver alkynyl clusters [19, 22]. Xie et al. also fabricated an extended two-dimensional infinite sheet formed by linking 1D silver cluster-based polymeric motifs, which consists of $Ag_{11}S_9$ cluster interconnected by thiolate ligands, and $[Mo_6O_{19}]^{2-}$ anions through intermolecular Ag$\cdots$O weak interactions [23]. The Zang group prepared two novel POM-based thiolate-protected silver coordination polymers, the one-dimensional Ag_{10}-Mo_6 and two-dimensional Ag_{18}-Mo_6, in high yields under essentially identical systems [24]. Ag_{10}-Mo_6 showed chain-like structures with an Ag_{10} core bridged by $[Mo_6O_{19}]^{2-}$ anions and exhibited bright green photoluminescence (Figure 16.3a). Ag_{18}-Mo_6 contain a unique 20-membered cycle-$Ag_{10}S_{10}$ with a diameter of approximately 11.382 Å resulting from the compact interlinking of alternating silver and sulfur atoms. Interconnecting the cycle-$Ag_{10}S_{10}$ and the Ag_3S^tBu and $AgCH_3CN$ motifs formed esthetic Ag–S sheets (Figure 16.3b). In addition, $[Mo_6O_{19}]^{2-}$ polyanions not only served as counter ions to fill in the large gaps of the cycle-$Ag_{10}S_{10}$ to balance the charges of the whole molecule, but also served to stabilize the structure by acting as inorganic ligands and forming Ag–O bonds.

16.2.1.3 Organic Linker

Metal–organic frameworks (MOFs) or porous coordination polymers (PCPs) with cluster-based nodes bridged by organic ligands via coordination bonds have become a prominent class of crystalline materials in separation, storage, catalysis, chemical sensing, and luminescent signaling devices. Cluster-based MOFs usually show virtues, including enhanced stability and larger channels, and conveniently integrate the functionality of cluster units and organic linkers. Inspired by the characteristic features of silver clusters, appropriate organic ligands might coordinately link and spatially separate silver clusters in certain orientations to form functional silver

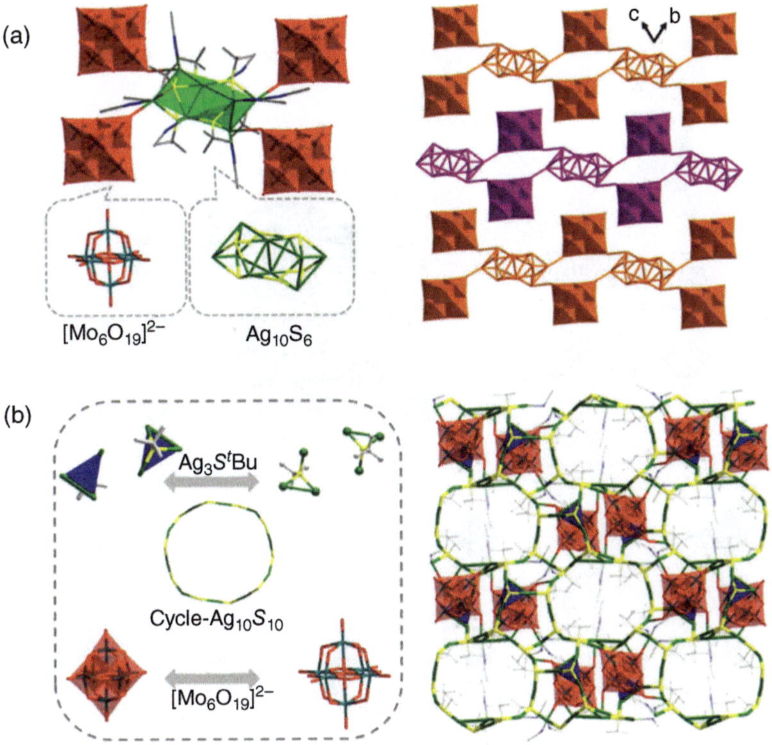

Figure 16.3 The crystal structures of Ag_{10}-Mo_6 (a) and Ag_{18}-Mo_6 (b). *Source:* Reproduced with permission from Ref. [24]. © 2020 The Royal Society of Chemistry.

cluster-assembled materials (SCAMs). In this section, we introduce some representative SCAMs categorized by organic linker types.

Pyridyl Linkers The coordination of pyridine to Ag^+ is common in metal complexes and MOFs, which follows the HSAB theory, as Ag^+ is a typical soft acid and pyridine is a borderline base. The medium coordination ability of Ag-pyridyl facilitates an optimized self-assembly rate during the formation of SCAMs, which consequently promises the single crystal growth for X-ray diffraction structure determination. Certainly, the size, geometrical configuration, and surface structure characteristics of the cluster nodes as well as the ligand type ultimately direct the structure of the SCAMs. In the past years, a series of SCAMs based on bi-/multi-pyridines have been synthesized and structurally characterized by X-ray diffraction. In this section, we discuss the SCAMs based on different silver cluster nodes.

In 2017, the Zang group first prepared a weakly luminescent dodecanuclear silver-chalcogenolate cluster (SCC) $[Ag_{12}(S^tBu)_6(CF_3COO)_6(CH_3CN)_6]\cdot CH_3CN$ (henceforth designated Ag_{12}) in low yield (5%), which is very unstable and completely deteriorated in air within 30 minutes. Each nanocluster has an $Ag_{12}S_6^{6-}$ core, which exhibits a Ag_5S_2–Ag_2S_2–Ag_5S_2 three-layer arrangement, with six Ag(I) vertices each bearing a volatile terminal CH_3CN ligand that can be substituted by a bridging N-linker to form porous MOFs. Thus, they used the linear bridging ligand, 4,4′-bipyridine (bpy), to replace the coordinated CH_3CN ligands of Ag_{12} to construct the rigid SCC-based MOF $[(Ag_{12}(S^tBu)_8(CF_3COO)_4(bpy)_4)]_n$ (Ag_{12}bpy) through both stepwise ligand-exchange method and one-pot synthetic strategy (Figure 16.4a), in which the secondary building units (SBUs) are

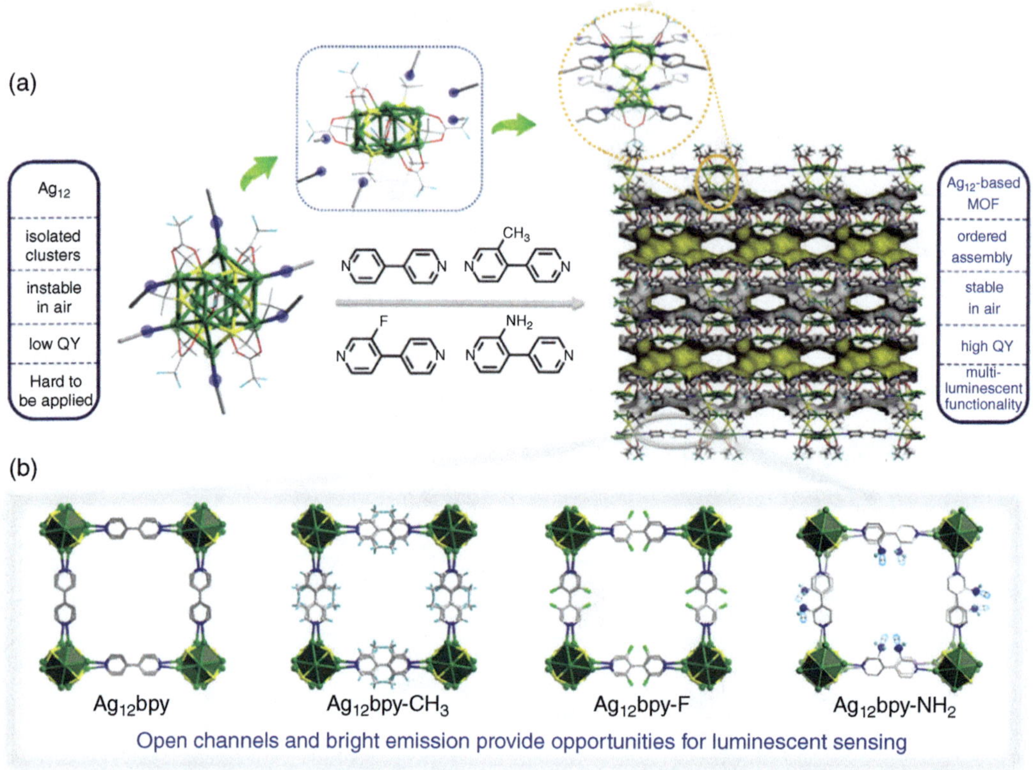

Figure 16.4 (a) Schematic representation of the ligand-exchange strategy used to obtain Ag_{12}bpy crystals. Interconnected channels of Ag_{12}bpy viewed along a axis, where the yellow surface represents the pore surface. The two rectangular boxes serve to compare the characteristics between Ag_{12} and Ag_{12}bpy crystals. The ligands bpy-F, bpy-NH_2, and bpy-CH_3 could also produce identical frameworks. (b) Perspective views of open channels of the isostructures of Ag_{12}bpy, Ag_{12}bpy-NH_2 Ag_{12}bpy-CH_3 and Ag_{12}bpy-F. Ag, green; C, gray; O, red; S, yellow; F, turquoise. H atoms are omitted for clarity. *Source:* Adapted with permission from Ref. [11]. © 2021 The Royal Society of Chemistry.

transformed quasi-isomers of Ag_{12} generated by induction of incoming bpy ligands [25]. Markedly enhanced stability (from minutes to a year) under ambient conditions and a ~ 60-fold increase in the RT luminescence quantum yield (QY 0.2–12.1%) were observed from Ag_{12} to Ag_{12}bpy.

This work enriches the cluster-based metal–organic framework portfolio, bridges the gap between silver chalcogenide/chalcogenolate clusters and metal–organic frameworks, and provides a foundation for further development of functional silver-cluster-based materials. Since then, SCAMs have gained an increasing interest and a series of such materials with atom-precise structures have been reported [11]. Meanwhile, the extensive studies on their properties have also been conducted [11].

The same group later investigated the effect of modification of the substituents on the bpy linker of Ag_{12}bpy on room-temperature emission [26]. $[Ag_{12}(S^tBu)_8(CF_3COO)_4(bpy\text{-}NH_2)_4]_n$ (abbreviated as Ag_{12}bpy-NH_2) having 3-amino-4,4′-bipyridine (bpy-NH_2) bearing an electron-donating $-NH_2$ group, $[Ag_{12}(S^tBu)_8(CF_3COO)_4(bpy\text{-}CH_3)_4]_n$, (abbreviated as Ag_{12}bpy-CH_3) possessing 3-methyl-4,4′-bipyridine (bpy-CH_3) with a weakly electron-donating methyl group, $[Ag_{12}(S^tBu)_8(CF_3COO)_4(bpy\text{-}F)_4]_n$ (abbreviated as Ag_{12}bpy-F) containing 3-fluoro-4,4′-bipyridine (bpy-F), which has a molecular framework isostructural with that of Ag_{12}bpy (Figure 16.4b).

The feasible modification of linkers in SCAMs paves the way to multifunctionalization and subsequent research based on the fundamental principles of structure-function relationships.

By tuning the synthesis methods, the cuboctahedra of $Ag_{12}S_6$ nodes could be further assembled into 2D and 3D network structures by bidentate ligands bpy, bpy-NH_2, bpa (1,2-bis(4-pyridyl) ethane), bpe (1,2-bis(4-pyridyl)ethylene), and CPPP (2,5-bis(4-cyanophenyl)-1,4-bis(4-(pyridine-4-yl)-phenyl)-1,4-dihydropyrrolo[3,2-b]pyrrole), tridentate TPPA (tris(4-pyridylphenyl)amine), quadridentate tppe (1,1,2,2-tetrakis(4-(pyridine-4-yl)phenyl)-ethene), and quadridentate TPyP (5,10,15,20-tetra(4-pyridyl)porphyrin) (Figure 16.5) [27–32]. In the cluster nodes of these SCAMs, the cuboctahedron consisting of 12 Ag atoms linked through Ag(I)-Ag(I) bonding often displays C_{3v} or quasi-C_{3v} symmetry, in which cuboctahedral or quasi-cuboctahedral Ag skeletons are connected via Ag–N bonds. Among them, Ag_{12}CPPP features a twofold self-interpenetrated 3D framework. Ag_{12}tppe is characterized by a highly symmetrical 3D network, in which the diameter of its cubic cage reaches 32 Å. $Ag_{12}Ag_8$tppe is notable because it incorporates two types of Ag_8 and Ag_{12} cluster nodes. The structural diversity of SCAMs (Figure 16.5) and the introduction of attractive AIE linkers could bring about a broader application range.

The fcc-Ag_{14} nanoclusters were first reported by the Zheng group in 2013. Each cluster holds 12 thiolate ligands on the edges and 8 triphenylphosphine ligands at the vertices [33]. In 2018, Wang et al. synthesized a similar superatomic fcc-Ag_{14} cluster via a "one-pot" self-reducing reaction by using carboranethiol, yet with the eight vertices anchored by volatile CH_3CN molecules [34]. The CH_3CN molecules could be easily substituted by pyridyl ligands in situ or step-by-step, which makes fcc-Ag_{14} an ideal eight-connected node in SCAMs. Accordingly, they adopted a series of length-variable bidentate *N*-heteroaromatic ligands [pyrazine (pz), di-pyridin-4-yl-diazene (dpz), 4,4′-bipyradine (bpy), 1,4-bis(4-pyridyl)benzene (dpbz)], and built up unprecedented 1D-to-3D superatomic SCAMs (Ag_{14}-pz, Ag_{14}-dpz, Ag_{14}-bpy, Ag_{14}-dpbz) via site-specific and directional

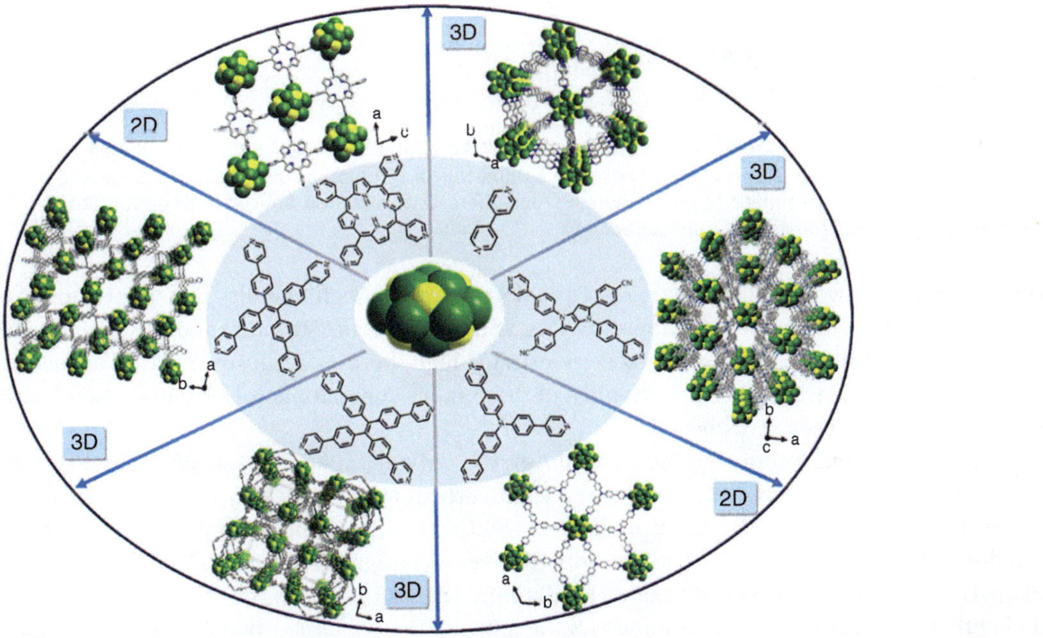

Figure 16.5 $Ag_{12}S_6$ nodes are assembled into two-dimensional and three-dimensional network structures by bidentate bpy, bpy-NH_2, and CPPP, the tridentate ligand TPPA, and the quadridentate ligands tppe and TPyP. *Source:* Adapted with permission from Ref. [11]. © 2021 The Royal Society of Chemistry.

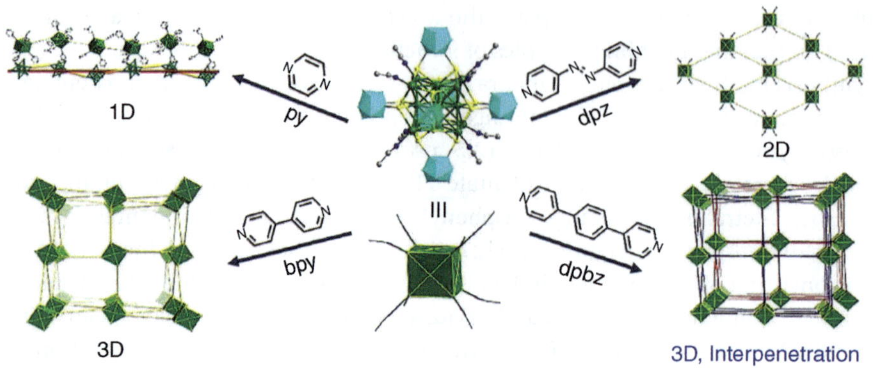

Figure 16.6 Bidentate N-containing ligands bridge mixed-valence (0/+1) Ag_{14} building blocks to form 1D-to-3D materials. *Source:* Adapted with permission from Ref. [11]. © 2021 The Royal Society of Chemistry.

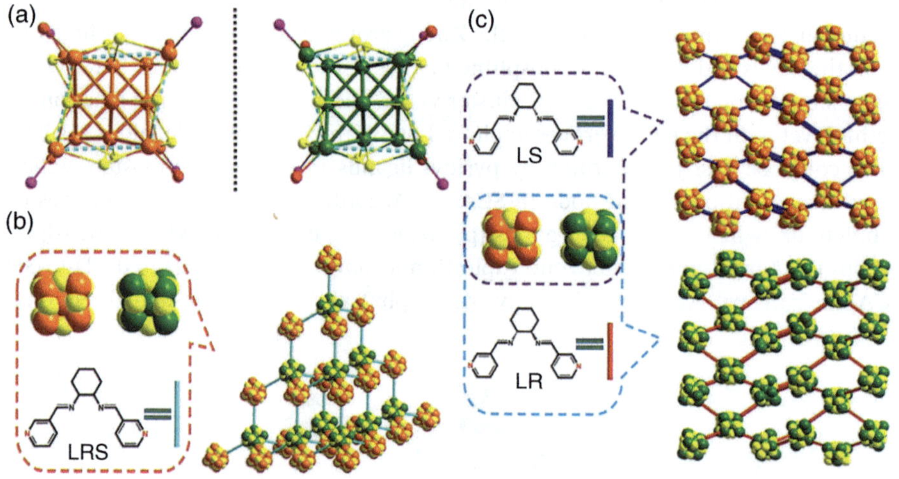

Figure 16.7 (a) Ag_{14} enantiomer pair (left, S-Ag_{14}; right, R-Ag_{14}). (b) Diamond-like topology of Ag_{14}-LRS. (c) Crystal structures of Ag_{14}-LS and Ag_{14}-LR, which exhibit mirror symmetry. Color legend: orange and green spheres, Ag; yellow sphere, S; pink sphere, P; red sphere, O. *Source:* Reproduced with permission from Ref. [35]. © 2021 American Chemical Society.

covalent-bond connection (Figure 16.6). Tunable patterning, high stability and photoluminescence were successfully achieved in these SCAMs, and the silver core structure and cluster emission were maintained all through. Ag_{14}-dpbz demonstrated high thermostability over 220 °C in air. The dual-emissive luminescent thermochromism of these stable superatomic silver nanocluster-based materials have also been investigated.

Recently, the Zheng group reported a chiral Ag_{14} cluster, $Ag_{14}(SPh(CF_3)_2)_{12}(PPh_3)_4(DMF)_4$, with a twisted fcc-$Ag_{14}$ structure [35]. Four of the eight vertices are occupied by triphenylphosphine, while others are occupied by DMF molecules (Figure 16.7a). Using racemic Ag_{14} clusters as building blocks and bidentate ligands with different configurations, (1*R*,2*R*,N^1*E*,N^2*E*)-N^1,N^2-bis(pyridin-3-ylmethylene)-cyclohexane-1,2-diamine (LR), (1*S*,2*S*,N^1*E*,N^2*E*)-N^1,N^2-bis-(pyridin-3-ylmethylene)cyclohexane-1,2-diamine (LS), as linkers, they obtained two chiral enantiomeric crystals, $[Ag_{14}(SPh(CF_3)_2)_{12}(PPh_3)_4(LR)_2]_n$ (Ag_{14}-LR) and $[Ag_{14}(SPh(CF_3)_2)_{12}(PPh_3)_4(LS)_2]_n$ (Ag_{14}-LS). Both Ag_{14}-LS and Ag_{14}-LR adopt a chiral 3D framework structure and exhibit perfect mirror symmetry (Figure 16.7c). Interestingly, structural analysis revealed that all the Ag_{14}

building blocks in Ag_{14}-LR are homochiral R-Ag_{14} and those in Ag_{14}-LS are homochiral S-Ag_{14}, which means that enantioconversion of Ag_{14} racemates to homochiral Ag_{14} building blocks took place in solution during the assembly process. The reaction of the racemic Ag_{14} clusters with an equimolar mixture of LR and LS (racemic LRS) afforded one achiral, $[Ag_{14}(SPh(CF_3)_2)_{12}(PPh_3)_4(LRS)_2]_n$ (Ag_{14}-LRS), 3D Ag_{14}-based assembly crystal. Ag_{14}-LRS contains both R-Ag_{14} and S-Ag_{14} building blocks in a ratio of 1 : 1. The two Ag_{14} building blocks connected on each LRS linkers possess different chiralities. The two enantiomers of LRS (LR and LS) are coordinated on the Ag_{14} building blocks with the same ratio. Both the building blocks and the linkers are racemic mixtures, which gives the obtained assembly material Ag_{14}-LRS an achiral diamond-like topology (Figure 16.7b).

Capsule-like $Ag_{10}S_6$ core with a parallel Ag_3-Ag_4-Ag_3 three-layer silver atom arrangement is also frequently seen in the SCAMs. Based on this Ag_{10} cluster node, the Zang group reported a hydrogen-bonded pillared-layered 2D SCC-MOF $[Ag_{10}(S^tBu)_6(CF_3COO)_2(PhPO_3H)_2(bpy)_2]_n$ (Ag_{10}bpy-2D) by introducing a multidentate phenylphosphonic acid ($PhPO_3H_2$) ligand as a functional hydrogen-bond donor and a 1D polymer $[Ag_{10}(S^tBu)_6(CF_3CF_2COO)_4(bpy)_2(CH_3CN)_2]_n$ (Ag_{10}bpy-CH_3CN) in succession [36, 37]. In Ag_{10}bpy-2D, each $[Ag_{10}(S^tBu)_6]$ cluster core is connected by four bpy linkers, resulting in a layered architecture with rhombic grids in the ab plane (Figure 16.8a). Two hydrogen-bonded $PhPO_3H$ linkers are located at the apex of the $Ag_{10}S_6$ capsule. The interlayer

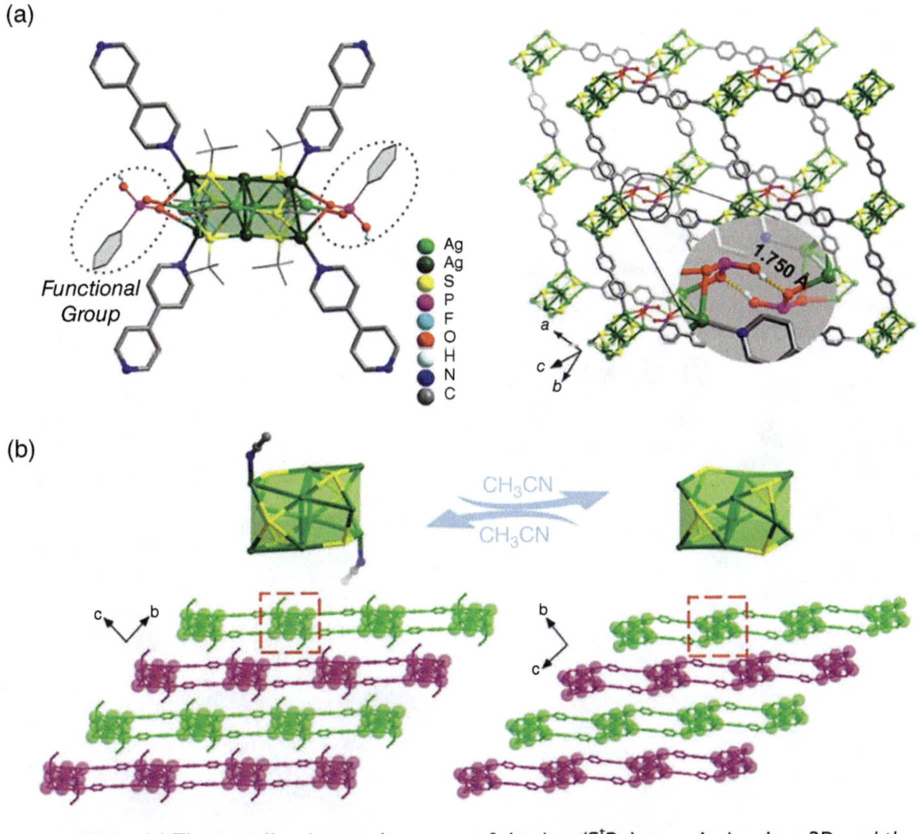

Figure 16.8 (a) The coordination environment of the $Ag_{10}(S^tBu)_6$ core in Ag_{10}bpy-2D, and the two-layer stack of the host framework of Ag_{10}bpy-2D with complementary hydrogen bonding (O–H···O, H···O distance is 1.750 Å) between interlayer $-PO_2OH$ moieties. *Source:* Adapted with permission from Ref. [36]. © 2018 American Chemical Society. (b) The reversible structure transformation between Ag_{10}bpy-CH_3CN and its desolvation form. Color labels: green, Ag; yellow, S; light gray, C; blue, N.

hydrogen-bonds lead to the layers stacking in a staggered A-B manner to form a three-dimensional (3D) supramolecular structure, where grids stack over each other in an offset fashion to yield the closed-porous phase with 9.6% void space (Figure 16.8a). Ag_{10}bpy-CH_3CN contains a highly related Ag_{10} node, which was also bridged by bpy yet into a 1D polymer (Figure 16.8b). The cluster nodes are also four-connected, but the linkers are oriented in the same direction. Two CH_3CN molecules are located in diagonal positions of the $Ag_{10}S_6$ core, the loss of which causes an elongation of the capsule $Ag_{10}S_6$ and accordingly the distortion of the supramolecular structure. Such a process is reversible in a CH_3CN vapor.

Other silver clusters with different size and geometrical structure are also reported as nodes in SCAMs. For instance, Ma et al. reported a novel 1D SCAM Ag_{18}bpy-NH_2 prepared by employing an intensely luminescent 3-amino-4,4′-bipyridine (bpy-NH_2) linker bridging the novel $Ag_{18}S_{10}$ clusters [38]. Each cluster node contains 18 silver ions co-stabilized by ten thiolates, two CF_3COO^-, four $PhPO_3H^-$ and one $PhPO_3^{2-}$. The overall cluster node is analogous to the drum-like Ag_{20} but with $PhPO_3^{2-}$ replacing the CO_3^{2-} template. The steric hindrance of the bulkier $PhPO_3^{2-}$ might account for the lower nuclearity.

The Bakr group reported a Ag_{16} cluster 0-D NC showing a distorted square gyrobicupola polyhedron templated by one enclosed Cl^- in the center [39]. By adding bpy and adjusting the ratio of the bpy linker to the Ag-thiolate complex in the synthesis, they obtained two distinct clusters form, Ag_{15} and Ag_{14}, each adopting a MOF-like structure with specific dimensionality (i.e. 1-D NCF and 2-D NCF, respectively) (Figure 16.9a). The cluster units in 1-D NCF and 2-D NCF share notable similarities. In particular, the units possess a similar geometrical structure in which the skeletons of cluster cores can be considered to be incomplete square gyrobicupolas. 0-D NC, 1-D NCF, and

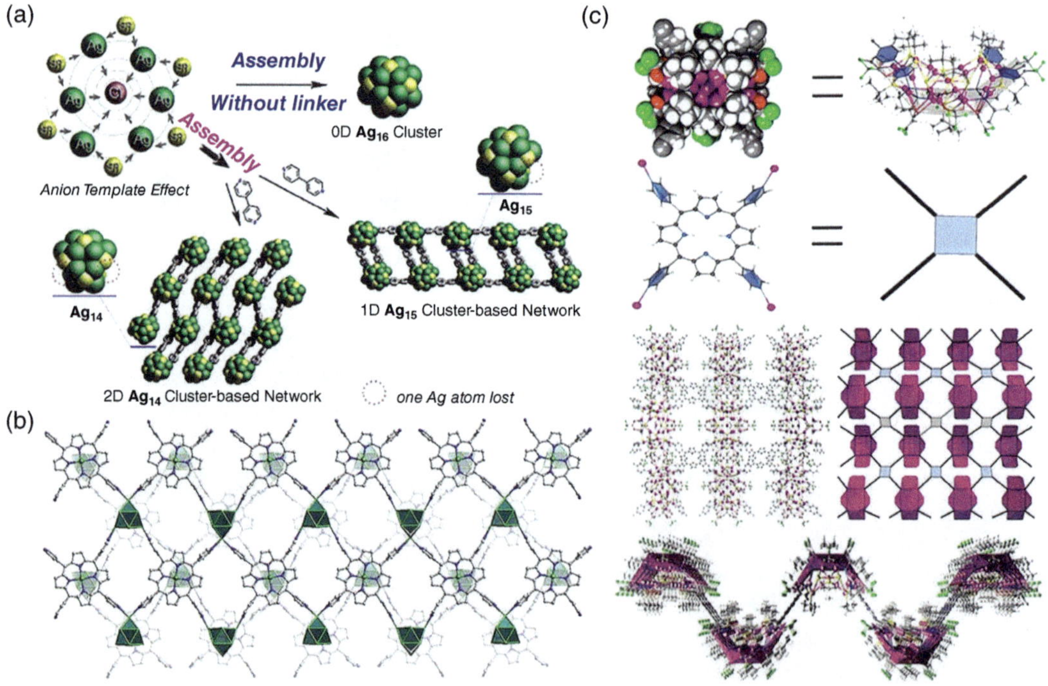

Figure 16.9 The assembly of 1-D NCF and 2-D NCF (a), Ag_9-AgTPyP (b), and Ag_{27}-MOF (c). *Source:* Panel a adapted with permission from Ref. [39]. © 2019 American Chemical Society; Panel b adapted with permission from Ref. [40]. © 2021 Wiley-VCH GmbH; Panel c adapted with permission from Ref. [41]. © 2020 Wiley-VCH GmbH.

2-D NCF exhibit qualitatively distinct characteristics in photoluminescence (PL) and thermal stability. Specifically, higher-dimensional structures show enhanced PL and higher thermal stability.

Recently, Zhao et al. presented a cationic 2D coordination structure Ag_{27}-MOF based on saddle-shaped, 27-nuclear silver clusters linked by a multitopic porphyrin-based ligand 5,10,15,20-tetra(4-pyridyl)porphyrin (TPyP-H_2) (Figure 16.9c) [41]. The saddle-shaped Ag_{27} cluster node is protected by fourteen $^tBuS^-$ and eight CF_3COO^- ligands, as well as templated by two enclosed S^{2-} ions. The overall cluster adopts a C_{2v} group symmetry. Due to the multitopic configuration of TPyP-H_2, it is bonded to four Ag_{27} clusters through four pyridyl groups, forming a square node. Ag_{27}-MOF adopts the 4-connected unimodal net with $\{4^4{\cdot}6^2\}$-*sql* net topology with both Ag_{27} cluster node and TPyP-H_2 treated as 4-connectors. With the same porphyrin-based tetrapyridine linker, Cao et al. synthesized a SCAM displaying a novel 2D framework, namely $[Ag_9(^tBuC{\equiv}C)_6(CF_3COO)_3(AgTPyP)]_n$ (Ag_9-AgTPyP), by a one-pot reaction of TPyP, CF_3COOAg, and $AgC{\equiv}C^tBu$ [40]. One Ag atom was spontaneously incorporated into the free-base TPyP ligand to form the metalloporphyrin AgTPyP. In the framework, the Ag_9 clusters are uniformly separated by Ag-centered porphyrin units (AgTPyP) (Figure 16.9b). The core of the Ag_9 cluster adopts a tower-like structure, capped by six $^tBuC{\equiv}C^-$ anionic ligands, and three CF_3COO^- ligands, which is further consolidated by numerous inner close Ag(I)···Ag(I) contacts. The 4-connected Ag_9 node is linked with μ_4-TPyP ligands to form a 2D framework that adopts an AB stacking mode.

The Mandal group fabricated two SCAMs, $[Ag_{11}(AdmS)_6(CF_3COO)_4(apy)_2(DMF)_2][NO_3]$ (Ag_{11}CAM) and $[Ag_{14}(S^tBu)_{10}(CF_3COO)_4$(4,4′-azopyridine)$_2]$ (Ag_{14}CAM), with a linear bipyridyl linker 4,4′-azopyridine [42, 43]. Ag_{11}CAM is composed of four-connected Ag_{11} cluster nodes, which adopt a unique distorted elongated square-bipyramid geometry, bridged by bipyridyl linkers to form a 1D polymeric structure. By contrast, the Ag_{14} cluster in Ag_{14}CAM is a distorted orthobielongated square pyramid in which two distorted elongated square pyramids (Johnson solids, J8) were fused by sharing a common square face. The silver clusters act as four-connected nodes linked by 4,4′-azopyridine to form a 2D layer-like CAM.

Dicarboxyl Linkers So far, the organic linkers employed to bridge silver clusters in SCAMs are primarily bitopic or multitopic pyridines. Exploiting new type linkers to develop novel SCAMs and investigate their structure property relationship is in sought. In addition, although in some cases the stability of silver clusters is significantly improved in pyridyl linker-based SCAMs, the universality is poor. The monodentate coordination mode of pyridine caused a lack of rigidity of the frameworks and might be responsible for the lower stability. In consideration of the common protective carboxylate groups in the ligand shell of the silver clusters, bi/multitopic carboxyl acids are promising linkers for SCAMs to further improve their stability.

The Liu group prepared two conductive 2D frameworks, SPc and SSc-2, based on an identical 14-center silver cluster bridged by fluorinated bicarboxylates, octafluoroadipic acid and tetradecafluoroazelaic acid, respectively, in the longitudinal and transverse directions through a solvothermal synthetic method [44, 45]. The Ag_{14} cluster nodes in both materials also adopt a drum-like geometry akin to the Ag_{16} node in 0-D NC, but with a Ag_3-Ag_8-Ag_3 arrangement instead. In SSc-2 the silver clusters serve as four-connected nodes (Figure 16.10a). The octafluoroadipic ligands are coordinated to the Ag centers by four O atoms through the mode of $\mu_4 - \eta^1, \eta^1, \eta^1, \eta^1$. In SPc, the Ag_{14} clusters act as an eight-connected node surrounded by eight DMF molecules and eight dicarboxylate groups of the eight tetradecafluoroazelaic acid ligands (Figure 16.10b). The tetradecafluoroazelate linkers are coordinated to the silver ion centers by four O atoms through two types of modes, $\mu_5 - \eta^1, \eta^2, \eta^1, \eta^1$ and $\mu_4 - \eta^1, \eta^1, \eta^1, \eta^1$. Both materials display great cycling stability, large capacity, and high-energy density without any modification, hence, acting as an excellent

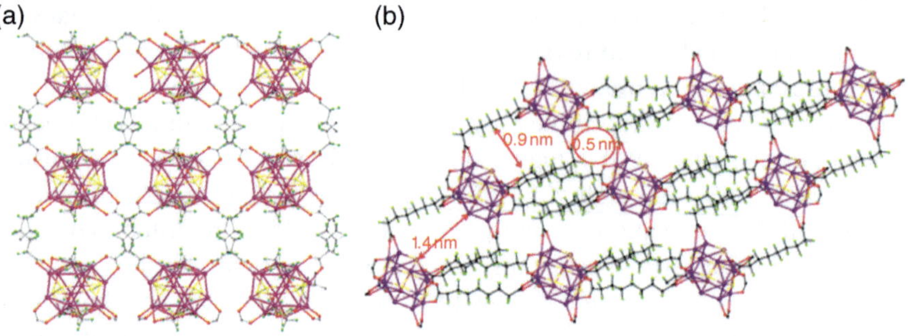

Figure 16.10 The 2D frameworks of SSc-2 (a) and SPc (b). *Source:* Panel a adapted with permission from Ref. [44]. © 2021 American Chemical Society; Panel b adapted with permission from Ref. [45]. © 2021 The Royal Society of Chemistry.

silver-based polymer supercapacitor, which benefits from the presence of fluorinated groups, 3D expansion of high-nucleus metallic clusters, and porosity.

The Zang group used linear 2,3,5,6-tetrafluoro-benzene-1,4-dicarboxylate (H_2L) to assemble silver-chalcogenide helical chains into an esthetic three-dimensional (3D) framework, namely $\{[Ag_6(S^tBu)_4(L)]\cdot guest\}_n$ (denoted as AgS-L) [46]. Wherein, 1D helices are arranged in a parallel manner and connected by two independent carboxyl groups from the H_2L ligands exclusively in a bidentate mode to generate a chiral open framework with channel openings of $9.2 \times 9.5\,Å^2$, which has a 3-c 3D network topology with a point symbol of $\{10^3\}$. It is noted that phenylphosphonic acid is a requisite in the synthesis that does not appear in the formula of AgS-L. It might act as a reagent regulating the pH of the reaction system and a competitor with H_2L in the formation of AgS-L. Due to the inherent water-resistant perfluorinated aromatic linkers and branched alkyl groups in the framework, the material exhibits excellent hydrophobic properties. Moreover, the ordered nanoarrays of $(-Ag-S-)_n$ chains facilitate electron transfer and enable semiconduction of AgS-L.

Of note, only fluorinated carboxylates were successfully introduced as linkers in SCAMs so far. Moreover, the preparation of SPc and SSc-2 adopted a solvothermal synthetic technique, while the synthesis of AgS-L required the phenylphosphonic acid as pH regulator and ligand competitor. It is known that the coordination reaction of bicarboxylates to silver cations is rapid and subsequently results in an insoluble precipitate to hinder the atom-precise structure determination. The introduction of electron-withdrawing fluorine in bicarboxylates, employment of competitive ligands, and adopting solvothermal synthetic method can efficiently regulate the reaction kinetics, which is beneficial for the crystal growth.

Thiourea Linkers Exploiting linkers with sulfur as donor atoms to fabricate SCAMs also face the same problem of fast precipitating due to the perfect match of the soft acid Ag and soft base S according to the HSAB theory. Thus, no thiolate-linker based SCAMs have been reported so far. As substitutes, the Duan group synthesized two tridentate ligands, 2,2′,2′′-((((1,3,5-Triazine-2,4,6-triyl)tris(azanediyl))tris(benzene-4,1-diyl))tris(ethan-1-yl-1-ylidene))tris(hydrazine carbothioamide) (TNS) and 2,2′,2′′-((((1,3,5-Triazine-2,4,6-triyl)tris(oxy))tris(benzene-4,1-diyl)) tris(methanylylidene))tris(hydrazinecarbothioamide) (TOS), by incorporating thiourea on the backbone of the tripodal ligands [47]. By slowly diffusing $AgBF_4$ into the solution of ligands, they obtained two silver-cluster-based frameworks TNS-Ag_8 and TOS-Ag_4 [47].

In TNS-Ag_8, the silver clusters exhibit D_{3d} symmetry with one of the silver atoms on the crystallographic threefold axis. Twelve ligands are coordinated to one cluster and can be divided into six pairs; the ligands in each pair stack together and function as a twofold linker to connect three different silver clusters. The silver clusters are regarded as 6-connection nodes that consolidate the three-dimensional frameworks with C_3-symmetric organic linkers. TOS-Ag_4 contains two asymmetric, independent, tetrahedral clusters, and each cluster acts as a 3-connection node for the construction of a twofold interpenetrated, three-dimensional framework. The clusters possess C_3 symmetry with one silver atom on the crystallographic threefold axis. The six ligands linked to one cluster are divided into three pairs, and the ligands in each pair are stacked together and function as twofold bridges for three different silver clusters. In this case, the silver clusters are considered 3-connection nodes that consolidate the 3D framework with a *srs*-type topology. Both frameworks are porous with large cross-sectional area of the channels to accommodate substrate molecules, which improves catalytic performance (see Section 16.3).

16.2.2 Gold Cluster-Assembled Materials (GCAMs)

Although gold shares many similarities with silver, particularly as soft Lewis acid in coordination chemistry, and the investigation of gold clusters has a longer history in comparison to silver clusters, the GCAMs are still rarely reported and the structure determination of GCAMs is still a big challenge. Actually, even the polymeric structures with simple gold cations as nodes are rare. The oxygen in carboxylate is a hard Lewis base, which does not match the soft gold atoms. Although nitrogen is a borderline Lewis base and the Au–N bonds are commonly seen in many coordination complexes, in addition, ligands with N donor atoms were successfully introduced in the protective layer of gold nanoclusters, there is still no progress in the exploration of GCAMs based on N-linkers.

Until now, the only high-dimensional GCAM with atom-precise structure was reported by the Wang group in 2014 [48]. First, by using (2-(3-methylpyrazinyl)-diphenylphosphine) (mdppz) as protecting ligands, they prepared a cluster $[(C)(Au\text{-mdppz})_6](BF_4)_2$, which bears six outward N donors that are available for coordination to metal ions in order to generate high-dimensional cluster-based polymers or frameworks (Figure 16.11a,b). Thus, the Au_6 cluster can be seen as a metal cluster linker. The cluster linker connects silver(I) ions with the outward N donors of its pyrazinyl groups to form a luminescent cluster-based MOF, $[(C)(Au\text{-mdppz})_6Ag_6](BF_4)_8$, with twofold interpenetrated NbO topology and hexagonal channels with a size of about 1.1 nm along the c axis (Figure 16.11c,d). Of note, although there are six outward N donors for extending connections, actually the Au_6 cluster only functions as a four-connected linker because of the terminal blocking of the silver ions by the H_2O molecules and BF_4^- counterions. The channels were available for solvent molecule exchanges along with corresponding luminescence responses.

Recently, the Yam group constructed an unprecedented gold(I) cluster cage $[Au_{24}\text{-Cl}]_C$ with $[(\mu_3\text{-S})Au_3]^+$ units as the nodes or vertices and V-shaped 1,3-bis(diphenylphosphino)benzene ligand with 2-phenyl substituent as the organic building block [49]. The complex cation of $[Au_{24}\text{-Cl}]_C$ displays a cubic structure, where eight $[(\mu_3\text{-S})Au_3]^+$ units occupy the vertices with the twelve diphosphine ligands spanning the edges of the cube. By virtue of the dynamic and reversible nature of the coordination bonds and aurophilic interactions, the gold(I) cluster cages exhibit interesting stimuli-responsive properties. A change of the counterions from Cl^- to PF_6^- or BF_4^- resulted in structural transformation from a cubic structure of $[Au_{24}\text{-PF}_6]_C$ and $[Au_{24}\text{-BF}_4]_C$ to a rhombic prism structure of $[Au_{24}\text{-PF}_6]_{RP}$ and $[Au_{24}\text{-BF}_4]_{RP}$ during the crystallization process, which could be recovered upon dissolution. The complex cation of $[Au_{24}\text{-PF}_6]_{RP}$ and $[Au_{24}\text{-BF}_4]_{RP}$ display rhombic prism structures with two rhomboidal and four square faces, and the bidentate phosphine ligands occupy the 12 edges and are linked by eight $[(\mu_3\text{-S})Au_3]^+$ vertices. Furthermore, upon heating the CD_3CN solution of $[Au_{24}\text{-PF}_6]_C$ at 353 K for around 0.5 hours, $[Au_{18}\text{-PF}_6]_{TP}$ was afforded, which displayed a triangular prism structure with six $[(\mu_3\text{-S})Au_3]^+$ units occupying the

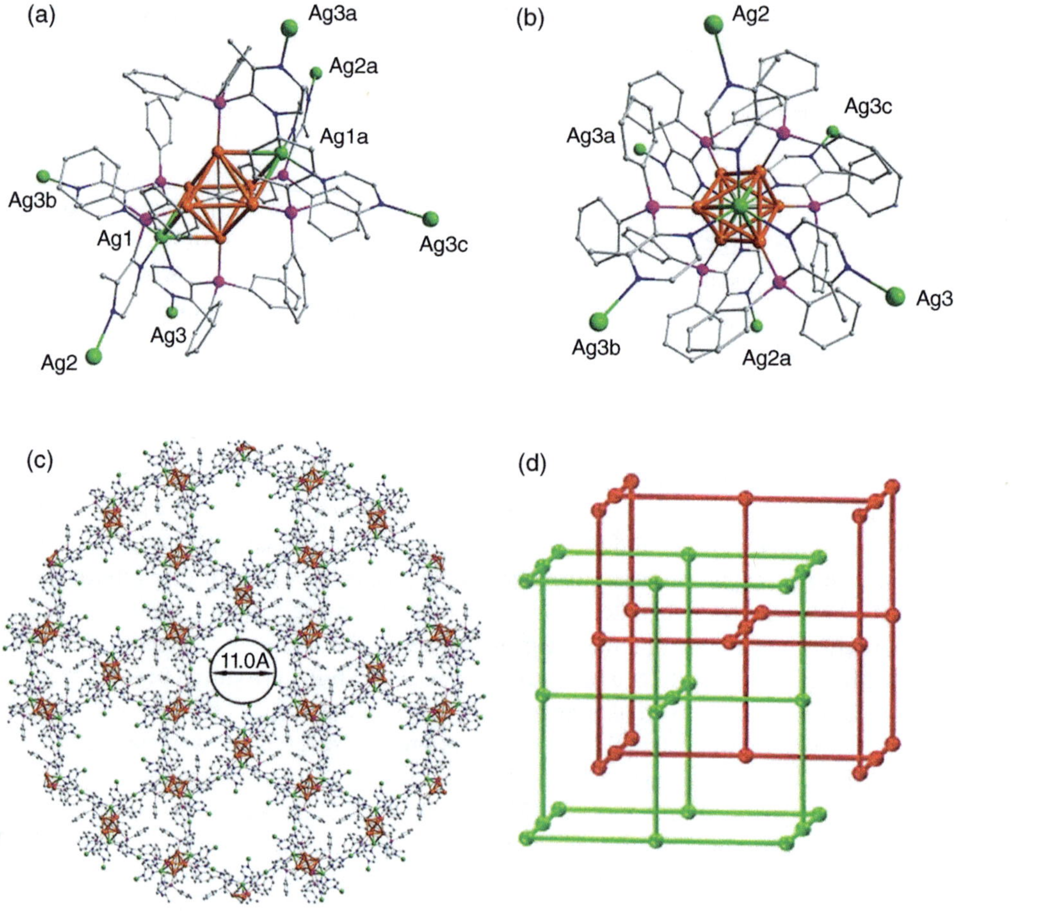

Figure 16.11 (a, b) Crystal structure of the C-$Au_6Ag_6(mdppz)_6$ nano-building block in different views. (c) Perspective view of Au_6Ag_6 cluster-based MOFs along the [0 0 1] direction. (d) Schematic illustration of the NbO topology in the twofold interpenetrated Au_6Ag_6 cluster-based MOFs. Color codes: orange sphere, Au; green sphere, Ag; purple sphere, P; gray sphere, C; blue sphere, N. *Source:* Redrawn with permission from Ref. [48]. © 2014 Wiley-VCH GmbH.

vertices and nine ligands spanning the edges (Figure 16.12). The reverse transformation from $[Au_{18}\text{-}PF_6]_{TP}$ to $[Au_{24}\text{-}PF_6]_C$ could be realized in DMSO-d_6, CD_3CN, and acetone-d_6 solutions at room temperature. This work suggests a new perspective to construct GCAMs by predesigning the ligands and cluster nodes with desired geometry and configuration.

16.2.3 Copper Cluster-Assembled Materials (CCAMs)

Cu^+ is also a soft Lewis acid, which is prone to coordinate to various donor atoms, including O, N, S, P, etc. as verified in coordination complexes, however, as copper clusters are susceptible to oxidation that results in precipitation, engineering their linkage with organic bridging groups to afford an infinite framework has rarely been reported and remains a technical challenge.

The Zang group successfully prepared a novel copper thiolate cluster-assembled materials, $Cu_{12}(S^tBu)_9(CF_3COO)_3(bpy)_4$ (abbreviated as Cu_{12}bpy), by bridging thiolate-protected Cu_{12} clusters with 4,4′-bipyridine (bpy) as the rigid organic linker [50]. Each Cu_{12} cluster is co-protected by nine

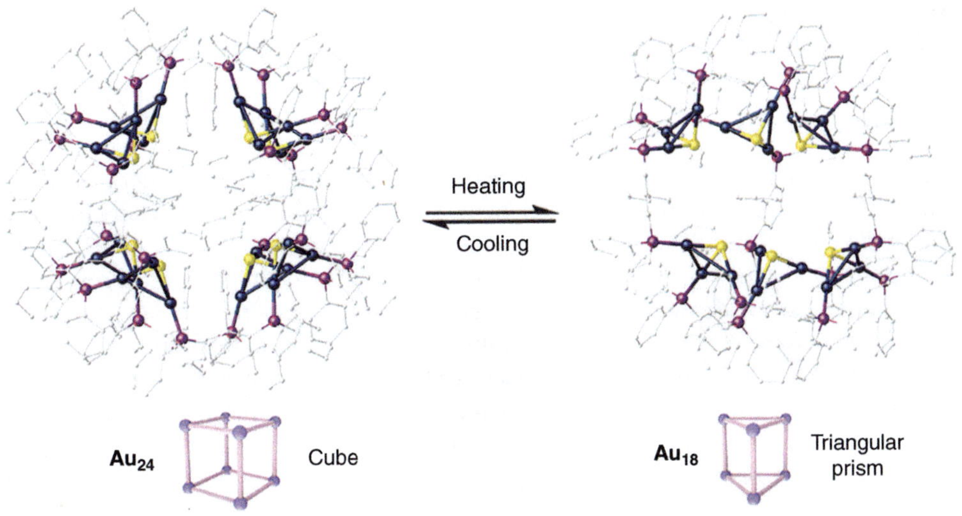

Figure 16.12 The structure transformation between the cubic Au_{24} cage and the triangular prism Au_{18} cage. *Source:* Adapted with permission from Ref. [49]. © 2021 American Chemical Society.

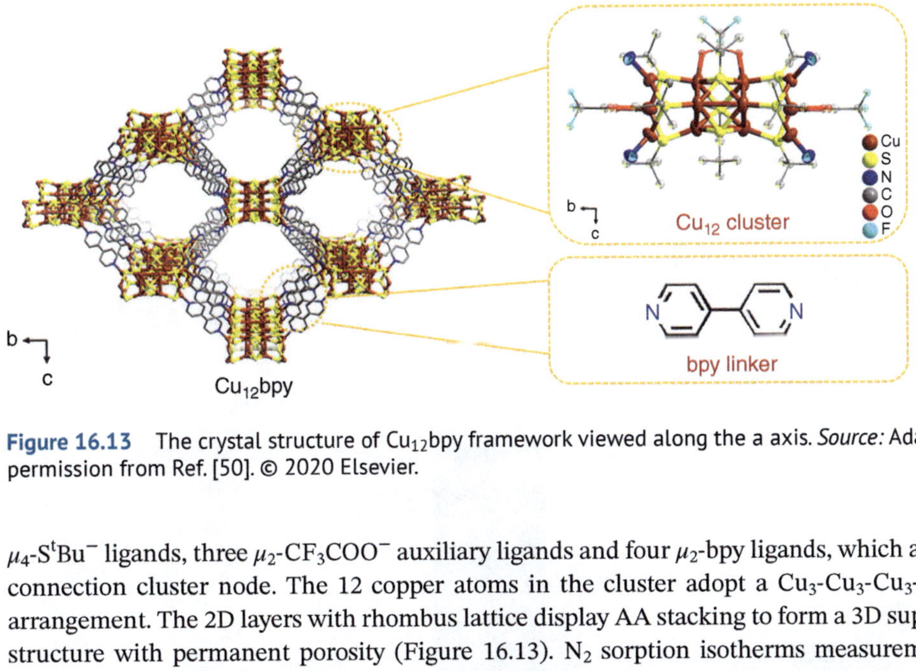

Figure 16.13 The crystal structure of Cu_{12}bpy framework viewed along the a axis. *Source:* Adapted with permission from Ref. [50]. © 2020 Elsevier.

μ_4-S^tBu^- ligands, three μ_2-CF_3COO^- auxiliary ligands and four μ_2-bpy ligands, which acts as a four-connection cluster node. The 12 copper atoms in the cluster adopt a Cu_3-Cu_3-Cu_3-Cu_3 rod-like arrangement. The 2D layers with rhombus lattice display AA stacking to form a 3D supramolecular structure with permanent porosity (Figure 16.13). N_2 sorption isotherms measurements at 77 K revealed a 141.5 $m^2\,g^{-1}$ Brunauer-Emmett-Teller (BET) surface area. The pore size distribution analysis based on nonlocal density functional theory (NL-DFT) calculation is centered at 1.2 nm.

16.3 Applications

Besides the inherent properties of the individual metal cluster and linker, the integration of them in SCAMs usually facilitates novel photophysical behaviors benefited from the various charge transfer pathways generated. In addition, the porous structure of SCAMs allows the adsorption of small molecules, which might endow SCAMs with luminescent sensing, molecular recognition, or catalytic properties.

16.3.1 Ratiometric Luminescent Temperature Sensing

Coinage metal clusters are known as potential temperature sensing luminescent materials mainly owing to the shrinking and elongation of metal–metal bonds upon temperature changing. The introduction of linkers could establish bi- or multi-emission centers in some SCAMs, which can be present in cluster and linkers, respectively. The distinct sensitivities to temperature of respective emission centers promise great potential in ratiometric temperature sensing. For Ag_{14}-dpbz, two emission peaks are observed when the temperature ranges from 83 K to room temperature (Figure 16.14a) [34]. When the temperature was decreased, the emission peak with a maximum at 518 nm grows much faster and becomes dominated at the temperature lower than 263 K, whereas the red emission band in the range of 580–700 nm becomes weaker. As a result, thermochromism of Ag_{14}-dpz was observed ranging widely from green to red (Figure 16.14a). For the crystals of

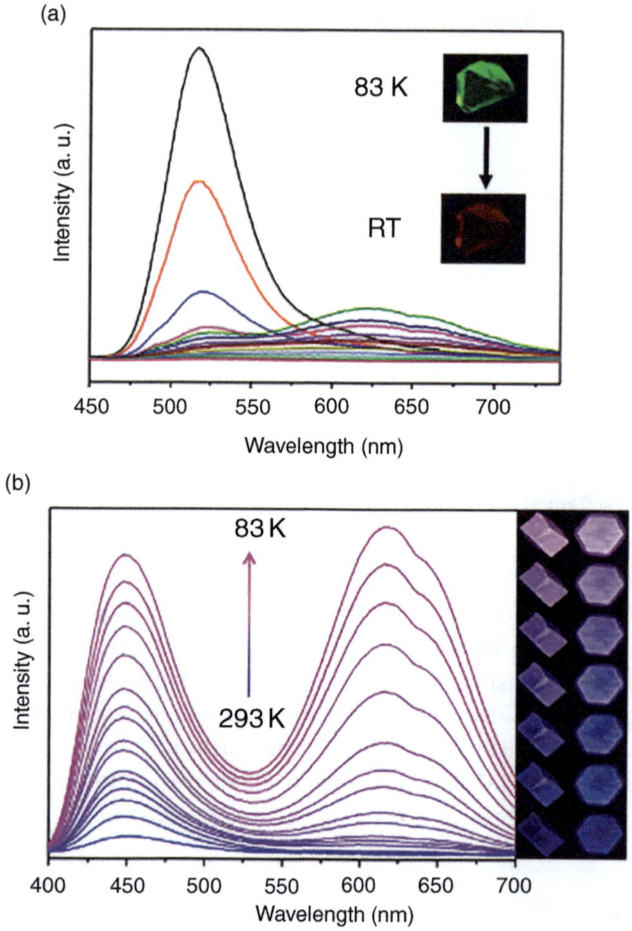

Figure 16.14 Variable-temperature emission spectra of Ag_{14}-dpbz (a) and Ag_{12}bpy-2/NH_2 (20:1) (b) and the corresponding crystal images. *Source:* Adapt with permission from Ref. [11]. © 2021 The Royal Society of Chemistry.

mixed-linker compound Ag_{12}bpy-2/NH_2 (20 : 1), temperature-dependent emission peaks and visual temperature sensing by color from purple pink to blue were observed (Figure 16.14b) [31]. A linear correlation of temperature (T) to the emission intensity ratio was found in the range 83–233 K (sensitivity 0.347% K^{-1}).

16.3.2 Luminescent Sensing and Identifying O_2 and VOCs

The molecules adsorbed in the SCAMs show different host-guest interactions with the frameworks, which usually result in synchronous variation of emissive parameters and also provides an excellent opportunity for luminescent sensing and identification. A series of SCAMs show potential applications in this field. In this section, only Ag_{12}bpy, Ag_{12}bpy-NH_2, and Ag_{10}bpy are discussed as representatives.

Ag_{12}bpy can function in ultrafast dual-function luminescence switching (<1 s), that is, turn-off by O_2 and multicolored turn-on by VOCs (Figure 16.15a) [25]. The switching mechanism was distinctly resolved by single-crystal X-ray diffraction (SCXRD) structural analysis of O_2/VOCs-included crystals, highlighting the importance of structural precision. Fluorescence (Fl) and phosphorescence (Ph) are two particular cases of luminescence. Fl originates from the first excited singlet state S_1, while Ph originates from the triplet state T_1. The amino-modified Ag_{12}bpy-NH_2 displayed two emissive peaks, one of which is a blue sub-nanosecond Fl component (0.37 ns under vacuum) at approximately 456 nm and the other is Ph at 556 nm (3.12 ms under vacuum) [26]. In comparison to the 200 ns Ph of Ag_{12}bpy, the lifetime of the Ph component of Ag_{12}bpy-NH_2 is prolonged by approximately 15000-fold, with an approximate 2% increase in QY. Thus, Ag_{12}bpy-NH_2 could detect trace oxygen gas in 0.3 s with a limit of detection (LOD) as low as 0.1 ppm in a ratiometric method, in which blue insensitive Fl functioned as a reference signal (Figure 16.15b). When the O_2 concentration varied in the range of 0–20 ppm, the emission color of Ag_{12}bpy-NH_2 contemporaneously varied, enabling the visible color response (Figure 16.15b). When the $-CH_3$ groups are anchored on the bpy linkers, they can interfere with the process by which the oxygen molecules interact with the emissive framework, and greatly decease the response sensitivity to molecular oxygen.

The flexible luminescent SCC-MOF Ag_{10}bpy exhibits a remarkable framework-breathing phenomenon upon adsorption of VOCs, accompanied by distinct responsive fluorescence [37]. Three chloro-derivatives of methane, namely, CH_2Cl_2, $CHCl_3$, and CCl_4, could be easily distinguished because of their striking difference of luminescent color in visible light (Figure 16.15c): CH_2Cl_2 displayed no shift (green), CCl_4 showed a 15 nm red shift (yellow-green), and $CHCl_3$ displayed the biggest red shift of 52 nm (yellow). The detailed investigation of $CHCl_3$- and CH_2Cl_2-induced structural dynamics and their structure-luminescence correlation, together with well-defined crystallographic characterization of the open- and closed-pore forms, well addressed such solvatochromism behaviors.

16.3.3 Catalytic Properties

The synergic effect between building blocks including silver clusters and organic linkers of SCAMs would produce the novel collective functionality. Ag_{12}TPyP has rich pores with a BET surface area of 234 $m^2\ g^{-1}$ and excellent chemical stability in air and various organic solvents [29]. Ag_{12}TPyP also has strong affinity for 2-chloroethyl ethyl sulfide (CEES) due to its unique pore structure and the interactions between the silver cluster and chlorine atoms of CEES, such that it shows rapid and high-dose adsorption of CEES. The porphyrin linkers in Ag_{12}TPyP have excellent light-harvesting capability and $^1O_2$ generation. Due to the synergistic efficiency between the silver

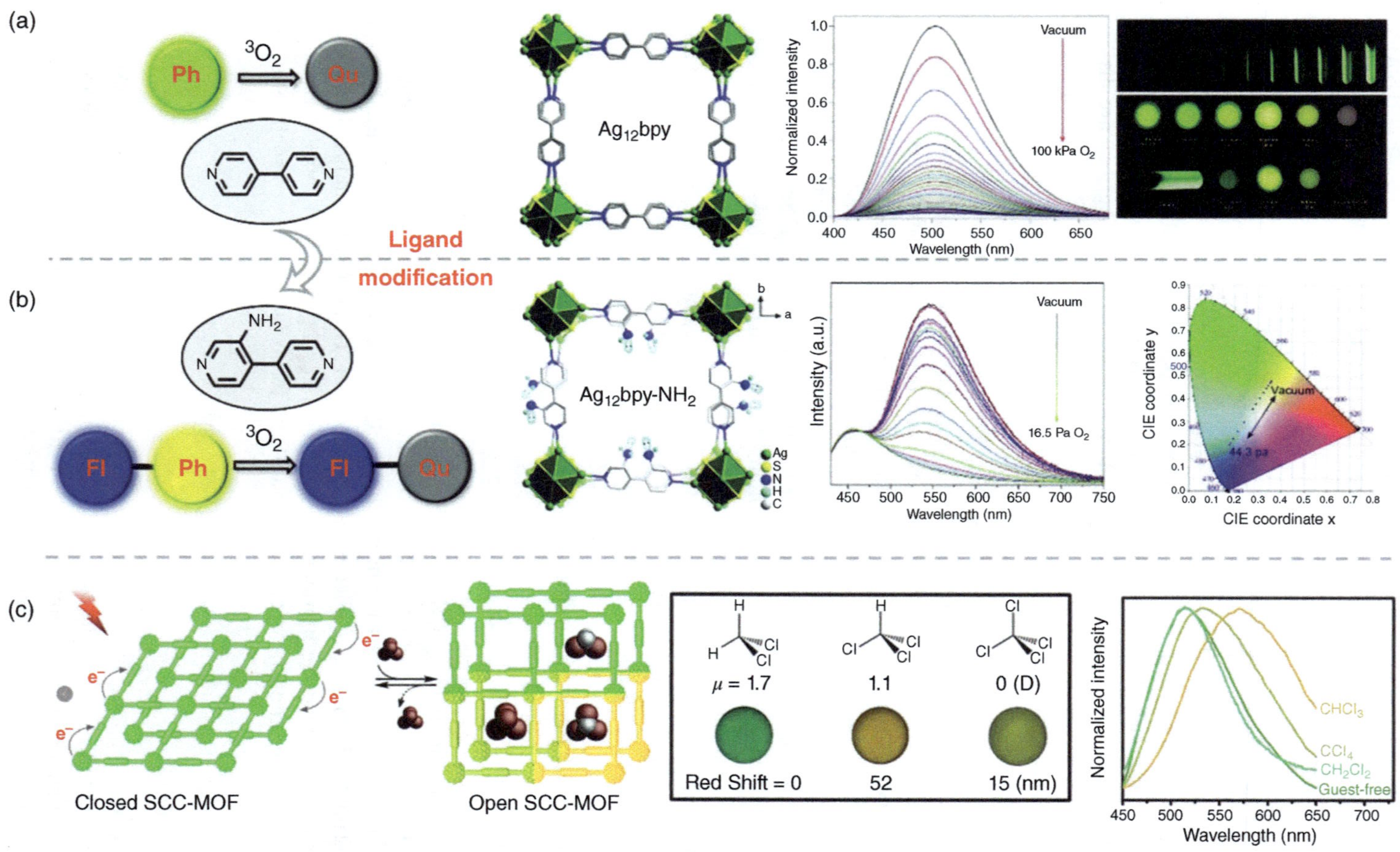

Figure 16.15 (a) Luminescence switching of Ag_{12}bpy responding to O_2 and VOCs. From left to right, the mechanism of O_2 quenching phosphorescence, structure of Ag_{12}bpy, O_2 partial pressure-dependent emission spectra, images of a glass tube containing Ag_{12}bpy powder samples from vacuum to air excited by UV light and the image of Ag_{12}bpy when responding to different VOCs under a UV light. (b) From left to right, the mechanism of ratiometric O_2 sensing based on dual-emission fluorescence and phosphorescence; structure of Ag_{12}bpy-NH_2; ratiometric emission spectra of Ag_{12}bpy-NH_2; the CIE coordinate diagram illustrating the color changing with O_2 pressure. Fl = Fluorescence, Ph = Phosphorescence and Qu = Quenched. (c) From left to right, the illustration of CH_2Cl_2, $CHCl_3$, and CCl_4 induced reversible pore-open/closed structural transformation and switchable solvatochromism; The structures of chloromethanes and their lumininescent responsive images and normalized PL spectra (excited at 365 nm). *Source:* Panels a, b adapted with permission from Ref. [11]. © 2021 The Royal Society of Chemistry; Panel c reproduced with permission from Ref. [37]. © 2018 American Chemical Society.

cluster, porphyrin-based linkers and the unique pores of SCAMs, $Ag_{12}TPyP$ efficiently degraded mustard-gas simulants with a half-life of 1.5 minutes with 100% selectivity by visible light (Figure 16.16a).

TOS-Ag_4 and TNS-Ag_8 both show channels with large enough cross-sectional area in their 3D structures for substrate molecules' adsorption [47]. The rich H-bonding donors and acceptors from the tripodal linker provide potential hydrogen-bonding interactions with substrates for synergistic activation and fixation during catalytic transformations. The trapped propargylamine molecules are then adjusted to suitable directions to allow the silver atoms to interact with the C–C bond through π coordination to further activate the substrates, which then react with CO_2 to form the product through nucleophilic attack. The synergistic interaction mode of the π-activation and hydrogen-bonding interactions accelerates the capture of the substrates around the catalytic sites and catalyzes the chemical transformation under mild reaction conditions. The total turnover number for TOS-Ag_4 was 930 per mole of silver atoms (760 per mole of Ag atoms for TNS-Ag_8,

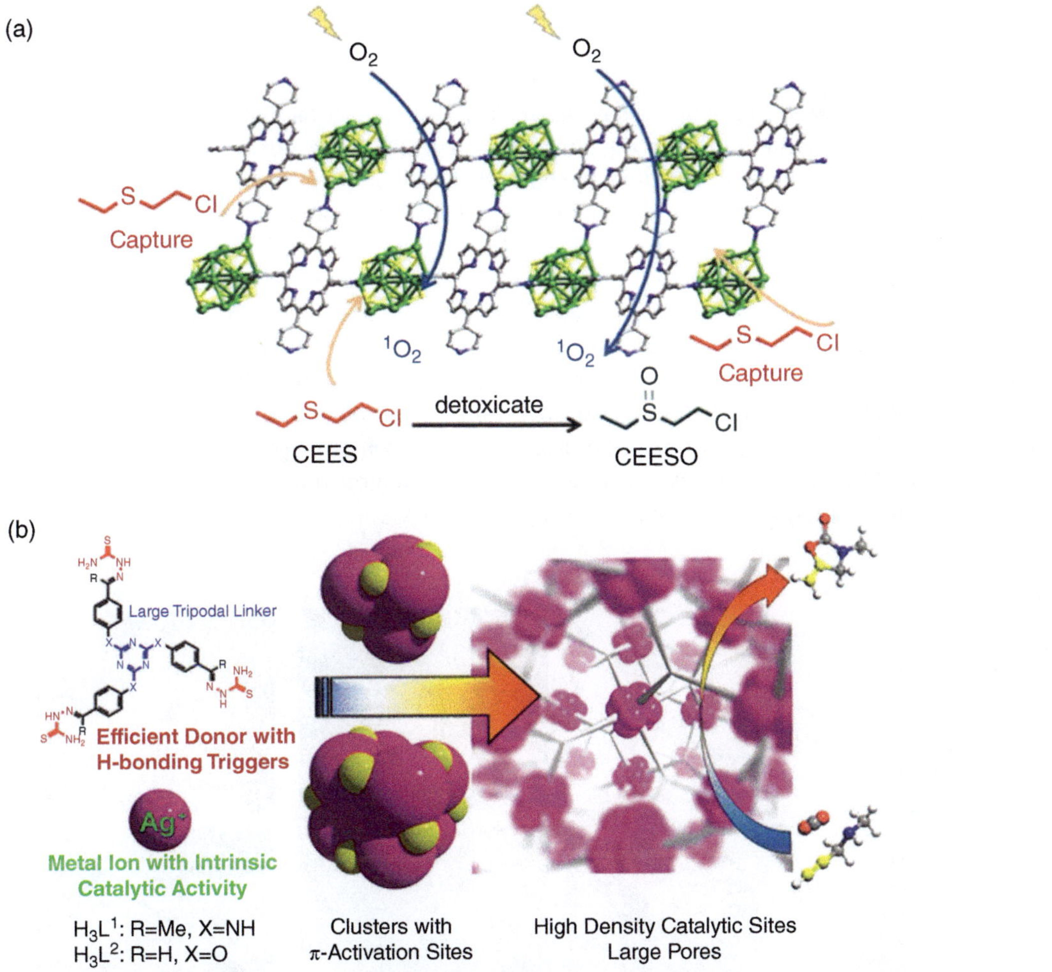

Figure 16.16 (a) Schematic illustration for the capture and photodetoxification of CEES by $Ag_{12}TPyP$. *Source:* Reproduced with permission from Ref. [29]. © 2019 American Chemical Society. (b) Perspective views of the construction of the porous coordination polymers and the catalytic conversion of carbon dioxide with propargylamines to oxazolidinones. *Source:* Reproduced with permission from Ref. [47]. © 2018 American Chemical Society.

Figure 16.16b). This catalytic system represents the highest catalytic efficiency reported, in comparison with other homogeneous or heterogeneous catalytic systems, for the cycloaddition of propargylamines. Furthermore, the strong silver–sulfur bonds of cluster nodes provide excellent chemical stability and recyclability for potential practical applications.

Later on, Ag_{27}-MOF was also studies as a high-performance catalyst toward CO_2-cycloadditions of N-propargylamines [41]. Ag_{27}-MOF shows high chemical stability and significant affinity toward CO_2. The amount of CO_2 uptakes is 14.5 and 10.3 $cm^3\,g^{-1}$ at 273 and 300 K, corresponding to about 3.8 and 2.7 CO_2 molecules per formula unit, respectively. These virtues facilitate its potential application in serving as a heterogeneous catalyst for CO_2 fixation. Moreover, the exposed silver nodes in the 2D framework could activate the C–C bonds though favorable Ag$\cdots\pi$ interactions. Accordingly, Ag_{27}-MOF shows excellent catalytic performance for carboxylative cyclization of propargylamines with CO_2. Benefited from the accessible catalytic platform of 2D MOF, both terminal and internal propargylamines with different sterically hindered substituents readily cyclize with CO_2 under ambient conditions.

16.3.4 Anti-Superbacteria

The Ag_9-AgTPyP integrated light harvesters, silver sites, and high surface areas are all in one complex with an orderly manner [40]. The Ag_9 cluster in Ag_9-AgTPyP has a preferential affinity to O_2. The porphyrinic unit behaves as an antenna to harvest visible light, leading to the formation of the excited state, which then transfers electrons to the catalytic sites around the Ag_9 cluster, enabling the activation of O_2 to various ROS, including singlet oxygen ($^1O_2$), superoxide anion ($\cdot O_2^-$), and hydrogen peroxide (H_2O_2) (Figure 16.17a). The unique 2D framework accelerates the ROS transportation. As a result, Ag_9-AgTPyP showed an outstanding photocatalytic inactivation efficiency against superbacteria methicillin-resistant *Staphylococcus aureus* (MRSA, >99.99999%) and *Pseudomonas aeruginosa* (*P. aeruginosa*, >99.999%) under visible light within 120 minutes at a catalyst dose of 50 $mg\,L^{-1}$. Moreover, Ag_9-AgTPyP film was fabricated as the bioprotective layer and incorporated into the face masks or bioprotective suits, which displayed intriguing photoactive antibacterial performance in both aerosol and liquid forms, demonstrating their great potential as bioprotective materials against superbacteria (Figure 16.17b).

Figure 16.17 (a) Schematic representation ofAg_9-AgTPyP photocatalytic inactivation of antibiotic-resistant bacteria. (b) Simulated bacterial aerosols and the interception test by N95 mask, and protective suit. *Source:* Adapted with permission from Ref. [40]. © 2021 Wiley-VCH GmbH.

16.4 Conclusion

This chapter summarized the coinage MCAMs, with a focus on those structurally well-defined systems determined by single-crystal X-ray diffraction. The MCAMs are sorted out in detail according to the metal type, linker type, and nuclearity of metal cluster nodes. We gave an outline of the crystal structures of some typical examples and analyzed the opportunities and challenges in the fabrication of such materials with different metal and linker components by following the HSAB theory. Benefiting from the novel structure features, the individual/synergy of the metal cluster node and selected linkers, MCAMs show potential applications in many fields, including catalysis, optical sensing, and biomedicine, to name a few. Several attractive representatives are introduced in the chapter. In future work, new types of linkers with strong coordination ability to coinage metal clusters should be explored to enhance the rigidity of the frameworks, and to improve the stability of the metal clusters accordingly. The rapid precipitation and oxidation in the synthesis of MCAMs, particularly the GCAMs or CCAMs should be overcome. Also, the applications of MCAMs should be further pursued according to their functionality and advantages.

Acknowledgments

S.Q.Z. thanks the financial support from the National Natural Science Foundation of China (No. 92061201, 21825106), the Program for Innovative Research Team (in Science and Technology) in Universities of Henan Province (19IRTSTHN022) and Zhengzhou University, Z.Y.W. acknowledges support from the National Natural Science Foundation of China (No. 21801228).

References

1 Chakraborty, I. and Pradeep, T. (2017). Atomically precise clusters of noble metals: emerging linke between atoms and nanoparticles. *Chem. Rev.* 117: 8208–8271.

2 Jin, R., Zeng, C., Zhou, M., and Chen, Y. (2016). Atomically precise colloidal metal nanoclusters and nanoparticles: fundamentals and opportunities. *Chem. Rev.* 116: 10346–10413.

3 Kang, X., Li, Y., Zhu, M., and Jin, R. (2020). Atomically precise alloy nanoclusters: syntheses, structures, and properties. *Chem. Soc. Rev.* 49: 6443–6514.

4 Zhang, M.-M., Li, K., and Zang, S.-Q. (2020). Progress in atomically precise coinage metal clusters with aggregation-induced emission and circularly polarized luminescence. *Adv. Opt. Mater.* 8: 1902152.

5 Goswami, N., Yao, Q., Luo, Z. et al. (2016). Luminescent metal nanoclusters with aggregation-induced emission. *J. Phys. Chem. Lett.* 7: 962–975.

6 Rival, J.V., Mymoona, P., Lakshmi, K.M. et al. (2021). Self-assembly of precision Noble metal nanoclusters: hierarchical structural complexity, colloidal superstructures, and applications. *Small* 17: 2005718.

7 Chakraborty, P., Nag, A., Chakraborty, A., and Pradeep, T. (2019). Approaching materials with atomic precision using supramolecular cluster assemblies. *Acc. Chem. Res.* 52: 2–11.

8 Bonacchi, S., Antonello, S., Dainese, T., and Maran, F. (2021). Atomically precise metal nanoclusters: novel building blocks for hierarchical structures. *Chem. Eur. J.* 27: 30–38.

9 Kang, X. and Zhu, M. (2019). Intra-cluster growth meets inter-cluster assembly: the molecular and supramolecular chemistry of atomically precise nanoclusters. *Coord. Chem. Rev.* 394: 1–38.

10 Luo, L. and Jin, R. (2021). Perspective atomically precise metal nanoclusters meet metal-organic frameworks. *Iscience* 24: 103206.

11 Jin, Y., Zhang, C., Dong, X.-Y. et al. (2021). Shell engineering to achieve modification and assembly of atomically-precise silver clusters. *Chem. Soc. Rev.* 50: 2297–2319.

12 Li, B., Huang, R.-W., Qin, J.-H. et al. (2014). Thermochromic luminescent nest-like silver thiolate cluster. *Chem. Eur. J.* 20: 12416–12420.

13 Wang, Z., Li, X.-Y., Liu, L.-W. et al. (2016). Beyond clusters: supramolecular networks self-assembled from nanosized silver clusters and inorganic anions. *Chem. Eur. J.* 22: 6830–6836.

14 Li, X.-Y., Su, H.-F., Kurmoo, M. et al. (2017). Structure, solution assembly, and electroconductivity of nanosized argento-organiccluster/framework templated by chromate. *Nanoscale* 9: 5305–5314.

15 Zhang, S.-S., Su, H.-F., Zhuang, G.-L. et al. (2018). A hexadecanuclear silver alkynyl cluster based NbO framework with triple emissions from the visible to near-infrared II region. *Chem. Commun.* 54: 11905–11908.

16 Chen, S., Du, W., Qin, C. et al. (2020). Assembly of the thiolated $Au_1Ag_{22}(S\text{-}Adm)_{12}{}^{3+}$ superatom complex into a framework material through direct linkage by SbF^{6-} anions. *Angew. Chem. Int. Ed.* 59: 7542–7547.

17 Wei, X., Kang, X., Zuo, Z. et al. (2021). Hierarchical structural complexity in atomically precise nanocluster frameworks. *Natl.Sci.Rev.* 8: nwaa077.

18 Wang, Q.-M., Lin, Y.-M., and Liu, K.-G. (2015). Role of anions associated with the formation and properties of silver clusters. *Acc. Chem. Res.* 48: 1570–1579.

19 Zhao, X.-L. and Mak, T.C.W. (2010). New polyoxometalate species stabilized in coordination networks constructed with the multinuclear silver(I) ethynediide aggregate $C_2@Ag_n$ ($n = 6$ and 7). *Inorg. Chem.* 49: 3676–3678.

20 Fan, X., Chen, S., Zhang, L., and Zhang, J. (2021). Protection of Ag clusters by metal-oxo modules. *Chem. Eur. J.* 27: 15563–15570.

21 Gruber, F. and Jansen, M. (2011). Supramolecular intercluster compounds consisting of 1D arrays of silver alkynyl clusters and Wells-Dawson anions, displaying ligand-free interfaces. *Z. Anorg. Allg. Chem.* 637: 1676–1679.

22 Hu, T., Li, Y.-H., Kuang, X.-F., and Lu, C.-Z. (2017). Synthesis and characterization of polyoxometalate-based silver(I) phenylethynide compounds with antibacterial and antifungal activities. *CrystEngComm* 19: 3445–3454.

23 Xie, A., Wang, Z., Cheng, L.-P. et al. (2019). An extended Ag-I cluster-based framework solid: silver-thiolate cluster linked polyoxometalate including Ag-I center dot center dot center dot H-C anagostic interactions. *Eur. J. Inorg. Chem.* 496–501.

24 Li, Y.-H., Wang, Z.-Y., Ma, B. et al. (2020). Self-assembly of thiolate-protected silver coordination polymers regulated by POMs. *Nanoscale* 12: 10944–10948.

25 Huang, R.-W., Wei, Y.-S., Dong, X.-Y. et al. (2017). Hypersensitive dual-function luminescence switching of a silver-chalcogenolate cluster-based metal-organic framework. *Nat. Chem.* 9: 689–697.

26 Dong, X.-Y., Si, Y., Yang, J.-S. et al. (2020). Ligand engineering to achieve enhanced ratiometric oxygen sensing in a silver cluster-based metal-organic framework. *Nat. Commun.* 11: 3678.

27 Wei, Z., Wu, X.-H., Luo, P. et al. (2019). Matrix coordination induced emission in a three-dimensional silver cluster-assembled material. *Chem. Eur. J.* 25: 2750–2756.

28 Wang, J.-Y., Huang, R.-W., Wei, Z. et al. (2019). Linker flexibility-dependent cluster transformations and cluster-controlled luminescence in isostructural silver cluster-assembled materials (SCAMs). *Chem. Eur. J.* 25: 3376–3381.

29 Cao, M., Pang, R., Wan, Q.-Y. et al. (2019). Porphyrinic silver cluster assembled material for simultaneous capture and photocatalysis of mustard-gas simulant. *J. Am. Chem. Soc.* 141: 14505–14509.

30 Wu, X.-H., Luo, P., Wei, Z. et al. (2019). Guest-triggered aggregation-induced emission in silver Chalcogenolate cluster metal-organic frameworks. *Adv. Sci.* 6: 1801304.

31 Huang, R.-W., Dong, X.-Y., Yan, B.-J. et al. (2018). Tandem silver cluster isomerism and mixed linkers to modulate the photoluminescence of cluster-assembled materials. *Angew. Chem. Int. Ed.* 57: 8560–8566.

32 Du, X.-S., Yan, B.-J., Wang, J.-Y. et al. (2018). Layer-sliding-driven crystal size and photoluminescence change in a novel SCC-MOF. *Chem. Commun.* 54: 5361–5364.

33 Yang, H., Lei, J., Wu, B. et al. (2013). Crystal structure of a luminescent thiolated Ag nanocluster with an octahedral Ag_6^{4+} core. *Chem. Commun.* 49: 300–302.

34 Wang, Z.-Y., Wang, M.-Q., Li, Y.-L. et al. (2018). Atomically precise site-specific tailoring and directional assembly of Superatomic silver nanoclusters. *J. Am. Chem. Soc.* 140: 1069–1076.

35 Deng, G., Teo, B.K., and Zheng, N. (2021). Assembly of chiral cluster-based metal-organic frameworks and the chirality memory effect during their disassembly. *J. Am. Chem. Soc.* 143: 10214–10220.

36 Dong, X.-Y., Huang, H.-L., Wang, J.-Y. et al. (2018). A flexible fluorescent SCC-MOF for switchable molecule identification and temperature display. *Chem. Mater.* 30: 2160–2167.

37 Li, Y.-H., Huang, R.-W., Luo, P. et al. (2019). 1D silver cluster-assembled materials act as a platform for selectively erasable photoluminescent switch of acetonitrile. *Sci. China Chem.* 62: 331–335.

38 Ma, X.-H., Wang, J.-Y., Guo, J.-J. et al. (2019). Reversible wide-range tuneable luminescence of a dual-stimuli- responsive silver cluster-assembled material. *Chin. J. Chem.* 37: 1120–1124.

39 Alhilaly, M.J., Huang, R.-W., Naphade, R. et al. (2019). Assembly of atomically precise silver nanoclusters into nanocluster-based frameworks. *J. Am. Chem. Soc.* 141: 9585–9592.

40 Cao, M., Wang, S., Hu, J.-H. et al. (2021). Silver cluster-porphyrin-assembled materials as advanced bioprotective materials for combating Superbacteria. *Adv. Sci.* e2103721.

41 Zhao, M., Huang, S., Fu, Q. et al. (2020). Ambient chemical fixation of CO_2 using a robust Ag_{27} cluster-based two-dimensional metal-organic framework. *Angew. Chem. Int. Ed.* 59: 20031–20036.

42 Das, A.K., Biswas, S., Thomas, A. et al. (2021). Switchable photon and phonon emission properties of an atomically precise Ag_{14} core-based two-dimensional silver cluster-assembled material. *Mater. Chem. Front.* 5: 8380–8386.

43 Das, A.K., Biswas, S., Manna, S.S. et al. (2021). Solvent-dependent photophysical properties of a semiconducting one-dimensional silver cluster-assembled material. *Inorg. Chem.* 60: 18234–18241.

44 Wu, Q.-S., Bigdeli, F., Rouhani, F. et al. (2021). New 3D porous silver nanopolycluster as a highly effective supercapacitor electrode: synthesis and study of the optical and electrochemical properties. *Inorg. Chem.* 60: 1523–1532.

45 Jing, Z., Rouhani, F., Bigdeli, F. et al. (2021). Stable supercapacitor electrode based on two-dimensional high nucleus silver nano-clusters as a green energy source. *Dalton Trans.* 50: 2606–2615.

46 Wang, J.-Y., Li, W.-H., Wei, Z. et al. (2020). A hydrophobic semiconducting metal-organic framework assembled from silver chalcogenide wires. *Chem. Commun.* 56: 2091–2094.

47 Chang, Z., Jing, X., He, C. et al. (2018). Silver clusters as robust nodes and π-activation sites for the construction of heterogeneous catalysts for the cycloaddition of propargylamines. *ACS Catal.* 8: 1384–1391.

48 Lei, Z., Pei, X.-L., Jiang, Z.-G., and Wang, Q.-M. (2014). Cluster linker approach: preparation of a luminescent porous framework with NbO topology by linking silver ions with Gold(I) clusters. *Angew. Chem. Int. Ed.* 53: 12771–12775.

49 Yan, L.-L., Yao, L.-Y., Ng, M., and Yam, V.W.-W. (2021). Stimuli-responsive and structure-adaptive three-dimensional Gold(I) cluster cages constructed via "De-aurophilic" interaction strategy. *J. Am. Chem. Soc.* 143: 19008–19017.

50 Wang, Q.-Y., Zhang, L., He, W.-M. et al. (2020). High-performance primary explosives derived from copper thiolate cluster-assembled materials for micro-initiating device. *Chem. Eng. J.* 389, 124455.

Index

Note: Page numbers in *italics* refer to Figures; those in **bold** to Tables.

Atomically Precise Nanochemistry, First Edition. Edited by Rongchao Jin and De-en Jiang.

d

e

f

g

h

m

n

Printed in the USA/Agawam, MA
September 16, 2026

910661.192